SOCIOLOGY

A Global Perspective

SOCIOLOGY
A Global Perspective

SECOND EDITION

Joan Ferrante
Northern Kentucky University

WADSWORTH PUBLISHING COMPANY
I(T)P™ An International Thomson Publishing Company

Belmont · Albany · Bonn · Boston · Cincinnati · Detroit ·
London · Madrid · Melbourne · Mexico City · New York · Paris ·
San Francisco · Singapore · Tokyo · Toronto · Washington

Contents

Developmental Editors: John Bergez and Maggie Murray
Editorial Assistant: Jason Moore
Production Editors: Angela Mann and Cathy Linberg
Text and Cover Designer: Andrew Ogus
Print Buyer: Karen Hunt
Art Editors: Kelly Murphy and Emma Nash
Permissions Editor: Jeanne Bosschart
Copy Editor: Robin Kelly
Proofreaders: Karen Stough and Kathy Lee
Indexer: Bernice Eisen
Photo Researcher: Stephen Forsling
Illustrators: Precision Graphics and Parrot Graphics
Cover Photographs: Credits appear on page A-39
Compositor: TSI Graphics
Printer: Quebecor Printing Book Group/Hawkins

Printed in the United States of America
   2 3 4 5 6 7 8 9 10—01 00 99 98 97 96 95

For more information, contact Wadsworth Publishing Company:

Wadsworth Publishing Company
10 Davis Drive
Belmont, California 94002, USA

International Thomson Editores
Campos Eliseos 385, Piso 7
Col. Polanco
11560 México D.F. México

International Thomson Publishing Europe
Berkshire House 168–173
High Holborn
London, WC1V 7AA, England

International Thomson Publishing GmbH
Königswinterer Strasse 418
53227 Bonn, Germany

Thomas Nelson Australia
102 Dodds Street
South Melbourne 3205
Victoria, Australia

International Thomson Publishing Asia
221 Henderson Road
#05-10 Henderson Building
Singapore 0315

Nelson Canada
1120 Birchmount Road
Scarborough, Ontario
Canada M1K 5G4

International Thomson Publishing Japan
Hirakawacho Kyowa Building, 3F
2-2-1 Hirakawacho
Chiyoda-ku, Tokyo 102, Japan

Library of Congress Cataloging-in-Publication Data

Ferrante-Wallace, Joan.
    Sociology : a global perspective / Joan Ferrante. — 2nd ed.
       p.   cm.
    Includes bibliographical references and index.
    ISBN 0-534-20976-9 (acid-free paper)
    1. Sociology.   2. Social history—Cross-cultural studies.
 I. Title.
 HM51.F47   1994   94-17688
 301—dc20

To my mother,
Annalee Taylor Ferrante

and in memory of my father,
Phillip S. Ferrante
(March 1, 1926–July 8, 1984)

BRIEF CONTENTS

CONTENTS

8

PREFACE

I wrote the first and second editions of *Sociology: A Global Perspective* with one overriding goal: to showcase the accomplishments of the discipline. I wanted to show the power, vitality, relevance, and versatility of sociological concepts and theories for framing and explaining a variety of social phenomena and issues. With this goal in mind I made some fundamental changes in the traditional textbook model, which sociologists (and other academics) typically have followed to organize, arrange, and present the knowledge of their discipline. The major changes are interrelated and include the following:

1. Introducing students to sociology's essential concepts and theoretical perspectives in a coordinated and integrated way, such that the final product is not an encyclopedia-style overview of the field

2. Demonstrating that sociological concepts and theories are not merely "definitions" but rather tools that can organize information, guide the way ideas are presented, and direct and clarify thinking about a broad range of topics

3. Presenting the sociological concepts and theories related to each chapter as *interdependent ideas* that have the power, when considered together, to drive a coherent discussion of almost any social phenomenon or issue

4. Expanding the application of sociological concepts and theories to events taking place not only within the United States but outside its borders and beyond any one country's boundaries.

With this new model in mind, I decided to structure each chapter so that it incorporated information about life in a particular country and applied sociological concepts and theories to an issue related to global interdependence. Consequently the topic of each chapter is presented with an emphasis on one of the following countries or geographical entities: Mexico, Japan, South Korea, Israel including the West Bank and Gaza, Zaire, India, the People's Republic of China, South Africa, Germany, the former Yugoslavia, Brazil, the United States, Lebanon, and the United States and Russia in the aftermath of the Cold War.

The choice of country for each topic is arbitrary in some ways because sociological concepts and theories can be applied to life in any society. Some countries, however, lend themselves to a particular chapter topic, in part because they speak to a particularly significant global issue. For example, Zaire is emphasized in Chapter 6 on "Social Interaction and the Social Construction of Reality" and the global issue of AIDS. I chose this central African country for two important reasons. First, focusing on Zaire points to evidence that HIV existed as early as 1959. This evidence has been traced to an unidentified blood sample frozen in that year and stored in a Zairean blood bank. Although this fact hardly proves that the AIDS virus originated in Zaire, the hypothesis of a Zairean connection has received considerable support from government and health officials in Western countries. Whether or not Zaire is actually the country where HIV originated, however, is irrelevant to the message of this chapter. Far more important is the idea that *reality is a social creation*. That is, people give meaning to events, traits, and objects. Second, focusing on Zaire and its relationship to other countries helps us connect the transmission of HIV to a complex set of intercontinental, international, and *intra*societal interactions. These interactions concern the unprecedented levels of international and intercontinental air travel of legal and illegal migrations from villages to cities and from country to country.

Similar considerations apply to each choice of which country to emphasize in a given chapter. Throughout the text, however, students are reminded that the concepts being discussed are not specific to the country being used to illustrate them. Frequent comparisons to U.S. society reinforce this idea.

Changes to Second Edition

The second edition includes some important changes. Among the most important changes are new chapters on race and ethnicity (with emphasis on Germany) and on gender (with emphasis on the former Yugoslavia) that allow an expanded discussion of issues related to social stratification. Of course the text has been thoroughly updated and approximately 25 percent of the book incorporates new material. I made the revision with an eye toward further clarifying the meaning of global interdependence, partly to demonstrate how the lives of people around the world are closely intertwined but especially to show even more clearly how the people in the United States are touched by, and touch the lives of, people from other countries.

I also revised the first edition to ease some reviewers' concerns that emphasizing life outside the United States sacrifices readers' understanding of life inside the United States. In this second edition I address this concern directly by incorporating material showing how the lives of people who live within the geographical boundaries of the United States are affected profoundly by factors that transcend those boundaries. In addition, I emphasize that insights about everyday behaviors and ways of organizing social life come from contrast. When people are exposed to different ways of thinking and behaving, they not only learn about another way of life but also gain important insights into their own ways. Each chapter ends with a focus essay that applies to one or more concepts introduced in that chapter to some aspect of life within the United States.

The tables, graphs, charts, photographs, and other illustrations have been revised to enhance the overall impact of sociology's concepts and theories. In addition I have worked to make the material presented in the text more cohesive and to make sociological concepts stand out more as the organizing tools driving the discussion. I have also added interconnections data with improved maps at the start of each chapter to give some sense of the extent to which the United States is connected to the world in general and to specific countries. I hope the overall effect of this change gives readers a clear idea of how the global perspective is a natural outgrowth of thinking sociologically.

Ancillary Materials

The Instructor's Manual contains a chapter outline in question and answer format, background notes on each country, overheads, and ideas and suggestions about how to present the material covered in each chapter of the textbook. Test Items written in collaboration with Robin Franck of Southwestern College (published separately) include all variety of questions (multiple-choice, true-false, and essay). We tried to create questions that test conceptual and application skills rather than the ability to memorize and recall small facts. The Study Guide contains study questions, concept applications, applied research questions, practice test questions, and a section related to continuing education. The latter was written with the hope that students will want to continue to learn about life in another country even after the course has ended.

Acknowledgments

I began to revise *Sociology: A Global Perspective* in February 1993 and completed the job in May 1994. The revision builds on the efforts of those who helped me with the first edition. One person to whom I am particularly indebted from that time is Sheryl Fullerton, the sociology editor who agreed to publish the book. I signed with Wadsworth because I was impressed with Sheryl's sensitivity to my vision of what I believed a textbook should be, and by her critique of the sample chapters I submitted to her.

For the second edition, I was fortunate to work with a team of professionals, some of whom had collaborated with me on the first edition. Many of their names are listed in a most unassuming manner on the copyright page of this text. My team included John Bergez (senior developmental editor), Stephen Forsling (photo researcher), Angela Mann (senior production editor), Kelly Murphy (art editor), Maggie Murray (developmental editor), Andrew Ogus (senior designer), and Robin Kelly (copy editor). Special thanks go to Cathy Linberg, who filled in for Angela Mann while she was on maternity leave (Congratulations Angela!). Cathy was an amazingly quick study. Although she was assigned the book midway through its production, she quickly familiarized herself with it and then proceeded to coordinate the various departments involved with its production and to attend to an enormous number of details in a highly efficient and conscientious manner.

Each person on my team played a major role in the production of this book. I cannot overstate the amount of support I received from each of them at all stages of the writing and production process. For the most part, my experiences do not support the image that many academics seem to hold about the author-publisher relationship—that if a choice must be made, academic integrity is sacrificed to profit (see special issue of *Teaching Sociology* on textbooks, July 1988). Our relationships were guided by my vision of what a textbook should be. In fact, I believe that my working relationships with the team members represent an ideal collaborative model demonstrating how ideas should be considered, exchanged, revised, and refined.

I give special thanks to Serina Beauparlant, with whom I have worked continuously since August 1988. During this time we established an excellent working relationship as well as a friendship. The usual adjectives ("sensitive," "dedicated," "competent," "conscientious," "bright," and so on) fall short of capturing her professional and personal qualities. Serina is one of a kind; she breaks the mold. I see her not merely as a publisher, but an editor who has a vision for the discipline, which she acquired through thoughtful conversations with a broad range of authors and other sociologists who teach at two-year, four-year, and graduate-level institutions. In June 1994 Serina left Wadsworth to become an editor at another company. I will miss her steady guidance and influence on this project.

I am also grateful to Susan Shook, recently promoted to assistant editor, for finding and contacting the following people, who reviewed the revision: John Brenner, Southwest Virgina Community College; Carol Copp, California State University, Fullerton; Joseph Drew, University of the District of Columbia; Lynn England, Brigham Young University; Juanita Firestone, University of Texas; Patricia Gadban, San Jose State University; T. Neal Garland, University of Akron; Kreutzer Garman, Dekalb College; Thomas Gold, University of California, Berkeley; Janet Hilowitz, University of Massachusetts; Jen Hlavacek, University of Colorado; Gary Hodge, Collin County Community College; Moon Jo, Lycoming College; Nancy Kleniewski, SUNY Geneso; Sunil Kukreja, University of Puget Sound; Martin Marger, Michigan State University; Michael Miller, University of Texas, San Antonio; Bronislaw Misztal, Indiana University/Purdue University at Fort Wayne; Senno Ogutu, Diablo Valley College; Kristin Park, Emporia State University; James Simms, Virginia State University; Behavan Sitaraman, University of Alabama in Huntsville; William Smith-Hinds, Lock Haven University; Richard Sweeny, Modesto Junior College; Susan Tiano, University of New Mexico; Walter Veit, Burlington Community College; Susan Walter, University of Central Oklahoma; Leslie Wang, University of Toledo; Richard Wood, De Anza College. Their thoughtful, constructive, and insightful comments are deeply appreciated.

I was fortunate to work on the revision with two developmental editors, John Bergez and Maggie Murray, from whom I received the highest quality feedback. Both John and Maggie considered and synthesized the reviewers' comments and devised a plan to revise the book in light of their critiques. Maggie was also my developmental editor for the first edition. I credit her for helping me to find the writing structure I needed to accomplish the goals outlined above. John coordinated the writing, production, photo, and art programs of the second edition in such a way that all the parts came together as a coherent whole so that every aspect of the book reflected the overall vision and goals. In addition, he wrote most, if not all, the captions that accompany the photographs included in each chapter. I learned from him that photographs function as more than breaks from the reading material. Rather, they can play an important pedagogical role if they are selected with care and the captions are written with the goal of synthesizing and enhancing the ideas covered.

Since February 1993 I have had the privilege of working with Renée D. Johnson (Northern Kentucky University, Class of 1994), my principal research assistant on this revision. Renée tracked down the research articles, books, and other documents I needed to write the revision. She did an outstanding job of finding the materials I needed, sometimes only with the vaguest of clues. She entered revisions into the computer and she maintained my files. I depended on Renée to comment on the effectiveness of my examples, reasoning, and writing from the viewpoint of the student reader. When I was under pressure to meet deadlines she worked willingly on weekends and evenings. Our relationship was truly collaborative because we learned and are still learning from each other. As a result of talking and working with her, and observing her in interaction with others, I was influenced in (and even changed) the ways I approached and framed some topics presented in this textbook.

I also appreciate the help of librarians at the University of Cincinnati, the Public Library of Cincinnati, and especially Northern Kentucky University. The NKU library faculty handled my endless requests for information in the most professional manner. For this I thank Nancy Campbell, Allen Ellis, Mary Ellen Elsbernd, Rebecca Kelm, Laura Sullivan, Sharon Taylor, Emily Werrell, and Theresa Wesley. Phil Yannerella the curator of government documents and his staff, Judy Brueggen and Elaine Richards, deserve special thanks because almost everyday they spent considerable time helping me to find the materials I needed.

I thank Karen Feinberg, on whom I relied to copyedit my manuscript before I sent it out for review. Karen and I have worked together in some capacity for about 10 years and I continue to be impressed by her unique style of editing. In addition, I thank my colleagues at Northern Kentucky University who contributed written materials to this text: Dr. Prince Brown for "Why Race Makes No Scientific Sense: The Case of African and Native Americans"; Dr. Sharlotte Neely for "Forced Culture Change of Native Americans by Bureau of Indian Affairs"; Dr. James Hopgood for "What Would Durkheim Say About James Dean?" and for several photographs included in Chapters 8 and 14; John Huss (Class of 1993) for "Peer Pressure as an Agent of Socialization"; Renée Johnson (Class of 1994) for "Journal Entry"; Yvonne Dupont and Karen Meyers for "The Problems of Value Rational

Action: The Case of the Hanford Nuclear Reservation;" and Lori Ling (Class of 1990) for "Tooning In to a Tiny Market." Kevin Steuart (Class of 1994) contributed "What is NAFTA?" and collaborated with me on writing "Telephone Calls as a Measure of Interconnection." In addition, Kevin found and compiled the data on interconnection that begin each chapter. As a result of his time and persistence, I was better able to illustrate the extent and essence of global interdependence.

My mother, Annalee Taylor Ferrante, handled the permissions, proofed quotes, and checked the accuracy of the references. As I mentioned in the preface to the first edition, this aspect of the research and writing process is considerably undervalued. While working on the first and the second editions I have come to recognize that my responsibility as a textbook author is to present the accomplishments of the discipline and to outline the ideas that bind together the people who call themselves sociologists. Therefore it is essential that the classic and contemporary writers—sociologists and nonsociologists alike—be quoted accurately and that references citing their work be accurate to aid readers who wish to find the original statements. Annalee Ferrante handled this job with the care and critical eye worthy of the great sociologists and other scholars from which I have drawn. If there are any inaccuracies with regard to the references and quotes it is because I failed to alert her to changes I made during my many drafts.

In closing I wish to acknowledge, as I did in the first edition, the tremendous influence of Dr. Horatio C Wood IV, M.D. (Federal Medical Center, Lexington, Kentucky), on my philosophy of education. I consider him my mentor. Finally, I wish to express my love for the most important person (still, no doubt about it) behind this project, my colleague and husband, Robert K. Wallace. I dedicate this book to the memory of my father, Phillip S. Ferrante, and to my mother and best friend, Annalee Taylor Ferrante.

SOCIOLOGY

A Global Perspective

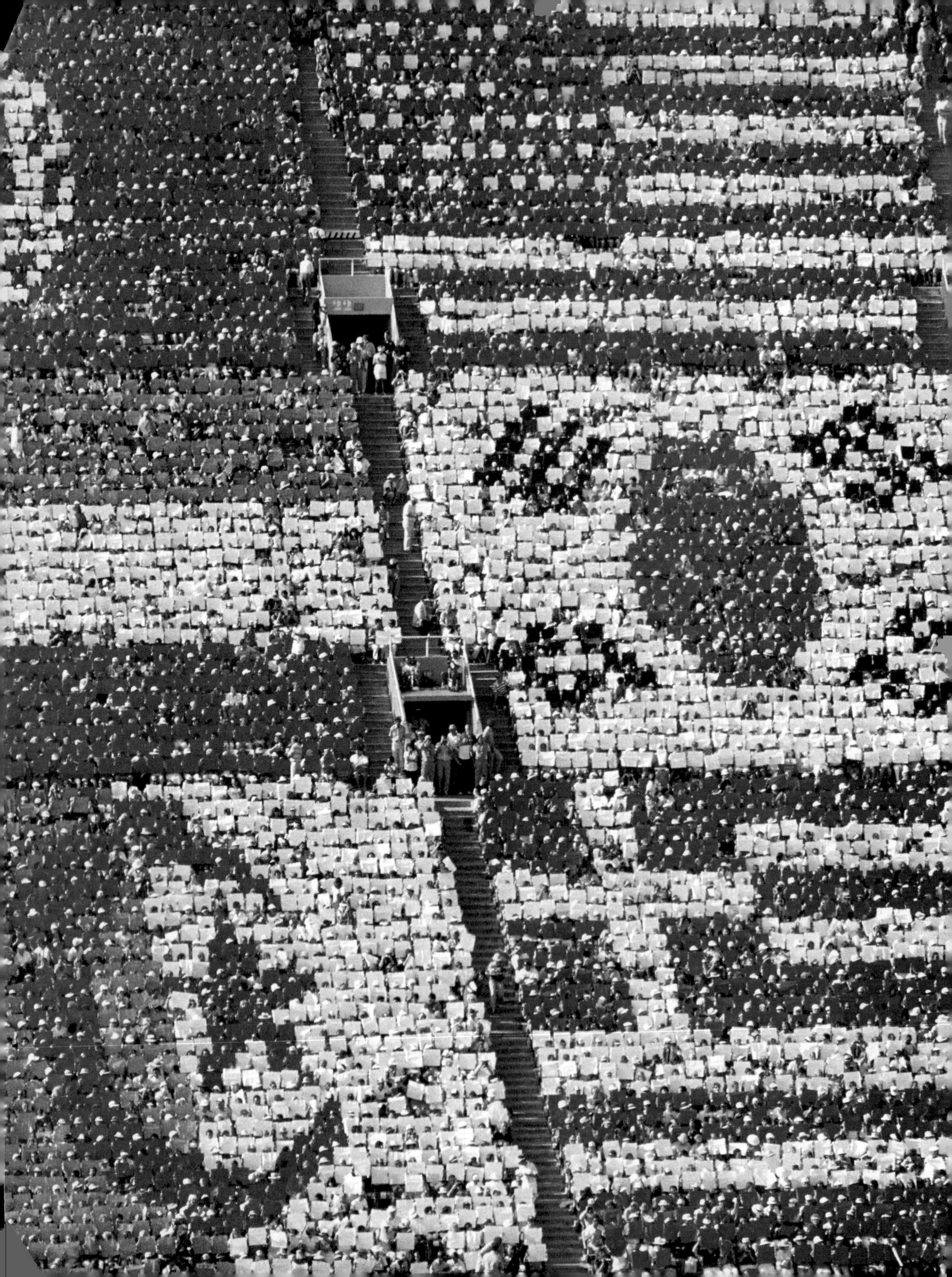

I THE SOCIOLOGICAL IMAGINATION

Opening ceremonies of the 1984 Olympics, Los Angeles.
David Burnett/Contact Press Images

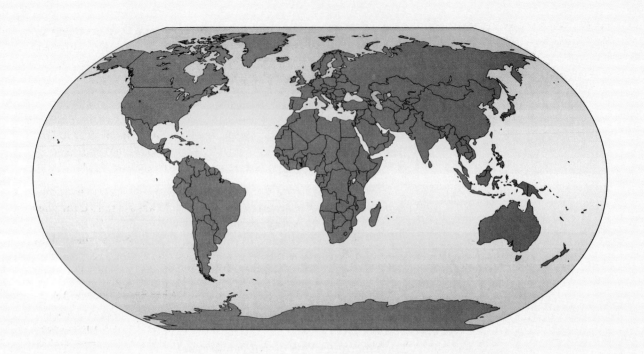

Foreign students enrolled in U.S. colleges for fall 1991	420,000
Doctoral degrees conferred by U.S. colleges to non-U.S. citizens in 1991	10,636
People living in the United States in 1990 who were born in other countries	21,632,000
Airline passengers flying in and out of the United States in 1991	68,750,337
People employed by foreign affiliates in the United States in 1991	4,705,300
Applications for utility patents filed in the United States in 1991 by inventors from foreign countries	76,351
Phone calls into and out of the United States in 1991	2,279,150,000
U.S. military personnel stationed in other countries, ashore and afloat, in 1993	324,864

Note: The most recent data available has been used in all chapters. See the Appendix for references for this data.

On the first day of class each semester, I routinely ask students in my Introduction to Sociology classes if they feel as if they have been touched personally in any way by global interdependence. Here are some of their responses:

"My grandmother worked at the Revlon factory in Cincinnati. About three years ago, Revlon moved all of its factories to Florida or Mexico where labor costs were cheaper. My grandmother had worked there for over 25 years. The only good thing that came out of this was that the company gave her retirement benefits."

"The company I work for—EG&G Technologies—has several companies throughout the world. Job listings are always posted for work in many of these foreign countries. In some cases the jobs require not only technical knowledge in a particular field but language skills and knowledge about how business operates (in other words, the business customs) in those countries."

"My father works for the government. He has been with the U.S. Postal Service for 22 years. He has won many management awards and has taught seminars about his specialties. He was turned down for a promotion that included traveling to different countries and inspecting other postal facilities—not because of his knowledge of postal operations, but because he didn't know a foreign language."

"One time I applied for an entry-level position with a major airline. I stated on my application that I had taken three years of French during high school. When a personnel manager contacted me, she asked if I was fluent in the language. After I told her with honesty that I didn't feel comfortable with the language training I had received, her views about my qualifications changed."

"At my job, I answer many phone inquiries from foreigners. I studied Spanish and thought I knew enough to get by. I had so many problems communicating with Spanish-speaking people over the phone. I couldn't understand a word they were saying. This was the first time I came into any contact with global interdependence. I realized that I didn't know the language as well as I thought I did."

"My entire family has been affected by global interdependence. My father is a General Motors employee. For over 20 years, he was employed at the Norwood plant, but three years ago it was shut down because of the lack of sales resulting from foreign competition. My father was laid off for two years. Then he had to transfer to the plant in Fort Wayne, Indiana. Although we've been very fortunate, I know many families that have suffered greatly."

"The construction company my father worked for is based in Ohio. The owners decided there were more job opportunities in Russia. They closed the construction company to go there."

In that particular sociology class, 60 of the 80 students (75 percent) answered the question about being touched by global interdependence in the affirmative—even though we hadn't yet discussed what global interdependence means or what it has to do with sociology.

Whenever I read the students' personal accounts of global interdependence, I am reminded of a distinction made by sociologist C. Wright Mills (1959) between troubles and issues. **Troubles** are private matters that can be explained in terms of personal characteristics (motivation level, mood, personality, or ability) or immediate relationships with others (family members, friends, acquaintances, or co-workers). The resolution of a trouble, if it can be resolved, lies in changing an individual's character or immediate relationships. Mills stated that "when, in a city of 100,000, only one man is unemployed, that is his personal trouble, and for its relief we properly look to the character of the man, his skills, and his immediate opportunities" (p. 9). **Issues,** on the other hand, are public matters that can be explained by factors outside an individual's control and immediate environment. Mills writes, "But when in a nation of 50 million employees, 15 million men are unemployed, that is an issue, and we may not hope to find its solution within the range of opportunities open to any one individual. The very structure of opportunities has collapsed" (p. 9).

The connection between troubles and issues is fundamental to what sociology is all about. As we will see, perceiving this connection is central to the quality of **sociological imagination**—the ability to link seemingly impersonal and remote historical forces to an individual's life. In today's world, having sociological imagination entails an understanding of global interdependence.

Troubles Versus Issues

My students' responses make clear that the situations they describe cannot be explained simply in terms of personal situation and character. No individual response—whether finding a new job, learning a foreign language, or transferring to another city—will control the historical force governing the particular situation to which the individual is responding. That force is **global interdependence.** In the most general sense, global interdependence is a state in which the lives of people around the world are intertwined closely and in which any one country's problems—the search for employment, drug abuse, environmental pollution, the acquisition and allocation of scarce resources—are part of a larger global situation (Cortes 1983). Although few people grasp completely the consequences of this trend (Craig 1987), newspaper and magazine headlines hint at this trend. Sample headlines read: "The Coming Global Boom," "Hollywood Takes to the Global Stage," "Global Goliath: Coke Conquers the World," "Students Need a Global View," "Small Companies Go Global," "A Global Program for the Environment," "Science Fiction News Reality: Pocket Phones for Global Calls," and so on. All of these headlines have a perspective of global interdependence.

Mills argues that people often feel "trapped" by issues because, even though they respond to their personal situations, they remain spectators to the larger forces that affect their lives. In other words,

The increasing number of advertisers who use the word "global" to describe the reach of their products and services is just one indication that global interdependence is already a reality for today's business.

Courtesy Sanwa Bank, Asian Advertisers, Inc.

they feel helpless because they have no control over these forces. When a war breaks out, for example, an ROTC student becomes a rocket launcher, a teenager buys and wears a sweatshirt that displays his or her support (or lack of support) for the war, an automobile owner thinks about the role of oil in the war while pumping gas, and a child whose parents are in the military has to live without them for a while. "The personal problem of war, when it occurs, may be how to survive it or how to die in it with honor; how to make money out of it; how to climb into the higher safety of the military apparatus; or how to contribute to the war's termination" (Mills 1959, p. 9). The *issues* of war, on the other hand, concern its causes: for example, a threat to

the resources (such as oil) on which a country depends, or a threat to a country's allies.

Likewise, when the CEOs of U.S.-based corporations decide to move some of their daily operations to a foreign country (see Figure 1.1) or to downsize their operations, the laid-off employees file for unemployment, return to college, open a business, take on third-shift work, or choose early retirement. The personal problems connected with this kind of economic restructuring may include finding a way to live on a reduced budget (see "How to Live with Less"), fighting drowsiness while driving home from third-shift work, or trying to stay cheerful while hunting for a job. Such responses do not change the issue of global interdependence,

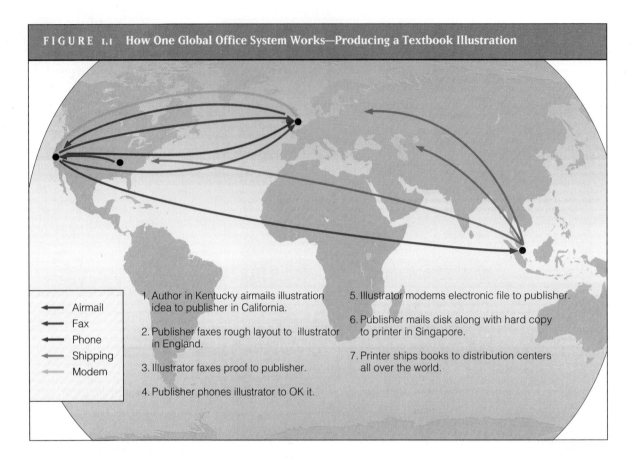

FIGURE 1.1   How One Global Office System Works—Producing a Textbook Illustration

Airmail
Fax
Phone
Shipping
Modem

1. Author in Kentucky airmails illustration idea to publisher in California.

2. Publisher faxes rough layout to illustrator in England.

3. Illustrator faxes proof to publisher.

4. Publisher phones illustrator to OK it.

5. Illustrator modems electronic file to publisher.

6. Publisher mails disk along with hard copy to printer in Singapore.

7. Printer ships books to distribution centers all over the world.

which is determined largely by an economic system that measures success in terms of profit. Profit is achieved by lowering production costs, hiring employees who will work for lower wages, introducing labor-saving technologies (computerization), and moving productions out of high-wage zones inside or outside the country. Such strategies leave workers vulnerable to unemployment.

Mills asks, "Is it any wonder that ordinary people feel they cannot cope with the larger worlds with which they are so suddenly confronted?" (pp. 4–5). Is it any wonder that they cannot understand the connection between these forces and their own lives, that they turn inward and try to protect their own sense of well-being, that they feel trapped? Mills believes that to gain some sense of control over their lives, people need more than information, because we live in a time when information dominates our attention and overwhelms our capacity to make sense of all we hear, see, and read in

a day. Thus, we may be exhausted by the struggle to learn from the information about the forces that shape our daily lives.

According to Mills, people need "a quality of mind that will help them to use information" in a way that they can think about "what is going on in the world and of what may be happening within themselves" (p. 5). Mills calls this quality of mind the sociological imagination. Those who possess the sociological imagination can view their inner life and human career in terms of larger historical forces. The payoff for those who possess this quality of mind is that they can understand their own experiences and "fate" by locating themselves in history; they can recognize the responses available to them by becoming aware of all of the individuals who share the same situation as themselves. Mills maintains that knowing one's place in the larger scheme of things is "In many ways . . . a terrible lesson; in many ways, a magnificent one" (p. 5).

How to Live with Less

The personal problems connected with unemployment and living on a reduced budget can be seen in the following example based on a real family's experience.

In March 1992, Jonathan Weber lost his $54,000-a-year-job as director of marketing at a software engineering company. He was paid for two weeks of unused vacation but got no severance pay and was ineligible for unemployment insurance. He has a wife and three children.

Mr. Weber, who has an MBA degree, has not been able to find work since leaving the company. The family has been forced to make tight cutbacks and live on Mrs. Weber's $29,000-a-year salary, plus $4,000 she earns from a weekend job she started in November 1992. This is how they manage while Mr. Weber continues to look for a job.

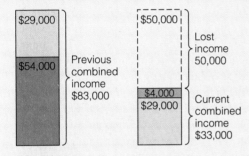

1992

March
Mr. Weber lost his $54,000-a-year-job.
Suspended immediately:
Long-distance telephone calls
Savings: $200 a month
Eating in restaurants
Savings: $400 a month
Weekend trips
Savings: $1,200 a year

May
Mr. Weber cashed in his $9,000 retirement fund.

June
Eliminated use of backyard pool.
Savings: $500 a year

July
Reduced minimum monthly payments with credit card holders.
Spread debt over four months with utility companies.

September
Cut two daughters' preschool attendance from five days a week to three half-days; Mr. Weber looks after them for the rest of the time.
Savings: $350 a month

November
Mrs. Weber took a part-time weekend job that paid $335 a month ($4,000 a year).

December
Reduced Christmas gifts from $1,600 the year before to $400.
Savings: $1,200

1993

January
Cancelled cable TV service.
Savings: $1,200 a year
Cancelled health club subscriptions.
Savings: $800 a year

June
Cancelled fall-term piano lessons for all three children.
Savings: $540 over six months

October
The Webers have managed to reduce spending by $16,800 a year but their income has shrunk even more. They keep about $28,500 of their $33,000 income after taxes and other deductions. They have kept mortgage payments current and still manage to buy food and other necessities by depleting their savings. But now they will either have to negotiate reduced or stretched out payments with their lender.

This knowledge can be a terrible lesson because we learn that human agony, hatred, selfishness, sadness, tragedy, and hopelessness have no limits. It can be a magnificent lesson because we learn that human dignity, pleasure, love, self-sacrifice, and effort also are unlimited. We will "come to know that the limits of 'human nature' are frighteningly broad." We will "come to know that every individual lives, from one generation to the next, in some society; that [people] live out a biography, and that [they live] it out within some historical sequence" (p. 6). Yet, if only by living in a society, people shape that society, however minutely, even as the society shapes them.

To understand this idea, consider the case of Philip Voss, a 48-year-old chief executive of Zatos International, an American subsidiary of the Japanese cosmetic giant, Shiseido. Since he joined the company in December 1989, Voss has taken weekly Japanese-language lessons. His decision to learn Japanese was a response to the "push and shove" of global interdependence, which he experienced on a personal level whenever he attended board meetings. "I'd be sitting with Shiseido executives in meetings, and they'd all be talking [in Japanese] among themselves and I'd have no idea what they were saying. You know they could have been saying, 'That Phil Voss, what a jerk,' for all I know." This response—the decision to learn Japanese—helped Voss feel less isolated, as he makes clear in his account of a speech he gave to top Japanese executives. "It was just a little speech—'I'm so glad to be here and meet you all'—that kind of thing. But I was nervous. And when I had finished it, they all rose from their seats and applauded me. I was thrilled" (Fanning 1990, p. F21). Yet, at the same time, Philip Voss's personal response became part of the "push and shove" of global interdependence. In other words, the sum of everyone's personal responses becomes what global interdependence is.

Those who possess the sociological imagination can grasp history and biography and the connections between the two. If we understand the forces shaping our lives, we are able to respond in ways that benefit our own lives as well as those of others. There is a striking difference between people who are able to locate their lives in history and those who are not. These who cannot make the connection are unlikely to know how to respond effectively to a world in which the lives of people around the globe are interconnected and in which one society's problems are part of a larger global problem. Almost by definition, such people lag behind those who are planning to study abroad, learn a foreign language, and/or read materials that enhance their understanding of other cultures.

Sociology offers a perspective that permits people to see the "push and shove" of history. It does not promise, however, that people can respond easily and in constructive ways if they understand the connection between their biographies and the larger forces, especially if they lack resources and the support of people who care. On the other hand, the sociological imagination decreases our chances of responding in inappropriate or misguided ways.

Consider the following example. The union members at a factory in Cincinnati allow only the owners of so-called American-made automobiles to park in the factory parking lot; members who drive foreign cars must park on the street. Similarly, in 1990, William Ford, a U.S. congressman from Michigan, decreed that some parking spaces near the capitol be reserved for staff members who own American automobiles (Newhouse News Service 1990; see "'Buy-American' Quandary"). Such responses are intended to encourage loyalty to U.S.-made products. But, in fact, this policy is misguided because there is no such thing as a car made in the United States. No country produces all of the parts needed to assemble an automobile. Furthermore, all of the major automakers in the world employ workers in countries other than their home countries. "One of the ironies of our age is that an American who buys a Ford automobile or an RCA television is likely to get *less* American workmanship than if he bought a Honda or a Matsushita TV" (Reich 1988, p. 78). As early as 1929, Ford Motor Company was a multinational corporation with 21 plants located in Europe. At that time General Motors had 16 plants in Europe (Marshall and Boys 1991). These historical facts offer a greater context for the misguided "buy American" argument.

The connection between biography and history becomes most evident to people when they can grasp two interrelated concepts—social relativity and the transformative powers of history.

"Buy-American" Quandary

With many products containing parts from around the world, it is often difficult to determine what is an "American-made" item. Los Angeles bureaucrats may have their hands full complying with ordinances being drafted under Charter Amendment G, which allows the city to give preference to products made in California or Los Angeles County and to require that a certain proportion of parts be made in America.

Electronic Typewriters

• **Brother:** Assembled in Tennessee from imported and U.S. parts by Nagoya, Japan-based Brother.
• **Smith Corona:** Made in New York and Singapore from U.S. and foreign parts. New York plant to be moved to Mexico. Smith Corona 47% owned by British firm Hanson.

Automobiles

• **Honda Accord:** Manufactured in Ohio from 75 percent U.S. parts.
• **Dodge Stealth:** Made in Japan by Mitsubishi.

Computers

• **Panasonic CF-370 notebook:** Manufactured in Texas by Ft. Worth–based Tandy for Osaka, Japan–based Matsushita, which sells in the United States under the Panasonic brand. Sixty percent of parts are imported, mostly from Japan.
• **Apple Powerbook:** One model made in San Diego by Tokyo-based Sony; another made by Apple in Ireland; another made in Colorado. All use substantial Japanese parts, but the company won't reveal the percentage.

SOURCE: "An American Dilemma: Typewriter Wars Point Up Problem of Defining U.S. Goods," by Susan Moffat. Copyright © 1992 by Susan Moffat. Reprinted by permission of the author and the *Los Angeles Times*.

When U.S. consumers buy an Apple Powerbook, they cannot be sure whether it was made by Irish workers, by U.S. workers in Colorado, or by U.S. workers employed at the Tokyo-based Sony plant in San Diego.
Courtesy Apple Computer, Inc.

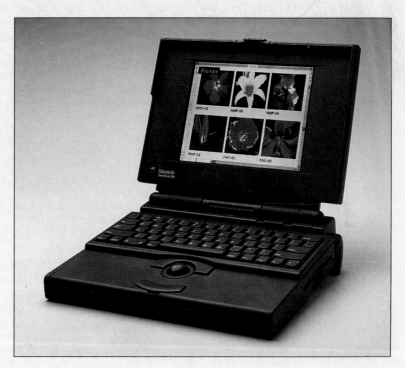

Social Relativity

Social relativity is the view that ideas, beliefs, and behavior vary according to time and place. People who have lived in more than one culture can teach us much about social relativity. Consider the excerpts on the next page from diaries kept by college students in my classes who were asked to interview foreign students about their adjustment to American life:

> I asked François (from France) what was the biggest adjustment he had to make coming to America. He answered in this way. When he came to America he took a job. The first day on the job he expected to take a lunch break around 12 o'clock. What seemed like several hours passed before his boss told him to take a half-hour lunch break. Lunch is not taken so lightly in France. François was accustomed to a larger meal and to taking as long as two hours to eat. (DIANA MATT)

> I asked Irma if she had any problems readjusting to her native culture when she visited Indonesia. She said that she did have problems. Irma mentioned that in Indonesia it is customary to drop in on a relative or friend without advance notice. In the United States, on the other hand, she had to learn to call people first. When she was home over Christmas break, she began dialing the phone number of a relative to let the person know that she and her mother were coming over. Her mother asked her why she was behaving so strangely. Irma replied that she wanted to check to see if this was a convenient time. Her mother thought Irma was going crazy. In Indonesia calling before a visit is equivalent to asking someone if they will go out of their way to make special preparations. Indonesians view such a request as rude. (KAREN MEYERS)

> I asked Hannah what were some of the hardest things for her to get used to since coming to the United States. She explained that in Jordan a girl would never introduce a boy as a boyfriend. If that were to happen, the girl would be disowned not only by the family but by the whole society. In general, the nature of family ties is very different in the two countries. In Jordan people do not always do what they would like. They have a deep respect for family, as do most people in Japan. This respect is so strong that family comes first and individual achievement second. Hannah described how marriage partners are chosen. The first and most important consideration is not the marriage partner per se, but the person's family. If someone was from a bad family, you would not marry the person for two reasons. First, the prevailing assumption is that if the family is bad, so is an individual raised in that family. Second, you would not disgrace the family by marrying a person of questionable character. (JIM EZELL)

These accounts show that ideas about eating, visiting relatives, and dating vary across regions and cultures. Furthermore, they help us to recognize and to understand that many of our ideas and behaviors are not personal responses but are products of the environment into which we were born.[1]

Transformative Powers of History

According to the concept of the **transformative powers of history**, the most significant historical events have dramatic consequences on people's thinking and behavior. This concept, like social relativity, sees events in the context of time and place. To understand the transformative power of an event, a person must have some idea of the way people thought or behaved before and after that event.

Consider the tremendous transformative power of the Industrial Revolution of the nineteenth century. Before the Industrial Revolution, people's lives varied little from one generation to the next. Thus, parents could assume that their children's daily lives would be much like their own and that their children would face the same challenges and problems that they faced. Consequently, societies embarked on long-term projects that extended to subsequent generations. For example, in the Middle Ages people built cathedrals that took several lifetimes to complete, "presumably never doubting that such edifices would be used and appreciated by their great-grandchildren when construction was complete" (Ornstein and Ehrlich 1989, p. 55).

During the Industrial Revolution, however, the pace of change increased to such a degree that parents could no longer assume that their children's lives would be like their own. "Imagine the reactions of an American today if asked to contribute to a building that would take 150 years to finish: 'We don't want to tie up our capital in something with no return for one hundred and fifty years!' 'Won't a new design and construction process make this one obsolete long before it's finished?'" (Ornstein and Ehrlich 1989, p. 55).

The sociological imagination permits those who adopt it to see how biography and history intersect—to understand that even some of the most personal experiences are shaped by time and place and are shaped by the transformative powers of history. This type of imagination is characteristic of all the great sociologists, including three of the most influential: Emile Durkheim, Max Weber, and Karl Marx. All had a tremendous impact on the discipline of sociology; all spent much of their professional careers in an attempt to understand the nature and consequences of one event—the Industrial Revolution.

The Industrial Revolution

Between 1492 and 1800 an interdependent world began to emerge. Europeans learned of, conquered, or colonized much of North America, South America, Asia, and Africa and set the tone of international relations for centuries to come. During this time colonists forced local populations to cultivate and harvest crops and to extract minerals and other raw materials for export to the colonists' home countries. When the Europeans' labor needs could not be met by indigenous populations, they imported slaves from Africa or indentured workers from Asia.

In light of colonization, one can argue that the world has been interdependent for at least 500 years. The scale of social and economic interdependence changed dramatically, however, with the Industrial Revolution, which gained momentum in England in about 1850 and soon spread to other European countries and to the United States. The Industrial Revolution drew people from even the most remote parts of the world into a process that produced unprecedented quantities of material goods, primarily for the benefit of the colonizing countries.

One fundamental feature of the Industrial Revolution was **mechanization,** the addition of external sources of power such as oil or steam to hand tools and to modes of transportation. This innovation turned the spinning wheel into a spinning machine, the hand loom into a power loom, and the black-smith's hammer into a power hammer. It replaced wind-powered sailboats with steamships, and horse-drawn carriages with trains. On a social level it changed the nature of work and the ways in which people interacted with one another.

The Nature of Work

Prior to the mechanization brought on by the Industrial Revolution, goods were produced and distributed at a human pace, as illustrated by the effort required to bake bread or to make glass:

- *Bakeries produced bread almost entirely by manual labor, the hardest operation being that of preparing the dough, "usually carried on in one dark corner of a cellar, by a man, stripped naked down to the waist, and painfully engaged in extricating his fingers from a gluey mass into which he furiously plunges alternately his clenched fists."*

- *In glassmaking, everything was done by hand, "the gatherers taking the metal from the furnace at the end of an iron rod, the blower shaping the body of the bottle with his breath, while the maker who finished the bottle off . . . tooled the neck with a light spring-handled pair of tongs. Each bottle was individually made no matter what household, shop, or tavern it was destined for."* (ZUBOFF 1988, PP. 37–38)

Workers paid a price for the reduction of the physical requirements necessary to produce goods, because skills and knowledge were intertwined with the physical effort. Before mechanization knowledge was

> inscribed in the laboring body—in hands, fingertips, wrists, feet, nose, eyes, ears, skin, muscles, shoulders, arms, and legs—as surely as it was inscribed in the brain. It was knowledge filled with intimate detail of materials and ambience—the color and consistency of metal as it was thrust into a blazing fire, the smooth finish of the clay as it gave up its moisture, the supple feel of the leather as it was beaten and stretched, the strength and delicacy of glass as it was filled with human breath. (ZUBOFF 1988, P. 40)

Industrialization transformed individual workshops into factories, craftspeople into machine operators, and hand production into machine production. Products previously designed as unique entities and assembled by a few people were now standardized and assembled by many workers, each performing only one function in the overall production process. This division and standardization of labor meant that no one person could say, "I made this; this is a unique product of my labor." The artisans yielded power over the production process to the factory owner because their skills were rendered obsolete by the machines. Now people with little or no skill could do the artisan's work, and at a faster pace.

The Nature of Interaction

Between 1820 and 1860 a series of developments—the railroad, the steamship, gas lighting, running water, central heating, stoves, iceboxes, the telegraph, and mass-circulation newspapers—transformed the ways in which people lived their daily lives. Although all of these developments were important, the railroad and the steamship were perhaps the most crucial. These new modes of travel connected people to one another in reliable, efficient, less time-consuming ways. These inventions caused people to believe they had "annihilated" time and space (Gordon 1989). They permitted

Lewis Hine's remarkable photograph, "The Steamfitter," exemplifies many people's concern about the molding of human beings to machines brought about by the Industrial Revolution.
International Museum of Photography at George Eastman House

people to travel day and night; in rain, snow, or sleet; across smooth and rough terrain.

Before the steam engine train was introduced, a trip by coach between Nashville, Tennessee, and Washington, D.C., took one month. Before the advent of the steamboat, a 150-mile trip on a canal took 32 hours (Gordon 1989). An overseas trip from Savannah, Georgia, to Liverpool, England, took months. Each year following the discovery of steam power, transportation became faster and more reliable. In 1819, the year of the first ocean crossing by a steam-driven vessel, the trip took one month. By 1881 the time needed to make that trip had been reduced to seven days (*The New Columbia Encyclopedia* 1975).

Railroads and steamships increased personal mobility as well as the freight traffic between previously remote areas. These modes of transportation facilitated an unprecedented degree of economic interdependence, competition, and upheaval. Now people in one area could be priced out of a livelihood if people in another area could provide goods and materials at a lower price (Gordon 1989).

Sociological Perspectives on Industrialization

Industrialization did more than change the nature of work and interactions; it affected virtually every aspect of daily life. "Within a few decades a social order which had existed for centuries vanished, and a new one, familiar in its outline to us in the late twentieth century, appeared" (Lengermann 1974, p. 28). The changes triggered by the Industrial Revolution, and especially by mechanization, are incalculable; they are still taking place.

Marx, Durkheim, and Weber all lived during the Industrial Revolution's most transformative decades. Although they wrote in the nineteenth and early twentieth centuries, their observations are still relevant today. In fact, many insights into the character of contemporary society can be gained by reading their writings, because those who witness and adjust to a significant event are intensely familiar with its consequences in daily life. Because most of us living today know only an industrialized life, we lack a contrast to help us understand how industrialization shapes contemporary existence. Consequently, the words of witnesses can be revealing, as the following anecdote suggests.

Recently a scientist was interviewed on the radio. He maintained that scientists were close to understanding the mechanisms that govern the aging process and that people might soon be able to live to be 150 years old. If the aging mechanisms in fact are controlled, the first people to witness the change will have to make the greatest adjustment. In contrast, people born after this discovery is made will know only a life in which they can expect to live 150 years. If these postdiscovery humans are curious, they may wish to understand how living to age 150 shapes their lives. To learn this, they will have to look to those who recorded life before the change and who described their adjustments to the so-called advancement.

So it is with industrialization: to understand how it shapes human life, we can look to some of the early sociologists, particularly Durkheim, Marx, and Weber. Because the Industrial Revolution (and all that it encompassed) was so important in the professional and personal lives of the early sociologists, the sociological perspective may be regarded as a "necessary tool for analyzing social life and the recurring issues that confront human beings caught up in the intensity and uncertainty of the modern era" (Boden, Giddens, and Molotch 1990, p. B1).

The Focus of Sociology

Because sociology emerged as an effort to understand the dramatic and almost incalculable effects of the Industrial Revolution on all phases of human life, the discipline is wide-ranging in its concerns. The value and the vastness of **sociology** are evident in the way we define the discipline (Bardis 1980). We might define sociology as the scientific study of the causes and consequences of human interaction.

Human interaction can include an encounter between strangers on an elevator, a conversation between a physician and a patient, the interrelations among teammates, and the relationships among people of a large organization. Although Marx, Durkheim, and Weber would agree with this definition, each thinker had distinct ideas about what sociologists should emphasize (Lengermann 1974).

Karl Marx (1818–1883)

Karl Marx was born in Germany but spent much of his professional life in London, working and writing in collaboration with Friedrich Engels. Two of Marx and Engels's most influential treatises are *The Communist Manifesto* and *Das Kapital. The Communist Manifesto,* a pamphlet issued in 1848, outlines the principles of communism, among other things. It begins, "A spectre is haunting Europe" and closes with these lines: "The workers have nothing to lose but their chains; they have a whole world to gain. Workers of all countries unite." *Das Kapital,* a massive multivolume work, is critical of the capitalist system and predicts its defeat by a more humane and more cooperative economic system, socialism. Marx's vision of the way to build a society as outlined in these treatises has profoundly influenced economic, social, and political life around the world, especially in the People's Republic of China, the former Soviet Union, and Eastern Europe—although Marx would not necessarily have approved of the effect.

According to Marx, the sociologist's task is to analyze and explain conflict, the major force that drives social change. The character of conflict is shaped directly and profoundly by the **means of production,** the resources (land, tools, equipment, factories, transportation, and labor) essential to the production and distribution of goods and services. Marx viewed every historical period as characterized by a system of production that gave rise to specific types of confrontation between an exploiting class and an exploited class. For Marx, class conflict was the vehicle that propelled people from one historical epoch to another. Over time, free men and slaves, nobles and commoners, barons and serfs, guildmasters and journeymen have confronted one another.

The appearance of machines as a means of production was accompanied by the rise of two distinct classes: the **bourgeoisie,** the owners of the means of production, and the **proletariat,** those who must sell their labor to the bourgeoisie. Marx expressed profound moral outrage over the plight of the proletariat, who at the time of his writings were unable to afford the products of their labor and whose living conditions were deplorable. In fact, every aspect of people's lives—their occupations, their religious beliefs, their level of education, their leisure activities—personifies the social class to which they belong. Marx devoted his life to documenting and understanding the causes and consequences of this inequality, which he connected to a fatal flaw in the organization of production (Lengermann 1974).

Emile Durkheim (1858–1918)

The Frenchman Emile Durkheim believed that the sociologist's task is to analyze and explain **solidarity,** the ties that bind people to one another, and the mechanisms through which solidarity is achieved. Just as Marx defined the means of production as a central concern to sociologists, Durkheim regarded solidarity as the essential concern. Durkheim observed that as society became industrialized, the nature of the ties that had bound people to one another changed in profound ways. Before the Industrial Revolution, people were drawn together because they were similar to one another: most people made their living by agriculture and "spent their whole lives within a few miles of where they had been born . . . [and] nearly all lived much as their parents and grandparents had lived before them" (Gordon 1989, p. 106). Order and stability existed in society because of this continuity across generations.

Durkheim maintained that a new form of solidarity based on differences and interdependence was characteristic of industrialized society. In such societies there is considerable specialization and the division of labor is complex. People are united by their inability to earn their livelihoods independent of others and by their need to cooperate with others to survive, even though they do not have emotional ties with most of the people they encounter in the course of a day.

Max Weber (1864–1920)

The German-born scholar Max Weber has had a monumental influence not only on sociology but also on political science, history, philosophy, economics, and anthropology. Although Weber recognized the significance of economic and material conditions in shaping history, he did not believe, as did Marx, that these conditions were the all-important forces in history. Weber regarded Marx's arguments as one-sided in that they neglected the interplay of economic forces with religious, psychological, social, political, and military pressures (Miller 1963).

According to Weber, the sociologist's task is to analyze and explain the course and the consequences of **social action**—actions that people take in response to others—with emphasis on the subjective meaning that the involved parties attach to their behavior. Certainly Weber's preoccupation with the forces that move people to action was influenced by the fact that he saw

> *the thrust behind sociological curiosity as residing in the endless variety of societies. Everywhere one looks one sees variety. Everywhere one finds [people] behaving differently. . . . Scrutiny of the facts can show endless ways of dealing with the problems of survival, and an infinite wealth of ideas. The sociological problem is to make sense of this variety. . . .*
> (LENGERMANN 1974, P. 96)

In view of this variety, Weber suggested that sociologists focus on the broad reasons why people pursue goals, whatever those goals may be (for example, to make a profit, to earn a college degree, or to be recognized by others). He believed that social actions could be classified as belonging to one of four important types: (1) traditional (a goal is pursued because it was pursued in the past), (2) affectional (a goal is pursued in response to an emotion such as revenge, love, or loyalty), (3) value-rational (a goal is pursued because it is valued, and it is pursued with no thought of foreseeable consequences and often without consideration of the appropriateness of the means chosen to achieve it), and (4) instrumental (a goal is pursued after it has been evaluated in relation to other goals and after thorough consideration of the various means to achieve it) (Abercrombie, Hill, and Turner 1988; Coser 1977; Freund 1968).

An example will help clarify the distinctions between the four types of action. Consider, for example, the goal of earning a college degree. If an individual pursues a college degree because everyone in her family for the past three ge[...] educated, the action can be classifie[...] If she pursues a degree for the love or p[...] of learning, the action is *affectional*. Her p[...] a college education is *value-rational* if she de[...] to attend college because potential employers valu[...] and demand a diploma. Sometimes the people who approach college in this way view it only as a ticket to apply for jobs; they lose sight of all but this narrow reason for earning a diploma. As a result, they may simply go through the motions in college and do only what they absolutely must do. They may take the easiest courses, have others write their papers, and skip classes while relying on others to take notes for them.

Instrumental action is more complex. Suppose the individual considers other goals before settling on the goal of earning a college degree. She might consider whether to travel and see the world, enlist in the military, or work a few years. Once she has chosen the goal, she might consider the various means to obtain it, including taking out a loan, living at home, enlisting in the army to obtain money for college, getting a job and attending school at night, and so on. The individual also may consider the best approach to learning in the context of an increasingly interdependent world. Such a strategy might include taking classes with the most challenging professors, enrolling in mathematics, science, and foreign language classes, and participating in campus activities. All of these behaviors illustrate instrumental action, a well-thought-out, careful approach to defining and achieving goals. With instrumental action, all angles and possibilities are considered.

Weber maintained that in the presence of industrialization, behavior was less likely to be guided by tradition or emotion and was more likely to be value-rational. (He believed that instrumental action was rare.) Weber was particularly concerned about the value-rational approach; he believed it could lead to **disenchantment**, a great spiritual void accompanied by a crisis of meaning. Disenchantment occurs when people focus so uncritically on the ways they go about achieving a valued goal that they lose sight of that goal.

The current system of education offers an example of disenchantment as a consequence of value-

al action. Multiple-choice and true-false tests, especially in large classes, are used often by teachers to measure how much knowledge students have acquired (the valued goal of education) by reading and attending classes. The danger is that students will come to associate learning with doing well on tests, or that teachers will teach students only the material they need to know to do well on tests. In the end, many students come to feel a spiritual void because they have lost sight of the real goals of education, such as personal empowerment and civic engagement.[2]

How the Discipline of Sociology Evolves

Even today, the ideas of Marx, Durkheim, and Weber continue to influence the discipline of sociology because their ideas have survived the test of time. Over many decades, people from a variety of backgrounds have found the ideas of these three scholars useful for thinking about a wide array of situations. (Ideas lose their usefulness if they cannot explain the situations and events that people consider important.)

Yet, despite the importance of Marx, Durkheim, and Weber, we must keep in mind that the discipline of sociology is always evolving and that tens of thousands of people have made it what it is today. The discipline evolves every time someone enters the profession. James M. Henslin, in a collection of essays titled *Down to Earth Sociology* (1993), asked the contributors to explain why they became sociologists. A sample of responses, which appear in the collection, shows how newcomers bring fresh insights and energy to the discipline:

> *I have always been interested in how individuals relate to society and how society relates to the individual. My interest in the social conditions that people experience—especially the marginality that so many blacks feel and how they relate to the wider social system—motivated me to go into sociology to look for some of the answers. I also had good teachers who inspired me. Later I found myself wanting to contribute in a meaningful way to correcting what I saw to be misrepresentations of reality in the academic literature about people who live in ghettos.* (ELIJAH ANDERSON, P. XXI)

> *A major reason I became a sociologist is that Everett C. Hughes, one of my teachers, used to observe that sociology continually shows us*

> *that "It could be otherwise." That is, social life—the way school, work, families, daily experience are organized—may feel permanent and given, but the arrangements are socially constructed, have changed over time, and can be changed.* (BARRIE THORNE, P. XXXI)

Harriet Martineau's Society in America was an important contribution to sociology, but it is only recently that her name and ideas have been included in sociology textbooks.
Archive Photos

I became a sociologist out of an interest in doing something about crime. I remained a sociologist because it became clear to me that until we have greater understanding of the political and economic conditions that lead some societies to have excessive amounts of crime we will never be able to do anything about the problem. Sociology is a beautiful discipline that affords an opportunity to investigate just about anything connected with human behavior and still claim an identity with a discipline. This is its strength, its promise, and why I find it thoroughly engaging, enjoyable, and fulfilling.

(WILLIAM J. CHAMBLISS, P. XXIII)

Unfortunately some voices and experiences have been left out of the record of sociology's history; still others were left out initially but were "discovered" later. Harriet Martineau (1802–1836) is one example. This Englishwoman began writing in 1825,[3] but only in recent years has her name been included in some sociology textbooks. From a sociological viewpoint, Martineau's most important book is *Society in America*.[4] Sociologists can learn much from the way she conducted her research on the United States. Martineau made it a point to see the country in all its diversity, and she believed it was important to hear "the casual conversation of all kinds of people" ([1837] 1968, p. 54). Martineau writes:

I visited almost every kind of institution. The prisons of Auburn, Philadelphia, and Nashville: the insane and other hospitals of almost every considerable place: the literary and scientific institutions, the factories of the north, the plantations of the south, the farms of the west. I lived in houses that might be called palaces, in log-houses, and in a farm house. I travelled much in wagons, as well as stages; also on horseback, and in some of the best and worst of steamboats. I saw weddings, and christenings; the gatherings of the richer at watering places, and of the humbler at country festivals. I was present at orations, at land sales, and in the slave market. I was in frequent attendance on the Supreme Court and the Senate; and witnessed some of the proceedings of state legislatures.

I travelled among several tribes of Indians; and spent months in the southern States, with negroes. (PP. 52–53)

Perhaps most useful are the methods that Martineau chose to make sense of all this information. First, she wanted to communicate her observations without expressing her judgments of the United

HAULING THE WHOLE WEEK'S PICKING

Artist William Henry Brown's depiction of slaves on a Mississippi plantation illustrates the kind of scene observed by Harriet Martineau, who made it a point to see the United States in all its diversity.

Detail from *Hauling the Whole Week's Picking* by William Henry Brown, c. 1842. The Historic New Orleans Collection, Museum/Research Center

In spite of his great athletic abilities, Josh Gibson ("the black Babe Ruth") could play professional baseball only in the Negro Leagues. Gibson was profoundly troubled by the fact that he never got to play with or against players like Joe DiMaggio.
UPI/Bettmann Newsphoto

States. Second, she gave a focus to her observations by asking the reader to compare the actual workings of the society with the principles on which the country was founded, thus testing the state of affairs against an ideal standard. Third, with this focus in mind, Martineau asked her reader "to judge for themselves, better than I can for them . . . how far the people of the United States lived up to, or fell below, their own theory" (pp. 48, 50).

Ignoring some people's ideas not only weakens the discipline; it also does a disservice to those who have been ignored. In *The Mismeasure of Man,* Stephen Jay Gould writes about the agony of being denied the opportunity to participate:

> *We pass through this world but once. Few tragedies can be more extensive than the stunting of life, few injustices deeper than the denial of an opportunity to strive or even to hope, by a limit imposed from without, but falsely identified as lying within.* (1981, PP. 28–29)

Gould's statements can be clarified by the case of Josh Gibson, an African-American baseball player whose talents can be equated to those of Babe Ruth. Legend has it that Gibson hit a thousand home runs

while in the Negro Leagues, as they were known then—leagues for players of color who were not permitted to play in "organized baseball." (Baseball was integrated with the arrival of Jackie Robinson in 1947.) Spectators said that Gibson's home runs were "quick smashing blows that flew off the bat and rushed out of the stadium" (Charyn 1978, p. 41). In 1943 Gibson suffered a series of nervous breakdowns and was institutionalized at St. Elizabeth's Hospital in Washington, D.C., leaving to play baseball on the weekends. He began to hear voices; his teammates found him "sitting alone engaged in a conversation with [Yankee star] Joe DiMaggio," a man he was never permitted to play with or against: "'C'mon, Joe, talk to me, why don't you talk to me? . . . Hey, Joe DiMaggio, it's me, you know me. Why don't you answer me? Huh, Joe? Huh? Why not?'" (p. 41). The case of Josh Gibson reminds us that people's physical characteristics (race, gender, age) cannot be the criteria for judging the worth of their ideas and contributions.

One of the most troubling aspects of writing this introduction to the discipline was deciding whose ideas to include. Over the course of my career as a sociologist I have read materials written by sociologists and nonsociologists that have enriched my understanding of the discipline. Yet, I cannot hear, see, or read all of the significant ideas that exist. Nor is there enough space to include all of the important ideas I have encountered. Please keep in mind that I am limiting your exposure to the discipline of sociology to what I have been able to see, experience, read, and write. Remember that thousands of meaningful contributions exist. I hope you will gain an appreciation of the sociological perspective from what I have included in this text and will come to believe that sociology offers a valuable way of looking at personal, local, national, international, and global events. As sociologist Peter Berger wrote in *Invitation to Sociology:*

> *The fascination of sociology lies in the fact that its perspective makes us see in a new light the very world in which we have lived all our lives.*
> *It can be said that the first wisdom of sociology is this—things are not what they seem.*
> (1963, PP. 21, 23)

The Importance of a Global Perspective

Marx, Durkheim, and Weber were wide-ranging and comparative in their outlook; they did not limit their observations to a single academic discipline, a single period in history, or a single society. They were particularly interested in the transformative powers of history, and they located the issues they studied according to time and place. All three lived at a time when Europe was colonizing much of Asia and Africa, and when Europeans were migrating to the United States, Canada, South Africa, Australia, New Zealand, and South America. Sociologist Patricia M. Lengermann believes that overseas expansion had significant consequences for the discipline of sociology. She writes:

> *Explorers, traders, missionaries, administrators, and anthropologists recorded and reported more or less accurately the details of life in the multitudes of new social groupings which they encountered. Westerners were deluged with the flood of ethnographic information. Never had man more evidence of the variety of answers which his species could produce in response to the problems of living. This knowledge was built into the foundations of sociology—indeed, one impulse behind the emergence of the field must surely have been Western man's need to interpret this evidence of cultural variation.*
> (1974, P. 37)

In part because of the specialization of knowledge into distinct disciplines, many sociologists abandoned the historical and comparative vision of the classic sociologists. Today the great majority of American sociologists focus disproportionately on contemporary life in the United States. Sociologists, however, are not the only individuals to disregard cultural and historical forces; such neglect is a shortcoming of most American educators. At all levels, education in the United States is structured in such a way that foreign countries, especially non-Western countries, are disregarded entirely. In addition, the United States is the only industrialized country in the world in which students are not required to learn a second language. This situation helps to explain why, in a recent United Nations study, Americans ranked next to last in their comprehension of foreign cultures (Ostar 1988).

The lack of a cross-cultural and historical perspective is particularly problematic because the United States no longer dominates the world economy. This does not necessarily mean that the United States is in decline; it means that Asian and European countries have recovered from World War II and other conflicts that left their people devastated and their economies in ruin. Immediately after the war the United States faced little or no competition. Now, more than 45 years later, the United States is only one of many economic powers in the world.

The case of the atomic bomb illustrates the changing role of the United States in world affairs since World War II. From 1945 to the mid-1950s the United States had a monopoly on atomic bombs and the means of delivery. During the mid- to late 1950s Russia developed its own atomic bombs.

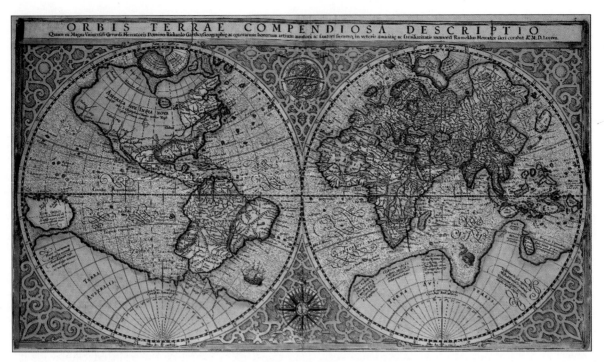

ORBIS TERRAE COMPENDIOSA DESCRIPTIO

*As this map (ca. 1595) illustrates, the age of exploration changed Western Europeans'
view of the world forever, but much remained unknown or only partially understood.
Today, a truly global perspective includes not only an accurate understanding of the
world's geography and diverse cultures, but an appreciation of the interdependence
of all who share the earth.*

Bridgeman Art Library

With the launching of the Sputnik satellites Russia introduced a means of delivery other than aircraft, namely the intercontinental ballistic missile. As we entered the 1980s, the United States was no longer one of two powers with nuclear weapons, but rather one of many countries with both the weapons and the means to deliver them (Ornstein and Ehrlich 1989). To date, 16 countries either have nuclear weapons, are suspected of having nuclear weapons, or have the potential to manufacture them in the near future (Edensword and Milhollin 1993; Marshall and Boys 1991).

To complicate matters, each of the world's approximately 190 countries, including those with nuclear capabilities, considers itself sovereign and recognizes no power as superior to itself (Harries 1990). Moreover, a country can purchase the technology and equipment it needs from corporations in other countries. For example, Iraq, a country with the potential to manufacture atomic bombs, purchased technology and equipment from corporations in 24 countries, including the United States. Since the Persian Gulf War, Western brokers have illegally shipped to Iraq the technology and equipment needed to restore its military arsenal to prewar levels (Jehl 1993).

In light of the steady decline in the United States's capacity to control international affairs since the end of World War II, we must adjust to global interdependence. To make this adjustment, Americans must change their point of view and increase their understanding of other peoples:

*We must see other nations not as 'aliens' but as
common partners on this planet—working toward common goals. We must see beyond our*

national borders and our cultural biases. . . . To do that we must develop a common understanding and a global perspective. (OSTAR 1988, P. 461)

This book represents one response to the call for a curriculum that reaches beyond the borders of the United States. The two most important goals of this text are: (1) to introduce the reader to essential sociological concepts, perspectives, and topics; and (2) to demonstrate the value of the sociological perspective for understanding the development of the self and for comprehending a wide array of contemporary issues, especially issues related to global interdependence. Each chapter covers a different sociological topic and emphasizes examples from one of the following countries or regions: Mexico, Japan, the Republic of Korea, Israel and the Occupied Territories, Zaire, India, the People's Republic of China, South Africa, Germany, the former Yugoslavia, Brazil, the United States, the Republic of Lebanon, and the United States and Russia in the aftermath of the Cold War.

In each chapter, I explain the rationale underlying the selection of a particular country and the pairing of that country with the sociological topic covered in that chapter. The selection of a particular country usually is related to one or more of these general considerations: (1) it offers insights into the workings of some aspect of U.S. society; (2) it illustrates that the lives of people in the United States are intertwined closely with those of people in other countries; (3) it helps us know and understand something of the world outside our borders; and (4) it enables us to clarify the issues we face as a country.

The selection of a particular country also is related to another consideration: that country is a vehicle for helping us think about a larger global trend. Each chapter addresses a global theme that demonstrates that "people live their lives in a sea of transactions that link them to worldwide systems of production, finance, communication, travel, education, military threats, and politics" (Alger and Harf 1985, p. 22). The global themes also help us understand that we live in an atmosphere in which one country's problems are part of a larger global problem. Finally, the global themes help us appreciate how an action taken in one part of the world can affect people in another (Isaacson 1992).

The interconnection data that begins each chapter gives us some sense of the extent to which the United States is connected to the world in general and to specific countries. Among other things, the data at the start of this chapter tells us that in 1991 (the last year for which data is available) there were more than 2,279,150,000 phone calls between people living in the United States and people in other countries. Think about what this number means. It represents the interactions of individuals in a very direct way. It does not matter who is talking or what is being said. What matters is that the conversation is between people inside and outside the United States.

It is easy to claim that the countries of the world are interconnected, but it is difficult to show what interconnection means or how it is played out on a day-to-day basis. The data helps us see how day-to-day, seemingly personal interactions (such as phoning or receiving a telephone call from someone in another country, attending college with foreign students, traveling to another country, working for a foreign employer, and performing military service outside the United States) add up to become global interdependence.

Perhaps the most important message is that sociological concepts and theories allow us to think about a wide array of global issues. In this textbook, these issues include the following:

- The transfer of labor-intensive manufacturing or assembly operations out of high-wage countries such as Germany, the United States, and Japan into low-wage countries such as China, Mexico, Taiwan, and South Korea

- The seemingly new outbreaks of conflicts between people who share a territory but differ from one another in ethnicity, race, language, history, or religion, and the mechanisms by which that conflict is passed from one generation to the next

- The rising concern about whether schools are educating a work force that can compete in an economic environment in which employers are drawing employees from a worldwide labor pool

- The unprecedented levels of international and intercontinental travel, the accompanying increase in interactions among people from different countries, and the relationship of these interactions to the development and spread of diseases

Although this book is not designed to provide a thorough overview of life in each of the countries discussed, I wrote it with the hope that the information presented here would stimulate readers to learn more about life outside the United States. I hope my readers will come to appreciate the richness and versatility of sociological concepts and perspectives for understanding life as we reach the end of the twentieth century and move into the twenty-first century. With this knowledge we can see beyond our immediate environment and relationships, and can place our biography in the context of the push and shove of history. At the same time we will come to recognize that our responses depend on a clear understanding of these forces.

A note about the adjective *American:* In the broadest sense of the word, the adjective applies to people living in North, South, and Central America, and all of the islands in the surrounding waters. Yet, it is almost always used in reference to the people, culture, government, history, economy, or other activities of the United States of America. Some critics argue that using *American* as synonymous with residents of the United States is an example of "American" arrogance. However, there is no succinct adjective such as Mexican, Canadian, Brazilian, Central American, South American, or North American that can be applied easily to life in the United States (*United Statesian* is not an accepted word and it is awkward). It is for this reason that I use the adjective *American* as synonymous with "pertaining to the United States."

Key Concepts

Bourgeoisie 16

Disenchantment 17

Global Interdependence 6

Issues 6 *compare w/troubles*

Means of Production 16

Mechanization 13

Proletariat 16

Social Action 17

Social Relativity 11

what is Sociological Imagination 6

Sociology 15

Solidarity 16

Transformative Powers of History 12

Troubles 6

Why did author say take a global perspective

Notes

1. Some additional excerpts from diaries kept by my students in my classes make this point even more clear:

   - When I asked Tomaz, from the Czech Republic, how life in the United States differed from life in his country, he said that high school classes begin at 8:00 A.M. and that they are 45 minutes long with a ten-minute break between classes. Classes change each week: one week he may have art and math, then the next week he may have science and social studies. Czech students can quit school after the eighth grade and go to work, or they can go on for four or more years of school. Every five months he gets a report card. Students do homework assignments but not for the grade. They do the assignments for their own personal growth and benefit. In order to graduate, Czech students take an oral exam in four subjects. This exam is administered as follows: the student picks out one question from a box containing 50 questions. He or she speaks on that question for 20 minutes in front of four teachers. (JODI DARLING)

- Alex, from Brazil, explained to me how he sees U.S. life as different from that in Brazil. Americans are more individual than Brazilians. He thinks this comes from how people are raised. For example, when U.S. parents buy a toy, it is usually for a particular child—that child does not have to share. In Brazil, when parents buy a toy, it belongs to all their children to share. Americans also compete with one another too much instead of trying to help everyone be equal. For example, in school everyone here competes to have the best grades. No one wants to help the people who are doing badly to do better. In Brazil we go out of our way to help those who aren't doing well do better. (ERIC HURST)

- Monika, a Swedish-born student, told me that in her country, *love* is a very strong word. Swedes use the word more selectively than Americans. People in the U.S. might say, "I *love* the outdoors." If a Swedish person said this, people would literally think something was wrong with the person. (TRACY KRAMER)

2. Personal empowerment means that people can critically examine the information they receive and can challenge assumptions. It is equivalent to possessing the sociological imagination. Civic engagement means that people take active roles in shaping the quality of life.

3. In her day, Harriet Martineau was a popular author. One measure of her popularity is that her first book sold more copies in a month than John Stuart Mill's *Principles of Economics* sold in four years. Martineau's success with the general public can be attributed in part to her style: she developed a plot, a setting, and a cast of characters and used them to illustrate economic principles at work in a community (Fletcher 1974).

4. Martineau receives the most credit for introducing sociology to England; she translated Auguste Comte's six-volume work *Positive Philosophy* into English. Many people consider Comte to be the father of sociology (Terry 1983; Webb 1960).

2 THEORETICAL PERSPECTIVES

with Emphasis on Mexico

Financial District, Reforma Avenue, Mexico City.
Larry Reider/Sipa Press

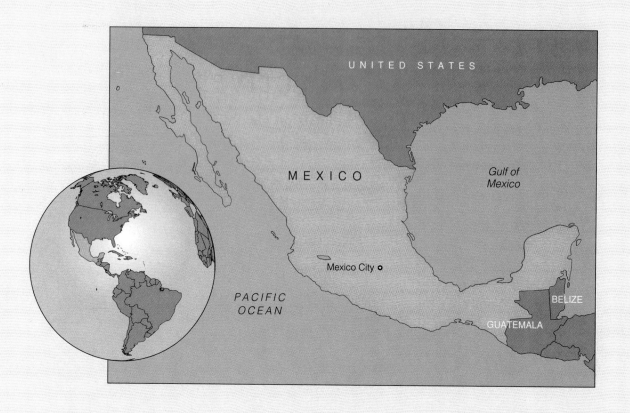

Students from Mexico enrolled in
college in the United States for fall 1991   7,000

Mexicans admitted into the United States in
fiscal year 1991 for temporary employment   19,418

People living in the United States in
1990 who were born in Mexico   4,447,000

Airline passengers flying between
the United States and Mexico in 1991   8,697,478

People employed by Mexican
affiliates in the United States in 1990   12,000

Applications for utility patents filed in the
United States in 1991 by inventors from Mexico   106

Phone calls made between
the United States and Mexico in 1991   240,581,000

Consider the following facts:

- There are approximately 2,000 foreign-owned manufacturing plants in Mexico. The great majority (over 90%) are U.S.-owned and are located along the northern border of Mexico.

- General Motors is the largest private employer in Mexico, with approximately 35,000 Mexican workers. GM-Mexico is expected to employ 70,000 persons by 1995.

- Approximately 4,000 freight trucks cross each day from the United States into Mexico to deliver equipment, components, and raw materials for assembly and to pick up assembled products (clothing, computers, automobiles) for return to the United States (Sanchez and Darling 1992).

- Each year about 60 million Americans cross into Mexico to do things like shop, dine, purchase over-the-counter and prescription drugs, visit physicians and dentists, and have their automobiles repaired (Oster 1989).

nyone can observe and verify the facts just presented. Yet, taken alone, facts such as these have no meaning; they give no explanation for why they are what they are. Facts are valuable and useful only when we begin to comprehend the reasons behind them. To make sense of facts, sociologists relate them to established and time-tested theories. In the most general sense, a **theory** is a framework that can be used to comprehend and explain events. In every science, theories serve to organize and explain events that can be observed through the senses (sight, taste, touch, hearing, and smell). A **sociological theory** is a set of principles and definitions that tell how societies operate and how people relate to one another. Three theories dominate the discipline of sociology: the functionalist, the conflict, and the symbolic interactionist perspectives.

This chapter outlines the basic assumptions and definitions of each of these founding theories and shows how each one offers a distinct framework that can be used to interpret any event. In this chapter we pay special attention to Mexico because, although the United States shares a 1,952-mile border with Mexico (which millions have crossed each year in order to shop and visit vacation spots), most Americans know little about what binds the two countries together. In truth, few people know the extent to which U.S. corporations have production operations in Mexico or understand their economic significance. In addition, most of us are unaware that Mexico is the United States's third-largest trading partner (after Japan and Can-

ada) or that the United States is Mexico's largest trading partner.[1] Lacking this knowledge, many Americans interpret the facts listed above to mean that Mexico poses an economic threat to the United States. They argue that Mexico is a major source of cheap labor, lures companies out of the United States, takes away jobs, and drives down workers' wages. Based on this interpretation, the natural impulse is to pass laws or to prevent the signing of trade agreements such as the North American Free Trade Agreement (NAFTA) so as to stop the flow of Mexicans into the United States, to prevent U.S. companies from investing in Mexico, and to limit the flow of Mexican-assembled products into the United States (see "What Is NAFTA?").

In light of this narrow, yet very common, interpretation[2] of events, the three major sociological perspectives offer us quite different ways to interpret these facts presented here and to assess the relationship between Mexico and the United States. In addition, the three perspectives offer us some constructive ways to think about not just this relationship but also a larger global trend: the transfer of labor-intensive manufacturing or assembly operations out of Western European countries, the United States, and Japan into low-wage, labor-abundant countries such as Mexico, Malaysia, Singapore, Philippines, Taiwan, South Korea, Haiti, Brazil, Dominican Republic, and Colombia (Grunwald 1985). In this chapter we use U.S. manufacturing operations in Mexico as a vehicle for thinking about this global trend. We begin with some background information on these manufacturing operations.

U.S. Manufacturing Operations in Mexico

General Motors (GM), the largest private employer in Mexico, represents one case among tens of thousands of foreign corporations that have transferred some of their manufacturing operations to low-wage, labor-abundant countries. In addition, GM represents one case among approximately 2,000 corporations that have transferred some of their manufacturing operations to Mexico.

In Mexico, a foreign-owned plant such as GM that purchases Mexican labor is called a *ma-*

quiladora (pronounced mah-kee-la-dora).[3] The *maquiladora* system works like this: foreign companies ship tools, machinery, raw materials, and components duty-free into low-wage, labor-abundant countries; workers use the tools and machinery to process and assemble the raw materials and components into finished or semifinished products, which are exported back to the country of the contracting company. Only the cost of labor to assemble the product is subject to tariff. Mexico, because of its

What Is NAFTA?

Many Americans have fiercely opposed NAFTA because they believe that companies will take advantage of the agreement to export U.S. jobs to Mexico.

Louis DeMatteis/JB Pictures

The North American Free Trade Agreement (NAFTA) is a 2,000-page document written with the purpose of eventually eliminating all trade barriers between the United States, Canada, and Mexico. The details of the agreement have been worked out and approved by the U.S. Congress. The agreement went into effect January 1, 1994. It sets guidelines for the gradual elimination of tariffs and duties on goods that are traded between the three countries. In its original form, it calls for the immediate elimination of many tariffs while others would be reduced over a period of up to 15 years. For example, half of U.S. farm goods exported to Mexico would immediately become duty-free. Throughout North America, all other tariffs on agricultural goods would be phased out over the next 15 years. Within five years of implementation of the agreement, 65 percent of U.S. goods would gain duty-free status (White and Maier 1992).

In general, the goals of the agreement are to eliminate or decrease trade barriers in order to increase trade between the three countries. Designers of the pact believe that this increase in trade will force the three economies to specialize in whatever they do best and, therefore, to become more efficient. The increased efficiency, they theorize, will lead to lower consumer prices, the creation of new jobs, and an increased standard of living for the people of all three countries.

However, one should not conclude that NAFTA will bring about these changes without causing problems to some segments of the three countries' economies. After all, the "North American Free Trade Agreement would create the largest market in the world: 360 million consumers with a total output of $6 trillion" (U.S. Department of State 1991, p. 4). It will unite three countries with very diverse cultures and levels of economic development. The United States has a population of 248 million and a per capita GNP of $21,110. Mexico, on the other hand, has a population of 85 million and a per capita GNP of $1,990.

Canada's population is only about a third of Mexico's, at 26 million with a per capita GNP of $19,020 (U.S. Department of State 1991).

NAFTA has the potential to "leave virtually no aspect of the economy untouched" (Risen 1992, A1). The significance and strength of the agreement's effects will vary depending on where the individual, corporation, or agency is positioned within the economy of its particular country. The effects the agreement will have on labor and the environment seem to have drawn the most attention. However, there is also a great deal of debate over the effects it will have on industries such as agriculture, automobile, finance, energy, and transportation.

Risen, James. 1992. "U.S., Mexico, Canada Agree to Form Huge Common Market." *Los Angeles Times* (August 13):A1+.

White, George and Andrea Maier. 1992. "A Closer Look at the Trade Agreement." *Los Angeles Times* (August 13):A7.

U.S. Department of State. 1991. "Background Information." *North American Free Trade Agreement* (May). Washington DC: U.S. Government Printing Office.

SOURCE: Kevin Steuart, Northern Kentucky University (Class of 1994).

proximity to the United States and because of its abundant, cheap labor force, is the United States's most important partner in off-shore assembly. Mexican workers perform a variety of labor-intensive tasks that range from assembling TV receivers to sorting U.S. retail store coupons. Assembly operations are concentrated, however, in the electronics, automotive, and textile sectors.

The *Maquiladora* Program was launched in 1965 after the *Bracero* Program[4] was terminated. The *Bracero* Program, which began in 1942, allowed Mexicans to work legally in the United States to relieve labor shortages in rural areas and to bolster the American work force during World War II. The *braceros* replaced American workers who were needed to work in defense plants or to serve in the armed forces. After World War II, the *Bracero* Program was extended several times to supplement the American work force during the Korean War and to provide laborers for agricultural and other low-wage work. With the termination of the *Bracero* Program, the Mexican government established the Border Industrialization Program.[5]

One purpose of the program was to encourage foreign investors to locate plants along the border so that employment opportunities could be created for returning *braceros*.[6] A second purpose was to increase the employment opportunities in border cities that were experiencing a high level of population growth due to a large influx of migrants from the interior of Mexico. (The termination of the *Bracero* Program increased the already high unemployment in the border region.) In addition, the program's creators envisioned that border industrialization would increase economic ties between the border region and the rest of the country, would provide foreign exchange, and would stimulate technology transfer from the United States to Mexico.

In 1965, Radio Corporation of America (RCA) became the first U.S. company to open a *maquila* plant. The first RCA products produced in Mexico were simple, relatively inexpensive parts that were not worth the cost of shipping overseas from Asia. The presence and success of the RCA plant gave credibility to the *Maquiladora* Program, making it "possible for Mexican promoters to point to the RCA plant as evidence that large and sophisticated

A Japanese-owned maquila.
Burroughs/Gamma Liaison

operations were viable and profitable on the south side of the U.S.–Mexican border" (Barrio 1988, p. 8).

As of 1993, approximately 2,000 U.S. companies had established *maquilas,* primarily along the approximately 2,000-mile border that divides Mexico from the United States (see Figure 2.1). Well-known companies with *maquila* operations include Xerox, Johnson & Johnson, GTE, Texas Instruments, Black and Decker, Uniroyal, Singer, Clark Equipment, Ford, General Motors, Chrysler, Honeywell, Kellogg, Foster Grant, and Westinghouse.[7] The great majority of *maquilas* are U.S.-owned, although Japanese, South Korean, Canadian, and a few European companies also have established *maquilas*. Mexico is an attractive location to corporations from these countries because it is a gateway to the U.S. consumer market. Also, the United

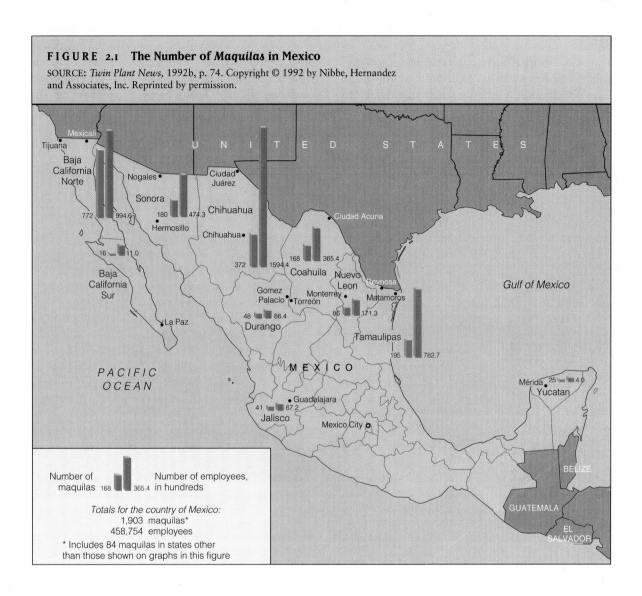

FIGURE 2.1 The Number of *Maquilas* in Mexico

SOURCE: *Twin Plant News*, 1992b, p. 74. Copyright © 1992 by Nibbe, Hernandez and Associates, Inc. Reprinted by permission.

States charges a relatively low duty on products assembled in Mexico and exported to the United States. The use of the Mexican work force for labor-intensive assembly by foreign corporations is expected to increase dramatically over the next decade.

This background information about *maquiladoras* raises a number of critical questions, including:

1. Why do *maquiladoras* exist, and what consequences do they have for the United States and Mexico?

2. Who benefits from and who loses with the *maquiladora* arrangement?

3. Does everyone in the United States and Mexico see *maquiladoras* in the same way?

These three questions are associated with the functionalist, conflict, and symbolic interactionist perspectives, respectively. These questions also can be asked about assembly operations in general and about virtually any event or situation. Each perspective not only offers us a central question to help guide our thinking about a particular event but also offers us a vocabulary for answering its central question. Keep in mind that one theoretical perspective is not necessarily better than the others. Each simply gives a different angle from which to analyze a particular event.

We turn first to an overview of the functionalist perspective. This overview is guided by the follow-ing questions: Why does a particular arrangement exist, and what are the consequences of this ar-rangement for society? Following this overview we will apply the functionalist perspective to the *maquiladora* arrangement.

The Functionalist Perspective

Functionalists focus on questions of order and sta-bility in society. They define society as a system of interrelated, interdependent parts. The parts are connected so closely that each one affects all of the others as well as the system as a whole. Functional-ists consider a **function** to be the contribution of a part to the larger system and its effect on other parts in the system.

To illustrate this complex concept, early func-tionalists used biological analogies such as: the hu-man body is made up of parts that include bones, cartilage, ligaments, muscles, a brain, a spinal cord, nerves, hormones, blood, blood vessels, a heart, a spleen, kidneys, lungs, and chemicals, all of which work together in an impressive harmony. Each part functions in a unique way to maintain the entire body, but it cannot be separated from other parts, which it affects and which in turn help it to func-tion. Consider eyelids: when they blink, they work in conjunction with tear fluid, tear ducts, the nasal cavity, and the brain to keep the corneas (the trans-parent coat over the eyes) from drying out and clouding over. Furthermore, some scientists specu-late that blinking functions in some way to activate the brain, which controls the body and performs thought processes (Rose 1988).

Functionalists see society, like the human body, as made up of parts, such as food-growing tech-niques, sports, medicine, bodily adornments, fu-neral rites, greetings, religious rituals, laws, lan-guage, modes of transportation, appliances, tools, and beliefs. Like the various body parts, each of the social parts functions to maintain the larger system. For example, one of the many functions of educa-tion in the United States is to transmit knowledge and skills from one generation to the next. Young people learn the previous generation's solutions for meeting various environmental challenges and are not forced to start from scratch. Other parts of the social system such as unwritten rules about appro-priate times to communicate with others, link peo-ple to one another in an orderly and predictable way. By extension, the absence of such rules can disrupt the society.

Anthropologist Edward T. Hall recounts an episode involving members of different cultures who did not share the same rules about appropriate times to communicate messages. As insignificant as this may sound, the different rules frustrated their interaction. Hall describes an interaction between some South Pacific island workers and an American manager after the islanders had come up with a so-lution to resolve a dispute with the manager over his hiring practices:

> When they finally arrived at a solution, they went en masse to see the plant manager and woke him up to tell him what had been decided. Unfortunately it was then between two and three o'clock in the morning. They did not know that it is a sign of extreme urgency to wake up Americans at this hour. As one might expect, the American plant manager, who un-derstood neither the local language nor the cul-ture nor what the hullabaloo was all about, thought he had a riot on his hands and called out the Marines. It simply never occurred to him that the parts of the day have a different meaning for these people than they have for [Americans]. (HALL 1959, P. 25)

Hall's example helps us see that all of the parts in a social system—no matter how seemingly routine

Why does poverty—a condition everyone agrees is problematic—persist in the middle of affluence? Functionalists have pointed to the ways in which the existence of poverty contributes to the stability of the social system.
Mary Ellen Mark/Library

and taken-for-granted—have functions that contribute to the smooth operation of the society.

In learning about functionalism, it is important to see that parts also affect and are affected by other parts. For example, the photocopying machine functions to connect people to one another by facilitating the sharing and distribution of information. Yet, the widespread use of this machine is linked to other parts of the system that connect people over great distances: interstate highways, jet travel, and communication satellites. The first commercial photocopier was introduced in the United States in the late 1950s, a period that coincides with the building of a national interstate highway system, the first domestic jet passenger service, and the launching of the Soviet communication satellite *Sputnik*. Although these inventions linked people with one another in different parts of the country and world, they also encouraged people to move apart and to interact in less spontaneous and less personal ways. Thus, photocopying documents (contracts, records of business transactions, policy changes, and so on) made it easier to share information with large numbers of people in diverse locations without calling face-to-face meetings.

In the most controversial form of this perspective, functionalists argue that all parts of society, even those that seem not to serve a purpose, make a contribution to the system's stability. Functionalists maintain that a part would cease to exist if it did not serve some function. Functionalists would argue that such parts of society as poverty, illegal immigration, and the transfer of labor from the United States into Mexico contribute to the stability of the social system.

Herbert Gans (1972) argued this point in his classic analysis of the functions of poverty. To his own question, "Why does poverty exist?" he answered that poverty performs at least 15 functions, five of which are described here.

- The poor have no choice but to do the unskilled, dangerous, temporary, dead-end, undignified, menial work of society at low pay. Hospitals, hotels, restaurants, factories, and farms draw their employees from a large pool of workers who are forced to work at minimum or below-minimum wages. This hiring policy keeps the costs of their services reasonable and increases profits.

Functions of Poverty

- Affluent persons contract out and pay low wages for many time-consuming activities such as housecleaning, yard work, child care, and grocery shopping. This practice gives them time for other, more "important" activities. This function of poverty was brought to national attention in 1993 when President Clinton's first two nominees for the U.S. Attorney General (Zoe Baird and Kimba Wood) disclosed that they had employed illegal immigrants to take care of their children.[8] This child-care arrangement is not confined to the most prominent and most affluent Americans. Many middle-class and even lower middle-class Americans make similar arrangements.

- The poor often volunteer for over-the-counter and prescription drug tests. Most new drugs must be tried on healthy subjects to determine potential side effects (for example, rashes, headaches, vomiting, constipation, drowsiness) and appropriate dosages. Money motivates subjects to volunteer. But because payment is relatively low, however, the tests attract a disproportionate share of poor people as subjects.

- The occupations of some middle-class workers—police officers, psychologists, social workers, border patrol guards, and so on—are intended to serve the needs or monitor poor people's behavior. For example, about 3,300 U.S. immigration agents patrol the U.S.–Mexico border (Kilborn 1992). Similarly, physicians and grocery store owners are reimbursed for serving poor people. The poor receive food stamps and medical cards; the providers receive the cash.

- Poor people purchase goods and services that otherwise would go unused. Day-old bread, used cars, and secondhand clothes are purchased by or donated to the poor.[9] In the realm of services, the labor of many less competent professionals (teachers, doctors, lawyers), who would not be hired in more affluent areas, is absorbed by low-income communities.

Gans outlined the economic usefulness of poverty to show how and why the parts of the society that everyone agrees are problematic and should be eliminated remain intact: these parts of the society contribute to the stability of the overall system.

Based on this reasoning, the economic system would be strained seriously if poverty were completely eliminated; industries, consumers, and occupational groups that benefit from poverty would be forced to adjust.

Although functionalists emphasize how parts contribute to the stability of the system, they also recognize that the system does not remain static: as one part changes, other parts adjust and change in ways that lessen or eliminate strain. In the United States, for example, when there are not enough qualified (or willing) workers to fill the available jobs, we increase the flow of immigrants willing or able to fill those jobs. In 1989 Congress passed legislation to increase the total number of legal immigrants by 100,000 to 400,000 per year and specifically those with advanced degrees, money to invest, or professional skills in short supply, such as engineering, nursing, and rural medical skills (Roberts 1990).

Critique of Functionalism

As you probably have realized by now, the functionalist perspective has a number of shortcomings. In the first place critics argue that functional theory is by nature conservative: it "is merely the orientation of the conservative social scientist who would defend the present order of things" (Merton 1967, p. 91). When functionalists identify how a problematic part of society such as poverty contributes to the system's stability, by definition they are justifying its existence and legitimating the status quo. *The New Yorker* cartoonist Roz Chast's illustration outlining the functions of oil spills (see Figure 2.2) shows how the functionalist argument can be twisted to justify even an ecological disaster. Critics argue that so-called system stability is more often than not achieved at a cost to some segment of the society such as those who are poor and powerless. Functionalists deny this criticism of their approach. When they define the function of controversial practices, they are simply illustrating why such parts continue to exist despite efforts to change or eliminate them.

Second, critics state that a part may not be functional when it is first introduced. Often, practices and inventions find their usefulness only after they

FIGURE 2.2   The Functionalist Argument Taken to an Extreme

SOURCE: From a drawing by R. Chast. Copyright © 1990 by *The New Yorker Magazine*, Inc. Reprinted by permission.

have come into existence. For example, the automobile, the photocopier, and the airplane link people with one another, but there was no real need for these inventions when they were first designed. Xerographic copying, for instance, was invented in 1938 but was not used commercially until 1958 (Andrews 1990). Likewise, when the automobile was invented there were no paved roads, so the automobile's use was limited. These examples complicate the functionalist argument that a part exists because it makes a positive contribution to society. They also suggest that necessity is not always the mother of invention but that invention often is the mother of necessity.

Third, critics assert that the part of the system that fills a function is not necessarily the only way or the most efficient way to meet that function. Although the automobile functions to connect people with one another, it is not an environmentally sound means of doing so. In fact, the widespread use of and reliance on the automobile hinders the development of more environmentally efficient modes of public transportation. Reliance on automobiles also reduces socialization, something that public transportation demands and reinforces.

Finally, because the functionalist perspective assumes that a part's function must support the smooth operation of society, it has difficulty accounting for the origins of social conflict or other forms of instability. This assumption leads functionalists to overlook the fact that inventions or practices may have negative consequences for the system or for certain groups within the society.

To address some of this criticism, sociologist Robert K. Merton (1967) introduced a few concepts that help us think about a part's overall effect on the social system and not just its contribution to stability. These concepts are manifest and latent functions, and manifest and latent dysfunctions.

Merton's Concepts

Merton distinguishes between two types of functions that contribute to the smooth operation of society: manifest functions and latent functions. **Manifest functions** are the intended, recognized, expected, or predictable consequences of a given part of the social system for the whole. **Latent functions** are the consequences that are unintended, unrecognized, unanticipated, or unpredicted. To illustrate this distinction let's consider the manifest and latent functions of the 1986 Immigration Reform and Control Act (IRCA). When Congress passed this act, it did so with the expectation of reducing the number of illegal immigrants entering the United States, encouraging illegal workers to apply for legal status,[10] and discouraging employers from hiring undocumented workers. (All of these consequences are manifest functions.) Prior to this act it was not against the law for employers to hire illegal immigrants (Sharp 1989).

On the other hand, Congress probably did not expect that the act would uncover 2.5 million undocumented workers who had used the social security numbers of relatives or friends, invented their own nine-digit numbers, or purchased fake social security cards. The act inadvertently functioned to straighten out social security accounts. When applicants filed for legal status, they received new accounts and past payments to other accounts were transferred accordingly (U.S. Immigration and Naturalization Service 1990).

Merton also points out that parts can have **dysfunctions** as well. That is, they can sometimes have disruptive consequences to the system or to some segment in society. Like functions, dysfunctions can be manifest or latent. **Manifest dysfunctions** are the expected or anticipated disruptions that a part causes in some segment of the system.

When the IRCA became effective, it took a while for U.S. employers to adjust to the many new hiring directives. There were many predictable disruptions (manifest dysfunctions) to the process. For example, the act required that employers check 2 of 29 possible documents that qualify as proof of citizenship. However, some of these documents, such as baptismal certificates or school report cards, have thousands of variations, making it harder for employers to distinguish between legitimate and fake documents (Behar 1990). Specifying a wide range of acceptable documentation was about the only fair way for illegal immigrants to prove that they had been in the United States long enough to qualify for amnesty. Yet, such liberal guidelines encouraged illegal immigrants to make or purchase fake documents. The policy therefore posed a predictable disruption to the amnesty process.

Latent dysfunctions are unintended, unanticipated negative consequences. The 1986 act has had several such consequences. Because employers had difficulty distinguishing between legitimate and fake documents, some employers discriminated against foreign-*looking* citizens in order to protect themselves against hiring illegal immigrants who presented fake documents. The U.S. General Accounting Office have estimated that 10 percent, or 461,000, employees were discriminated against in this way (U.S. General Accounting Office 1990).

The everyday reality of illegal immigration is highlighted by this highway sign in the border city of San Ysidro, California. The sign warns motorists to watch for illegal immigrants running across the highways and at the same time warns the migrants that crossing illegally is prohibited.
Steve Rubin/JB Pictures

Sociologists Katharine M. Donato, Jorge Durand, and Douglas S. Massey (1992) have identified another latent dysfunction: the IRCA legislation did not deter undocumented migration from Mexico to the United States. They estimated that "the probability of taking a first illegal trip, the probability of repeat migration, the probability of apprehension by the Border Patrol, the probability of using a border smuggler, and the costs of illegal border crossing" have not changed since the IRCA took effect in 1986.

Another latent dysfunction involves the three million illegal immigrants who were granted amnesty under the 1986 act. Workers who gained legal status were then able to make demands on employers for higher wages and better working conditions. However, this had the unexpected consequence of causing employers to risk fines by hiring illegals, who would make no demands and would accept low wages and poor working conditions in order to remain in the United States.

You can see from just this brief analysis of the Immigration Reform and Control Act that the concepts of manifest and latent functions and dysfunctions make for a more balanced perspective than does the concept of function alone. The addition of these concepts eliminates many of the criticisms leveled at the functionalist perspective. On the other hand, this broader functionalist approach introduces a new problem: it gives us "no techniques . . . for adding up the pluses and minuses and coming out with some meaningful over-all calculation or quotient of net effect" (Tumin 1964, p. 385). In regard to the Immigration Reform and Control Act, we are left with the question of whether the *overall* impact of this act has had a positive or negative effect on society.

The Functionalist Perspective on the Maquiladora Program

To see how the functionalist perspective can be applied to a specific event, let's consider how functionalists would view the *Maquiladora* Program. From the functionalist viewpoint this program can be defined as a response to restore order to both Mexican and American societies. When the *Bracero* Program ended in 1964, Mexico needed to create jobs for returning *braceros*. U.S. corporations needed a low-wage labor pool to lower the cost of production and compete in the global economy. If U.S. companies such as General Motors had not moved labor-intensive assembly operations to countries such as Mexico to save on labor costs, including employee benefits, they could not have survived the new competition from companies like Nissan, Toyota, and Hyundai. We can use Merton's concepts to assess the overall consequences of the labor-transfer program on the United States and Mexico.

Manifest Functions of the Maquiladora Program

One way that some advocates of the *Maquiladora* Program assess its contribution to stability is to estimate the effect on Mexico and the United States if the *Maquiladora* Program were eliminated. They project that under such a scenario Mexico's gross national product (GNP) would drop by $3.1 billion because *maquilas* are directly and indirectly responsible for an estimated 593,000 jobs. The U.S. economy would be affected similarly: prices for American goods would increase by a projected 36 percent and the GNP would drop by $3.7 billion. The higher prices would mean a reduced demand for U.S. products and a corresponding loss of manufacturing, management, marketing, and retail jobs (Cornejo 1988).[11]

Proponents argue that this projected loss would occur because *maquilas* function in two important ways to save and create U.S. jobs. First, American companies supply raw materials and components for assembly by Mexican *maquila* workers. *Maquilas* in Juárez, Mexico, alone buy components and raw materials from 4,600 suppliers in 42 states (Sanders 1986). The U.S. Department of Commerce estimates that 75,000 American workers produce and ship the materials used by *maquila* workers (Applebome 1986). Second, the low-cost, labor-intensive *maquila* output is sent to sister plants in the United States for final assembly. Thus, the Mexican contribution saves manufacturing jobs and permits American workers to receive higher wages for their work. If the *Maquiladora* Program were eliminated, it is extremely unlikely that jobs would return to the United States; jobs simply would move to another off-shore location (Pranis 1990). For example, jobs might be transferred to any of 40 export processing zones in Asia (Warr 1989).

This economic arrangement is typified well by the *maquila* plant Delredo, a division of Delco, one of the largest producers of magnets. Delredo is located in Nuevo Laredo, Mexico. Iron components needed to make magnets are purchased from companies in Oklahoma. The magnets are produced at Delredo and then are shipped to a sister plant in Rochester, New York. There they are placed in electric motors that operate power windows and windshield wipers (Jacobson 1988).

Latent Functions of the Maquiladora Program

On a latent level the growth of *maquila* plants along the border functions to further integrate Mexico with the United States socially, economically, and politically.[12] People from both sides of the border cross freely to work, shop, eat, or go to school. This integration is symbolized by the bridge and border crossing that connect border cities with their respective sister cities across the Rio Grande. These bridges increase not only the flow of goods but also the exchange of persons, information, and services between the United States and Mexico.

The increased interaction among border peoples is blurring the culture at the border. The culture is neither Mexican nor American; it is a blend (Langley 1988). This blurred line of demarcation at the border is functional in that it eases the strains usually caused by differences in language, outlook, and customs that inevitably arise when people from distinct cultures interact. This buffer zone may prove essential for smooth relations between Mexico and the United States as the two countries become more dependent on each other.

There are border residents who understand both worlds and act to mediate, facilitate, and smooth exchanges between the two countries. Bilingual consultants who are comfortable with both cultures help U.S. businesses establish assembly operations in Mexico. There are also academic centers that facilitate the blending of cultures. They include the American Center for Mexican Studies at the University of California in San Diego, the Program on Mexico at the University of California in Los Angeles, the Center for Frontier Studies at the University of Texas in El Paso, and Mexico's Colegio de la Frontera Norte in Tijuana (Sanders 1987).

The latent functions just described are positive, but you should not take that to mean that there are no cultural frictions at the border. In *Days of Obligation: An Argument with My Mexican Father,* Richard Rodriguez (1992) reminds us that on the U.S. side of the border "are petitions to declare English the official language of the United States [and] the Ku Klux Klan nativists posing as environmentalists, blaming illegal immigration for freeway congestion" (p. 84). (See "A Common Border Culture?") Rodriguez's observations suggest that there are some dysfunctional consequences to Mexican–United States economic and social interdependence.

Manifest Dysfunctions of the Maquiladora ***Program*** Although the *Maquiladora* Program contributes to stable relations between Mexico and the United States, it has several expected (manifest) and unexpected (latent) dysfunctions. Job displacement is one obvious manifest dysfunction: whenever an event alters the way in which significant numbers of people earn their livelihoods, we can anticipate that some adjustments will have to be made. In the United States, for example, many workers have been laid off or fired because their jobs were transferred to Mexico. The disappearance of entry-level manufacturing jobs has left many unskilled workers without jobs and lacking the qualifications to fill newly created jobs. Workers in cities that specialized in handling only routine manufacturing jobs were particularly affected. Thirty counties in the United States have a 25-percent unemployment rate; more than 100 counties have 15-percent unemployment (Murray 1988).

Flint, Michigan, along with the surrounding Genesee County, a region of 430,000 people, is a notable example. During the late 1970s General Motors employed 80,000 people in the Flint area. In a 10-year span 32,000 jobs disappeared permanently. Over the course of the 1990s it is expected that GM will cut down to only 35,000 workers in the Flint area (Jacobson 1988). Rural counties are particularly vulnerable to this form of job displacement because manufacturing jobs in these areas tend to be concentrated in routine manufacturing industries (food, textiles, furniture production, apparel). These industries require less skilled labor (such as garment inspection, repetitive assembly, and simple machine operation) and can be shifted to foreign-based plants (O'Hare 1988).

It would be very misleading, however, if we were to attribute job displacement and loss in the United States to just the *Maquiladora* Program. Mexico is one of many locations in the world to which U.S.-based corporations have moved assembly and other operations. Another often overlooked reason for job displacement and loss is the failure of U.S. corporations to invest in the technologies that would allow them to compete in a global economy.

As you might expect, the U.S. decision to terminate the *Bracero* Program compounded unemployment problems along the border, which the *Maquiladora* Program was expected to alleviate. Although we have little information about the fate of the *braceros,* we know from numbers alone that the Border Industrialization Program did not function to absorb them into the Mexican economy. More than four million permits to work in the United States were issued to Mexican *braceros* between 1951 and 1964 (Garcia 1980), whereas the *maquila* industry has generated about 600,000 jobs over a 25-year period.

Latent Dysfunctions of the Maquiladoras ***Program*** Several latent dysfunctions are also associated with the growth of *maquilas* along the border. First, because the Border Industrial Program has increased economic interdependence between the United States and Mexico, problems in one country affect the economic life of the other. For example, the 1974–75 and 1981–82 recessions in the United States contributed to *maquila* plant layoffs and closings. When the $100 billion debt crisis hit Mexico in the early 1980s and the peso subsequently was devalued, retail businesses on the U.S. side of the border were devastated. (It is estimated that, before this crisis, 40 to 70 percent of the spending by border Mexicans took place on the U.S. side.) In Brownsville, Texas, for example, retail sales dropped by 68 percent in 1981 (Hansen 1985). Although *maquila* owners benefit from a weak Mexican economy and a weak peso, the adjustments that the Mexican government imposes on its people—forced austerity, currency devaluation, curtailment of social programs, decline of real wages—make life miserable for the masses and could lead to political upheaval that might devastate both the Mexican and the American economies.

Another latent dysfunction is the transfer of significant numbers of white-collar jobs out of the United States into Mexico. Signs of this trend are

A Common Border Culture?

Tijuana is only one of several major Mexican cities along the U.S. frontier, composing a highly urbanized society in which 85 percent of the population live in 12 cities. The largest in 1980 were Juárez (567,365), Mexicali (510,664), Tijuana (461,257), Matamoros (238,840), Reynosa (211,412), and Nuevo Laredo (203,286). With Juárez and Tijuana now the largest of them, one must know these cities to understand the frontier. All of them are among Mexico's largest cities, and they all lie directly across from American cities with which, to varying degrees, they have symbiotic relationships. The smaller American counterpart cities depend more closely on specifically frontier activities than the larger ones like San Diego and El Paso, which are more likely to have significant autonomous economic activities, as well.

The profound differences in overall standard of living and way of life between the United States and Mexico make it difficult to speak casually of a common border economy or culture. The paradox of integration with separation is reflected in the following comment: "On the Mexican–U.S. border two economic and social systems are articulated in the most dramatic and categorical coincidence between underdevelopment and development, creating with their fusion a complex social structure."*

The most casual observation of San Diego and Tijuana confirms that one is an American (albeit California) city, the other Mexican (though a northern frontier one). This is so even though they have many similar economic activities, their middle classes live comparably, and they share a common population stratum, since much of San Diego's population is Mexican in background. Though culture varies *within* each country (Tijuana differs from Guadalajara, San Diego from Boston), the sharp economic and cultural differences of the two countries make even their frontier cities distinctly Mexican or American in appearance and style. Perhaps most important, those on each side are conscious of their national identity and have a certain pride in their cities. Tijuana may seem somewhat poor and chaotic to Americans, but the *tijuanense* perceives his city as one of the most desirable ones in Mexico.

The fact that two countries are involved explains why cooperation between the cities on common problems has been difficult, involving complex negotiations between authorities not only from San Diego and Tijuana, but from the state and federal levels in both countries. Water, sewage, air pollution, land use, and economic development affect both communities, but disputes have often dragged on with only partial or at times unsuccessful outcomes. A specialist on border relations points out that "despite the attractiveness of this vision of binational planning and environmental management, little has been accomplished at the micro-regional level. One obstacle is the sense members have of their political community. . . . Second, there is a welter of planning issues that may be skewed by distinct cultural understandings and priorities, project pacing, matching of counterparts, and allocation of costs, as well as the proper language for justifying cross-border cooperation. Third, the proper involvement of federal, state, and local-level agencies has, in many instances, an asymmetrical emphasis on both sides of the border."†

Although several public, private, and mixed associations of San Diego and Tijuana leaders hold frequent meetings and have achieved agreement on a number of issues, the San Diego–Tijuana border area is not truly integrated into a serious structure of common policy and action. They have, however, collaborated in areas like flood control and sewage disposal for the Tijuana River (which lies mostly in Mexico, but empties into the Pacific in the United States), drug control, regional transport, water supply, cultural exchanges, and expositions.

The manifestations of a common border culture are found less in formal agreements than in the activities of thousands of people who divide their lives between both sides. On the map, the border appears as something fixed, but to individuals who cross frequently, it is a minor aspect of their personal and economic activities. The 1987 demographic surveys of Baja California indicate that about eight percent of the economically active population of Tijuana works in California. Some hold "green cards"—alien permanent residence permits which allow them to work in the United States. Others have "blue cards"— border crossing permits for residents of the Mexican frontier cities allow-

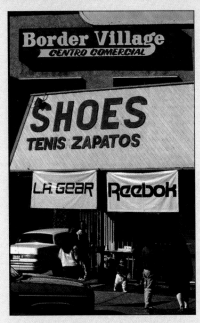

Two distinct cultures and economies are intermingled in the towns on either side of the U.S./Mexican border.
Les Stone/Sygma

ing them to stay up to 72 hours within 25 miles of the border. Though blue card holders are not supposed to work, many do. Some Tijuana residents hold jobs on both sides. Several thousand American citizens also commute every day, working in California, but living in the Tijuana area where costs are lower.

The economic interchange is even more pronounced. Dollars are as common as pesos in Tijuana, serving as legal tender not only in tourist shops, but everywhere else. Tijuana's exchange houses are major conduits for flight of capital from

Mexico and for the entrance of earnings by legal and illegal workers, as well as remittances to relatives from California. Tens of thousands of people cross the border in each direction every day with the intention of spending money. The chief groups going to Tijuana have already been discussed. Those going to California include workers, ordinary Mexicans on shopping trips, wholesale and retail merchants, importers of inputs for Tijuana factories.

Across the border from Tijuana much of the population is composed of migrants from Mexico who came within the past two generations. They brought with them aspects of their culture like music, food, and religion, but especially a tenacious retention of Spanish language that is unusual among immigrant groups. While "americanization" has created in them a Mexican-American culture different from the Mexican, the heritage is unmistakable.

The culture of the border is in a state of continuous change. The San Diego–Tijuana area is quite different now than it was in the 1920s, when both were small cities and Tijuana's economy rested on vice. Now they are huge, complex cities that will exceed a million inhabitants well before the year 2000 (San Diego, which had 875,504 in 1980, has probably already done so). Their location in different countries will continue to distinguish them, but increased formal bonds are likely to be added to the already existent human and economic ones. In time, Tijuana will exceed San Diego in population, becoming the second

largest city on the Pacific Coast north of the equator. Presumably, by then its significance as a commercial and industrial center will have overcome completely its anachronistic reputation as a center of vice.

SOURCE: "Tijuana, Mexico's Pacific Coast Metropolis," by Thomas G. Sanders. Pp. 7–8 in *UFSI Reports* No. 38. Copyright © 1987 by Universities Field Staff International, Inc.

*Ramón de Jesús Ramírez Acosta and Victor Castillo Rodríguez, "La frontera México-Estados Unidos: Estudio de las economiás de Baja California y California," *Cuadernos de Economía*, série 1 (1985). 13–14.

†Joseph Nalven, "Transboundary Environmental Problem Solving: Social Process, Cultural Perception," *Natural Resource Journal*, XXVI (fall, 1986), 801–02. On specific problems of land use planning, see Lawrence A. Herzog, "The Cross Cultural Dimensions of Urban Land Use Policy in the US-Mexico Border: A San Diego-Tijuana Case Study," *The Social Science Journal*, XXII (July 1985), 29–46. Herzog shows that separate decisions on land use in the Otay Mesa led to poor site planning and inadequate environmental controls. A major obstacle was different land-use policies in the two countries.

TABLE 2.1   Latent and Manifest Functions and Dysfunctions of the *Maquiladora* Program

The discussion of latent and manifest functions and dysfunctions associated with the transfer of labor-intensive manufacturing jobs from the United States into Mexico shows the functionalists' approach in analyzing any "part." This chart lists the various functions and dysfunctions associated with the *Maquiladora* Program.

	Manifest	**Latent**
Functions	*Maquilas* are a source of jobs for Mexicans. GNP of Mexico increases. U.S. GNP increases. *Maquilas* save and create jobs in the United States.	The increased exchange of goods, services, persons, and information between cities on both sides of the border has created a "buffer" zone. There is a transfer of white-collar jobs from the United States into Mexico (a plus for Mexico).
Dysfunctions	There is a mismatch between skills needed for jobs lost and jobs created in the United States. Many workers lost jobs in the United States. The returning *braceros* were not hired.	Problems in one economy affect the other economy. There is a transfer of white-collar jobs out of the United States into Mexico (a minus for the United States). Rapid growth of *maquila* cities contributes to problems on both sides of the border.

---

just beginning to emerge. When the *maquila* industry started in 1965, most of the white-collar work that was necessary for *maquila* operations was done by workers in the United States. Since then, however, many Mexican citizens have participated in high-school equivalency programs offered by *maquilas* or have graduated from one of Mexico's public universities or technical schools whose missions are to prepare young people for skilled work (Uchitelle 1993). Engineers, technicians, administrators, computer operators, accountants, personnel officers, and lower and middle managers now can be drawn from the Mexican population. Because salaries for Mexican technicians average $8,400 a year, compared with $25,000 or more earned by their American counterparts, this alternative is attractive for *maquila* owners (Jacobson 1988). Although this situation is functional for Mexico, it is a latent dysfunction for the white-collar segment of the American labor force.

The populations of the border cities, where most *maquila* plants are located, have increased rapidly due to migrations of persons seeking employment from the interior of Mexico. This rapid and unregulated population growth has resulted in another latent dysfunction: the establishment of large human settlements without proper sanitation, transportation, water, electricity, or social services. The negative consequences of such rapid and unplanned growth for Tijuana became evident in January 1993 after 12 days of heavy rain. Squatter settlements and other housing developments built in the 30 or so canyons around the Tijuana River blocked the natural course of the water, causing flooding and mud slides in low-lying areas. As a result, more than 5,000 people were left homeless (Golden 1993). Again it is important to point out that the *maquila* industry is one factor in a complex series of factors contributing to these problems.

Problems such as pollution, traffic congestion, flooding, and disease affect community life on both sides of the border (Herzog 1985). Hence, these problems cannot be solved by one side alone. Problems do not stop at the border, as illustrated by the

TABLE 2.2 Principal Activities of the U.S. Border Patrol, 1989

Persons		Conveyances	
Questioned	19,806,036	Examined	7,806,782
Apprehended	906,535	Seized	10,789
Deportable aliens located	891,147	Value of seizures (merchandise)	
(non-Mexican)	60,162	Narcotics	$1.2 billion
		Other	$21.2 million
Smugglers of aliens located	13,794		

SOURCE: U.S. Immigration and Naturalization Service (1990), p. 130.

automobile and truck exhaust generated by traffic slowdowns and delays on the many bridges and other border crossings between Mexico and the United States. It is estimated that vehicles waiting to cross the Cordova Bridge (Bridge of the Americas), which joins El Paso, Texas, and Ciudad Juárez, emit 1,280 tons of carbon monoxide, hydrocarbons, and oxides of nitrogen annually (Roderick and Villalobos 1992). Obviously, emissions from commercial vehicles affect both sides of the border.

Table 2.1 summarizes the manifest and latent functions and dysfunctions associated with the *maquiladora* industry. These concepts help us answer the questions: Why do *maquiladoras* exist, and what consequences do *maquilas* have on the United States and Mexico? The strength of the functionalist perspective is that it gives us a balanced overview of a part's contribution—negative, positive, intended, and unintended. The weakness of this perspective is that it still leaves us wondering whether the overall impact of *maquilas* is positive or negative for the United States and Mexico. (As a case in point see "Women in the *Maquiladoras:* Latent Function or Dysfunction?") On the other hand, the functionalist perspective does help us see that some segments in each country benefit more than others from this labor-transfer arrangement. At the same time, it leaves us believing that the negative consequences are simply the costs for overall order and stability. As we will learn in the next section, such a conclusion would receive no support from a conflict theorist.

The Conflict Perspective

In contrast to functionalists, who emphasize order and stability, conflict theorists focus on conflict as an inevitable fact of social life. These sociologists emphasize the role of competition in producing conflict. Dominant and subordinate groups in society compete for scarce and valued resources (such as wealth, education, power, prestige, leisure, and health care). Those who gain control of these resources have the power to protect their own interests.

Conflict between groups can take many forms besides outright physical confrontation, including subtle manipulation, disagreement, dominance, tension, and hostility. Table 2.2 shows various forms of conflict in encounters between border patrol agents and people suspected of being illegals, including

Women in the Maquiladoras:
Latent Function or Dysfunction?

The . . . debate over the North American Free Trade Agreement (NAFTA) . . . focused public attention on the Mexican *maquiladora* industry. Although it has never been demonstrated that the movement of U.S. firms to Mexico and elsewhere is the main cause of rising unemployment in the United States, many Americans see Mexican workers as direct competitors with U.S. labor for manufacturing jobs. The knowledge that most *maquiladora* workers are women has heightened the controversy, for many Americans believe that the women hired by U.S.-based firms have less need for wages than the U.S. workers, many of whom are men, whose jobs are being eliminated. The *maquiladoras'* employment practices are also a focus of concern for many Mexican citizens, who lament that U.S. companies' preference for female labor is disrupting Mexican society. Whether or not one agrees with these criticisms, it is difficult to escape the conclusion that women's work in *maquiladoras* is generating considerable debate.

Women have always been the preferred work force for the unskilled, low-wage jobs that characterize the *maquiladoras*. Although most managers and skilled technicians are men, most of the workers who assemble clothing, electronics components, and other products are women. When the *maquiladoras* were first established, most assemblers were young, single, childless women who worked for a few years before dropping out of the labor force to marry and raise families. Over time, as more *maquiladoras*

came to the border and the demand for female labor began to exceed the supply of young, unmarried women willing to take assembly jobs, firms changed their recruitment practices to encompass more married women and more women with children. Regardless of their age or family arrangements, most women assemblers make essential economic contributions to households in which men are often unemployed or underemployed.

The large-scale incorporation of women into the manufacturing work force was a latent function of the *Maquiladora Program*. When the Border Industrialization Program (BIP) was established in the mid-1960s, it was intended to alleviate the unemployment and underemployment troubling the border region. Because Mexican culture defines wage work as a male activity and views women in terms of their roles as wives and mothers, it was assumed that men were the primary victims of unemployment and that men would be the key beneficiaries of the new jobs created within the *maquiladoras*. The founders of the BIP did not anticipate that the U.S. firms establishing *maquiladoras* would focus their recruitment efforts on women. Their hiring practices in Mexico were not unique; transnational corporations investing in various developing countries employ women for assembly jobs. Although *maquiladora* managers attribute their preference for female labor to women's manual dexterity, patience, and tolerance for monotonous work, feminist critics counter that these companies hire women

because they are cheaper to employ than men, and because their gender-role socialization makes them more docile, and thus less likely to organize unions, than men. As a result of these employment practices, the *Maquiladora Program* has not reduced male unemployment, but instead has drawn a previously non-employed sector of the population into the labor force.

Was this unintended consequence a latent function or dysfunction? Has the employment of women been beneficial or detrimental to Mexican women, their families, and the society generally? Critics of the program emphasize its negative consequences. Some argue that the *maquiladoras* have harmed women by subjecting them to exploitative, menial, low-wage jobs that impair their physical and mental health. Others claim that women's *maquiladora* employment has damaged the Mexican family. In Mexico, conventional gender roles assign women to the sphere of home and family and men to the world of waged employment, leading to a clear-cut gender division of labor within and outside the household. In encouraging women to take jobs outside the home, critics claim, the *Maquiladora Program* has disrupted family equilibrium by giving women new statuses that weaken their adherence to traditional feminine roles, and by giving women resources that enable them to challenge men's traditional roles as sole family wage-earners. Critics claim that male alcoholism, divorce, spousal abuse, and other indicators of family disruption are on the rise, and that these dislocations are a direct result of women's employment in the *maquiladoras*. Finally, critics argue that the *maquiladoras'* preference for female labor has contributed to the border region's unemployment problems. In their view, rather than providing jobs for those who most need them—men with families to support—the *maquiladoras* have employed women who previously were supported by husbands or fathers. Once women join the economically active population by taking jobs in *maquilas*, they develop a life style requiring continual waged employment. If, as is often the case, their factory lays them off or they quit their jobs because they cannot tolerate the drudgery and monotony of assembly work, they will look for another, different kind of job. When displaced *maquiladora* workers take jobs that could have been filled by men, they contribute to male unemployment by displacing men from the labor force.

Advocates of the *Maquiladora Program*, by contrast, emphasize its positive consequences: in their view, the *maquiladoras* provide jobs for women who must earn a wage to support themselves and their families. Many Mexican women are single household heads supporting their children without assistance from male partners. Those who live in stable partnerships often have little economic security, because many Mexican men are unemployed, and others earn a wage that is insufficient to support their households. Throughout the last decade, Mexico has suffered from a severe economic crisis marked by spiraling inflation

and cutbacks on government services, which have made it difficult for households to subsist on a single wage-earner's income. In most households, the earnings of all adult members, whether male or female, have been essential to the family's economic survival. Without the employment opportunities provided by the *maquiladoras*, many border region families might have faced economic destitution. Not only has the *Maquiladora Program* benefited many families by providing women with needed income, but it has offered women jobs that increase their status in society. According to this view, wage work gives women resources that increase their personal autonomy and enhance their bargaining power relative to male partners or fathers. Women who earn an income tend to have more input into household decisions than nonemployed women who are completely dependent on male members to support them. Not only do they have some say-so about how their wages will be spent, but they have the economic wherewithal to leave abusive marriages when necessary to protect themselves and their children. Their employment in the *maquiladoras* thus provides the economic foundation for their liberation from repressive family relationships.

Clearly, whether one views the women's employment in the *maquiladoras* as a latent function or dysfunction for Mexican society depends upon his or her theoretical perspective. Those who believe that women benefit from paid employment are apt to stress the positive

consequences of the *Maquiladora Program*, while those who see the *maquiladoras* as vehicles of capitalist exploitation or those who view women's roles solely in domestic terms are apt to stress the *maquiladoras'* negative impacts on society. Regardless of one's position on these issues, it is clear that for the time being the *maquiladoras* will continue to employ women in the bulk of their assembly jobs. Policies such as raising the federal minimum wage to guarantee women workers a decent standard of living, and regulating working conditions within firms to protect workers' physical and mental health, are possible strategies for ensuring that the negative consequences of women's employment do not outweigh the benefits.

SOURCE: Susan Tiano, University of New Mexico (1994).

questioning, arrests, deportation, and property and vehicle searches and seizures. Notice that in 1989 almost eight million vehicles were searched and approximately 20 million people were questioned, resulting in more than 900,000 apprehensions and 10,000 seizures. In fact, the conflict is not just confined to encounters with illegal immigrants. Agents obviously confront or question "illegal-looking" people as well, because there are many more searches and questionings than apprehensions.

Conflict theorists draw their inspiration from Karl Marx, who focused on class conflict. As you saw in Chapter 1, Marx maintained that there are two major classes and that class membership is determined by relationship to the means of production. The more powerful class is the bourgeoisie, or those who own the means of production (land, machinery, buildings, tools) and purchase labor. The bourgeoisie, motivated by the desire for profit, need constantly to expand markets for their products. In addition, they search for ways to make production more efficient and less dependent on human labor (using machines, robots, and automation, for example), and they strive to find the cheapest labor and raw materials. These needs spread "the bourgeoisie over the whole surface of the globe. It must nestle everywhere, settle everywhere, establish connexions everywhere" (Marx [1888] 1961, p. 531). According to Marx, in less than 100 years of existence, the bourgeoisie "has created more massive and more colossal productive forces than have all preceding generations together" (Marx [1888] 1961, p. 531).

The less powerful class, the proletariat, consists of the workers who own nothing of the production process except their labor. The proletariat's labor is a commodity no different from machines and raw materials. Mechanization combined with the specialization of labor has left the worker with no skills; the worker is an "appendage of the machine and it is only the most simple, most monotonous, and most easily acquired knack that is required of him" (Marx, p. 532). As a result, workers produce goods that have no individual character and no sentimental value to either the worker or the consumer.

Conflict exists between the two classes because those who own the means of production exploit workers by "stealing" the value of their labor. They do so by paying workers only a fraction of the profits they make from the workers' labor and by pushing them to increase output. Increased output without a commensurate pay raise reduces wages to an ever smaller fraction of the profit.

The exploitation of the proletariat by the bourgeoisie is disguised by a **facade of legitimacy**—an explanation that members in dominant groups give to justify their actions—or by a justifying ideology. Conflict theorists define **ideologies** as fundamental ideas that support the interests of dominant groups. (The notion that poor people are poor because they are lazy rather than because they are paid low wages is an example of an ideology.) Because ideologies are backed by those in power, they are taken as accurate explanations of why things are the way they are.

On closer analysis, however, ideologies are at best half truths, based on "misleading arguments, incomplete analyses, unsupported assertions, and implausible premises. . . . [A]ll ideologies foster illusions and cast a veil over clear thinking . . . that enable class divisions in society to persist" (Carver 1987, pp. 89–90). The capitalists' exploitation of the proletariat is justified by the argument that the proletariat are free to take their labor elsewhere if they don't like the arrangement. Upon closer analysis, however, this is not the case. On the most basic level employers have considerably more leverage over workers than vice versa: if workers make too many demands, are unreliable, or do not produce— or if business is slow—employers can fire their workers. Workers have no comparable leverage against unreliable and overdemanding employers. Furthermore, many workers must take what employers offer, because hundreds of other workers may be waiting to fill the jobs if they refuse. For the most part, workers are an "incoherent mass scattered over the whole country, and broken up by their mutual competition" (Marx, p. 533).

Several examples show how the owners of production hold the advantage over workers who are competing against one another for jobs. In 1989 there were three farm workers for every job available in Florida. In this situation employers could offer low wages. If workers refused to accept the wages, there were always others who would.

A "FRONTLINE" (1990) interview with a migrant worker named Mr. Silva, whose home base is in Florida, reveals that often workers are angry (rightly or wrongly), not with farm owners but with other migrant workers who accept the low wages: "They're running us out of there. Those Guatemalans, they say, 'If you pay him $3.35 an hour, you can pay me $3.00 an hour.'"

This scenario is hardly different from the coal miners' strike against Pittston Coal Company or the bus drivers' strike against Greyhound that occurred in 1989 and 1990. In both cases company executives hired permanent replacements for striking workers. By hiring new workers, Pittston managed to keep 25 of its 30 mines and plants open in Virginia and West Virginia (Hinds 1989, 1990); likewise, Greyhound trained 2,600 new drivers in 13 days to replace striking bus drivers (Bearden 1990). This strike lasted three years, from April 1990 until May 1993. A U.S. Federal judge subsequently ruled that Greyhound officials had provoked the bus drivers to strike (Myerson 1994). Replacement workers accepted the jobs because, as one worker put it, "I got to make a living too. My family comes first; it's every man for himself now" (Dupont 1989, p. 16y). In some cases the jobs provide the replacement workers with a better income: "They [the strikers] say that they're not makin' enough money. Well, they can come down here and make what I was makin' and see how they live" (Madden 1990, p. 15). These cases show that the widespread availability of replacements weakens the power of the workers (and of the unions that represent them) to make demands of employers (Kilborn 1990a). As a result, the effectiveness of strikes has diminished in recent years (see Figure 2.3).

The ideas of Marx inspired most conflict theorists, and the fundamental theory has many subtle variations. Yet, despite these variations, most conflict theorists ask this basic question: Who benefits from a particular pattern or social arrangement, and at whose expense? In answering this question, conflict theorists strive to identify practices that the dominant groups have established, consciously or unconsciously, to promote and protect their interests. Exposing these practices helps to explain why there is unequal access to valued and scarce resources.

Most conflict theorists also examine the facade of legitimacy that supports existing practices. They observe how exploitive practices are justified logically by those in power. The most common methods of justification are (1) blaming the victims by proposing that character flaws impede their chances of success and (2) emphasizing that the less successful benefit from the system established by the powerful. Consider the argument a Denver woman gives "MacNeil/Lehrer Newshour" correspondent Tom Bearden (1993) for hiring illegal immigrants to care for her children.

MR. BEARDEN: Does the employer of an undocumented worker have too much power over that person? There are some who believe that people who hire undocumented aliens gain an unfair power over them, it gives them influence over them because they're, in a sense, collaborating in something that's against the law. Do you agree with that, or have any thoughts about that?

DENVER WOMAN: I guess I would disagree with that. The one thing that you get in undocumented child care or the biggest thing that you probably get, my woman from Mexico was available to me 24 hours a day. I mean, her cost of living in Mexico and quality of life in Mexico compared to what she got in my household were two extremes. When we hired her, she said, "I'll get up with the baby, I'll be available all hours of the day, I'll clean the house, I'll cook." They do everything. And if you hire someone from here in the States, all they're going to do is take care of your children. So not only do you have a differentiation in price, you have a differentiation in services in your household. I have to admit that was, at that point, with a newborn infant, was wonderful to have someone who was so available. . . .

MR. BEARDEN: And it's not like indentured servitude?

DENVER WOMAN: That crossed my mind, and after she had been here for six months or so, we went to a schedule where she finished at 6 or 7 o'clock at night. And I don't think I ever really took advantage of her. Once a week I'd have her get up with the baby, so I didn't. . . . [S]he was available to me, but I don't feel like I really took

FIGURE 2.3 On Strike in the United States: Number of Strikes Involving 1,000 or More Employees

Since the 1980s the number of strikes by workers declined to an all-time low. The decline is related to the practice of hiring permanent replacement workers, which gained national attention when President Reagan replaced striking air traffic controllers, and to the large number of young people (baby boomers) who entered the labor force at a time when the number of high-paying industrial jobs was in decline.

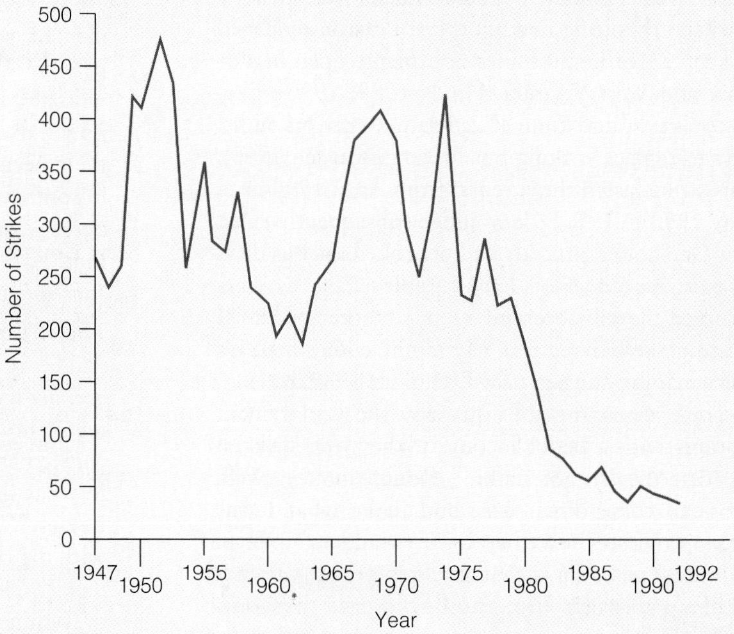

SOURCE: U.S. Bureau of Labor Statistics (1993).

advantage of her, other than the fact that I paid her less and she was certainly more available. But she got paid more here than she would have gotten paid if she'd stayed where she was. (p. 8)

Conflict theorists would take issue with the logic that this Denver woman uses to justify hiring an undocumented worker at a low salary. When it comes right down to it, the Denver woman is protecting and promoting her interests (having someone available all hours of the day to cook, clean,

and do child care) at the expense of the Mexican worker.

As a final example of how capitalists use a facade of legitimacy to justify exploitive practices, let's look at how they explain the fate of jobless workers. More than 12 million American factory workers have lost jobs because of plant relocations and shutdowns. Many of the companies from which factory workers have been laid off, such as General Electric and General Motors, have established job re-entry programs. An article by Peter

Kilborn in *The New York Times* highlighted the plight of 1,200 union workers laid off after General Electric moved refrigerator production from Cicero, Illinois, to Decatur, Alabama. There, nonunion workers are paid $9.50 per hour, $4 less than the company paid the Cicero employees. In Kilborn's article, Robert Jones, then the U.S. assistant secretary of labor for employment and training, described the laid-off workers as being "as dysfunctional as you can get." Others who were interviewed described the workers as "not able to meet entry-level requirements of other jobs" and as "relatively old and unskilled." Many were described as functioning at fifth- or sixth-grade levels, and some Hispanics were said to "have never had to communicate in English" (Kilborn 1990b).

At face value, these comments suggest that companies like General Electric, in conjunction with the U.S. Department of Labor, are concerned enough to be spending $4 billion in efforts to retrain laid-off employees who, employers believe, lack the intelligence, motivation, skills, and so on, to take advantage of the potential employment opportunities available to them. Conflict theorists, however, would point out that the fate of these jobless employees represents the tragic outcome of a production process that requires so little from its workers (other than repetitive manual labor) that the workers have acquired no skills and have nothing marketable to show for decades of employment. From a conflict perspective, the laid-off workers are not to blame for their fate. Those who used them as if they were machines to do mindless, repetitive work are to blame.

Critique of Conflict Theory

Like the functionalist perspective, conflict theory has its shortcomings. A major criticism is that it overemphasizes the tensions and divisions between dominant and subordinate groups and underemphasizes the stability and order that exist within societies. It tends to assume that those who own the means of production are all-powerful and impose their will on workers who have nothing to offer except their labor. The theory also assumes that the owners exploit the natural resources and the cheap labor of poor countries at will and without resistance. This is a somewhat simplistic view of the employer-employee relationship and of relationships between corporations and host countries. It also tends to ignore the real contributions of industrialization in improving people's standard of living.

Moreover, the owners of production do not always strip workers of all of their skills. For example, the IBM typewriter factory in Lexington, Kentucky, has automated its plant and retrained employees who once performed simple tasks to do more demanding and less monotonous work. Some employers recognize that, although machines are cheaper to maintain than people, they must still invest in people who are needed for complex tasks (Pranis 1990). Many companies deserve attention for their favorable employee policies. Xerox, Johnson & Johnson, and AT&T are just three companies praised by *Working Mother* magazine in its article, "100 Best Companies for Working Mothers." Noteworthy companies also are named in *The 100 Best Companies to Work for in America* (Flanigan 1993; Silverstein 1993).

Finally, conflict theorists tend to neglect situations in which consumers, citizen groups, or workers use economic incentives to modify or control the way capitalists pursue profit. For example, the Environmental Defense Fund negotiated successfully with McDonald's to ban polystyrene packaging at its 8,500 restaurants across the United States. The Earth Island Institute Dolphin Project used lawsuits, court injunctions, and letter-writing campaigns to persuade Starkist to change its fishing techniques so as to make oceans safer for dolphins (Koenenn 1992).

The Conflict Perspective on the Maquiladora Program

From a conflict perspective, *maquilas* represent the pursuit of profit. The means of production (machinery, raw materials, tools, and other components) are owned by capitalists in the United States and other foreign countries. The Mexican workers own only their labor and must sell it to the owners of production at a low price. They cannot demand higher wages and better working conditions because the employers can easily replace them from a large pool of Mexican workers. If labor problems emerge, foreign companies can move their operations to the interior of Mexico or to another country.

As for the facade of legitimacy, there are those who maintain that the *Maquiladoras* Program benefits both the United States and Mexico. Mexico gains because its growing work force gets jobs and the United States gains because its industries become competitive in the world market. The loss of some American jobs is more than compensated for, considering the alternatives of bankruptcy and widespread unemployment.

However, from a conflict point of view, the facade of legitimacy masks the real purpose of the *maquila* industry, which is to increase profits by exploiting the most vulnerable and least expensive labor. U.S. businesspeople describe this labor arrangement as being of mutual benefit. But, in reality, they do not want Mexico to prosper because their success is tied to a fragile Mexican economy: the weaker the peso, the lower the wages that company owners have to pay workers. As long as Mexican labor is a bargain in the world labor market, companies in the United States and other countries will continue to locate assembly plant operations there.

The importance of the Mexican plant to General Motors' overall operations was made plain in a conversation between Bill Moyers and a GM plant manager, presented in the CBS documentary "One River, One Country: The U.S.–Mexican Border." An excerpt of their conversation follows. The plant manager's evasiveness on the question of wages suggests that the owners of production are fully aware that they exploit Mexican workers.

GM MANAGER: We are constructing the rear body wiring harness that goes into all GM autos for assembly in the United States.

MOYERS: These are the electrical wires that run to backup lights and turn signals . . .

GM MANAGER: [interrupts] . . . backup lights and turn signals and to license plate lamps.

MOYERS: How important is this plant to General Motors?

GM MANAGER: If one of these wiring harnesses does not get built, an automobile goes across the assembly line without a wiring harness in it.

MOYERS: If the wiring doesn't get done, then no car gets built?

GM MANAGER: That's right; we are a single-source plant, as all our plants are.

MOYERS: And what about their wages? What do these young people get paid?

GM MANAGER: Well, their wages obviously are on the Mexican wage scale, and we follow those standards as a *maquila* industry.

MOYERS: So what are those?

GM MANAGER: [pause] The numbers are ever-changing. The minimum wage has just undergone a new adjustment. It will be a 25 percent increase [to] 1,640 pesos a day for the minimum wage.

MOYERS: So that's about—under the new devaluation—about three dollars?

GM MANAGER: I believe that's right.

MOYERS: Is that per hour or per day?

GM MANAGER: Per day. (MOYERS 1986)

The Moyers interview shows why Mexico is an attractive place for investment. Mexico's low-cost labor pool is virtually guaranteed for decades to come. Nearly one million new jobs must be created each year in Mexico's job market just to absorb those who reach working age. (Sixty-four percent of Mexico's population is under 23 years of age.) Yet, despite low-cost labor along the border, there is a movement among outside investors to find even cheaper labor pools within Mexico's near and deep interiors.

Compared to the employer, the Mexican worker gains very little (see Table 2.3). *Maquila* jobs are characterized by insecurity, lack of advancement, and exceedingly low wages. Most of the work is mind-numbing and outrageously repetitive. In electronics plants, for example, workers peer "all day through a microscope, bonding hair-thin gold wires to a silicon chip destined to end up inside a pocket calculator" (Ehrenreich and Fuentes 1985, p. 373). In the *maquilas* where manufacturers' coupons are sorted, workers sort up to 1,300 coupons each per hour. Ten of thousands of times per day, workers drag coupons across an electronic bar code scanner, which flashes a numeric code that identifies the slot where the worker is to file the coupon (Glionna 1992).

TABLE 2.3   The Cost of Everyday Items

According to the conflict theorist Karl Marx, workers exchange their labor for a wage that goes to purchase what they need to survive. A comparison of the labor-time Mexican and American workers "pay" for a number of everyday items is shown below. Prices of various items in Mexico and metropolitan San Francisco (converted at N$ 3.27 to the dollar) are based on recent random samplings and adjusted for different packaging sizes and weights. The time worked to earn these goods is based on a 40-hour workweek and 1993 U.S. Labor Department statistics for the average wage, including fringe benefits, of manufacturing workers in the United States and Mexico ($10.58 and $2.17 per hour, respectively).

Item	Mexican Price ($US)	Time to Earn (hours)	U.S. Price ($US)	Time to Earn (hours)
Nintendo Action Set	$134.18	61.83	$99.99	9.45
Goldstar Microwave	$207.38	95.56	$179.95	17.08
Shampoo (440 ml)	$2.60	1.20	$3.79	.36
Gillette Razors (10 pack)	$4.86	2.24	$8.99	.85
Black Beans (lb)	$.36	.17	$.89	.08
White Cheese (lb)	$1.73	.80	$3.99	.38
Ham (lb)	$1.99	.92	$1.18	.11
Rice (lb)	$.36	.17	$.47	.04
Dark Rum (750 ml)	$4.86	2.24	$5.88	.53

SOURCE: Compiled by Jason Moore, *Excelsior* (1994).

Along with exploiting Mexican labor, multinational corporations pollute the environment. The Mexican equivalent of the U.S. Environmental Protection Agency found that approximately 1,000 American-owned *maquila* plants generate hazardous waste. Only about one-third comply with Mexican laws requiring them to file reports on how hazardous waste is handled. Only one in five could document that they dispose of hazardous wastes properly[13] (Suro 1991). In other words, *maquilas* take from the Mexican people and give little in return:

> U.S. corporations come right across the line. They take our energy but leave nothing. The maquiladoras *haven't donated buildings, parks, or anything else. General Motors squeezes Mexican brains and pays vassal wages for the privilege of inhaling paint thinner. Women go crazy attaching gray polyester sleeves over and over, never learning how to make the whole blouse. Fluoride factories that no city in the U.S. will* accept come here. Companies pull out overnight if they feel like it. The people have no say in what happens to them. (WEISMAN 1986, P. 12)

Generally speaking, in the context of Mexico's overall employment problems, the *maquilas* do little to alleviate unemployment. In 1988 about 500,000 persons worked at *maquila* plants, which have been in existence since 1965; to stave off unemployment, though, Mexico would have to generate about one million new jobs a year.

American and other foreign employers have the upper hand not only because Mexico needs a seemingly endless number of jobs but also because the country faces intense pressures to generate foreign currency to pay back $100 billion in loans. Lending institutions in countries like the United States make profits from the high interest rates they charge. *Maquila* owners benefit from the conditions of the loans, which stipulate that Mexico must promote export manufacturing as a means of generating foreign currency for loan repayments.[14]

In the meantime, American companies with assembly plants in Mexico gain another measure of control over workers and communities in the United States. In the face of a threatened relocation, employees take pay cuts, work harder, and complain less. Absenteeism declines and productivity rises. Many local, state, and federal governments give land, special tax breaks, and wage concessions to keep operations in their region or to entice companies to locate new operations in their areas (Lekachman 1985).

In response to all of this, conflict theorists ask: Who benefits from the transfer of labor-intensive manufacturing or assembly operations from the United States into low-wage, labor-abundant countries such as Mexico? The answer is the owners of production, or the capitalists:

> But if you stretch it a little, we're all a part of it—anyone who has anything to do with Ford, GM, GE, or if you own anything electronic

that's not Asian. Anyone in Wisconsin or Michigan working for these firms or using their products does so at the expense of Latin America. (WEISMAN 1986, P. 133)

Unlike the functionalist perspective, which is unclear about the overall effect that an event or arrangement has on society, the conflict perspective zeros in on its exploitive consequences. Thus, for conflict theorists the overall effect of the *maquiladoras* is clear: *maquiladoras* benefit a small segment—the capitalists. The clear losers are the workers. It is not an arrangement that benefits workers or strengthens their role in the production process.

We turn now to the third theoretical perspective—symbolic interaction. It is distinct from the functionalist and conflict perspectives, which focus on the way social systems are organized. Instead, symbolic interactionists focus on how people experience and understand the social world.

The Symbolic Interactionist Perspective

In contrast to functionalists, who ask how parts contribute to order and stability, and conflict theorists, who ask who benefits from a particular social arrangement, symbolic interactionists ask how people define reality. In particular, they focus on how people make sense of the world, on how they experience and define what they and others are doing, and on how they influence and are influenced by one another. These theorists argue that something very important is overlooked if an analysis does not consider these issues.

Symbolic interactionists have drawn much of their inspiration from American sociologist George Herbert Mead (1863–1931). Mead was concerned with how the self develops, with how people attach meanings to their own and other people's actions, with how people learn these meanings, and with how meanings evolve. Consequently he focused on people and their relationships with one another. He maintained that we learn meanings from others, that we organize our lives around those meanings, and that meanings are subject to change (Mead 1934).

According to symbolic interactionists, symbols play a central role in human life. A **symbol** is any kind of physical phenomenon—a word, an object, a color, a sound, a feeling, an odor, a movement, a taste—to which people assign a meaning or value (White 1949). The meaning or value is not evident from the physical phenomenon alone. This is a deceptively simple idea that suggests that people *decide* what something means. In order to grasp this idea, you might think about how young children question the meaning of everything. As a parent once told me about children, "They don't understand anything. Everything is learned." Meaning, for the child, evolves through interactions with others.

Let's look at another example. Consider the colors black and white. *The Synonym Finder* (Rodale, 1986) lists 160 synonyms for *black* including *threatening, menacing, treacherous,* and *sinful*. For the most part, the color has no positive associations. On the other hand, 75 of the synonyms listed for *white* have positive associations, including *pure, stainless, innocent,* and *immaculate*. Although the color white has some negative associations (for example, colorless, dull, dingy, bloodless, bland, nondescript), it is not generally associated with ideas of danger.

Where do these symbolic associations come from? Some of my students have argued that those for black are related to a universal fear of the dark. In other words, the black of night is naturally threatening to all people. However, this interpretation may not be so universal. A Melanesian story, "Finding Night," from *In the Beginning: Creation Stories from Around the World*, shows that "blackness" can symbolize positive ideas such as rest, self-renewal, and rejuvenation. The following excerpt from this story illustrates the point. In this passage, the god Quat is asked to do something to put an end to daylight.

In the beginning, there was light. It never dimmed, this light was over everything. It was bright all-light everywhere, and there was no rest from it. . . .

"It's too light. . . . Quat, do something. We don't like the world so bright all the time. Make something to stop it, please, Quat."

Quat looked everywhere for something. Something that was not light. He could find nothing. Light was everywhere. He'd heard about such a place at the far edge of the sky, and it was called Qong, *Night. Quat . . . sailed over the sea toward the far edge.*

He sailed and sailed. Finally, Quat reached the edge where the sky came down and he could touch it. There lived Qong.

Night was dark. It had no light anywhere in it. It touched Quat about his eyes. . . . It taught him sleep, as well. And the great darkness, Night, gave him another piece of itself.

So Quat went home, taking the piece of Night in his hand. . . . (HAMILTON 1988, PP. 10–13)

Consider a final example: the various meanings assigned to a suntan. In the United States, a tan has at various times represented quite different ideas about social class, youth, and health. Around the turn of the century, wealthy persons purposely avoided tanning to distinguish themselves from members of the working class (farmers and laborers). Pale complexions showed that they did not have to make their livings outdoors, laboring under the sun. Then, as the basis of the U.S. economy changed from agriculture to manufacturing, a large portion of the population moved indoors to work. The meaning attached to a pale complexion changed accordingly to represent unrelieved indoor labor; a tan came to symbolize abundant leisure time (Tuleja 1987).

The presence or absence of a suntan also has stood for ideas about health and youth. Many Americans describe a tan as something that makes them look good and feel better. But this meaning is likely to change with increasing reports about the connection between exposure to the sun and premature aging and skin cancer. In the face of such evidence, a tanned complexion symbolizes the skin's "desperate attempt to protect the body from radiation."[15] These changes in the meaning of a suntan underscore the fact that a physical form becomes a symbol because people agree on its meaning. Likewise, they demonstrate that meanings of symbols can change as conditions change.

Symbolic interactionists maintain that people must share a symbol system if they are to communicate with one another. Without some degree of mutual understanding, encounters with others would be ambiguous and confusing. The importance of shared symbols frequently is overlooked unless a misunderstanding occurs. Several situation-comedy television shows—including "Mork and Mindy," "Perfect Strangers," and "Beverly Hillbillies"—have depended on such misunderstandings. They feature characters who do not share the same symbols as do other characters.

On a more serious level, diplomatic relations between countries can be strained when their respective leaders and citizens do not define events in the same way. For example, Mexicans and Americans interpreted very differently the arrival of 24,000 U.S. Marines to El Chorrillo, a Panama City barrio, to arrest General Manuel Noriega in 1989.[16] David Gergen, editor of *U.S. News & World Report* at the time, describes these differences:

We understood in the American press that Noriega had declared war against us and this is what we all reported. Their understanding here in Mexico was that Noriega had said we are in a state of war with the United States. Not that he had declared war. We see this as a surgical strike to grab Noriega and restore democracy. They see it as an occupation by American forces—yet another example of Yankee imperialism. We see this in terms of the number of American soldiers being killed. . . . Here in Mexico they talk almost entirely about the number of Panamanian civilians being killed and the newspapers display bodies of civilians and they have pictures of buildings that appear to be civilian buildings with great clouds of smoke over them and people in the street. And the [Mexican] government talks about the great numbers of civilian deaths. (GERGEN 1989, P. 12)

This example shows that problems arise when involved parties place different interpretations on the same event. It also shows that during interaction the parties involved do not respond directly to the surroundings and to each other's actions, words, and gestures. Instead they make interpretations first and then respond on the basis of those interpretations (Blumer 1962). This interpretation-response process is taken for granted. Usually we are not conscious that the meanings that we assign to objects, people, and settings make encounters understandable and shape our reactions. To make us aware of this fact, something must happen that challenges our interpretations.

Anthropologist Edward T. Hall describes the interpretation that guides how American white males shake hands: "The emphasis is on a firm strong handshake with direct and unblinking eye contact. One must demonstrate mutual respect, equality of status (for the moment at least), strength, sincerity, and dependability" (Hall 1992, p. 105).

This familiar interpretation, however, is challenged by the Navajo idea of a handshake. For the Navajo,

the emphasis is on proper feelings rather than image. One does not look the other in the eye (to do so signals anger or displeasure). All that is necessary is to hold the other human being in one's peripheral visual field while grasping his hand gently, so as not to disturb the natural flow of feeling between his state of being and yours. These handshakes could be protracted because the Navajo . . . like to ease into things and are jarred by abrupt transitions. (PP. 105–106)

In order to understand what makes the symbolic interactionist perspective distinct from that of the functionalist and conflict theorists, consider how the various theorists might answer the question: Why did the United States send troops to the Persian Gulf region after President Saddam Hussein sent Iraqi troops into Kuwait?

From a conflict perspective the American military buildup in the Persian Gulf between August 1990 and January 1991 can be analyzed: (1) in terms of a conflict over a scarce and valued resource (oil) and (2) in terms of the minority or working-class background of U.S. troops, especially as compared to the upper-class background of most government leaders who decide whether military action is necessary.

The buildup also can be analyzed in terms of its unifying function. That is, the Gulf situation functioned to turn the American people's attention away from divisive issues at home (for example, the savings and loan crisis, the huge budget deficit, relatively high levels of unemployment, and an impending recession). A functionalist would acknowledge the class differences between military personnel and government leaders but would emphasize that the military functions to offer working-class people, the unemployed and the poor, and those who want to serve their country in a military capacity a chance to earn money to go to college, to learn a skill, to gain on-the-job experience, or to fulfill a moral obligation.

Symbolic interactionists, in contrast, would focus on how the involved parties interpret the situation and construe each other's words and actions. In August 1990, for example, President Saddam Hussein "took issue with the way the West responded to his explanation of why he was detaining Westerners" who were described in the West as hostages (Gault 1990, p. 7). Western media reported that Iraqi leaders were using Americans and other foreigners as shields to protect important military weapons plants. Saddam Hussein (1990) claims that he was using foreigners to prevent war.

Critique of Symbolic Interaction

Symbolic interactionists inquire into factors that influence how we interpret what we say and do, especially those factors that promote the same interpre-

Functionalists, conflict theorists, and symbolic interactionists differ in their analysis of events such as the Gulf War. What might each theorist focus on in this photograph of a Colorado Springs rally?
Matt Slothower/Impact Visuals

tations from significant numbers of people. Related topics of interest include the origins of symbolic meanings; how meanings persist; and under what circumstances people question, challenge, criticize, or reconstruct meanings. Although symbolic interactionists are interested in these topics, they have established no systematic frameworks for predicting what symbolic meanings will be generated, for determining how meanings persist, or for understanding how meanings change. These are the shortcomings of the symbolic interactionist perspective.

Because of these shortcomings, the symbolic interactionist perspective does not give precise guidelines about where to focus one's attention. For example, whose interpretation should we focus on when we analyze an event such as the arrest of General Manuel Noriega or the U.S. involvement in the Persian Gulf? The cast of characters involved in these events is virtually endless. Even if we were able to consider every interpretation, we would still be left with the questions What really happened, and whose interpretation best captures the reality of the situation?

The Symbolic Interactionist Perspective on the Maquiladora Program

As we have seen, symbolic interactionists explore meanings that people assign to words, objects, ac-

tions, and human characteristics. Such a broad focus allows considerable flexibility when it comes to applying it to foreign manufacturing operations in Mexico. We can use this perspective to examine the different meanings assigned to *maquila* operations by people living on opposite sides of the U.S.–Mexico border, by policymakers in the United States and Mexico, by American and Mexican businesspeople, by union leaders in both countries, or by displaced workers. We can also use it to analyze why U.S. corporations initially were slow to move assembly operations to Mexico. When the value-added system (in which only the labor costs associated with assembling the product are subject to tariff) first became available in the mid-1960s, Hong Kong, not Mexico, grew to be the most important off-shore assembly partner of the United States. Long-standing images associated with Mexican workers and Asian workers may help explain why this happened.

Americans tend to stereotype Asians as hardworking, intelligent, and obedient (Yim 1989). In sharp contrast, Mexicans are often stereotyped as unambitious and lazy. The unflattering image that many people hold is that of the "lazy peon asleep in the sun, his sombrero tipped forward over his eyes" (Dodge 1988, p. 48). Many Americans also tend to associate Mexico with the afternoon siesta and believe that the country closes shop for a couple of hours after a big noontime meal. In view of these contrasting conceptions, it is understandable that a label bearing "Assembled in Mexico" evokes a much different meaning for consumers and businesspeople than a label bearing "Assembled in Hong Kong."

These images of the siesta and the sombrero, although prevalent, are at odds with the fact that the annual rate of growth in Mexico before the 1983 oil crisis was one of the highest in the world. In a country short on capital and arable land, that achievement required considerable human effort (Dodge 1988). Further, many who equate the afternoon siesta with lack of ambition and unwillingness to work probably do not know that for many Mexicans the workday begins at 5:30 A.M. and ends at 7:00 or 8:00 P.M. When we consider that Mexico is now the United States's most important off-shore assembly partner and that Japan is moving rapidly to capitalize on the strengths of the Mexican labor force, we might expect such negative images to fade. Yet, the results of a recent survey of 2,800 Japanese youths by the Mexican government suggest that these conceptions are still strong and are shared by people other than Americans.[17] When Japanese youths were asked, "What comes to mind when you hear the word *Mexico?*" some 30 percent answered *sombrero*, 20 percent said *dirt*, 10 percent said *crime*, and 30 percent said *desert* (Pearce 1987).

Although the origins of these images of lazy Mexicans are not clear, they are rooted in part in American views toward poverty and its causes and are perpetuated by the mass media. There is much poverty among the Mexican masses, and Americans typically define the poor as a drain on society. They tend to attribute poverty to inferior traits of the people themselves—laziness, lack of discipline, lack of skill, and so on. For some reason, many Americans find it difficult to envision people who work hard and yet remain poor.

In spite of the negative images of Mexico, this country has emerged over the past decade as a world leader in off-shore assembly operations. To see Mexico as a viable production site, owners of American companies had to redefine Mexican attributes in such a way that the Mexican labor force became a viable alternative to the Asian labor force, especially now that Asian countries were becoming industrial powers in their own right. Together, Mexican communities trying to attract American investors, corporations offering shelter plans, and U.S. border communities that might benefit from *maquila* operations advertised a new image of Mexico. What attributes did they emphasize to offset the negative stereotypes of Mexicans? Let's take a look at some of the strategies they used (and continue to use) to market (or define) Mexico.

First, advertisers list the names of U.S. companies with established plants in Mexico to assure potential clients that reputable companies have succeeded there. Second, they emphasize the proximity of *maquila* cities to the United States to remind potential clients that the border is "close to home" and that transportation and communication costs are cheaper than in Asia. Third, they describe the size and cost of the Mexican labor force relative to that of other countries. Finally, they show pictures of happy, middle-class-type workers to counter images of exploited foreign workers and to suggest that the workers' standard of living benefits from *maquila* employment. Here are some examples of advertisements from *Twin Plant News* (1990a, 1990b, 1992a), which illustrate these strategies:

- "What Industrial City Is 3 Times the Size of Dallas and 45 Times Closer to You Than Taiwan?" (The advertisement promotes Monterrey, Nuevo León.)

- "Shopping the Interior [of Mexico]? Picture This . . . ABUNDANT LABOR . . . Worker Housing for 180,000 is in close proximity to the park."

- "More than 1,400 U.S. owned and operated factories called 'Twin Plants' or '*Maquiladoras*' are now operating along the U. S.–Mexican border."

The symbolic interactionist perspective adds yet another dimension to the way we approach social events. It asks us to consider how the involved parties interpret an event. In this sense, it enhances the functionalist and conflict perspectives, which focus on the origins (societal need versus profit) and consequences (stability and disruption versus exploitation). Each perspective offers a unique set of questions and concepts to answer those questions and alerts us to avoid making simple statements or generalizations about events. The North American Free Trade Agreement represents an event that supporters and opponents alike evaluated with simple generalizations.

Discussion

We began this chapter with some facts that describe ways in which Mexico and the United States are interconnected. Taken alone, the facts tell us little about the relationship between the two countries. At first glance, many Americans see these facts as proof that Mexico threatens their economic well-being. The three theoretical perspectives give us a strategy for thinking about these facts and tempering hasty and oversimplistic reactions. The strategy is reflected in the questions and vocabulary of each perspective (see Table 2.4).

No single sociological perspective, of course, can give us a complete picture of social events. Each of the three basic perspectives offers only one way of looking at this phenomenon. With regard to *maquilas,* functionalists emphasize how the arrangements between the United States and Mexico create a pattern that contributes to the overall economic stability and the well-being of both countries. Conflict theorists examine how these arrangements benefit the owners of production and exploit workers in both Mexico and the United States. Symbolic interactionists consider the meanings assigned by various groups to *maquila* industries and explore how these meanings affect relationships between Mexicans and Americans. Significantly, despite the differences among these theories, none can be used to support the interpretation that Mexico is a drain on the American economy. In fact, all three support the notion that, for better or worse, the two countries are dependent upon each other.

GOOD POINT

Very few sociologists adhere to only one perspective and maintain that it should be adopted, to the neglect of the other two perspectives. Ideally, the three perspectives should not be viewed as clashing or incompatible. In fact, as you have seen, they overlap considerably. The conflict perspective's focus on exploitation overlaps the functionalist perspective's focus on manifest and latent dysfunctions. Both perspectives recognize the loss of American jobs, the low wages paid to Mexican workers, and the vulnerability of communities that rely on

This maquila worker is paid about $5.00 per day to sew pants for export to the United States.
Louis DeMatteis/JB Pictures

TABLE 2.4 Overview of the Three Theoretical Perspectives			
	Functionalist Perspective	**Conflict Perspective**	**Symbolic Interactionist Perspective**
Focus	order and stability	conflict over scarce and valued resources	shared meaning
Vision of Society	system of interrelated parts	dominant and subordinate groups in conflict over scarce and valued resources	interaction is dependent on shared symbols
Key Terms	function, dysfunction, manifest, and latent	means of production, facade of legitimacy	symbols
Central Question	How does a part contribute to overall stability of a society?	Who benefits from a particular pattern or social arrangement, and at whose expense?	How are symbolic meanings generated?
Major Criticisms	defends existing social arrangements; offers no technique to establish a part's "net effect"	exaggerates tension and divisions in society	no systematic framework for predicting which symbolic meanings will be generated or for how meanings persist or change

assembly jobs. But their emphasis differs. Conflict theorists make exploitation the focus. Functionalists view the negative consequences as unfortunate but probably necessary side effects if both countries are to survive and to prosper economically. Yet, despite these differences, the functionalists' understanding of the benefits of exploitive practices complements the conflict theorists' need to define and eliminate such practices. Understanding the function of these practices helps explain why they persist. Knowing why they persist helps in defining policies to eliminate them.

Symbolic interactionists can benefit from the insights of functionalists and conflict theorists as they attempt to understand the various meanings of *maquilas* for different segments of the world population. The functionalist and conflict perspectives can be used to explain the origins of the various symbolic meanings assigned to *maquilas*. Among those segments of society that benefit or profit, *maquilas* are more likely to evoke positive images. Conversely, *maquilas* probably evoke negative im-

ages among exploited segments. Interestingly, however, the people who benefit from *maquilas* are often unaware that they do so and thus define Mexico as a burden on the American economy. Similarly, persons whom conflict theorists would define as exploited are often unaware that they are exploited, and may define the *maquilas* in positive terms. Both symbolic interaction and conflict theory (specifically the concept *facade of legitimacy*) could be used to explain this incongruence.

Because no one perspective can capture all aspects of a situation, we can know more of a given situation if we apply more viewpoints. All three perspectives are useful in that each makes a distinct contribution to understanding. In addition to analyzing *maquilas*, these perspectives can be used to examine other arrangements among human beings. In fact, most sociological analyses contain elements of more than one perspective; thus it is difficult to associate an analysis with one theory.

In the chapters that follow, we will examine sociological explanations, definitions, and research on a number of important topics: culture, socialization, social interaction and the construction of reality, deviance, social organizations, education, religion, social stratification, race and ethnicity, gender, family, and social change. Usually, we will be unable to identify one of the theoretical perspectives underlying a sociological analysis; the analysis often will be based on a combination of perspectives. Sometimes, however, one particular framework will be especially well suited to a topic, or a sociologist will be aligned clearly with a particular framework. In those cases we will identify the sociologist with a perspective or will state the perspective that governs an analysis.

FOCUS
Evaluating NAFTA

The *Maquiladora* Program is not the same thing as NAFTA. This program does not contribute to the free movement of goods and services between Mexico and the United States. The *Maquiladora* Program simply allows U.S.-based corporations to purchase Mexican labor to assemble and process goods and services, which are exported back to the United States, not sold to the Mexican people. NAFTA, on the other hand, is an agreement that seeks to remove barriers to trade and to the movement of goods and services between countries.

These differences notwithstanding, our discussion of the *maquila* industry is relevant to NAFTA. We learned in this chapter that the *Maquiladora* Program cannot be analyzed easily or in simple terms. Its effects on the economies of Mexico and the United States are mixed. The same can also be said about NAFTA: its effects are complex and cannot be summarized by simple generalizations. The comments of NPR commentator James Fallows, which aired a few days before the U.S. Congress voted, illustrate this point.

NAFTA Not a Major Factor Regarding U.S. Economy

James Fallows

Those in favor of NAFTA often talk as if the future of world trade hangs on this document. Two days ago, President Clinton said that NAFTA was essential to America's economic hopes. Those opposed talk instead about a future of tarpaper shacks for American workers and what Ross Perot calls the "giant sucking sound" of jobs going south to Mexico. They're all wrong. This issue is not as clear-cut, nor even as important as either side now says. It's not clear-cut because NAFTA's effects, by any realistic estimate, would be a complicated mixture of good and bad. Some people would gain in each country and others would lose. The overall benefits and losses are impossible to predict because there's no historic evidence of what happens when two countries with such different living standards and political systems attempt to integrate. The one thing that is certain about NAFTA's effect on the United States is that for the foreseeable future it wouldn't be very much. The Mexican economy is about five percent as large as America's. The normal growth of the U.S. economy creates the equivalent of another whole Mexico about every 20 months.

Whether you think of Mexico as a market for U.S. products, or as a threat to U.S. jobs, it's not economically big enough to be a major factor either way. Nor would NAFTA's effect on U.S. policies be dramatic since most tariffs and other barriers to Mexican goods are already gone. Right now, with no NAFTA, the average U.S. tariff on Mexican products is under four percent. And even if NAFTA is rejected, businesses will keep looking around the world for the least expensive source of labor. To me, the pluses of this agreement slightly outweigh the minuses. The strongest pro-NAFTA argument is not strictly economic, but strategic. These two countries are stuck with each other as neighbors, and the United States will be richer and calmer if its

SOURCE: National Public Radio/"Morning Edition." Commentary by James Fallows (September 17). Copyright © 1993 by National Public Radio®. Reprinted by permission.

neighbors are too. To Mexico, NAFTA is economically crucial, and its defeat would be a big political blow. At a time when Europeans and Asians are tightening their regional economic blocks, we might as well think about ours.

Other people might disagree about the balance of good and bad in NAFTA, but we should stop pretending that it's an all-or-nothing deal. When I hear the exaggerated claims for and against this agreement, I start thinking of . . . [m]odern political talk shows . . . [training] us to think that each week's issue is urgently important, until there's a new issue next week, and that to demonstrate importance you should always overstate and shout. NAFTA is more suitable to a discussion in a monotone, or even a whisper. It's a medium-important issue. Why not treat it that way?

Key Concepts

(handwritten annotation: W' — crossed out — KNOW What)

Dysfunctions 38	Latent Dysfunctions 38	Sociological Theory 30
Facade of Legitimacy 48	Latent Functions 38	Symbol 54
Function 34	Manifest Dysfunctions 38	Theory 30
Ideologies 48	Manifest Functions 38	

(handwritten annotation: 3 Theoretical Perspectives: conflict, symbolic Interaction, Functionalism (conservative))

Notes

1. Mexico is the United States's third-largest trading partner (after Japan and Canada). The United States is Mexico's largest trading partner, accounting for two-thirds of Mexico's foreign trade. In 1992, the United States had a $6 billion trade surplus with Mexico (*The New York Times* 1993).

2. Even top government officials view Mexico in narrow terms. According to Gary Jacobs, a businessman with an interest in seeing the border develop:

   There is no constituency for the border. There has been absolutely no recognition [in Washington] of what the U.S.–Mexican border is all about. . . . Here we have this huge neighbor to the south . . . [and] the United States does not have a foreign policy toward Mexico. . . . It takes Mexico for granted. (JACOBS 1986)

3. *Maquila* is a derivative of the verb *maquilar,* "to do work for another." Originally, *maquila* signified the toll that a farmer paid to a miller for processing grain (Magaziner and Patinkin 1989, p. 319).

4. The AFL-CIO also had an interest in seeing the *Bracero* Program end and lobbied for its termination. In particular, union leaders maintained that if the program ended, the size of the American labor force would shrink and wages would rise (Pranis 1989).

5. A foreign company can take part in the Border Industrialization Program in several ways: it can (1) establish its own plant, (2) subcontract with an existing firm in Mexico, or (3) participate in a shelter plan. A shelter plan is an arrangement with a liaison company that provides facilities (buildings and warehouses), contracts with a work force, and handles paperwork and red tape. The contracting company provides raw materials and components, production equipment, and on-site management.

6. *Bracero* is derived from the Spanish *brazo,* which means "arm." The word refers to manual laborers (those who use their arms).

7. The *Maquiladora* Program as such did not create a new opportunity; it merely gave firms another off-shore option. In the 1960s, U.S. firms already were using labor in Puerto Rico (in the economic

development project Operation Boot Strap) and were subcontracting to Japanese and Hong Kong firms (Pranis 1990).

8. Before November 6, 1986—the day the Immigration Reform and Control Act went into effect—it was not against the law to hire undocumented workers. Kimba Wood had employed an illegal immigrant before this date; her actions therefore were not illegal. Zoe Baird, on the other hand, hired illegal immigrants after this date.

9. Mexican workers and border residents contribute to the smooth functioning of the American economy. For example, residents of Tijuana, a Mexican border city, purchase a considerable number of secondhand products, such as automobiles and electrical appliances, from the United States (Sanders 1987). In the United States, documented and undocumented Mexican workers make up a significant segment of the migrant workers who plant, cultivate, and harvest fruits, nuts, and vegetables. They also work in other low-paying jobs in the hotel, restaurant, and child-care sectors.

10. The 1986 act permitted illegal immigrants to apply for temporary legal status if they could establish continuous residence in the United States since January 1, 1982. The act also allowed illegal immigrants who could establish that they had performed seasonal agricultural services in the United States for at least 90 days between May 1, 1985, and May 1, 1986, to apply for legal temporary residence status (U.S. Immigration Reform and Control Act of 1986).

11. Most analysts agree that there is no way to know for sure the impact the elimination of the *Maquiladora* Program might have on the U.S. or Mexican economies. For example, advocates cite considerable job loss and proponents cite increases in employment. As with the NAFTA debate both sides probably exaggerate their positions.

12. We must remember that, although it is clear the *maquila* industry has increased the social and economic integration, especially along the border separating the two countries, the two countries have always been interconnected. This fact is obvious when we simply consider that in 1848 the U.S. government annexed (in the name of manifest destiny) Mexican territory along with many of its inhabitants (who lived in what is now California, Nevada, Texas, Utah, and parts of Arizona, Colorado, New Mexico, and Wyoming). Just because a line was drawn separating the two countries politically does not mean that the social and economic ties between people on either side ceased to exist.

13. An agreement between Mexico and the United States, known as Annex III, specifies that hazardous waste and materials (HWM) produced at *maquila* plants must be returned to the country of origin. An application must be filed 45 working days before every HWM shipment. The application specifies, among other things, a detailed description of the HWM, the route and final destination, and emergency measures to be taken in case of accidental spill (Partida and Ochoa 1990).

14. Foreign companies cannot pay workers in dollars. They must exchange dollars for pesos to ensure that the dollars go to the government to pay off debts, rather than to the workers.

15. I heard a dermatologist use this expression on a "MacNeil/Lehrer Newshour" program.

16. Estimates of Panamanian civilian and military casualties range between 300 and 4,000; the most common estimates cited are between 300 and 700 (Uhlig 1990).

17. The Mexican government expends considerable energy in assessing foreigners' view of its society, as evidenced by its survey of Japanese youths. In response to the Japanese people's largely negative view of Mexico, the Mexican government has developed programs to attract Japanese tourists and to promote student and cultural exchanges. The programs are aimed at changing the attitudes toward Mexico by encouraging Japanese people to learn about Mexico firsthand. Approximately 100,000 Japanese tourists visit Mexico each year.

3 RESEARCH METHODS IN THE CONTEXT OF THE INFORMATION EXPLOSION

with Emphasis on Japan

A bullet train with Mount Fuji in the background.
Tony Stone Images

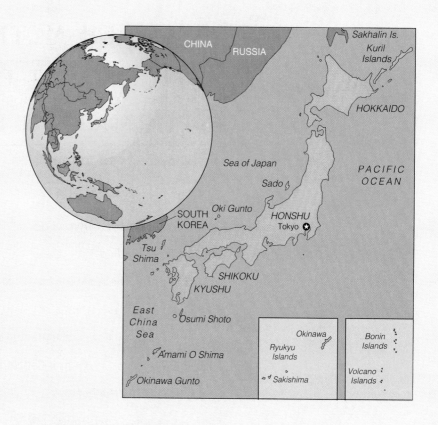

CHINA
RUSSIA
Sakhalin Is.
Kuril Islands
HOKKAIDO
PACIFIC OCEAN
Sea of Japan
Sado
Oki Gunto
HONSHU
Tokyo ✪
SOUTH KOREA
Tsu Shima
SHIKOKU
KYUSHU
East China Sea
Osumi Shoto
A'mami O Shima
Okinawa Gunto
Okinawa
Ryukyu Islands
Sakishima
Bonin Islands
Volcano Islands

Students from Japan enrolled in
college in the United States for fall 1991 — 42,000

Japanese admitted into the United
States in fiscal year 1991 for temporary employment — 37,813

People living in the United States in
1990 who were born in Japan — 422,000

Airline passengers flying
between the United States and Japan in 1991 — 8,941,126

U.S. military in Japan in 1993 — 46,821

People employed in 1990 by
Japanese affiliates in the United States — 616,700

Applications for utility patents filed in the
United States in 1991 by inventors from Japan — 36,846

Phone calls made between
the United States and Japan in 1991 — 97,785,000

Observing how little people overseas know about what is going on in Japan makes me adamant about increasing the channels through which information about Japan is transmitted to the world. We absolutely must make ourselves better understood. Information is overflowing within Japan, but a great deal of the most important information of wider interest does not ever get out of the Japanese language, let alone into international media. In my view, about 40 percent of the distrust in Japan from abroad comes from misunderstanding and lack of information.

(RYŪZŌ 1991, P. 281)

We learned in Chapter 2, which explored theoretical perspectives, that facts are meaningless without theory and that theories give us the tools (questions and vocabulary) to explain facts. In that chapter we learned that there are approximately 2,000 foreign-owned manufacturing plants in Mexico. We used three sociological theories to comprehend and explain this fact. Although sociologists use theories to explain facts (which must be accurate and observable), they also use facts to test a theory's explanatory power. A theory's usefulness is challenged if it cannot explain accurate facts. Research is the means by which we gather facts and explain facts.

This chapter discusses the methods used by sociologists to conduct research. **Research** is a fact-gathering and fact-explaining enterprise governed by strict rules (Hagan 1989). **Research methods** are the various techniques that sociologists and other investigators use to formulate meaningful research questions and to collect, analyze, and interpret facts in ways that allow other researchers to check the results.

We need to possess a working knowledge of research methods even if we do not plan to become sociologists or to do research of our own. One important reason is connected with a relatively new global phenomenon—the **information explosion**. This dramatic term describes an unprecedented rate of increase in the volume of data due to the development of the computer and of telecommunications.[1] In this chapter we make a distinction between data and information. **Data** consists of printed, visual, and spoken materials. Data becomes **information** after someone reads it, listens to it, or views it.

In addition to coping with large quantities of data, people also have to consider the data's quality. Most of the data that we hear, read, and see has been created by others. Therefore we can never be sure that it is accurate.

In "Too Much of a Good Thing?: Dilemmas of an Information Society," Donald Michael (1984) argues that we cannot assume that more data will lessen uncertainty and increase feelings of control and security. Instead the opposite may be true: more data can overwhelm us to the point that we conclude we cannot believe anything we hear, read,

The backdrop to the annual ceremony recalling the bombing of Hiroshima, held at Heiwa Koen (Peace Park), dramatizes how Japan rose from the ashes of World War II to become one of the world's economic powers.
Charlie Cole/Sipa Press

or see. We need not accept Michael's gloomy assessment, however, if we possess a working knowledge of social research methods. Such knowledge gives us the skills to identify and create high-quality data.

In this chapter, we give special attention to Japan for several reasons. First, a great deal has been written about Japanese culture, business practices, educational system, and lifestyle in general. The amount of printed material alone on Japan is overwhelming. Much of this material explores Japan's economic and technological successes over the past five decades. It addresses how Japan rose

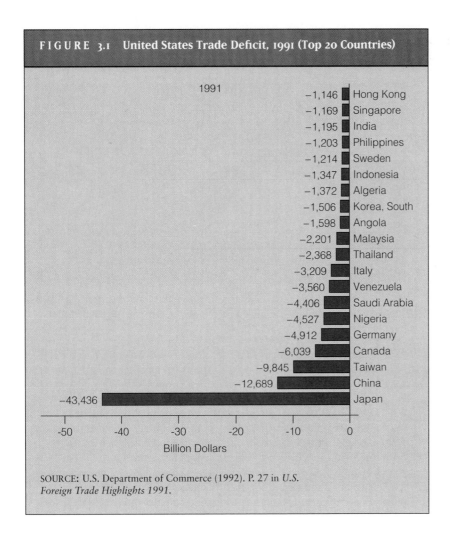

FIGURE 3.1   United States Trade Deficit, 1991 (Top 20 Countries)

1991

−1,146	Hong Kong
−1,169	Singapore
−1,195	India
−1,203	Philippines
−1,214	Sweden
−1,347	Indonesia
−1,372	Algeria
−1,506	Korea, South
−1,598	Angola
−2,201	Malaysia
−2,368	Thailand
−3,209	Italy
−3,560	Venezuela
−4,406	Saudi Arabia
−4,527	Nigeria
−4,912	Germany
−6,039	Canada
−9,845	Taiwan
−12,689	China
−43,436	Japan

Billion Dollars

SOURCE: U.S. Department of Commerce (1992). P. 27 in *U.S. Foreign Trade Highlights 1991*.

from the ashes of World War II (1939–1945) to become one of the top three exporters in the world (Saxonhouse 1991). This feat is even more amazing because Japan is a resource-poor country: it must import all of its raw materials and much of its food. The United States has played an important role in Japan's rise because it gave economic assistance and helped the country rebuild after the war. In addition, the United States is linked to Japan as its number-one trading partner.

Most Americans know that of the 179 countries with which the United States trades, the largest trade deficit is with Japan (see Figure 3.1). In fact, the trade imbalance with Japan is the single largest contributor to the record trade deficits that the United States incurred in the 1980s and 1990s (U.S. Department of Commerce 1985, 1989a, 1989b). Although there is considerable debate over what this large deficit means, it has increased public awareness about the interdependence between the United States and Japan (see "Figuring the Trade Deficit with Japan").

Politicians, the media, and many social critics do not challenge the way the trade deficit is calculated. And they use the trade deficit as a measure of the overall health of the American economy, Japanese aggressiveness, the patriotism of American consumers, the competitiveness of American corporations, and the overall quality of American workmanship. The massive attention given to the trade

Figuring the Trade Deficit with Japan

Consider the trade balance between two countries. It is calculated by adding up the dollar value of the goods and services that each country exports to the other and then subtracting the smaller amount from the larger. On the basis of this formula, the United States has sustained a trade deficit with Japan for approximately a decade and a half.

Some critics argue that this formula is not an accurate or valid measure of the amount of trade between the United States and Japan. It does not consider the dollar value of goods and services produced by American and Japanese companies located in the other's country. A more valid measure would be to add up the goods and services that each country exports to the other, *plus* what each country's firms produce and sell within the other country (Robinson 1985).

Richard D. Robinson, a professor of international management, figured the U.S. trade deficit with Japan on the basis of each formula. According to the formula that counts annual American exports to Japan ($25.6 billion) and Japanese exports to the United States ($56.8 billion), the trade deficit is $31.2 billion for the United States. According to the second formula, which includes sales by Japanese-owned businesses in the United States ($12.6 billion) and sales by American-owned businesses in Japan ($43.9 billion), the trade surplus is $100 million for the United States (significantly different from the deficit figure of $31.2 billion). To evaluate this difference ($31.2 billion versus $100 million), it is important to understand the difference between a million and a billion.

How does this worker in a Honda factory in Marysville, Ohio, illustrate the difficulty of calculating the trade balance between two countries in a global economy?
Andy Snow/Saba

"For example, knowing that it takes only about eleven and a half days for a million seconds to tick away, whereas almost thirty two years are required for a billion seconds to pass, gives one a better grasp of the relative magnitude of these two common numbers" (Paulos 1988, p. 10).

According to the second formula, then, the United States has a slight trade surplus with Japan. Even this formula, however, is not free of problems. For one thing, it does not take into account that countries export goods and services *indirectly* to one another. For example, 70 percent of Japanese males use Schick razors. Schick, with headquarters in Connecticut, exports razors to Japan via Hong Kong (Totten 1990). Similarly, Japan exports cars to the United States from Canada, Mexico, and various countries in Asia.

On the basis of this information we can argue that the best operational definition is a third formula: for each country, add up the dollar value of goods and services it exports to the other country, *plus* what that country's firms produce and sell within the other country, *plus* what that country exports indirectly to the other country. Then figure the difference between the two totals. Although this third formula is now accepted as a more valid measure of trade, it is virtually impossible to find complete and accurate figures on indirect exports. In addition, anyone who tries to calculate indirect exports faces the challenge of determining the national identity or origin of the exported products. For example, should the approximately 164,000 vehicles produced in Japan and exported to the United States as Chryslers, Dodges, and Chevrolets be considered Japanese exports (Sanger 1992)? Similarly, should the 50,000 Nissan Quest Mini-Vans produced at Ford's Avon, Ohio, production plant be considered Japanese exports? As you can see, the issue of trade balance is not simple or clear-cut.

deficit has caused Americans at almost every level to question why they are not selling as many American products abroad as they are purchasing from foreign countries (particularly Japan) and to question what quality in the Japanese has made them so "successful" in such a short time. In search of an answer to this question, American politicians, journalists, and social critics have pointed to such things as the special bond between Japanese mothers and children, the Japanese work ethic (which is said to be so extreme for men that Japanese children grow up "fatherless"), the value that the Japanese place on the group, and the general unwillingness of Japanese consumers to purchase U.S.-manufactured products. A basic understanding of research methods enables us not only to evaluate the accuracy of the things we read, hear, and view about Japan, but to cope with the consequences of the information explosion.

The Information Explosion

At least two technological innovations are responsible for the information explosion: computers and telecommunications. Both technologies help people create, store, retrieve, and distribute large quantities of data at mind-boggling speeds. Comparing the size and capabilities of the first computers to those of the present suggests why the volume of data has increased so rapidly within the past 50 years. The computers of the 1940s weighed five tons, stood eight feet tall, were 51 feet long, and contained 17,468 vacuum tubes and 5,000 miles of wiring (Joseph 1982). They performed simple calculations in a few seconds but tended to overheat and break down. The first computers used so much power that the lights in nearby towns failed when they were turned on. Because of their size and cost, the only people who used computers worked for the U.S. Defense Department and the Census Bureau. Today, in contrast, a single silicon chip a quarter-inch thick can process millions of bits of information in a second. The chip has reduced computer size and cost and has made possible the widespread use of the personal computer.

Similarly, telecommunications have increased our ability to send data quickly across space. Although the telephone, radio, and television have existed in some form for as long as 100 years, methods of transmitting clear and fast signals have changed considerably. Fiber optics cables have replaced wire cables as the means of transmitting images, voices, and data.

In 1923 the cable connecting Britain and the United States contained 80,000 miles of iron and steel wire (enough to circle the earth three times) and four million pounds of copper. It could transmit the equivalent of 1,200 letters of the alphabet per minute across the ocean. In contrast, when the capabilities of fiber optics are exploited fully, a single fiber the diameter of a human hair can carry the entire telephone voice traffic of the United States and can transmit the contents of the Library of Congress to any place in the world in a few seconds (Lucky 1985). Transmitting the contents of the Library of Congress is no small feat if we stop to consider that this library houses 100 million items on 532 miles of shelves (Thomas 1992).

Such innovations as the personal computer and fiber optics put data-creation technologies into the hands of the general public and increase the speed of the following processes (Branscomb 1981; Compaine 1981):

Data entry (from typing to scan and voice entry)

Editing (from white-out and retyping to word processing)

Duplication (from carbon copying and offset printing to photocopying, computer display, and laser printing)

Storage (from the shelf, catalog, and file drawer to digital mass storage)

Retrieval (from card catalogs to computer searches)

Distribution (from mail and freight to fax and satellite transmission)

Complex calculations (from pencil and paper to computer)

Sociologist Orrin Klapp wrote about the information explosion in *Overload and Boredom: Essays on the Quality of Life in the Information Society*. He used a vivid metaphor to describe the dilemma of sorting through and keeping up with the massive amounts of data generated. Klapp envisioned a person "seated at a table fitting [together] pieces of a gigantic jigsaw puzzle. From a funnel overhead, pieces are pouring onto the table faster than one can fit them. Most of these pieces do not match up. Indeed, they do not all belong to the same puzzle" (Klapp 1986, p. 110). The pieces falling from overhead represent research accumulating at a pace that interferes with people's ability to organize it into a comprehensible pattern. Klapp used this analogy to show that the speed with which data is produced and distributed overwhelms the brain's capacity to organize and evaluate it.

When we want information on any subject, we not only must select from a large quantity of data but also must often sift through distorted, exaggerated presentations. Klapp gives some reasons for these added burdens. New technologies permit large numbers of magazines, newspapers, radio stations, and television channels to exist. As a result, message senders must compete for our attention. Reporters, producers, and others in the media often devise ways to entice us to read and listen. Common strategies include eye-catching headlines, misleading titles, and shocking stories (murders, car accidents, plane crashes). Some eye-catching book titles and news headlines related to Japan, for example, include *The Coming War with Japan, Agents of Influence, The End of the American Century,* "The Chip Wars with Japan: A Struggle for Survival," "U.S. High-Tech Lead over Japan in Danger," and "Battle Against Japan for the Future."

Too often such exaggerated headlines lure people into reading and listening to material that turns out to be trivial, repetitive, contradictory, and ultimately uninformative in that it does not add to understanding. The titles and headlines cited above, for example, might catch our attention but they mask at least one important idea—that the U.S. relationship with Japan is not a simple matter of "us" against "them." Consider those headlines about competition in technology. In fact, although the Japanese dominate the world market in memory chips, the United States dominates the world market in software. In addition, some American companies are cooperating with Japanese companies: Texas Instruments and Hitachi are working jointly to design more advanced memory chips; Motorola produces memory chips jointly with Toshiba (Sanger 1989, 1990).

Klapp also blames **dearth of feedback** as a factor in creating data that is poor in quality. What does he mean by "dearth of feedback"? Much of the data that is televised and published is not subjected to honest, constructive feedback because there are too many messages and not enough critical readers and listeners to evaluate the data before it is released or picked up by the popular media. Without feedback the creators cannot correct their mistakes; thus, the data they produce becomes poor in quality. Klapp believes that data overload, coupled with distortion, exaggeration, and triviality, is as problematic to readers as a lack of data. Anyone who reads professional journals; listens to the radio; watches television; or reads newspapers, magazines, and books widely and critically must realize that there are few barriers to media attention or publication:

> *There seems to be no study too fragmented, no hypothesis too trivial, no literature citation too biased or too egotistical, no design too warped, no methodology too bungled, no presentation of results too inaccurate, too obscure, and too contradictory, no analysis too self-serving, no argument too circular, no conclusions too trifling or too unjustified, and no grammar and syntax too offensive for a paper to end up in print.* (RENNIE 1986, P. 2391)

It is discouraging to confront the tremendous amount of data that is released before being evaluated for accuracy or worth. And it is easy to become pessimistic about the human ability to organize, evaluate, comprehend, and trust or question the growing quantity of data. Yet, the situation has another side. Although the ease with which data is produced seems to have reduced the overall quality of the material, it also can increase the chances that

Technological innovations like optical cables enable people to transmit vast volumes of information with unprecedented speed. The social consequences of the revolution in information technology include new jobs, new demands for lifelong learning, and the problem of coping with information overload.

Bob Thomason/Tony Stone Images

good and useful ideas will receive some exposure. In addition, the variety means that there is something for everyone. Even though the data is not particularly well organized, researchers still can draw from it and organize it in new and unexpected ways. Moreover, the data explosion does not negate the need to be informed; it simply increases the need to be able to create, identify, and synthesize data and turn it into useful and worthwhile information. Decisions still must be made, actions still must be taken, and policies still must be formed. No constructive decision, action, or policy can be founded on haphazard, misleading, or inadequate data. That situation would be equivalent to a physician's decision to perform heart surgery based only on the intuition that such action will solve the patient's health problems and without or-

dering medical tests, reviewing the patient's history, or interviewing the patient beforehand.

The larger point is that we live in a society in which people need to be more than computer literate (able to operate a computer and use it to input, access, and output data). People also need to be **research methods literate**—that is, they must know how to collect data that is worth putting into the computer and they must know how to interpret the data that comes out of it. Unless computer literate people also have research methods literacy, they merely possess the skills to enter data. We turn now to basic techniques and strategies that sociologists (and all other researchers) use to evaluate and gather reliable data. We start with the guiding principle—the scientific method.

The Scientific Method

Sociologists are guided by the scientific method when they investigate human behavior; in this sense they are scientists. The **scientific method** is an approach to data collection guided by two assumptions: (1) that knowledge about the world is ac-

quired through observation, and (2) that the truth of the knowledge is confirmed by verification—by others making the same observations. Researchers collect data that they and others can see, hear, taste, touch, and smell. They must report the process by

scientific method

which they make their observations and must present conclusions so that interested parties can duplicate that process. If observations cannot be duplicated, or if upon duplication the results differ substantially from those of the original study, the study is considered suspect. Findings endure as long as they can withstand continued re-examination and duplication by the scientific community. When researchers know that others are thoughtfully critiquing and checking their work, the result is reinforcement of careful, thoughtful, honest, and conscientious behavior. Moreover, this "checking" encourages researchers to maintain **objectivity**—that is, to not let personal and subjective views about the topic influence the outcome of the research.

Because of continued re-examination and revision, research is both a process and a dialogue. It is a process because findings and conclusions never are considered final. It is a dialogue because a critical conversation between researchers and readers leads to more questions and additional research.

This description of the scientific method is an ideal one, because it outlines how researchers and reviewers *should* behave. In practice, though, questionable behaviors on both sides do arise sometimes. It sometimes happens that some research is dismissed as unimportant (often even before it is read) and as unworthy of examination simply because the topic is controversial, departs from mainstream thinking, or is written by someone from a group that is considered "inferior." Moreover, the scientific method works on the assumption that researchers are honest—that they do not manipulate data to support personal, economic, and political agendas. But there is no certain way to know the extent to which researchers actually are honest. In one survey sponsored by the American Association for the Advancement of Science (AAAS), 25 percent of the 1,500 people surveyed reported that in the past 10 years they had witnessed someone faking, falsifying, or plagiarizing data (Marsa 1992).

The standards encouraged by the scientific method make research more than a fact-generating enterprise. As we are about to see, research is a carefully planned, multistep, fact-gathering and fact-explaining enterprise (Rossi 1988) that involves a number of interdependent steps:

1. Defining the topic of investigation
2. Reviewing the literature
3. Identifying core concepts and forming hypotheses
4. Choosing a research design and collecting data
5. Analyzing the data
6. Drawing conclusions

Researchers do not always follow the steps in sequence, however. Sometimes they do not define the topic (step 1) until they have familiarized themselves with the literature (step 2). Sometimes an opportunity arises to gather information about a group (step 4), and a project is defined to fit the opportunity (step 1). Although the six steps need not be followed in sequence, all need to be completed to ensure the quality of the project.

In the sections that follow we will examine each stage individually, making reference to a variety of research projects that examine some aspect of Japanese society. We give particular attention throughout the chapter to *Communicative Styles of Japanese and Americans: Images and Realities*, a research project conducted by social psychologist Dean Barnlund (1989). One purpose of Barnlund's study, which took more than a decade to complete, was to compare and contrast the many ways in which Japanese and Americans communicate with strangers. According to Barnlund, the popular literature and media suggest that "the Japanese are likely to be startled at the ease with which Americans approach and enter into intense conversations with people they scarcely know. . . . The Americans, in turn, confront Japanese who appear reluctant to meet strangers and slow to get to know them" (pp. 190–91). Barnlund spent 10 years researching this topic because he believed that first-time encounters between representatives of each country could be strained if neither side were familiar with the other's communicative style.

This focus on Barnlund's study serves two purposes. First, it allows us to follow a single study through all six stages of the research process and to share in the decision making that accompanies each stage. Second, it gives us an opportunity to learn what it means to be critical consumers of research. That is, we will pay attention to how Barnlund gathered and explained his facts and will seek to identify any shortcomings. If we know the short-comings, then we will be better able to evaluate the significance of the findings.

Step 1: Defining the Topic for Investigation

The first step of a research project is to choose a topic. It would be impossible to compile a comprehensive list of the topics that sociologists study, because almost any subject involving human beings is open to investigation. Sociology is distinguished from other disciplines not by the topics it covers but by the perspectives it uses to study topics (see Chapters 1 and 2).

Good researchers explain to their readers *why* their chosen topic is significant. Explanation is vital because it clarifies the purpose and significance of the project and also clarifies the motivation for doing the work. If you don't know why you are conducting a project, it is unlikely to generate much personal or public interest.

Researchers choose their topics for a number of reasons. Personal interest is a common and often underestimated motive. It is perhaps the most significant reason why someone chooses a specific topic to study. This is especially true if we consider how a researcher eventually chooses one topic from a virtually infinite set of possibilities. Consider the reasons why researcher Anne Allison (1991) studied *obentō*, the lunch boxes that Japanese mothers make up for their children when they are in nursery school. An *obentō* is a "small box packaged with a five or six course miniaturized meal whose pieces and parts are artistically arranged, perfectly cut, and neatly arranged" (Allison 1991, p. 196). Allison became interested in the *obentō*s while she was living in Japan and taking her child to a Japanese nursery school. Her son's nursery school teacher talked with Allison "daily about the progress he was making finishing his *obentō*s" (p. 200). As Allison tells it:

The intensity of these talks struck me at the time as curious. We had just settled in Japan and David, a highly verbal child, was attending a foreign school in a foreign language he had not yet mastered; he was the only non-Japanese child in the school. Many of his behaviors dur-ing this time were disruptive: for example, he went up and down the line of children during morning exercises hitting each child on the head. Hamada-sensei [the teacher], however, chose to discuss the obentōs. I thought surely David's survival in and adjustment to this environment depended much more on other factors, such as learning Japanese. Yet it was the obentō that was discussed with such recall of detail ("David ate all his peas today, but not a single carrot until I asked him to do so three times") and seriousness that I assumed her attention was being misplaced. (PP. 200–201)

On a personal level, Allison undertook this study in order to learn more about her son's new environment so she could help him adapt to it. Realizing that the *obentō* was somehow significant to his success in a Japanese preschool, Allison chose it as a research topic. The choice of a research topic, however, usually has some further significance that goes beyond personal interest. Allison's research on *obentō*s, for example, examines why Japanese mothers are expected to make such elaborate lunches for their children and why Japanese teachers pay so much attention to how well children eat lunch. In addition, Allison's research offers readers some insights into the Japanese system of preschool education and allows us to see the educational significance of eating lunch—a seemingly routine activity. According to Allison's research, the elaborately prepared lunch is a sign of a Japanese woman's commitment as a mother, which inspires her child to be similarly committed as a student.

Barnlund's Study: Step 1

Dean Barnlund chose to do research on Japanese and American styles of communication because he wanted first to understand two communication styles and then to speculate on the international significance of any differences that he found. His

The preschool obentōs that inspired Anne Allison's study are a good example of how sociologists' curiosity may be aroused by almost any facet of social life.
Ulrike Welsch

research is timely and important to U.S.–Japan business, social, and political relations. Understanding these differences helps both sides identify potential strains, meet challenges in communications, and maintain constructive relationships (see Table 3.1). Barnlund explained how first-time encounters between representatives of each country could be strained if neither side were familiar with the other's communication styles. He stated that this strain could be reduced when each party realized that the other's behavior was not a negative reaction but was merely different.

Step 2: *Reviewing the Literature*

All good researchers take existing research into account. They read what knowledgeable authorities have written on the chosen topic if only to avoid repeating what has already been done. Even if researchers believe that they have revolutionary ideas, they must consider the works of past thinkers and show how they advance their ideas or correct errors or oversights. More important, reading the relevant literature can generate insights that the researcher may not have considered.

Researchers have no simple formula for selecting articles and books to read on a chosen topic. There are some general strategies to follow when reviewing literature. First, researchers do not rely solely on popular sources such as *Time, Psychology Today,*

Reader's Digest, Newsweek, and *USA Today* as background information. Such sources are written for general audiences and are not designed to give readers detailed information on a topic. Often they include only the most sensational elements of a topic because the overriding goal is to sell copies and make a profit rather than contribute to a field of knowledge (Knepper 1992).

Instead of relying on popular sources, sociologists rely on research and articles published in scholarly journals. There are hundreds of journals (some that are sponsored by sociological associations and others that are not) that publish research conducted by sociologists. The American Sociological Association, for example, sponsors the

TABLE 3.1   Differences in Negotiating Behavior Between Japanese and American Officials

In the foreword to *National Negotiating Styles*, a publication of the U.S. Department of State, editor Hans Binnendijk comments that "a better understanding of each nation's particular style can strengthen the ability of the United States to negotiate a better deal" (Thayer and Weiss 1987).

American Style	Japanese Style
Americans sometimes overstate the first position to allow for retreat. Economic positions are often cast in harsh, challenging language.	Japanese rarely overstate the first position, though sometimes they are vague. Japanese like to regard their position as reasonable for both sides.
Americans keep final formulation of the first position secret until the first negotiating session.	Japanese usually leak their position to some American before it is formally revealed.
Americans respond to the press on an "if-asked" basis. There is an adversarial relationship between officials and reporters. Officials favor the domestic press only slightly over the foreign press.	Japanese initiate encounters with the press. They expect, and often get, editorial sympathy from the domestic press, at least in foreign economic negotiations. Officials isolate the foreign press from the domestic press.
Americans try to maintain secrecy over the course of the negotiations until the end of a negotiating session.	Japanese usually reveal the tenor and substance of the negotiations and sometimes the details as the negotiations go along.
Americans like to establish a principle and then search out a solution based on that principle.	Japanese like to talk about practical solutions, resolving matters case-by-case. They allow the solution to precede the principle.
Americans tend to compromise too soon, particularly if Japanese negotiators recognize the American principle.	Japanese find compromise difficult. They often create a fictive principle or offer meaningless concessions.
Americans place great value on winning an argument.	Japanese try to stress areas of agreement.
Americans are adversarial.	Japanese try to avoid contention.
Americans cast negotiations in terms of victory and defeat.	Japanese negotiate to avoid failure.
Americans tend to conduct their business in the negotiating hall, though they are aware that activities outside can be important.	Japanese like to conduct real negotiations away from the formal negotiating hall, using the formal session to announce agreements reached elsewhere.
Americans see the negotiated solution as final and implementation naturally flowing therefrom.	Japanese see the negotiated solution as one more stage and implementation as a subject for further negotiation.

Note: The description of the American style was derived in part from a document prepared by an anonymous Japanese diplomat that circulated in Tokyo when Michael Blaker's book, *Japanese International Negotiating Style*, was published in 1977.

SOURCE: Adapted from "The Changing Logic of a Former Minor Power" by Nathaniel B. Thayer and Stephen E. Weiss. Pp. 45–74 in *National Negotiating Styles*, edited by H. Binnendijk. U.S. Department of State (1987).

following journals: *American Sociological Review, The Journal of Health and Social Behavior, Sociology of Education, Teaching Sociology,* and *Contemporary Sociology: A Journal of Reviews.* Although researchers might search specific journals for related articles, they are more likely to use a computerized database such as Sociofile, which contains titles, abstracts (summaries of the articles), and sources of articles appearing in thousands of journals.

Second, the author's and publisher's reputation is an important criterion in selecting a book or article. (Unfortunately, reputation can be difficult to determine.) Third, the date of publication can be important if a researcher is interested in the most recent material on a topic. Some of the most insightful work on a topic, however, may have been written a century or more ago. Recall that in Chapter 1 we discussed the importance of Emile Durkheim, Karl Marx, and Max Weber to the field of sociology. All of these thinkers wrote in the nineteenth or early twentieth century.

A final strategy that good researchers follow when they review the literature is to make a point of selecting several articles that present an opposing viewpoint on the topic. Such an approach opens a researcher's mind to other ways of thinking. Let's return to Barnlund's study of U.S.–Japanese communication patterns and consider his approach to reviewing the sociological literature.

Barnlund's Study: Step 2

Barnlund read more than 200 books and articles in conjunction with his study. He read the following works by sociologists to gain general insight into thinking about relations between strangers:

Emile Durkheim, *Sociology and Philosophy* ([1924] 1953)

Margaret Wood, *The Stranger* (1934)

Erving Goffman, *Behavior in Public Places* (1963)

Lyn Lofland, *A World of Strangers* (1973)

Barnlund learned a number of significant things about strangers while reviewing the literature. Throughout most of history, nearly all of the people an individual encountered each day were familiar. As societies industrialized, however, this pattern changed. The familiar world was transformed into one populated by strangers who occupied the same space but remained personally uninvolved. Under this arrangement we do not notice or talk to most of the people around us. We avoid physical, verbal, and even visual contact with most strangers as we go about the activities of daily life. Yet, we know that a stranger eventually can become an acquaintance, a friend, or a spouse, or can affect us personally in other ways.

The rules that govern this passage from stranger to nonstranger vary from person to person and from culture to culture. Apart from individual differences, cultural constraints determine the line that separates strangers from acquaintances and the ease with which that line can be crossed. To gain an appreciation of Japanese and American constraints regarding strangers, Barnlund read Alexis de Tocqueville ([1835] 1956), Kurt Lewin (1948), Kano Tsutomu (1976), and Robert Ozaki (1978). These readings helped him construct a profile of Japanese and American behaviors toward strangers.

Barnlund learned that in both countries strangers are regarded as nonacquaintances or unknown persons, yet, Americans and Japanese behave quite differently toward strangers. In general, the Japanese are uncomfortable with unfamiliar settings and unfamiliar people. Therefore, they tend to avoid strangers.

The literature suggests that the Japanese treat strangers as inconsequential or unessential beings for at least two reasons. First, the Japanese language is structured to consider status; therefore, it is very awkward to address persons of unknown status. Second, the sense of group orientation is strong among the Japanese and sharpens the distinction between the group and outsiders. Thus, one might argue that a stranger's presence threatens group cohesion.

Although Americans sometimes negatively associate strangers with trespassers and outsiders, they also sometimes see them in a positive light as loners, mavericks, newcomers, and immigrants willing to leave their homeland in search of "the American dream." Compared with the Japanese, Americans are more open to strangers; they are more willing to start conversations and to share personal information. For one thing, the English language is flexible enough to accommodate persons of unknown status. In addition, the American orientation toward the individual places less value on group identification and thus eases contact with outsiders.

Barnlund's familiarity with the literature on strangers helped him to define his project clearly. Although he found the literature fascinating and provocative, it told him only in the vaguest and most impressionistic ways what distinguishes Japanese from American behavior. Barnlund decided to address this lack of specific knowledge by asking Japanese and Americans how they view and react to strangers. He planned to use his findings to support or contradict the generalizations he had found in the existing literature.

Step 3: Identifying Core Concepts and Forming Hypotheses

After deciding on a topic and reading the relevant literature (not necessarily in that order), sociologists typically state their core concepts. Concepts are powerful thinking and communication tools that enable us to give and receive complex information in an efficient manner. When we hear or read a concept and are familiar with its meaning, it brings to our consciousness a host of associations stored in our brains. For example, we encountered the concept *symbol* in Chapter 2. For someone who comprehends this concept, it should evoke appropriate images, ideas, and background information (Hirsch 1988), including the notion that a symbol is a physical phenomenon to which people assign meaning. This concept *symbol* should also trigger the idea that the meanings attached to physical phenomena vary according to time and place.

Scientists from different disciplines are distinguished from one another by the concepts on which they draw to frame and explain observations. Concepts help distinguish sociology from other disciplines such as anthropology, psychology, biology, chemistry, and physics. Even when scientists from different disciplines study the same topics, the concepts governing their sciences prompt them to look at different features. To illustrate this point, we can look at the concept of suicide.

Psychologists define suicide as an intentional act of aggression turned inward. When psychologists study suicide they focus on the unique life of the suicidal individual to understand motives. The concept of suicide does not trigger sociologists to think in this way, however, because they think of suicide as "the severing of relationships" (Durkheim 1951).

Sociologists maintain that the structure of relationships provides important clues for understanding the act of suicide. Thus, they are interested in which groups tend to have unusually high or unusually low suicide rates. For example, suicide rates are unusually high among the suddenly unemployed, the elderly, and adolescents. Sociologists ask themselves, What is it about being a member of that group that contributes to unusually high rates of suicide? Specifically how does being an elderly person, an unemployed person, or a teenager strain relationships with family, peers, and employers to the point that a person might choose to commit suicide in order to end the relationships?

Durkheim maintained that one's relationship to the group is the key to understanding suicide rates. Those who find themselves in any of the following situations are likely to have higher suicide rates than others who have more secure and stable attachments to a group: (1) excessively isolated from the group (the chronically ill or elderly), (2) excessively attached to the group to the point that the self cannot be separated from the group (recruits undergoing basic training), (3) suddenly thrown out of a group (the suddenly unemployed, lottery winners, widows and widowers), and (4) hopelessly locked into a group (a prisoner on death row, those without resources to escape a confining environment such as an impoverished rural community). Keep in mind that the four categories are not clearcut. Prisoners on death row are in a situation that isolates them from family and friends. At the same time, prisoners are hopelessly confined to living in a highly structured environment with other inmates.

A clear statement of core concepts enables researchers to focus their investigations. In clarifying the concept of suicide as the severing of relationships, for example, sociologists direct their attention to group membership and how membership in a particular group affects relationships with other people. This conceptual focus helps sociological researchers identify the variables they want to study. A **variable** is any trait or characteristic that can change under different conditions or that consists of more than one category. Sex, for example, is a variable; it is generally divided into two categories: male and female. The variable marital status is often separated into six categories: single, living together, married, separated, divorced, and widowed. The variable suicide rate can vary from zero suicides per 1,000 persons in a population to 1,000 suicides per 1,000 persons. This latter rate of course is very unusual and is limited to cases of mass suicide.

Researchers strive to explain particular behaviors. The behavior to be explained is the **dependent variable.** The variable that explains the dependent variable is the **independent variable.** Thus, a change in the independent variable brings about a change in the dependent variable. A **hypothesis** or trial explanation put forward as the focus of research predicts how independent and dependent variables are related. This trial idea specifies what outcomes will occur as the independent variable varies.

Drawing on Durkheim's theory of suicide, sociologist Satomi Kurosu (1991), for example, hypothesized that rural populations would have a higher suicide rate than urban populations in Japan. His independent variable was geographical location, which has two broad categories: urban and rural. His dependent variable was suicide rates. Kurosu reasoned that geographical location would affect suicide rates because rural areas in Japan are characterized by conditions of sparse populations with more males than females, stagnant economies, and a disproportionately large percentage of elderly residents. (Kurosu's research supported his hypothesis.)

Barnlund's Study: Step 3

Two of the major guiding concepts in Barnlund's research are culture and strangers. Culture is a society's distinctive and complete design for living; it

Modes of greeting illustrate Barnlund's belief that people raised in the same culture share tendencies to behave in similar ways in the face of similarly perceived situations. How would this scene look if the businessmen had been born and raised in the United States?

Charles Gupton/Tony Stone Images

comprises "the things people have, the things they do, and what they think" (Herskovits 1948, p. 625). Chapter 4 explores more fully the concept of culture.

Barnlund argues that culture is reflected in people's behavior and that people born and raised in the same culture share tendencies "to act in similar ways in the face of similarly perceived situations" (Barnlund and Araki 1985, p. 9). For the purpose of his study, Barnlund determined that strangers are unknown, "unintroduced" persons. "Unintroduced" is a particularly important feature of the concept of stranger because it prompts us to consider how people who do not know each other make contact without the aid of a third party.

Determining a clear statement of core concepts enabled Barnlund (1989) to focus his investigation. Of all the subjects on which he could have focused with regard to Japan and the United States, he chose to compare the meanings that natives of two countries assign to strangers, and to examine the behavior toward strangers that follows from these meanings. This conceptual focus helped him identify independent and dependent variables for his study. In Barnlund's study the independent variable is cultural background and the dependent variable is behavior toward strangers. The variable cultural

Questions Used to Operationalize Behavior Toward Strangers

Below are some of the questions Barnlund asked Japanese and American students. In parentheses is the variable that each question is intended to operational-ize. (Note: These are approximations of Barnlund's questions because he did not include the question-naire in his book.)

1. (*Predisposition to notice*) During the past month, how often did you notice and think about a stranger's personal characteris-tics?
   __ Never
   __ Seldom
   __ Occasionally
   __ Frequently
   __ Very often

2. (*Predisposition to consider inter-acting with*) During the past month, how often did you con-sider talking to a stranger?
   __ Never
   __ Seldom
   __ Occasionally
   __ Frequently
   __ Very often

3. (*Predisposition to initiate conver-sations*) During the past month, how often did you actually start a conversation with a stranger?
   __ Never
   __ Seldom
   __ Occasionally
   __ Frequently
   __ Very often

4. (*Predisposition to respond*) Dur-ing the past month, how often did you respond favorably to strangers who started a conversa-tion with you?
   __ Never
   __ Seldom
   __ Occasionally
   __ Frequently
   __ Very often

5. (*Outcome*) Identify three strangers to whom you talked re-cently. Check what you learned about each stranger.

   #1 #2 #3
   __ __ __ I learned the person's name.
   __ __ __ I learned where the person lives.
   __ __ __ I learned the person's phone number.
   __ __ __ I learned the person's occupation.
   __ __ __ I learned the person's marital status.
   __ __ __ I learned about the person's activities and interests.

SOURCE: Adapted from Barnlund (1989).

background has two categories: Japanese and Amer-ican. Because behavior toward strangers covers such a broad range, Barnlund decided to break it down into a number of specific variables, including:

- Predisposition to notice strangers
- Predisposition to consider interacting with strangers
- Predisposition to initiate conversation with strangers
- Predisposition to respond to strangers
- Frequency of encounters with strangers
- Outcomes of talking to strangers (specifically what people learn from each other)

In Barnlund's study the first five variables in the list can be answered in one of five ways. Thus they have five categories: never, seldom, occasionally, frequently, and very often. (See "Questions Used to Operationalize Behavior Toward Strangers.") The last variable in the list—outcome of talking to strangers—has at least six categories:

I learned the person's name

I learned where the person lives

I learned the person's phone number

I learned the person's occupation

I learned the person's marital status

I learned about the person's activities and interests

Barnlund maintains that if we know the culture that people are from, we can predict with reason-able accuracy their behavior toward strangers.

From this belief, Barnlund generated several hypotheses about Japanese and American behavior toward strangers:

- Japanese are less likely than Americans to notice strangers.

- Japanese are less likely than Americans to consider interacting with strangers.

- Japanese are less likely to initiate conversation with strangers.

- Japanese are less likely than Americans to respond to strangers.

- Japanese are less likely than Americans to consider interacting with strangers.

- When contact is made with strangers, Japanese learn less than Americans learn about them.

Step 4: Choosing a Design and Collecting the Data

Once researchers have clarified core concepts and have stated hypotheses, they must decide on a **research design,** a plan for gathering data to test the hypotheses. A research design specifies the **unit of analysis** (who or what is to be studied) and the **method of data collection** (the procedures used to gather relevant data). One research design in itself is not better than another. The unit of analysis and the data-gathering procedures that are chosen for one study may not be appropriate for another study. Thus researchers choose a design to fit the circumstances of each study (Smith 1991).

Units of Analysis

Peter Rossi (1988) identified a number of units that sociologists study, other than individuals. This list includes traces, documents, territories, households, and small groups.

Traces Traces are materials or situations that yield information about human activity, such as the items that people throw away, the number of lights on in a house,[2] or changes in water pressure.[3] One example of research that examines traces is the Garbage Project, a University of Arizona program directed by William L. Rathje (Rathje and Murphy 1992). Since the 1970s, Rathje and his team of researchers have been collecting garbage from landfills and selected neighborhood garbage cans. Among other things, Rathje has found that the hazardous waste generated by households across the United States equals the hazardous waste generated by commercial establishments.

Documents Documents are written or printed materials, such as magazines, books, calendars, graffiti, birth certificates, and traffic tickets. For her research on *obentō*s, Anne Allison (1991) examined a variety of documents. She did a content analysis of *obentō* magazines and cookbooks and of *obentō* guidelines that are issued biweekly by Japanese nursery schools. From examining the documents, Allison realized the cultural importance of the *obentō*. She found that the magazines, cookbooks, and guidelines were filled with recipes, hints, pictures, and ideas about how to prepare food so that children will eat it. One common suggestion was that the mother should design the presentation of food items and the bag in which the food is carried. In other words, the mother should not let someone else prepare the *obentō*.

Territories Territories are settings that have borders or that are set aside for particular activities. Examples include countries, states, counties, cities, streets, neighborhoods, classrooms, and buildings. William H. Whyte's *City: Rediscovering the Center* (1988) is an example of research that examines a territory. For more than 16 years Whyte visited and observed New York City and other cities across the United States and around the world. He was interested in identifying the conditions under which people use urban spaces. Whyte found that some city spaces simply are not designed to be used: "Steps too steep, doors too tough to open, ledges you cannot sit on because they are too high or too low, or have spikes on them" (p. 1). Often city streets are simply not very interesting places.

Among other things, Whyte also found that Tokyo's streets are consistently more interesting than most city streets in the United States. This is because the Japanese "do not use zoning to enforce a rigid separation of uses. They encourage a mixture, not only side by side, but upwards. In the buildings you will see showrooms, shops, pachinko parlors, offices, all mixed together and with glass-walled restaurants rising one on top of the other—three, four, and five stories up" (p. 89).[4]

Households **Households** include all related and unrelated persons who share the same dwelling. Economist Raymond A. Jussaume, Jr., and sociologist Yoshiharu Yamada surveyed households in Seattle and in Kobe, Japan, and found some important differences between the two. The very composition of Kobe and Seattle households is different: 3 percent of Kobe households have no children or elderly residents, in contrast to 15.7 percent of Seattle households. Also, although virtually every Kobe and Seattle household has at least one telephone, only 2.1 percent of Kobe households have unlisted telephone numbers, compared to approximately 20 percent of Seattle households (Jussaume and Yamada 1990).

Small Groups **Small groups** are defined as two to about twenty people who interact with one another in meaningful ways (Shotola 1992). Examples include father-child pairs, doctor-patient pairs, families, sports teams, circles of friends, and committees. An example of research conducted at this level of analysis is the work of sociologist Masako Ishii-Kuntz (1992). She studied father-child relationships in three countries (Japan, Germany, and the United States) in order to evaluate the popular belief that Japanese fathers are largely absent from the home and uninvolved in their children's lives. The results of Ishii-Kuntz's study are described later in this chapter.

Population and Samples

Because of time constraints alone, researchers cannot study entire **populations**—the total number of individuals, traces, documents, territories, households, or groups that could be studied. Instead they study a **sample**, or a portion of the cases from a larger population. For example, the Garbage Project researchers have sampled approximately 250,000 pounds of garbage since 1973. Ishii-Kuntz (1992) sampled "1,149 Japanese, 1,000 American, and 1,003 German father-child pairs" (p. 106) for her comparative study of Japanese, German, and American father-child relationships.

Ideally, a sample should be a **random sample:** every case in the population should have an equal chance of being selected. The classic, if inefficient, way of selecting a random sample is to assign every case a number, place the cards or slips of paper on which the numbers are written into a container, thoroughly mix the cards, and pull out one card at a time until the desired sample size is achieved. Most researchers do not use this tedious system but use computer programs to generate their samples. If every case has an equal chance of becoming part of the sample, then theoretically the sample should be a **representative sample**—that is, one that has the same distribution of characteristics (such as age, gender, and ethnic composition) as the population from which it is selected. Thus, if the population from which a sample is drawn is 12 percent Latino, then 12 percent of a representative sample will be Latino. Sometimes just by chance researchers draw samples that do not represent the population with regard to specified characteristics. In that case, researchers may draw additional cases from the categories that are underrepresented in the sample. In theory, if the sample is representative, then whatever is true for the sample is also true for the larger population.

Obtaining a random sample is not as easy as it might appear. For one thing researchers must begin with a **sampling frame**—a complete list of every case in the population—and each member of the population must have an equal chance of being selected. Securing such a complete list can be difficult. Campus, city, and telephone directories are easy to acquire, but lists of American citizens,[5] of adopted children in the United States,[6] of American-owned companies in Japan, or of Japanese-owned companies in the United States are not so easily obtained.[7] Almost all lists omit some people (persons with unlisted numbers, members too new to be listed, or between-semester transfer students) and include some people who no longer belong (those who have moved, died, or dropped out). The important point is that the researcher consider the extent to which the list is incomplete and update it before drawing a sample. Even if the list is complete, the researcher also must think of the cost and time required to take random samples and consider the problems of inducing all sampled persons to participate.

Researchers sometimes select samples to study that they know are not representative because the subjects are accessible. For example, researchers often study high school and college students because they are a captive audience. In addition, researchers may choose their subjects in spite of the fact that little is known about them, or because the subjects have special characteristics or their experiences clarify important social issues. The point is that when a sample is not representative it is important to justify why it was chosen.

Basic Methods of Data Collection

In addition to identifying who or what is to be studied, the design also must include a plan for collecting information. Researchers can choose from a variety of data-gathering methods including surveys, interviews, observations, and secondary data analysis.

Self-Administered Questionnaire   A **self-administered questionnaire** is a set of questions given (or mailed) to respondents, who read the instructions and fill in the answers themselves. This method of data collection is very common. The questionnaires found in magazines or books, displayed on tables or racks in service-oriented establishments (hospitals, garages, restaurants, groceries, physicians' offices), and mailed to households are all self-administered questionnaires. This method of data collection has a number of advantages: no interviewers are needed to ask respondents questions; it can be given to large numbers of people at one time; and respondents are not influenced by an interviewer's facial expressions or body language, so they feel more free to give unpopular or controversial responses.

At the same time, there are some disadvantages to such questionnaires. Respondents are often self-selected in the sense that they choose to ignore or fill out the questionnaire. This leaves researchers wondering whether the people who do not fill out a questionnaire have different opinions than those who do. The results of a questionnaire depend not only on respondents' decisions to fill out and return it but on the quality of the survey questions asked and a host of other considerations. (See "Differences in Construction of U.S. and Japanese Versions of the Self-Administered Questionnaire.")

Interviews   In comparison to questionnaires, **interviews** are more personal. They are face-to-face sessions or telephone conversations between an interviewer and a respondent, in which the interviewer asks questions and records respondents' answers. As respondents give answers, interviewers must avoid pauses, expressions of surprise, or body language that reflect value judgments. Refraining from such conduct helps respondents feel comfortable and encourages them to give honest answers.

Interviews can be structured or unstructured. In a **structured interview,** the wording and the sequence of questions are set in advance and cannot be altered during the course of the interview. In one kind of structured interview respondents are free to answer the questions as they see fit, although the interviewer may ask them to clarify or explain answers in more detail. In another kind of structured interview the respondents choose answers from a response list that the interviewer reads to them.

Sociologist Hisako Matsuo's (1992) study of second- and third-generation Japanese-Americans made use of face-to-face structured interviews. As part of the interview, Japanese-American respondents were presented the following three situations designed to determine the extent to which they identify with Japan or the United States:

1. Japan became an economic power after World War II. On a scale of 1 to 4, with 1 being "very much" and 4 being "not at all," rate how proud you are of Japan's economic accomplishments.

2. Suppose that you are watching a volleyball game of the Olympics, the United States versus Japan. On a scale of 1 to 5, with 1 being "absolutely Japan" and 5 being "absolutely the U.S.," which team do you support?

3. On a scale of 1 to 10, with 1 being "a complete Japanese identity" and 10 being "a complete

American identity," indicate the number that most clearly describes your identity.

These three interview items illustrate the directedness of a structured interview.

In contrast, an **unstructured interview** is flexible and open-ended. The question-answer sequence is spontaneous and like a conversation: the questions are not set in advance and are not asked in a set order. The interviewer allows the respondent to take the conversation in directions the respondent defines as crucial. The interviewer's role is to give focus to the interview, to ask for further explanation or clarification, and to probe and follow up interesting ideas expressed by the respondent. The interviewer appraises the meaning of answers to questions and uses what was learned to ask follow-up questions. Talk show hosts often use an unstructured format to interview their guests. However, sociologists have much different goals than talk show hosts. For one thing, sociologists do not formulate questions with the goal of entertaining an audience. In addition, sociologists strive to ask questions in a neutral way, and there is no audience reaction to influence how respondents answer the questions.

Observation Observation is watching, listening to, and recording behavior and conversations as they happen. This technique sounds easy, but observation is more than seeing and listening. The challenge of observation lies in knowing what to look for, while still remaining open to other considerations; success results from identifying what is worth observing. "It is a crucial choice, often determining the success or failure of months of work, often differentiating the brilliant observer from the . . . plodder" (Gregg 1989, p. 53). Good observation techniques must be learned and then developed through practice. Practice is needed to know what is worth observing, to be alert to unusual features, to take detailed notes, and to make associations between observed behaviors.

If observers come from a culture different from the one they are observing, they must be careful not to misinterpret or misrepresent what is happening. Imagine for a moment how an uninformed, naive observer might describe a sumō wrestling match: "One big, fat guy tries to ground another big, fat guy or force him out of the ring in a match that can last as little as three seconds" (Schonberg 1981, p. B9). Actually, for those who understand it, sumō wrestling is "a sport rich with tradition, pageantry, and elegance and filled with action, excitement, and heroes—dedicated to an almost impossible standard of excellence down to the last detail" (Thayer 1983, p. 271).

Observational techniques are especially useful for studying behavior as it occurs, for learning things that cannot be surveyed easily, and for acquiring the viewpoint of the persons under observation. Observation can take two forms: participant and nonparticipant. **Nonparticipant observation** is detached watching and listening: the researcher only observes and does not interact or become involved in the daily life of those being studied.

A good example of nonparticipant observation is Catherine C. Lewis's (1988) research on Japanese first-grade classrooms. Lewis observed 15 Japanese classrooms each for a full day, "from morning greetings through the children's departure from school" (Lewis, p. 161). Lewis observed that

> it was common for two [student] monitors to stand at the front of the class, ask the class to be quiet, announce what subject was about to be studied, and ask the children to rise and greet the teacher. In some classrooms, the monitors picked which group showed the best order, cautioned groups to improve their posture, granted check marks to groups that were quiet and ready to begin the class period, or chose which groups would receive lunch first. Often, the monitors assembled and quieted the class when the teacher was not yet present. Typically, monitors rotated each day, and all children became monitors in turn. (PP. 162–63)

In contrast to nonparticipant observation, researchers engage in **participant observation** when they join a group and assume the role of a group member, when they interact directly with those whom they are studying, when they assume a position critical to the outcome of the study, or when they live in a community under study. Anne Allison's research on *obentō*s described earlier in this chapter is an excellent example of participant observation. Her research revolved around conversations she had with other mothers, daily

Differences in Construction of U.S. and Japanese Versions of the Self-Administered Questionnaire

When constructing a self-administered questionnaire, researchers must make sure that the questions are clear to respondents so that they will not be misunderstood. If respondents are from different cultures, researchers may have to adjust the questions and the format of the questionnaire and accompanying letter so that potential respondents view the researcher's request for them to participate in a study as legitimate and worthy of their time and attention. In the excerpt below, we learn some ways that researchers R. A. Jussaume, Jr., and Y. Yamada adjusted the questionnaire to accommodate Japanese and American respondents.

Design Component	United States	Japan
Size of cover letter	8½″ × 11″	8¼″ x 11½″
Return envelope	U.S. standard #9 (white)	Japanese standard #4 (off-white)
Outgoing envelope	U.S. standard #10 (white)	Japanese standard #3 (off-white)
Address style	Left-to-right/printed	Top-to-bottom/handwritten
Personalization	Signature	Personal seal
Cover letter context	Bilateral trade	Understanding Japan

Many of the adjustments required little more than common sense. For example, standard letter size in Japan is not 8½ by 11 inches, but approximately 8¼ by 11½. Envelope sizes are also different and open on the narrow as opposed to the wide side. For the Kobe survey, Japanese standard number 3 and number 4 size envelopes of an off-white color were used. In addition, addresses were written on envelopes from the top down, instead of from left to right, in accordance with the prevailing cultural practice.

In the TDM [total design method], trust is established with the respondent by having the researcher personally sign each cover letter. In Japan, signatures have no legal validity and little cultural significance, although it is understood by Japanese that this is *not* the case in "Western" cultures. Therefore, while the American researcher signed his name to the Japanese cover letter, the Japanese counterpart impressed his *Inkan* (personal seal) to the letter. This conveyed to the respondent the importance which both researchers placed on the project.

Another alteration was in how the respondent's name was affixed to the envelope and the cover letter. With the Seattle sample, the common practice of printing names and addresses on the envelope and the letter was followed. However, with the Kobe sample, the names of the respondents were written on the envelope and on the top of the letter in longhand. While Japanese-language word processors are commonplace, correspondence in Japan still is generally written in longhand. It was

conversations with the teacher about her son's eating habits, Mothers' Association meetings, and other school-related events.

In both participant and nonparticipant observation, researchers must decide whether to hide or to announce their identity and purpose. One of the primary reasons for choosing concealment is to avoid the **Hawthorne effect**,[8] a phenomenon whereby observed persons alter their behavior when they learn

they are being observed. If researchers announce their identity and purpose, they must give participants time to adjust to their presence. Usually, if researchers are present for a long enough time, the subjects eventually will display natural, uninhibited behaviors.

Secondary Sources Another strategy for gathering data is to use **secondary sources**. This is data that

thought that Japanese respondents would react more positively to the survey if they saw that the effort had been taken to write out their names in longhand.

Other adaptations were made in the style and content of the cover letters and postcard. In the opening paragraphs of the Japanese letter, mention was made of how the results of the study would help promote international understanding of Japanese food consumption patterns. When animal proteins produced from imported feed grains are included, Japan is said to import more than half of its base food calories. The Japanese people are very sensitive to the issue of international trade in agricultural commodities and want exporting nations to be more accommodating in producing the kinds of products they wish to consume. On the American side, the point was made that the results of the survey would be potentially useful in helping American producers sell more goods to Japan. As the bilateral trade imbalance with Japan has become a major political issue in the United States, it was thought that American respondents would react positively to participating in

a project which would have potential positive ramifications on the problem.

In both the English and Japanese letters, an equally strong emphasis was placed on three essential features of the TDM. First, respondents were told how their names were selected, that their responses would represent those of many other people, and that their participation was invaluable. Second, the confidentiality of the survey was stressed and participants were requested not to write their name anywhere on the survey instrument. Finally, respondents in both countries were offered a report of the comparative results of the study as an incentive for participation.

The Japanese cover letter was written independently of the English, however. The main points to be covered in the Japanese letter were discussed, with reference made to the themes covered in the English letter. The Japanese member of the research team then composed an original letter for the Kobe survey. This letter was pretested on a small number of Japanese women to ensure that the tone of the letter was neither too condescending nor overly

polite. This was an important procedure, for written Japanese is much more formal and complex than English is. Arguably, the style of Japanese *personal* correspondence is more formal than most American business correspondence.

The first mailing for the Seattle sample was sent out on 7 June 1988 and that for the Kobe sample on 16 June 1988. Exactly one week after each initial mailing, a follow-up postcard requesting those who had not yet responded to the survey to do so was sent out to *all* respondents. Two weeks after the postcard, a second mailing was distributed to those from whom responses had not been received.

SOURCE: "A Comparison of the Viability of Mail Surveys in Japan and the United States of America," by R. A. Jussaume, Jr., and Y. Yamada. Pp. 222–23 in *Public Opinion Quarterly*. Copyright © 1990 by the American Association for Public Opinion Research. Published by the University of Chicago Press. Reprinted by permission.

has been collected by other researchers for some other purpose. Government researchers, for example, collect and publish data on many areas of life including births, deaths, marriages, divorces, crime, education, travel, and trade.

Sociologists Lynn K. White and Yuko Matsumato (1992) conducted comparative research on divorce in Japan and the United States. Their work is an example of a study that uses this kind of sec-

ondary data. White and Matsumato sought to explain why Japan's divorce rate is one-fifth that of the United States: Japan's annual divorce rate is 4.8 per 1,000 married women, whereas the rate in the United States is 22.6 per 1,000 married women. The researchers used census data from each country to obtain divorce rates for Japan's 47 prefectures and for the 50 states. They also obtained data on the following variables: (1) the percentage of

Japanese and of American women in the labor force, (2) the percentage of Japanese women and American women who are self-employed, (3) the percentage of Japanese women and American women who work in family-owned businesses, and (4) the ratio of females per 100 males in professional and administrative occupations. White and Matsumato found that these four variables helped explain divorce rates in Japan and in the United States. But they did not help explain why Japan's divorce rate is one-fifth that of the United States.

A second kind of secondary data source consists of materials that people have written, recorded, or created for reasons other than research (Singleton, Straits, and Straits 1993). Examples include television commercials and other advertisements, letters, diaries, home videos, poems, photographs, art work, graffiti, movies, and song lyrics.

Operational Definition: Reliability and Validity

We have looked at four methods of collecting data—self-administered surveys, interviews, observation, and secondary data analysis. There are other ways to collect data, such as experimental design,[9] but the four methods discussed in this chapter are the ones sociologists are most likely to use. No matter which method researchers choose, they should collect data to test hypotheses and, in order for the findings to matter, other researchers must be able to replicate the study. For this reason it is necessary that researchers give clear and precise definitions and instructions about how to observe and measure the variables being studied. In the language of research, such definitions and accompanying instructions are called **operational definitions**. An analogy can be drawn between an operational definition and a recipe. Just as anyone with basic cooking skills can follow a recipe to achieve a desired end, anyone with basic research skills should be able to replicate a study if he or she knows the operational definitions (Katzer, Cook, and Crouch 1991). For example an operational definition of education is the number of years of formal schooling the person has completed.

If the operational definitions are not clear or do not indicate accurately the behaviors they were de-

signed to represent, they are of questionable value. Good operational definitions are reliable and valid. **Reliability** is the extent to which the operational definition gives consistent results. The question How many magazines do you read each month? may not yield reliable answers because respondents may forget some magazines. Thus if you asked the question at two different times the likelihood that the respondent might give two different answers is high. One way to increase the reliability of this question is to ask respondents to list the magazines that they have read in the past week. The act of listing forces respondents to think harder about the question, and shortening the amount of time to think back on makes it easier to remember. In the case of survey questions, reliability can be checked by asking the question at two different times to learn whether the answer remains the same.

Validity is the degree to which an operational definition measures what it claims to measure. Professors give tests to measure students' knowledge of a particular subject as covered in class lectures, discussions, reading assignments, and other projects. Students may question the validity of this measure if the questions on a test reflect only the material covered in the textbook or only the material covered in lectures. In such instances students may argue that the test does not measure what it claims to measure.

Similarly, when the trade deficit with Japan is used as an operational definition of the openness of the Japanese market, of Japanese aggressiveness, or of the patriotism of American consumers, we must question whether it is a valid measure of such behaviors. For example, we know that the trade deficit as currently calculated cannot be used as a measure of the openness of Japanese markets to American goods because it does not consider (1) the dollar value of goods and services produced by American and Japanese companies located in the other's country or (2) the dollar value of the indirect exports that each country sends to the other. Some examples of more valid measures of market openness are (1) the number of trade barriers each country puts up against the other, (2) the number of suggestions each country offers the other for opening its markets, and (3) per capita purchases of imports (see Figure 3.2).

FIGURE 3.2 Alternative Measures of the Relative Openness of U.S. and Japanese Markets

One Measure: U.S. Trade Deficit with Japan

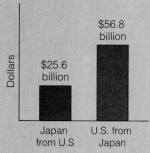

Value of Imports

Possible interpretation:
U.S. market is far more open than Japan's; U.S. is at severe disadvantage in terms of trade.

Alternative #1

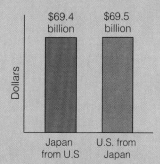

Value of Imports Plus Sales by Japanese-Owned Businesses in U.S. and U.S.-Owned Businesses in Japan

Possible interpretation:
U.S. enjoys slight trade surplus with Japan; the markets are about equally open.

Alternative #2

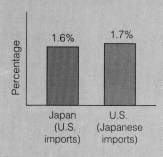

Imports as Percentage of GNP

Possible interpretation:
U.S. / Japanese imports are about equal.

Alternative #3

Per Capita Purchases of Imports (Annual)

Possible interpretation:
U.S. consumers spend more on Japanese goods than Japanese consumers spend on U.S. goods.

SOURCES: Adapted from Sanger (1992) and Totten (1990).

Barnlund's Study: Step 4

In step 4, researchers must decide on a research design or the plan for gathering data to test hypotheses. They have to specify the unit of analysis (individuals, traces, documents, territories, households, or small groups) and identify their sample. In addition, researchers must decide on a method of data collection (self-administered questionnaire, interviews, observation, or secondary source analysis). Finally, they must provide clear operational definitions for their independent and dependent variables.

Barnlund (1989) chose to study individuals born and raised in the United States or in Japan as his unit of analysis. Obviously he could not study everyone born and raised in each country. Consequently Barnlund decided to use a sample of college students who met this criterion. His sample included 423 Japanese and 444 American college students between the ages of 18 and 24, born and raised in Japan or in the United States, respectively. The Japanese participants were enrolled in public and private universities in Fukuoka, Kyoto, Matsue, Tokyo, and Chiba City. The American participants were enrolled in public and private universities in New York, Minneapolis, Denver, Reno, and San Francisco. Although Barnlund did not state how the students were selected, such a sample clearly is not representative of all of the people born and raised in Japan and the United States.

Barnlund acknowledged that college students constitute a group that is not representative of American or Japanese populations, but he suggested that there are some advantages to studying them: "While they are not representative of their cultures as a whole, they constitute a large and important part of it" (p. 49). More significantly, future Japanese and American leaders are likely to come from the ranks of the college-educated, and, "since we seek the shape of the future rather than the past, their patterns of communicating with one another may suggest its direction" (p. 49).

Barnlund chose the self-administered questionnaire as his method of gathering data. He used a series of questions to operationalize the various dimensions of behavior toward strangers. All questions concerning predisposition were preceded by a definition of *stranger* (an unknown, unintroduced person) and by a short introduction. (Refer back to "Questions Used to Operationalize Behavior Toward Strangers.")

The questions that Barnlund used to operationalize his dependent variable (behavior toward strangers) may not be reliable for at least one reason. Asking respondents to think about encounters with strangers over the past month is problematic because a month is a long time; most encounters with strangers are likely to be vague in the respondents' memories. Thus, if respondents were asked these questions at two different times, they may give different answers. Barnlund could increase the reliability of his questions if he asked respondents to think about encounters with strangers over the past week or past 24 hours.

Barnlund's questions have some validity problems as well. Specifically, critics argue that his definition of a stranger as an unknown, unintroduced person is too vague: it needs to be made more specific and should include some qualifications with regard to setting and time. For example, if the setting is such that a person must interact with a stranger (such as a store clerk), should that interaction count as a situation that taps people's disposition toward strangers? Critics believe Barnlund should have instructed the respondent that the stranger is an unknown, unintroduced person whom the respondent is not required to approach in order to purchase goods and services or to fill a requirement imposed by a third party. In addition, the instructions should have included some time limit beyond which a person ceases to be a stranger. For example, if a respondent says hello or waves to another person several times before initiating a conversation, is that other person still a stranger? It might have been wise to also specify that a stranger is someone whom you have encountered for the first time. Because Barnlund's definition of strangers is so lenient, we must ask if he is really measuring behavior toward strangers.[10]

These criticisms do not necessarily diminish the value of Barnlund's efforts. Any researcher who tries to assign operational definitions to complex concepts such as predisposition to strangers will encounter problems such as these. The challenge is to anticipate the difficulties and to strive to construct the best possible operational definitions.

Step 5: Analyzing the Data

When researchers get to the stage of analyzing collected data, they search for themes, meaningful patterns, and/or links between variables. They present their findings in a form that allows others to see the themes, patterns, and/or links as well. Researchers may make use of graphs, frequency tables, photographs, statistical data, and so on. The choice of presentation depends on what results are significant and how best to show them.

Researchers William T. Bailey and Wade C. Mackey used tables to summarize the data they collected in their 1989 nonparticipant observation study of 18,272 child-adult interactions in public settings in Japan and in the United States. These researchers constructed tables that show the percentages of child-adult interactions in Japan and in the United States involving women only, men only, and men and women. The tables show that, contrary to the popular image of Japanese men as people who work long hours and spend little or no time with their children, Japanese men are observed as frequently as American men interacting with their children in public settings (see Figure 3.3).

Barnlund's Study: Step 5

When it came time to analyze data, Barnlund looked for meaningful links between the independent variable (cultural background) and the dependent variable (behavior toward strangers). According to his analysis, Americans notice, consider interacting with, initiate conversation with, and respond to strangers more frequently than do the Japanese. This observation is supported by the fact that Americans answered "frequently" and "very often" more often than "never" and "seldom" to the predisposition questions (that is, How often did you notice, consider interacting with, start a conversation with, or respond favorably to a stranger?). Conversely, the Japanese answered "never" and "seldom" more often than "frequently" and "very often."

Barnlund found that the most pronounced difference between Japanese and Americans was in their predisposition to consider interacting and in their actual interaction with strangers. When interaction did take place with a stranger, the Japanese most frequently learned the stranger's occupation (an important indicator of status). Americans, on the other hand, most frequently learned the stranger's name, followed closely by activities and interests. One-third of the American respondents acquired phone numbers and addresses (important facts for further contact), compared to one-fifth of the Japanese. One shortcoming in Barnlund's analysis of the data is that he did not use tables to present his findings. He presented his findings in sentence form (that is, "One third of American respondents . . . while one-fifth of Japanese respondents . . . " and so on). This is a shortcoming because readers know only the number that Barnlund chose to highlight.

Contrary to a common stereotype, the results of Bailey and Mackey's comparative study of parent-child interactions indicated that Japanese fathers spend about as much time with their children as U.S. fathers do with theirs.

Sonia Katchian/Photo Shuttle: Japan

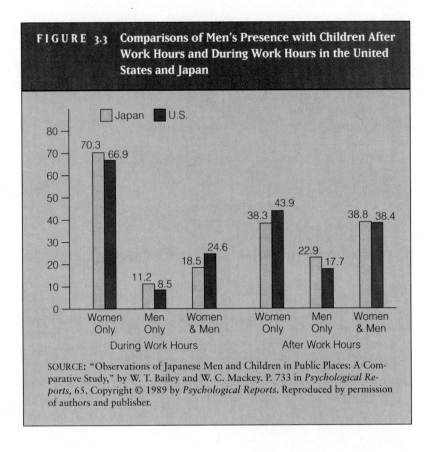

FIGURE 3.3 Comparisons of Men's Presence with Children After Work Hours and During Work Hours in the United States and Japan

SOURCE: "Observations of Japanese Men and Children in Public Places: A Comparative Study," by W. T. Bailey and W. C. Mackey. P. 733 in *Psychological Reports*, 65. Copyright © 1989 by *Psychological Reports*. Reproduced by permission of authors and publisher.

Step 6: Drawing Conclusions

In the final stage of the research process, sociologists comment on the **generalizability** of findings, the extent to which the findings can be applied to the larger population from which the sample is drawn. The sample used and the response rate are both important when it comes to generalizability. If a sample is randomly selected, subjects all agree to participate, and the response rate is high, we can say it is representative of the population and theoretically the findings are generalizable to that population. If a sample is chosen for other reasons—perhaps because it is especially accessible or interesting—then the findings cannot be generalized to the larger population. For example, Lee Iacocca, chairman of Chrysler Corporation, reported in a series of commercials that "by a margin of 4 to 1, the test drivers preferred the Plymouths and

Dodges to the Hondas" (Levin 1990, p. C3). This finding can hardly be generalized to all American car buyers, however, because the "sample" involved 200 new-car shoppers in Los Angeles who currently owned domestic models and were older-than-average drivers (Levin 1990). The fact that the findings based on a sample cannot be generalized to the larger population does not necessarily diminish the value of the study, however. The value of the study depends on the reasons a nonrandom sample was used and the insights that it generates. In the case of the Chrysler Corporation study, the sample was selected to bias responses in favor of Chrysler products.

Even though one goal of drawing conclusions is to make generalizations about the larger population, generalizations are not statements of certainty that apply to everyone. Consider the comparative research on father-child interaction in three cultures conducted by sociologist Ishii-Kuntz (1992). In this case, for example, knowing a respondent's culture did not mean that the researchers could predict the frequency of father-child interaction for any one pair. The study data shows that Japanese fathers are significantly less likely than American and German fathers to eat dinner with their children every day. Clearly, this does not mean that all individual Japanese fathers behave in this way but that as a group they interact with their children less frequently than do American and German fathers.

At least three conditions must be met before a researcher can claim that an independent variable contributes significantly toward explaining a dependent variable. The first condition is that the independent variable must precede the dependent variable in time. Time sequence can be established easily when the independent variable is a predetermined factor such as sex, ethnicity, or birth date. These kinds of factors are fixed before a person is capable of any kind of behavior. Usually, however, time order cannot be established so easily.

A second condition that must be met before a researcher can claim that an independent variable contributes toward explaining a dependent variable is that the two variables must be correlated. A **correlation** is simply a mathematical representation of the extent to which a change in one variable is associated with a change in another (Cameron 1963). Establishing a correlation is a necessary step but is not in itself sufficient to prove cause. A correlation shows only that the variables are related; it does not mean that one variable causes the other. For one thing, a correlation can be spurious. A **spurious correlation** is one that is coincidental or accidental; in reality, some third variable is related to both the independent and the dependent variables. The presence of the third variable makes it seem that those two variables are related. A good example is the seeming correlation between the number of fire trucks sent to the scene of a fire and the amount of damage done in dollars. (The more fire trucks, the greater the damage.) However, common sense tells us that this is a spurious correlation. A third variable—the size of the fire—is responsible for both the number of fire trucks sent and the amount of damage done.

Another example of a spurious relationship is the seeming correlation between physical features and academic achievement, especially in mathematics and science. In the United States students with coarse, straight, black hair and dark eyes with epicanthic folds characteristic of Japanese and other Asian people are thought to be higher academic achievers than students with blue eyes and blond hair (white) or students with dark eyes and tightly curled black hair (African-American). Common sense tells us that physical features do not correlate with academic achievement; the difference in achievement is due to several other variables such as parental involvement, time devoted to schoolwork, and amount of after-school employment. It happens that the students with straight black hair and dark eyes with epicanthic folds tend to devote more time to homework, to work less at after-school jobs, and to watch less television after school, as well as to have parents who supervise them as they do their homework. These behaviors are tied to academic achievement, not to students' physical appearance.

Finally, if a researcher is to claim that an independent variable helps to explain a dependent variable, there must be no evidence that another variable is responsible for a spurious correlation between the independent and the dependent variables. To check this possibility, sociologists identify **control variables**, variables suspected of causing spurious correlations. Researchers determine whether a control variable is responsible by holding it constant and re-examining the relationship between the independent and the dependent variables.

For example, one variable suspected of causing the spurious correlation between academic achievement and physical traits is after-school employment. To control for after-school employment, researchers classify Americans of Japanese descent, Americans of European descent, and Americans of African descent into two separate groups each: those who work after school and those who do not. Then the relationship between physical traits and achievement is re-examined separately within each of the two groups. If the correlation between academic achievement and physical traits disappears in the presence of employment, we know that the relationship is spurious and that a third variable— amount of after-school employment—is actually

Are Japanese students (shown here being congratulated by colleagues on passing their exams) more studious than their American counterparts? If so, the difference is not due to genetic predispositions, but to the high value that Japanese society places on academic success and examination scores.
Sonia Katchian/Photo Shuttle: Japan

responsible for differences in academic achievement and for making it seem that differences in physical attributes cause differences in academic achievement.

Thinking about the possibility of a spurious relationship is especially important when the independent variable is an **ascribed characteristic:** any trait (age, sex, ethnicity, culture at birth) that is biological in origin and/or cannot be changed. Such findings can be used to stereotype and stigmatize some groups of people. For example, many Americans believe that Japanese workers are more loyal than Americans, that Japanese students are more studious, that Japanese criminals are more contrite, that the Japanese are less litigious than Americans, and that Japanese people in general are more group-oriented. Actually, however, such observed differences are beyond the control of individual Japanese and can be explained as due to some other factors:

The "loyalty" of white-collar workers to their company, in contrast to the constant movement of employees in other countries, is one clear example. [In Japan] the major corporations tacitly agree never to hire someone who has left another firm. Japanese children are studious in large part because admission to the University of Tokyo, which is based on examination

scores, is essentially their only hope for having an influential place in society.

Japanese are "nonlitigious," not just because of their alleged love of consensus but also because of the acute shortage of lawyers. The Ministry of Justice controls the Legal Training and Research Institute, where future lawyers and judges must train, and it admits only 2 percent of those who apply. (Of the 23,855 who took the entrance examination in 1985, 486 were admitted.)

Most criminals arrested by the police confess partly out of a sense of remorse but also because they know what a trial would mean: in 99 percent of criminal trials, the verdict is guilty.
(FALLOWS 1989, P. 28)

Even if researchers are able to establish the three conditions mentioned above—that (1) the independent variable precedes the dependent variable in time, (2) there is a correlation between the independent and dependent variables, and (3) there is no evidence of a spurious correlation—it is extremely difficult for them to claim that the independent variable causes the dependent variable. Researchers use a **probabilistic model** of cause in which the hypothesized effect does not always result from a hypothesized cause. In other words, knowing a person's culture (the hypothesized cause or independent variable) does not mean that we can predict his or her behavior toward strangers (the hypothesized effect or dependent variable) with 100 percent accuracy. Under a probabilistic model if researchers meet the three conditions they can say the independent variable makes a significant contribution toward explaining the dependent variable.

Barnlund's Study: Step 6

Barnlund's study adds to what we know about Japanese and American communication patterns, but it is difficult to generalize his findings to non-college segments of each society. (A general hazard of using college students is that it is difficult to apply the findings to other segments of the society.) Recall that Barnlund justified his decision to survey American and Japanese college students by arguing that Japanese and American leaders are likely to come from the ranks of the college-educated.

Barnlund also considered a number of things that could be responsible for causing a spurious relationship between culture and behavior toward strangers. He checked to make sure that the Japanese and the American students in the survey were similar in all traits except inherited culture. When Barnlund compared American and Japanese students he found no differences in age, proportion of males to females, years of education, or family size. He concluded that "these samples are sufficiently alike with regard to age, sex, home environment, education, and family size to attribute any differences in their communicative styles mainly to the influence of culture" (1989, p. 50). This conclusion

may have been premature, however. Barnlund acknowledged that there were some differences with regard to place of residence: "More Japanese lived at home with their parents while a greater proportion of Americans lived in apartments with peers" (p. 49). Place of residence could be related to behavior toward strangers in that students who live away from home are more open to new experiences (and, by extension, to meeting strangers) than are those who live at home.

Upon reviewing Barnlund's findings we cannot say with certainty that culture is the variable that explains differences in the way Japanese and Americans behave toward strangers. In addition, Barnlund's results cannot be generalized to first-time encounters between representatives of each country. It is difficult to apply his findings to settings in which the strangers belong to another culture and are required to talk with each other in order to negotiate a business transaction.

These shortcomings may leave you feeling that reviewing Barnlund's study is a waste of time. Yet, one could argue that in the process of critiquing Barnlund's study we have become more open-minded about the Japanese and are less likely to act toward them on the basis of faulty data. For example, although Barnlund's data shows that the Japanese tend to notice, consider interacting with, and initiate conversation with strangers less frequently than Americans, we cannot classify them as unfriendly. The fact that some Japanese answer "very often" and "frequently" to questions about strangers indicates that the Japanese do not all behave in uniform ways toward strangers.

Discussion

In this chapter we have identified the technological forces that allow people to produce large quantities of data, and we have considered the quality control problems—dearth of feedback and exaggerated, distorted headlines—that are associated with the information explosion. As a result we have to think critically about the research we encounter because

in the workaday world it is clear that most of what is proffered as research is secondhand and dubious. The magazine researchers find one reference and think a fact or judgment [is] confirmed. Opinion research is continually shown up by its results. Gallup gives one figure, Harris another. And in the stream of reports about

A quality contol meeting at a Japanese television factory.

Charles Gupton/Tony Stone Images

divorce, disease, and diet, about education, ecology, economics, and government, indeed about all subjects of practical concern, the chaos is permanent and daunting.

It is safe to say that the steady stream of "studies" does not really add to the public's knowledge. One report cancels another; details do not stick in the mind; and interest is too often aroused by what goes against common sense. (BARZUN 1987, P. 20)

Despite such problems, a basic knowledge of sound research techniques helps us sort through data in a way that allows us to distinguish media hype from balanced research. We have learned that there are numerous reasons why we should be skeptical of research.

First, all scientists—chemists, biologists, sociologists, economists—make choices about what they will study based on their values and personal interests. Sociologists may be interested in communication patterns of Japanese and Americans because they believe that Americans are falling short in trade negotiations with the Japanese or are unsophisticated about other cultures and need to learn about them. Likewise, chemists and biologists often choose a research topic on the basis of their values about what kinds of problems need to be solved. Because researchers have opinions and values that influence their choices of topics, some mechanism

must be in place to prevent personal values from influencing the outcome of the research. That mechanism is the scientific method, which requires researchers to report the means by which they arrive at their findings so that interested parties can scrutinize and verify the process.

Adhering to the principles of the scientific method requires discipline, patience, and constructive feedback. Researchers, however, often violate these principles. For example, operational definitions (such as trade balances) are often constructed without careful consideration of their reliability and validity, samples are selected carelessly, and interpretations are sloppy and inaccurate.[11] The likelihood that sloppy research will be published increases as the amount of research to be evaluated increases, simply because there are more researchers than critical readers. To compound the matter, when sloppy research is published, it becomes open to the public; at that point it is difficult to predict how findings will be used.

All disciplines face quality assurance problems. Social scientists, however, face an additional problem that physical scientists usually do not encounter: they study people who can refuse to cooperate, conceal information, or give answers to please or frustrate researchers. As sociologist Fritz Machlup put it, a rock does not say to a geologist, "'I am a beast,' nor does it say: 'I am here because I did not like it up there near the glaciers, where I used to live; here I like it fine, especially this nice view of the valley'" (1988, p. 60). People, on the other hand, are less open to scrutiny.

It is difficult to learn the extent to which people conceal information from researchers. For example, a Gallup poll surveyed 1,239 American households to determine how many had mailed the completed forms back to the Census Bureau. According to the findings of this poll, 80 percent had returned the forms, which contradicts U.S. Census Bureau figures of 63 percent (Kagay 1990). It could be that 17 percent of those surveyed in the Gallup poll lied. It also could be that the sample drawn was not representative of the population. If the sample had been selected through a random-digit-dialing method, only those with working telephones would have been surveyed. Perhaps people who have telephones are more likely to return the form than those who do not have telephones.

It is important to realize that a person cannot expect to conduct the perfect research project, one with no flaws or shortcomings. For this reason it is essential that the reader be able to assess how well the researcher worked to meet the challenges of doing research. For example, although Barnlund's research contains a number of conceptual and other methodological problems, critiquing it has at least one valuable outcome: it forces readers to think through the study's flaws. In identifying the flaws we actually gain information. In the case of Barnlund's study we learn how difficult it is to make statements about how people born and raised in a particular country behave. Such a discovery encourages us to question findings that make broad and unqualified generalizations. In the end this leaves us more broad-minded about human behavior and with fewer stereotypes about people.

FOCUS

Number of Telephone Messages as a Measure of Interdependence

One of the most critical steps in the research process is the operationalization of concepts. As stated in this chapter, operational definitions are specific instructions for observing and recording core concepts and variables. The criteria must be precise and clear so there is no question about how they are to be measured. If operational definitions are not clear or do not indicate accurately the behaviors they were designed to represent, their value must be questioned. Good operational definitions must be reliable and valid.

If operational definitions are not reliable and valid, it does not matter how much time, money, or effort a person puts into the project; the findings will be considered suspect. The following is an example that illustrates the careful thought that must go into constructing good operational definitions. It concerns messages between countries sent in the

form of telephone messages (which includes conversations, faxes, telexes, and telegrams). Can we use the number of telephone messages between countries as a measure of interdependence or interconnection? This is an important question, because people throw the word *interdependence* around but fail to be specific about its meaning.

The Telephone Message Deficit

Kevin Steuart and Joan Ferrante

Consider the Federal Communications data on the annual phone message deficit that the United States has with the countries covered in this textbook. The data shows that in all cases the United States sends more phone messages than it receives. (See Table 3.2.)

Just as the trade deficit between the United States and another country is used as a measure of interdependence, suppose that the phone message deficit is used as an operational definition of interdependence. (In this case, interdependence is taken

SOURCE: Joan Ferrante and Kevin Steuart, Northern Kentucky University (1994).

to mean a state in which the lives of people in two countries are interconnected.) Is the phone message deficit a good measure of interconnection or interdependence? (As you read the critique below, think about whether there are comparable criticisms for the trade deficit.)

Actually, there are a number of problems with the message deficit as a measure of interconnection.

1. The message deficit does not tell us the total number of phone exchanges made between two countries, nor does it tell us how many messages people in the United States send to a particular country or how many people in a particular country send messages to the United States. For example, a phone message deficit with Mexico of 53,707,000 calls could result from an almost endless number of scenarios, including:

    a. 93,437,000 phone messages originating from Mexico with 147,144,000 messages originating from the United States for a total of 240,581,000 messages (147,144,000 − 93,437,000 = 53,707,000).

    b. 5,682,000 messages originating from Mexico and 59,389,000 messages originating from the United States for a total of 65,071,000 messages (59,389,000 − 5,682,000 = 53,707,000).

2. A large message deficit with a particular country does not necessarily mean that the interconnection between the United States and that country is stronger than an interconnection characterized by a small message deficit. For example, the total number of messages sent between India and the United States is 22,327,000. The deficit is 793,000. The total number of messages sent between Israel and the United States is 22,001,000. The deficit is 12,581,000.

For these two reasons, the total number of phone messages is a better operational definition of interdependence than the phone message deficit.

Now suppose that we decide to go with the total number of messages sent between the United States and another country as a measure of interconnection. (See Table 3.3.) Are total phone messages a good measure of interdependence? There is at least

TABLE 3.2 The Phone Message Deficit	
Country	**Calls from the U.S. Less Calls to the U.S.**
Brazil	4,488,000
China	2,032,000
Germany	19,247,000
India	793,000
Israel	12,581,000
Japan	5,579,900
South Korea	7,222,000
Lebanon	270,000
Mexico	53,707,000
South Africa	1,178,000
Zaire	58,000
TOTAL	107,155,000

one problem: the population size of a country can distort the size of the interconnection. For example, Israel has a population of 4,037,620, whereas Japan's population is 123,611,167. The total number of messages between Israel and the United States is 22,001,000. The total number of messages between Japan and the United States is 97,785,000. On the basis of just this information, one might conclude that Japan is more interconnected than Israel with the United States. If we figure per capita messages for Israel (22,001,000/4,037,620 or 5.45) and for Japan (97,785,000/123,611,167 or 0.79), we can see that Israel sends almost seven times as many phone messages per person as Japan.

So, if we decide to stay with total number of phone messages as a measure of interconnection, then we will have to standardize for population size. That is, we must divide the total number of phone messages by the population size of the country. (See Table 3.4.)

Another problem with the message deficit as a measure of interconnection, one might argue, is that we don't know who is making the phone calls. For example:

1. U.S. students attending school in Japan, U.S. businesspeople in Japan, or U.S. vacationers in

TABLE 3.3   Average Annual Telephone Messages to and from Selected Countries

Country	Messages to	Messages from	Total Messages
Brazil	17,717,000	13,229,000	30,946,000
China	6,094,000	4,062,000	10,156,000
Germany	67,669,000	43,422,000	111,091,000
India	11,560,000	10,767,000	22,327,000
Israel	17,291,000	4,710,000	22,001,000
Japan	51,682,000	46,103,000	97,785,000
South Korea	25,841,000	18,619,000	44,460,000
Lebanon	270,000	0	270,000
Mexico	147,144,000	93,437,000	240,581,000
South Africa	4,300,000	3,122,000	7,422,000
Zaire	58,000	0	58,000
TOTAL	349,626,000	237,471,000	587,097,000

TABLE 3.4   Average Annual Telephone Messages and Population (By Country)

Country	Total Messages	Population Size	Per Capita Messages
Brazil	30,946,000	153,322,000	0.20
China	10,156,000	1,133,682,501	0.01
Germany	111,091,000	79,753,227	1.39
India	22,327,000	846,302,688	0.03
Israel	22,001,000	4,037,620	5.45
Japan	97,785,000	123,611,167	0.79
South Korea	44,460,000	43,410,899	1.02
Lebanon	270,000	2,126,325	0.13
Mexico	240,581,000	81,249,645	2.96
South Africa	7,422,000	26,288,390	0.28
Zaire	58,000	29,671,407	0.002
TOTAL	587,097,000	2,523,455,869	0.23

Japan may be making calls to the U.S. from Japan.

2. German students attending school in the United States, German businesspeople in the United States, or German vacationers may be making calls to Germany from the United States.

3. A third party (say, someone from India visiting Japan) may be calling someone in the United States.

4. U.S. students attending school in China, or U.S. businesspeople or vacationers in China, as well as Chinese residents, may be calling collect to the United States. The U.S. Federal Communications Department counts collect calls as originating in the country paying the bill.

Do these possibilities affect the validity of this measure of interconnectedness? We would argue no, because even when the messages are between people from the same countries, the callers are still phoning each other from two distinct countries.

Even if some of the people calling from Mexico are U.S. visitors, they are in Mexico for some reason and they are interacting with people in Mexico in some capacity. Therefore, a phone message documents, in a rough way, interconnection. In other words, total messages is not a perfect measure of interdependence but it is one of the best indicators we have. Can you think of a better measure of interconnection? If you can, please write the Sociology Program at Northern Kentucky University, Highland Heights, KY 41099.

Key Concepts

Ascribed Characteristic 94

Concepts 79

Control Variables 93

Correlation 93

Data 68

Dearth of Feedback 72

Dependent Variable 80

Documents 82

Generalizability 92

Hawthorne Effect 86

Household 83

Hypothesis 80

Independent Variable 80

Information 68

Information Explosion 68

Interviews 84

Method of Data Collection 82

Nonparticipant Observation 85

Objectivity 74

Observation 85

Operational Definitions 88

Participant Observation 85

Population 83

Probabilistic Model 94

Random Sample 83

Reliability 88

Representative Sample 83

Research 68

Research Design 82

Research Methods 68

Research Methods Literate 73

Sample 83

Sampling Frame 83

Scientific Method 73

Secondary Sources 86

Self-Administered Questionnaire 84

Small Groups 83

Spurious Correlation 93

Structured Interview 84

Territories 82

Traces 82

Unit of Analysis 82

Unstructured Interview 85

Validity 88

Variable 80

Notes

1. Telecommunications technologies include radio, television, and modem. These technologies convert voice, data, or images into electronic impulses to transmit them. There have been four distinct revolutions in human communication: the development of speech, writing, printing, and telecommunications (Bell 1979).

2. Sociologists Robert and Helen Lynd ([1929] 1956) observed the times at which lights were turned on during winter mornings to determine whether people from working-class households started their days earlier than middle-class people.

3. Researchers have studied reductions in municipal water pressure to obtain estimates of the number of people watching commercials shown at breaks in prime-time programs. The assumption is that when commercials are interesting, fewer people

go to the bathroom than when commercials are uninteresting (Rossi 1988).

4. Pachinko is a type of pinball game. Pachinko parlors are typically located near train stations and commercial districts (Kōji 1983).

5. Even the U.S. government does not have a list of all citizens. Census Bureau officials encountered enormous problems in trying to distribute the 1990 census forms to every household (and to the homeless) and in trying to induce everyone (citizens and noncitizens) to participate. Almost 5 million forms sent to household addresses on the Census Bureau address list were returned by the post office as undeliverable (Barringer 1990). Almost 40 percent of households did not return the forms. As a result, 210,000 Census Bureau workers were initially hired nationwide and 56,000 more had been requested to visit the household addresses from which no forms were returned (Barringer 1990). "Since they only carry addresses, the workers often try to get names from a mailbox or a building superintendent, then try to conduct a telephone interview. As a last resort, they are to use a neighbor, a mail carrier or other knowledgeable source to find out vital information about occupants, like gender and race" (Navarro 1990, p. E4).

6. In the United States, 48 states keep adoption records closed. Because of this practice, there is no way of knowing the number of adoptees in this country. Most adoptees and their families whom researchers have studied come from the population of people who use mental health services (Caplan 1990).

7. For example, the U.S. government does not keep records of which U.S. companies are sold to foreigners. In fact, officials from the Office of Trade and Investment Analysis had to examine newspapers, magazines, and business and trade journals to estimate the number of foreign purchases of U.S. companies (U.S. Department of Commerce 1989b).

8. The term "Hawthorne effect" originates from a series of studies of workers' productivity conducted in the 1920s and 1930s, which involved female employees of the Hawthorne, Illinois, plant of Western Electric. Researchers found that no matter how they varied working conditions—

bright versus dim lighting, long versus short breaks, frequent versus no breaks, piece rate pay versus fixed salary—workers' productivity increased. One explanation for these findings is that workers were responding positively to the fact that they had been singled out for study (Roethlisberger and Dickson 1939).

9. The classic experimental design consists of two groups matched with respect to all characteristics (age, marital status, ethnicity, attitudes, health status) except that one group is exposed to a hypothesized independent variable and the other group is not. The group exposed to the independent variable is called the *experimental group;* the group not exposed to the independent variable is called the *control group.* The two groups then are compared to determine whether there are any significant differences between them. Any such differences are assumed to be caused by the independent variable.

In a well-known type of experimental design, investigators test a particular drug's effectiveness by giving a placebo—a sugar pill or an equally ineffective substance—to those in the control group and the actual drug to those in the experimental group. The researchers then compare the two groups with regard to health outcomes. If the experimental group has a better rate of recovery than the control group, the researchers can safely conclude that it is due to the drug.

10. In the spring semester of 1989, I used Barnlund's study as a text in a social psychology course that I taught. These criticisms were offered by the students in that class.

11. This anecdote, which describes the career of major league umpire Bill Clem, can be applied to researchers as well:

> *At a testimonial dinner for Bill Clem, . . . a speaker rose to praise Clem by saying, "He always calls them the way he sees them." At this, a second speaker rose to disagree. "No, he always calls them the way they are." Clem then got up and replied, "You're both wrong. Until I call 'em, they ain't nothin'."* (LEWONTIN 1990, P. 6)

Similarly, observations "ain't nothin'" until researchers construct measures and apply them.

4 CULTURE

with Emphasis on the Republic of Korea[1]

Demilitarized zone between North and South Korea.

Nathan Benn/Woodfin Camp & Associates

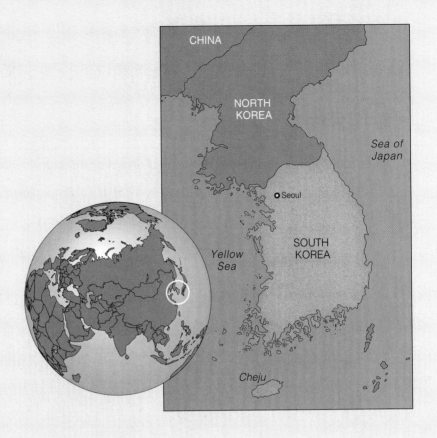

Students from South Korea enrolled in college in the United States for fall 1991	25,720
South Koreans admitted into the United States in fiscal year 1991 for temporary employment	7,721
People living in the United States in 1990 who were born in Korea	663,000
Airline passengers flying between the United States and South Korea in 1991	1,586,068
U.S. military personnel in South Korea in 1993	34,800
People employed in 1990 by South Korean affiliates in the United States	8,100
Applications for utility patents filed in 1991 in the United States by inventors from South Korea	1,321
Phone calls made between the United States and South Korea in 1991	44,460,000

They came to Seoul not for pleasure but for their future. They were nervous on arrival, for by and large they were small-town Midwesterners, owners of modest companies, and many of them had not traveled very much. Now not only were they in a distant and strange land but they had arrived there vulnerable, almost beholden, needing to make a deal, and they might have to give away part of their company in the process. They came because they feared they could no longer compete at home, and they had been told by their most important customers, the giant American assembly companies, to get their costs down. Korea was to them, like it or not, their best hope. Korea, they had been told, was the new Japan and for them a way of holding off the Japanese challenge.

What had once been a trickle of them had become by the spring of 1986 a torrent. In Seoul they met their new partners, men with whom they could not communicate at all. William Vaughan, a Chrysler representative in Korea, had watched them come and witnessed their desperation, knowing they were there to survive, for if they could not work something out, they were convinced, they would soon be out of business. Vaughan had an unusual job; he represented Chrysler in its continuing negotiations to complete the massive deal with Samsung, the vast South Korean manufacturing combine, but in addition to that he was a kind of matchmaker between the American parts manufacturers used by Chrysler and the Koreans, trying to find Korean companies who could succeed in this alien new world of autos.
(HALBERSTAM 1986, P. 697)

During the 1980s and 1990s, tens of thousands of partnerships, joint ventures, and production agreements were formed between U.S. and foreign industries and corporations. As you might imagine, there is considerable debate about the meaning of this trend for the overall economic well-being of the United States. But one implication is clear: Americans are part of a more interconnected and more competitive world economic climate than they were even a decade ago. The new global competition and accompanying concerns about world position have caused many American business leaders, educators, government officials, and social critics to compare the United States to other countries on a host of attributes including geographic and scientific knowledge, unemployment rates, savings rates, literacy, high school graduation rates, productivity, divorce rates, health costs, military expenditures, and credit card use. The assumption underlying such comparisons is that there is "a connection between the kinds of everyday behaviors a society encourages and its [economic] stability and prosperity" (Fallows 1990, p. 14).

In addition, the new global competition has caused U.S. government, business, and educational leaders to emphasize the need to train citizens, employees, and students to be more sensitive to cultural differences. For example, Honeywell, a large Minnesota-based high-technology firm with annual foreign revenues of more than $1.2 billion, has designed a seminar to raise cross-cultural awareness among its employees. The training program focuses on general knowledge about (1) foreign countries and particularly about the countries in which a Honeywell employee is entering; (2) values, practices, and assumptions in countries other than the United States; and (3) knowledge of one's own cultural framework including its values, assumptions, and perceptions about other cultures (Kim and Mauborgne 1987).

Sociologists define **culture** as a society's distinctive and complete design for living (Kluckhohn 1964). Culture includes everything tangible and intangible that people of a society create, acquire from other societies, and transmit to subsequent generations. "In the final analysis [culture] comprises the things that people have, the things that they do, and what they think" (Herskovits 1948,

p. 625). Culture is a fundamental and extremely complex concept. Its importance in explaining behavior is comparable to that of both genes and personality.

Sociologists distinguish between culture and society. A **society** is a group of people living in a given territory who share a culture and who interact with people of that territory more than with people of another territory. The broad nature of this definition suggests that a society can be a group of countries (Western, Eastern, or African societies), a single country (South Korea, North Korea, or the United States), or a community within a country (Old Order Amish, Koreatown, or Chinatown). Again, the important features of a society are common territory, shared culture, and interaction.

In this chapter we explore eight essential principles of the concept of culture and compare South Korean and American cultures. South Korea was chosen for several reasons. First, South Korea is one of 12 countries and city-states in the Pacific Rim, a region that is expected to become a formidable economic force in the twenty-first century. The countries and city-states of the Pacific Rim are Japan, South Korea, China, Taiwan, Hong Kong, the Philippines, Vietnam, Thailand, Malaysia, Singapore, Indonesia, and Australia. Most Americans are unfamiliar with Pacific Rim countries, with the exception of Japan, and are generally unaware of the rapid economic development taking place in that region. In fact, South Korea, Taiwan, Hong Kong, and Singapore are known popularly as "the next Japans" or "the four dragons." Along with the People's Republic of China, they are among the most rapidly developing economies in the world. All four are credited with achieving economic development at an unprecedented speed in a region of the world characterized by relatively few natural resources (Kim 1993). Along with Japan, South Korea, Taiwan, Hong Kong, and Singapore are the major players in the Asian Economic Community, one of three regional trading blocks expected to dominate the global economy over the next decade (Welch 1991). The Asian Economic Community will compete with North America and the European Union (see Figure 4.1).

Another reason for focusing on South Korea is that 35 years after the Korean War, which divided

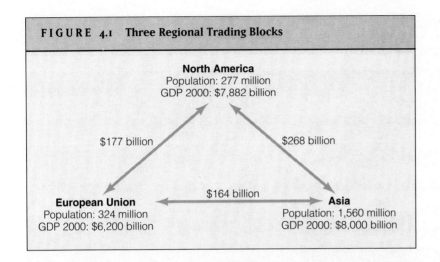

FIGURE 4.1 **Three Regional Trading Blocks**

North America
Population: 277 million
GDP 2000: $7,882 billion

$177 billion

$268 billion

$164 billion

European Union
Population: 324 million
GDP 2000: $6,200 billion

Asia
Population: 1,560 million
GDP 2000: $8,000 billion

Since the end of World War II, Korea has been divided by a demilitarized zone into North and South Korea.
T. Matsumoto/Sygma

the country into North Korea and South Korea, approximately 35,000 U.S. military personnel are stationed in South Korea. Yet, despite this long involvement in Korean affairs, Americans have interacted with Koreans mainly on the basis of narrowly conceived national security or military interests. Little attention has been paid to how Koreans live or how they view the United States.

Now that South Korea has become integrated into the world economy and is intertwined economically with the United States, Americans must learn to interact with Koreans as equal partners. To do so, American officials, business leaders, and educators must learn to appreciate, understand, and respect Korean culture.

Although this chapter focuses on South Korea and the United States, we can apply the concepts we discuss here to understand *any* culture and to frame other cross-cultural comparisons. As you read about culture, remember that South Korea is referred to broadly as a country possessing an Eastern or Asian culture and that the United States is regarded as a country possessing a Western culture. Therefore, many of the patterns described here are

Placing Ourselves on a Cultural Map

We all know that *East* and *West* signify geographical as well as cultural notions. We speak of the Eastern and the Western worlds and cultures. The question explored here is this: Because East and West are concrete geographical notions, are these notions culturally concrete as well? In other words, if we portray a society as a Western society, must this society be located in the Western Hemisphere? Conversely, if we portray a society as an Eastern society, must this society be located in the Eastern Hemisphere? My answer is that (1) the modern world is too closely intertwined culturally for this kind of dichotomization, (2) East and West are culturally more imagined than real, and (3) we live in a world that is becoming more heterogeneous.

Let me ask you this simple question: What does a map of the United States show? It shows the location of different states, counties, cities, roads, and national parks, as well as mountain ranges, rivers, and lakes. Now, try to imagine a cultural map of the United States. Can you visualize this map as clearly as the previous one? Perhaps not, especially if you remember that sociologists define culture as the culmination of all the material and nonmaterial products and aspects of a society. Such a definition makes visualization of this map much more problematic. For example, apart from geography, what does it mean to be a Kentuckian? How different is a Kentuckian from a Californian? At first glance, not many differences may come to mind. After all, residents of both states are Americans and they share cultural norms: love for the flag,

Scenes like this photo of a Korean-American on parade are not uncommon in Los Angeles—one of the most racially and ethnically diverse communities in the United States. How would you draw a "cultural map" of a place like Los Angeles?
Alon Reininger/Contact Press Images

football, apple pie, fast cars, open space, Hollywood, and many other things. You may conclude that both Kentuckians and Californians are "Americans" by the fact of location within the boundaries of the United States.

But there are also differences between the two states. California has a large immigrant population that lives in and around its metropolitan areas (San Francisco, Los Angeles, San Diego, Fresno, Sacramento, and so on). California is culturally more heterogeneous than Kentucky. Therefore, in California, the cultural notions of East and West become more nuanced, or vanish altogether.

Because of California's large Asian, Hispanic, and Middle East-ern populations, drawing a cultural map of California would require more time, energy, and imagination than a cultural map of Kentucky. We need to know the many countries from which these people emigrated; we need to know the languages they speak, the food they eat, the music they listen to, the religions they follow, the sports they like to watch, and a host of other elements.

SOURCE: Hamid Kusha, Northern Kentucky University (revised by J. Ferrante, 1993).

not necessarily unique to Korea or to the United States, but are shared with other Eastern or Western societies. At the same time, do not overestimate the similarity among countries that share a broad cultural tradition. Do not assume, for example, that South Korea (or any other Pacific Rim country or city-state) is just like Japan. As we will see, much of Korean identity is tied up with being "not Japanese" (Fallows 1988). To assume that South Korea is like Japan is equivalent to assuming that the United States is just like a Western European country, such as Germany or England. As we know, the United States is a country that celebrates its independence from European influence.

Finally, do not assume that all people who live within the borders of a specific country share the same culture. In every country, even ethnically homogeneous countries such as South Korea, some differences always exist among groups and individuals living in that country. (See "Placing Ourselves on a Cultural Map.")

Because the term *culture* includes everything that a group of people has, thinks, and does, we should not be surprised to learn that social scientists do not agree on a specific definition (Segall 1984). Despite their disagreements over definitions, sociologists do agree about the essential principles of culture. They believe that culture consists of tangible and intangible components; that people inherit and learn a culture; that biological, environmental, and historical forces shape and change culture; and that culture is a tool that people use to evaluate other societies and to adapt to problems of living. A basic familiarity with these principles gives us a framework for thinking about any culture.

Material and Nonmaterial Components of Culture

Material Culture

Principle 1: Culture consists of material and nonmaterial components. In sociological terms, **material culture** consists of all the objects or physical substances available to the people of a society (Horton and Hunt 1984). Examples of such objects include the telephone, the shovel, the writing system, paper money, the computer, the microwave oven, the automobile, and writing paper. All of these items began as resources[2] (plants, trees, minerals, or ores) and were converted into other forms for a purpose—to transport people, animals, and goods; to cook food; to make calculations; and so on. Sociologists refer to **technology** as the knowledge, skills, and tools used to transform resources into forms with specific purposes, and the skills and knowledge required to use them.

According to sociologists, the significant features of material culture are the purpose for which the culture is designed, the value placed on it, and the fact that some people are unhappy without it and direct their energies toward acquiring it

(Kluckhohn 1949). In other words, sociologists strive to learn what the objects of material culture mean to the people who use them (Rohner 1984). Consider the automobile, for example. It "is neither good nor bad. It all depends on the use to which it is put. It can serve doctors and patients, bank robbers, truck drivers, and kidnappers" (Braunschweig 1990, p. 23). Along these lines, sociologists would find it interesting that Americans, especially teenagers, define the automobile as the mechanism by which they gain independence. A student in one of my Introduction to Sociology classes gave the following written response to the question Describe an event that has affected your life in a significant way. This response shows what owning an automobile can mean to some young adults:

Getting my driver's license and purchasing an automobile is one event that has had a profound effect on my life. Now I am free to do what I want when I want; I do not have to rely on someone else to get where I want to go. On

the other hand, in order to pay for my car and related expenses I work 32–40 hours a week. I would not have to work as much if I did not own a car. One reason my car costs me so much is my insurance premiums are very high and have become even higher because I was in an accident.

Nonmaterial Culture

In sociological terms, **nonmaterial culture** includes the intangible human creations (things that cannot be identified directly through the senses) that exert considerable influence over people's behavior. Four of the most important of these creations are beliefs, values, norms, and symbols. (See Chapter 2 for a discussion of symbols.)

Beliefs Conceptions that people accept as true about how the world operates and where individuals fit in it are known as **beliefs**. Beliefs can be rooted in blind faith, experience, tradition, or the scientific method. (For a discussion of the scientific method, see Chapter 3.) Whatever the origin, beliefs can exert powerful influences on behavior.

To learn about the connection between people's beliefs and behaviors, sociologists Evon Z. Vogt and John M. Roberts (1960) studied beliefs about nature held by two groups living in Gallup, New Mexico. One group consisted of Protestant-American homesteaders from Texas. The other group consisted of Zuni, a pueblo-dwelling group of Native Americans. The researchers found that homesteaders see themselves as separate from nature. They think of nature as "something to be controlled and exploited by man for his own ends and material comfort" (Vogt and Roberts 1960, p. 115). The Zuni, on the other hand, believe themselves to be neither masters nor victims of nature, but to have an integral and essentially cooperative role in the cosmic scheme of things.

The differing belief systems of these two groups are reflected in how each relates to the environment, particularly in how each group responds to droughts. The homesteaders equip themselves with tractors, practice modern farming methods, and, when droughts occur, seed the clouds with silver iodide. The Zuni, on the other hand, have developed

a body of knowledge about how to live in a difficult environment. In addition, they increase their ceremonial activity during hard times. From the Zuni perspective people need only do their part and "the gods will do the rest. With centuries of summer rains to testify to the soundness of this view, Zuni [are] deeply opposed to rainmaking with airplanes and silver iodide" (Vogt and Roberts 1960, p. 116).

Values General and shared conceptions of what is good, right, appropriate, worthwhile, and important with regard to modes of conduct (for example, self-reliance and obedience) and states of existence (for example, freedom of choice and equal opportunity) are known as **values**.[3] Whereas beliefs are conceptions about how the world and people in it operate, values are conceptions about how the world and people *should* be. Sociologist Robin Williams (1970) has identified nine core values that people in the United States consider important even if they are not always realized:

1. equal opportunity (not equality of outcome but equal opportunity to compete)
2. achievement (especially occupational)
3. activity (active role in manipulating the environment to achieve desired ends)
4. results (what is actually achieved rather than the process of obtaining desired results)
5. future outlook (as opposed to looking to the past and the present)
6. science (as a means of solving problems)
7. material comfort (consumption and a high standard of living)
8. democracy (right of political participation)
9. survival of the fittest

We distinguish between societies not on the basis of which values are present, but rather according to which values people in that culture choose, when they are in situations in which they must choose. Americans, for example, place considerable value on the individual as an individual; they stress personal achievement and unique style (free choice). In contrast, Koreans value the individual in relationship to the group (particularly the family); they stress self-discipline and respect toward those who are older.

In sports, for example, the value placed by Americans on the individual is evidenced in the fact that they single out the most valuable player of a game, a season, a league, or a tournament. Furthermore, when Americans view an outstanding athletic feat, they tend to give more credit to the individual's talent or desire to win than to disciplined practice. In addition, American athletes work to find the style that is right for them and are willing to change this style if it does not bring success. Koreans, on the other hand, do spotlight individual achievement, but, in doing so, place considerable value on discipline, particularly on form (that is, adhering to time-tested and efficient methods of accomplishing goals). From the Korean point of view, athletic achievement does not occur simply because a person wants to excel or because he or she possesses raw talent. Athletic competence develops over time, after the individual masters and appreciates the steps that combine to produce the intended result. Compared to the American system, the Korean system minimizes individual achievement because the achiever owes success to the mastery of technique.

Values, we should note, transcend any particular situation. For example, the American emphasis on individual achievement and unique style—and the Korean emphasis on the group, form, and discipline—are not confined to one single area of life, such as sports. As we will see later in this chapter, these cultural values permeate many areas of life.

Norms All societies have guidelines that govern moral standards and even the most routine aspects of life. Sociologists call the written and unwritten rules that specify the behaviors appropriate to specific situations **norms**. Some norms are considered more important than other norms. In this regard, sociologist William Graham Sumner (1907) distinguished between folkways and mores. **Folkways** are norms that apply to the routine matters—eating, sleeping, appearance, posture, use of appliances, and relationships with various people, animals, and the environment.

Consider the folkways that govern how a meal typically is eaten at Korean and American dinner tables. In Korea, diners sit at low tables with their legs crossed. They do not pass items to one another, except to small children. Instead, they reach and

The Korean table has no clear place settings, a practice that reflects values about the relationship of the individual to the group. In addition, Korean diners use the same utensils for serving and eating.
David Bartruff/Artistry International

stretch across one another and use their chopsticks to lift small portions from serving bowls to individual rice bowls or directly to their mouths. The Korean norms of table etiquette—reaching across instead of passing, having no clear place settings, and using the same utensils to eat and to serve—de-emphasize the individual and reinforce the greater importance of the group.

Americans follow different dining folkways. They sit in chairs at tables approximately waist high. They have individual place settings, marked clearly by place mats or blocked off by eating utensils. It is considered impolite to reach across another person's space and to use personal utensils to take food from the serving bowls. Diners pass items around the table and use special serving utensils. The fact that Americans have clearly marked eating spaces, do not trespass into other diners' spaces, and use separate utensils to take food reinforces values about the importance of the individual.

Koreans and Americans even have different folkways about how they should use resources such as notebook paper and the electricity needed to keep refrigerators cold.[4] Koreans open the refrigerator door only as wide as necessary to remove an item, blocking the opening to minimize the amount of cold air that escapes. Americans open the refrigerator door wide and often leave it open while they decide what they want or until they move the desired item to a stove or countertop. As for notebook paper, Koreans fill every possible space on a sheet of paper before throwing it away. Sometimes they use a ruler to draw extra lines between those already on the paper in order to double the writing space. Americans will sometimes throw away a sheet of paper with only one line of writing because they do not like what they wrote.

Mores are norms that people define as pivotal to the well-being of the group. Obvious examples of mores are norms that prohibit cannibalism or the unjust and deliberate taking of another person's life. People who violate mores usually are severely punished (imprisoned, institutionalized, or executed); mores are considered to be final and un-

changeable. In contrast, there is considerable tolerance toward nonconformity to a folkway, and the consequences of violating folkways are usually minor—a disapproving stare, whispers behind one's back, or laughter. Thus, people can conceive of changing folkways that govern the number of meals they eat each day, for example, but they cannot conceive of changing mores.

One way to explain differences in behavior is to point to differences in Korean and American values and norms. One could argue that Koreans value conservation and Americans value consumption and that they have devised standards of appropriate behavior that reflect these values. Yet, it is not very satisfying simply to say that values and norms guide behavior. We must investigate the geographical and historical circumstances that gave rise to specific norms and values.

The Role of Geographical and Historical Forces

Principle 2: Geographical and historical forces shape the character of culture. Sociologists operate under the assumption that culture is "a buffer between [people] and [their] habitat" (Herskovits 1948, p. 630). That is, material and nonmaterial aspects of culture represent the solutions that people of a society have worked out over time to meet their distinctive historical and geographical challenges:

> All mankind shares a unique ability to adapt to circumstances and resolve the problems of survival. It was this talent which carried successive generations of people into the many niches of environmental opportunity that the world has to offer—from forest, to grassland, desert, seashore, and icecap. And in each case, people developed ways of life appropriate to the particular habitats and circumstances they encountered. (READER 1988, P. 7)

Part of the reason that Koreans and Americans use refrigerators and paper differently has to do with the amount of natural resources in each country. Korea has no oil, only moderate supplies of

coal, and depleted forests. Relative to Korea, the United States possesses abundant supplies of oil, wood, and coal. Although Koreans can import these resources, they face pressures unknown to most people in the United States, even as Americans come to realize that resources are dwindling. Because Koreans depend on other nations for most resources, they are vulnerable to any world event that might disrupt the flow of resources into their country. This vulnerability reinforces the need to use resources sparingly and to not take them for granted.

In sum, conservation and consumption-oriented values are rooted in circumstances of shortage and abundance. To understand this connection, recall a time when your electricity or water was turned off. Think about the inconvenience you experienced after a few minutes and how it increased after a few hours. The idea that one must conserve available resources takes root. People take care to minimize the number of times they open the refrigerator door.

Imagine how a permanent resource shortage or the dependence on other countries for resources

These Taiwanese women are sorting through electronic trash at a government-run yard—trash imported from the United States. Resource-poor countries such as Taiwan and South Korea are often more conservation-oriented than resource-rich countries such as the United States.

Jodi Cobb/National Geographic Society

could affect people's lives. A long-term resource shortage affected Californians when their state experienced a six-year drought (1987–1992). In some water districts, Californians cut their use of water by 25 to 48 percent. In Contra Costa County, for example, customers cut water consumption below the 280 gallons of water per day recommended under the voluntary rationing plan to an average of 165 gallons (Ingram 1992).

In contrast, you can imagine how the abundance of resources breaks down conservation-oriented behaviors. After spending some time in the United States, many Koreans stop double-lining their paper because there are few incentives to do so. Supplies of paper are abundant and no one else is conserving in this way.

For the most part people do not question the origin of the values they follow and the norms to which they conform "any more than a baby analyzes the atmosphere before it begins to breathe it" (Sumner 1907, p. 76), nor are they aware of alternatives. This is because many values and norms that people believe in and adhere to were established before they were born. Thus, people behave as they do simply because they know of no other way. And, because these behaviors seem so natural, we lose sight of the fact that culture is learned.

Culture Is Learned

Principle 3: Culture is learned. Humans are born "with two endowments, or, more properly stated, with one and into one" (Lidz 1976, p. 5). We are born *with* a genetic endowment and *into* a culture. Our parents transmit by way of their genes a biological heritage, common to all humans but uniquely individual. The genetic heritage that we share with all human beings gives us a capacity for language development, an upright stance, four movable fingers and an opposable thumb on each hand, as well as other characteristics. If these traits seem too obvious to mention, consider that they allow humans to speak innumerable languages, to perform countless movements, and to devise and use many inventions and objects. In fact, "most people are shaped to the form of their culture, because of the enormous malleability of their genetic endowment" (Benedict 1976, p. 14). Regardless of their physical traits (for example, eye shape and color, hair texture and color, skin color), babies are destined to learn the ways of the culture into which they are born and raised.

Anthropologist Clyde Kluckhohn's acquaintance with a blue-eyed, blond-haired man born in the United States but raised in China illustrates this point:

Some years ago I met in New York City a young man who did not speak a word of English and was obviously bewildered by American ways.

By "blood" he was as American as you or I, for his parents had gone from Indiana to China as missionaries. Orphaned in infancy, he was reared by a Chinese family in a remote village. All who met him found him more Chinese than American. The facts of his blue eyes and light hair were less impressive than a Chinese style of gait, Chinese arm and hand movements, Chinese facial expression, and Chinese modes of thought. The biological heritage was American, but the cultural training had been Chinese. He returned to China. (KLUCKHOHN 1949, P. 19)

The point is that our genetic endowment gives us human and physical characteristics, not our cultural characteristics. We cannot assume that someone comes from a particular culture simply because he or she looks like a person whom we expect to come from that culture.

An excerpt from a letter written by a first-generation Taiwanese-American mother to her daughter shows that even parents have a hard time accepting this idea when their children are raised in a country and a culture different from their own:

To you, Taiwan is just a fun but humid place that you visited one summer, and your grandparents are just fuzzy voices over a telephone. To me, there is a lifetime that sits thousands of miles away, tucked inside the navy blue suit you see on me now. And to me, Taiwan is still home. HOME. Yes, our white stucco house with the orange door is home, too, but part of my blood still flows toward Taiwan. Can you understand that? Maybe that's why sometimes I expect you to understand Chinese culture without having experienced any of it firsthand. I think that you have the same blood, and that it pulses to the same beat. You ARE Chinese still, and I know that you have some interest and even some pride in it, but there's so much you don't know. It is your right to know your family's experiences, even if you don't care about them. Maybe someday, you will care. (YEH 1991, P. 2)

The development of language illustrates the relationship between genetic and cultural heritages. Human genetic endowment gives us a brain that is flexible enough to allow us to learn the language(s) that we hear spoken by the people around us.

Language is an important nonmaterial component of culture. As young children learn words and the meanings of words, they learn about their culture. They also acquire a tool that enables them to think about the world, to interpret their experiences, to establish and maintain relationships, and to convey information. Anyone who speaks only one language might not realize this property of language until he or she learns another language. To become fluent in another language is not merely a matter of reading and conversing in that language but of actually being able to think in that language. Similarly, when young children learn the language of their culture, they acquire a thinking tool.[5]

The following characteristics of language show the relationship between learning the meaning of words and learning the ways of a culture:

- *Language conveys important messages above and beyond the actual meaning of words.* Words have two levels of meaning—denotative and connotative. Denotation is literal definition, whereas connotation is the set of associations that a word evokes. The connotation of a word is as important as the literal definition.

  The importance of connotation can be seen with the Korean and English words for *individual*, which in both languages literally means "a single human being." In the United States, the connotations associated with *individual* are generally positive: a distinct person, one of a kind, or someone capable of independent action. In Korea, the word for individual (*kaein*) has connotations of selfishness or of one interested only in his or her own affairs (Underwood 1977).

- *Words refer to more than things: they also describe relationships.* The word *adoption*, for example, as in the adoption of a child, refers to more than the child; it involves more than taking in and raising that child. It also implies the presence of biological and adoptive parents and evokes assumptions about the relationship between the child, the biological parents, and the adoptive parents. The norms that guide these relationships are important features of the word. In Korea, most people think of adoption as "a system whereby a sonless couple may receive a son from one of the husband's brothers or male cousins" (Peterson 1977, p. 28). An adopted son cannot come from the wife's family or from a sister of the husband. In the United States, adoption usually is a situation in which a child's biological parents release him or her to responsible adults who agree to raise the child as their own. The adoptive parents usually have no connection with the biological parents.

- *Words mirror cultural values.* Language embodies features considered important to the culture. In Korean society, age is an exceedingly important measure of status: the older a person is, the more status, or recognition, he or she has in the society. Korean language acknowledges the importance of age by its use of special age-based hierarchical titles for everyone. In fact, it is nearly impossible to carry on a conversation, even among siblings, without taking age into consideration. Every word referring to one's brother or sister acknowledges his or her age in relation to the speaker. Even twins are not equal; one twin was born first. Furthermore, norms that guide Korean forms of address do not allow the speaker to refer to brothers or sisters by first name.

  The manner of addressing siblings according to age is only one example of the emphasis given to age in the Korean language. Variations are found even in the responses to a question like "Where are you going?" between persons of the same age. If the friends are teenagers, a response such as "to school" would be *"hakkyo-ey ka-un-ta"*; if adults, *"hakkyo-ey ka-a"*; if mature adults, *"hakkyo-ey ka-ney"*; and so on (Park 1979, p. 68).

  Among Koreans, the importance of the group over the individual is reflected in rules governing the writing and speaking of one's name. Koreans tend to identify themselves by stating the family name first and then the given name. In effect, the family is more important than the individual. Likewise, a letter is addressed to the country, the province, the city, the street, the house number, and, finally, the recipient.

- *Common expressions embody the preoccupations of the culture.* Frequently used phrases and words serve as indicators of cultural preoccupations—stresses, strains, and values. For instance, Americans use the word *my* to express "ownership" of persons or things over which they do not have exclusive rights: *my* mother, *my* school, *my* country. The use of *my* reflects the American preoccupation with the needs of the individual over those of the group. In contrast, Koreans express possession as shared: *our* mother, *our* school, *our* country. The use of the plural possessive reflects the Korean preoccupation with the needs of the group over individual interests.

  Another example of cultural preoccupations is the Korean response to a full moon—"Isn't that sad?"—or to singing birds—"They are weeping!" These comments represent patterns of response rooted in centuries of invasions and wars. In part because of its geographical location, Korea has had a continuous history of invasions—by the Japanese, the Chinese, and the Russians—that caused severe hardship, substantial loss of life, and widespread devastation. Generations of tragedy have created in the Korean people a sadness that is reflected in their responses to many natural phenomena.

These four characteristics of language demonstrate that children do more than learn words and meanings. They also acquire a perspective that reflects what is important to the culture.

Culture as a Tool for Problems of Living

Principle 4: Culture is the tool that enables the individual to adjust to the problems of living. Although our biological heritage is flexible, it presents all of us with a number of challenges. Humans are dependent on others for a relatively long time; they feel emotion; they experience organic drives such as hunger, thirst, and sex; and they age and eventually die. These biological inevitabilities have given rise to functional requisites, arrangements, or "formulas" necessary for the survival of all societies. There

The Importance of Corn

The driving wheel of the supermarket is not always visible: it is not the business of a driving wheel to be ostentatious. But it is there—everywhere. It is American corn, or maize. You cannot buy anything at all in a North American supermarket which has been untouched by corn, with the occasional and single exception of fresh fish—and even that has almost certainly been delivered to the store in cartons or wrappings which are partially created out of corn. Meat *is* largely corn. So is milk: American livestock and poultry are fed and fattened on corn and cornstalks. Frozen meat and fish has a light corn starch coating on it to prevent excessive drying. The brown and golden colouring which constitutes the visual appeal of many soft drinks and puddings comes from corn. All canned foods are bathed in liquid containing corn. Every carton, every wrapping, every plastic container depends on corn products—indeed all modern paper and cardboard, with the exception of newspaper and tissue, is coated in corn.

One primary product of the maize plant is corn oil, which is not

The Corn Palace in Mitchell, South Dakota—a major corn-producing state—symbolizes the importance of corn to that community. The United States produces more than 440 billion pounds of corn every year, as much as the rest of the world put together.
Peter Essick

only a cooking fat but is important in margarine (butter, remember, is also corn). Corn oil is an essential ingredient in soap, in insecticides (all vegetables and fruits in a supermarket have been treated with insecticides), and of course in such factory-made products as mayonnaise and salad dressings. The taste-bud sensitizer, monosodium glutamate or MSG, is commonly made of corn protein.

---

are formulas for caring for children; for satisfying the need for food, drink, and sex; for channeling and displaying emotions; for segmenting the stages and activities of the life cycle; and, eventually, for departing this world. In the remainder of this section we will focus on the differing cultural formulas for dealing with two other biological events—hunger and the social emotions.

Cultural Formulas for Hunger

All people become hungry, but the factors that stimulate and satisfy appetite vary considerably across cultures. One indicator of a culture's influence is that people define only a portion of the potential food available to them as edible. Culture determines what is defined as edible, who prepares the food, how the food is served and eaten, what the relationship is among those eating together, how many meals are eaten in a day, and when during the day meals are eaten.

Dogs and snakes are among the foods defined by many Koreans as edible, but they are not defined as such by Americans. These differences are rooted in historical and environmental factors. Whereas the United States has an abundance of fertile, flat land

Corn syrup—viscous, cheap, not too sweet—is the very basis of candy, ketchup, and commercial ice cream. It is used in processed meats, condensed milk, soft drinks, many modern beers, gin, and vodka. It even goes into the purple marks stamped on meat and other foods. Corn syrup provides body where "body" is lacking, in sauces and soups for instance (the trade says it adds "mouth-feel"). It prevents crystallization and discolouring; it makes foods hold their shape, prevents ingredients from separating, and stabilizes moisture content. It is extremely useful when long shelf-life is the goal.

Corn starch is to be found in baby foods, jams, pickles, vinegar, yeast. It serves as a carrier for the bubbling agents in baking powder; is mixed in with table salt, sugar (especially icing sugar), and many instant coffees in order to promote easy pouring. It is essential in anything dehydrated, such as milk (already corn, of course) or instant potato flakes. Corn starch is white, odourless, tasteless, and easily moulded. It is the invisible coating and the universal neutral carrier for the active ingredients in thousands of products, from headache tablets, toothpastes, and cosmetics to detergents, dog food, match heads, and charcoal briquettes.

All textiles, all leathers are covered in corn. Corn is used when making things stick (adhesives all contain corn)—and also whenever it is necessary that things should *not* stick: candy is dusted or coated with corn, all kinds of metal and plastic moulds use corn. North Americans eat only one-tenth of the corn their countries produce, but that tenth amounts to one and a third kilograms (3 lb.) of corn—in milk, poultry, cheese, meat, butter, and the rest—per person per day.

The supermarket does not by any means represent all the uses of corn in our culture. If you live in North America—and even very possibly if you do not—the house you live in and the furniture in it, the car you drive, even the road you drive on, all depend for their very existence on corn. Modern corn production "grew up" with the industrial and technological revolutions, and the makers of those revolutions were often North Americans. They turned their problem-solving attention to the most readily available raw materials and made whatever they wanted to make—antibiotics or deep-drilling oil-well mud or ceramic spark-plug insulators or embalming fluids—out of the material at hand. And that material was the hardy and obliging fruit of the grass which the Indians called *maïs*.

In English, the word *corn* denotes the staple grain of a country. Wheat is "corn" to the people of a country where wheaten bread is the staple. Oats is "corn" to people who eat oats; rye is "corn" if the staple is rye. When Europeans arrived in America they saw that, for the Indians, maize was the basic food, so the English-speaking newcomers called it "Indian corn." We continue in North America to recognize the primacy of maize in our culture by calling it "corn."

SOURCE: Pp. 22–23 in *Much Depends on Dinner*, by Margaret Visser. Copyright © 1988 by Grove Press, New York. Reprinted with permission

for grazing and food production,[6] Korea is comparatively mountainous. Consequently, the available land is used to grow crops, not to graze cattle. The few existing cattle in Korea are important to the agricultural system as a source of labor—to pull plows. Even today, cattle are more efficient than tractors in tilling steep inclines. The agricultural importance of cattle, combined with the lack of land to support a cattle industry, discourages the widespread practice of eating beef and encourages using dogs and snakes as alternative food sources.[7]

Another interesting cultural difference is that rice is the staple of the Korean diet, whereas corn is the staple of the American diet. Very little in the American diet is not affected by corn, though few Americans realize this. In addition to being a vegetable, corn (in one form or another) is found in soft drinks, canned foods, candy, condensed milk, baby food, jams, instant coffee, instant potatoes, and soup, among other things (see "The Importance of Corn").

Corn is a "gift" from Native Americans to all people who settled in the United States. The Native Americans recognized the significance of corn to life by referring to it as "our mother," "our life," or "she who sustains us" (Visser 1988). It is also worth noting that, among the many colors of corn, only the yellow and the white varieties are defined as edible by the dominant American culture; the more exotic colors (blue, green, orange, black, and red) are considered fit only for decoration at Thanksgiving time (although blue corn chips and other blue corn products have become available in gourmet food stores and trendy restaurants). One might speculate that this arbitrary selection is tied to the early American immigrants' rejection of the exotic elements of Native American cultures.

Koreans have no such lack of awareness of rice as a staple. They recognize and appreciate the significance of rice in their lives and eat it at all meals—breakfast, lunch, and dinner. In fact, the Korean word for food is *rice*. Most Koreans are aware of the many uses of rice and the by-products of rice plants: to feed livestock; to make soap, margarine, beer, wine, cosmetics, paper, and laundry starch; to warm houses; to provide inexpensive fuel for steam engines; to make bricks, plaster, hats, sandals, and raincoats; and to use as packing material to prevent items from breaking in shipping.

Cultural Formulas for Social Emotions

Culture also influences the expression of emotion, just as it influences people's response to food needs. **Social emotions** are internal bodily sensations that we experience in relationships with other people. Grief, love, guilt, jealousy, and embarrassment are a few examples of social emotions. Grief, for instance, is felt at the loss of a relationship. Love reflects the strong attachment that one person feels for another person. Jealousy can arise from fear of losing the affection of another (Gordon 1981). People do not simply express social emotions directly; they interpret, evaluate, and modify their internal bodily sensations upon considering "feeling rules" (Hochschild 1976, 1979).

Feeling rules are norms that specify appropriate ways to express the internal sensations. They define sensations that one should feel toward another person. In the dominant culture of the United States, for example, same-sex friends are supposed to like each other but not feel romantic love toward each other. The process by which we come to learn feeling rules is complex; it evolves through interactions with others.

In the novel *Rubyfruit Jungle*, author Rita Mae Brown describes a situation in which feeling rules are articulated. The central character, Molly, who is about seven years old at the time, wonders whether girls can marry each other. She approaches Leota, a girlfriend whom she likes very much, about this possibility.

> *"Leota, you thought about getting married?"*
> *"Yeah, I'll get married and have six children and wear an apron like my mother, only my husband will be handsome."*
> *"Who you gonna marry?"*
> *"I don't know yet."*
> *"Why don't you marry me? I'm not handsome, but I'm pretty."*
> *"Girls can't get married."*
> *"Says who?"*
> *"It's a rule."*
> *"It's a dumb rule. Anyway, you like me better than anybody, don't you? I like you better than anybody."*
> *"I like you best, but I still think girls can't get married."* (BROWN 1988, P. 49)

In another scene in the same novel, Molly walks in on her father (Carl) while he is comforting his friend Ep, whose wife has just died. In the passage that follows, Molly reflects on the feeling rules that apply to men:

> *I was planning to hotfoot it out on the porch and watch the stars but I never made it because Ep and Carl were in the living room and Carl was holding Ep. He had both arms around him and every now and then he'd smooth down Ep's hair or put his cheek next to his head. Ep was crying just like Leroy. I couldn't make out what they were saying to each other. A couple of times I could hear Carl telling Ep he had to hang on, that's all anybody can do is hang on. I was afraid they were going to get up and see me so I hurried back to my room. I'd never seen men hold each other. I thought the only things they were allowed to do was shake hands or fight. But if Carl was holding Ep maybe it wasn't against the rules. Since I wasn't sure, I thought I'd keep it to myself and never tell. I was glad they could touch each other. Maybe all men did that after everyone went to bed so no*

one would know the toughness was for show.
Or maybe they only did it when someone died.
I wasn't sure at all and it bothered me. (P. 28)

These examples show that people learn norms that specify how, when, where, and to whom to display emotions.

Feeling rules apply to male-female relationships as well as to same-sex friends. Different norms, for example, govern the body language that men and women use to show affection toward each other. A Korean husband and wife almost never touch, hug, or kiss in public; instead they express affection inwardly. Indeed, Koreans tend to value the concealment of such emotions. When they see American couples express love overtly by touching, caressing, and kissing in public, they regard that behavior as a sign of insecurity. Korean couples have no need to express love overtly; they know they love each other (Park 1979).

Sociologist Choong Soon Kim (1989) found that these feeling rules applied even to Korean couples separated from one another for 30 years or more. Choong observed the "Family Reunion" program, which took place in South Korea in June 1983. The program was a television campaign designed to reunite relatives living throughout South Korea who had been separated from one another as a result of the Korean War (Jun and Dayan 1986). Kim noted that no husband and wife kissed when reunited and most couples did not hug. No Korean onlookers commented that they found this behavior unusual. In the United States, by contrast, if reunited couples did not display affection, onlookers would wonder about the quality of the relationship.

Another example of the role culture plays in channeling expressions of emotion has to do with laughter. Laughter can be an expression of emotions such as anxiety, sadness, nervousness, happi-ness, or despair. No matter what emotion laughter releases or defuses, however, it occurs when something is out of place in a given situation or when behavior in a specific circumstance is different from what is expected. Because culture provides the guidelines for what is expected in a specific set of circumstances, this observation about laughter implies that what is funny in one culture may not be funny in another. In fact, communication specialists generally agree that jokes do not translate well. The following anecdote illustrates why:

> *I began my speech with a joke that took me about two minutes to tell. Then my interpreter translated my story, and about 30 seconds later the Japanese audience laughed loudly. I continued with my talk, which seemed well received . . . but at the end, just to make sure, I asked the interpreter, "How did you translate my joke so quickly?" The interpreter replied: "Oh, I didn't translate your story at all. I didn't understand it. I simply said, 'Our foreign speaker has just told a joke, so would you all please laugh.'"*
> (MORAN 1987, P. 74)

To this point we have discussed a number of principles about culture. We have emphasized that culture is a tool that people learn and draw from to meet the challenges of living. The discussion so far may have led you to believe that the nonmaterial components of culture have greater influence over behavior than do the material components, but, in fact, there is a debate about which element is more influential. There is no simple answer to this question because material and nonmaterial components of culture are interrelated: the nonmaterial shapes the material, and the material—particularly the introduction of some new object or technological advance—shapes the nonmaterial.

The Relationship Between
Material and Nonmaterial Culture

Principle 5: It is difficult to separate the effects of nonmaterial and material cultures on behavior. To see how nonmaterial components of culture shape the material, consider the flags of the United States and of South Korea. (The South Korean flag was the flag of Korea before the country was divided at the 38th parallel into North and South Korea.) A flag can be designed in any number of ways, so why did Korean and American designers settle on these particular designs? One could argue that designers drew on symbols that reflected important themes in each country's history.

The South Korean flag has a circle in the middle that is divided into two comma-shaped halves, the red half for yin (feminine) and the blue half for yang (masculine). The yin/yang design represents contributing, yet opposing, forces in the universe and depicts balance and harmony as a solution to contradictory or conflicting forces. The four trigrams (sets of three lines) around the circle represent heaven (≡) opposite earth (☷) and water (☵) opposite fire (☲). The message conveyed in the design is that humans have the "responsibility to balance these forces for optimum social harmony and human progress" (Reid 1988, p. 171). The use of the yin/yang symbol and the trigrams, taken from the *I Ching,* or *Book of Changes,* an ancient Chinese book (1122 B.C.), acknowledges the value of the past and China's influence on Korean culture. The flag's design also reflects a belief in the interdependence between forces rather than the dominance of one force over another. Finally, it represents an important norm by which Koreans are guided: people have the responsibility to balance opposing or contradictory forces to ensure human progress (Yoo 1987).

Similarly, the flag of the United States reflects its nonmaterial culture. Each of the 13 alternating red and white stripes represents one of the 13 original colonies. Each of the 50 stars on the blue field represents a state. Although the United States began as a British colony, the flag does not acknowledge that influence (except in the red, white, and blue colors). The stars and the stripes do not overlap. Separateness reflects the value placed on independence or freedom from a strong central power. The omission of a symbol that represents Europe corresponds with the importance of shaking off the influence of the past to make a clean start. It matches the norm that rejects the role of the past in defining a person.

We can see another influence of nonmaterial culture on material culture in the different burial customs of Koreans and Americans. Most Americans from a Christian background have the deceased placed in a wooden coffin and then in a concrete-lined metal vault for burial. The double enclosure keeps the dead body separate from the earth as it

The use of the yin/yang symbol and trigrams on South Korea's flag reflects a belief in the interdependence among natural forces.

decays. This method ensures that the deceased will not become "confused" with the earth; it reflects the belief that humans and their environment are independent. In contrast, the Korean practice of burying the dead in wooden coffins that decay corresponds with the belief that humans and their environment are interdependent and that decay is part of that interdependence.

Just as nonmaterial components of culture shape flags and coffins, material components shape the nonmaterial. For example, the microwave oven is an important material influence on American values and norms, specifically values that emphasize the individual over the group and norms that govern when the members of a family should eat.

Approximately 70 percent of American households have microwave ovens, which have eliminated one of the incentives for families to eat together. Before the microwave was available, it was more efficient for one member of the household to cook for everyone because of the time—whether calculated in person time or in oven time—required to cook a meal. Now that meals take only minutes to prepare, it is no longer considered inefficient for family members to cook and eat separately. The microwave also enhances independence in another way: a person is not tied to a particular meal schedule or even family group:

The old dining room table required each individual to give up some personal autonomy and

A traditional Korean bier.
Dave Bartruff/Artistry International

A Hasidic Jewish burial procession in New York.
Nathan Benn/Woodfin Camp & Associates

A cremation in northern Bali.
R. Ian Lloyd Productions

Food left at a gravesite in a Japanese cemetery to nourish the deceased spirits when they visit the material world during the annual Obon Festival.
Bill Bachman/Photo Researchers, Inc.

Visitors to a grave at Arlington National Cemetery during Memorial Day Week.
Art Stein/Photoresearchers

One expression of material and nonmaterial culture is the norms, values, and beliefs various societies have developed in response to the biological inevitability of death.

bow to the dictates of the group and the social system. If we stop eating together, we shall save time for ourselves and achieve mealtime self-sufficiency. . . . The communal meal is our primary ritual for encouraging the family to gather together every day. If it is lost to us, we shall have to invent new ways to be a family. It is worth considering whether the shared joy that food can provide is worth giving up. (VISSER 1989, P. 42)

In Korean society the writing system of *hangul,* introduced by King Sejong in 1446, has had far-reaching effects on the culture. This writing system was developed over a 20-year period with the explicit purpose of encouraging literacy. What makes *hangul* unique is that anyone who speaks Korean can learn to read in a matter of days, because the simple and consistent phonetic rules make it "one of the clearest and most logical systems of writing ever devised" (Iyer 1988, p. 48).

The efficiency of *hangul* is evident when it is contrasted with the English alphabet, whose phonetic rules are inconsistent. Although each of the 26 letters theoretically is associated with a distinct sound, some letters symbolize more than one sound, especially when combined with other letters. Combinations of vowels can be quite problematic. Consider the vowel sounds in the following words, comparing the spellings: *break* versus *freak, sew* versus *few, food* versus *good, paid* versus *said,* and *shoe* versus *foe* ("Letters" 1988).

Hangul has had an immense impact on Korean society. First, it freed Koreans from the writing and reading system that was previously in use—a system based on the complex ideograms of the Chinese system. The relative simplicity of *hangul* is responsible for the nearly 100 percent literacy rate in

Korea. Second, *hangul* supports the strong Korean sense of and desire for national unity. Because virtually everybody can read, Koreans have a common base of knowledge. Such shared knowledge encourages unity by breaking down barriers due to differences in literacy levels. Finally, the widespread literacy has contributed to South Korea's rapid rise as an economic force in the world. Higher literacy rates lead to greater employability, lower error rates at work, and fewer industrial accidents—all of which give Korea an advantage over countries with lower rates of literacy. The United States is learning only now that a literate work force, even in entry-level positions, is essential to the country's ability to compete in the worldwide marketplace.

The examples discussed in this subsection show that inventions often have consequences that go beyond their intended purpose. The microwave oven not only changed the time and effort it takes to cook a meal; it also changed how family members relate to one another. Likewise, the purpose of a writing system is to record ideas, but the design of the Korean writing system promoted widespread literacy and equality. These examples not only show how material components affect values and norms, but also demonstrate that the introduction of a new invention can have profound consequences on the way in which people in a society relate to one another.

Most people tend to think that their material and nonmaterial culture is self-created. They underestimate the extent to which ideas, materials, products, and other inventions are connected in some way to foreign sources or borrowed outright from those sources (see "The West's Debt to China").

Cultural Diffusion

Principle 6: People borrow ideas, materials, products, and other inventions from other societies. The process by which an idea, an invention, or some other cultural item is borrowed from a foreign source is called **diffusion**. The term *borrow* is used in the broadest sense and can mean to usurp, pirate, steal, imitate, plagiarize, purchase, or copy.

Basketball, a U.S. invention, has been borrowed by people in 75 countries, including South Korea, where 21 clubs are registered with the *Federation Internationale de Basketball*. Baseball, another U.S. invention, has been borrowed by people in more than 90 countries (*World Monitor* 1992, 1993). See Figure 4.2.

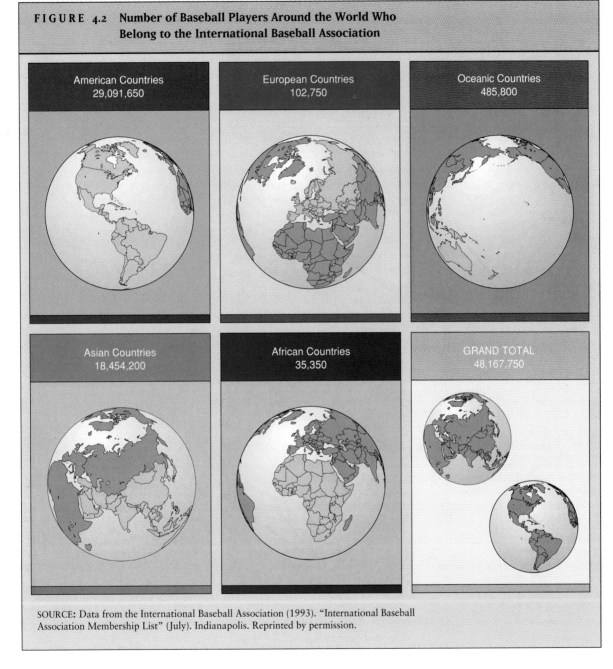

FIGURE 4.2 **Number of Baseball Players Around the World Who Belong to the International Baseball Association**

American Countries
29,091,650

European Countries
102,750

Oceanic Countries
485,800

Asian Countries
18,454,200

African Countries
35,350

GRAND TOTAL
48,167,750

SOURCE: Data from the International Baseball Association (1993). "International Baseball Association Membership List" (July). Indianapolis. Reprinted by permission.

The opportunity to borrow occurs every time two people from different cultures make contact, whether it be by phone, as paths cross during travel, in the course of attending school, or because an inventor from one country files a patent in another country. Opportunities for cultural diffusion between the United States and South Korea exist as a result of the 45-year presence of U.S. troops, international trade, and the arrival and return of Koreans who have attended colleges and universities in the United States. (The data at the beginning of the chapter shows that more than 25,000 South Koreans enrolled in colleges and universities in the United States in 1991.)

The West's Debt to China

One of the greatest untold secrets of history is that the "modern world" in which we live is a unique synthesis of Chinese and Western ingredients. Possibly more than half of the basic inventions and discoveries upon which the "modern world" rests come from China. And yet few people know this. Why?

The Chinese themselves are as ignorant of this fact as Westerners. From the seventeenth century onwards, the Chinese became increasingly dazzled by European technological expertise, having experienced a period of amnesia regarding their own achievements. When the Chinese were shown a mechanical clock by Jesuit missionaries, they were awestruck. They had forgotten that it was they who had invented mechanical clocks in the first place!

It is just as much a surprise for the Chinese as for Westerners to realize that *modern* agriculture, *modern* shipping, the *modern* oil industry, *modern* astronomical observatories, *modern* music, decimal mathematics, paper money, umbrellas, fishing reels, wheelbarrows, multi-stage rockets, guns, underwater mines, poison gas, parachutes, hot-air balloons, manned flight, brandy, whisky, the game of chess, printing, and even the essential design of the steam engine, all came from China.

Without the importation from China of nautical and navigational improvements such as ships' rudders, the compass and multiple masts, the great European Voyages of Discovery could never have been undertaken, Columbus would not have sailed to America, and Europeans would never have established colonial empires.

Without the importation from China of the stirrup, to enable them to stay on horseback, knights of old would never have ridden in their shining armor to aid damsels in distress; there would have been no Age of Chivalry. And without the importation from China of guns and gunpowder, the knights would not have been knocked from their horses by bullets which pierced the armor, bringing the Age of Chivalry to an end.

Without the importation from China of paper and printing, Europe would have continued for much longer to copy books by hand. Literacy would not have become so widespread.

Johann Gutenberg did *not* invent movable type. It was invented in China. William Harvey did *not* discover the circulation of the blood in the body. It was discovered—or rather, always assumed—in China. Isaac Newton was *not* the first to discover his First Law of Motion. It was discovered in China.

These myths and many others are shattered by our discovery of the true Chinese origins of many of the things, all around us, which we take for granted. Some of our greatest achievements turn out to have been not achievements at all, but simple borrowings. Yet there is no reason for us to feel inferior or downcast at the realization that much of the genius of mankind's advance was Chinese rather than European. For it is exciting to realize that the East and the West are not as far apart in spirit or in fact as most of us have been led, by appearances, to believe, and that the East and the West *are already combined* in a synthesis so powerful and so profound that it is all-pervading. Within this synthesis we live our daily lives, and from it there is no escape. The modern world *is* a combination of Eastern and Western ingredients which are inextricably fused. The fact that we are largely unaware of it is perhaps one of the greatest cases of historical blindness in the existence of the human race.

Why are we ignorant of this gigantic, obvious truth? The main reason is surely that the Chinese themselves lost sight of it. If the very originators of the inventions and discoveries no longer claim them, and if even their memory of them has faded, why should their inheritors trouble to resurrect their lost claims? Until our own time, it is questionable whether many Westerners even wanted to know the truth. It is always more satisfying to the ego to think that we have reached our present position alone and unaided, that we are the proud masters of all abilities and all crafts. . . .

We need to set this matter right, from both ends. And I can think of no better single illustration of the folly of Western complacency and self-satisfaction than the lesson to be drawn from the history of agriculture. Today, a handful of Western nations have grain surpluses and feed the world. When Asia starves, the West sends grain. We assume that Western agriculture is the very pinnacle of what is possible in the productive use of soil for the growth of food. But we should take to heart the astonishing and disturbing fact that the European agricultural revolution, which laid the basis for the Industrial Revolution, came about only because of the importation of

The stirrup—a device reminiscent of jousting knights in shining armor and countless cowboys of the American West—was actually invented in China. Its invention led to the development of the game of polo.

Detail of water color painting on silk by Li-Lin, c. 1635. Courtesy Victoria and Albert Museum

Chinese ideas and inventions. The growing of crops in rows, intensive hoeing of weeds, the "modern" seed drill, the iron plow, the moldboard to turn the plowed soil, and efficient harnesses were all imported from China. Before the arrival from China of the trace harness and collar harness, Westerners choked their horses with straps round their throats. Although ancient Italy could produce plenty of grain, it could not be transported overland to Rome for lack of satisfactory harnesses. Rome depended on shipments of grain by sea from places like Egypt. As for sowing methods—probably over half of Europe's seed was wasted every year before the Chinese idea of the seed drill came to the attention of Europeans. Countless millions of farmers throughout European history broke their backs and their spirits by plowing with ridiculously poor plows, while for two thousand years the Chinese were enjoying their relatively effortless method. Indeed, until two centuries ago, the West was so backward in agriculture compared to China, that the West was the Underdeveloped World in comparison to the Chinese Developed World. The tables have now turned. But for how long? And what an uncomfortable realization it is that the West owes its very ability to eat today to the adoption of Chinese inventions two centuries ago.

It would be better if the nations and the peoples of the world had a clearer understanding of each other, allowing the mental chasm between East and West to be bridged. After all they are, and have been for several centuries, intimate partners in the business of building a world civilization. The technological world of today is a product of both East and West to an extent which until recently no one had ever imagined. It is now time for the Chinese contribution to be recognized and acknowledged, by East and West alike. And, above all, let this be recognized by today's schoolchildren, who will be the generation to absorb it into their most fundamental conceptions about the world. When that happens, Chinese and Westerners will be able to look each other in the eye, knowing themselves to be true and full partners.

SOURCE: From "The West's Debt to China." Pp. 9–12 in *The Genius of China: 3,000 Years of Science, Discovery, and Inventions,* by Robert Temple. Copyright © 1986 by Multimedia Publications (UK) Ltd. Reprinted by permission of the publisher.

For example, the presence of the American military in South Korea since 1950 has encouraged diffusion in a number of ways. First, the American Forces in Korea Network (AFKN) exposed Koreans to American television and music and (by extension) to American values and norms. Second, job opportunities opened up for Koreans on U.S. bases, and a service industry (bars, gift shops, brothels, tailor shops, laundries) grew to meet the soldiers' needs. Companies such as Hyundai and Daelim got their start by filling contracts for the U.S. military. Third, American soldiers, in collaboration with Korean black marketeers, leaked PX goods into Korean society. At one point 60 percent of the available PX materials flowed into Korean society. This situation stimulated a desire for these products among the Korean people (Bok 1987).

People of one society do not borrow ideas, materials, or inventions indiscriminately from another society. Borrowing is almost always selective: ideas and inventions are accepted or rejected depending on how useful they are to the receiving culture. They must be useful enough to make their acceptance worth the trouble.

Selective borrowing has an important implication for anyone trying to sell products to people who live in foreign markets: if each market has its own tastes, "the producer who doesn't tailor is the producer who will fail" (Magaziner and Patinkin 1989, p. 36). Many Americans have been slow to grasp this concept as they try to convince foreigners to purchase their products:

> *American firms rarely customize. "They just send us products made for Americans and say 'Why don't you Koreans buy them?'" [explains Kim, a Korean engineer working for Samsung]. Sometimes American exports are almost unusable. Although Korean households have 220-volt electricity, U.S. firms have been known to try selling refrigerators built for 110 volts, with only rudimentary converters. "Even the chocolate," says Kim. "My kids like chocolate very much, but they don't like American chocolate. Too sweet. You want to sell chocolate here, you have to know our taste."* (MAGAZINER AND PATINKIN 1989, PP. 37–38)

Even if people in one society accept a foreign idea or invention, they are still selective about which features of the item they adopt. Even the simplest invention is really a complex of elements, including various associations and ideas of how it should be used. Thus, not surprisingly, people borrow the most concrete and most tangible elements and then develop new associations and shape the item to serve new ends (Linton 1936). For example, the Japanese borrowed the game of baseball from Americans but modified it considerably to fit important Japanese cultural values. In a 1988 National Public Television "Frontline" documentary, "American Game, Japanese Rules," reporters asked American athletes playing baseball in Japan about the differences and the similarities in how the game is played in the two countries. One American athlete said, "Well, they play nine innings. That's about the only thing they have in common [with us in the game]. After we put on the uniform I'm not sure what we're doing out there" ("Frontline" 1988, p. 2).

During this documentary, American athletes described a number of striking differences:

> *I think that the biggest difference is, they play for ties. You know, that's unheard of back home. You'll play all day and all night to break a tie. But over here, you play for ties and you have a time limit on a game, that is three hours and twenty minutes.*

> *A tie is a wonderful thing in Japan. In the U.S., being a professional athlete when I played it, a tie is like kissing your sister. I mean, I'd just as soon not have that happen, I'd just as soon not play the game. In Japan they're ecstatic when you have a tie because everybody came out, everybody had the big fight, they had the big confrontation. They all fought together but nobody lost—no face was lost. So that is a perfect day.*

> *When they win the championship they only win by a couple of games. [Say] they're up by ten games [at some point in the season], by the end of the year maybe they'll be up by two. That looks okay, it doesn't look like one team blew out the other team and embarrassed them.*

> *What they feel is that, for instance, if a foreigner comes over and he's just hitting a lot of home runs, a great deal of home runs, the umpire—I wouldn't say the umpire—the league and people themselves would expect the umpire*

to expand the strike zone to balance things out. They feel that this foreigner has to be very strong and if the pitchers would keep throwing the ball over the plate and he's going to hit a home run every time up and that's not fair. So what they do is they'll start calling pitches this far outside a strike, that far inside a strike. And their philosophy is they're being fair because they're balancing it out. ("FRONTLINE" 1988, PP. 2–3, 8)

A particularly significant global trend is the diffusion of technology from already developed countries to the newly industrializing countries such as Korea and other Pacific Rim countries. From the mid-1960s to early 1970s, these nations moved to produce goods such as calculators, computers, cars, microwave ovens, televisions, radios, shoes, and clothing for already developed markets in the United States, Western Europe, and Canada. After they learned how to make these products, they perfected them by making them more functional, more efficient, higher in quality, and lower in price.

Two international business consultants, Ira Magaziner and Mark Patinkin, visited Korea in 1977 and witnessed the beginnings of the borrowing and perfecting process there. They describe this process in *The Silent War:*

The factory floors were bare concrete, and people were hand-wheeling parts to and from the production line. I moved on to Samsung's research lab, which reminded me of a dilapidated high-school science classroom. But the work going on there intrigued me. They'd gathered color televisions from every major company in the world—RCA, GE, Hitachi—and were using them to design a model of their own. They were working on refrigerators and other appliances as well. The chief engineer was young, well-trained, a recent graduate of an American university. I asked him about Samsung's color-television strategy, telling him I presumed the company planned to buy parts from overseas, only doing assembly in Korea. Not at all, he said. They were going to make everything themselves—even the color picture tube. They'd already picked the best foreign models, he said, and signed agreements for technical assistance. (MAGAZINER AND PATINKIN 1989, PP. 23–24)

Microwave oven production is a case in point. Since 1979 the conglomerate Samsung has been the world's leading producer of this product, invented 40 years earlier in the United States. Samsung workers assemble about 80,000 ovens per week; and about one in every three microwave ovens sold in the United States, although not necessarily under the Samsung brand name, is made in Korea (Magaziner and Patinkin 1989). Although Koreans borrowed American technology, they improved on the production process. Cost-efficient Korean production is a major reason why 70 percent of American households own at least one microwave oven.

One of the reasons why people are selective with regard to the items they borrow from another culture is that they evaluate those items in terms of the standards of their home culture.

The Home Culture as the Standard

Principle 7: The home culture is usually the standard that people use to make judgments about the material and nonmaterial cultures of another society. Most people come to learn and accept the ways of their culture as natural. When they encounter other cultures they can experience mental and physical strain. Sociologists use the term **culture shock** to describe the strain that people from one culture experience when they must reorient themselves to the ways of a new culture. In particular, they must adjust to a new language and to the idea that the behaviors and responses they learned in their home culture and have come to take for granted do not apply in the foreign setting. The intensity of culture shock depends on several factors: (1) the extent to which the home and foreign cultures are different, (2) the level of preparation or knowledge about the new culture, and (3) the circumstances (vacation, job transfer, or war) surrounding the encounter. Some cases of culture

A higher proportion of Korean immigrants own small businesses, especially food stores, than any other immigrant group in the United States. What cultural issues might these merchants confront as they interact with their customers?

Chuck Fishman/Woodfin Camp & Associates

shock are so intense and so unsettling that people become ill. Among the symptoms are "obsessive concern with cleanliness, depression, compulsive eating and drinking, excessive sleeping, irritability, lack of self confidence, fits of weeping, nausea" (Lamb 1987, p. 270).

In his book *Communication Styles in Two Different Cultures*, Myung-Seok Park, a professor of communication, describes the stress encountered when someone enters a foreign society. These experiences are typical of the adjustments that Koreans must make when they come to the United States. Taken as separate incidents, the two encounters are not especially stressful. It is the cumulative effect of a series of such encounters that causes culture shock.

> When I studied at the University of Hawaii, my academic advisor was an old, retired English professor, Dr. Elizabeth Carr. All of the participants were struck by her enthusiasm, deep devotion and her unfailing health. So at the end of the fall semester I said to her, "I would like to extend my sincere thanks to you for the enormous help and enlightening guidance you gave us in spite of your great age." Suddenly she put on a serious look, and I saw a portion of her mouth twisting. I had an inkling that she

> seemed unhappy about the way I expressed my thanks to her. Understandably enough, I was not a little embarrassed. A few hours later she told me that my remark "in spite of your great age" had reminded her suddenly that she was very old. I felt as if I had committed a big crime. I restrained myself from commenting on age any more. (PARK 1979, P. 66)

> One Saturday afternoon I was drinking in an American drinking establishment with some of my new American friends. What surprised me at that moment was that whenever the bar girl brought some bottles of beer, she made change and took away a certain amount out of the money in front of each drinker. It was only I who did not put money on the table. (We Korean people pay the total amount for drinks when we leave.) When the money placed in front of each person ran out, the girl proceeded to take another person's money. What was still more surprising to me was that each one filled his own glass and drank without passing the glass to his friend and without asking him to drink. I was somewhat bewildered because I had never poured my own glass before. . . . I realized that there was something cold and unfriendly in the American way of drinking. (PP. 38–39)

Some accounts from a Japanese physician visiting the United States for the first time provide additional illustrations of culture shock:

Another thing that made me nervous was the custom whereby the American host will ask a guest, before the meal, whether he would prefer a strong or a soft drink. Then, if the guest asks for liquor, he will ask him whether, for example, he prefers scotch or bourbon. When the guest has made this decision, he next has to give instructions as to how much he wishes to drink, and how he wants it served. With the main meal, fortunately, one has only to eat what one is served, but once it is over one has to choose whether to take coffee or tea, and—in even greater detail—whether one wants it with sugar, and milk, and so on. I soon realized that this was only the American's way of showing politeness to his guest, but in my own mind I had a strong feeling that I couldn't care less. What a lot of trivial choices they were obliging one to make—I sometimes felt—almost as though they were doing it to reassure themselves of their own freedom. My perplexity, of course, undoubtedly came from my unfamiliarity with American social customs, and I would perhaps have done better to accept it as it stood, as an American custom. (DOI 1986, P. 12)

The "please help yourself" that Americans use so often had a rather unpleasant ring in my ears before I became used to English conversation. The meaning, of course, is simply "please take what you want without hesitation," but literally translated it has somehow the flavor of "nobody else will help you," and I could not see how it came to be an expression of good will. The Japanese sensibility would demand that, in entertaining, a host should show sensitivity in detecting what was required and should himself "help" his guests. To leave a guest unfamiliar with the house to "help himself" would seem excessively lacking in consideration. This increased still further my feeling that Americans were a people who did not show the same consideration and sensitivity towards others as the Japanese. As a result, my early days in America, which would have been lonely at any rate, so far from home, were made lonelier still. (P. 13)

Americans in Korea or Japan would be equally confused and bewildered by the Korean habits of pouring and drinking from each other's glasses, by their constant references to age (especially mature age), and by Japanese norms about hospitality. This is because most people take an ethnocentric view toward alien cultures. People who hold the viewpoint of ethnocentrism use their home culture as the standard for judging the worth of foreign ways. From this viewpoint, "one's group is the center of everything, and all others are scaled and rated with reference to it" (Sumner 1907, p. 13). Thus, other cultures are seen as "strange" or, worse, as "inferior."

Ethnocentrism

There are different levels and consequences of ethnocentrism. The most harmless is simply defining foreign ways as peculiar, as did some American visitors who attended the 1988 Summer Olympic Games in Seoul. Learning that some Koreans eat dog meat, some people made jokes about it. People speculated about the consequences of asking for a doggy bag, and they made puns about dog-oriented dishes—Great Danish, fettuccine Alfido, and Greyhound, the favorite fast food (Henry 1988).

The most extreme and most destructive form of ethnocentrism is **cultural genocide,** wherein the people of one society define the culture of another society not only as offensive but as so intolerable that they attempt to destroy it. There is overwhelming evidence, for example, that the Japanese tried to exterminate Korean culture between 1910 and 1945. After the Japanese annexed Korea in 1910, Japanese became the official language, *hangul* was banned, Koreans were given Japanese names, Korean children were taught by Japanese teachers, Korean literature and history were abandoned, ancient temples—important symbols of Korean heritage—were razed and bulldozed, and the Korean flag could not be flown. The Japanese brutally suppressed all resistance on the part of the Korean people. When Koreans tried to declare their right to self-determination in March 1919, thousands of people were injured or killed in clashes with the Japanese military.[8] Unfortunately, American history is filled with instances of this type of ethnocentrism, as when the U.S. Bureau of Indian Affairs

Forced Culture Change of Native Americans by Bureau of Indian Affairs Schools

The Eastern band of Cherokee Indians lives on a federal reservation in the mountains of North Carolina and today numbers more than 8,000 people. In the fall of 1990, the North Carolina Cherokees took direct control of their school system. For almost a century, however, dating back to 1892, Cherokee schools, like schools for most other Native Americans, were operated directly by the Bureau of Indian Affairs (BIA) and financed by federal taxes.

For much of the twentieth century, formal education was a tool to wipe out Indian culture and transform Indians into "red-skinned whites." The process of acculturation was accomplished over the objections of parents and tribal leaders and often at the expense of the physical and emotional well-being of Indian students. For most Native Americans the most oppressive phase of BIA education occurred during the first three decades of the twentieth century. Even today, however, there are lingering effects.

The hallmark of BIA education as it evolved around the turn of the century was the boarding school. In 1920 the BIA superintendent on the Cherokee reservation wrote about the importance of "taking the most promising young men and women away from reservation influences" (Eastern Cherokee Agency File 1920). The "bad influences" of the reservation were not crime or other social problems, but parents who spoke Cherokee or wore moccasins.

All over the United States, Indian students were shuffled around to maximize the distance between themselves and their families. Thus, although a boarding school existed on the Cherokee reservation, many of the students were non-Cherokee Indians from distant reservations; in turn, many Cherokees, even as young as age six, were shipped to Chilocco Boarding School in Oklahoma, to Haskell in Kansas, to Carlisle in Pennsylvania, or to Hampton in Virginia. Even where day schools were available for elementary school–age students, the situation often changed with junior high or high school where boarding schools became the norm.

In 1918 the Cherokee BIA superintendent commented on one small boy whose whole family "have been known to stand him [the superintendent] off with drawn knives" (Annual Report of the Commissioner of Indian Affairs 1918). For many Native Americans, whether Navajo, Cheyenne, Cherokee, or others, the months of August and September were spent trying to hide their children from the truant officers. Captured children were taken crying off to the out-of-state boarding schools. If they were lucky, they got to return home for a visit the following summer. In many cases, however, it was years before children saw their parents again. It was not uncommon for a weeping 6-year-old to be taken away and return to his parents only at age 16 when he could legally drop out of school.

To make things even worse, the boarding school was often an alien environment where only English could be spoken and uniforms had to be worn. The schools stressed vocational subjects: boys learned to farm—ironic for Cherokee boys who were removed from their family farms—and girls were trained to be servants. Girls at the Cherokee Boarding School were taught to do laundry and to cook ("the demand for them as house servants in Asheville and other neighboring towns, could not be supplied") (Young 1894, p. 172).

Children caught speaking their native language had their mouths washed out with soap or were

---

forced Native Americans to attend boarding schools (see "Forced Culture Change of Native Americans by Bureau of Indian Affairs Schools").

Sociologist Everett Hughes identifies yet another type of ethnocentrism:

One can think so exclusively in terms of his own social world that he simply has no set of concepts for comparing one social world with another. He can believe so deeply in the ways and the ideas of his own world that he has no point of reference for discussing those of other peoples, times, and places. Or he can be so engrossed in his own world that he lacks curiosity about any other; others simply do not concern him. (HUGHES 1984, P. 474)

This Cherokee school was one of the special schools established by the U.S. Bureau of Indian Affairs to "Americanize" the native people of the United States. To further this goal, the school required pupils to speak English and to wear uniforms.
Courtesy Museum of the Cherokee Indian

beaten if they were repeat offenders. At the Cherokee Boarding School children who attempted to run away were chained to their dormitory beds at night.

Perhaps the best description of the BIA boarding school era came from anthropologist John Collier who, as commissioner of Indian affairs under Franklin Roosevelt's New Deal, attempted to repeal many of the oppressive policies. In 1941, Collier, writing in a magazine called *Indians at Work*, reflected on a night in the 1920s when he was allowed to sleep over at the boys' dormitory at the Cherokee Boarding School. Collier had gotten lost hiking in the mountains and asked to stay over the night until he could get transportation off the reservation:

> *The little boy inmates could not or would not talk English to each other, and they dared not talk Cherokee. . . . This was an Oliver Twist place but with every light and shade of imagination disbarred. . . . Horror itself gave up the fight.*
> (COLLIER 1941, P. 3)

References

Collier, John (1941) Editorial. *Indians at Work* (September 1–8).

Commissioner of Indian Affairs (1918) *Annual Report of the Commissioner of Indian Affairs*. Interior Department Report, vol. 2. Indian Affairs and Territories, 65th Congress, 3rd Session, House Document 1455.

Eastern Cherokee Agency File (1892–1958) Eastern Cherokee Agency Correspondence. Federal Archives and Records Center, East Point, Georgia.

Young, Virginia (1894) "A Sketch of the Cherokee People on the Indian Reservation of North Carolina." *Woman's Progress*: 171–72.

SOURCE: Sharlotte Neely, Northern Kentucky University (1991).

James F. Larson and Nancy Rivenburgh, professors of communication, offer a vivid example of this type of ethnocentrism. They list a number of comments made by television commentators during the opening ceremonies of the 1988 Summer Olympic Games held in Seoul, Korea. The comments are characterized by simple, sweeping generalizations that provide little insight into the people who live in the countries mentioned:

- Oman is probably the "hottest nation in the world."
- Mexico is "one of the most highly emotional countries in the world."
- Americans are "cool cats," "superstars" from "the most famous of all Olympic nations."
- Ireland, the "home of the leprechaun and four-leaf clover."

- Japan once occupied Korea but the countries are now "making their peace" (LARSON AND RIVENBURGH 1991, PP. 85–86).

Since the end of the Korean War, millions of Americans—military personnel, technicians, social workers, and educators—have spent time in South Korea as advisors but not as learners. For this reason, "surprisingly few Americans have come away from their Korean experience with much understanding of the country, its people, language, or culture" (Hurst 1984a, p. 1).

A conversation between author Simon Winchester (1988) and the Korean owner of a bar frequented by American service personnel shows that from the Korean viewpoint many American military people display the kind of ethnocentrism Everett Hughes described:

Just then two burly and unshaven airmen walked past. One wore a patch on his jacket that said "Munitions Storage—We tell you where to stick it!" The other had a T-shirt with the words: "Kill 'em All: Let God Sort the Bastards Out." Both were sporting newly stitched shoulder patches showing what appeared to be a small plane—the fuselage looking remarkably phallic—beneath the rubric "One Hundred Successful Missions to A-Town." Mr. Kwong shook his head with distaste.

"That's what I can't take. Don't they ever learn? We need to be respected here, and they're not respecting us. They treat us like we're some backward Third World country, and you know we're not. We're proud, we've got good reason to be. But this. . . ." He gestured with despair. I said I hadn't found anything very offensive about the two passing airmen. "Maybe not, maybe I react too much," he said. "I've worked for twenty-five years trying to bring the two communities together. I organize them to go out to meet families. I try to persuade them to learn a bit of Korean, to eat some of the food, to understand why they're here. But they don't want to know. And it's the way some of them treat our women, and our men too. Some of them just have no respect for us. The way they see it, they're top of the pile, and everyone else is

nothing. It makes me mad." (WINCHESTER 1988, P. 144)

Cultural Relativism

A perspective that runs counter to ethnocentrism is cultural relativism. **Cultural relativism** means two things: **(1)** that a foreign culture should not be judged by the standards of a home culture, and **(2)** that a behavior or way of thinking must be examined in its cultural context—in terms of that society's values, norms, beliefs, environmental challenges, and history. Cultural relativism is a perspective that aims to understand, not to condone or to discredit, foreign behavior and thinking.

For example, the Korean practice of defining infants as one year of age at birth cannot be understood if it is evaluated according to dominant American values, which emphasize the future over the past. These values support the idea that the birth of a baby represents a fresh start with unlimited possibilities no matter what the social class, ethnicity, or educational level of the baby's parents. From this point of view, it makes sense to define age at birth as zero. The Korean practice, however, must be considered in light of Korean values. Defining age in this way corresponds with Korean values about the importance of the past and with beliefs about the relationship of the individual to the group. The message is that generations are interdependent and that past events (even before conception) are important to a person's life. The Korean way of defining age ties together past, present, and future generations.

Similarly, most Americans cannot understand why some Koreans eat dog meat. This reaction should not be surprising when we consider that more than one-third of American households include at least one dog. In general, dogs are valued pets. The United States has more than 10,000 pet shops, 19,000 dog food vendors, 11,000 grooming shops, 7,000 kennels, and 300 pet cemeteries, as well as 200 products for dogs, ranging from feeding dishes to raincoats and sunglasses (Rosenfeld 1987). On the other hand, most Koreans are equally appalled that Americans often let dogs live in their homes, allow them to lick their faces, and spend so much money on them when there are many poor and homeless people in the United States. When we consider the historical and environmental challenges surrounding the Korean decision to eat dogs, the practice might not seem so unreasonable.

Reverse Ethnocentrism

A position of cultural relativism not only neutralizes a we-are-superior-and-you-have-to-do-everything-like-us attitude but also counteracts **reverse ethnocentrism**—the tendency to see the home culture as inferior to a foreign culture. The sentiment "How wonderful Asia is; we have to become more like them" expresses this viewpoint (Berger 1989, p. 493). For example, although most Americans know little about Korean culture, an overwhelming number (98.8 percent) believe that Koreans are more industrious than Americans; further, 96.8 percent believe Koreans are more self-disciplined, and 88.8 percent believe Koreans are more serious than Americans (Yim 1989).

Americans have this image not only of Koreans but of Asian peoples in general, particularly the Japanese. Education is perhaps the most highly publicized area in which Japanese are believed to be superior to Americans. "The image is one of an obedient, hard-working mass of students rigorously preparing for grueling examinations under the firm supervision of dedicated teachers who still retain the respect of society at large" (Hurst 1984b, p. 2). This popular and powerful image leads Americans to believe that the Japanese educational system is superior in every way to the American system.

This image is only part of the picture, however. In reality, both countries have serious educational problems, and each looks to the other in hopes of finding solutions: ironically, each group "criticize[s] in its own system the qualities the other finds attractive there" (Hurst 1984b, p. 6). Americans generally recognize that their system offers too much choice and too little direction; they call for tougher standards, more tests, more homework, and a uniform curriculum that includes greater emphasis on mathematics and science and less on electives. The Japanese, on the other hand, recognize that their system offers students too few choices and too much direction. The Japanese students spend almost all of their free time studying, because they are cramming for examinations, the results of which determine their academic future (what kind of high school they can attend, whether they will attend college), future employment, and social status. As a result, the Japanese are calling for less emphasis on tests and more choice in courses, as well as greater emphasis on social activities. The point is that every culture has its strengths and weaknesses; to assume that any culture is perfect is misguided thinking.

In addition to countering reverse ethnocentrism, the viewpoint of cultural relativism, with its emphasis on context, cautions against assuming that people in a society behave as a spontaneous, coordinated unit. No single statement about behavior or thinking can apply to everyone in the society.

Subcultures

Principle 8: In every society there are groups that possess distinctive traits that set them apart from the main culture. Groups that share in some parts of the dominant culture but have their own distinctive values, norms, language, or material culture are called **subcultures**. A subculture that conspicuously challenges, rejects, or clashes with the central norms and values of the dominant culture is referred to as a **counterculture**. All of the cultural principles discussed thus far apply to subcultures (which include countercultures).

Often we think we can identify subcultures on the basis of physical traits, ethnicity, religious background, geographic region, age, gender, socioeconomic or occupational status, dress, or behavior defined as deviant by society. However, determining which people constitute a subculture is a complex task that requires careful thought; it must go beyond simply including everyone who shares a particular trait. For example, using broad ethnic categories as a criterion for identifying the various subcultures within the United States makes little sense. Realistically, a biologically and culturally in-termixed population numbering hundreds of mil-lions cannot be divided neatly into white, African-American, Hispanic, Native American, and Asian subcultures (Clifton 1989). The broad categoriza-tion of *Native American,* for example, ignores the fact that the early residents of North America "practiced a multiplicity of customs and lifestyles, held an enormous variety of values and beliefs, spoke numerous languages mutually unintelligible to the many speakers, and did not conceive of themselves as a single people—if they knew about each other at all" (Berkhofer 1978, p. 3). The point is that the presence or absence of a single trait, es-pecially a physical trait, cannot be the only crite-rion for classifying someone as part of a subculture. Sociologists determine whether a group of people constitutes a subculture by learning whether they share a language, values, norms, or a territory, and whether they interact with one another more than with people outside the group.

One characteristic central to all subcultures is that their members are isolated in some way from other people in the society. This isolation may be voluntary or it may result from an accident of ge-ography or it may be imposed consciously or un-consciously by a dominant group. Or it may be a combination of all three. Whatever the reason, sub-cultures are cut off in some way. The cutoff may be total or it may be limited to selected segments of life such as work, school, recreation and leisure, dating and marriage, friendships, religion, medical care, or housing.

The Old Order Amish are an example of a sub-culture within the United States that is cut off vol-untarily from the mainstream culture. One domi-nant characteristic of the Old Order Amish is their value for self-sufficiency and the corresponding rejection of material culture that promotes de-pendence: their homes have no furnaces, running water, indoor toilets, television sets, radios, electric refrigerators, or doorbells, and they have no cars (Kephart 1987).

A geographically isolated subculture within the United States is that of the Aleuts, natives of the Aleutian Islands, which stretch in a chain from Alaska across the northern Pacific Ocean. Their

language, an Eskimo-Aleut dialect, is incomprehen-sible to other American Eskimo groups. The plight of other Native American groups is an example of isolation imposed by a dominant group. These groups represent numerous subcultures within the United States that were forced onto reservations by the U.S. government. About one-third live on res-ervations and are geographically, economically, and socially isolated from mainstream American culture.

Some subcultures within the United States are isolated in different ways or to different degrees; their members integrate into mainstream culture when possible but voluntarily or involuntarily ac-cept a segregated role in other areas of life. In gen-eral, African-Americans who work or attend school primarily with whites are often excluded from per-sonal and social relationships with white col-leagues. This exclusion forces them to form their own fraternities, study groups, support groups, and other organizations.

The number and variety of affiliations between members of a subculture and people outside their group are a rough indicator of the subculture's rela-tionship to mainstream culture. In general, the greater the number and the more diverse the affilia-tions with "outsiders," the more likely it is that the subculture shares a significant portion of the main-stream culture. Some subcultures are **institutionally complete** (Breton 1967); their members do not in-teract with anyone outside their subculture to shop for food, attend school, receive medical care, or find companionship because these needs are pro-vided for by the subculture. In the United States, re-tirement communities are a typical example of a setting in which residents' needs to affiliate with the mainstream culture are minimized. In South Korea, where the total number of non-Koreans is less than one percent of the population, the largest minority group—composed of 50,000 Chinese—lives in in-stitutionally complete ethnic communities.

Despite the ethnic similarity among the people, Korea has been divided since 1945 into two insti-tutionally complete societies—North Korea and South Korea. The people of the two Koreas have no relationship with one another (including no corre-spondence by mail, phone, or travel) and technically remain at war with one another. The line that separates the two people at the 38th parallel is the most heavily militarized region of the world; each side is prepared to stop the other from invading. Both sides recognize their common language, ethnicity, history, and culture, and both believe in unification. Yet, despite these similarities, neither side has been able to compromise on economic or government structure, the elective process, or the appointive process for government offices.

In view of the recent unification of Germany, we might speculate that the two Koreas might also reunite soon. South Korean officials predict, however, that it may take 20 years before they eventually reunite. One important difference is that West Germany, in comparison to South Korea, was far more wealthy and thus better able to absorb the staggering economic costs ($250 billion) of reunification. Recently, South Korea's assistant minister for reunification commented: "After seeing what happened in Germany, especially in the economic area, it has changed attitudes toward unification in Korea. The biggest lesson we learned is that we need to prepare steadily, and we need to have realistic expectations. We cannot just be ruled by sentiment" (Protzman 1991; Sterngold 1991, p. C1).

Often there is a clear association between institutional completeness and language differences.

Persons who cannot speak the language of the dominant culture are likely to live in institutionally complete ethnic communities (for example, Little Italy, Chinatown, Koreatown, Mexican barrios). Of the 750,000 Koreans living in the United States, approximately 300,000 live in Southern California. A large portion of these 300,000 live in an institutionally complete Koreatown west of downtown Los Angeles.

Whether an immigrant chooses to belong to a community of like persons depends upon a variety of factors: circumstances of migration (forced or voluntary), age, gender, status of mother country relative to host country, knowledge of the host country's language, level of skills, education, and social class. Korean immigrants who cannot obtain bank loans to start businesses may become members of *kyes*, informal financing systems of 20 or so would-be entrepreneurs to which each member promises to contribute several hundred dollars a month for 20 months. Each month one member of the *kye* receives $10,000 and continues to pay into the system until all 20 members have received $10,000 each (Reinhold 1989; Sanchez 1987). The presence of *kyes* in some Korean-American communities means that those Koreans do not have to interact with banks. In this sense *kyes* contribute to institutional completeness.

Sociological Perspectives on Culture

Although sociologists tend to agree about the eight essential principles of culture, they have quite different views about how culture affects people's lives. These different views correspond to the three major sociological perspectives: functionalism, conflict theory, and symbolic interaction. (See Chapter 2 for a detailed discussion of these perspectives.)

The Functionalist View of Culture

According to the functionalist perspective, societies can operate smoothly only if their members are able to meet the demands and challenges of the en-

vironment in effective, coordinated ways. In this vein, functionalists emphasize that culture serves as a buffer between people and their environment. Culture represents the solutions that the people of a society have worked out over time to meet distinct environmental and historical challenges and have passed on to the next generation "so that each new generation does not have to start out from scratch" (Critchfield 1985, p. 3). Language, for example, functions to give people a sense of unity and to link them to one another so that cooperative action is possible. If you can recall times when the people around you were speaking another language (for example, as on an elevator), the unifying and cooperative function of language should become obvious.

The functionalist perspective on culture has at least one major limitation: it tends to ignore dysfunctions (negative and harmful aspects of the material and nonmaterial components of culture) and overemphasize culture's integrative qualities, on the assumption that if something exists, it is by definition useful for society. Conflict theorists, on the other hand, focus on dysfunctional aspects of culture.

The Conflict View of Culture

From a conflict perspective, society is not held together because everyone learns and shares the same cultural values. Conflict theorists are concerned with how the groups that control the means of material production impose their products, values, and norms on other groups. In this regard they focus on how culture (material and nonmaterial) is shaped by industrialization. For example, food, fashion, and leisure are shaped by the means of production. (In the case of industrialized societies, the capitalist system makes it easier for people to abandon making clothes, preparing food, and entertaining themselves, because they can consume what machines have made for them.)

A major shortcoming of the conflict perspective is that it sees the owners of production as imposing a culture (products, values, and norms) on other, less powerful groups. It hardly acknowledges that members of these less powerful groups often want to purchase what industrialization has to offer, thus rejecting (consciously or unconsciously) aspects of their culture.

The Symbolic Interactionist View of Culture

Symbolic interactionists are not concerned with the functions of culture or with questions of how a dominant culture is imposed on others. The symbolic interactionist is more concerned with the symbolic properties of culture. Symbolic interactionists define the symbol as essential to civilization because (1) cultures emerge and are perpetuated as a result of symbols, (2) interaction between people could not take place without symbols, and (3) infants are transformed into human beings when they acquire symbols. A symbol may be defined as any kind of physical form—a material object, a color, a sound, a word, an odor, a movement, a taste—that receives its value or meaning from those who use it (White 1949).

The meaning of a symbol is not inherent in the form itself.[9] The color appropriate to mourning, for example, varies across cultures; it may be black, white, red, yellow, green, or any other color. The meaning associated with a word is not intrinsic to the combination of letters or strokes that make it that word: the Korean word *kaein* and the English word *human* both refer to a member of the species *homo sapiens*. A raised eyebrow may signify surprise, interest, attraction, or fear. Colors, words, and gestures become symbols because people have agreed upon their meanings over time. Hence, symbolic interactionists conceive of culture as an elaborate and complex symbol system that has been constructed and is passed on from one generation to another.

Symbolic interactionists also are interested in how meanings are constructed and how they change over time and across cultures. The symbolic interaction perspective makes us aware that the key to understanding and interacting with someone from another culture is to learn the symbol system of that culture. At present, business-oriented magazines and journals published in the United States print columns and articles that alert readers to the differences between American and foreign symbol systems and that caution against misinterpretations.

The symbolic interaction perspective has at least two weaknesses. First, although it is relatively easy to trace why a particular symbol is associated with a physical form after the fact, there is no systematic framework for predicting what symbolic meanings will arise. Moreover, it is not clear how people come to agree on meanings.

Discussion

In this chapter we examined the eight essential sociological principles of culture and three viewpoints about how culture affects people's lives. Although culture is an important and complex concept that defies a simple definition, people take it for granted. As anthropologist Ralph Linton (1936) said, "The last thing a fish would ever notice would be water." Similarly, people do not notice culture except in unusual circumstances (Henslin 1985). Because people inherit a culture at birth and begin to assimilate it very early in life, the role of culture is comparable to the role of gravity. We rarely think about the presence of gravity, but if it were to disappear suddenly, life as we know it would be thrown out of balance, to say the least. The influence of culture is equally pervasive. Although we rarely think about culture, it provides a blueprint, or a set of guidelines, for thinking and behaving with regard to other people, animate beings (animals, plants, and so on), and inanimate objects (for example, microwave ovens, automobiles, and works of art).

Comparing the contrasting cultural practices should not lead to regarding the ways of one culture as better than those of another. Every culture has shortcomings and destructive features as well as virtues and positive features. "We must recognize the unhappiness, the degradation, the misery, the incredible brutality, cruelty, and human wastage in all cultures which each tends to ignore while stressing [the virtue]" (Frank 1948, p. 394).

There is virtue, for example, in the American emphasis on the future rather than the past. Such a viewpoint is optimistic; it allows a person to make a fresh start despite past mistakes. On the other hand, it has negative implications. If the past is considered irrelevant and not worth understanding, then (1) mistakes are repeated, (2) old people feel useless because their accomplishments are in the past, and (3) people lack the feeling of continuity between generations because each generation perceives itself as starting anew. Similarly, the Korean emphasis on the past and on tradition has its virtues: (1) perhaps fewer mistakes from the past tend to be repeated, (2) the old feel useful, and (3) the generations feel a common bond. Yet, this viewpoint also contains some destructive elements, as when the ways of the past do not allow for change, when the old do not learn from the young, and when an individual is trapped in the mistakes of the past.

In a little over twelve years' time, the United States has become part of a more interdependent, more competitive world economic order. The current level of global interdependence and competition reflects a multitude of interactions between Americans and people of different cultures. Such encounters make for the unusual circumstances that have forced Americans to acknowledge other ways of doing things. Although American sociologists and anthropologists have written about culture for more than 100 years, only in the past

A Korean artist doing fine calligraphy.
Kim Newton/Woodfin Camp & Associates

decade have nonacademics (particularly business-persons) taken an interest in culture with the understanding that knowledge about other cultures will help to negotiate deals and so sell products to foreign consumers.

From a sociological point of view, there are other reasons for learning about foreign cultures. For example, comparing one way of thinking and behaving with other ways teaches that human behavior is constructed, not determined; that choice exists; and that people need not be prisoners of their culture. In addition, nations are interdependent in other than economic ways. Environmental pollution, for example, affects every nation and cannot be solved by one nation acting alone. Nations can learn conservation techniques from one

another; if Americans simply know that Koreans double-line their notebook paper and close their refrigerator doors quickly, they will be more aware of their own wasteful practices. Americans also can study programs enacted by other countries to reduce pollution. In the case of recycling, many city officials in the United States studied the Japanese recycling system before starting their own programs. Most important, learning about other cultures offers insights into the workings of American society, helps us know and understand the world outside our borders, enables us to clarify issues we face as a country, and enlightens us and gives us an appreciation of what we have borrowed from other cultures and heretofore considered uniquely American.

FOCUS
The Insight of Edward T. Hall

This chapter may give you the impression that the call for cross-cultural awareness is a fad—something that emerged in the early 1990s. Actually, sociologist Edward T. Hall made a case, about 35 years ago, in *The Silent Language* (1959), for why Americans should learn more about other cultures. The introduction to this now classic book is reprinted here because it gives us insights about the behavior Americans must change if they are to interact with people from other cultures as equal parties.

A Case for Multicultural Awareness

Edward T. Hall

Though the United States has spent billions of dollars on foreign aid programs, it has captured neither the affection nor esteem of the rest of the world. In many countries today Americans are cordially disliked; in others merely tolerated. The rea-

sons for this sad state of affairs are many and varied, and some of them are beyond the control of anything this country might do to try to correct them. But harsh as it may seem to the ordinary citizen, filled as he is with good intentions and natural generosity, much of the foreigners' animosity has been generated by the way Americans behave.

As a country we are apt to be guilty of great ethnocentrism. In many of our foreign aid programs we employ a heavy-handed technique in dealing with local nationals. We insist that everyone else do things our way. Consequently we manage to convey the impression that we simply regard foreign nationals as "underdeveloped Americans." Most of our behavior does not spring from malice but from ignorance, which is as grievous a sin in international relations. We are not only almost totally ignorant of what is expected in other countries, we are equally ignorant of what we are communicating to other people by our own normal behavior.

It is not my thesis that Americans should be universally loved. But I take no consolation in the remark of a government official who stated that "we don't have to be liked just so long as we are respected." In most countries we are neither liked nor respected. It is time that Americans learned how to communicate effectively with foreign nationals. It is time that we stop alienating people with whom we are trying to work.

For many years I have been concerned with the selection and training of Americans to work in foreign countries for both government and business. I am convinced that much of our difficulty with people in other countries stems from the fact that so little is known about cross-cultural communication. Because of this lack, much of the good will and great effort of the nation has been wasted in its overseas programs. When Americans are sent abroad to deal with other peoples they should first be carefully selected as to their suitability to work in a foreign culture. They should also be taught to speak and read the language of the country of assignment and thoroughly trained in the culture of the country. All of this takes time and costs money. However, unless we are willing to select and train personnel, we simply sell ourselves short overseas.

Yet this formal training in the language, history, government, and customs of another nation is only the first step in a comprehensive program. Of equal importance is an introduction to the non-verbal language which exists in every country of the world and among the various groups within each country. Most Americans are only dimly aware of this silent language even though they use it every day. They are not conscious of the elaborate patterning of behavior which prescribes our handling of time, our spatial relationships, our attitudes toward work, play, and learning. In addition to what we say with our verbal language we are constantly communicating our real feelings in our silent language—the language of behavior. Sometimes this is correctly interpreted by other nationalities, but more often it is not.

Difficulties in intercultural communication are seldom seen for what they are. When it becomes apparent to people of different countries that they are not understanding one another, each tends to blame it on "those foreigners," on their stupidity, deceit, or craziness. The following examples will illuminate some of these cross-cultural cross-purposes at their most poignant.

Despite a host of favorable auspices an American mission in Greece was having great difficulty working out an agreement with Greek officials. Efforts to negotiate met with resistance and suspicion on the part of the Greeks. The Americans were unable to conclude the agreements needed to start new projects. Upon later examination of this exasperating situation two unsuspected reasons were found for the stalemate: First, Americans pride themselves on being outspoken and forthright. These qualities are regarded as a liability by the Greeks. They are taken to indicate a lack of finesse which the Greeks deplore. The American directness immediately prejudiced the Greeks. Second, when the Americans arranged meetings with the Greeks they tried to limit the length of the meetings and to reach agreements on general principles first, delegating the drafting of details to subcommittees. The Greeks regarded this practice as a device to pull the wool over their eyes. The Greek practice is to work out details in front of all concerned and continue meetings for as long as is necessary. The result of this misunderstanding was a series of unproductive meetings with each side deploring the other's behavior.

In the Middle East, Americans usually have a difficult time with the Arabs. I remember an American agriculturalist who went to Egypt to teach modern agricultural methods to the Egyptian farmers. At one point in his work he asked his interpreter to ask a farmer how much he expected his field to yield that year. The farmer responded by becoming very excited and angry. In an obvious attempt to soften the reply the interpreter said, "He says he doesn't know." The American realized something had gone wrong, but he had no way of knowing what. Later I learned that the Arabs regard anyone who tries to look into the future as slightly insane. When the American asked him about his future yield, the Egyptian was highly insulted since he thought the American considered him crazy. To the Arab only God knows the future, and it is presumptuous even to talk about it.

In Japan I once interviewed an American scholar who was sent to Japan to teach American history to Japanese university professors. The course was well under way when the American began to doubt if the Japanese understood his lectures. Since he did not speak Japanese, he asked for an interpreter. After a few lectures with the interpreter translating for him, the American asked the group to meet without him and make a report on what they were learning from the course. The next time the American met with the class the interpreter told him that the class understood only about 50 percent of what had been going on. The American was discouraged and upset. What he didn't know was that he had inadvertently insulted the group by requesting an interpreter. In Japan a sign of an educated man is his ability to speak English. The Japanese professors felt that the American had caused them to lose face by implying that they were uneducated when he requested the interpreter.

Americans often do so badly in their jobs overseas that military officers have a real fear of being assigned to some countries. I once heard a retired admiral talking to an army general about a mutual acquaintance. "Poor old Charley," lamented the admiral, "he got mixed up with those Orientals in the Far East and it ruined his career." Periodically, after an incident like the Girard case, which was a tragedy of errors by both American military and diplomatic personnel in Japan, there is a brief flurry of interest in "doing something about better selection of personnel for foreign assignment." As one Pentagon aide remarked, "At least we ought to be able to select them so they don't shoot the local civilians."

Obviously there are always going to be unavoidable incidents when our forces are stationed in foreign countries. Many of the incidents are made worse, however, by the inept way the Americans handle the consequences. When incidents do occur, the Americans rarely know how to act in such a way as to avoid adding fuel to the fire. They are usually blind to the fact that what passes as ordinary, acceptable American behavior is often interpreted in such a way by foreigners that it distorts our true sentiments or our intentions.

If this book does nothing more than to plant this idea, it will have served its purpose. Yet in writing these pages I have had a more ambitious goal. This book was written for the layman, the person who is at times perplexed by the life in which he finds himself, who often feels driven here and there by forces he does not understand, who may see others doing things that genuinely mystify him at home in America as well as overseas in another culture. I hope that I can show the reader that behind the apparent mystery, confusion, and disorganization of life there is order; and that this understanding will set him to re-examine the human world around him. I hope that it will also interest the reader in the subject of culture and lead him to follow his own interests and make his own observations.

Key Concepts

Beliefs 110

Counterculture 133

Cultural Genocide 129

Cultural Relativism 132

Culture 106

Culture Shock 127

Diffusion 122

Ethnocentrism 129

Feeling Rules 118

Folkways 111

Institutionally Complete 134

Material Culture 109

Mores 112

Nonmaterial Culture 110

Norms 111

Reverse Ethnocentrism 133

Social Emotions 118

Society 106

Subcultures 133

Technology 109

Values 110

8Principles of culture

Notes

1. Republic of Korea is the official name for South Korea. In this chapter the terms *South Korea* and *Korea* are used interchangeably. Any reference to North Korea specifically is noted as such.

2. Raw resources become usable resources only when people learn to see and use them as such. Until people learned how to use oil for light and fuel, for example, charlatans sold it as medicine.

Now, the many uses to which oil has been put have "created an umbilical dependency. Modern transportation, factories, and food production feed on oil. Almost nothing . . . can work without oil" (Gelb 1990, p. 34). Similarly, uranium existed only in an academic context until people discovered its radioactive properties, which make it the essential ingredient of the atomic bomb and nuclear fuels. And waterfalls were simply navigational obstacles until people learned to convert the force of the water into energy (Chinoy 1961).

3. Because values cannot be observed directly, sociologists take one of two approaches to determine what values people regard as important: (1) they assume that values are reflected in behavior and can be determined by observing actual behavior, or (2) they ask people to tell them what their values are. Both methods have their drawbacks. With the first approach, researchers may make wrong interpretations because people don't always behave according to their professed values ("Do as I say, not as I do"). With the second approach, people may not be willing or able to state what values are important to them.

4. In most American homes, the refrigerator is the single largest energy user. About 1,200 kilowatt-hours a year are needed to keep a refrigerator running, about twice the amount that Japanese refrigerators use. The difference is due to size (Japanese refrigerators are smaller) and to energy-efficient engineering (Wald 1989).

5. In the 1920s, linguist Edward Sapir wrote a highly influential book, *Language: An Introduction to the Study of Speech*. Sapir, who is responsible for alerting us to the social uses and structure of language, believed that "No two languages are ever sufficiently similar to be considered as representing the same social reality. The worlds in which different societies live are distinct worlds, not merely the same world with different labels attached" (Sapir 1949, p. 162). These assumptions underlie the linguistic relativity hypothesis advanced by Sapir: languages are so different that it is nearly impossible to make translations in which words produce approximately the same effects in the speaker of language *X* as in the speaker of language *Y*. Today, most social scientists reject this position and argue that, although considerable work may be required, it is possible for people to discover one another's meanings and to translate from one language to another.

6. Even in the United States cattle are relatively new as a major source of meat. The widespread use of beef as food began only after the westward expansion of the 1870s and the development of the tractor. Before that time the population was largely confined to the densely wooded eastern states, where pigs rather than cows thrived. Thus, pork was the major source of meat. The settlement of the West opened an abundance of rich flatland for grazing and food production (Tuleja 1987).

7. After the city of Seoul was awarded the 1988 Summer Olympic Games, the South Korean government took steps to rid the city of snake and dog shops. Several forces caused this move, including Korean sensitivity to foreign opinion and the power of the global mass media. The history of foreign involvement in Korean affairs has made Koreans self-conscious about what people from other countries think about Korean culture. Thus, to avoid ridicule and embarrassment, the South Korean government moved to close the dog and snake shops so they would not be present for the cameras.

8. The Japanese maintained that their purpose in destroying Korean culture was to help modernize Korea. Even today, many Japanese argue that under their occupation Korean agricultural and industrial productivity improved and that highways, railroads, ports, communication systems, and industrial plants were built. The Koreans' earnings and output, however, went to support Japanese society and left most Koreans impoverished.

   As Japan became more deeply involved in World War II, Koreans were forced to serve in the military or to fill the factory positions of Japanese workers fighting in the war. During this 35-year period the Koreans were subservient to the Japanese in every way. In essence, the Japanese attempted to "Japanize" the Koreans by taking away their language, ideas, and material culture.

9. There are some exceptions to the rule. The meanings assigned to symbols sometimes may be linked to qualities inherent in them (such as green as a symbol of nature). Usually, however, the symbols chosen to represent meanings are arbitrary.

5 SOCIALIZATION

with Emphasis on Israel, the West Bank, and Gaza

A view of Jerusalem, with the Wailing Wall in the foreground.
James Stanfield/National Geographic Society

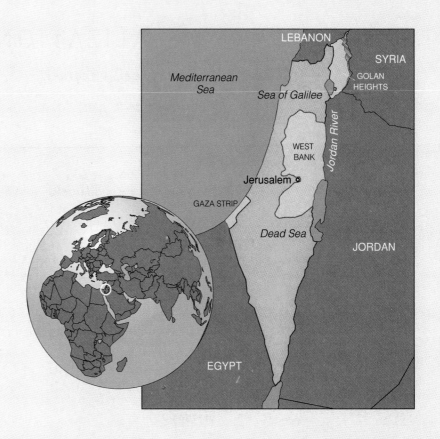

Israelis admitted into the United States in fiscal year 1991 for temporary employment	5,867
People living in the United States in 1990 who were born in Israel	97,000
Airline passengers flying between the United States and Israel in 1991	450,624
U.S. military personnel in Israel in 1993	62
People employed in 1990 by Israeli affiliates in the United States	3,100
Students from Israel enrolled in college in the United States for fall 1991	3,130
Phone calls made between the United States and Israel in 1991	22,001,000
Applications for utility patents filed in 1991 in the United States by inventors from Israel	633

Let me say to you, the Palestinians, we are destined to live together on the same soil in the same land.

We, the soldiers who have returned from battles stained with blood; we who have seen our relatives and friends killed before our eyes; we who have attended their funerals and cannot look into the eyes of their parents; we who have come from a land where parents bury their children; we who have fought against you, the Palestinians, we say to you today in a loud and a clear voice: Enough of blood and tears.

Enough! We have no desire for revenge; we have—we harbor no hatred toward you. We, like you, are people—people who want to build a home, to plant a tree, to love, live side by side with you in dignity, in affinity, as human beings, as free men.

We are today giving peace a chance and saying to you—and saying again to you—enough. Let us pray that a day will come when we all say farewell to the arms.

We wish to open a new chapter in the sad book of our lives together, a chapter of mutual recognition, of good neighborliness, of mutual respect, of understanding. We hope to embark on a new era in the history of the Middle East.

PRIME MINISTER YITZHAK RABIN
September 13, 1993
Middle East Peace Signing Ceremony

At birth, the human cerebral cortex (the seat of complex thought) is not sufficiently developed to permit a sophisticated awareness of self and others or reflection on the rules of social life, which we call *norms*.[1] These capacities develop, however, as children mature biologically and as they interact with others.

To illustrate, most two-year-olds are biologically ready to show concern for what adults regard as the rules of life. They are bothered when rules are violated: paint peeling from a table, broken toys, small holes in clothing, and persons in distress raise troubling questions. From a young child's point of view, when something is "broken," then somebody somewhere has done something very wrong (Kagan 1988a, 1988b, 1989).

To show this kind of concern with standards, two-year-olds must first be exposed to information that leads them to expect behavior, people, and objects to be a certain way (Kagan 1989).[2] They develop these expectations in the course of their social relations with adults. Children go to adults with their questions and needs. Adults respond in different ways: they may offer explanations, express concern, try to help, show no concern, or pay no attention. Through many such simple exchanges, children learn how to think about objects and people. They learn about the social group to which they primarily belong, and about other groups to which they do not belong. In addition, through these exchanges children are acquiring basic skills such as the ability to talk, walk upright, and reason. This learning from others about the social world is part of a complex lifelong process called **socialization.**

Socialization begins immediately after birth and continues throughout life. It is a process by which newcomers develop their human capacities and acquire a unique personality and identity. This process also enables **internalization,** in which people take as their own and accept as binding the norms, values, beliefs, and language needed to participate in the larger community. Socialization is also the process by which culture is passed on from generation to generation.

This chapter explores the socialization process and its significance for both societies and individuals. We examine how nature (biology) and nurture (experience) interact to produce individuals who are like others of their society and group, yet are uniquely themselves. We discuss the influence of collective memory and group membership on social identity. We also consider different theories of early socialization and how it differs from re-socialization.

Our emphasis in this chapter is on the fierce century-long conflict between Jews and Arabs in Israel and the West Bank and Gaza. This conflict is one of an estimated 60 internal conflicts going on in the world between people who share a territory but who differ from one another in ethnicity, race, language, or religion (Kotowitz and Moody 1993). Most of these conflicts have long histories, which means that the conflict has been passed on from one generation to the next. We concentrate on Israel and the West Bank and Gaza because several times in the past (notably, in September 1993) the United States has been a major sponsor of meetings between Israeli and Palestinian representatives. Achieving peace will not be easy, because everyone involved has been affected in some way by the conflict. For most people, the conflict has been a part of their lives since they were born. In this chapter we draw on socialization concepts and theories to help us understand how the conflict has been "passed down" from one generation to the next.

Keep in mind that socialization is just one factor of many that help us understand why the Israeli-Palestinian conflict has lasted at least a century. (Chapters 9, 10, and 13 address other factors that fuel ongoing internal conflicts.) Although we concentrate on these two groups (Palestinians and Israelis), we must keep in mind that the issues raised here are relevant to the strategies people everywhere use to teach newcomers how to participate in the society in which they are born. The newcomers, however, do not become carbon copies of their teachers. They learn about the environment they inherit and then come to terms with it in unique ways.

Coming to Terms:
The Palestinians and Israelis

[handwritten: Hebrews = Israelites = Jews / Palestinians = Arabs]

In December 1987, Palestinians in the West Bank and Gaza launched a series of violent demonstrations against Israeli police, military, and other authorities who have occupied the territories since 1967. The *intifada* (or *intifadah*) is now into its seventh year: "It is the most sustained, coherent and comprehensive opposition to date" (Lesch 1989, p. 89). See "The Intifadah: Organization and Phases." The following excerpt describes the vicious cycle of Palestinian protest and Israeli attempts to suppress that protest:

> *One of my friends recently served in the [occupied] territories. He kept on talking about his inner conflict. During the day, he said, you join in everything—beatings, shootings, driving [Palestinian] women out of houses about to be blown up—and you have no problem with any of it. You're ready to maim the first stone-thrower you come across without a qualm. But later, back at home, you feel miserable thinking about it, and toss about in bed unable to sleep. But while it's happening you feel nothing. There you stand as a soldier with your moral claims, surrounded by hundreds spitting at you, calling you names, throwing bottles and stones, calling your mother a whore. And as you walk through a village, and out of every window somebody is shouting, you begin to be afraid. An incredible fear takes hold of you, and then comes the moment when you can't stand it any longer and you hit out at the next person you see. For a fleeting moment you simply have to feel you're still stronger and able to defend yourself or you'll go crazy. Most of us, he told me, then just hit out, without looking, and you see these terrible explosions of rage. In the evening or on weekends they sit around and talk about what's happened. Feeling desperate, they must keep talking. He himself, he said, had talked about it so often, and almost everyone he knows feels tormented.* (SICHROVSKY 1991, PP. 61–62)

The situation that this Israeli soldier finds himself in grows out of a century-long dispute between Palestinian Arabs and Israeli Jews over the land between the Jordan River and the Mediterranean Sea. (See the map at the beginning of this chapter.) Both sides call this land "home." The Palestinians, descendants of Canaanites, Muslims, and Christians, had lived on this land for more than 2,000 years. The Hebrews (Israelites), exiled from Egypt, arrived in the area around 1200 B.C. and established a kingdom (Eretz Israel) with Jerusalem as its capital. The Romans conquered this land around 70 B.C. They treated the Jews harshly and suppressed all expression of Jewish culture, including religion and language. The Jews actively resisted and rebelled against Roman rule. The Romans responded by expelling most of the Jews. Only a very small number of determined Jews remained through successive occupations by Persians, Arabs, European Crusaders, Turks, and, ultimately, the British. The dispersed Jews, victims of discrimination abroad, never forgot their "home" in Palestine.

In the late nineteenth century, in a growing climate of anti-Semitism throughout Europe and Russia, Theodor Herzl founded the modern Zionist movement.[3] Herzl, a Jewish émigré living in England, believed that the only way to combat European anti-Semitism was to establish a Jewish state. He formulated a plan to return dispersed Jews to Palestine, which he thought belonged rightfully to Jews everywhere. Young European and Russian Jews followed Herzl's lead and emigrated to Palestine to buy land and build settlements.

Shortly after the Jewish return movement began, World War I came to an end. The British, who had defeated the Turks, were given control over Palestine by the Allies. Although the British set limits on the number of returning Jews, the Nazi Holocaust during World War II increased the flow of Jewish refugees. It gave a decisive push and a desperate urgency to the Jewish return movement. Haunted by the Holocaust, which had claimed more than six million lives (approximately one-third of all European Jews), the Jews were determined to establish their own state and their own army to protect themselves from future aggression, and to reduce

Haunted by the Holocaust and widespread persecution throughout Europe, Jewish refugees fled to the territory they called Israel and the Palestinian people called Palestine.
Robert Capa/Magnum

their dependency on other countries for food, shelter, jobs, and passports when they assumed refugee status.[4]

The steady stream of Jewish colonists notwithstanding, Palestine was then home to approximately 1.2 million Palestinian Arabs (Smooha 1980).[5] From the time that Jews began to return, Arabs and Jews fought local battles over the land. Members of each group burned crops, destroyed trees, stole animals, sabotaged irrigation systems, and destroyed agricultural equipment belonging to the other. As the British prepared to withdraw following World War II, they asked the United Nations to act as an outside mediator in the growing dispute. On November 29, 1947, the United Nations voted to partition Palestine into two independent states, one Jewish and the other Arab. Under this plan 400,000 Palestinians were living in the Jewish half and an unknown number of Jews were living in the Palestinian side (Rubinstein 1991). The Palestinians could not tolerate this arrangement. From their point of view it was inconceivable that representatives from other countries could vote to divide and give away their land. As the British gradually withdrew troops and military equipment, the struggle between Palestinians and Israelis over the now-evacuated territory escalated.

On May 14, 1948, Jewish leaders declared Israel an independent state. The next day, Palestinians and Arab armies from Egypt, Syria, Jordan, Lebanon, and Iraq attacked from all sides. When the Jews defeated the Arab armies, the country of Palestine ceased to exist, and about one million Palestinian Arabs fled or were driven out to refugee camps controlled by Jordan, Syria, Egypt, and the United Nations.[6] Approximately 160,000 Palestinians remained (Elon 1993). A second large Palestinian displacement occurred after Israel defeated armies from Jordan, Syria, Iraq, and Egypt in the Six-Day War in 1967. Approximately 700,000 Palestinian Arabs [7] fled to neighboring Arab countries when Israeli troops took control of the West Bank of the Jordan River, East Jerusalem, Gaza, and Golan Heights seized from Syria (see Figure 5.1, p. 151). Arab governments responded by expelling their Jewish citizens, many of whom found

With the defeat of Arab armies in 1948, the country of Palestine ceased to exist and about 1 million Palestinians became refugees.
UPI/Bettmann

asylum in Israel. Today, approximately two million Palestinian refugees and their descendants live in Jordan, Syria, and Lebanon (Broder and Kempster 1993).

Currently, one million Palestinians, descendants of those who remained in 1948, live in Israel (Elon 1993). The West Bank, Gaza, and East Jerusalem are home to about 1.8 million Palestinians and 230,000 Israelis who have built settlements there (Charney 1988; T. Friedman 1989; Hull 1991). It is estimated that Israeli settlers and military personnel use 34 percent of the Gaza Strip and 52 percent of the West Bank land (R. Friedman 1992; Shadid and Seltzer 1988). Settlements and military installations have been constructed with two goals in mind: "One, to surround the major Palestinian towns and prevent their expansion; and two, to cut off the Arab towns from one another" (Matar 1983, p. 124).

West Bank and Gaza Palestinians have now been under Israeli military rule for almost three decades. In May 1994 Israeli soldiers withdrew from Gaza and from the town of Jericho on the West Bank, giving limited self-rule to the Palestinians. Negotiations continue on extending self-rule beyond Jericho.

The Palestinians have exerted systematic and persistent pressure aimed at ending the Israeli occupation since December 1987 with their uprising, called the *intifada*. Translated literally, the word *intifada* means "a tremor, shudder or shiver." The word is derived from the Arabic *Nafada*, which means "to shake, to shake off, shake out, dust off, to shake off one's laziness, to have reached the end of, be finished with, to rid oneself of something, to refuse to have anything to do with . . . someone" (T. Friedman 1989, p. 375). The *intifada* is being waged most visibly by the "children of the stones"—young Palestinians who throw rocks and bottles at Israeli settlers and soldiers.[8]

The Palestinian-Israeli conflict has an additional dimension of complexity because the two communities are intertwined economically.

On any given day one could find the Israeli army arresting all Palestinian males ages eighteen and over in one West Bank village, while in the next village an Israeli contractor would be hiring all Palestinian males eighteen and over to build a new Jewish town. (T. FRIEDMAN 1989, P. 360)

The Intifadah: Organization and Phases

Demonstrations began in December 1987 in the refugee camps, which had always been a key locus of protest, and spread to the towns and villages. Although Gazan villages had grown to the point that they had merged over the years with the camps and towns, the 500 villages on the West Bank were scattered widely and had not been closely involved in the nationalist protests in the past. This time, however, West Bank villagers were keenly motivated to join the *intifadah* because of the massive seizures of agricultural and grazing land over the past decade which had threatened, disrupted or destroyed their very livelihood.

At the local level, the overall leadership is linked to myriad political committees in neighborhoods in Palestinian refugee camps, villages and towns. Other committees handle health, agricultural, educational and women's issues. In each section of a town, residents have elected a committee to coordinate their resistance efforts and handle emergencies. Residents donate small sums of money so that food can be stockpiled for emergencies. A team checks and cleans old wells and cisterns for use if Israel cuts off the water lines. A census is taken of all the residents so that each person's skills are known; each is then assigned a role to play in case of emergency. The neighborhood committee also supervises the planting of vegetable gardens and provides aid to the needy.

Communication on organizational and political issues takes place partly by word of mouth, but its main articulation comes from the *bayanat,* the official statements of the Unified Leadership.

The *bayanat* set out specific instructions for the coming week or fortnight, detailing the days on which there would be general strikes and demonstrations and the hours at which shops should open and close. They also conveyed requests for certain officials to resign or extended congratulations to particular towns or villages for their resistance efforts. For example, *bayanat* have requested that policemen and appointed mayors resign, have set business shop hours from nine o'clock to noon daily, and have reminded people to boycott Israeli goods when there are Palestinian-made substitutes. *Bayanat* have also urged workers to work inside Israel only in cases of dire necessity and have praised the efforts of Jewish peace forces inside Israel in support of the Palestinian cause.

The methods utilized in the *intifadah* have evolved over the course of time. The early period was marked by an emphasis on mass protests, with large numbers of people pouring into the streets to confront Israeli soldiers with stones, burning tires and barricades. Such large-scale demonstrations are difficult to sustain for many months, which explains why, over time, they evolved into cat-and-mouse tactics. A group of youths would set up a makeshift barricade, which would attract a military patrol. The soldiers would shoot at the youngsters, who would hurl stones and then scatter into the alleys. Alternatively, a jeep with soldiers would station itself in the middle of the vegetable market in a refugee camp, flaunting its authority while the residents hurried to shop. After a while, its presence would inflame the situation to the point that boys would start pelting it with stones. The soldiers would leap out, shooting in all directions as they chased the children, beating them with truncheons and shooting volleys of tear gas into the market area. For most of the early period, the youths' aim was to keep the military off-balance, to harass them and try to minimize their own casualties.

SOURCE: Adapted from "Anatomy of an Uprising: The Palestinian *Intifadah,*" by Ann Mosely Lesch. Pp. 94–100 in *Palestinians Under Occupation.* Copyright © 1989 by the Center for Contemporary Arab Studies. Reprinted by permission.

Israelis and Palestinians have little in common, other than the economic relationship they share. (In this relationship, Palestinians are concentrated in low-status jobs.) These two groups live segregated lives, for the most part, and do not share a language, religion, schools, residence, or military service.

The conflict between Jews and Arabs has lasted more than a century and its meaning has passed from generation to generation.[9] Only the form of the conflict has changed. Before 1948, both sides engaged in terrorist activities as well as land and property destruction. From 1948 through the early 1970s (and again in the early 1980s), the conflict

FIGURE 5.1   Palestinian Refugee Camps in the West Bank, Gaza, and the Surrounding Nations of Jordan, Syria, and Lebanon

SOURCES: U.S. Bureau for Refugee Programs (1988) and "Where the Israeli Settlements Are" (1993a). Copyright © 1993 by The New York Times Company. Reprinted by permission.

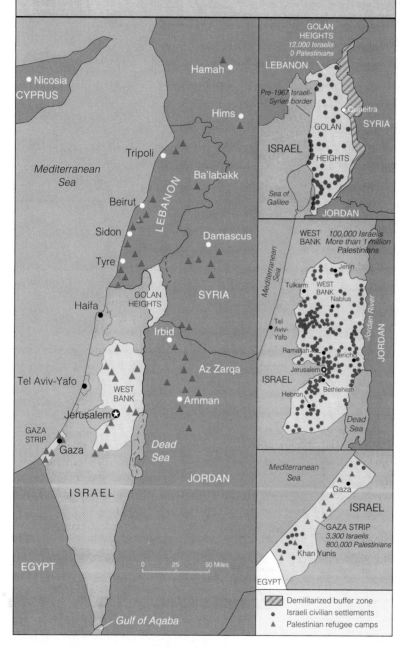

A historic handshake: The peace accord signed in Washington, D.C. by lifelong enemies Yitzhak Rabin and Yasir Arafat marked the first step toward possible settlement of the century-long conflict between Jewish Israelis and Palestinians.
Cynthia Johnson/Time Magazine

involved international armies, guerrilla fighters, and terrorist infiltrators. For almost two decades the Palestinians lived under military occupation, waiting impatiently for a resolution to the "Palestinian question." Now a new generation fights the occupation with knives, Molotov cocktails, flags, rocks, graffiti, strikes, boycotts, and barricades. The Israelis respond with gunfire, deportations, imprisonments, curfews, and school closures.

This long sequence of historical events has brought us to the present, where the "children of the stones," who have known life only under Israeli military occupation, throw stones at Israeli settlers, visitors, and young soldiers born long after Israel became an independent state. The *intifada,* in conjunction with the end to the Cold War, pushed Arabs and Israelis to participate in peace talks jointly sponsored by Moscow and Washington. This cooperation between Moscow and Washington would not have been possible before the Cold War ended. Before the Cold War ended, foreign policy in the Middle East was dominated by U.S. and Soviet efforts to contain the spread of the other's economic and political systems.

On September 13, 1993, the Prime Minister of Israel, Yitzhak Rabin, and the Chairman of the Palestinian Liberation Organization, Yasir Arafat, after several years of negotiations (including 18 months of secret negotiations held in and mediated by leaders in Norway) met on the U.S. White House lawn to sign a peace accord. For the first time, each side acknowledged the other's right to exist and took the first steps toward a permanent settlement of their conflict. However, there are still many unresolved issues. To complicate matters, there are various factions on each side that are attempting to influence the course of the settlement and the outcome of subsequent talks to match their own vision of the future (Schiff and Ya'ari 1991). Factions on both sides plan to initiate violent confrontations between Palestinians and Israelis with the hope that such actions will put an end to a possible peace settlement and future talks.

In studying this conflict sociologists ask these questions: (1) how do members of a new generation learn about and come to terms with the environment they have inherited, and (2) how is conflict passed down from one generation to another? For sociologists, part of the answer lies with socialization, a process that involves nature and nurture.

Nature and Nurture

No discussion of socialization can ignore the importance of nature and nurture to physical, intellectual, social, and personality development. **Nature** is the term for human genetic makeup or biological inheritance. **Nurture** refers to the environment or the interaction experiences that make up every individual's life. Some scientists debate the relative importance of genes and environment, arguing that one is substantially more important than the other to all phases of human development. But most consider such a debate futile; it is impossible to separate the influence of the two factors or to say that one is more important. Both are essential to socialization. Trying to distinguish the separate contributions of nature and nurture is analogous to examining a tape player and a cassette tape separately to determine what is recorded on the tape rather than studying how the two work together to produce the sound (Ornstein and Thompson 1984).

The development of the human brain illustrates rather dramatically the inseparable qualities of genes and environment. By our human genetic makeup, we possess a cerebral cortex—the thinking part of the brain—which allows us to organize, remember, communicate, understand, and create.[10] The cortex is made up of at least 100 billion neurons or nerve cells (Montgomery 1989); the fibers of the cell form thousands of synapses, which make connections with other cells. The number of interconnections is nearly infinite; it would take 32 million years, counting one synapse per second, to establish the number (Hellerstein 1988, Montgomery 1989).

Perhaps the most outstanding feature of the human brain is its flexibility. Scientists believe that the brain may be "set up" to learn any of the more than 5,000 known human languages. In the first months of life, babies are able to babble the sounds needed to speak all of these languages, but this enormous potential is reduced by the language (or languages) that the baby hears and eventually learns. Evidence suggests that the brain's language flexibility begins to diminish when the child reaches one year of age (Ornstein and Thompson 1984; Restak 1988). The larger implication is that genetic makeup provides essential raw materials but that these materials can be shaped by the environment in many different ways.

The human genetic makeup is flexible enough to enable a person to learn the values, beliefs, norms, behavior, and language of any culture. A multitude of experiences must combine with genetic makeup, however, to create a Palestinian who desires a homeland, believes that he or she should not "pay" for the Holocaust, and protests actively against the occupation. Likewise, nature and nurture combine to create an Israeli who values a homeland, believes the Jews have a legitimate right to land they were forced to leave 2,000 years ago, and fights to preserve the boundaries of his or her country. These ideas, however, are learned in interaction with others. If there is no contact with others, a person cannot ever become a normally functioning human being, let alone learn to become part of society.

The Importance of Social Contact

Cases of children raised in extreme isolation or in restrictive and unstimulating environments show the importance of social contact (nurture) to normal development. Some of the earliest and most systematic work in this area was done by sociologist Kingsley Davis, psychiatrists Anna Freud (with Sophie Dann) and Rene Spitz, and sociologist Peter Townsend. Their work shows how neglect and lack of socialization influence emotional, mental, and even physical development.

Cases of Extreme Isolation

In two classic articles, "Extreme Isolation of a Child" and "Final Note on a Case of Extreme Isolation," sociologist Kingsley Davis (1940, 1947) documented and compared the separate yet similar lives of two girls, Anna and Isabelle. Each girl had received a minimum of human care during the first six years of her life. Both were illegitimate children and for that reason were rejected and forced into seclusion. When authorities discovered them, they were living in dark, attic-like rooms, shut off from the rest of the family and from daily activities. Although both girls were six years old when authorities intervened, they exhibited behavior comparable to that of six-month-old children. Anna "had no glimmering of speech, absolutely no ability to walk, no sense of gesture, not the least capacity to feed herself even when the food was put in front of her, and no comprehension of cleanliness. She was so apathetic that it was hard to tell whether or not she could hear" (Davis 1947, p. 434).

Like Anna, Isabelle had not developed speech; she communicated with gestures and croaks. Because of a lack of sunshine and a poor diet, she had developed rickets: "Her legs in particular were affected; they 'were so bowed that as she stood erect the soles of her shoes came nearly flat together, and she got about with a skittering gait'" (Davis 1947, p. 436). Isabelle also exhibited extreme fear and hostility toward strangers.

Anna was placed in a private home for retarded children until she died four years later. At the time of her death, she behaved and thought at the level of a two-year-old child. Isabelle, on the other hand, was placed in an intensive and systematic program designed to help her master speech, reading, and other important skills. After two years in the program, Isabelle had achieved a level of thought and behavior normal for someone her age.

On the basis of Anna and Isabelle's case histories, Davis concluded that extreme isolation has a profound and negative effect on mental and physical development. On the other hand, Davis concluded, Isabelle's case demonstrates that extreme "isolation up to the age of six, with failure to acquire any form of speech and hence failure to grasp nearly the whole world of cultural meaning, does not preclude the subsequent acquisition of these" (Davis 1947, p. 437).

In addition, Davis speculated on the question of why Isabelle did so much better than Anna. He offered two possible explanations. First, Anna may

have inherited a physical and mental constitution that was less hardy than Isabelle's. Second, Anna's condition may have been a result of not having received the intensive and systematic therapy that Isabelle received. The case history comparisons are inconclusive, though. Because Anna died at age 10, researchers will never know whether she eventually could have achieved a state of normal development if she had lived longer.

In drawing these conclusions, Davis overlooked one important factor. Although Isabelle had spent her early childhood years in a dark room, shut off from the rest of her mother's family, she spent most of this time with her deaf-mute mother. Davis and the medical staff who treated Isabelle seemed to equate deafness and muteness with feeble-mindedness (not an unusual association in the 1940s). In addition, they seemed to assume that being in a room with a deaf-mute is equivalent to a state of isolation. A considerable body of evidence, however, now suggests that those who are deaf and mute have rich symbolic capacities (see Sacks 1989). The fact that Isabelle was able to communicate through gestures and croaks suggests that she had established an important and meaningful bond with another human being. Although the bond was less than ideal, it gave her an advantage over Anna. In view of this possibility, we must question Davis's conclusion that children may be able to overcome the effects of extreme isolation during the first six years of life, provided they have a "good enough" constitution and systematic training. We use the phrase "good enough" because researchers do not know the exact profile of a person or training regime that would allow someone to overcome such effects.

Children of the Holocaust

Anna Freud and Sophie Dann (1958) studied six German-Jewish children whose parents had been killed in the gas chambers of Nazi Germany. The children were shunted from one foster home to another for a year before they were sent to the ward for motherless children at the concentration camp at Tereszin. The ward was staffed by malnourished and overworked nurses, themselves concentration camp inmates. After the war the six children were housed in three different institutionlike environments. Eventually, they were sent to a country cottage where they received intensive social and emotional care.

During their short lives, these children had been deprived of stable emotional ties and relationships with caring adults. Freud and Dann found that the children were ignorant of the meaning of family and grew excessively upset when they were separated from one another, even for a few seconds. In addition,

> they showed no pleasure in the arrangements which had been made for them and behaved in a wild, restless, and uncontrollably noisy manner. During the first days after arrival they destroyed all the toys and damaged much of the furniture. Toward the staff they behaved either with cold indifference or with active hostility, making no exception for the young assistant Maureen who had accompanied them from Windermere and was their only link with the immediate past. At times they ignored the adults so completely that they would not look up when one of them entered the room. They would turn to an adult when in some immediate need, but treat the same person as nonexistent once more when the need was fulfilled. In anger, they would hit the adults, bite or spit. (FREUD AND DANN 1958, P. 130)

Orphanages and Nursing Homes

Other evidence of the importance of social contact comes from less extreme cases of neglect. Rene Spitz (1951) studied 91 infants who were raised by their parents during their first three to four months, but who later were placed in orphanages because of unfortunate circumstances. At the time of their entry into the institution, the infants were physically and emotionally normal. At the orphanages, they received adequate care with regard to bodily needs—good food, clothing, diaper changes, clean nurseries—but little personal attention. Because there was only one nurse for every 8 to 12 children, the children were starved emotionally. The emotional starvation caused by the lack of social contact resulted in such rapid physical and developmental deterioration that a significant number of the children died. Others became completely passive, lying on their backs in their cots. Many were unable to stand, walk, or talk (Spitz 1951).

Such cases teach us that children need close contact and stimulation from others in order to develop normally. Adequate stimulation means the existence of a strong tie with a caring adult. The tie must be characterized by a bond of mutual expectation between caregiver and baby. In other words, there must be at least one person who knows the baby well enough to understand his or her needs and feelings and who will act to satisfy those needs and desires. Under such conditions the child learns that certain actions on his or her part elicit predictable responses: getting excited may cause Dad to be equally excited; crying may get Mother to soothe the child. When researchers set up experimental situations in which a parent fails to respond to his or her infant in expected ways (even for a few moments), the baby suffers considerable tension and distress ("Nova" 1986).

Meaningful social contact and stimulation from others are important at any age. Strong social ties with caring people are linked to overall social, psychological, and physical well-being. British sociologist Peter Townsend (1962) describes the effects of minimal interaction that can characterize life for the elderly in nursing homes. The consequences for the institutionalized elderly are strikingly similar to those described by Spitz in his studies of institutionalized children:

> In the institution people live communally with a minimum of privacy, and yet their relationships with each other are slender. Many subsist in a kind of defensive shell of isolation. Their mobility is restricted, and they have little access to general society. Their social experiences are limited, and the staff leads a rather separate existence from them. They are subtly oriented toward a system in which they submit to orderly routine and lack creative occupation, and cannot exercise much self-determination. They are deprived of intimate family relationships. . . . The result for the individual seems to be a gradual process of depersonalization. He may become resigned and depressed and may display

no interest in the future or in things not imme-diately personal. He sometimes becomes apa-thetic, talks little, and lacks initiative. His per-sonal and toilet habits may deteriorate.
(TOWNSEND 1962, PP. 146–47)

The works of Kingsley Davis, Anna Freud, Rene Spitz, and Peter Townsend support the idea that a person's overall well-being depends on meaningful interaction experiences with others. On a more fun-damental level social interaction is essential to a de-veloping sense of self. Sociologists, psychologists, and biologists agree that "it is impossible to con-ceive of a self arising outside of social experience" (Mead 1934, p. 135). Yet, if the biological mecha-nisms involved in remembering or learning and re-calling names, faces, words, and the meaning of sig-nificant symbols were not present, people could not interact with one another in meaningful ways.

You have to begin to lose your memory, if only in bits and pieces, to realize that memory is what makes our lives. Life without memory is no life at all. . . . Our memory is our coherence, our reason, our feeling, even our action. With-out it we are nothing. (BUNUEL 1985, P. 22)

Individual and Collective Memory

Memory, the capacity to retain and recall past ex-periences, is easily overlooked as we explore social-ization. Yet, without memory, individuals and even whole societies would be cut off from the past. On the individual level, memory is what allows people to retain their experiences. On the societal level, memory preserves the cultural past.

How memory works is still largely a mystery. Apparently some physical trace remains in the brain after new learning takes place. The latest neu-rological evidence suggests that the physical trace is stored in an anatomical entity called an **engram.** Engrams, or *memory traces,* as they are sometimes called, are formed by chemicals produced in the brain. They store in physical form the recollections of experiences—a mass of information, impres-sions, and images unique to each person:

It may have been a time of listening to music, a time of looking in at the door of a dance hall, a time of imagining the action of robbers from a comic strip, a time of waking from a vivid dream, a time of laughing conversation with friends, a time of listening to a little son to make sure he was safe, a time of watching illu-minated signs, a time of lying in the delivery room at childbirth, a time of being frightened by a menacing man, a time of watching people enter the room with snow on their clothes.
(PENFIELD AND PEROT 1963, P. 687)

Scientists do not believe that engrams store actual records of past events, like films stored on videocas-settes. More likely, engrams store edited or consoli-dated versions of experiences and events, which are edited further each time they are recalled.

As we noted earlier, memory has more than an individual quality; it is strongly social. First, no one can participate in society without an ability to re-member and recall such things as names, faces, places, words, symbols, and norms. Second, most "newcomers" easily learn the language, norms, val-ues, and beliefs of their surrounding culture. We take it for granted that people have this informa-tion stored in memory. Third, people born at ap-proximately the same time and place are likely to have lived through many of the same events. These experiences, each uniquely personal and yet similar to one another, remain in memory long after the event has passed. We will use the phrase **collective memory** to describe the experiences shared and re-called by significant numbers of people (Coser 1992; Halbwachs 1980). Such memories are re-vived, preserved, shared, passed on, and recast in many forms, such as stories, holidays, and museums.[11]

Israel is alive with memorials and reminders of the past. For example, the *Jerusalem Post* runs ret-rospective columns with headlines such as "No School for Jewish Children in Poland." And the roadsides in Israel are littered with remnants of

Palestinian jeeps, cars, and trucks from the 1948 War of Independence (Bourne 1990).

Palestinians displaced by the 1948 and 1967 wars retain memories of their former homeland, and, like the Jews, they pass them down to those who lack personal memories. Some name their children after the cities and towns in which they had lived before the 1948 war (Al-Batrawi and Rabbani 1991). Most tell their children about the places where they used to live; they teach them to call those places home, and even to respond with the name of that land if they are asked where they come from. When author David Grossman (1988) asked a group of Palestinian children in a West Bank refugee camp to tell him their birthplace, each replied with the name of a former Arab town:

> *Everyone I spoke to in the camp is trained—almost from birth—to live this double life: they sit here, very much here . . . but they are also there. . . . I ask a five-year-old boy where he is from, and he immediately answers, "Jaffa," which is today part of Tel Aviv.*
>
> *"Have you ever seen Jaffa?"*
>
> *"No, but my grandfather saw it." His father, apparently, was born here, but his grandfather came from Jaffa.*
>
> *"And is it beautiful, Jaffa?"*
>
> *"Yes. It has orchards and vineyards and the sea."*
>
> *And farther down, . . . I meet a young girl sitting on a cement wall, reading an illustrated magazine. . . . She is from Lod, not far from Ben-Gurion International Airport, forty years ago an Arab town. She is sixteen. She tells me, giggling, of the beauty of Lod. Of its houses, which were big as palaces. "And in every room a hand-painted carpet. And the land was wonderful, and the sky was always blue. . . .*
>
> *And the tomatoes there were red and big, and everything came to us from the earth, and the earth gave us and gave us more."*
>
> *"Have you visited there, Lod?"*
>
> *"Of course not."*
>
> *"Aren't you curious to see it now?"*
>
> *"Only when we return."* (GROSSMAN 1988, PP. 6–7)

The Jews who went to Palestine to escape persecution and who fought to establish the state of Israel hold memories that are both parallel to and different from Palestinian memories. Although members of both groups participated in many of the same historical events, their memories differ because they witnessed these events from different viewpoints. Because Israel has participated in six wars with neighboring countries, has occupied the West Bank and Gaza for more than 27 years, and is a refuge for persecuted Jews from more than 80 countries around the world, virtually everyone in Israel has memories of war and persecution. The memories may concern personal involvement in war, waiting for a loved one to return, or fleeing places where they were deemed unfit to exist.

The point is that socialization is not possible without memory. Memory is the mechanism by which group expectations become internalized into the self and, by extension, into the whole society, and by which the past remains an integral, living part of the present. Both Israelis and Palestinians, when asked about the state of affairs, pull from memory things said and done in past times and in other places. The picture that any one individual holds cannot be a complete record of the past. Yet, memories include the perceived reasons why things are the way they are. The Israelis built a strong defense force and rule the Occupied Territories with an iron hand because they never want their existence to be threatened again. The Palestinians, on the other hand, resist Israeli control and are determined to establish a homeland to replace the one they remember losing (see "Memory and Identity Across the Generations").

Memory of past experiences allows individuals to participate in society and shapes their viewpoint. In his essay "The Problem of Generations," sociologist Karl Mannheim (1952) maintained that first impressions or early childhood experiences are fundamental to a person's view of the world. In fact, Mannheim believed that an event has more biographical significance if it is experienced early in life than if it is experienced later in life. Many of our most decisive early experiences take place in groups, some of which leave a powerful and lasting impression.

These charred Hebrew scriptures are preserved today as a powerful reminder of Nazi Germany's attempt to destroy Jewish culture in the Holocaust. Just as sacred books represent one way in which culture is transmitted across generations, so too do traumatic events help to shape the collective memory of a society and its people.

Holocaust Memorial Council

The Role of Groups

In the most general sense a **group** is two or more persons who share a distinct identity (the biological children of a specific couple; members of a gymnastics team, military unit, club, or organization; or persons sharing a common cultural tradition), feel a sense of belonging, and interact with each other in direct and/or indirect, but broadly predictable, ways. Interaction is broadly predictable because norms govern the behavior expected of members depending on their position (mother, coach, teammate, brother, sister, employee) within the group. Groups vary according to a whole host of characteristics including size, degree of intimacy among members, member characteristics, purpose, duration, and the extent to which the members socialize newcomers and each other. However, sociologists identify primary groups and ingroups and outgroups as particularly powerful socialization agents.

Primary Groups

Primary groups, such as the family or a high school sports team, are characterized by face-to-face contact and strong ties among members. Primary groups are not always united by harmony and love; they can be united by hatred for another group. But in either case the ties are emotional. A primary group is made up of members who strive to achieve "some desired place in the thoughts of [the] others" and who feel allegiance to the others (Cooley 1909, p. 24). A person may never achieve the desired place but may still be preoccupied with that goal. In this sense primary groups are "fundamental in forming the social nature and ideals of the individual" (Cooley 1909, p. 23). The family is an important primary group because it gives the individual his or her deepest and earliest experiences with relationships and because it gives newcomers their first exposure to the "rule of life." In addition, the family can serve to buffer its members against the effects of negative circumstances, or it can exacerbate these effects.

Sociologists Amith Ben-David and Yoav Lavee (1992) interviewed 64 Israelis to learn how members of their families behaved toward one another during the SCUD missile attacks, launched by Iraq

Primary groups like the family provide meaningful interactions with others that are vital to our physical, mental, and social well-being. In the process, primary groups have an immense influence on our socialization and on how we see the world.
Esaias Baitel/Gamma-Liaison

during the Persian Gulf War. During these attacks, families gathered in sealed rooms and put on gas masks. The researchers found that families varied in their response to this life-threatening situation. Some respondents reported that interaction was positive and supportive: "We laughed and we took pictures of each other with the gas masks on" or "We talked about different things, about the war, we told jokes, we heard the announcements on the radio" (p. 39).

Other respondents reported that interaction was minimal but that a feeling of togetherness prevailed: "I was quiet, immersed in my thoughts. We were all around the radio . . . nobody talked much. We all sat there and we were trying to listen to what was happening outside" (p. 40).

Finally, some respondents reported that interaction among family members was tense: "We fought with the kids about putting on their masks, and also between us about whether the kids should put on their masks. There was much shouting and noise" (p. 39). The point is that even under extremely stressful circumstances such as war, the family can respond in ways that increase or decrease that stress.

We do not know the extent to which Israeli and Palestinian populations are composed of people who suffered the loss of important primary group figures in their lives. The historical experiences, especially those of Israelis, suggest that the proportion may be very high. For example, approximately one-third of the 8,000 Ethiopian Jews airlifted to Israel in 1984 came from one-parent families. In many cases, the other parent had died from war-related causes or from starvation (*Encyclopedia Judaica Yearbook* 1987).

It seems that children whose primary group remains intact emerge in relatively good psychological condition despite widespread turmoil, violence, and destruction around them (Freud and Burlingham 1943). This finding gains indirect support from what we know about groups that turn to violence to achieve their ends. Such groups tend to draw recruits most often from populations that have suffered extreme humiliation and brutality at the hands of uniformed representatives of authority and justice (Fields 1979). For example, a high proportion of Palestinians have had their homes blown up, their villages destroyed, close family members or friends imprisoned and beaten by Israeli soldiers

Memory and Identity Across the Generations

Sociologists Donald E. Miller and Lorna Touryan Miller interviewed more than 100 survivors of the Armenian genocide, which wiped out at least 1.5 million Armenians in Ottoman Turkey in the early 1900s. They also interviewed the grandsons of four survivors.

From 1915–23, approximately 1.5 million[*] Armenians died in Ottoman Turkey in what has been called the first genocide of the twentieth century. As a proportion of the total population of Armenians, the events of 1915 claimed the lives of about one-half the population living in Turkey and approximately one-third of the world-wide population of Armenians. The male population was killed principally through direct massacre. The women and children suffered a more excruciating death as they were deported from their historic towns and villages to the deserts of Syria. During the deportation marches, women were raped,

---

[*]The figure 1.5 million is subject to debate. Published estimates range from 600,000 to 1.5 million.

children were abducted, and the few goods they carried were plundered. Even more sadistic, however, was death through attrition—entire caravans died from a combination of starvation, dehydration, exhaustion, and disease. Deportation was a rubric for extermination. (MILLER AND MILLER 1991, P. 13)

Based on these interviews, Miller and Miller offered some thoughts about how collective memory is transmitted from one generation to another and how it affects a person's identity with regard to ancestry. They present six main points:

First, traumatic events such as genocides—but one might also include natural disasters such as floods, earthquakes, and famines, or lost wars and uncontrolled epidemics—potentially serve as the axial point for group and generational self-understanding. Conversations that link generations radiate around these events. These traumatic events become the template through which generations relate to each other and through which group self-understanding evolves. These

events are the object of interpretation, reinterpretation, dispute, rejection, embrace, and/or denial. But whatever else they may be, they are not ignored. They define the parameters of communal conversations, thus providing the components from which collective identity is built, even if the ordering and interpretation of these events are highly idiosyncratic.

Secondly, grandparents are primary carriers and transmitters of collective group memories. Grandparents symbolize the past to grandchildren, and in this regard their task is to embody the heritage, the collective memory of the family or group. For all of their protests, grandchildren hunger for someplace to stand, for roots that define from whence they have come. In this regard, the relationship between grandchildren and grandparents is potentially a spiritual one, if one means by this term that grandparents possess the stories which may serve as the loom for weaving a personal identity and meaning system. Identity is the story we tell others—as well as ourselves—about who we are.

---

or parents or siblings killed. Furthermore, research shows that both the Palestinians still living in refugee camps and the most recent Jewish émigrés to Israel are least likely to believe that the conflict can be solved by negotiation. We might speculate that living in refugee camps or the experience of being forced to migrate to Israel socialized some of the members in these groups to believe that armed struggle represents the most viable solution (Shadid and Seltzer 1988; Yishai 1985).

One very clear example of a primary group is a military unit. A unit's success in battle depends on strong ties among its members. Soldiers in the primary group, it seems, fight for one another, rather than for victory per se, in the stress of battle (Dyer 1985). Military units train their recruits always to think of the group before the self. In fact, the paramount goal of military training is to make the individual feel inseparable from his or her unit. Some common strategies to achieve this goal include

And these stories are inevitably rooted, at some level, in the soil of experience offered to grandchildren by grandparents.

Third, when our stories of origin include moral contradictions, we may reject them because they paralyze and threaten us, or we may seek to correct our past in an effort to achieve personal wholeness and healing. What may appear to be fanatical or frenzied behavior, may in fact be based in battles with demons whose origins are rooted in the injustices (and failures) which surround the stories of our predecessors. The privileged position of youth offers grandchildren a unique perspective for interpreting the successes, failures and injustices of parents and grandparents. Indeed, it is this reinterpretation of events by each generation that provides the dynamic of cultural evolution and changing group self-definition.

Fourth, parents establish the context for mediation between grandparents and grandchildren. Parents control the frequency of exposure of grandchildren to grandparents. Parents communicate in direct and indirect ways whether the stories told by grandparents are to be valued. Specifically, the affective response of parents to the stories of grandparents establish the parameters for the respect that grandchildren feel toward grandparents, and therefore the attitude they have toward the generational inheritance which grandparents offer.

Fifth, parents and grandparents together bear primary responsibility for creating and maintaining the institutions that preserve cultural and group values. The schools they build, the clubs they support, the organizations they found—and the money and volunteer hours that go into sustaining these programs—these are the institutions that preserve cultural memory and socialize each new generation of children and grandchildren. While these institutions do not replace the important interaction that occurs among children, parents, and grandparents, they reinforce values learned in the home as well as potentially universalize these values by locating them in a broader frame of reference.

Sixth, there is nothing mechanical about the transmission of memory from one generation to another. Within a given family, different children appropriate the cultural inheritance that is available to them in potentially quite different ways. While cultural appropriation always occurs within a particular historical context which may predispose a child to be more or less open to the heritage represented by parents and grandparents, nevertheless it appears that children exercise considerable control in what they accept and what they reject. The struggle for identity is potentially very individual and idiosyncratic, which is another way of saying that individuals exercise considerable freedom in constructing their personal sense of meaning. (PP. 35–37)

SOURCE: From "Memory and Identity Across the Generations: A Case Study of Armenian Survivors and Their Progeny," by Donald E. Miller and Lorna T. Miller. Pp. 13, 35–37 in *Qualitative Sociology.* Copyright © 1991 Human Science Press, Inc. Reprinted by permission.

ordering recruits to wear uniforms, to shave their heads, to march in unison, to sleep and eat together, to live in isolation from the larger society, and to perform tasks that require the successful participation of all unit members: if one member fails, the entire unit fails. Another key strategy is to focus the unit's attention on fighting together against a common enemy. An external enemy gives a group singular direction and thus increases its internal cohesiveness.

Almost every Israeli can claim membership in this type of primary group because virtually every Israeli citizen, male and female, serves in the military for three years and two years, respectively. Men must serve on active duty for at least one month every year until they are 64 years old. Military training is similarly an important experience for many Palestinians, although it is less formal. Palestinian youths, especially those living in Syrian, Lebanese, Egyptian, and Jordanian refugee camps (which are outside of Israeli control), join youth clubs and train to protect the camps from attack. The focus on a common enemy helps to establish and maintain the boundaries of the military unit. All types of primary groups, however, have boundaries—a sense of who is in the group and who is outside of the group.

Ingroups and Outgroups

Sociologists use the term ingroup to describe those groups with which people identify and feel closely attached, particularly when that attachment is founded on hatred for or opposition toward another group. Ingroups exert their influence on our social identity in conjunction with outgroups. An outgroup is a group of individuals toward which members of an ingroup feel separateness, opposition, or even hatred. Outgroups, which exist in relation to an ingroup, also can make us conscious of where we belong. Obviously, one person's ingroup is another person's outgroup. The very existence of an outgroup heightens loyalty among ingroup members and magnifies characteristics that distinguish the ingroup from the outgroup. An outgroup can unify an ingroup even when the ingroup members are extremely different from one another. For example, one could argue from a functionalist viewpoint that Israel benefits from the Palestinian presence in Israel and the Occupied Territories. The presence of non-Israelis provides a thread of unification among Israelis, who are themselves culturally, linguistically, religiously, and politically diverse. One reason for Israel's diverse population is that, since 1948, Jews from 80 different countries have settled there (see Figure 5.2). To ease communication problems caused by diversity, Israeli law requires that everyone learn Hebrew. In addition to a common language, the unifying threads are the desire for a homeland free of persecution and the ongoing conflict with an outgroup—Palestinians and Arabs in surrounding states.

Similarly, the presence of Israelis acts to unite an equally diverse Palestinian society. Palestinians come from different ethnic and religious groups, clans, and political orientations. They may be Muslim, Christian, or Druze.

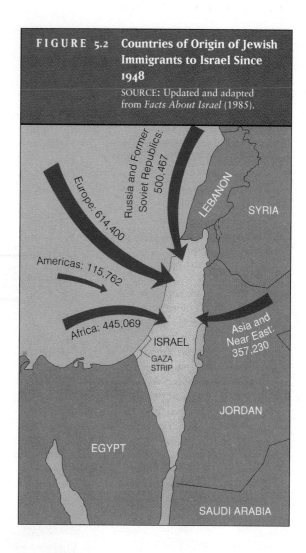

FIGURE 5.2 Countries of Origin of Jewish Immigrants to Israel Since 1948

SOURCE: Updated and adapted from *Facts About Israel* (1985).

Loyalty to an ingroup and opposition to an outgroup are accompanied by an us-versus-them consciousness. In Israel, the West Bank, and Gaza, this consciousness is reinforced by the fact that the work lives and personal lives of most Palestinians and Israelis are interrupted time and again by wars and military obligations. Israeli men must leave their jobs and families at least once a year to serve in the army. (Since the *intifada*, the annual length of service has increased from 30 to 60 days.) Palestinians frequently sacrifice work and family when they protest the occupation with strikes, work stoppages, and business closures. Israeli officials also

Often members of ingroups and outgroups clash over symbols. Before the 1993 peace accord, it was illegal in Israel to display the Palestinian flag. The flag thus became the focus of many conflicts between Israelis and Palestinians.
Stephane Compoint/Sygma

disrupt Palestinian work and family life when they arrest dissenters or when they close Palestinian businesses as punishment for protesting the occupation. Each group holds the other responsible for the state of affairs, and each seeks to control the other.

Boundaries between the two groups are sharp; they are reinforced by residential, occupational, educational, or religious segregation.[12] The 19 Israeli settlements in the Gaza Strip are protected by Israeli soldiers as well as 10-foot barbed-wire fences and land mines (R. Friedman 1992). Because there is little interaction between ingroup and outgroup members, they know little about one another. This lack of firsthand experience deepens and reinforces misrepresentations, mistrust, and misunderstandings between members of the two groups. Members of one group tend to view members of the other in the most stereotypical of terms. Often one of the groups has superior status, material conditions, and facilities. In this case, Israelis have the superior economic and political status. Palestinians, many of whom have graduated from technical colleges or universities, can obtain only manual labor jobs and low-status service jobs.

Dr. Yorum Bilu at Hebrew University of Jerusalem designed and conducted a particularly creative research study to examine the consequences of ingroup-outgroup relations on Israel's West Bank. Dr. Bilu and two of his students asked youths ages 11 to 13 from Palestinian refugee camps and Israeli settlements on the West Bank to keep a journal of their dreams over a specified period. Seventeen percent of Israeli children wrote that they dreamed about encounters with Arabs; 30 percent of the Palestinian children dreamed about meeting Jews:

> *Among 328 dreams of meetings (Jews and Arabs) there is not one character identified by name. There is not a single figure defined by a personal, individual appearance. All the descriptions, without exception, are completely stereotyped; the characters defined only by their ethnic identification (Jew, Arab, Zionist, etc.) or by value-laden terms with negative connotations (the terrorists, the oppressors, etc.). . . .*
>
> *The majority of the interactions in the dreams indicate a hard and threatening reality, a fragile world with no defense. . . .*
>
> *An Arab child dreams: "The Zionist Army surrounds our house and breaks in. My big brother is taken to prison and is tortured there. The soldiers continue to search the house. They throw everything around, but do not find the person they want [the dreamer himself]. They leave the house, but return, helped by a treacherous neighbor. This time they find me, and my relatives, after we have all hidden in the closet in fright."*

*A Jewish child dreams: ". . . suddenly some-
one grabs me, and I see that it is happening in
my house, but my family went away, and Arab
children are walking through our rooms, and
their father holds me, he has a kaffiyeh and his
face is cruel, and I am not surprised that it is
happening, that these Arabs now live in my
house."* (GROSSMAN 1988, PP. 30, 32–33)

Often an ingroup and an outgroup clash over
symbols—objects or gestures that are clearly associated with and valued by one group. These objects
can be defined by members of the other group as so
threatening that they seek to eliminate them: destroying the objects becomes a way of destroying
the group. Danny Rubinstein, an Israeli reporter
covering the West Bank, observed that most of the
clashes between Palestinian youths and Israeli soldiers are over symbols:

*Of the hundreds of clashes I have witnessed pit-
ting Palestinian youths against Israeli military
and administrative authorities in the West Bank
and Gaza, most have involved symbols. Thus,
for example, an ongoing battle is being waged
over the Palestinian flag. Arabs hoist the flag
(which is very much like the Jordanian flag),
while Israeli soldiers bring it down and attempt
to catch and punish the perpetrators. At times
the situation takes a ridiculous turn. Some time
ago, in Bethlehem, I heard an Israeli officer
issue an order to close down for a week all
shops on a certain street where a Palestinian
flag had been hoisted on the corner utility pole
the night before. I saw schoolgirls in Hebron*

*knitting satchels modeled on the Palestinian flag
and clothing stores with window displays
arranged to fit its color and pattern.*

*The flag is but one example. Military cen-
sors crack down on anti-Israeli expression in
Arabic language newspapers, textbooks and
plays. Every day Israeli soldiers remove pro-
Palestinian graffiti from walls in Arab towns
and villages. The warfare over symbols is accompanied by demonstrations of young people,
rock throwing, protest marches and the burning
of automobile tires—routine incidents that
reached a new height in early December 1987.*
(RUBINSTEIN 1988, P. 24)

To this point we have examined how socialization is a product of nature and nurture. We have
discussed how genetic makeup provides each individual with potentials that are developed to the extent made possible by the environment. We also
have considered the importance of stimulation
from caregivers in developing our genetic potential
and the connection between group membership
and self-awareness. Even groups to which we do
not belong can have powerful influences on our
sense of self. For example, an outgroup makes us
clearly aware of who "they" are, and this in turn
reminds us of who "we" are. Next we will examine
the theories of two symbolic interactionists, George
Herbert Mead and Charles Horton Cooley, regarding some specific ways in which the self develops and in which information is transmitted to
newcomers.

Symbolic Interactionism and Self-Development

Humans are not born with a sense of self; it evolves
through regular interaction with others. The emergence of a sense of self depends on our physiological capacity for **reflexive thinking**—stepping outside the self and observing and evaluating it from
another's viewpoint. Reflexive thinking allows a
person to learn what he or she is like in others'
eyes and to adjust and direct behavior in ways that
meet others' expectations. In essence, self-aware-

ness emerges hand in hand with awareness of others and of their evaluations of one's behavior and
appearance.

The Emergence of Self-Awareness

According to George Herbert Mead, significant
symbols and gestures are the mechanisms that
allow an individual to interact with others and, in

Symbolic interactionists emphasize that learning to communicate in symbols, especially language, plays a key role in the development of the self.

James Stanfield/National Geographic Society

the process, to learn about the self. (For a discussion of symbolic interaction theory, see Chapter 2.) A **significant symbol** is a word, gesture, or other learned sign that is used "to convey a meaning from one person to another, and that has the same meaning for the person transmitting it as for the person receiving it" (Theodorson and Theodorson 1979, p. 430). Language is a particularly important significant symbol because "it is only through language that we enter fully into our human estate and culture, communicate freely with our fellows, acquire and share information. If we cannot do this, we will be bizarrely disabled and cut off" (Sacks 1989, p. 8).

Symbolic gestures or signs are nonverbal cues, such as tone of voice, inflection, facial expression, posture, and other body movements or positions that convey meaning from one person to another. For example, Palestinians use the V-sign to symbolize solidarity: "We are here; we endure; we exist; we will not give up" (Morrow 1988, p. 36).

As people learn significant symbols—language and symbolic gestures—they also acquire the ability to do reflexive thinking and to adjust the presentation of self to meet other people's expectations. George Herbert Mead believed, however, that humans do not adhere mechanically to others' expectations. Instead, a dialogue goes on continuously between two phases of the self—the *I* and the *me*.

The *me* is Mead's term for the self as the internalized expectations of others. Before an individual acts, the *me* takes others into account by assessing the appropriateness of the act and anticipating the responses. The *I* is the spontaneous, autonomous, creative self, capable of rejecting expectations and acting in unconventional, inappropriate, or unexpected ways. For example, a student recently expressed disappointment at receiving a failing grade on an examination. Upon seeing her score she blurted out, "A 50! I skipped two classes to study for this stupid test!" Before making this comment, she failed to anticipate its effect on the professor. Presumably, the unexpected grade came as such a shock that her spontaneous *I* overwhelmed her calculating *me*.

Although Mead does not specify how the *I* emerges, we know that a spontaneous, creative self must exist; otherwise human life would never change and would stagnate. Mead was more specific about how the *me* develops; it develops through imitation, play, and games, all three of which give the developing child practice with role-taking.

Role-Taking

Mead assumed that the self is a product of interaction experiences. He maintained that children acquire a sense of self when they become objects to themselves. That is, they are able to imagine the effect of their words and actions on other people. According to Mead, a person can see him- or herself

① Roletakingstage

as an object after learning to role-take. **Role-taking** involves stepping outside the self and viewing its appearance and behavior imaginatively from an outsider's perspective.

Researchers have devised an ingenious method for determining when a child is developmentally capable of role-taking. A researcher puts a spot of rouge on the child's nose and then places the child in front of a mirror.[13] If the child shows no concern with the rouge, it is assumed that he or she has not yet acquired a set of standards about how he or she ought to look; that is, the child cannot role-take or see him- or herself from another person's viewpoint. If, on the other hand, the child shows concern over the rouge, it is assumed that he or she has formed some notion of self-appearance and therefore can role-take (Kagan 1989).

Mead hypothesized that children learn to take the role of others (1) through imitation, (2) through play, and (3) through games. Each of these methods involves a progressively sophisticated level of role-taking. Mead developed the stages of his theory along these lines. The three stages are as follows.

② *The Preparatory Stage* In this stage children have not yet developed the mental capabilities that allow them to role-take. Although they mimic or imitate people in their environment, they have almost no understanding of the behaviors they are imitating. Children may imitate spontaneously (by mimicking a parent writing, cooking, reading the newspaper, and so on) or they may repeat things that adults encourage them to say and reward them for saying. In the process of imitating, children learn to function symbolically; that is, they learn that particular actions and words arouse predictable responses from others. For example, Israeli children may be taught to respond "I will defend my homeland, Israel," to the question "What will you do when you grow up?" Similarly, Palestinian parents teach their children where they come from, even before the children learn notions of geography and understand the historical circumstances of their living arrangements. A typical exchange between a Palestinian parent and child would go something like this:

PARENT: Where are you from?

CHILD: I am from Palestine—from the city of Hebron.

PARENT: What is Israel?

CHILD: The real name for Israel is Palestine.

Both Palestinian and Israeli children, like children in nearly every culture, learn to sing patriotic songs and say prayers before they can understand the meaning of the words. Jenny Bourne, a political activist and a member of a delegation that visited the Occupied Territories, was struck by the fact that as soon as some two-year-old Palestinian children "saw the cameras come out, they were up and alert, hands outstretched as taut fingers made in unison the victory sign for our photos. No [Palestinian] child we met anywhere wanted to be photographed without that sign" (Bourne 1990, p. 70).

 The Play Stage Mead saw children's play as the mechanism by which they practice role-taking. **Play** is a voluntary and often spontaneous activity with few or no formal rules, which is not subject to constraints of time (for example, 20-minute halves, 15-minute quarters) or place (for example, a gymnasium, a regulation-size field). Children, in particular, play whenever and wherever the urge strikes. If there are rules, they are not imposed upon participants by higher authorities (for example, rule books, officials). Participants undertake play for their amusement, entertainment, or relaxation. These characteristics make play less socially complicated than organized games (Corsaro 1985; Figler and Whitaker 1991).

In the play stage, children pretend to be **significant others**: people or characters who are important in their lives—important in that they have considerable influence on a child's self-evaluation and encourage the child to behave in a particular manner. Children recognize behavior patterns characteristic of these significant persons and incorporate them into their play. When a little girl plays with a doll and pretends she is the doll's mother, she talks and acts toward the doll as her mother does toward her. By pretending to be the mother, she gains a sense of the mother's expectations and perspective and learns to see herself as an object. Similarly, two children playing doctor and patient are learning to see the world from viewpoints other than their own and to understand how a patient acts in relation to a doctor (and vice versa).

Children's role-taking can come only from what they see and hear. For the most part, Palestinian children in the West Bank and Gaza have never seen an adult male Israeli without a gun. Palestinian children's play reflects their experiences: the children pretend to be Israeli soldiers arresting and beating other Palestinian children who are pretending to be stone throwers. They use sticks and cola cans as if they were guns and tear-gas canisters (Usher 1991). One evening ABC news featured a segment on Palestinian children engaged in this type of play. When asked by the reporter which they preferred to be, soldiers or stone throwers, the children replied, "Soldiers, because they have more power and can kill."

Israeli children have had little experience with Palestinians except as manual laborers or "terrorists." Thus, it is hardly surprising that some Israeli kindergartners pretend that Israelis are Smurfs (good guys) and Palestinians portray Gargamel (a bad guy in a TV program). Israeli children, like Palestinian children, pretend to be soldiers because both men and women must serve in the Israeli military beginning at age 18. Israeli children pick up their fathers' guns and declare, "I'm going to kill the Arabs with this" or "Arabs are bad and must be killed." Through this type of play, children learn how they think the "enemy" views them and how they should view the enemy. Similarly, American children were conditioned during the Cold War to regard characters with foreign accents, like Boris and Natasha from "Rocky and Bullwinkle," as evil.

 The Game Stage In Mead's theory, the play stage is followed by the game stage. **Games** are structured and organized activities that almost always involve more than one person. They are characterized by a number of constraints, including one or more of the following: established roles and rules, an outcome toward which all activity is directed, and an agreed-upon starting time and place. Through games, children learn (1) to follow established rules, (2) to take simultaneously the role of all participants involved in the game, and (3) to see how their position fits in relation to all other positions.

When children first take part in organized sports, their efforts seem chaotic. Instead of making an organized response to a ball hit to the infield, for example, everyone tries to retrieve the ball, leaving nobody to catch it when a throw is needed to stop an advancing runner. This chaos exists because children have not developed to the point where they can see how their role fits with the roles of everyone else in the game. Without such knowledge a game cannot have order. Through playing games, children learn to organize their behavior around the **generalized other,** that is, around a system of expected behaviors, meanings, and points of view that transcend those of the people participating. "The attitude of the generalized other is the attitude of the whole community. Thus, for example, in the case of such a social group as a baseball team, the team is the generalized other insofar as it enters—as an organized process or activity—into the experience of [those participating]" (Mead 1934, p. 119). In other words, when children play organized sports, they practice fitting their behavior into an established behavior system.

In view of this information, not surprisingly, games are the tools used in programs designed to break down barriers between Palestinian and Jewish children and adolescents. The games involve activities like

> *throwing an orange into the air, calling a person's name to catch it, throwing it again with another's name, and again and again as the whoops of laughter fill the room. Then they all crowd together, take each other's hands, and turn around until they are enmeshed in a tangle of arms. Intertwined with each other, they try to unravel themselves without letting go. They talk to each other, giving advice, crouching so another can step over an arm, stooping so others can swing arms over heads, spinning around, trying to turn the snarled mess of Arab and Jewish bodies into a clean circle.* (SHIPLER 1986, P. 537)

Although these games seem merely fun, sociologists contend that participants are learning to see things from another perspective and to play their parts successfully in a shared activity. The participants cannot be effective unless they understand their own roles in relation to everyone else's. Although the children trying to untangle themselves may not be fully aware of it, they are learning that a Palestinian (or an Israeli) can be in positions like their own. If anyone can become untangled, participants must be able to understand everyone else's situation.

As we have learned, George Herbert Mead assumed that the self develops through interaction with others. Mead identified the interaction that occurs in play and games as important to children's self-development. When children participate in play and games, they practice at seeing the world from the point of view of others and they gain a sense of how others expect them to behave. Sociologist Charles Horton Cooley offers a more general theory about how the self develops.

The Looking-Glass Self

Like Mead, Charles Horton Cooley assumed that the self is a product of interaction experiences. Cooley coined the phrase **looking-glass self** to describe the way in which a sense of self develops: people act as mirrors for one another. We see ourselves reflected in others' reactions to our appearance and behaviors. We acquire a sense of self by being sensitive to the appraisals of ourselves that we perceive others to have: "Each to each a looking glass, / Reflects the other that [does] pass" (Cooley 1961, p. 824). As we interact, we visualize how we appear to others, we imagine a judgment of that appearance, and we develop a feeling somewhere between pride and shame: "The thing that moves us to pride or shame is not the mere mechanical reflection of ourselves but . . . the imagined effect of this reflection upon another's mind" (Cooley 1961, p. 824).

Cooley goes so far as to argue that "the solid facts of social life are the facts of the imagination." According to this logic, one person's effect on another is defined most accurately as what one person

imagines the other will do and say on a particular occasion (Faris 1964). Because Cooley defines the imagining or interpreting of others' reactions as critical to self-awareness, he believes that people are affected deeply even when the image they see reflected is exaggerated or distorted. One responds to the perceived reaction rather than to the actual reaction.

On the other hand, we cannot overlook the fact that more often than not our imaginations of how other people will react and behave rests on past experiences with others. In the case of Palestinians and Israelis, for example, each group aims a number of powerful images at the other (see "The Complexity of the Israeli-Palestinian Conflict"). For example, Palestinians call the Israelis Nazis and equate the occupation of their country and the accompanying imprisonments, beatings, and identification checks with the concentration camps. These labels are quite painful to Israelis, who see little similarity between concentration camps and the Palestinian situation. Israelis, on the other hand, react by defining the Palestinians as culturally primitive and as incapable of managing their own affairs. They tell the Palestinians that the Israelis are responsible for turning the worthless desert land occupied previously by a backward Palestinian people into a modern, high-technology state.

Both Mead and Cooley's theories suggest that self-awareness derives from an ability to think—to step outside oneself and view the self from another's perspective. Although Cooley and Mead describe the mechanisms (imitation, play, games, and other people) by which people learn about themselves, neither theorist addresses how a person acquires this level of cognitive sophistication. To answer this question we must turn to the work of Swiss psychologist Jean Piaget.

Cognitive Development

Piaget is the author of many influential and provocative books about how children think, reason, and learn. The titles of some of his many books—*The Language and Thought of the Child* (1923),

The Child's Conception of the World (1929), *The Moral Judgement of the Child* (1932), *The Child's Conception of Time* (1946), and *On the Development of Memory and Identity* (1967)—give some clues about the many categories of childhood thinking that Piaget investigated.

Piaget's influence reaches across many disciplines: biology, education, sociology, psychiatry, psychology, and philosophy. His ideas about how children develop increasingly sophisticated levels of reasoning stem from his study of water snails (*Limnaea stagnalis*), which spend their early life in stagnant waters. When transferred to tidal water, these lazy snails engage in motor activity that develops the size and shape of the shell to help them remain on the rocks and avoid being swept away (Satterly 1987).

Building upon this observation, Piaget arrived at the concept of **active adaptation,** a biologically based tendency to adjust to and resolve environmental challenges. The theme of active adaptation runs through almost all of Piaget's writings. He believed that learning and reasoning are rooted in active adaptation. He defined logical thought, another biologically based human attribute, as an important tool for meeting and resolving environmental challenges. Logical thought emerges according to a gradually unfolding genetic timetable. This unfolding must be accompanied by direct experiences with persons and objects; otherwise, a child will not realize his or her potential ability. On the basis of cumulative experiences, a child constructs and reconstructs his or her conceptions of the world.

Piaget's model of cognitive development includes four broad stages, each characterized by a progressively more sophisticated reasoning level. A child cannot proceed from one stage to another until the reasoning challenges of earlier stages are mastered. Piaget maintained that reasoning abilities cannot be hurried; a more sophisticated level of understanding will not show itself until the brain is ready.

- *Sensorimotor stage (from birth to about age 2).* In this stage children explore the world with their senses (taste, touch, sight, hearing, and smell). The cognitive accomplishments of this stage include an understanding of the self as separate from other persons and the notion of object permanence, the realization that objects and persons exist even when they are out of sight. Before this notion takes hold, very young children act as if an object does not exist when they can no longer see it.

- *Preoperational stage (from about ages 2 to about 7).* Piaget focused most of his attention on this stage. Children in this stage typically demonstrate three characteristic types of thinking. They think anthropomorphically; that is, they assign human feelings to inanimate objects. They believe that objects such as the sun, the moon, nails, marbles, trees, and clouds have motives, feelings, and intentions (for example, dark clouds are angry; a nail that sinks to the bottom of a glass filled with water is tired). They think nonconservatively, a term Piaget used to signify an inability to appreciate that matter can change form but still remain the same in quantity. They think egocentrically in that they cannot conceive how the world looks from another's point of view. Thus, if a child facing a room full of people (all of whom are looking in his or her direction) is asked to draw a picture of how a person in the back of the room sees the people, the child will draw the picture as he or she sees the people. Related to egocentric thinking is centration, a tendency to center attention on one detail of an event. As a result of this tendency, the child fails to process other features of a situation (see Figure 5.3).

- *Concrete operational stage (from about ages 7 to 12).* By the time children enter this stage, they have mastered these preoperational tasks but have difficulty in thinking hypothetically or abstractly without reference to a concrete event or image. For example, a child in this stage has difficulty in understanding a life without him or herself in it. One 12-year-old struggling to grasp this idea said to me, "I am the beginning and the end; the world begins with me and ends with me."

- *Formal operational stage (from the onset of adolescence onward).* At this point people are able to think abstractly. For example, they can conceptualize their existence as a part of a much larger historical continuum and a larger context.

As far as we know, this progression by stages toward increasingly sophisticated levels of reasoning

FIGURE 5.3   Centration in a Child's Drawing

As they draw, young children can focus on only one detail at a time. For example, they draw the passengers and the automobile separately. When they think about the passengers or a car, they fail to consider them as a unit.

SOURCE: From "Children's Drawings of Human Figures," by Norman H. Freeman. P. 138 in *The Oxford Companion to the Mind*. Copyright © 1987 Oxford University Press. Reprinted by permission.

Piaget's theory suggests that the emergence of social concern in many young people may be explained in part by the development of their ability to think abstractly, and consequently to identify with issues and causes.
Bruce de Lis/Picture Group

is universal, but the *content* of people's thinking varies across cultures. As an example, all Palestinian and Israeli children learn the following rule: "If you ever see an unattended package or bag on a street or bus, don't touch it. Notify an adult immediately." Knowing the rule is a matter of safety because the package might contain a bomb. American children typically are not exposed to the same dangers and hence have no need to consider situations in which this rule can be relaxed.

The New York Times reporter David Shipler, in his book *Arab and Jew,* describes his frustrations in explaining to his young children that they did not have to follow this rule when they returned to the United States. When he set some bags of newspapers on the curb to be picked up, his children reported suspicious packages outside:

> *Michael [age 7] ran in another day to report a plastic cup of some sort in the street. I had seen it and asked him to throw it away. He adamantly refused to go near it, and he remained solidly unmoved by my extravagant as-*

surances that we didn't have to worry about bombs on a quiet, tree-lined suburban street in America. (SHIPLER 1986, P. 83)

From the perspective of Piaget's theory, Michael centered all of his attention on one detail (the rule) and could not respond to other aspects of the situation that made the rule irrelevant, such as geographic location.

The theories of Mead, Cooley, and Piaget all suggest that the process of social development is multifaceted and continues over time. It is important to realize that socialization takes place throughout the life cycle and does not cease at a particular age. One reason that socialization is a lifelong process is that people make any number of transitions over a lifetime: from single to married, from married to divorced or widowed, from childless to parent, from healthy to disabled, from one career to another, from civilian status to military status, and from employed to retired. In making such transitions, people undergo resocialization.

Resocialization

Resocialization is the process of being socialized over again. In particular it is a process of discarding values and behaviors unsuited to new circumstances and replacing them with new, more appro-

priate values and norms (standards of appearance and behavior). A considerable amount of resocialization happens naturally over a lifetime and involves no formal training; people simply learn as

The Complexity of the Israeli-Palestinian Conflict

This written exchange between two University of Cincinnati students shows the complexity of the Israeli-Palestinian conflict. The exchange began after the editorial "Most in U.S. Don't Grasp Complex Israeli Situation" appeared in the student newspaper *The News Record* (Jacobs 1989).

U.S. Should Realize Truth About Israel

To the Editor:

I want to thank Dan Jacobs for his Jan. 30 article. He stated the truth about the complicated situation in my little country, Israel.

The issue of security for Israel (which is the size of Massachusetts) is a matter of survival. This is the reason for mandatory service in the IDF (Israeli defense forces) for high school graduates in Israel, both males and females. Females serve for a period of 24 months and males for 36 months.

Since the day our country was given to us by the United Nations we have been surrounded by enemies. We have had six major wars and a long, bitter history of terrorism attacks.

Among our enemies was—and still is—Mr. Arafat, who established the PLO terror organization to destroy the Israeli state physically and morally, trying to drive the Israelis out by attacking and killing civilians.

The only Palestinian leader that I respect—at least for his honesty—is Abu-Nidal, who claims he wants the whole state of Israel and not only the West Bank, which is the hidden agenda that people fail to see in Arafat's plan.

My American friends, I ask you not to forget the American citizens who were terror victims (TWA in Greece, The Italian Cruise boat, 241 marines and lately, Pan-Am). The American government had a good reason to count on and trust Israel. Israel is one of the few countries in the world that welcomes and loves American people.

Please, before you pick a side, listen to both Israelis and Palestinians. Go to the literature, inquire and learn the historical truth.

Mira Jacob, Senior
Social Work

Israel a Military Power Violates Human Rights

To the Editor:

In the Feb. 6 *News Record*, Mira Jacob responded to Dan Jacobs' article, which she found accurate in its portrayal of her "little country."

Her "little country" is the fourth military power in the world; it owns and operates nuclear plants and nuclear weapons (the only such country in the Middle East!). Her "little country" is also the seventh-largest arms producer in the world and has fought five wars which generally resulted in the occupation of neighboring countries.

May I point out that, according to the State Department report on Human Rights released two days ago, this "little country" killed 360 civilian Palestinians, 70% of whom were under the age of 17? The "little country" deported 48 Palestinians, demolished 150 homes, made 200 families homeless, injured 7,000 women and children (disabling 1,500 for life), put 25,000 Palestinians in jail, and placed 3,500 under administrative detention (up to six months imprisonment possible without legal procedure).

This "little country" confiscated 52% of the land in the West Bank and 42% in the Gaza Strip. The "little country" turned the West Bank and Gaza Strip into experimental labs, using beatings, tear gas, rubber

---

they go. For example, people marry, change jobs, become parents, change religions, and retire without formal preparation or training. However, some resocialization requires that, to occupy new positions, people must undergo formal and systematic training and demonstrate that they have internalized appropriate knowledge, suitable values, and correct codes of conduct.

Such systematic resocialization can be voluntary or imposed (Rose, Glazer, and Glazer 1979). It is voluntary when people choose to participate in a process or program designed to "remake" them. Examples of voluntary resocialization are wide-ranging—the unemployed youth who enlists in the army to acquire a technical skill, the college graduate who pursues medical education, the drug addict

and plastic bullets, metal-plastic bullets, Iron Fist policy, bone breaking, the burials of live Palestinians and collective punishments.

Palestinians in that "little country" cannot even plant a tree without permission from the military government! How much sympathy does the world owe this "little country"?

Nasser Atta, Graduate Student
Department of Sociology

Israel Modernized Country Deserving Trust, Sympathy

To the Editor:
This is to respond to Nasser Atta's letter on Monday, Feb. 13.*

It seems Atta is asking Israel to justify or defend its achievements.

Yes—Israel is 40 years old, and during this short era, in spite of wars, it has been turned from a desert into a beautiful, modern westernized country.

If our neighbors wouldn't invade us and be so hostile toward us—we probably wouldn't have the fourth best military in the world. Unfortunately, people like myself, my family members and friends of mine had to serve in the Israeli Defense Force, to protect our dream and our children from becoming victims of a second holocaust. (If I may recall—one of

the goals of the P.L.O.)

The Israelis have turned the west bank into a kind of "experimental lab" introducing the Palestinian people to a much better quality of life.

Stop and think: Why don't 17-year-olds work on their graduation diplomas, and children on their studies like they did prior to the uprising? Instead they are engaging in delinquent, violent behavior against the 18 to 21 year-old Israeli soldiers that were never taught to hate.

As far as permission to "plant a tree"—maybe you should learn more about zoning laws applied in the United States, Israel and every civilized country.

As far as the sympathy issue to Israel, it comes from trust. Israel earned trust by being a faithful ally to the United States.

If Yassar Arafat's terrorist gang, or other Arab terror organizations wouldn't engage in terror activity all over the world, maybe Arafat wouldn't have such a hard time trying to convince the United States and the rest of the world to trust him.

As an Israeli, I will be very hard to convince that those Palestinians (who can be watched on TV everyday) throwing stones and fire bombs are accepting my existence.

There was no and will never be any justification for terror—so please stop twisting the truth. Concentrate on proving the P.L.O. innocent instead of trying to prove my people's guilt!

Mira Jacob, Senior
Social Work

---

*Nasser Atta did not respond to Mira Jacob's critique of his letter. He maintained that his position remains unchanged and that to respond would be the start of a never-ending series of letters.

SOURCE: *The News Record,* University of Cincinnati. Reprinted by permission.

---

who seeks treatment, the alcoholic who joins Alcoholics Anonymous. Resocialization is imposed when people are forced to undergo a program designed to rehabilitate them or to correct some supposed deficiency in their earlier socialization. Military boot camp (when a draft exists), prisons, mental institutions, and schools (when the law forces citizens to attend school for a specified length of time) are ex-

amples of environments that are designed to resocialize but that people also enter involuntarily.

In *Asylums: Essays on the Social Situation of Mental Patients and Other Inmates,* sociologist Erving Goffman writes about a setting—total institutions (with particular focus on mental institutions)—where people undergo systematic socialization. **Total institutions** are settings in which people surrender control of their lives, voluntarily or involuntarily, to an administrative staff and in which they (as inmates) carry out daily activities (eating, sleeping, recreation) in the "immediate company of a large batch of others, all of whom are [theoretically] treated alike and required to do the same thing together" (Goffman 1961, p. 6). Total institutions include homes for the blind, the elderly, the orphaned, and the indigent; mental hospitals; jails; penitentiaries; prisoner-of-war camps; concentration camps; army barracks; boarding schools; and monasteries and convents. Their total character is symbolized by barriers to social interaction, "such as locked doors, high walls, barbed wire, cliffs, water, forests, or moors" (p. 4).

Goffman was able to identify the general and standard mechanisms, despite their wide range, that the staffs of all total institutions employ to resocialize "inmates." When the inmates arrive, the staff strips them of their possessions and their usual appearances (and the equipment and services by which their appearances are maintained). In addition, the staff sharply limits interactions with people outside the institution in order to establish a "deep initial break with past roles" (p. 14).

We very generally find staff employing what are called admission procedures, such as taking a life history, photographing, weighing, fingerprinting, assigning numbers, searching, listing personal possessions for storage, undressing, bathing, disinfecting, haircutting, issuing institutional clothing, instructing as to rules, and assigning to quarters. The new arrival allows himself to be shaped and coded into an object that can be fed into the administrative machinery of the establishment. (P. 16)

Goffman maintains that the admission procedures function to prepare inmates to leave off past roles and to take on new roles. The inmates learn their new roles by participating in the various enforced activities that staff members have designed, to fulfill the official aims of the total institution, whether the aims are to care for the incapable, to keep inmates out of the community, or to teach people new roles (for example, to be a soldier, priest, or nun).

In general, it is easier to resocialize people when they want to be resocialized than when they are forced to abandon old values and behaviors. Furthermore, resocialization is likely to be easier if acquiring new values and behaviors requires competence rather than subservience (Rose, Glazer, and Glazer 1979). A case in point is the resocialization that takes place in medical school. Theoretically, medical students learn (among other things) to be emotionally detached in their attitudes toward patients, to not prefer one type of patient over another (that is, patients of a particular ethnicity, gender, or age, or even cooperative patients over noncooperative patients), and to provide medical care whenever it is required (Merton 1976). These attitudes are necessary for proper diagnosis and treatment, as the following account makes clear.

An emergency room physician, Elisabeth Rosenthal, reported on an unkempt man who had been brought into the emergency room. He was uncooperative, his speech did not make sense, he did not know the date or the name of the president of the United States, and he could not decide whether he was in a hotel or a hospital. On the basis of his appearance and behavior, he seemed to be drunk; the best treatment seemed to be to let him sleep it off. In theory, the physician is trained not to be influenced by stereotypes and to look beyond commonplace interpretations associated with certain physical traits.

In this man's case subsequent tests revealed that he had a massive kidney infection: "His incoherence and his lack of cooperation were caused not by intoxication but by a metabolic disturbance resulting from his infection" (Rosenthal 1989, p. 82). For the physician, professional competence is demonstrated by learning to abandon or hold in check widely shared misconceptions about the meaning of behavior, especially the behavior of persons in different ethnic, gender, and age groups. Similarly, Israeli physicians who treat Palestinian patients must learn to not let the larger conflict between Israelis and Palestinians interfere with treatment plans.

Michael Gorkin (1986) believes that many problems can develop between Israeli psychiatrists and Palestinian patients during therapy. (The same could be said of Palestinian psychiatrists and Israeli patients, but there are very few Palestinian psychiatrists in Israel. Consequently, most Palestinians who go to psychiatrists go to Israeli ones.) If Israeli psychiatrists do not learn to manage their stereotypes and prejudices with regard to Palestinians and to familiarize themselves with Arab culture, they may treat the client in counterproductive ways. In addition, the patient must trust the psychiatrist if therapy is to be successful.

Establishing trust is made difficult by the fact that almost all Israeli psychiatrists serve in army reserve units and must cancel therapy hours several times a year for several weeks. Many are reluctant to tell their Palestinian clients the reason for their absence; most hope that their patients will not inquire. Palestinians, of course, are aware of this commitment. Gorkin maintains that the Israeli psychiatrist must address these strains constructively with the Palestinian client in the initial sessions. He does not believe that the discussion will resolve these differences but thinks that it "sets the stage

for openness in the therapeutic interaction and conveys the message that this crucial issue is not taboo" (Griffith 1977, p. 38). The Israeli psychiatrist is likely to come to terms with prejudices and stereotypes because he or she has chosen to become a physician and because coming to terms with these beliefs demonstrates professional competence achieved through resocialization during the medical training period.

Therein lies the dilemma in finding a resolution to the Palestinian-Israeli conflict. Both sides use what they hope are resocialization measures that attempt to force the other side to change its position about land rights. Israelis deport, imprison, impose curfews, close schools, level houses, and kill. Palestinians throw stones, strike, boycott Israeli products, and kill. Both Palestinians and Israelis seem to believe that if they make life miserable enough for the other, each will gain a homeland. The problem is that if one side wins through intimidation, the other side by definition assumes a subservient position.

Discussion

Socialization is a multifaceted, lifelong process. Among other things, it is a process through which newcomers acquire a sense of self and internalize the norms, values, beliefs, and language needed to participate in the larger community. Many of the important socialization theorists—George Herbert Mead, Charles Horton Cooley, and Erving Goffman—are considered to be symbolic interactionists. Consequently, this chapter centers around the process by which newcomers acquire the skills to become social beings, including the ability to understand the symbol system of their culture, to internalize the pattern of expected behaviors, and to interpret how they fit into it.

On the other hand, the material describing the effects of extreme social isolation and limited association with people demonstrates that socialization goes beyond the needs of individuals. It functions to link people to one another in orderly and predictable ways. Without the benefits of social interaction, newcomers fail to thrive physically and to

learn the skills they need to achieve meaningful connections with others. In addition, without meaningful contact between the generations, culture (solutions to the problems of living) cannot be passed on from one generation to the next. Without this link a culture ceases to exist.

In addition to being a process by which newcomers acquire a symbol system, solutions to problems of living, and the skills to interact in orderly and predictable ways, socialization is also a process by which newcomers learn to think and behave in ways that reflect the interests of the teachers and to accept and fit into a system that benefits some groups more than others.

Whatever its consequences, socialization consists of a gradual unfolding of one's genetic endowment and an extraordinary number of social experiences. We are just beginning to discover how experiences (nurture) and genes (nature) combine to enlarge human potential, to bestow an identity, and to shape thought and behavior. One point is clear: it is impossible to separate the influences of nature and nurture or to say that one is more important than the other for development:

> *Genetic determination is like the blueprint of a beautiful house. But the house itself is not there; you can't sleep in a blueprint. The kind of building you eventually have will depend on the choice of which bricks, which wood, which glass are used—just as the virgin brain will be shaped by what is given to it from the environment.* (DELGADO 1970, P. 170)

Without stimulation from other people, we cannot learn language, appropriate behaviors, and the skills and information we need to live with other people. Without the operation of biological mechanisms, we cannot learn or develop, no matter how great the stimulation. Yet, biology is not destiny, and people are more than the sum of their experiences. Our biological materials can be shaped in innumerable ways. For example, at birth we have the flexibility to learn any of the world's more than 5,000 languages and to learn the ways of any culture, but we learn the language or languages we hear spoken and the culture into which we are born. One might argue that, although we are flexible biologically, we have little control over how the raw materials are shaped—that we are products of our upbringing and other circumstances beyond our control. Yet, our biological makeup gives us a brain that is capable of thinking and creating and that allows us to think about our environment and to act to change it. In this vein, George Herbert Mead believed that humans do not adhere mechanically to others' expectations. He believed, rather, that there is a continuous dialogue between two phases of the self: the *I* and the *me*.

The material in this chapter may leave you with the impression that human beings are largely products of circumstances beyond their control; after all, Israelis and Palestinians have been engaged in armed and unarmed conflict for almost half a century without reaching a solution. Hindsight, however, always makes events seem more predictable than they are.

Noted biologist Stephen Jay Gould studies patterns of life's history on the earth over billions of years. One important theme in his research is that

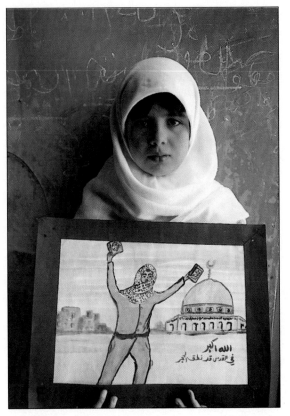

Through socialization this Palestinian girl has learned to dream of living someday in a glorious Jerusalem—a city she has never seen.
Joanna B. Pinneo/Aurora

the "complexities of history overwhelm simple predictable law-like behavior and we have to understand patterns as a complex, unrepeatable, unpredictable historical result—not a random one, not an accidental one, [because] it is explainable after it happens but if we wind the tape of life back to the beginning [and start over] it may not happen again" (Gould 1990).

What we know about the socialization and resocialization processes suggests that life does not have to unfold in a seemingly predictable fashion. Human genetic and social makeup contains considerable potential for change. First, people are not born with preconceived notions about standards of appearance and behavior. To develop standards, people must be exposed to information that leads them to expect people, behavior, and objects to be a certain way.

Second, the cerebral cortex (which makes us capable of thinking and creating) allows people to think reflexively—to step outside the self and observe and evaluate it from another viewpoint. A significant number of Israeli Jews, American Jews, Palestinian Arabs, and Palestinian-Americans have written books and articles with this purpose in mind—to understand others' viewpoints. Many of these books have been best-sellers in the United States and in Israel, which suggests there is considerable interest in reaching solutions that consider the well-being of all involved.

Third, the mechanisms that teach prejudice and hatred for another group—mechanisms such as imitation, play, and games—also teach respect and understanding. The problem is that they often are used to teach children respect and understanding after the children already have learned prejudice and hatred through the same mechanisms.

Fourth, people can be resocialized to abandon one way of thinking and behaving for another. Preferably, however, people will choose to abandon old habits and, in doing so, will gain a feeling of competence.

Finally, as noted early in this chapter, each generation learns about the environment it inherits and then comes to terms with it in unique ways. Although the older generations may share their personally acquired memories with later generations, these memories cannot affect the behavior of the younger generations in the same way. The continuous emergence of new generations "serves the necessary social purpose of enabling us to forget" or to at least rethink an approach to a situation (Mannheim 1952, p. 294). On the other hand, we cannot assume that people of one generation cannot change the way they think about and approach their environment. Israeli Prime Minister Yitzhak Rabin and P.L.O. Chairman Yasir Arafat, lifelong bitter enemies, recognized the other's right to exist. They did so in an exchange of letters before the peace accord ceremony.

Mr. Prime Minister . . .

The P.L.O. recognizes the right of the State of Israel to exist in peace and security . . . [and] renounces the use of terrorism and other acts of violence. . . .

> *Sincerely,*
> *Yasir Arafat*
> *Chairman*
> *The Palestinian Liberation Organization*

Mr. Chairman . . .

The Government of Israel has decided to recognize the P.L.O. as the representative of the Palestinian people. . . .

> *Yitzhak Rabin*
> *Prime Minister of Israel*

(*THE NEW YORK TIMES* 1993b, P. A1)

FOCUS
Peer Pressure as an Agent of Socialization

John Huss

On August 28, 1993, *The New York Times* published "Appearances Are Destructive," an editorial by the South African–born author Mark Mathabane in which he describes and comments on the peer pressure his sisters experienced in a U.S. public school to dress the "right way."

They were constantly taunted for their homely outfits. A couple of times they came home in tears. In South Africa students were required to wear uniforms, so my sisters had never been preoccupied with clothes and jewelry.

SOURCE: John Huss, Northern Kentucky University, Class of 1993.

They became so distraught that they insisted on transferring to different schools despite my reassurances that there was nothing wrong with them because of what they wore.

I have visited enough public schools around the country to know that my sisters' experiences are not unique. In schools in many areas, Nike, Calvin Klein, Adidas, Reebok and Gucci are more familiar names to students than Zora Neale Hurston, Shakespeare, and Faulkner. Many students seem to pay more attention to what's on their bodies than in their minds.

Teachers have shared their frustrations with me at being unable to teach those students willing to learn because classes are frequently disrupted by other students ogling themselves in mirrors, painting their fingernails, combing their hair, shining their gigantic shoes or comparing designer labels on jackets, caps and jewelry (P. A11). *

Mathabane believes that American students compete most fiercely with each other over who dresses the best, not over academic achievements. In light of his observations, we might ask: How is it that peer pressure exerts such influence over many students' lives? Why do many students need the "right" pair of jeans, the "right" sports cap, or the "right" pair of camp moccasins? The functionalist, conflict, and symbolic interactionist perspectives provide some insights on these questions.

Recall that the functionalist perspective is guided by the following questions: Why does a particular arrangement exist? What are the consequences of this arrangement for society? The functionalist conclusion is simply that *some* inequality is necessary for the maintenance of society as we know it. How can this be?

In general, functionalists argue that the desire for money and material possessions—such as designer clothing—motivates people to work hard and meet the standards of excellence that are required in many important jobs. Without equality or reward, the most capable people would not be mo-

tivated to train for or perform the demanding tasks that are essential to the economic system. If necessary, it seems, parents can live vicariously through their well-dressed offspring. From the functionalist standpoint, teenagers who reap these benefits will naturally want to continue dwelling in this fashion mecca and will seek high-level, high-paying occupations in order to rightfully maintain their spot in the fitting room of life.

On the other hand, teenagers with little financial backing from parents will also strive to obtain these spoils of materialism. After-school jobs have become a major force in teen life. A study by University of Michigan professor Harold Stevenson reveals that 74 percent of high school juniors in Minneapolis worked. The goal or significance of work seems to diminish for seniors, though. About 47 percent of male seniors and 36 percent of female seniors put in no more than 20 hours per week at their places of employment. In 1991, only 10 percent of high school seniors surveyed said they were saving most of their earnings for college, and just 6 percent said they used most of their earnings to help pay family living expenses. Instead, teens admitted, most earnings were funneled into a "luxury label" lifestyle.

Let us now turn our attention to the central question concerning the conflict perspective: Who benefits and who loses from this arrangement? First, conflict theorists assert that inequality exists because the middle and upper classes benefit from it. The working poor are exploited; they are paid low wages so that their employers can make fatter profits and live more affluent lives. While admittedly this process falls short of a great conspiracy, there can be little doubt that exclusive clothing and ornamentation are being used by members of a common social background as show of superiority and class rank. Instead of rejecting the ideology that promotes a "clothes-make-the-person" attitude, the "have nots" spend much of their time mimicking the successful dressers and often lose sight of other more important priorities and long-range goals.

In analyzing social inequality, conflict theorists argue that it exists *not* because it is functional for society as a whole but because some people have been able to achieve political and economic power and have managed to pass on these advantages to

---

*From "Appearances Are Destructive," by Mark Mathabane. Copyright © 1993 by The New York Times Company. Reprinted by permission.

their descendants. These people have learned that expensive clothing can highlight the advantages and separate the wheat from the chaff. Clothing brands then act as a brand of a different sort.

The interactionist perspective gives a different slant on this puzzling dilemma. To begin with, many symbolic interactionists consider major components of society—such as the economy—to be abstractions. After all, these components cannot act or exist by themselves. It is people who exist and act, and it is only through the social behavior of individuals that society can come into being at all. Society is ultimately created, maintained, and changed by the social interaction of its members. Symbolic interactionists say, Let's face it, clothes are symbols. As a society we share common ground as to what these symbols mean to us. By the same token, designer labels and logos represent something to the members of our society—especially, but not exclusively, in the teenage population. The right clothes mean you're "it." The wrong clothes mean second choice, second rate, and second class.

The functionalists seek to understand why a situation exists, what function it serves, and what consequences it presents. Peer pressure to dress the right way functions to encourage individuals to work harder to obtain the things they believe they need or deserve (in this case, clothes). Offering rewards (peer admiration or even envy) for success persuades members of society to work harder to achieve what the society values.

Conflict theorists are concerned with who benefits and who loses from an arrangement. Designer clothing helps identify those members in society who have achieved a particular level of affluence. From this viewpoint, those who can afford exclusive items lord them over the lower classes who cannot. This creates a class envy and perpetuates the "us" and "them" mentality and a false consciousness as students strive to acquire the "right" clothes.

Symbolic interactionists see clothing as a symbol that delivers a message. Clothes send a message to peers about where we fit in, where we want to fit in, or whether we fit in at all. In light of what these perspectives say about the origins of peer pressure to dress the right way, do you think requiring students to wear uniforms will significantly diminish this peer pressure and encourage students to apply pressure on others to achieve academic success?

Key Concepts

whats the role of a group

Notes

1. Two major developments occur in the second year of life, but developmental researchers from Western countries place disproportionate emphasis on one of these—the emergence of the self and self-awareness. In contrast, researchers from non-Western societies emphasize the two-year-old's ability to appreciate standards and the difference between right and wrong (Kagan 1988b, 1989).

2. Kagan (1989) reports the results of a study in which 14- and 19-month-old children were given a set of 22 toys to play with. Of the 22 toys, 10 were unflawed, 10 were flawed (for example, a boat with holes in the bottom, a doll with black streaks on its face), and 2 were meaningless wooden forms. No 14-month-old child showed concern with the flawed toys, but 60 percent of the 19-month-old children showed clear concern with the flawed toys: they brought the toys to their mothers and said things like "broke," "bad," or "yucky." According to Kagan, this "moral sense is one of the most profound accomplishments. We should view it just as we view singing and speaking and walking. It is a maturational milestone that will be acquired by every child as long as they live in a world of human beings" (Kagan 1988a, p. 62).

3. Zionism is the plan and the movement to establish a homeland in Palestine for Jews scattered across the globe. After the establishment of Israel, "Zionism was widened to include material and moral support for Israel" (Patai 1971, p. 1262).

4. One of the last messages received by the outside world from the Warsaw ghetto as it was being crushed by the Nazis was: "The world is silent. The world *knows* (it is inconceivable that it should not) and stays silent. God's vicar in the Vatican is silent; there is silence in London and Washington; the American Jews are silent. This silence is astonishing and horrifying" (Steiner 1967, p. 160).

5. In this biography of Theodor Herzl, Ernst Pawel (1989) writes that Herzl never acknowledged the presence of large Arab villages, contrary to the findings of Leo Motzkin, the man he commissioned to survey the land. Herzl formulated his plan on the basis of two misconceptions: first, that the land was unpopulated, and, second, that any people who lived there would welcome the modernization the Jews would bring.

6. The figures on the number of Palestinians displaced because of the 1948 war vary from 500,000 to 1.3 million.

7. The figures on the number of Palestinians displaced by the 1967 war vary from 168,000 to 800,000.

8. Although Palestinian youth play a critical and very confrontational role in the *intifada*, a complex structure of committees and leadership directs the uprising. The *intifada* is supported through boycotts, business closures, and strikes.

9. Since Israel declared statehood in 1948, Israelis and Palestinians have fought six major wars (1948, 1956, 1967, 1968–71, 1973, and 1982); hardly a day has passed without acts of retaliation by both sides.

10. Ornstein and Thompson (1984) provide an excellent metaphor for visualizing the cortex:

    Form your hands into fists. Each is about the size of one of the brain's hemispheres, and when both fists are joined at the heel of the hand they describe not only the approximate size and shape of the entire brain but also its symmetrical structure. Next, put on a pair of thick gloves, preferably light gray. These represent the cortex (Latin for "bark"—the newest part of the brain and the area whose functioning results in the most characteristically human creations, such as language and art. (PP. 21–22)

11. A case in point is the United States Holocaust Memorial Museum in Washington, DC. The museum was founded with the realization that surviving witnesses to the Holocaust will be dead in about 10 to 20 years. The museum will help preserve Jewish collective memory of this event by housing "object survivors"—letters, diaries, identity papers, armbands, clothes—that document life in the concentration camps and ghettos (Goldman 1989).

12. Approximately 8,000 Ethiopian Jews were airlifted to Israel through Operation Moses. Israeli officials debated the consequences of settling the people as a unit or scattering them throughout Israel. The Israelis feared that if the group were settled as a unit, the people would retain traits (language and culture) that would turn them into an outgroup (Nesvisky 1986).

13. In an interview published in *Omni* magazine, culture historian Morris Berman stated:

The most traumatic event in life is the moment [we] become aware, usually in the third year of life, of [our] specular or . . . mirror image and realize that the image is what people mean when they say [our name]. Before this overwhelming discovery, all of us feel one with the external environment. Afterward there is a tear in the fabric: I am "in here" and "that" is "out there."

The event is traumatic because you realize you can be an other for other others; you can be interpreted from the outside in a way [that is consistent] or antagonistic to the way you experience yourself. [It's the beginning of alienation.] (BERMAN 1991, P. 64)

6 SOCIAL INTERACTION AND THE SOCIAL CONSTRUCTION OF REALITY

with Emphasis on Zaire

Children canoeing to school on the Ngiri River.
Jacques Jangoux/Tony Stone Images

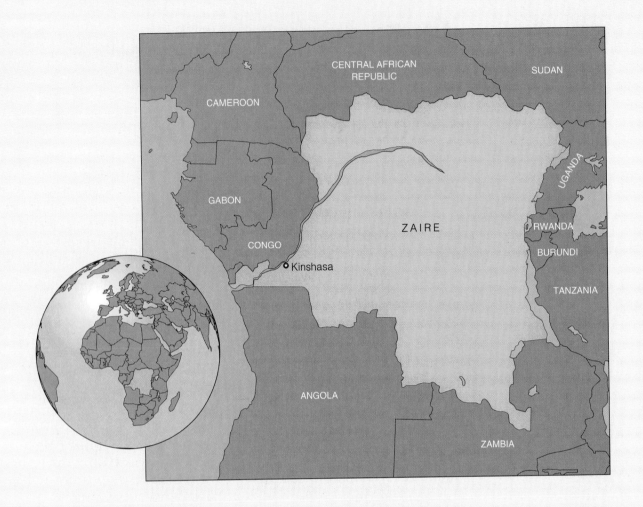

U.S. military personnel in Zaire in 1993   8

Phone calls made from
the United States to Zaire in 1991   58,000

Students from Zaire enrolled in
college in the United States for fall 1991   383

ZAÏRE 2ᴷ

MALACHITE

Margrethe Rask, a Danish surgeon, was exposed to the virus now known as human immunodeficiency virus, or HIV, while working in a small village clinic in Zaire. The excerpt that follows highlights some of the final events in Dr. Rask's life, her last interactions with colleagues and close friends.

Grethe Rask gasped her short, sparse breaths from an oxygen bottle. . . . "I'd better go home to die," Grethe had told [her friend] Ib Bygbjerg matter-of-factly. The only thing her doctors could agree on was the woman's terminal prognosis. All else was mystery. Also newly returned from Africa, Bygbjerg pondered the compounding mysteries of Grethe's health. None of it made sense. In early 1977, it appeared that she might be getting better; at least the swelling in her lymph nodes had gone down, even as she became more fatigued. But she had continued working, finally taking a brief vacation in South Africa in early July.

Suddenly, she could not breathe. Terrified, Grethe flew to Copenhagen, sustained on the flight by bottled oxygen. For months now, the top medical specialists of Denmark had tested and studied the surgeon. None, however, could fathom why the woman should, for no apparent reason, be dying. There was also the curious array of health problems that suddenly appeared. Her mouth became covered with yeast infections. Staph infections spread in her blood. Serum tests showed that something had gone awry in her immune system; her body lacked T-cells, the [essential parts of] the body's defensive line against disease. But biopsies showed she was not suffering from a lymph cancer that might explain not only the T-cell deficiency but her body's apparent inability to stave off infection. The doctors could only gravely tell her that she was suffering from progressive lung disease of unknown cause. And, yes, in answer to her blunt questions, she would die.

Finally, tired of the poking and endless testing by the Copenhagen doctors, Grethe Rask retreated to her cottage near Thisted. A local doctor fitted out her bedroom with oxygen bottles. Grethe's longtime female companion, who was a nurse in a nearby hospital, tended her. Grethe lay in the lonely whitewashed farmhouse and remembered her years in Africa while the North Sea winds piled the first winter snows across Jutland.

In Copenhagen, Ib Bygbjerg, now at the State University Hospital, fretted continually about his friend. Certainly, there must be an answer to the mysteries of her medical charts. Maybe if they ran more tests. . . . It could be some common tropical culprit they had overlooked, he argued. She would be cured, and they would all chuckle over how easily the problem had been solved when they sipped wine and ate goose on the Feast of the Hearts. Bygbjerg pleaded with the doctors, and the doctors pleaded with Grethe Rask, and reluctantly the wan surgeon returned to the old *Rigshospitalet* in Copenhagen for one last chance. On December 12, 1977, just twelve days before the Feast of the Hearts, Margrethe P. Rask died. She was forty-seven years old.

(RANDY SHILTS 1987, PP. 6–7)

We can visualize some of the interactions between Grethe Rask and her friends and colleagues. For example, we can visualize Dr. Rask telling her friend Ib Bygbjerg, "I'd better go home to die," or asking Copenhagen doctors whether she would die, after they tell her that she is "suffering from progressive lung disease of unknown cause." Finally, when Dr. Rask decides to leave the hospital and die at home, we can imagine Ib Bygbjerg pleading, "Please come back to the hospital for more tests; maybe there is still hope."

Sociologists looking at this situation would agree that Dr. Rask's illness is the obvious and immediate reason for these **social interactions**—everyday events in which at least two people communicate and respond through language and symbolic gestures to affect one another's behavior and thinking. In the process, the parties involved define, interpret, and attach meaning to the encounter (see the section in Chapter 5 on self-development). Sociologists also assume that any social interaction reflects forces beyond the obvious and immediate. Hence they strive to locate the interaction according to time (history) and place (culture). From a sociological viewpoint, history and culture limit the range of potential experience and point people "towards certain definite modes of behaviour, feeling, and thought" (Mannheim 1952, p. 291).

When sociologists study social interaction, they seek to understand and explain the forces of context and content. **Context** consists of the larger historical circumstances that bring people together. **Content** includes the cultural frameworks (norms, values, beliefs, material culture) that guide behavior, dialogue, and interpretations of events. In the case of Dr. Rask, sociologists would determine the context by asking what historical events brought Dr. Rask to Africa in the first place and what further events put her in direct contact with a deadly virus. To understand the content of her interactions, sociologists would ask how the parties involved are influenced by their cultural frameworks as they strive to define, interpret, and respond meaningfully to her condition.

In this chapter we explore the sociological theories and concepts that sociologists use to analyze any social interaction in terms of context and content. We give particular focus to social interaction as it relates to the transmission of HIV and the treatment of AIDS (see "What Is the Difference Between AIDS and HIV?"). In doing so, we look closely at the central African country of Zaire (formerly the Belgian Congo) for two important reasons. First, focusing on Zaire and its relationship to other countries helps us connect the transmission of HIV to a complex set of intercontinental, international, and *intra*societal interactions. Specifically, these interactions involve unprecedented levels of international and intercontinental air travel of the privileged for pleasure and business, as well as legal and illegal migrations of the underprivileged from villages to cities and from country to country (Sontag 1989).

Second, focusing on Zaire points to evidence that HIV existed as early as 1959. The evidence has been traced to an unidentified blood sample frozen in that year and stored in a Zairean blood bank. Although this hardly proves that the AIDS virus originated in Zaire, this hypothesis has received considerable support from government and health officials in Western nations.

Whether or not Zaire is actually the country of origin of HIV is irrelevant to the message of this chapter. Far more important is the idea that reality is a social construction. That is, people give meaning to phenomena (events, traits, and objects). The meanings almost always emphasize some aspect of a phenomenon and ignore other aspects. For exam-

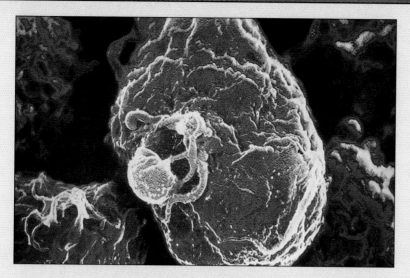

AIDS virus budding from a human lymphocyte, the white blood cells that form in lymphoid tissues.
Bill Longcore/Science Source/Photoresearchers

Acquired immunodeficiency syndrome (AIDS) is a fatal disease that severely compromises the human body's ability to fight infections and is caused by the Human Immunodeficiency Virus (HIV). HIV infection can be transmitted by sexual intercourse between men and between men and women; by exposure to contaminated blood or blood products; by sharing or reusing contaminated needles; and during pregnancy, childbirth, and possibly breastfeeding, from woman to child.

There is no evidence that HIV infection is transmitted through casual contact, water, air, or insects.

Although persons infected with HIV may not show any clinical symptoms of AIDS for months or years, they may never become free of the virus and may infect others without realizing it. An individual is considered to have AIDS if a blood test indicates the presence of antibodies to the [HIV] virus and if he or she has one or more debilitating and potentially fatal cancers, neurological disorders, or bacterial, protozoal, or fungal infections that are characteristic of the syndrome.

SOURCE: U.S. General Accounting Office (1987), p. 7.

ple, to say that HIV traveled from Zaire to the United States ignores the possibility that it traveled from the United States to Zaire. When we compare Western values and beliefs related to the origin of the virus and the treatment of AIDS with non-Western, African values and beliefs, we realize that the Western framework is only one way of viewing the AIDS phenomenon. Moreover, by comparing the two frameworks, we can see that the meanings people give to an event have enormous consequences for the individuals involved (medical personnel, infected persons and those close to them, and noninfected persons). These meanings influence how people interact with one another and what decisions they make and actions they take to deal with HIV infection and AIDS.

Continue to think about Dr. Rask as you read this chapter. As we explore issues of context and content, we will see that an individual's seemingly unique and personal interactions are affected by history and culture, just as they are affected by genetics (Tuchman 1981). We begin by exploring the context of Dr. Rask's interactions—the unprecedented mixing of the world's peoples and the large-scale social disruptions that have accompanied the emergence of worldwide economic and social interdependence. If we can understand these social forces, we can understand more about the transmission of viruses in general and the transmission of HIV in particular.

The Context of Social Interaction

Emile Durkheim was one of the first sociologists to provide insights into the social forces that contributed to the rise of a "global village." In *The Division of Labor in Society*, Durkheim gives us a general framework for understanding both global interdependence and conditions that can cause large-scale social upheaval, leaving people vulnerable to phenomena like AIDS. More specifically, Durkheim's ideas provide a framework for understanding how Zaire was transformed, in less than 200 years, from a land of isolated and independent nations to a country characterized by immense social disruptions and participation in the world economy. (In this chapter, *nation* is a geographical area occupied by people who share a culture and a history. By contrast, a country is a political entity, recognized by foreign governments, with a civilian and military bureaucracy to enforce its rules.)

Durkheim observed that an increase in population size and density intensifies the demand for resources. This in turn stimulates the development of more efficient methods for producing goods and services. As population size and density increase, society "advances steadily towards powerful machines, towards great concentrations of forces and capital, and consequently to the extreme division of labor" (Durkheim [1933] 1964, p. 39). As Durkheim described it, **division of labor** refers to work that is broken down into specialized tasks, with each task performed by a different set of persons. Not only are the tasks themselves specialized, but the parts and materials needed to manufacture products come from many geographical regions.

The West vigorously colonized much of Asia, Africa, and the Pacific in the late nineteenth and early twentieth centuries because of the growing demand for resources (see Figure 6.1). Western countries forced local populations to cultivate and harvest crops and to extract minerals and ores for export. The Belgian government claimed territory in central Africa, named it the Belgian Congo, and forced the people living there to extract rubber and mine copper. As industrialization proceeded in Europe, so did the demand for various raw materials. Over time the world grew to depend on Zaire as a source of cobalt (needed to manufacture jet en-

gines), industrial diamonds, zinc, silver, gold, manganese (needed to make steel and aluminum dry-cell batteries), and uranium (needed to generate atomic energy and to fuel the atomic bomb). The worldwide division of labor now included the indigenous people of Zaire, who mined the raw materials needed for products in distant parts of the world.

Durkheim noted that as the division of labor becomes more specialized and as the sources of materials for products become more geographically diverse, a new kind of solidarity or moral force emerges. Durkheim used the term **solidarity** to describe the ties that bind people to one another in a society. He referred to the solidarity that characterizes preindustrial society as mechanical, and the solidarity that characterizes industrial societies as organic.

Mechanical Solidarity

Mechanical solidarity is social order and cohesion based on a common conscience or uniform thinking and behavior. In this situation, everyone views the world in much the same way. A person's "first duty is to resemble everybody else, [and] not to have anything personal about one's beliefs and actions" (Durkheim [1933] 1964, p. 396). Such uniformity derives from a simple division of labor and the corresponding lack of specialization. (In other words, everyone is a jack-of-all-trades.) When everyone in a society does the same thing, they have common experiences, possess similar skills, and hold similar beliefs, attitudes, and thoughts. Therefore, a simple division of labor means that people are more alike than different. People are bound together because similarity gives rise to consensus. In societies characterized by mechanical solidarity, the ties that bind people to one another are based primarily on kinship and religion.

As one of about 200 nations in Zaire, each of which has a distinct language and belief system, the Mbuti pygmies, a hunting-and-gathering people who live in the Ituri Forest (an equatorial rain forest) of northeastern Zaire, exhibit this type of solidarity. Their culture has changed little over the past

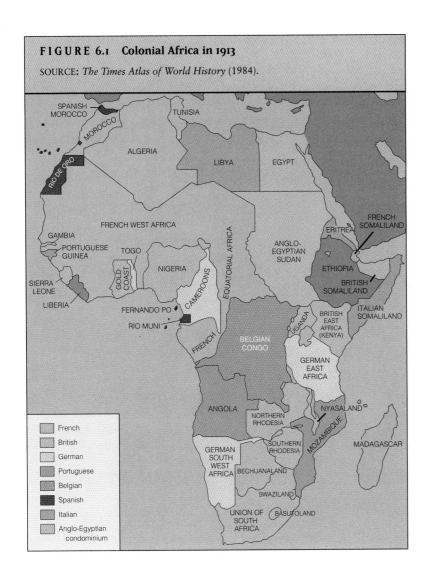

FIGURE 6.1   Colonial Africa in 1913

SOURCE: *The Times Atlas of World History* (1984).

thousand years. The Mbuti reflect how people in Zaire used to relate to one another before colonization. Their society represents the way of life that many people were forced to abandon after colonization began.

The Mbuti share a forest-oriented value system. Their common conscience derives from the fact that the forest gives them food, firewood, and materials for shelter and clothing. Anthropologist Colin Turnbull has written extensively about the Mbuti and their value system in three fine books, *The Forest People* (1961), *Wayward Servants* (1965), and *The Human Cycle* (1983). Excerpts

from these books show the extent to which Mbuti forest-centered values permeate their life:

> *For them the forest is sacred, it is the very source of their existence, of all goodness. . . . Young or old, male or female . . . the Mbuti talk, shout, whisper, and sing to the forest, addressing it as mother or father or both.* (1983, P. 30)

> *It is not surprising that the Mbuti recognize their dependence upon the forest and refer to it as "Father" or "Mother" because as they say it gives them food, warmth, shelter, and clothing*

just like their parents. What is perhaps surprising is that the Mbuti say that the forest also, like their parents, gives them affection. . . . The forest is more than mere environment to the Mbuti. It is a living, conscious thing, both natural and supernatural, something that has to be depended upon, respected, trusted, obeyed, and loved. The love demanded of the Mbuti is no romanticism, and perhaps it might be better included under "respect." It is their world, and in return for their affection and trust it supplies them with all their needs. (1965, P. 19)

Turnbull provides several examples of the intimacy between the Mbuti and the forest. In one instance, he came upon a youth dancing and singing by himself in the forest under the moonlight: "He was adorned with a forest flower in his hair and with forest leaves in his belt of vines and his loin cloth of forest bark. Alone with his inner world he danced and sang in evident ecstasy" (1983, p. 32). When questioned as to why he was dancing alone, he answered, "'I am not dancing alone, I am dancing with the forest . . .'" (1965, p. 253).

In a second instance, Turnbull asked a Mbuti pygmy if he would like to see a part of the world outside the forest. The pygmy hesitated a long time before asking how far they would go beyond the forest. Not more than a day's drive from the last of the trees, replied Turnbull, to which the Mbuti pygmy responded with disbelief, "No trees? No trees at all?" He was highly disturbed about this, and asked if it was a good country. From the Mbuti perspective, people living without trees must be very bad to deserve that punishment. In the end he agreed to go if they took enough food to last them until they returned to the forest. "He was going to have nothing to do with 'savages' who lived in a land without trees" (1961, p. 248).

Finally, Turnbull notes that the pygmies are aware of the ongoing destruction of the rain forest by companies that push them farther into the forest's interior. By consensus the pygmies do not wish to leave the forest and become part of the modern world: "The forest is our home; when we leave the forest, or when the forest dies, we shall die. We are the people of the forest" (1961, p. 260).

The Mbuti, a hunting and gathering people, exemplify Durkheim's concept of mechanical solidarity. The core of their "common conscience" is a value system centered on the forest, which provides them with all the necessities of life.

Sarah Errington/Hutchison Library

Organic Solidarity

A society with a complex division of labor is characterized by **organic solidarity**—social order based on interdependence and cooperation among people performing a wide range of diverse and specialized tasks. A complex division of labor increases differences among people, and common conscience in turn decreases. Yet, Durkheim argues that the ties that bind people to one another can be very strong nonetheless. In societies characterized by a complex division of labor, the ties that bind people to one another are no longer based on similarity and common conscience but on differences and interdependence. When the division of labor is complex and when the materials for products are geographically scattered, few individuals possess the knowledge, skills, and materials to permit self-sufficiency. Consequently, people find themselves dependent on others. Social ties are strong because people need each other to survive.

Specialization and interdependence mean that every individual contributes a small part in creating a product or delivering a service. Because of specialization, relationships between people take on a transitory, limited, impersonal, and abstract character. We relate to one another in terms of our specialized roles. We buy tires from a dealer; we interact with a sales clerk by telephone, computer, and fax; we fly from city to city in a matter of hours and are served by flight attendants; we pay a supermarket cashier for coffee; and we deal with a lab technician for only a few minutes when we give blood. We do not need to know these people personally to interact with them. Nor do we need to know that the rubber in the tires, the cobalt from which the jet engine is built, and the coffee we purchase come from Zaire. Similarly, we do not need to know whether the blood we give is kept in the United States or is exported elsewhere.

When we interact in this manner, we can ignore personal differences and can treat those who perform the same tasks as interchangeable. Yet, members of society

> *are united by ties which extend deeper and far beyond the short moments during which the exchange is made. Each of the functions that they exercise is, in a fixed way, dependent upon others. . . . [W]e are involved in a complex of obligations from which we have no right to free ourselves.* (DURKHEIM [1933] 1964, P. 227)

In other words, because everyone is dependent on everyone else, each individual has a stake in preserving the system.

A curious feature of organic solidarity is that, although people live in a state of interdependence with others, they maintain little awareness of it, possibly because of the fleeting and impersonal nature of the relationships. Because the ties with most of the people with whom we come in contact during a day are instrumental (we interact with them for a specific reason) rather than emotional, we seem to live independently of one another.

Durkheim hypothesized that societies become more vulnerable as the division of labor becomes more complex and more specialized. He was particularly concerned with the kinds of events that break down the ability of individuals to connect with one another in meaningful ways through their labor, a process we take for granted until something disrupts those connections. Such events include (1) industrial and commercial crises caused by such occurrences as plant closings, massive layoffs, crop failures, and war; (2) workers' strikes; (3) job specialization, insofar as workers are so isolated that few people grasp the workings and consequences of the overall enterprise; (4) forced division of labor to such limits that occupations are based on inherited traits (race, sex) rather than on ability, in which case the "lower" groups aspire to the positions that are closed to them and seek to dispossess those who occupy such positions; and (5) inefficient management and development of workers' talents and abilities so that work for them is nonexistent, irregular, intermittent, or subject to high turnover. For example, a country might not develop enough workers (teachers, scientists, nurses) or too many workers (athletes and entertainers) for available positions, or it might fail to retrain people whose positions are vulnerable to layoff or obsolescence. These events are particularly disruptive when they arise suddenly.

Zaire in Transition

From 1883 to the present, at least one of these five disruptive situations has existed in Zaire. A brief summary of Zaire's history in the past 100 years shows the extent to which these disruptions have created a social order that severs the connections that people have to one another.

King Leopold's Presence in Africa

Most people associate the name Mark Twain with the novels *Huckleberry Finn* and *Tom Sawyer*. They do not think of him as an avid critic of American and European imperialism who wrote essays and pamphlets expressing his outrage at the "Great Powers for the way they exercised their 'unwilling' missions in South Africa, China, and the Philippines" (Meltzer 1960, p. 256). When the Congo Reform Association approached Twain in early 1905 to "lend his voice 'for the cause of the Congo natives'" (p. 257), Twain responded by writing "King Leopold's Soliloquy." In the essay, he presents a report filed by the Reverend H. E. Scrivener, a British missionary, on the plight of the people of the Belgian Congo under King Leopold's rule. Since no magazine editor in the United States would agree to publish this essay, Congo reform groups issued it as a pamphlet and sold it for 25 cents, with proceeds going toward the relief of the Congo people. The following is excerpted from that report.

Soon we began talking, and without any encouragement on my part the natives began the tales I had become so accustomed to. They were living in peace and quietness when the white men came in from the lake with all sorts of requests to do this and that, and they thought it meant slavery. So they attempted to keep the white men out of their country but without avail. The rifles were too much for them. So they submitted and made up their minds to do the best they could under the altered circumstances.

First came the command to build houses for the soldiers, and this was done without a murmur. Then they had to feed the soldiers and all the men and women—hangers on—who accompanied them. Then they were told to bring in rubber. This was quite a new thing for them to do. There was rubber in the forest several days away from their home, but that it was worth anything was news to them. A small reward was offered and a rush was made for the rubber. "What strange white men, to give us cloth and beads for the sap of a wild vine." They rejoiced in what they thought their good fortune. But soon the reward was reduced until at last they were told to bring in the rubber for nothing. To this they tried to demur; but to their great surprise several were shot by the soldiers, and the rest were told, with many curses and blows, to go at once or more would be killed. Terrified, they began to prepare their food for the fortnight's absence from the village which the collection of rubber entailed. The soldiers discovered them sitting about. "What,

Belgian Imperialism (1883–1960)

Before 1883, inhabitants of what is now Zaire lived in villages characterized by common conscience and a simple division of labor. In 1883, however, King Leopold II of Belgium claimed the land as his private property and millions of people were forced from their villages to work the land and mine raw materials. Leopold's personal hold over the land was formally legitimized in 1885 by leaders of 14 European countries attending the Berlin West Africa Conference. The purpose of this conference was to carve Africa into colonies. The continent was divided without regard to preexisting national boundaries, so that friendly nations were split apart and hostile nations were thrown together.

For 23 years Leopold capitalized on the world's growing demand for rubber. His reign over Zaire was the "vilest scramble for loot that ever disfigured the history of human conscience and geographical location" (Conrad 1971, p. 118). The methods he used to extract rubber for his own personal gain involved atrocities so ghastly that in 1908 international outrage forced the Belgian government to assume administration of the Belgian Congo (see "King Leopold's Presence in Africa"). The Belgian government operated more humanely than Leopold; but it, too, forced the indigenous people to build roads so that metals and crops could be transported from mines and fields across the country for export. Africans were forced to leave their villages to work the mines, to cultivate

not gone yet?" Bang! bang! bang! and down fell one and another, dead, in the midst of wives and companions. There is a terrible wail and an attempt made to prepare the dead for burial, but this is not allowed. All must go at once to the forest. Without food? Yes, without food. And off the poor wretches had to go without even their tinder boxes to make fires. Many died in the forests of hunger and exposure, and still more from the rifles of the ferocious soldiers in charge of the post. In spite of all their efforts the amount fell off and more and more were killed. I was shown around the place, and the sites of former big chiefs' settlements were pointed out. A careful estimate made the population of, say, seven years ago, to be 2,000 people in and about the post, within a radius of, say, a quarter of a mile. All told, they would not muster 200 now, and there is so much sadness and gloom about them that they are fast decreasing.

We stayed there all day on Monday and had many talks with the people. On the Sunday some of the boys had told me of some bones which they had seen, so on the Monday I asked to be shown these bones. Lying about on the grass, within a few yards of the house I was occupying, were numbers of human skulls, bones, in some cases complete skeletons. I counted thirty-six skulls, and saw many sets of bones from which the skulls were missing. I called one of the men and asked the meaning of it. "When the rubber palaver began," said he, "the soldiers shot so many we grew tired of burying, and very often we were not allowed to bury; and so just dragged the bodies out into the grass and left them. There are hundreds all around if you would like to see them." But I had seen more than enough, and was sickened by the stories that came from men and women alike of the awful time they had passed through. The Bulgarian

atrocities might be considered as mildness itself when compared with what was done here. How the people submitted I don't know, and even now I wonder as I think of their patience. That some of them managed to run away is some cause for thankfulness. I stayed there two days and the one thing that impressed itself upon me was the collection of rubber. I saw long files of men come in, as at Bongo, with their little baskets under their arms; saw them paid their milk tin full of salt, and the two yards of calico flung to the headmen; saw their trembling timidity, and in fact a great deal that all went to prove the state of terrorism that exists and the virtual slavery in which the people are held.

SOURCE: From "King Leopold's Soliloquy on the Belgian Congo," in *Mark Twain and the Three R's*, by Mark Twain, pp. 47–48. Copyright © 1973 by Maxwell Geismar. Reprinted with permission of International Publishers.

and harvest the crops, and to live alongside the roads and maintain them. The Belgians introduced a cash economy, imported goods from Europe that eventually became essential to native life, established a government, and built schools, hospitals, and roads. Under this system the African people could acquire cash in one of two ways: growing cash crops or selling their labor. They were no longer allowed to be self-sufficient, as they were before European colonization.

The introduction of European goods and a cash economy pulled the people who inhabited the Belgian Congo into a worldwide division of labor and created a migrant labor system within the country. Since colonization people have moved continuously from the villages to the cities, mining camps, and

plantations (Watson 1970). In addition, family members have endured prolonged separations as a result of the migrant labor system.

The migrant labor system affected Africans' lives in many fundamental ways. . . . [M]ale workers were typically recruited from designated labour supply areas great distances from the centres of economic activity. This entailed prolonged family separations which had serious physical and psychological repercussions for all concerned. The populations of African towns 'recruited by migration' were characterized by a heavy preponderance of men living in intolerably insecure and depressing conditions and lacking the benefits of family life or other customary supports. (DOYAL 1981, P. 114)

Under the migrant labor system, men travel to cities, mining camps, and plantations to find work. In the absence of men, women grow food and run the village markets.

Margaret Gowan/Tony Stone Images

The Belgians did not anticipate the Africans' anger about the exploitation of their land, minerals, and people, and were not prepared for the violent confrontations that took place in the late 1950s. Those in power termed the revolutions "savage" and claimed that the Africans did not appreciate the "benefits" of colonialism. When the Belgians pulled out suddenly in 1960, the Belgian Congo became an independent country without trained military officers, businesspeople, teachers, doctors, or civil servants. In fact, there were only 120 medical doctors in a country of 33 million people (Fox 1988).

Independence of Zaire (1960–present)

In the vacuum left by the Belgians, the various African ethnic groups that had been forced together to form the Belgian Congo now fought each other to obtain power. Civil wars raged until 1965, when Sese Mobutu took power with the help of the United States. Since that time several power struggles have occurred between various nations within Zaire, especially in the later 1970s. To stop the rebellions, Mobutu called on mercenary forces from Morocco, Belgium, and France. To make up for the lack of skilled personnel, Mobutu invited French, Danes, Haitians, Portuguese, Greeks, Arabs, Lebanese, Pakistanis, and Indians to work as civil servants, teachers, doctors, traders, businesspeople,

and researchers. It was through this invitation that Dr. Rask arrived in Zaire.

In addition to problems posed by civil war, several other major problems have made life difficult for the people of Zaire since independence. First, many cash crops were priced out of competition in a growing world economy, and a technological revolution in synthetic products reduced the demand for African raw materials. Second, civil wars raging in neighboring countries have caused thousands of refugees from Sudan, Rwanda, Angola, Uganda, Congo, and Burundi to flee to Zaire. Meanwhile, Zaireans suffering from their own civil wars and economic problems sought refuge in those same countries (Brooke 1988a; U.S. Bureau for Refugee Programs 1988). Finally, Mobutu has diverted much of Zaire's wealth to European banks and has invested it in property outside of Africa (R. Kramer 1993; Brooke 1988b). Some people estimate that Mobutu's personal fortune is worth as much as $5 billion (R. Kramer 1993).[1]

The local populations have suffered greatly and are still suffering from the ongoing massive upheaval in Zaire. These events disrupted the division of labor and people lost an important social connection to one another. Such "change of existence, whether it be sudden or prepared, always brings forth a painful crisis" (Durkheim [1933] 1964, p. 241). Out of economic necessity, a desire for a higher standard of living, and a need to escape war,

The massive social disruption produced by civil wars in Zaire includes the destruction of entire villages, such as this one in Kivu.
David Orr

many people have left their villages for the cities and for industrial, plantation, and mining sites. Women, children, and the elderly left behind in the villages have had little choice but to change to higher-yield and less labor-intensive crops to survive. Unfortunately, the new crops, such as cassava, are low in protein and high in carbohydrates. (Low-protein diets compromise the human immune system, making people more vulnerable to infection.) To further complicate matters, significant numbers of single women with no means of supporting themselves in the villages and rural areas migrated to labor sites in search of employment. Because there are few employment opportunities for women at the labor sites, many are forced to survive through prostitution. In the meantime, when the men and women who had migrated out of the villages to find employment became sick and could no longer work, they returned home to their villages and infected an already vulnerable people with whatever diseases they carried (Hunt 1989).

The magnitude of these migrations is reflected in the population increase of Kinshasa, the capital city, which grew from 390,000 in 1950 to 3.5 million in 1988. A large portion of its population (almost 60 percent) lives in squatter slums, the largest of which is named the Cite. Here, the

streets [are] stuffed with children and families living under cardboard roofs held down by

rocks. Kinshasa [is] a wasteland of flooded streets and cracked sidewalks, smoldering garbage and bars catering to whores and lonely white men. There [is] an end-of-civilization atmosphere, with survivors finding shelter in the rubble. (CLARKE 1988, PP. 175, 178)

As a result of this upheaval and mismanagement, Zaire fell from its position as one of the wealthiest colonies in Africa to become the poorest independent country and one of the 12 poorest countries in the world. Its annual per capita income is $220. Another indicator of how dire the situation has become, in 1960 Zaire had 90,000 miles of passable highway; in 1988 the country had only 6,000 miles of dependable roads. The poor roads affect the ability to transport medical supplies, food, and fertilizer to villages that have become dependent on these commodities (Noble 1992).

These historical events are the contextual forces that brought Dr. Margrethe Rask to a small village clinic in Zaire. Amid such chaos and poverty, Dr. Rask had to perform operations on less than a shoe-string budget, with only minimal supplies. "Even a favored clinic would never have such basics as sterile rubber gloves or disposable needles. You just used needles again and again until they wore out; once gloves had worn through you risked dipping your hands in your patient's blood because that was what needed to be done" (Shilts 1987, p. 4).

Civil wars and decades of economic deterioration in Zaire have created conditions in which the sick and injured must be transported over long distances to receive needed medical care.
David Orr

The Global Context of AIDS and HIV

Somewhere in this unprecedented mixing of people from all over the world are the conditions that facilitated both the activation and the transmission of HIV (see "Global Patterns of HIV and AIDS"). The importing and exporting of blood (which brings people into contact with one another in indirect ways), the development of wide-body jet aircraft, and large-scale migrations are cited as events that increased opportunities for large numbers of people from different countries and from different regions of the same country to interact with one another and to transmit the HIV infection through unprotected sexual intercourse, needle sharing, and other activities that involve blood and blood products (De Cock and McCormick 1988).

The importance of considering these factors is supported by the fact that disease patterns historically are affected by changes in population density and transportation patterns, both of which bring together previously isolated groups (McNeill 1976). Leading AIDS researchers believe that the transmission of HIV infection is indeed linked to changes in population density and transportation. Interestingly, "the medical condition which was later to be called AIDS began to be noticed in the late 1970s and early 1980s in several widely separated locations, including Belgium, France, Haiti, the United States, Zaire, and Zambia" (The Panos Institute 1989, p. 72). Before this time, the virus may have survived in a dormant state in an isolated population with a tolerance to the virus but may have been activated when this population came into contact with another population with no tolerance. Or two harmless retroviruses, each existing in a previously isolated population, may have interacted to produce a third, lethal virus. The point is that

> *a major dislocation in the social structure—love, hate, peace, war, urbanization, overpopulation, economic depression, people having so much leisure [or having no alternative source of income] they sleep with five different people a night—whatever it is that puts a stress on the ecological system, can alter the equilibrium between [people] and microbes. Such great dislocations can lead to plagues and epidemics.*
> (KRAUSE 1993, P. XII)

In view of this information about the global context of interaction, it is difficult to say who is responsible for triggering and transmitting the virus that can cause AIDS. Clearly, "the foreigners

Global Patterns of HIV and AIDS

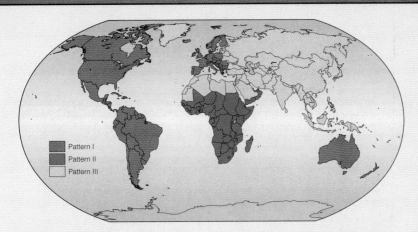

Pattern I
Pattern II
Pattern III

Pattern I countries are typically industrialized countries and include North America, many Western European countries, Australia, New Zealand, and parts of Latin America. Most cases of HIV infection occur among homosexual males and intravenous drug users. In contrast, heterosexual transmission is responsible for the majority of cases of HIV infection in the pattern II areas of Central, Eastern, and Southern Africa, and increasingly in some areas of Latin America, especially the Caribbean. HIV has only recently been introduced into the pattern III areas of Eastern Europe, North Africa, the Middle East, Asia, and the Pacific (excluding Australia and New Zealand). These countries have reported relatively few AIDS cases and of these, the majority are as a result of infected imported blood products or sexual contact with individuals from pattern I or II areas. (The designations employed and the presentation of material on this map do not imply the expression of any opinion whatsoever on the part of the World Health Organization concerning the legal status of any country, territory, city, or area or of its authorities, or concerning the delimitation of its frontiers or boundaries.)

SOURCE: "The Global Patterns and Prevalence of AIDS and HIV Infection," by James Chin and Jonathan M. Mann. P. S250 in *AIDS* 2. Copyright © 1988 by Gower Academic Journals. Reprinted by permission of the authors and *Current Science*.

introduced hitherto unknown diseases and probably aggravated some previously endemic diseases to epidemic proportions by the facilitation of transportation, forced migration of rural populations to work sites, and the creation of congested cities" (Lasker 1977, p. 280). For example, the destruction of the Ituri rain forest in Zaire by companies from around the world brings developers into contact with previously isolated populations (such as the Mbuti) and forces many people to migrate to the city because they have lost their homes. In addition to disrupting people's lives, the commercial activities in the rain forest cause climatic changes (such as the greenhouse effect) that can alter the structure of viruses. The point is not for us to determine who started the transmission (because that is impossible to ascertain), but to recognize the extent to which the world's people are interacting with one another and to become aware that the actions of one group can affect other groups (see "The Implications of Mad Cow Disease").

Placing Dr. Rask's interactions in this global context helps us see how historical events bring people into interaction with one another. The opportunity for Dr. Rask to go to Zaire and practice medicine, the circumstances that placed her in direct contact with patients' blood, and her subsequent need to seek medical treatment arose from a unique sequence of historical events.

In the next section we will focus on the concepts used by sociologists when they examine the *content* of social interaction. We will see that when people interact, they identify the social status of the people they interact with. Once they determine another person's status in relation to their own, people proceed to interact on the basis of role expectations.

The Implications of Mad Cow Disease

First the cattle started going wild and dying. A dairy farmer in southern England offered the earliest reports, in 1986, of cows exhibiting weird behavior: agitation, fearfulness, aggression, and eventually the inability to stand. Within months other farmers reported the same strange symptoms, and it became clear that some infection was ravaging the herds of Great Britain and threatening the safety of the nation's meat supply.

But what truly captured the attention of the British people—even more than the eventual loss of thousands of cattle to infection or slaughter—was the death in Bristol of a Siamese cat. This was the spring of 1990; soon two more cat deaths were confirmed. Before they died, the cats had gone crazy just the way the cows had. They became aggressive or extremely withdrawn, highly agitated or abnormally limp.

Very frightening indeed. Here was a disease with no respect for species boundaries, capable of felling even the kittens that shared people's [apartments]. Could human cases be far behind?

By the time the cats died, the London tabloids were proclaiming "mad cow disease" the next great scourge. Hysteria notwithstanding, that prophecy has turned out to have a ring of truth. At last count, more than five years after the disease was first detected in cows, it had led to the death of eighteen thousand cattle. And it had spread not only to cats but also to antelopes, an Arabian oryx, and other residents of the London Zoo. So far, no cases have been reported in peo-

ple. The disorder is thought to be caused by a strange "slow" virus, one that grows unnoticed in the nervous system for years. Ominously, the virus persists in the environment too. An American scientist who buried a pot of virus-laden soil in his backyard in Maryland recently dug it up after three years—and found that the virus was still infectious.

Mad cow disease is an example of the ways in which human actions—in this case, a single change in a manufacturing process that changed the purity of cattle feed—can create breeding grounds for new viruses. The pattern occurs again and again in the development of what are now known as "emerging viruses." An emerging virus, of which the most notorious is the one that causes AIDS, is a virus that crosses species or geographic boundaries and shows up with unprecedented virulence in unexpected places.

Viral emergence was until recently thought to be a rare event. And, until recently, it was thought to be accounted for by mutations—bizarre, totally unpredictable changes in a virus's genetic arrangement. But as scientists began to look systematically at how and why viruses emerge, they were surprised to find how rare it was that mutations were responsible. Much more important in terms of the emergence of new viruses—which turned out to be far more common than was once suspected—was human behavior.

Knowing this casts a new light on some of the most important social issues of our time. Every deci-

sion made anywhere in any sphere of life—environmental, political, demographic, economic, military—carries with it implications about disease that reverberate halfway around the world. When the Aswân High Dam was built in Egypt, the new body of still water allowed mosquitoes to thrive, and the viruses they carried became a new threat. When used tires were shipped from Japan to Texas, the mosquitoes that hitched a ride in the wet rims eventually found their way to a new continent; because the mosquitoes can carry viruses never before seen by Texans, their presence was also a new threat to public health. And when the city borders of Seoul were pushed farther into the countryside, urban Koreans were exposed to the virus that field mice had been carrying for centuries—and many contracted a raging hemorrhagic fever that kills at least 10 percent of its victims.

SOURCE: *A Dancing Matrix: Voyages Along the Viral Frontier,* by Robin Marantz Henig, pp. ix–x. Copyright © 1993 by Robin Marantz Henig. Reprinted by permission of Alfred A. Knopf, Inc.

The Content of Social Interaction

As the division of labor has become more specialized and the sources of labor and raw materials have become more geographically diverse, the ties that connect people to one another have shifted in character from mechanical to organic, and the number of interactions people have with strangers has increased. How can people interact smoothly with people they know nothing about? They eliminate "strangeness" by identifying the social positions or social status of the strangers with whom they interact. Knowing a person's social status gives us some idea of the behaviors we can expect from someone in that status. It also affects how we will interact with that person. To grasp this principle, think about the times you meet someone for the first time. How do you start the interaction? You ask the stranger a question to determine his or her social status. You might ask, "What do you do?" That question sets the interaction into motion.

Social Status

In everyday language, people use the term *social status* to mean rank or prestige. To sociologists, **social status** refers to a position in a social structure. A **social structure** consists of two or more people interacting and interrelating in specific expected ways, regardless of the unique personalities involved. For example, a social structure can consist of the two statuses of doctor and patient and their relationship. Other familiar examples of statuses in a two-person social structure are husband and wife, professor and student, sister and brother, and employer and employee.

Examples of multiple-status social structures are a family, an athletic team, a school, a large corporation, and a government. Again, the common characteristic of all social structures is that it is possible to generalize about the behavior of people in each of the statuses, no matter who occupies them. Just as the behavior of a person occupying the status of a football quarterback is broadly predictable, so too is the behavior of a person occupying the status of nurse, secretary, mechanic, patient, or physician. Once we know a person's status we think we know how to interact with the person.

Types of Status Statuses can be of two kinds—ascribed or achieved. An **ascribed status** is a position that people acquire through no fault or virtue of their own. Examples include gender, age, ethnic, and health statuses; these include male, female, African-American, Native American, adolescent, senior citizen, retired, son, daughter, and disabled. An **achieved status** is a position earned (or lost) by a person's own actions or abilities. Occupational, educational, parental, and marital statuses include athlete, senator, secretary, physician, high school dropout, divorcé, and single parent.

The distinction between ascribed and achieved statuses is not always clear-cut. It is often the case that achieved statuses such as financial, occupational, and educational positions are related to ascribed statuses such as gender, ethnicity, and age. For example, in the United States, physicians are disproportionately male (80 percent) and white (92 percent), whereas registered nurses are overwhelmingly female (86.2 percent) and white (90 percent) (U.S. Bureau of the Census 1992b).

Stigmas Some ascribed and achieved statuses are such that they overshadow all other statuses that a person occupies. Sociologist Erving Goffman calls such statuses **stigmas** and classifies them into three broad varieties: (1) physical deformities; (2) character blemishes due to factors such as sexual orientation, mental hospitalization, or imprisonment; and (3) stigmas of ethnicity, nationality, or religion. When a person possesses a stigma, he or she is reduced in the eyes of others from a multifaceted person to a person with one tainted status. To illustrate this point Goffman opens *Stigma: Notes on the Management of Spoiled Identity* (1963) with a letter written by a sixteen-year-old girl born without a nose. Although she is a good student, has a good figure, and is a good dancer, no one she meets can get past the fact that she has no nose. (See Chapter 10 for more on social stigma.)

Every person occupies a number of statuses. For example, Dr. Rask was middle-aged, a female, a physician, a patient, and Danish. A status has meaning, however, only in relation to other statuses. For example, the status of a physician takes on quite different meanings depending on whether the interaction is with someone who occupies the same status or a different status such as patient, spouse, or nurse. This is because a physician's role varies according to the status of the person with whom he or she interacts.

Social Roles

Sociologists use the term **role** to describe the behavior expected of a status in relationship to another status (for example, professor to student). The distinction between role and status is subtle but noteworthy: people *occupy* statuses and *enact* roles.

Associated with every status is a **role set,** or an array of roles. For example, the status of physician entails, among other roles, the role of physician in relationship to patient, to nurse, to other doctors, and to a patient's family members. The sociological significance of statuses and roles is that they make it possible for us to interact with other people without knowing them. Once we determine another person's status in relation to our own, we interact on the basis of role expectations attached to that status relationship.

Role expectations are socially prescribed and include both rights and obligations. The **rights** associated with a role define what a person assuming that role can demand or expect from others depending on their status. Teachers have the right to demand and expect that students will come prepared for class. The **obligations** associated with a role define the appropriate relationship and behavior that the person enacting that role must assume toward others occupying a particular status. Teachers have an obligation to their students to come to class prepared.

One of the best-known descriptions of a role and its accompanying rights and obligations was given by sociologist Talcott Parsons (1975). According to Parsons we assume a **sick role** when we are sick. Ideally, sick persons have an obligation to try to get well, to seek technically competent help, and to cooperate with a treatment plan. Sick persons also have certain rights: they are exempt from "normal" social obligations and are not held responsible for their illness.

Social Roles and Individual Behavior Roles set general limits on how we think and act but do not imply that behavior is totally predictable. Sometimes people do not meet their role obligations, as when professors come to class unprepared or when patients do not cooperate with their physicians' treatment plans or when physicians blame their patients for getting sick (for example, thinking they have AIDS because they engaged in promiscuous and perverse sexual activity, or thinking they have lung cancer because they don't have the will to quit smoking). By definition, when people fail to meet their role obligations, other people are not accorded their role rights. When professors are unprepared, students' rights are violated; when patients do not follow treatment plans, physicians' rights are violated; when physicians blame their patients, patients' rights are violated. Moreover, the idea of role does not imply that all people occupying the same status enact the roles of that status in exactly the same way. Roles are enacted differently because of individual personalities and interpretations of how the role should be carried out. Finally, roles are enacted differently because people resolve role strain and role conflict differently.

Role strain is a predicament in which contradictory or conflicting expectations are associated with the role that a person is occupying. For example, military physicians, as physicians, have an obligation to preserve life. On the other hand, they are employed to care for patients who have been placed deliberately in situations that threaten their health and lives. **Role conflict** is a predicament in which the expectations associated with two or more roles in a role set are contradictory. For example, a sick person has an obligation to want to get better and to comply with treatment plans. This obligation, however, can interfere with other roles that the person holds, as in the case of a woman who finds that the side effects of a prescribed drug prevent her from being alert enough to work or care for her children.

Socially prescribed rights and obligations notwithstanding, there is always room for improvisation and personal style. Yet, despite variations in how people enact roles, role is still a useful concept because, for the most part, an appreciable degree of

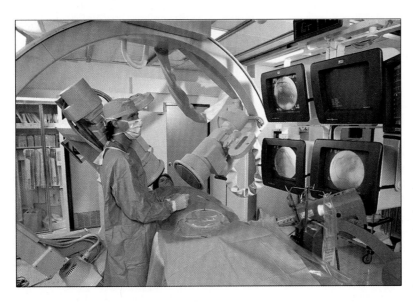

The Western approach to healing is rooted in cultural beliefs about the nature of disease and the power of science and technology to combat it.
Bob Daemmrich/Stock, Boston

predictability exists, "sufficient to enable most of the people, most of the time to go about [the] business of social life without having to improvise judgments anew in each newly confronted situation" (Merton 1957, p. 370). In other words, the variations usually fall within "a certain range of culturally acceptable behavior—if the performance of a role deviates very much from the expected range of behavior the individual will be negatively sanctioned" (p. 370).

Cultural Variations in Roles: The Patient-Physician Interaction The behaviors expected of one status in relation to another status vary across cultures. Role expectations are intertwined with norms, values, beliefs, and nonmaterial culture (see Chapter 4). In the United States the major objective of the patient-physician interaction is to determine the exact physiological malfunction and use the available material culture and technology (tests, equipment, machines, drugs, surgery, transfusions) to treat it. This objective is shaped by a profound cultural belief in the ability of science to solve problems. Practitioners of Western medicine are expected to use all of the tools of science at hand to establish the cause, combat the disease for as long as possible, and return the body to a healthy state. Even when a person is close to death, most Ameri-

cans tend to believe that a scientific cure could emerge at any moment. When physicians violate this norm, the violation is almost always accompanied by intense public debate. In view of this cultural orientation, it is not surprising that in the United States, physicians and their patients rely heavily on technological elements such as X-rays and CAT scans to diagnose the condition and on vaccines, drugs, and surgery to cure it. This reliance is reflected in the fact that the United States, with a population of 226 million—5 percent of the world's population—consumes 23 percent of the world's pharmaceutical supply (Peretz 1984).

We can contrast the U.S. physician-patient relationship with the traditional African healer-patient relationship. Although Zaire has modern health care facilities, the majority of people go to *curandeiros,* or traditional healers.[2] The social interaction between *curandeiro* and patient is very different from the U.S. physician-patient social interaction.

Just as Western physicians do, traditional healers recognize the organic and physical aspects of disease. But they also attach considerable importance to other factors—supernatural causes, social relationships (hostilities, stress, family strain), and psychological distress. This holistic perspective allows for a more personal relationship between healer and patient. Another significant difference from Western medicine is that healers concentrate on providing symptom relief instead of searching for a total cure. African *curandeiros,* for example, focus on treating the wasting symptoms of AIDS such as diarrhea, headaches, fevers, and weight loss, and they employ remedies with few side effects so as not to make people sicker (Hilts 1988). Obviously, each system has its attractions and its weaknesses, as evidenced by the fact that Westerners suffering from incurable diseases sometimes turn to traditional, non–science-based medicine.[3] Likewise, when traditional methods fail Africans, they may make a trip to a hospital or clinic to consult with a doctor of Western medicine.

Sociologist Ruth Kornfield observed Western-trained physicians working in Zaire's urban hospitals and found that success in treating patients was linked to the foreign physician's ability to tolerate and respect other models of illness and include them in a treatment plan. For example, among some Zairean ethnic groups, when a person becomes ill, the patient's kin form a therapy management group and make decisions about administering treatments. Because many people in Zaire believe that illnesses are caused by disturbances in social relationships, the cure must involve a "reorganization of the social relations of the sick person

that [is] satisfactory for those involved" (Kornfield 1986, p. 369; also see Kaptchuk and Croucher 1986, pp. 106–08).

The point of this discussion is not to evaluate the quality or outcome of either culture's patient-practitioner interaction. Rather, by comparing the two cultures, we can see more clearly that people think and behave in largely automatic ways because they are influenced by norms, values, beliefs, and nonmaterial culture. Those involved with Dr. Rask automatically assumed a scientific framework to define the origin and treatment of her condition. Such a conceptual framework defines illness as "a state of disease and dysfunction 'impersonally' caused by microorganisms, inborn metabolic disturbances, or physical or psychic stress" (Fox 1988, p. 505). On this basis, we can begin to understand why Dr. Rask, other physicians, and her close friends all defined her illness as a condition contracted in Africa, requiring hospitalization, having biological origins, and related to direct contact with a patient's blood. The interactions and dialogue between Dr. Rask, other medical personnel, and friends would have been quite different if they had believed that her conditions were caused by spiritual beings, conflict with family, failure to observe religious practice, or the forces of nature, gods, ancestors, or other beings.

The Dramaturgical Model of Social Interaction

A number of sociologists have compared roles attached to statuses with dramatic roles played by actors. Erving Goffman is a sociologist associated with the **dramaturgical model** of social interaction. In this model, social interaction is viewed as though it were theater, people as though they were actors, and roles as though they were performances presented before an audience in a particular setting. People in social situations resemble actors in that they must be convincing to others and must demonstrate who they are and what they want through verbal and nonverbal cues. In social situations, as on a stage, people manage the setting, their dress,

their words, and their gestures to correspond to the impressions they are trying to make or the image they are trying to project. This process is called **impression management.**

On the surface, the process of impression management may strike us as manipulative and deceitful. Most of the time, however, people are not even aware that they are engaged in impression management because they are simply behaving in ways they regard as natural. Women engage in impression management when they remove hair from their faces, legs, armpits, and other areas of their bodies and present themselves as hairless in these

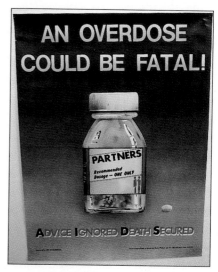

Almost every country in the world produces AIDS-prevention messages tailored to the values and beliefs of people in that country. These posters are from Uganda (left) and Barbados (right).

areas. From Goffman's perspective, even if people are aware that they are engaged in impression management, it can be both a constructive and a normal feature of social interaction because smooth interactions depend on everyone's behaving in socially expected and appropriate ways. If people spoke and behaved entirely as they pleased, civilization would break down. Goffman (1959) also recognizes the dark side of impression management that occurs when people manipulate the audience in deliberately deceitful and hurtful ways.

Impression management often presents us with a dilemma. If we do not conceal inappropriate and unexpected thoughts and behavior, we risk offending or losing our audience. Yet, if we conceal our true reactions, we may feel that we are being deceitful, insincere, or dishonest or that we are "selling out." According to Goffman, in most social interactions the people involved weigh the costs of losing their audience against the costs of losing their integrity. If keeping the audience is important, concealment is necessary; if showing our true reactions is important, we may risk losing the audience.

In the United States people who test positive for HIV antibodies face the dilemma of impression management. If they disclose the test results, they risk discrimination and the loss of their jobs, insurance coverage, friends, and family. If they keep this information to themselves, they may feel they are being untrue to themselves and to others who care about them. Rarely are interaction situations "either-or." Usually people compromise between the two extremes.

The dramaturgical model and the specific idea of impression management are useful concepts for understanding the dilemma that one partner may face if he or she suggests using a condom as a precautionary condition of sexual intercourse. This dilemma is quite different in the United States than in Zaire. In both countries health officials recommend condom use as a way to reduce substantially the risk of HIV infection. Risk of infection extends to new sexual partners, members of high-risk groups, and even currently monogamous partners if they were sexually active in the past, used intravenous drugs, or received a blood transfusion in the past 10 years. (At the time of this writing, the interval between HIV infection and the onset of AIDS is known to be as long as 10 years. In essence, an individual is going to bed not only with the other per-

Point of View: Women Are More Than Mothers

Kathryn Carovano maintains that those who design prevention programs must consider the facts that, for many women around the world, "motherhood brings status and security," and men, not women, usually decide whether to use condoms. Her article, which follows, describes how these facts affect the spread or prevention of HIV infection.

The vast majority of women with AIDS and HIV infection are aged 15 to 45. Yet to date, women in this age group—if not involved in prostitution—have probably received the least attention in AIDS prevention programmes.

The main reason for this is the false notion that these women are not at risk. Most of them are confronted with the issue of AIDS for the first time when pregnant or considering pregnancy. Generally they receive information at that time, because of society's concern for their unborn children. Many HIV-positive women, in both industrialised and developing countries, only learn that they are infected when one of their children is diagnosed with AIDS.

These women are frequently referred to as "women of reproductive age." But ironically, the only advice they receive on reducing their risk of HIV transmission is to rely on condoms or non-penetrative sex—both of which prevent conception. While this may be an added advantage for some, it is a critical flaw for those wishing to conceive.

Motherhood brings status and security to many women's lives. In many cultures, women are told that the purpose of their existence is to bring forth new life, especially male, and that their value depends on this. Often there is no social place for women who are unable, or choose not, to have children.

While this situation continues, measures to prevent HIV infection which ignore the importance of motherhood will have only limited impact. There is an urgent need for research to try to develop a virucide which would prevent HIV transmission, but not conception.

Women need to be able to protect themselves from HIV without being forced to give up the option of having children. In an ideal world, mutual faithful monogamy would allow women to protect themselves and have healthy children. But this is not the reality of relationships for much of the world's population.

Women are more than mothers, however, and the core issue is not about women deciding whether or not to have children, but about women's right to separate sexuality from procreation and to be in control of sexual decision-making.

Unlike most other contraceptive methods available today, condoms and non-penetrative sex both require male co-operation. From family planning experience we know that, while many men might agree on the importance of limiting the size of their families, few accept primary responsibility for the prevention of pregnancy.

Alternative, preventive technology controlled by women is essential. In the meantime, prevention education programmes are also needed that make all sexually active women aware that they are at risk from AIDS. The first woman to be diagnosed with AIDS in Mexico in 1985 was a 52-year-old housewife living in Mexico City; her only known "risk" behaviour was having unprotected sexual intercourse with her husband.

SOURCE: "Point of View: Women Are More Than Mothers," by Kathryn Carovano (1990). P. 2 in WorldAIDS (8). Reprinted by courtesy of The Panos Institute, Arlington, VA and the Bureau of Hygiene and Tropical Diseases, London.

son but also with that person's sexual partners of the past 10 years.)

In both Zaire and the United States, the subject of condom use is a sensitive one because of the message that the condom conveys to potential sexual partners. In Zaire condom use is associated with birth control, not with disease prevention. The taboo against birth control is strong because many Africans measure their spiritual and material wealth by the number of offspring. If children survive, they become economic assets to the family as well as links to ancestors (Whitaker 1988).[4] In view of the strong pressures to have children, Zaireans believe that condom use virtually deprives them of the approval of their families and ancestors (see "Point of View: Women Are More Than Mothers").

Researcher Kathleen Irwin (1991) and 15 colleagues interviewed healthy factory workers and their wives from Kinshasa, Zaire. They found that, although many respondents had heard of condoms, few actually used them. The researchers also found that, among these respondents, it is the men who decide whether to use and purchase condoms.

For many people in the United States the contraceptive value of condoms is a secondary concern at best. The primary concern is to have recreational sex that is also safe. In the United States if a person suggests using a condom during sex, he or she is implying that the partner's sexual orientation or sexual history is suspect. Advertisers try to package and market condoms in ways that counteract this message. Condoms come in all colors, are designed and manufactured in forms that stimulate greater sexual sensation, and show sexually appealing scenes on outer packages. The point is that in both Zaire and the United States, "sexual behavior is based on long-standing cultural traditions and social values and may be very difficult to change" (U.S. General Accounting Office 1987). This fact in turn makes it difficult for any person to manipulate the meaning of a condom without offending his or her sexual partner. Therefore, many people resist using condoms, even in high-risk situations (Giese 1987). Ironically, raising questions of personal vulnerability does not appear to be an effective strategy for convincing people to protect themselves from the risk of HIV because most people believe that bad things happen to other people, not to themselves.

The dramaturgical model of social interaction is useful because it helps us see how the need to convey the right impressions can work to discourage people from behaving in ways that are not in their best (or others' best) interest. Goffman uses another theater analogy—staging behavior—to identify situations in which people are most likely to engage in impression management.

Staging Behavior

Just as the theater has a front stage and a back stage, so does everyday life. The **front stage** is the area visible to the audience, where people take care to create and maintain expected images and behavior. The **back stage** is the area out of the audience's sight, where individuals can "let their hair down" and do things that would be inappropriate or unexpected on the front stage. Because backstage behavior frequently contradicts frontstage behavior, we take great care to conceal it from the audience. Goffman uses the restaurant as an example. Restaurant employees do things in the kitchen and pantry (back stage) that they would not do in the dining areas (front stage), such as eating from customers' plates, dropping food on the floor and putting it back on a plate, and yelling at each other. Once they enter the dining area, however, such behavior stops. How often have you, as a customer, seen a server eat a scallop or a french fry from a customer's plate while in the dining area? If you have ever worked in a restaurant, however, you know that this is fairly common backstage behavior.

The division between front stage and back stage is hardly unique to restaurants but is found in nearly every social setting. In relation to the AIDS crisis, we can name a host of environments that have a front stage and a back stage, such as hospitals, doctors' offices, and blood banks. Much as restaurant personnel shield diners from backstage behavior, medical personnel shield patients from backstage behavior. For example, most people know little about the blood bank industry beyond what they see when they donate, sell, or receive blood. Although the public can research such industries and learn about their inner workings, most people do not have the time to study every industry or do not know what questions to ask about the industry that provides them with goods and services. For example, most people probably give little thought about whom the blood industry collects blood from, and would never think to ask the question of where does blood come from. Thus, they would be surprised to find out that Mexicans cross the border into the United States to give blood. In 1980, a few years before HIV was discovered, *Newsweek* magazine ran a story on the thousands of poor Mexicans who earn $10.00 by undergoing a procedure called plasmapheresis (technicians take blood, separate the red cells from the plasma and reinject the donor with his or her own cells minus the plasma). Because red cells are returned to donors they may undergo the procedure up to twice a week. The article highlighted the international nature of blood collection (Clark and McGuire 1980).

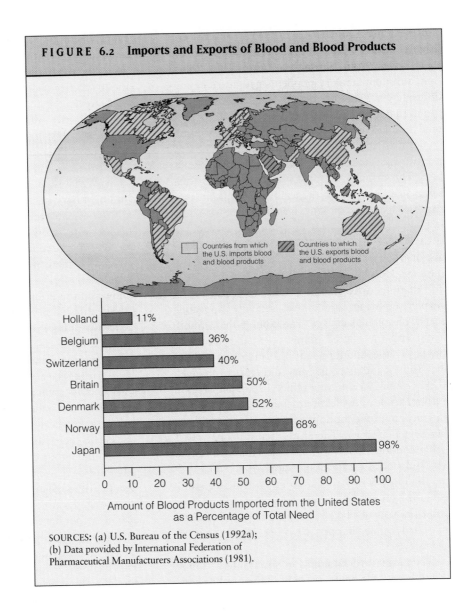

FIGURE 6.2   Imports and Exports of Blood and Blood Products

Countries from which the U.S. imports blood and blood products

Countries to which the U.S. exports blood and blood products

Holland 11%
Belgium 36%
Switzerland 40%
Britain 50%
Denmark 52%
Norway 68%
Japan 98%

Amount of Blood Products Imported from the United States
as a Percentage of Total Need

SOURCES: (a) U.S. Bureau of the Census (1992a);
(b) Data provided by International Federation of
Pharmaceutical Manufacturers Associations (1981).

The Back Stage of Blood Banks

Blood bank officials found themselves in a dilemma in 1981, when officials at the Centers for Disease Control made known their suspicion that the HIV virus was being transmitted through blood products. This revelation meant that not only the United States' blood supply but also the world's blood supply was contaminated, because the United States

supplies about 30 percent of the world's demand for blood and blood products (U.S. Bureau of the Census 1992a; *The Economist,* 1981, 1983; also see Figure 6.2).

Until spring of 1985, U.S. blood bank officials continued to affirm publicly their faith in the safety of the country's blood supply, insisting that there was no need to screen donors. Yet, these officials never revealed to the public the many shortcomings in production methods that could jeopardize the safety of blood. (By *shortcomings* I do not mean negligence, but rather deficiencies in medical knowledge and in the level of technology.) Looking back on their policies, blood bank officials later argued that they practiced this concealment to prevent a worldwide panic. (In Goffman's terminology, they did not want to lose their audience.) Such a panic would have brought chaos to the medical system, which depends on blood products.[5] Still, the delay in implementing screening exposed many people to infection.

Here lies a potential connection to Zaire. Malaria is a common disease in Zaire, especially among children. The disease leaves its victims vulnerable to severe anemia, and blood transfusions are used to treat the anemia. As mentioned earlier, the United States is a large exporter of blood products. Many countries that import this blood re-export it, and the Red Cross delivers blood products to countries in need. Therefore, some of the blood used in Zaire is likely to have originated in the United States.

Although blood bank officials announced publicly that the risk of HIV infection from blood products was one in a million, knowledge of the backstage collection, production, and distribution of blood products leaves no doubt that the risks were in fact higher, especially to hemophiliacs, whose plasma lacks the substance Factor VIII, which aids in clotting, or whose plasma contains an excess of anticlotting material (U.S. Department of Health and Human Services 1990). In fact, we now know that 50 percent of hemophiliacs were HIV-infected from Factor VIII treatments before the first case of AIDS appeared in this group ("Frontline" 1993).

Even after blood bank officials agreed to start screening blood for HIV infection in spring 1985 (after years of debate over whether to test), they still pronounced the blood supply to be safe. They did not, however, announce the shortcomings of the screening tests: (1) the antibodies that the test measures may not appear in the blood for 11 months after infection with the virus, and (2) a small but unknown percentage of HIV carriers never develops detectable antibodies (Kolata 1989).[6,7]

In hindsight, it is easy to criticize blood bank officials' response to the situation. Yet, in fairness to the blood industry, we must acknowledge the legitimacy of their wish not to induce worldwide panic, especially very early on when no tests were available to screen blood for the infection. As the head of the New York Blood Center argued, "You shouldn't yell *fire* in a crowded theater, even if there is a fire, because the resulting panic can cause more deaths than the threat" (Grmek 1990, p. 162). In addition, we must recognize that it is impossible to eliminate every element of risk. Even so, one troubling fact suggests that their decision may have been motivated by profit: although U.S. companies tested new blood for the domestic market, they did not test blood already stored in their inventories and they continued to supply untested blood to foreign countries for at least six months after the tests were available ("Frontline" 1993; Hiatt 1988; Johnson and Murray 1988).

To this point we have examined a number of concepts that help us analyze the content of any social interaction. Specifically, when examining content sociologists identify the statuses of those people involved. Once statuses have been identified, they focus on the roles expected of each status in relation to the other, with special emphasis on rights and obligations. Sociologists use a dramaturgical model to think about how people enact roles. Thus they focus on impression management (how people manage the setting, their dress, their words, and their gestures) to correspond to role expectations. Impression management is most important when people are front stage as opposed to back stage.

Knowing about status, roles, impression management, and front/back stages helps sociologists predict much of the content of social interaction. However, when we interact we do more than identify the statuses of the people involved and act to behave in ways consistent with role expectations. We also try to assign causes to our own and others' behaviors. That is, we posit explanations for behavior, and we may act differently depending on what explanations we come up with. A theoretical approach that helps us understand how we arrive at our everyday explanations of behavior is attribution theory.

Attribution Theory

Social life is complex. As we have seen, people need a great deal of historical, cultural, and biographical information if they are to understand the causes of even the most routine behaviors. Unfortunately, it is nearly impossible for people to have this information at hand every time they want to understand the causes of behavior. For one thing, "the real environment is altogether too big, too complex, and too fleeting for direct acquaintance" (Lippmann 1976, p. 178).

Yet, despite our limitations in understanding causes of behavior, most people do attempt to determine a cause anyway, even if they rarely stop to examine critically the accuracy of their explanations. As most of us know very well, ill-defined, incorrect, and inaccurate perceptions of cause do not keep people from forming opinions and taking action. Such perceptions, however, result in actions that have real consequences. Sociologists William and Dorothy Thomas described this process very simply: "If [people] define situations as real they are real in their consequences" (Thomas and Thomas [1928] 1970, p. 572).

Attribution theory rests on the assumption that people assign (or attribute) a cause to behavior in order to make sense of it. People usually attribute cause to either dispositional traits or situational factors. **Dispositional traits** include personal or group traits such as motivation level, mood, and inherent ability. **Situational factors** include forces outside an individual's control such as environmental conditions or bad luck. When evaluating the causes of their own behavior, people tend to favor situational factors. When evaluating the causes of another's behavior, on the other hand, people tend to point to dispositional traits.[8]

A memorable example of attributing cause to dispositional rather than situational characteristics appears in Colin Turnbull's *The Lonely African*. Turnbull describes how a European farm owner reacts to an African farmhand who wears a tie without a shirt: "'I've actually got a farmhand who wears a tie—but the stupid bastard doesn't realize you don't wear a tie without a shirt!'" (Turnbull 1962, p. 21). The owner attributes the behavior to a dispositional factor—the farmhand is simply stupid. If the farm owner had thought about the behavior in terms of situational factors, he would have found that the farmhand wears the tie "because it makes a bright splash of color, and is useful for tying up bundles, and refuses to wear the shirt that collects dirt and sweat and makes the Europeans smell so bad" (p. 21). Right or wrong, the attributions that people make shape the content of social interaction. In other words, the way in which people construct reality affects how they treat and deal with different individuals and groups.

Throughout history, whenever medical professionals have lacked the knowledge or technology to combat a disease, especially one of epidemic proportions, the general population tends to hold some groups of people within or outside of the society responsible for causing the disease (Swenson 1988). In the sixteenth and seventeenth centuries, for example, the English called syphilis "the French disease," and the French called it "the German disease." In 1918 the worldwide influenza epidemic, which infected more than one billion people and killed more than 25 million, was called "the Spanish flu" even though there was no evidence to support the idea that it originated in Spain.

In a similar vein, U.S. medical researchers trying to map the geographical origin and spread of AIDS have hypothesized various interaction scenarios between specific groups of people who are inferred to behave in careless, irresponsible, or immoral ways to explain how AIDS has spread transcontinentally. The hypotheses usually assume that the disease originated in Zaire and then spread to the United States via Europe, Haiti, or Cuba. Yet, there is no evidence to support the possibility that it did not spread transcontinentally from the United States. Two of the most prevalent hypotheses are as follows:

- The virus traveled from Zaire to Haiti to the United States. In the mid-1960s a large number of Haitians went to Zaire to fill middle-management positions in the newly independent state. Mobutu sent them home in the 1970s. American homosexuals vacationing in Port-au-Prince brought the virus back to New York and San Francisco.

when we look for a cause in ourselves, it is seen as situational

when we look for a cause in others, it is seen as dispositional

HAVING HIV
REMINDS US
OF WHAT'S
IMPORTANT
IN LIFE:

LOVE, FAMILY.
A good Marriage.
HUMOR &
A good
Doctor.

Be Here for The Cure. Get Early Treatment for HIV.
Call 1 800-367-2437

Public-health messages like this one are specifically designed to counter many people's mistaken belief that only certain high-risk groups are vulnerable to AIDS. These beliefs often stem from dispositional attributions people make about the causes of HIV infection.

Courtesy San Francisco AIDS Foundation; photograph by Annie Leibovitz

- The virus traveled from Zaire to Cuba to the United States. Cubans brought it back from Angola, which shares a long border with Zaire. In the late 1970s the Cuban government purged the army of undesirables, including some homosexual veterans who had served in Angola. Many of these Cubans migrated to Miami.

Attributing cause to dispositional factors seems to reduce uncertainty about the source and spread of the disease. The rules are clear: if we do not interact with members of that group or behave like members of that group, we can avoid the disease. Such rules provide us with the secure feeling that the disease cannot affect *us* because it affects *them* (Grover 1987).

In the United States, many people believe that HIV and AIDS is confined to members of a few high-risk groups, notably hemophiliacs, male homosexuals, and intravenous (IV) drug users. Dispositional explanations for the high risk among members of these groups imply that these members earned their disease as a penalty for perverse, indulgent, and illegal behaviors (Sontag 1989). As late as 1985, medical and government officials referred to AIDS as the "gay plague," even though there was overwhelming evidence that the HIV infection was transmitted through heterosexual intercourse, needle sharing, and other exchanges of blood and blood products or some other bodily fluids. In his book *And the Band Played On*, journalist Randy Shilts (1987) argues that the public health and medical research response to AIDS was delayed several years because policymakers believed that casualties were limited to unpopular and socially powerless groups such as homosexuals, IV drug users, and Haitians.

Similarly, dispositional thinkers might explain the high incidence of HIV infection and AIDS among both males and females in Zaire (or other African countries) in terms of excessive sexual promiscuity or polygynous marriages,[9] or in terms of rituals involving monkeys, female circumcision, and scarification (see "The Myth of African Promiscuity"). Actually, HIV infection in Zaire is attributable to four major modes of infection: heterosexual contact, blood transfusions, reuse of needles for medical purposes, and mother-to-child transmission.

Whereas Americans tend to regard HIV infection as originating in Africa and as being transmitted through bizarre and indulgent behaviors, Zaireans tend to view AIDS as an American disease:

Westerners had brought AIDS to Africa with their "weird sexual propositions"—a view echoed by La Gazette *in July 1987, which referred to Westerners coming to Africa with their "sexual perversions." "Many of the venereal diseases now found in Kenya," said an editorial in the* Kenyan Standard, *"were brought into the country by the same foreigners who are now waging a smear campaign against us."* (THE PANOS INSTITUTE 1989, P. 75)

The Myth of African Promiscuity

Many phenomena have led outsiders to consider Africans as promiscuous; polygamy is one of them. In pre-colonial times the men who had more than one wife were the elders, distinguished warriors, priests and successful traders who could pay the bride price demanded or were worthy of the dowry to be received. Marriage was about alliances between families and therefore involved kinship connections in which the emphasis was not on mutual sexual attraction or physical bonds. Not all men could marry more than one wife, and some had no wives at all.

Other forms of marriage in Africa which appear unusual to outsiders, including the inheritance of a dead sister's or dead brother's spouse, often have very limited physical sexual basis and are based on access to property and rights of inheritance. They reflect the unequal position of women in these societies, who may have no legal rights of existence except through a husband and who must be seen as wives even if this does not involve the obligations of physical sex.

The prevalence of semi-nudity may suggest to Europeans sexual availability but it is more a case of comfort and convenience. Relaxed attitudes to the body are often accompanied by rigid rules about sexual interaction and eligibility for marriage. Fertility dances and dances that celebrate puberty, maidenhood and bridal status indeed express appreciation of the human body, but are no more suggestive in the sexual sense than some European folk-dances, ice-skating and ballet in which women throw up their legs in front of a paying audience.

Although there is tremendous diversity in cultural groupings in Africa, throughout the continent sexual life is surrounded by rites and rules that tie up with age and entry into adult life. These include female and male circumcision, child marriage, early betrothal, the control of age-grades over such sources of brideprice and dowry as cattle, the various tests of manhood which certain groups subject their young men to before they can take wives, and the importance, in fact near fetishism, of virginity among some groups. All of these limit the access or the rights of young men to sexual interaction with young girls, while other sanctions against the breaching of sexual protocols range from fines to ostracism and in certain cases corporal punishment and homicide. Along with the more mundane sanctions are mystical ones such as oath-taking, putting spells and curses on culprits and invoking the displeasure of ancestral spirits.

Various arrangements in some pre-colonial urban centres and among certain warrior communities came very close to prostitution, but the sale of sex for money only came with monetisation and, of course, the European presence. Colonial policies needed cheap African labour for such activities as mining and construction, while denying the workers permission to bring their

---

In Zaire and other African countries, people often refer to the United States as the "United States of AIDS." Condoms are called "American socks," and aid in the form of condoms is known as "foreign AIDS" (Brooke 1987; Hilts 1988). People commonly believe that (1) AIDS arrived via rich American sports fans who came to Kinshasa to watch the Ali-Foreman fight in 1977, (2) the virus was manufactured in an American laboratory for military germ warfare and was unleashed deliberately on African people,[10] (3) the disease came from American canned goods sent as foreign aid, and (4) the AIDS epidemic can be traced to the way in which the polio vaccine was manufactured and administered to African people 30 years ago. In each scenario, one can infer that those giving the disease to African people are morally suspect, evil plotters, or careless; these are all dispositional characteristics.

Dispositional theories such as the ones just listed, whether American or African, are alike in that they define a clear culprit or scapegoat. A **scapegoat** is a person or a group that is assigned blame for conditions that cannot be controlled, that threaten a community's sense of well-being, or that shake the foundations of a trusted institution.

families to live with them, so they would not have any claims for urban land.

Africans could only live in towns if they had employment and could show proof of it, a system of control of population movement of which current South African pass laws are only a perfection. With men housed in barrack-like accommodation for periods of up to a year, the authorities encouraged or turned a blind eye to the growth and spread of brothels. Prostitution also spread wherever there were centres of trade and administration, and sometimes led to the emergence of ethnic specialisation as women from one area were recruited into brothels in cities elsewhere. In this sense prostitutes were also a type of migrant labour.

Another effect of colonial urbanisation was the breakdown of traditional rules and protocols of marriage, leading to unprecedented inter-ethnic marriages and sexual contacts. The new anonymity offered by cities made their inhabitants accessible to foreign and alien contacts both for transient sexual and more lasting relationships. The pluralism of urban life leaves individuals without the strict regulation of sex and marriage which the ethnic culture once provided, with no formal rules applicable to liaisons between partners from different ethnic groups or between Africans and outsiders.

Westernisation has brought with it not only new airports, roads and institutions but also the adoption of Western lifestyles, tastes and fashions. Modern young Africans look to Europe and North America for fashion, music and dance. They accept Levi jeans along with Coca-Cola, rock music, heroin and free sex.

In the immediate post-independence affluence of countries such as Nigeria or Kenya, the urban young have both the money and the leisure to enjoy a lifestyle which includes extensive and varied sexual activities. Their prey might be called the "new poor": the massive pool of young women living in the most deprived conditions in shanty towns and slums across Africa, who are available for the promise of a meal, new clothes, or a few pounds. The same poor, whether as prostitutes or as clerks and petty officials predisposed to taking bribes, are vulnerable to the hard currency offered by foreign visitors. Their sexual availability is born out of real material needs and reinforced by imported norms that advocate free sex.

This is contemporary sexual life in many African cities and towns, where many people take their cue from London, Paris, New York and the rest of the "civilised" world. Their attitudes may reinforce a myth of African promiscuity, but the worldwide breakdown in sexual restraints and control is not a specifically African problem.

SOURCE: "The Myth of African Promiscuity," by Tade Aina. Pp. 78–80 in *Blaming Others: Prejudice, Race and Worldwide AIDS.* Copyright © 1988 by The Panos Institute. Reprinted by permission of New Society Publishers, Philadelphia, PA.

Usually, the scapegoat belongs to a group whose members are already vulnerable, hated, powerless, or viewed as different.

The public identification of scapegoats gives the appearance that something is being done to protect the so-called general public; at the same time, it diverts public attention from those who have the power to assign labels. In the United States, identifying AIDS as the "gay plague" diverted attention from the blood banks and the risks associated with medical treatments involving blood products. Blood bank officials maintained that the supply was safe as long as homosexuals abstained from giving blood. In Zaire, identifying AIDS as an American disease diverted attention away from corrupt officials who were funneling money out of Zaire, leaving its people poor and malnourished and (by extension) vulnerable to HIV infection.

From a sociological perspective, dispositional explanations that point to a group—or to characteristics supposedly inherent in members of that group—are simplistic and potentially destructive not only to the group but to the search for solutions. When the focus is a specific group and that group's behavior, the solution is framed in terms of controlling that group. In the meantime, the

Scapegoating is a life-or-death issue for members of ACT-UP, a group formed to aggressively protest what its members see as a slow and insufficient response by public officials to the AIDS epidemic. They believe that policymakers would act more decisively if AIDS were not seen as primarily a problem for unpopular or socially powerless groups.

Steven Rubin/JB Pictures

problem can spread to members of other groups who believe they are not at risk because they do not share the problematic attribute of the groups identified as high-risk. This kind of misguided thinking about risk applies even to physicians and medical researchers.

Medical sociologist Michael Bloor (1991) and colleagues argue that the official statistics on the modes of transmission are influenced by researchers' beliefs about the relative riskiness of behaviors. For example, an HIV-positive man who has received a blood transfusion and who has had sexual relations with another man is placed in the transmission category "homosexual" rather than "blood recipient." This approach to classification ignores the possibility that a homosexual male can become HIV-infected through a blood transfusion.

Similarly, until 1993 the official definition of AIDS did not include HIV-related gynecological disorders such as cervical cancer as one of the conditions that constituted a diagnosis of AIDS. (Also not included under the official definition of AIDS until 1993 were HIV-related pulmonary tu-

berculosis and recurrent pneumonia.) Under the old definition, HIV-positive persons were said to have AIDS only if they developed one of 23 illnesses, many of which were peculiar to the gay population with AIDS. Under the new, revised definition, HIV-positive women with cervical cancer are officially diagnosed as having AIDS. Before this change, physicians did not ask women with cervical cancer to be tested for HIV and did not give HIV-positive women with this condition the opportunity to be treated for AIDS (Barr 1990; Stolberg 1992). These two examples show that attributions about who "should" have AIDS affect the way in which the condition of AIDS is defined and influence the statistics about who has AIDS. They also show that such attributions affect the content of the physician-patient interaction.

In addition to acknowledging the shortcomings related to AIDS classification and diagnosis, we also have to acknowledge that we simply do not know how many people are HIV-infected worldwide, or who is actually infected.

Determining Who Is HIV-Infected

To obtain information on who is actually infected with HIV, every country in the world would have to administer blood tests to a random sample of its population. Unfortunately, (but perhaps not surprisingly) people resist being tested.

A planned random sampling of the U.S. population sponsored by the Centers for Disease Control (CDC) was aborted after 31 percent of the people in the pilot study refused to participate, despite assurances of confidentiality (Johnson and Murray

1988). Two researchers involved with this project concluded that "it does not seem likely that studies using data on HIV risk or infection status, even with complete protection of individual identity, will be practical until the stigma of AIDS diminishes" (Hurley and Pinder 1992, p. 625).

The United States is not the only country whose people do not want to be tested. This resistance seems to be universal. For example, Zairean officials were reluctant to disclose the number of AIDS cases and infection rates to United Nations officials or to allow foreign medical researchers to test Zairean citizens because of their sensitivity to the unsubstantiated but widely held belief that Zaire was the cradle of AIDS (Noble 1989).

Random sampling, however, is the most dependable method we have of determining the number of HIV-infected persons. Until we have such information, we cannot know what factors cause a person with HIV to contract AIDS. Random blood samples would permit comparisons between the lifestyles of infected but symptom-free people and of infected people who have developed AIDS or AIDS-related complex (ARC).[11] From such comparisons we could learn which cofactors cause a healthy carrier to develop ARC or AIDS. Such cofactors might include diet, exposure to hazardous materials, or prolonged exposure to the sun or to tanning booth rays—anything that might compromise the immune system in such a way as to activate a dormant infection. For example, why do some people remain HIV-infected for years without developing AIDS? Why do other people develop AIDS shortly after exposure to HIV? Why is HIV absent from the bodies of some patients with AIDS-like symptoms (Liversidge 1993)? Finally, how is it

that at one time HIV may have been a harmless virus? How did the virus change to become the causative pathogen of AIDS? (Some scientists speculate that HIV has been around in an inactive state for at least a century.)[12]

The point is that a person may develop AIDS as a result of other factors besides the behavior that causes a person to contract HIV infection. As long as there is no systematic and random sampling of populations, people will continue to speculate on these factors, either overestimating or underestimating the prevalence of infection and the projected numbers of AIDS cases worldwide.

This situation leaves people with a dilemma about how to deal with a complex health problem such as AIDS when there are so many unanswered questions. Most people do not have the time to inform themselves about all of the contextual forces underlying HIV and AIDS. Yet, this does not stop people from acting or attributing cause on the basis of limited information. Even if people don't have the time to inform themselves, they do have the option of at least being critical of their information sources; for most people, the source is television.

Public health officials believe that the media need to deliver at least one important message with regard to HIV: "It is not who you are; it is how you live and what you do" (S. Kramer 1988, p. 43). For the most part, however, this has not been the message transmitted to the American public. Because television news and information shows are an important source of information for most Americans (98 percent of American households have television sets), we will examine how television producers present information in general and the AIDS phenomenon in particular.

Television: A Special Case of Reality Construction

When sociologists say that reality is constructed, they mean that people assign meaning to interaction or to some other event. When people assign meaning they almost always emphasize some aspects of that event and ignore others. In this chapter we have looked at some of the strategies that people use to construct reality. Consider the following points:

1. When people assign meaning, they tend to ignore the larger context.
2. When they interact with others, they assign meaning first by determining their own social status in relation to the other parties and then by drawing upon learned expectations of how people in some social statuses are to behave.

3. People create reality when they engage in impression management. That is, they manage the setting, their dress, their words, and their gestures to correspond to impressions they are trying to make.

4. People control access to the back stage so that outsiders to the back stage form opinions on the basis of the front stage.

5. People attribute cause to dispositional traits when evaluating others' behavior, and they attribute cause to situational factors when evaluating their own behavior.

People also assign meaning to events based on first-hand experiences. Television gives people access to events that people would never have the chance to experience if left to their own resources. Thus, our analysis of how people come to construct the reality they do would not be complete if we ignored the format that television, especially television news, uses to present information about what is going on in the world.

Television conquers time and space: it allows us to see what is going on in the world as soon as it happens. However, consider the following features of national and local news—television at its most serious and most informative:

- The average length of a camera shot is 3.5 seconds.

- Every three or four news items, no matter how serious, are followed by three or four commercials.

- The news of the day is often presented as a series of sensationalized images.

- Approximately 15 news items are presented in a 30-minute news program with approximately 12 different commercials.

- Most news coverage of events focuses on the moment; each item is presented without a context.

In *Amusing Ourselves to Death*, Neil Postman (1985) examines the format of news programs and asks how it affects how people think about the world. Overall, he believes, this format gives viewers the impression that the world is unmanageable and that events just seem to happen. More to the point, it gives the public only the most superficial

facts about events in the world. For example, most people in the United States know that AIDS exists and they know basic facts about how the virus is transmitted and about how to reduce risk of transmission. On the other hand, a large percentage (71 percent) also admit that they do not know a lot about AIDS (U.S. Department of Health and Human Services 1992).

What characteristics of the news format produce this consequence? The brevity of camera shots is probably the main problem. Postman argues that with the average length of a shot being just 3.5 seconds, facts are pushed into and out of consciousness in rapid succession and viewers do not have sufficient time to reflect on what they have seen and to evaluate it properly. Furthermore, commercials defuse the effects of any news event; they give the following message: I can't do anything about what is happening in the world, but I can do something to feel good about myself if I eat the right cereal, own the right car, and use the right hair spray.

Television is image-oriented; a picture is a moment in time and, by definition, is removed from any context. This quality often causes television news to be sensationalistic. The highly publicized case of Ryan White (who died in April 1990 at age 18 from AIDS-related respiratory failure) illustrates just how sensationalistic news reports can be. As a child, White had received HIV-infected Factor VIII while being treated for hemophilia and later was diagnosed as having AIDS. White was barred from attending Western Middle School in Kokomo, Indiana in August 1985 after school officials learned that he had AIDS (Kerr 1990). When White returned to school in April 1986, reporters and camera crews covered the event in what the school principal termed a sensationalistic manner:

> *I understand that the media has a job to do, but I think there is a fine line between informing the public and creating controversy in order for a story to keep continuing.*
>
> *It seems like the problems were brought out by those who jumped on the sensational. These were published and displayed on television and it created an excited atmosphere in what was really a pretty calm school situation. . . .*
>
> *When Ryan did come back, there were 30 kids whose parents took them out of school.*

The story of Ryan White, a hemophiliac boy who contracted HIV, was widely publicized in the news media. How does this kind of selective coverage change people's view of a phenomenon like AIDS?

Mary Ann Carter/Sipa Press

That's what made the news, but there were 365 other children who stayed in the school. (COLBY 1986, P. 19)

An image-oriented format tends to ignore those historical, social, cultural, or political contexts that would make the event more understandable. Viewers are left with vivid, sensationalistic images of enraged parents, an emaciated gay AIDS patient, a skid row drug addict, a prostitute, a family home destroyed by fearful neighbors, Africans walking to an AIDS clinic, or gays protesting discrimination.

Rarely, however, do television producers present AIDS as a chronic but often manageable condition or portray those with AIDS as leading responsible lives. Imagine what people's initial attitudes toward AIDS might have been if the first discussion of the disease had centered around the life of someone like Dr. Rask instead of a small group of homosexual males. Trying to understand how Dr. Rask contracted HIV would certainly have told us more about the complex social origin of the disease than focusing on homosexual practices.

This is not to deny that television is an important tool for informing large audiences. Despite the sensationalistic coverage most people in the United States know that AIDS exists and that it is related in some way to a virus, but some confusion remains about how HIV is transmitted. For example, about 29 percent of people in the United States believe that a person runs the risk of becoming infected from donating blood, and 10 percent don't know whether donating blood puts them at risk (U.S. Department of Health and Human Services 1992).

To illustrate how little context television provides, you might list the things you have learned about AIDS from this chapter that you did not learn from the media. One might argue that there is not enough time to learn about AIDS from television because so many other important events are competing for attention in the news. Still, this point does not eliminate the need for context in understanding events.

Another way to illustrate how little context television provides is to count how many events (other than sports and weather items) you remember from last night's news. If you can't remember many events, then the information may have been presented so quickly that you could not reflect long enough to absorb it. To remedy this problem, newscasters might reduce the number of stories, increase the time given to context, and cover stories in less reactive and more reflective ways. In the case of AIDS, for example, producers could show segments explaining how HIV differs from AIDS or how the television image of AIDS is different from the experience of AIDS. With regard to this last suggestion, television producers are quite good at covering and discussing presidential campaigns as media events. They could make a similar analysis with regard to AIDS (Treichler 1989).

Diamond miners in Zaire.
J. Abbas/Leo de Wys Inc.

Discussion

Dr. Rask represents one person with AIDS caught up in the historical events that led to worldwide economic interdependence that, in turn, brought people from all over the world into contact with one another. A greater knowledge of this context helps us see what larger circumstances put her at risk for HIV infection.

The importance of considering interactions on a global scale is supported by the fact that disease patterns historically are affected by changes in population density and changes in transportation, both of which bring previously isolated groups into contact with one another. We focused on Zaire and on the interconnections between its people and people from other countries to show that a complex set of interactions lies behind the transmission of HIV infection. This complexity makes it impossible to state conclusively that HIV originated in Zaire. Even if a previously isolated group in Zaire such as the Mbuti was identified as the group that harbored the virus in a dormant state, could we in good conscience define the Mbuti as the source? What about the many forces that brought them out of isolation into contact with groups having no immunity to the virus?

Context is only one dimension of the social interaction process that sociologists examine. They also seek to understand the content of social interaction. Sociologists use the dramaturgical and the attribution models along with the concepts of status and role, rights, obligations, impression management, front stage, and back stage to analyze the dialogue, actions, and reactions that take place when people interact, as well as to understand how the parties involved in social interaction strive to define, interpret, and attach meaning to the encounter. People associated with Dr. Rask, for example, assumed a scientific framework to define the origin, significance, and treatment of her condition. As a result, Dr. Rask experienced not only a physical state of sickness but a social state in which her behavior and the behavior of the involved parties reflected the assumption that her physiological malfunction could be understood and corrected with the available medical technologies (tests, drugs, machines, surgery, and so on).

This chapter delivers several other important lessons. First, dispositional explanations that point to a group or to characteristics supposedly inherent in members of that group are simplistic and potentially destructive. When the focus is the group, the solutions are framed in terms of controlling that group rather than understanding the problem and finding solutions. Situational explanations, on the other hand, make people feel less secure over the short term but in the long run offer the best hope of finding a lasting solution.

Second, although television is an important tool for informing large audiences (for example, that AIDS exists and that preventive measures can be taken), it has some limitations. Because television is image-oriented, newscasters tend not to present the historical, social, political, or cultural contexts that would make news events more understandable. The emphasis on image rather than content creates viewers who know, for instance, that AIDS exists but little else about AIDS. This point suggests that television information must be supplemented with information from alternative sources.

Third, even when people understand the risks of a disease such as AIDS and know the ways to prevent infection, it does not necessarily follow that they will behave accordingly. Goffman's concept of impression management reminds us that social meanings associated with preventive measures are powerful and can cause people to resist using those measures even in clearly high-risk situations.

Most important, the global transmission of HIV infection illustrates a point about interdependence on a global scale: when something goes wrong in one part of the world, other parts of the world are affected. Our discussion of the world's reliance on blood products from the United States illustrates

this point; so do many other phenomena, such as global warming and illegal drug trade. Interdependence is not necessarily a negative situation. It can also lead to greater efforts to solve problems and keep the system running. As futurist John Naisbitt suggests, "If we get sufficiently interlaced economically, we will probably *not* bomb each other off the face of the planet" (1984, p. 79). Such interdependence also means that "AIDS cannot be stopped in any country unless it is stopped in all countries" (Mahler, quoted in Sontag 1989, p. 91); "it cannot be mastered in the West unless it is overcome everywhere" (Rozenbaum, quoted in Sontag 1989, p. 91). These comments suggest that if AIDS policies are to be effective, they must involve worldwide effort.

Such an effort is underway. The World Health Organization (WHO) is sponsoring, directing, and coordinating a global strategy to prevent HIV and control its transmission. The Global Programme on AIDS supports national AIDS programs in 150 countries that include health education, prevention information, blood transfusion services, and cross-cultural research on human behavior and effective communication. WHO officials maintain that this program offers health benefits beyond the prevention of AIDS: "The global response to AIDS offers a great opportunity to accelerate the strengthening of our health care infrastructures" in general (World Health Organization 1988, p. 15).

FOCUS
The Effect of Skin Color and Gender on Social Interaction

The concepts introduced in this chapter can be applied to understanding any social interaction. A good example can be found in the book *Streetwise* by sociologist Elijah Anderson (1990). In this book Anderson describes some all-too-common interaction scenarios between black males (an ascribed status) and police officers. He describes how in the eyes of many law enforcers skin color overshadows all other attributes the black male may possess. In addition he describes the pressures that such a perspective puts on black males. Finally, Anderson describes some impression management strategies that black males use in the presence of police officers.

The Police and the Black Male
Elijah Anderson

The police, in the Village-Northton as elsewhere, represent society's formal, legitimate means of social control. Their role includes protecting law-

SOURCE: Adapted from "The Police and the Black Male," pp. 190, 194–97 in *Streetwise: Race, Class, and Change in an Urban Community* by Elijah Anderson. Copyright © 1990 by the University of Chicago Press. Reprinted by permission.

abiding citizens from those who are not law-abiding, by preventing crime and by apprehending likely criminals. Precisely how the police fulfill the public's expectations is strongly related to how they view the neighborhood and the people who live there. On the streets, color-coding often works to confuse race, age, class, gender, incivility, and criminality, and it expresses itself most concretely in the person of the anonymous black male. In doing their job, the police often become willing parties to this general color-coding of the public environment, and related distinctions, particularly those of skin color and gender, come to convey definite meanings. Although such coding may make the work of the police more manageable, it may also fit well with their own presuppositions regarding race and class relations, thus shaping officers' perceptions of crime "in the city." Moreover, the anonymous black male is usually an ambiguous figure who arouses the utmost caution and is generally considered dangerous until he proves he is not. . . .

There are some who charge— . . . perhaps with good reason—that the police are primarily agents of the middle class who are working to make the area more hospitable to middle-class people at the expense of the lower classes. It is obvious that the police assume whites in the community are at least middle class and are trustworthy on the streets. Hence the police may be seen primarily as protecting "law-abiding" middle-class whites against anonymous "criminal" black males.

To be white is to be seen by the police—at least superficially—as an ally, eligible for consideration and for much more deferential treatment than that accorded blacks in general. This attitude may be grounded in the backgrounds of the police themselves. Many have grown up in Eastern City's "ethnic" neighborhoods. They may serve what they perceive as their own class and neighborhood interests, which often translates as keeping blacks "in their place"—away from neighborhoods that are socially defined as "white." In trying to do their job, the police appear to engage in an informal policy of monitoring young black men as a means of controlling crime, and often they seem to go beyond the bounds of duty. The following field note shows what pressures and racism young black men in the Village may endure at the hands of the police:

At 8:30 on a Thursday evening in June I saw a police car stopped on a side street near the Village. Beside the car stood a policeman with a young black man. I pulled up behind the police car and waited to see what would happen. When the policeman released the young man, I got out of my car and asked the youth for an interview.

"So what did he say to you when they stopped you? What was the problem?" I asked. "I was just coming around the corner, and he stopped me, asked me what was my name, and all that. And what I had in my bag. And where I was coming from. Where I lived, you know, all the basic stuff, I guess. Then he searched me down and, you know, asked me who were the supposedly tough guys around here? That's about it. I couldn't tell him who they are. How do I know? Other gang members could, but I'm not from a gang, you know. But he tried to put me in a gang bag, though." "How old are you?" I asked. "I'm seventeen, I'll be eighteen next month." "Did he give any reason for stopping you?" "No, he didn't. He just wanted my address, where I lived, where I was coming from, that kind of thing. I don't have no police record or nothin'. I guess he stopped me on principle, 'cause I'm black." "How does that make you feel?" I asked. "Well, it doesn't bother me too much, you know, as long as I know that I hadn't done nothin', but I guess it just happens around here. They just stop young black guys and ask 'em questions, you know. What can you do?"

On the streets late at night, the average young black man is suspicious of others he encounters, and he is particularly wary of the police. If he is dressed in the uniform of the "gangster," such as a black leather jacket, sneakers, and a "gangster cap," if he is carrying a radio or a suspicious bag (which may be confiscated), or if he is moving too fast or too slow, the police may stop him. As part of the routine, they search him and make him sit in the police car while they run a check to see whether there is a "detainer" on him. If there is nothing, he is allowed to go on his way. After this ordeal the youth is often left afraid, sometimes shaking, and uncertain about the area he had previously taken for granted. He is upset in part because he is painfully aware of how close he has come to being in "big trouble." He knows of other youths who have gotten into a "world of trouble" simply by being on the streets at the wrong time or when the police were pursuing a criminal. In these circumstances, particularly at night, it is relatively easy for one black man to be mistaken for another. Over the years, while walking through the neighborhood I have on occasion been stopped and questioned by police chasing a mugger, but after explaining myself I was released.

Many youths, however, have reason to fear such mistaken identity or harassment, since they might be jailed, if only for a short time, and would have to post bail money and pay legal fees to extricate themselves from the mess. When law-abiding blacks are ensnared by the criminal justice system, the scenario may proceed as follows. A young man is arbitrarily stopped by the police and questioned. If he cannot effectively negotiate with the officer(s), he may be accused of a crime and arrested. To resolve this situation he needs financial resources, which for him are in short supply. If he does not have money for any attorney, which often happens, he is left to a public defender who may be more interested in going along with the court system than in fighting for a poor black person. Without legal support, he may well wind up "doing time" even if he is innocent of the charges brought against him. The next time he is stopped for questioning he will have a record, which will make detention all the more likely.

Because the young black man is aware of many cases when an "innocent" black person was wrongly accused and detained, he develops an "attitude" toward the police. The street word for police is "the man," signifying a certain machismo, power, and authority. He becomes concerned when he notices "the man" in the community or when the police focus on him because he is outside his own neighborhood. The youth knows, or soon finds out, that he exists in a legally precarious state. Hence he is motivated to avoid the police, and his public life becomes severely circumscribed.

To obtain fair treatment when confronted by the police, the young man may wage a campaign for social regard so intense that at times it borders on obsequiousness. As one streetwise black youth said:

"If you show a cop that you nice and not a smart-ass, they be nice to you. They talk to you like the man you are. You gonna get ignorant like a little kid, they gonna get ignorant with you." Young black males often are particularly deferential toward the police even when they are completely within their rights and have done nothing wrong. Most often this is not out of blind acceptance or respect for the "law," but because they know the police can cause them hardship. When confronted or arrested, they adopt a particular style of behavior to get on the policeman's good side. Some simply "go limp" or politely ask, "What seems to be the trouble, officer?" This pose requires a deference that is in sharp contrast with the youth's more usual image, but many seem to take it in stride or not even to realize it. Because they are concerned primarily with staying out of trouble, and because they perceive the police as arbitrary in their use of power, many defer in an equally arbitrary way. Because of these pressures, however, black youths tend to be especially mindful of the police and, when they are around, to watch their own behavior in public. Many have come to expect harassment and are inured to it; they simply tolerate it as part of living in the Village-Northton.

After a certain age, say twenty-four, a black man may no longer be stopped so often, but he continues to be the object of police scrutiny. As one twenty-seven-year-old black college graduate speculated:

I think they see me with my little bag with papers in it. They see me with penny loafers on. I have a tie on, some days. They don't stop me so much now. See, it depends on the circumstances. If something goes down, and they hear that the guy had on a big black coat, I may be the one. But when I was younger, they could just stop me, carte blanche, any old time. Name taken, searched, and this went on endlessly. From the time I was about twelve until I was sixteen or seventeen, endlessly, endlessly. And I come from a lower-middle-class black neighborhood, OK, that borders a white neighborhood. One neighborhood is all black, and one is all white. OK, just because we were so close to that neighborhood, we were stopped endlessly. And it happened even more when we went up into a suburban community. When we would ride up and out to the suburbs, we were stopped every time we did it.

If it happened today, now that I'm older, I would really be upset. In the old days when I was younger, I didn't know any better. You just expected it, you knew it was gonna happen. Cops would come up, "What you doing, where you coming from?" Say things to you. They might even call you nigger.

Such scrutiny and harassment by local police makes black youths see them as a problem to get beyond, to deal with, and their attempts affect their overall behavior. To avoid encounters with the man, some streetwise young men camouflage themselves, giving up the urban uniform and emblems that identify them as "legitimate" objects of police attention. They may adopt a more conventional presentation of self, wearing chinos, sweat suits, and generally more conservative dress. Some youths have been known to "ditch" a favorite jacket if they see others wearing one like it, because wearing it increases their chances of being mistaken for someone else who may have committed a crime.

But such strategies do not always work over the long run and must be constantly modified. For instance, because so many young ghetto blacks have begun to wear Fila and Adidas sweat suits as status symbols, such dress has become incorporated into the public image generally associated with young black males. These athletic suits, particularly the more expensive and colorful ones, along with high-priced sneakers, have become the leisure dress of successful drug dealers, and other youths will often mimic their wardrobe to "go for bad" in the quest for local esteem. Hence what was once a "square" mark of distinction approximating the conventions of the wider culture has been adopted by a neighborhood group devalued by that same culture. As we saw earlier, the young black male enjoys a certain power over fashion: whatever the collective peer group embraces can become "hip" in a manner the wider society may not desire. These same styles then attract the attention of the agents of social control.

Key Concepts

Achieved Status 199	Impression Management 202	Scapegoat 210
Ascribed Status 199	Mechanical Solidarity 188	Sick Role 200
Back Stage 205	Obligations 200	Situational Factors 208
Content 186	Organic Solidarity 190	Social Interactions 186
Context 186	Rights 200	Social Status 199
Dispositional Traits 208	Role 200	Social Structure 199
Division of Labor 188	Role Conflict 200	Solidarity 188
Dramaturgical Model 202	Role Set 200	Stigmas 199
Front Stage 205	Role Strain 200	

Notes

1. Although Mobutu's latest term in office officially ended in 1991, presidential elections have yet to be held.

2. In Zaire the chief "physicians" are traditional healers. Only a small proportion are like the witch doctors of legend. (In fact, many wear business suits.) Traditional healers operate from a variety of beliefs and practices, but most approach health care from a holistic perspective.

3. Currently, many Western doctors are showing considerable interest in traditional African medicine. There seems to be a movement toward in-

corporating elements of both scientific and traditional medicine into health care and toward learning more about the curative properties of the herbs and remedies used by healers before the plants that supply them become extinct (because of the destruction of the rain forests). Apparently, traditional medicines work faster and better for some diseases (such as hepatitis) than prescribed pharmaceuticals (Lamb 1987).

4. In Zaire the infant mortality rate is 130 per thousand; one-half of all children in Zaire die before they reach the age of five.

5. Many medical treatments depend on blood products: "Blood transfusions save the lives of premature infants and are crucial for children with Cooley's anemia and other hereditary blood disorders. A vaccine against hepatitis B is derived from blood. Injections of gamma globulin, prepared from blood, are effective in helping prevent hepatitis A, chicken pox, rabies, and other ailments. Platelet transfusions are a key to some cancer treatments" (Altman 1986, p. A1).

6. Today there are three checks to prevent HIV-infected people from donating blood (Zuck 1988):
   - Voluntary self-exclusion by high-risk donors.
   - Confidential exclusion when donors feel pressured to give blood. (After giving blood they can tell the technician drawing blood that they wish their blood to be used for research only.)
   - HIV antibody testing to detect infected units.

7. When health officials comment on the safety of the blood supply, they usually place the risk of receiving contaminated blood products in the context of other events. The risk of receiving contaminated blood is estimated to be 1 in 40,000. The odds of death from influenza are 1 in 5,000 cases and from an automobile accident, 1 in 5,000 per year (Zuck 1988).

8. Generally, people use different criteria when they attribute cause to their own behavior than when they attribute cause to another's behavior. With regard to other people's failures or shortcomings, we tend to overestimate the extent to which they are caused by dispositional traits. In contrast, we overestimate the extent to which our own failures are due to situational factors. With regard to success, we tend to exaggerate the extent to which others' successes are caused by situational factors (a lucky break; they were in the right place at the right time). We tend to explain our own successes as resulting from dispositional factors such as personal effort and sacrifice.

9. African officials become outraged when the Western media equate polygynous marriage arrangements with promiscuity. One villager responded as follows to this charge: "'We have several wives, and we are faithful to them all, and we care for all their children until we die. You people can not even be faithful to one wife, and your children are such a nuisance to you that you send them away from home as soon as they can walk'" (Turnbull 1962, p. 28).

10. This theory has some basis in fact. In 1969 the U.S. Defense Department discussed the theoretical possibility of the following development:
    Within the next 5 to 10 years it would probably be possible to make a new infective micro-organism which could differ in certain important respects from any known disease-causing organisms. Most important of these is that it might be refractory to the immunological and therapeutic processes upon which we depend to maintain our relative freedom from infectious disease. (HARRIS AND PAXMAN 1982, P. 241)

11. AIDS-related complex (ARC) is a term applied to HIV-infected persons whose symptoms do not meet the definition of AIDS set forth by the CDC. ARC encompasses a wide range of symptoms, including fever, rash, and bacterial or viral infections, most of which point to a diminished ability of the body's immune system.

12. Actually, the HIV-AIDS connection is not a clear-cut one. In recognition of this, more than 100 biologists from around the world have formed an organization (the Group for the Scientific Reappraisal of the HIV/AIDS Hypothesis) with the purpose of rethinking this connection. This group also issues a newsletter titled *Rethinking AIDS* (Liversidge 1993).

7 SOCIAL ORGANIZATIONS
with Emphasis on the Multinational Corporation in India

Golden Temple, Varanasi, India.
Herbert Lanks/Monkmeyer Press

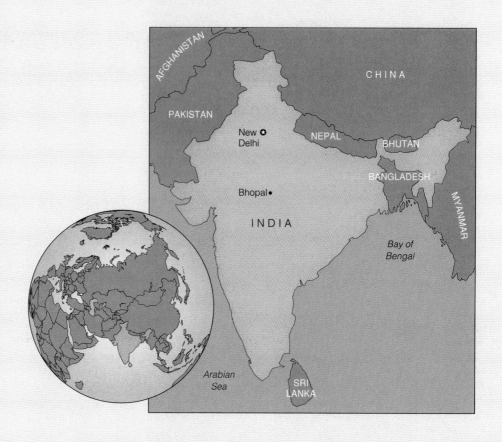

Students from India enrolled in college in the United States for fall 1991	29,000
Indians admitted into the United States in fiscal year 1991 for temporary employment	8,747
People living in the United States in 1990 who were born in India	463,000
Airline passengers flying between the United States and India in 1991	100,237
U.S. military personnel in India in 1993	30
Applications for utility patents filed in the United States in 1991 by inventors from India	51
Phone calls made between the United States and India in 1991	22,327,000

भारत
INDIA
25

उद्योग
Industries

On December 3, 1984, approximately 40 tons of methylisocyanate (MIC), a highly toxic, volatile, flammable chemical used in making pesticides, escaped from a Union Carbide storage tank and blanketed the densely populated city of Bhopal, India. Investigators determined that between 120 and 140 gallons of water somehow had entered the storage tank containing MIC. The combination triggered a violent chemical reaction that could not be contained. As a result, approximately 800,000 residents awoke coughing, vomiting, and with eyes burning and watering. They opened their doors and joined the "largest unplanned human exodus of the industrial age":

> Those able to board a bicycle, moped, bullock car, bus, or vehicle of any kind did. But for most of the poor, their feet were the only form of transportation available. Many dropped along the way, gasping for breath, choking on their own vomit and, finally, drowning in their own fluids. Families were separated; whole groups were wiped out at a time. Those strong enough to keep going ran 3, 6, up to 12 miles before they stopped. Most ran until they dropped.
> (WEIR 1987, P. 17)

Although exact numbers are not known, the most conservative estimates are that the chemical accident killed at least 2,500 people and injured another 250,000 (see Table 7.1). Many of the injured live with the long-term and chronic side effects of their exposure, which include lung and kidney damage, visual impairment, skin diseases and eruptions, neurological disorders, and gynecological damage (Everest 1986). In 1989 Union Carbide paid $470 million to the Indian government as compensation to the victims and their families. This breaks down to about $3,000 per family affected. However, legal complications have delayed these payments. As of 1993 only about 700 people had received some payment (Hazarika 1993).

TABLE 7.1 Estimates of Casualties Caused by Bhopal Accident			
	Deaths	**Injuries**	**Source of Data**
Indian government	1,754	200,000	Lawsuit filed in New York
Indian newspapers	2,500	200,000–300,000	*Times of India, India Today*
U.S. newspapers	2,000+	200,000	*The New York Times, Washington Post*
Voluntary organizations	3,000–10,000	300,000	[estimates]
Delhi Science Forum	5,000	250,000	*Social Scientist*
Eyewitness interviews	6,000–15,000	300,000	Personal interviews*
Circumstantial evidence of death	10,000	—	Shrouds sold Cremation wood used Missing persons estimated
Claims against Union Carbide substantiated by medical records	4,000	3,500	*The New York Times*
Total claims against Union Carbide	13,000	330,000	*The New York Times*

*Personal interviews with over 100 residents of Bhopal were conducted beginning on December 3, 1984. Even on the morning of December 3, local residents claimed that nearly 6,000 people had died.
SOURCES: From *Bhopal: Anatomy of a Crisis*, by Paul Shrivastava, p. 55. Copyright © 1992 by Paul Shrivastava. Reprinted by permission of the author and Paul Chapman Publishing, Ltd., London. Also adapted from *The New York Times* (1992) and Hazarika (1993).

Whenever there is a technology-related disaster involving the loss of human life, investigators search the scene and interview people, looking to uncover explanations. In the case of the Bhopal tragedy, some explanations that investigators found for what happened at Bhopal include technology failures, managerial and operating flaws, training deficiencies, insufficient staff, inadequate safety procedures, and widespread inattention to known risks. In the final analysis it seemed to investigators that almost everyone involved with the Bhopal plant had in some way avoided responsibility for correcting known risks or anticipating dangers. Investigators also concluded that the disaster could have been prevented had officials made public the available knowledge about the extreme toxicity and reactivity of MIC, and had they considered emergency plans to cope with an accident involving MIC (Jasanoff 1988).

These explanations and conclusions are not unique to the incident at Bhopal. They underlie other disasters such as those at Three Mile Island and Chernobyl and the explosion of the space shuttle *Challenger*. Many industrial accidents occurred in the 1980s, posing dangers to both humans and the environment (see Table 7.2). In fact, the dominant theme of most disaster investigations is that workers at virtually every level—from maintenance workers to the chief executive officer—somehow ignore, do not receive, fail to act on, fail to enforce, or fail to pass along information that could prevent the disaster from occurring. This finding suggests that the causes of so-called accidents go beyond individual mistakes or technological failures. Rather, it is something about the organizations themselves that halts the spread of important information and prevents responsibility for problems from falling on any one individual.

Baldev/Sygma

Morvan/Sipa Press

The chemical accident at Bhopal killed at least 2,500 people and injured another 250,000. Sociologists ask how the nature of organizations may contribute to tragedies like this one.

In this chapter we examine concepts that sociologists use to analyze an **organization,** defined as a coordinating mechanism created by people to achieve stated objectives. Those objectives may be to maintain order (for example, a police department); to challenge an established order (for example, the Consumers Union of the United States[1]); to keep track of people (a census bureau); to grow, harvest, or process food (Pepsico); to produce goods (Sony); to make pesticides (Union Carbide); or to provide a service (a hospital) (Aldrich and Marsden 1988).

From a sociological perspective, organizations can be studied apart from the people who make them up. This is because organizations have a life that extends to some degree beyond the people who constitute them. This idea is supported by the simple fact that organizations continue on even as their members die, quit, or retire, or get fired, promoted, or transferred. Organizations are a taken-for-granted aspect of life:

> *Consider, however, that much of an individual's biography could be written in terms of encounters with [them]: born in a hospital, educated in a school system, licensed to drive by a state agency, loaned money by a financial institution, employed by a corporation, cared for by a hospital and/or nursing home, and at death served by as many as five organizations—a law firm, a probate court, a religious organization, a mortician, and a florist.* (ALDRICH AND MARSDEN 1988, P. 362)

Because organizations are so much a part of our lives, we rarely consider how they operate, how much power they have, and how much social responsibility they assume.

227

TABLE 7.2   Selected Industrial Accidents of Environmental Significance, World, 1981–1993*

Date	Country and Location	Origin of Accident	Products Involved	Number of Deaths[†]	Number of Injured	Number of Evacuated
1981	Puerto Rico, San Juan	Rupture in factory	Chlorine	—	200	2,000
	USA, Gelsmar	Plant	Chlorine	—	140	—
	Mexico, Montana	Derailment	Chlorine	29	1,000	5,000
	USA, Castalc	Plant	Propylene	—	100	—
	Venezuela, Iacoa	Explosion	Oil	145	1,000	—
1982	USA, Livingston	Rail accident	Chemicals	—	9	3,000
	Venezuela, Caracas	Tank explosion	Explosives	101	1,000	—
	USA, Vernon	Plant	Methyl acrylate	—	355	—
	USA, Taft	Explosion	Acrolein	—	—	17,000
1983	USA, Denver	Rail accident	Nitric acid	—	43	2,000
	Nicaragua, Corinto	Tank explosion	Oil	—	—	23,000
1984	USA, Sauget	Plant	Phosph. oxychloride	—	125	—
	Brazil, Sao Paulo	Pipeline explosion	Gasoline	508	—	—
	USA, Peabody	Plant	Benzene	1	125	—
	USA, Linden	Plant	Malathion	—	161	—
	USA, Middleport	Plant	Methyl isocyanate	—	110	—
	Mexico, St. J. Ixhuatopec	Tank explosion	Gas	452	4,248	31,000
	India, Bhopal	Leakage	Methyl isocyanate	2,500	50,000	200,000
	Pakistan, Gahri Bhoda	Gas-pipe explosion	Natural gas	60	—	—
	Peru, Callao	Pipeline explosion	Tetraethyl	—	—	3,000
	Mexico, Matamoras	Fertilizer factory	Ammonia	—	200	3,000
1985	USA, Institute	Fire	Aldicarbe oxide	—	140	—
1986	USSR, Chernobyl	Reactor explosion	Nuclear	29	300	135,000
	Switzerland, Basel	Warehouse fire	Chemicals	—	—	—
1988	USA, Commerce	Chemical reaction	Chlorine	—	—	20,000
1990	Russia, Ufa	Explosion	Phenol	—	110	4,000+
1991	Thailand, Bangkok	Explosion	Methyl bromide and others	3	—	6,000
1993	USA, Milwaukee	Water supply	Protozoan cryptosporidium	40	400,000	—

*This is only a partial listing.
[†]Inclusion criteria are more than 50 dead or more than 100 injured or more than 2,000 evacuated or more than $50 million in damages.

SOURCE: From the Organization for Economic Cooperation and Development 1987; *The New York Times Index* 1988, 1990, 1991, 1993. Reprinted by permission.

The concepts that sociologists use to study organizations can help us understand how they can be powerful coordinating mechanisms that channel individual effort into achieving goals that benefit the lives of many people. At the same time we can use these concepts to help us see how these coordinating mechanisms can contribute to ignorance, misinformation, and failure to take responsibility for known risks. These issues of accountability are particularly relevant when they involve technologies that have the potential to cause irreparable damage to people and to the environment. For this reason we will focus on Union Carbide, a major chemical corporation with headquarters in the United States; specifically, we will focus on Union Carbide's plant in Bhopal, India.

This focus allows us to consider a special kind of organization, the **multinational corporation,** which owns or controls production or service facilities in countries outside the one in which it is headquartered (U.S. General Accounting Office 1978). The emphasis on a multinational corporation in a country such as India (see "A Traveler's View of India") allows us to consider the special issues of multinationals in countries that are poor and that have large populations.[2] These issues include (1) cultural differences between workers who come from one country and members of management who come from another country, (2) the economic conditions of the host country, and (3) the fact that the total operations of a multinational corporation are not subject to laws of one government. This last problem—regulating the multinational corporation—is a "terribly complex and thorny issue" yet to be resolved (Keller 1986, p. 12).

The Multinational Corporation: Agent of Colonialism or Progress?

As stated earlier, multinational corporations (or just "multinationals") are enterprises that own or control production and service facilities in countries other than the one in which their headquarters are located. The United Nations estimates that there are at least 35,000 multinationals worldwide with 150,000 foreign affiliates (Clark 1993). Multinationals are headquartered disproportionately in the United States, Japan, and Western Europe (see Table 7.3). Multinationals compete against rival corporations for global market share, and they plan, produce, and sell on a multicountry and even a global scale. In addition, they recruit employees, extract resources, acquire capital, and borrow technology on a multicountry or worldwide scale (Kennedy 1993; Khan 1986; U.S. General Accounting Office 1978).

Multinationals establish operations in foreign countries in order to obtain raw materials or to make use of an inexpensive labor force (for example, a *maquila* assembly plant in Mexico, a lumber company in Brazil, a mining company in South Africa). They also establish subsidiary companies in foreign countries and employ their citizens to manufacture goods or provide services that are marketed to customers in those countries (for example, IBM Japan).

Hewlett-Packard is one example of an organization that epitomizes the far-flung nature of the multinational corporation that plans on a global scale. It has 95,000 employees working in 475 sales offices in 40 countries and at 55 factory sites in 15 countries. The offices and the factory sites are connected by computer technology that integrates voice, video, and information sharing and generates a staggering eight million pages of text each day (National Public Radio 1990).

Critics maintain that multinational corporations are engines of destruction. That is, they exploit people and resources in order to manufacture products inexpensively. They take advantage of cheap and desperately poor labor forces, lenient environmental regulations, and sometimes nonexistent worker safety standards. According to these critics, multinational corporations represent another kind of colonialism.[3] Advocates of multinational corporations, on the other hand, maintain that these corporations are agents of progress. They praise the

TABLE 7.3   The World's Largest Industrial Corporations, 1991

Company	Headquarters in	Profits ($ millions)	Employees (Number)
1 General Motors	U.S.	(4,452.8)	756,300
2 Royal Dutch/Shell Group	Britain/Netherlands	4,249.3	133,000
3 Exxon	U.S.	5,600.0	101,000
4 Ford Motor	U.S.	(2,258.0)	332,700
5 Toyota Motor	Japan	3,143.2	102,423
6 Intl. Business Machines	U.S.	(2,827.0)	344,553
7 IRI	Italy	(254.1)	407,169
8 General Electric	U.S.	2,636.0	284,000
9 British Petroleum	Britain	802.8	111,900
10 Daimler-Benz	Germany	1,129.4	379,252
11 Mobil	U.S.	1,920.0	67,500
12 Hitachi	Japan	1,629.2	309,757
13 Matsushita Electric Industrial	Japan	1,832.5	210,848
14 Philip Morris	U.S.	3,006.0	166,000
15 Fiat	Italy	898.7	287,957
16 Volkswagen	Germany	665.5	265,566
17 Siemens	Germany	1,135.2	402,000
18 Samsung Group	South Korea	347.3	187,377
19 Nissan Motor	Japan	340.9	138,326
20 Unilever	Britain/Netherlands	1,842.6	298,000
21 ENI	Italy	872.0	N.A.
22 E.I. Du Pont de Nemours	U.S.	1,403.0	133,000
23 Texaco	U.S.	1,294.0	40,181
24 Chevron	U.S.	1,293.0	55,123
25 Elf Aquitaine	France	1,737.1	86,900

SOURCE: From "The World's Largest Multinational Industrial Corporations," *Fortune.* Copyright © 1992 by *Time,* Inc. Reprinted by permission.

multinationals' ability to transcend political hostilities, to transfer technology, and to promote cultural understanding.

In reality, no simple evaluation can be made that would apply to all multinationals (see Table 7.4). Obviously, at some level they "do spread goods, capital, and technology around the globe. They do contribute to a rise in overall economic activity. They do employ hundreds of thousands of workers around the world, often paying more than the prevailing wage" (Barnet and Müller 1974, p. 151). George Keller, chairman of the board and chief executive officer of Chevron Corporation, explains:

I don't want to sound like Pollyanna. But I'm proud that U.S. companies have a well-deserved reputation for contributing to the communities where they operate overseas. They don't do it solely out of charity. They do it because . . . an improvement in the local economic and social infrastructure is essential if they are to operate effectively. In some developing nations, the multinationals may virtually create the basis of a modern economy.

Chevron faced that situation in Saudi Arabia in the mid-thirties. Our people had come to search for oil in what was then a sparsely inhabited desert.

To conduct our operations, we had to help create the necessary environment. We drilled water wells, built roads, and developed electrical

---

TABLE 7.4 Pros and Cons of Multinational Corporations (MNCs)

Frequently Heard MNC Claims of Benefits for Host Nations

Provide new products
Introduce and develop new technical skills
Introduce new managerial and organizational techniques
Promote higher employment
Yield higher productivity
Provide greater access to international markets
Provide for greater accumulation of foreign exchange
Supplement foreign aid objectives and programs of home countries directed toward the host
Serve as a point of contact for host country businesspeople and officials in the home country
Encourage the development of new ancillary or spin-off industries
Assume investment risks that might not have been undertaken by others
Mobilize capital for productive purposes that might have gone to other, less fruitful uses

Frequently Heard Criticisms of MNCs by Host Nations

Lead to a loss of cultural identity and traditions with the creation of new consumer tastes and demands
Be used as channels for foreign (especially U.S.) political influence
Possess a competitive advantage over local industries
Create inflationary pressures
Misapply host country resources
Exploit host country wealth for the primary benefit of the citizens of other nations
Lead to loss of control by hosts over their own economies
Possess neither sufficient understanding nor concern for the local economy, labor conditions, and national se-
    curity requirements
Dominate key industries
Divert local savings from investment by nationals
Restrict access to modern technology by centralizing research and development facilities in the home country
    and by employing home country nationals in key management positions

SOURCE: Adapted from U.S. General Accounting Office (1978).

---

power. *As we made progress, the local communities grew and developed into prosperous cities. Schools and hospitals were built, and local industries emerged.* (KELLER 1986, PP. 125–26)

Even so, the means employed to create the environments that multinational companies need to achieve the maximum profit for owners and stockholders (the valued goal) are not necessarily those that alleviate a host country's problems of mass starvation, mass unemployment, and gross inequality. Critics argue that, if anything, multinationals aggravate these problems because their pursuit of profit is closely related to gross social and ecological imbalances that would be obvious to anyone visiting a country such as India:

What a curious contradiction of rags and riches. One out of every ten thousand persons lives in a palace with high walls and gardens and a Cadillac in the driveway. A few blocks away hundreds are sleeping in the streets, which they share with beggars, chewing gum hawkers, prostitutes, and shoeshine boys. Around the corner tens of thousands are jammed in huts without electricity or plumbing. Outside the city most of the population scratches out a bare subsistence on small plots, many owned by the few who lived behind the high walls. Even

A Traveler's View of India

Poor and *densely populated* are two adjectives that often come to mind when we think of India. As the information in the excerpt below makes clear, this is only one of the many sides of Indian society.

Street scene in Rajasthan.
Pablo Bartholomew/Gamma-Liaison

An exquisite stone screen from a mosque in Ahmedabad.
Lindsay Hebberd/Woodfin Camp & Associates

Travel in India is a total experience, one which no visitor ever forgets. A vast subcontinent of over 800 million people, it covers 22 states, stretches some 2000 miles north to south, and about 1700 miles east to west, and contains more different languages, religions, races and cultures than any other country in the world. It is also one of the oldest civilisations known, dating back some 5000 years, and was in its heyday one of the richest, known as 'The Golden Bird of the East'.

Whatever you want from a holiday, it's here in India. Royal palaces, desert fortresses, beach resorts, hill-stations, temples, mountains and lakes—you'll find them all, and a lot more besides. And India is still one of the cheapest tourist destinations in the world. Where else could you stay at a five-star hotel for less than £40 (US$22) per night, or dine out on the most sumptuous cuisine for less than £10 (US$5.5), or find such good shopping bargains—top-quality silks and brocades, furnishings and paintings, carvings and carpets, jewellery and gems? If you're into sports and recreation, there are good facilities in many cities for golf and fishing, for horse-riding and swimming, for tennis and squash, while in outlying areas you can explore wildlife parks on elephant-back or cross the remote Thar Desert on a camel safari. Finally, don't forget India's rich cultural heritage. Here you can enjoy some of the best music, dance and theatre in the East.

But this is only one side of India. Yes, this is a place of exotic enchantment, of religious mystique, of great natural beauty. But it is also a place of incredible noise, squalor and poverty. As one tourist officer remarked: 'India is a large country, with large problems. Everything here is on a grand scale—both good and bad.' This powerful sense of contrast, this alternation between luxury and poverty, beauty and ugliness, efficiency and chaos, is the key to understanding India. It's something which produces a very ambivalent reaction from foreign travellers, and many develop a 'love-hate' relationship. As one visitor commented:

> In one week in India, I had seen enough hunger and poverty and deprivation to haunt me for a lifetime. But I had also seen a land rich in colour and vitality, where time had stood still for centuries, where the strong culture of the people seemed less corrupted by Western standards than any other country in Asia.

India is a very personal experience, and every traveller's impressions are different. But nobody returns unaffected or unchanged. It's an amazing and contradictory place—very black and white, seldom peaceful, never private . . . and often frustrating. But once it's in your bloodstream, you'll never get it out. You'll never fully understand it and you'll never see it all, but the compulsion to keep on trying will probably send you back time and again.

Perhaps the most absorbing part of India is not her 'sights' but her street-life. The typical Indian street is a living theatre of people shaving, hawking goods, gossiping, making clothes and preparing betel, of sa-cred cows, camels and dogs jostling for position with taxis, bicycles, rickshaws and pilgrims, of heaped mountains of colourful spices, fruits, vegetables and incense. It never seems to stop, and if you like 'action', you'll never want to leave. Everywhere you look, every corner you turn, every person you meet, will make an indelible impression. As one person put it: 'The very next thing you see will probably be the most amazing sight of your life.' If you're into photography, you'll just love it. Even if you aren't, you'll still be hypnotized by it. In India, every glance is a picture.

SOURCE: *Cadogan Guides: India*, by Frank Kusy. Pp. 1–2 in 1989 rev. ed. Copyright © 1987 by Frank Kusy. Reprinted by permission of Globe Pequot Press.

Temples on the shore of a lake in the heart of the Indian Desert.
Lee Day/Black Star

where the soil is rich and the climate agreeable most people go to sleep hungry. The stock market is booming, but babies die and children with distended bellies and spindly legs are everywhere. There are luxurious restaurants and stinking open sewers. The capital boasts late-model computers and receives jumbo jets every day, but more than half of the people cannot read. (BARNET AND MÜLLER 1974, PP. 133–34)

How are multinationals connected with this kind of social imbalance when it seems that "without the technologies and the capital that multinationals help to introduce, developing countries would have little hope of eradicating poverty and hunger" (*Union Carbide Annual Report* 1984, p. 107)?

Features of Modern Organizations

Sociologist Max Weber gives us one framework for understanding the two faces of organizations— organizations capable of (1) efficiently managing people, information, goods, and services on a worldwide scale and (2) promoting inefficient, irresponsible, and destructive actions that can affect the well-being of the entire planet. Weber's ideas about organizations are built on an understanding of rationalization and its significance in modern life.

Rationalization as a Tool in Modern Organizations

In Chapter 1 we learned that, according to Max Weber, the sociologist's main task is to analyze and explain the course and consequences of social action (actions influenced by other people, including the thoughts and feelings that lead to particular actions). Weber recognized the endless variety of social action and maintained that the sociologist's task is to make sense of it, not just to observe it (Lengermann 1974). In light of this task, Weber classified social action into four types, according to the reasons why people pursue a goal: traditional, affectional, value-rational, and instrumental. He also contended that ever since the onset of the Industrial Revolution, an individual's actions are less likely to be guided by tradition or emotion and more likely to be value-rational. Weber was particularly concerned about the value-rational action because, as we learned in Chapter 1, the valued goal can become so all-important that people lose

sight of the negative consequences that can arise from the methods used to reach that goal.

According to Weber, rationalization is a product of human technological and organizational ingenuity and proficiency and it coincides with the specialization, the division of labor, and the mechanization that revolutionized the production process. Weber defined **rationalization** as a process whereby thought and action rooted in emotion (love, hatred, revenge, joy), in superstition, in respect for mysterious forces, and in tradition are replaced by thought and action grounded in the logical assessment of cause and effect or the means to achieve a particular end (Freund 1968).

The thought and action guided by tradition, superstition, and emotion is different from the thought and action guided by the means-to-an-end logic of value-rational action. We can illustrate these differences by comparing two distinctly different meanings applied to trees. One meaning is applied by science and industry. The other is held by the Bonda, a small tribe that lives in the Orissa Mountains in India.

The Bonda reflect a culture rooted in emotion, superstition, and respect for mysterious forces. The Bonda believe that spirits inhabit the earth, the sky, and the water; they believe that sickness, death, and poor harvests are caused by evil spirits. They are particularly respectful of spirits who live in trees and plants. Bonda priests specify which trees can and cannot be cut down. The people do not touch those trees considered to be the homes of gods and genies. Some trees are left standing if the priests

Superstition VS Rational Emotion

believe that their removal would displease phantoms or demons and would cause them to send poor harvests, sickness, and deaths in order to avenge crimes against trees (Chenevière 1987). In essence, the belief that trees, plants, and animals possess souls leads people to feel reverence and respect for nature and to behave accordingly toward it.

Rationalization discredits the idea that plants and trees have spirits. Science and technology have enabled us to break down trees and plants into various components and assign them precise functions in the larger production process. In contemporary society, for example, trees are thought of as a means to an end—a source of food, wood, rubber, quinine (a drug used to combat malaria), turpentine (an ingredient of paint thinner and solvents), cellulose (used to produce paper, textiles, and explosives), and resins (used in lacquers, varnishes, inks, adhesives, plastics, and pharmaceuticals). From a value-rational point of view, nature is something to use: "Rivers are something to dam; swamps are something to drain; oaks are something to cut; mountains are something to sell and lakes are sewers to use for corporate waste" (Young 1975, p. 29).

It is not that people in "rational" environments do not value nature on some level; rather, they place greater value on the goals of profit, employment, convenience, and global competition. For example, *The New York Times* reported on a study published in a leading scientific journal, *Nature*, that justified forest conservation because "revenues generated by harvesting edible fruits, rubber, oils, and cocoa from 2.5 acres of tropical rain forest are nearly two times greater than the return on timber or the value of the land if used for grazing cattle" (1989, p. 24Y). Presumably, if the return on timber were greater than that from harvesting the forest's products, there would be no support for conservation.

One can argue that rationalization has released people from the bondage of superstition and tradition and has given people unprecedented control over nature. One major negative side effect of rationalization, however, is what Weber called the **disenchantment of the world**—a great spiritual void accompanied by a crisis of meaning. As we learned in Chapter 1, disenchantment occurs when the very

process of achieving a valued goal is such that it leaves people with a great spiritual void (as when a college student takes the easiest courses to achieve the goal of getting a college diploma).

Weber made several important qualifications regarding value-rational thought and action. First, he used the term *rationalization* to refer to the way in which daily life is organized socially to accommodate large numbers of people, not necessarily to the way individuals actually think (Freund 1968). For example, a large chemical industry makes the products that enable millions of people to use them. The companies that make pesticides advertise them as a rational means for the quick and efficient killing of bugs in the house, in the garden, or on pets. Yet, most people who buy and use these products have no idea how the chemicals work, where they come from, or what consequences they bring except that they kill bugs. Thus, on a personal level, people deal with pesticides as if they were magic.

Second, rationalization does not assume better understanding or greater knowledge. People who live in a value-rational environment typically know little about their surroundings (nature, technology, the economy). "The consumer buys any number of products in the grocery without knowing what substances they are made of. By contrast, 'primitive' man in the bush knows infinitely more about the conditions under which he lives, the tools he uses and the food he consumes" (Freund 1968, p. 20). Most people are not troubled by such ignorance but are content to let specialists or experts know how things work and how to make corrections when something goes wrong. People assume that if they ever need this information, they can consult an expert or go to the library and look it up.

Finally, instrumental action is rare. When people are determining a goal and deciding on the means (actions) to be employed, they seldom consider and evaluate competing goals or other, more appropriate but less expedient means of reaching the stated goal. For example, people often turn to technology as the means of solving problems that they define as important. Rarely does anyone evaluate the overall consequences of a technology for the quality of life on the planet. Scientist Klaus-Heinrich Standke offers four criteria by which to evaluate a prospective technology:

Ratioalization = disenchantment in environments — of the world

Which technology is "appropriate"? A new school of thought attempts to optimize the parameters by which a prospective technology is to be judged. An "optimum technology" is accordingly that which

(1) is directed towards the highest possible human goals;

(2) uses mineral and energy resources most efficiently and preserves or enhances the environment;

(3) preserves or enhances "good work" for the maximum number of human beings;

(4) uses the very best scientific and technical information and combines them with the wisdom and highest values of the culture.

(STANDKE 1986, P. 66)

More often than not, people set valued goals without first considering the possible disruptive or destructive social consequences of the means or strategies used to reach them. This failure to consider such consequences is at the heart of the "destructive side" of organizations in general and of multinational corporations in particular.

Value-Rational Action: Chemical Companies in India

In the case of India, during the 1960s and 1970s the government encouraged chemical companies such as Union Carbide to locate in India (see Figure 7.1). They were supposed to be part of the Green Revolution, a plan to relieve chronic food shortages and help the country become self-sufficient in food production through agricultural technologies, including treated seeds, pesticides, and fertilizers (Derdak 1988). In addition, the chemical companies used local labor and regional raw materials and thus provided employment opportunities. The manufactured products were used to prevent malaria and other insect-borne diseases and to protect crops and harvests from insects, rodents, and diseases.

The short-term agricultural yields were indeed impressive, but the means chosen to achieve those goals have had negative long-term consequences. For example, only the wealthiest Indian farmers

were able to purchase the chemical technologies. In conjunction with mechanization, chemical technologies allowed the wealthier farmers to farm more efficiently. As a result, they pushed the poor, small farmer, who could not compete, off the land and out of business. In the end, millions of poor farmers migrated to the cities in search of work, but the cities were unable to absorb them.

Another long-term negative consequence is related more directly to the use of chemical substances. Although pesticides succeed in killing most bugs, the few bugs genetically able to tolerate the pesticide survive to reproduce even more resilient offspring. Scientists estimate that there are now 504 insects, or "superbugs," resistant to one or more pesticides (Holmes 1992).

Perhaps one of the most vivid illustrations of the long-term negative consequences of chemical use is the widespread and damaging flooding in northeast India and neighboring Bangladesh as a consequence of the extensive deforestation of the Himalayas. Areas such as the Tarai Forest in the Himalayan foothills were uninhabitable in the past because of the danger of malaria-carrying mosquitoes. Pesticides such as DDT eradicated the mosquitoes and opened forests to lumbering, farming, grazing, and limestone quarrying, thus clearing the land of trees and plants. Now water from rain and melting snow runs unimpeded from the denuded slopes of the Himalayas, carrying chemical-laden soil that clogs and pollutes the rivers. The September 1988 flood put three-quarters of Bangladesh under water and left approximately 30 million people homeless and 1,300 dead from drowning. To compound an already hazardous condition, the malaria-carrying mosquito has developed a resistance to the chemicals and drugs used to prevent the spread of malaria and now threatens more lives than ever. The point is that long-term massive chemical use, although profitable for chemical companies, cannot help to achieve true development goals because the countries in question will pay tremendous externality costs—costs that are not figured into the price of a product but that are nevertheless a price we pay for using or creating a product. In this case, the cost of restoring contaminated and barren environments and assisting people to cope is an externality cost (Lepkowski 1985).

FIGURE 7.1 American Companies in India That Manufacture Chemicals

American Cyanamid Co.
Ashland Chemical Co.
Dow Chemical Co.
Drew Chemical Corp.
Eastman Kodak Co.
Ferro Corporation
Gamlen Chemical Co.
Hercules Inc.

E. F. Houghton & Co.
Lubrizol Corp.
Merck Sharp & Dohme Intl.
Monsanto Co.
Occidental Petroleum Corp.
Pennwalt Corp.
Pfizer Inc.
Rohm & Hass Co.

G. D. Searle & Co.
Selas Corp. of America
Sybron Corp.
Union Carbide Corp.
United Catalysts Inc.
Velsicol Chemical Corp.
Wheelabrator-Frye Inc.

The map below shows the number of American companies and American chemical manufacturers located in each city.

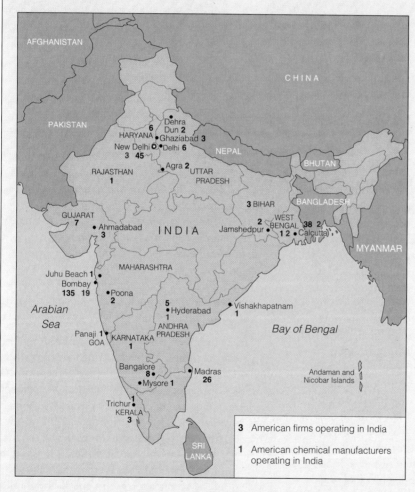

SOURCE: List compiled from the *Directory of American Firms Operating in Foreign Countries*, 11th ed., Vol. 3, pp. 807–30. Copyright © 1987 by World Trade Academy Press. Reprinted by permission.

Dan Habib/Impact Visuals

Robert Wallis/JB Pictures

These cigarette advertisements for Marlboro (downtown Shanghai, left) and Lark (Tokyo, right) do not carry the kind of health warning that is legally required in the United States. When profit-making organizations are not operating under legal constraints, do they have any obligation to protect consumers from the hazards associated with their products?

Multinationals are not charities, of course, and one could argue that they are not responsible for how the people who purchase their products use them. Nevertheless, many people question whether the multinationals and other corporations should have the right to ignore the larger long-term effects of their products on people and on the environment. We can also ask whether it is acceptable for the executives and stockholders of multinational corporations to point to increases in a "poor" country's gross national product or that country's increased ability to export as evidence that multinationals are agents of progress when, by most social measures, their presence has exacerbated world poverty, world unemployment, and world inequality (Barnet and Müller 1974).

In this regard, consider the statement of Keith Richardson, the public affairs chief of B.A.T Industries,[4] the world's largest manufacturer of cigarettes. When asked

whether the company didn't feel some obligation to put on their cigarette packs sold in the Third World the kind of health warnings they are required to carry in Britain, the United States, and other countries [he replied,] "These are sovereign countries, and they are unenthusiastic about being told what to do by pressure groups in the UK and the United States. We fit in with what the government wants in each country. We let the marketplace decide. We do not try to impose." (MOSKOWITZ 1987, PP. 40–41)

Closer to home, consider the role of automobile makers in creating American dependence on foreign oil. (Americans account for five percent of the world's population; yet, they consume 40 percent of the gasoline in the world, two-thirds of which goes toward fuel for transportation.) Consider American consumers' preferences regarding the fuel efficiency of automobiles. Chevrolet offered two versions of its 1990 Corsica. From a standstill, one model took 14 seconds to reach a speed of 60 miles per hour; the other took 10 seconds and used five more gallons of gas per mile (Wald 1990). Most consumers chose the faster, less fuel-efficient model.

In view of the various world oil crises—most recently the Persian Gulf crisis—should corporations produce and market products that are not fuel-efficient, even if those products satisfy public wants? Regardless of the position we take in response to such questions, our views will be more realistic and well informed if we understand how organizations in general operate.

One problem with many business organizations is their apparent unwillingness to accept responsibility for the ways in which people use the products they manufacture and the effects of marketing harmful products. Next we turn to another problem of business organizations—the systematic procedures that they follow to produce and distribute goods and services in the most efficient (especially the most cost-efficient) manner. A key concept to understanding these workings is that of the bureaucracy.

The Concept of Bureaucracy

Weber defined **bureaucracy**, in theory, as a completely rational organization—one that uses the most efficient means to achieve a valued goal, whether that goal is making money, recruiting soldiers, counting people, or collecting taxes. The following are some major characteristics of a bureaucracy that allow it to coordinate people so all of their actions center on achieving the goals of the organization.

- There is a clear-cut division of labor: each office or position is assigned a specific task toward accomplishing the organizational goals.

- Authority is hierarchical: each lower office is under the control and supervision of a higher office.

- Written rules and regulations specify the exact nature of relationships among personnel and describe the way in which tasks should be carried out.

- Positions are filled on the basis of qualifications determined by objective criteria (academic degree, seniority, merit points, or test results) and not on the basis of emotional considerations such as family ties or friendship.

- Administrative decisions, rules, regulations, procedures, and activities are recorded in a standardized format and are preserved in permanent files.

- Authority belongs to the position and not to the particular person who fills the position or office. The implication is that one person can have authority over another on the job because he or she holds a higher position, but those in higher positions can have no authority over another's personal life away from the job.

- Organizational personnel treat clients as "cases" and "without hatred or passion, and hence without affection or enthusiasm" (Weber 1947, p. 340). This approach is necessary because emotion and special circumstances can interfere with the efficient delivery of goods and services.

Taken together, these characteristics describe a bureaucracy as an **ideal type**—ideal not in the sense of being desirable but as a standard against which real cases can be compared. Anyone involved with an organization realizes that the actual behavior departs from the ideal type. Thus one might ask, What is the use of listing essential characteristics if no organization exemplifies them? This list is a useful tool because it identifies important organizational features. (Note, however, that having these traits does not guarantee that things run perfectly; the rules and policies themselves can cause problems.) Rather, by comparing the actual operation to the ideal, one can determine the extent to which an organization departs from these traits or adheres to them too rigidly.

In the case of the Bhopal disaster, serious problems can be linked to either rigid adherence to or blatant departures from official rules, regulations, and procedures. The following list summarizes some of the problems:

- A refrigeration unit designed to keep MIC cool and to inhibit chemical reactions had been turned off for several months.

- Two of the plant's three main safety systems were not working.

- On the eve of the disaster, an employee who did not meet the plant's training requirements was

Residents living around the Institute, West Virginia, Union Carbide plant responded in much the same way as Bhopal residents did to a chemical leak at the plant.

A. Tannenbaum/Sygma

assigned the task of cleaning out an improperly sealed pipe leading to the MIC tank. (Workers suspect that this is probably how water entered the tank.)

• The problem tank containing MIC was filled to 73 percent of capacity rather than the recommended 50 percent. The restriction was imposed so that, in the event of a reaction, there would be more time for corrective action, because a less full tank would cause pressure in the tank to rise less quickly.

• Plant operators failed to move some of the MIC in the problem tank to a spare tank, as required, because they said the spare was not empty, as it should have been.

• Training, experience, and educational qualifications of employees at the Bhopal plant were reduced sharply after the plant began to lose money.

• Instruments monitoring the chemical were unreliable.

• The Bhopal plant did not have the computer system that was present in sister plants in the United States to monitor the chemicals and to alert the staff quickly to problems.

• There were no effective public warnings of the potentially disastrous effects of a chemical accident. Officials did not provide contingency plans or shelters, nor did they distribute brochures or other materials to the people in the area surrounding the plant.

• Most employees panicked as the gas escaped. They failed to behave in recommended ways that could have helped to evacuate residents in company buses (see Everest 1986; Jasanoff 1988; Shabecoff 1988a, 1988b, 1989; Weir 1987 for details on all of these problems).[5]

This list of problems clearly shows that employees at all levels—from managers who did not hire qualified people, inspect equipment, or draft contingency plans in case of emergency to the actions of employees who routinely violated standard operating procedures—did not behave in a responsible manner. In light of this fact, sociologists ask what factors within organizations might be behind these problems.

Factors That Influence Behavior in Organizations

On paper, the job descriptions, the relationships among personnel, and the procedures for performing work-related tasks are well defined and predictable. However, the actual workings of organizations are not as predictable, because the people involved with organizations vary in the extent to which they adhere to rules and regulations. Three factors that influence how people act are the informal relationships they form with others, the way they are trained to do their jobs, and the way their performances are evaluated.

Formal Versus Informal Dimensions of Organizations

Sociologists distinguish between formal and informal aspects of organizations. The **formal dimension** consists of the official, written guidelines, rules, regulations, and policies that define the goals of the organization and its relationship to other organizations and with integral parties (for example, the government or the stockholders). This term also applies to the roles, the nature of the relationships among roles, and the way in which tasks should be carried out to realize the goals. The **informal dimension** includes worker-generated norms that evade, bypass, do not correspond with, or are not systematically stated in official policies, rules, and regulations. In the most general sense, this term applies to behavior that does not correspond to written plans (Sekulic 1978). Examples of worker-generated norms include unwritten rules about standards of interaction, the appropriate content of conversations between employees of different ranks, and the pace at which people should work.

Worker-generated norms are so much a part of daily life that people rarely think about them. Consequently, they are best illustrated by situations in which members of an organization come from different cultures (see "Valuing Cultural Diversity"). *Twin Plant News,* a magazine that covers the *maquila* industry, printed the details of an encounter between an American manufacturing manager (Frank) and a Mexican personnel director (Pablo). It shows that "misunderstandings of [norms governing] each other's methods and motives can disrupt organizational harmony" (McIntosh-Fletcher 1990, p. 32). In the following account, Frank has just arrived at the Mexican plant and is meeting Pablo for the first time:

> *Frank wants to impress upon Pablo that he will be an effective employer. He instinctively takes on his most self-assured, logical, and factual approach, as this has worked in the past. He has prepared to talk about turnover figures, salaries, current staffing problems and other concerns they had in Chicago as related to the Mexican operation.*
>
> *Desiring to anticipate what Frank would expect from a person in his position, Pablo has asked his two assistants to join them and told the entire staff he would introduce Frank. When together, Pablo tries to be gracious, maintain good eye contact and ask about Frank's family. He also refers most of Frank's questions to his assistants to give them recognition and an opportunity to establish a relationship with Frank.*
>
> *Departing an hour later, Frank is determined to speak to the plant manager about getting a Personnel Director he can trust with the facts. Pablo seemed uneasy with his questions and kept looking to his two assistants for the answers. He also expected a person in Pablo's position to have a better command of the English language. [Never mind that Frank doesn't know Spanish.]*
>
> *Once Frank left, Pablo asked his assistants whether they thought anyone would get along with this "gringo." He noted that Frank appeared in a hurry, was impatient with introductions, seemed uncomfortable with all the handshaking among the department staff and even appeared angry.* (MCINTOSH-FLETCHER 1990, P. 32)

One area of informal organization that sociologists have studied quite extensively is worker-generated norms that govern output or physical effort.

Valuing Cultural Diversity

In 1992, the U.S. Department of Labor published a self-instructional package about cultural diversity in the workplace. The package includes a series of case studies in which the reader is given a scenario, asked to answer some questions about that scenario, and then asked to check his or her responses against the "correct" responses. A case study from this manual is included here.

What Is Diversity?

Experiencing diversity in the workplace can be a shock. Diversity means that people of different cultural backgrounds are not always going to act the way you would expect. If you are unprepared for this it can lead to:

- Misunderstanding

- Frustration

- Bad decisions

Diversity in the Workplace: A Case Study

Tran, a Vietnamese by birth, is applying for a supervisory position for which he is well qualified at a large corporation. Marie, the human resource director, conducts the initial interview. After examining Tran's meticulously prepared resume, she is impressed with his qualifications and asks Tran to elaborate on his professional experience.

When Tran speaks, Marie is surprised that he doesn't seem to have much to say, but repeatedly points out that it's all on the resume. Marie becomes confused, thinking that Tran must be lacking in self-confidence if he is so reluctant to mention his professional strengths. She also is disturbed to hear the thickness of his accent and concludes that not only is Tran lacking in self-esteem, but that he must be uneducated to have such a poor grasp of English. Marie decides that despite the strengths of Tran's resume, it would be unwise to recommend hiring him.

Marie has just faced a common challenge confronting companies today: the difficulty to accurately assess the qualifications of foreign-born applicants and workers. Language and cultural differences result in the underestimation of applicants' abilities and poor employee appraisals.

Tran's reluctance to praise himself is a frustrating behavior for American managers. Tran reflects an Asian social value opposed to the values of the majority of Americans. To an Asian (and many Europeans), proper behavior during an employment interview means appearing modest and letting the resume speak for itself. As a job seeker, to call inordinate attention to oneself would be extremely rude. Contrast this with the American interview practice of calling attention to oneself and to one's achievements in a professional setting, which is considered a virtue and a sign of self-respect.

Questions   Please write out your answers to the following two questions. (Answer sheet reads: There is no one correct answer. However, consider the following points.)

1. Was Marie correct in assuming that Tran lacked confidence and was uneducated?

   Answer: She was probably not correct. She failed to detect that Tran and she were following different rules of behavior.

2. What are the consequences of Marie's action?

   Answer: The major consequence for the company is that it may have missed a chance to hire a potentially hard-working and valuable employee. The consequences for Tran may be even more severe.

SOURCE: U.S. Department of Labor (1992), pp. 4–6.

These include informal norms against working too hard (those who do so are often called "rate busters"), working too slowly, or slacking off, as well as norms about the number and length of coffee breaks and the length of the lunch break.

Both positive and negative consequences are associated with informal norms. Informal norms about bending the rules, cutting through red tape, and handling unusual cases or problems, for example, can increase organizational efficiency and effectiveness. Informal norms about after-work activities, friendships, and unofficial communication channels can promote loyalty and work satisfaction. On the negative side, informal norms that put the worker and public safety at risk can have destructive consequences.

At the Bhopal plant, the informal methods that workers used to monitor leaks explain, in part, how the chemical reaction went out of control at the plant on December 3, 1984. Workers could not rely on the alarm systems and gauges that monitored chemical pressure, temperature, level, flow, and composition because these monitoring devices were notoriously unreliable. Usually, the gauges did not show correct readings or did not work. When they were working, they were inadequate because the gauge range was too limited to show critical danger levels. By all accounts, the management was generally unresponsive to workers' complaints about plant equipment. Consequently, workers operated on the assumption that the instruments were inaccurate and monitored the MIC and other chemical leaks according to whether their eyes watered or burned. Their eyes burned and watered frequently, however, so they often ignored even this signal. This method of leak detection was a direct violation of official plant procedures. The Union Carbide technical manual on MIC states: "Although the tear gas effects of the vapor are extremely unpleasant, this property cannot be used as a means to alert personnel." The worker-generated norm for monitoring leaks was also used by Union Carbide workers at the West Virginia chemical production plant (Weir 1987).

The operator monitoring the gauges on the evening of the Bhopal crisis wasn't concerned when pressure gauges connected to the MIC tank rose substantially between 10:30 and 11:00 P.M., and he felt no need to report this change to the shift supervisor. Other operators smelled MIC and noticed that their eyes were irritated, but they maintained that this was no cause for alarm. When they reported the possibility of a leak to the plant supervisor, he decided to look into it after a tea break (Kurzman 1987). Once again, it is important to remember that these kinds of problems are not unique to Bhopal; investigations show that they also existed at other disaster sites such as Three Mile Island and Chernobyl.

There are two other factors that affect the way people behave in organizations—specifically, whether they behave in flexible or rigid ways. One factor is how people are trained to do their jobs. The other is how the organization evaluates worker performance.

Trained Incapacity

If an organization is to operate in a safe, creditable, predictable, and efficient manner, its members need to follow rules, guidelines, regulations, and procedures. Organizations train workers to perform their jobs a certain way and reward them for good performances. However, when workers are trained to respond mechanically to the dictates of the job, they risk developing what economist and social critic Thorstein Veblen (1933) called **trained incapacity**, the inability to respond to new and unusual circumstances or to recognize when official rules and procedures are outmoded or no longer applicable. In other words, workers are trained to do their jobs only under normal circumstances and in a certain way; they are not trained to respond in imaginative and creative ways or to anticipate *what-if* scenarios so that they can perform under a variety of changing circumstances.

In her 1988 book, *In the Age of the Smart Machine*, social psychologist Shoshana Zuboff distinguishes between work environments that promote trained incapacity and those that promote empowering behavior. Zuboff's conclusions are the result of more than a decade of field research in various work environments including pulp mills, a telecommunications company, a dental insurance claims office, a large pharmaceutical company, and the Brazilian offices of a global bank. All of these workplaces had one trait in common: the workers were learning to use computers. We focus here on the experiences of pulp mill workers.

Zuboff found that pulp mill employees who had worked in conventional mills all of their lives were overwhelmed at first by the new condition of having to run the mill while seated at computer screens. These comments illustrate their reactions:

With computerization I am further away from my job than I have ever been before. I used to listen to the sounds the boiler makes and know just how it was running. I could look at the fire in the furnace and tell by its color how it was burning. I knew what kinds of adjustments were needed by the shades of color I saw. A lot of the men also said that there were smells that told you different things about how it was running. I feel uncomfortable being away from these sights and smells. Now I only have numbers to go by. I am scared of that boiler, and I feel that I should be closer to it in order to control it. (P. 63)

When I go out and touch something, I know what will happen. There is a fear of not being out on the floor watching things. It is like turning your back in a dark alley. You don't know what is behind you; you don't know what might be happening. It all becomes remote from you, and it makes you feel vulnerable. It was like being a new operator all over again. Today I push buttons instead of opening valves on the digester. If I push the wrong button, will I screw up? Will anything happen? (PP. 63–64)

With the change to the computer it's like driving down the highway with your lights out and someone else pushing the accelerator. (P. 64)

What strikes me as most strange, hardest to get used to, is the idea of touching a button and making a motor run. It's the remoteness. I can start it from up here, and that is hard to conceive. I can be up in the control room and touch the keyboard, and something very far away in that process will be affected. It takes a while to gain confidence that it will be OK, that what you do through the terminal actually will have the right effects. . . . It's hard to imagine that I am sitting down here in front of this terminal and running a whole piece of that plant outside. The buttons do all the work. (P. 82)

Zuboff believes that management can choose to use computers as automating tools or informing tools. To automate means to use the computer to increase workers' speed and consistency, as a source of surveillance (e.g., by checking up on workers or keeping precise records on the number of keystrokes per minute), and to maintain divisions of knowledge and thus a hierarchical arrangement between management and workers. The pulp

mill workers' comments show that this choice has resulted in trained incapacity:

Currently, managers make all the decisions. . . . Operators don't want to hear about alternatives. They have been trained to do, not to think. There is a fear of being punished if you think. This translates into a fear of the new technology. (P. 74)

Sometimes I am amazed when I realize that we stare at the screen even when it has gone down. You get in the habit and you just keep staring even if there is nothing there. (P. 66)

We had another experience with the feedwater pumps, which supply water to the boiler to make steam. There was a power outage. Something in the computer canceled the alarm. The operator had a lot of trouble and did not look at the readout of the water level and never got an alarm. The tank ran empty, the pumps tripped. The pump finally tore up because there was no water feeding it. (P. 69)

We have so much data from the computer. . . . Operators are tempted not to tour the plant. They just sit at the computer and watch for alarms. One weekend I found a tank overflowing in digesting. I went to the operator and told him, and he said, "It can't be; the computer says my level is fine." I am afraid of what happens if we trust the computer too much. (P. 69)

On the other hand, management can choose to use computers as informating tools. To informate means to empower workers with knowledge of the overall production process, with the expectation that they will make critical and collaborative judgments about production tasks. The pulp mill workers who use the computer as an informating tool experience work very differently than those who use the computer as an automating tool. The following quotes illustrate this point:

Each number is telling you about something, and you draw a picture in your own mind. Each number has a picture, and each number is connected to another number, et cetera, . . . and you get a map. You see the number, the equipment, and all the pieces of the equipment related to it. (P. 87)

Computers can be powerful informating tools that allow workers such as air traffic controllers to manage huge amounts of information in their decision making. But computers can also be used merely as automating devices —or to monitor workers' performances.
Richard Pasley/Stock, Boston

To do the job well now you need to understand this part of the mill and how it relates to the rest of the plant. You need a concept of what you are doing. Now you can't just look around you and know what is happening; you can't just see it. You have to check through the data on the computer to see your effects. And if you don't know what to look for in the data, you won't know what's happening. (P. 94)

[Before automation we] never expected them to understand how the plant works, just to operate it. But now if they don't know the theory behind how the plant works, how can we expect them to understand all of the variables in the new computer system and how these variables interact? (P. 95)

If something is happening, if something is going wrong, you don't go down and fix it. Instead, you stay up here and think about the sequence. . . . You get it done through your thinking. But dealing with information instead of things is very . . . well, very intriguing. I am very aware of the need for my mental involvement now. I am always wondering: Where am I at? It all occurs in your mind now. (P. 75)

Things occur to me now that never would have occurred to me before. With all of this information in front of me, I begin to think about how

to do the job better. And, being freed from all that manual activity, you really have time to look at things, to think about them, and to anticipate. (P. 75)

The computer makes your job easier . . . but it also makes things more complicated. You have to know how to read it and what it means. That is the biggest problem. What does that number actually mean? You have to know this if you want to really learn how to trust the technology. (P. 81)

Virtually all investigative reports of the Bhopal disaster point to trained incapacity as an important contributing factor in causing the runaway chemical reaction. They especially point to the training supplied after the plant began to lose money and was put up for sale. In the early days of operation, well-educated Indian supervisors who had been trained at company headquarters in the United States frequently reminded workers to wear masks, goggles, and protective clothing while handling chemicals. Over time, however, this vigilance weakened, and protective equipment fell into disuse or was not replaced when it wore out. Worker training was reduced from an intensive one-year course to a four-month course and subsequently to a 30-day crash course. Most workers later reported that they were trained to master certain steps but not to handle the chemical in all its conditions. They had no knowledge of the production process as a whole or of the rationale behind many of the rules and procedures, as the following comments from several Bhopal operators reveal:

> *I was trained for one particular area and one particular job, I don't know about other jobs. During training they just said "These are the valves you are supposed to turn, this is the system in which you work, here are the instruments and what they indicate. That's it."* (DIAMOND 1985A, P. A7)

> *[My three months of instrument training and two weeks of theoretical work taught me to operate only one of several MIC systems.] If there was a problem in another MIC system, I don't know how to deal with it.* (DIAMOND 1985A, P. A7)

> *[I] knew the pipe [leading to the MIC storage tank] was unsealed but "it was not my job" to do anything about it.* (1985B, P. A6)

> *The management said MIC could give you a rash on your skin or irritate your eyes. They never said it could kill you.*

> *No one at this plant thought MIC could kill more than one or two people.* (THE NEW YORK TIMES 1985A, P. A6)

As further evidence that workers were trained to do their jobs in a rote manner, most plant workers at all levels had little understanding of the chemicals they worked with, especially of how the chemicals might react under unusual circumstances. Union Carbide published an MIC manual stating that MIC "may cause skin and eye burns on contact. Vapors are extremely irritating and cause chest pain, coughing and choking. May cause fatal pulmonary edema. Repeated exposure may cause asthma" (*The New York Times* 1985a, p. A6). The manual warned that the chemical was toxic, volatile, and flammable. Yet, many workers seemed genuinely surprised that MIC was so extremely dangerous. In fact, many did not receive the manual, which was written in technical English and which even English-speaking workers found difficult to read and understand.

Many newspaper, magazine, and network news reporters suggested that trained incapacity is a problem unique to underdeveloped countries such as India and that, in the case of Bhopal, the lack of *what-if* thinking is rooted in Hindu beliefs. Some reports suggested that inferior Indian labor and resources were behind the tragedy. Many accounts of the Bhopal crisis cite the fact that Indian law required Union Carbide to design, engineer, build, operate, and staff its Bhopal chemical plant with local labor, materials, and machines unless the company could show that local resources were not available. These accounts insinuate that Indian laws required Union Carbide to compromise safety standards. In a similar vein, many of these reports point out that foreign companies were encouraged to use manual production systems to create more jobs for the large unemployed Indian population.

Before we accept such conclusions, we need to ask how many American chemical workers are issued manuals and trained to understand the properties of the chemicals with which they work. The result of a recent Environmental Protection Agency (EPA) study suggests that many people in the United States who work with chemicals do not understand the hazards of these chemicals or know how to handle them except under routine conditions. Between 1980 and 1985, 6,928 chemical accidents occurred in the United States, causing 139 deaths and 1,500 injuries (Diamond 1985c). The large number of accidents reflects that in the United States, as elsewhere, standardized training programs and intensive worker training for runaway chemical reactions are rare. "The best company programs . . . include a month of classes in safety principles and two-week refresher courses yearly. The worst have no classroom or refresher training. . . . They say here's your safety gear—we wear it when things go wrong" (Diamond 1985c, p. D11).

Another EPA report showed that between 1961 and 1989 there were 17 industrial accidents in the United States in which the chemicals released exceeded the amount and toxicity of those released at Bhopal. Fortunately, these chemical releases did not have the deadly consequences of the Bhopal accident because they occurred in remote regions, the wind blew the chemicals away from heavily populated areas, or the chemicals leaked in a liquid state. Liquid chemicals, although highly toxic,

The explosion that ripped through this petrochemical plant in Louisiana was one of almost 7,000 such chemical accidents that occurred in the United States between 1980 and 1985, around the time of the Bhopal incident.
Laura Elliot/Sygma

diffuse more slowly, so that workers have time to contain them or to allow people to be evacuated (Shabecoff 1989). The point is that if people in the United States view the Bhopal crisis as something unique to countries populated by so-called illiterate peasants,[6] they are discouraged from analyzing their own industries and the shortcomings in the ways in which workers are trained.

So far we have looked at informal relationships between people in organizations and at the ways in which people are trained to do their jobs to understand the factors that determine how closely they adhere to organizational rules and regulations. Now we turn to a third factor: how organizations evaluate job performances.

Statistical Records of Performance

In large organizations, supervisors often compile statistics on absenteeism, profits, losses, customer satisfaction, total sales, and production quotas as a way to measure individual, departmental, and overall organizational performance. Such measures can be convenient and useful management tools because they are considered to be objective and precise and because they permit systematic comparison of individuals across time and departments. On the basis of numbers, management can reward good performances through salary increases and promotions and can take action to correct poor performances. We learned in Chapter 3, however, that operational definitions of key variables such as performance are problematic when they are not reliable or valid. In his 1974 book *On the Nature of Organizations,* sociologist Peter Blau examines the problems that can occur when managers use faulty measures without taking their shortcomings into consideration.

One problem with statistical measures of performance is that a chosen measure may not be a valid indicator of what it is intended to measure, or it may measure performance by too narrow a criterion. For example, occupational safety is often measured by the number of accidents that occur on the job. On the basis of this indicator, the chemical industry has one of the lowest accident rates of all industries. This indicator, however, has been criticized as too narrow and lacking validity: chemical workers may be less likely to suffer physical injury on the job than to suffer illnesses whose symptoms go unrecognized as related to chemical exposures. Furthermore, exposure-related illnesses may take years to develop.

Another example of a narrow indicator is one used commonly to regulate workers' exposure to chemicals: this measure is the acceptable number of deaths per 1,000 workers. The Occupational Safety and Health Administration (OSHA) defines as acceptable 6.2 cases of cancer per every 1,000 people who work with formaldehyde. For people who work with arsenic and benzene, 8 cases per 1,000 and 152 per 1,000, respectively, are considered acceptable (Shabecoff 1985). This definition means that after 152 out of 1,000 people who work with benzene die, the risk becomes unacceptable and working conditions must be investigated. Such a measure is problematic because it requires that a certain number of deaths be reached before a situation is considered unusual and before corrective action can be taken.

A second problem with statistical measures of performance is that they encourage employees to concentrate on achieving good scores and to ignore problems generated by their drive to score well. If quarterly profit is used as an indicator of corporate performance, management may do as Bhopal management did—cut employees by 25 percent and cut costs in such critical areas as worker safety, plant maintenance, and employee training (Wexler 1989). Such single-minded efforts are most prevalent in organizations in which employee, departmental, or company performance is evaluated according to rigid measures and where strong sanctions are applied if workers do not meet target figures.

A third problem with statistical measures of performance is that people tend to pay attention only to those areas that are being measured and to overlook those for which no measures exist. For example, ecological economists criticize traditional economists because the latter ignore resource depletion and loss of human skills (through death, injury, or lack of training) when they calculate outcomes such as profit and productivity. Ecological economist Robert Respetto cites the case of Indonesia to demonstrate how traditional measures of agricultural output are affected when resource depletion is considered. According to Respetto, losses from soil erosion "reduce the net value of crop production by about 40 percent. And net losses of forest resources actually exceed timber harvests. Moreover, from 1980 to 1984 depletion of oil fields reduced the value of Indonesia's reserves by about $10 billion annually" (quoted in Passell 1990, p. B6). If these factors were considered in calculating productivity

and net profit, imagine the conservation efforts that would have to be expended to increase profits and productivity.

In the case of Bhopal, no measure existed to monitor how often equipment such as refrigeration units, safety systems, and gauges failed to work. Because no measure was in place, responsibility for this dimension of operational safety obviously was never assigned to a position in the ranks of Union Carbide or in any of the many Indian government agencies.[7] As a result, everyone who was interviewed, from the president of Union Carbide to the Bhopal police, maintained that they were not responsible for operational safety. The following quotes illustrate this point:

> *But safety was the responsibility of the Indian personnel. It was "a local issue."* [*Warren M. Anderson, chairman, Union Carbide Corporation*] (ENGLER 1985, P. 498)

> *We expected that Union Carbide Corporation would try and palm off the blame on Union Carbide, India. But UCC cannot escape responsibility. They should have ensured that such* [*safety*] *lapses could not occur.* [*Kamal K. Pareek, senior project engineer, Bhopal Union Carbide plant*] (DIAMOND 1985C, P. A1)

> *Responsibility for plant maintenance, hiring and training of employees, establishing levels of training and determining proper staffing levels rests with plant management.* [*Union Carbide spokesperson*] (DIAMOND 1985A, P. A6)

> *We do not design, maintain and operate plants. We only check to see that there are enough protective masks and safety guards.* [*factory inspector, India Labor Department*]

> *It is the basic responsibility of the company to make the community aware of* [*the hazards*]. *I have not seen anything so far to show that this was done.* [*Arjun Singh, official, Madhya Pradesh government*] (REINHOLD 1985, P. A8)

As we have seen, many potential problems are associated with statistical measures of performance. To ensure that important conditions such as occupational safety are monitored, it is advisable to develop thoughtful and accurate indicators to measure them, to assign responsibility to a definite position, and to tie the measures to the evaluation of performance by persons occupying that position. When no such system is in place, responsibility for accident prevention never rests squarely with specific people. This diffusion of responsibility also can occur when decision makers rely on experts for advice or when the power to make decisions is concentrated in the hands of a few people at the top.

Obstacles to Good Decision Making

In his writings about bureaucracy, Weber emphasized that power was not located in the person but in the position that a person occupied in the division of labor. The kind of power described by Weber is clear-cut and familiar: a superior gives orders to subordinates, who are required to carry out those orders. The superior's power is supported by the threat of sanctions: demotions, layoffs, firings. Sociologists Peter Blau and Richard Schoenherr recognize the importance of this form of power but identify a second, more ambiguous type—expert power—which they believe is "more dangerous than seems evident for democracy and . . . is not readily identifiable as power" (1973, p. 19).

Expert Knowledge and Responsibility

According to Blau and Schoenherr, expert power is connected to the fact that organizations are becoming increasingly professionalized. **Professionalization** is a trend whereby organizations hire experts (such as chemists, physicists, accountants, lawyers, engineers, psychologists, or sociologists) who have formal training in a particular subject or activity that is essential to achieving organizational goals. Experts are not trained by the organization, however; they receive their training in colleges and universities. Theoretically, they are allowed to be self-directed and are not subjected to narrow job descriptions or direct supervision. Experts use the frameworks of their chosen profession to analyze situations, solve problems, or invent new technologies. From the experts' viewpoints, the information, service, or innovation they provide to the organization is technical and neutral. They do not necessarily think about or have control over the application of that information, service, or invention.

Blau and Schoenherr regard this arrangement between experts and organizations as problematic because it leaves nobody accountable for the ac-

tions of powerful corporations and because it complicates attempts to find individuals "whose judgments [are] the ultimate source of a given action" (pp. 20–21). This situation is complicated for two reasons. First, the recommendations and judgments of experts rest on specialized knowledge and training. The experts may understand principles of accounting, physics, biology, chemistry, or sociology, but their training for the most part is compartmentalized; they know one subject very well but they do not know other subjects. For example, a chemist may be able to design a pesticide, but he or she has not been trained to consider the limitations of the people who use it. Similarly, a sociologist may understand the abilities and limitations of people who use a pesticide but may not understand the technology. Because the sociologist does not understand the chemistry, he or she cannot design the details of a program to educate consumers.

In addition, decision making in large organizations is complex because no single person provides all of the input that goes into a decision. A decision is a joint product of information and judgments by a variety of experts. Often the decision maker does not understand the principles underlying an expert's recommendations and judgments. The problem is that when something goes wrong, the experts claim that they only provided information, suggestions, and recommendations, whereas management claims that it cannot predict the consequences of an invention or a service that only the experts understand.

Blau and Schoenherr emphasize that the men and women who give expert advice or make decisions based on expert advice are decent people but that their training and point of view make them unable to anticipate and plan for the unintended consequences. The chemists who created pesticides and fertilizers believed that their inventions would help feed the world. Little did they know that these chemicals would be misused and overused to the point of ecological crisis. This brings us to another issue related to decision making—oligarchy.

The Problems with Oligarchy

Oligarchy is rule by the few, or the concentration of decision-making power in the hands of a few persons who hold the top positions in a hierarchy.

> *One of the most bizarre features of any advanced industrial society in our time is that the cardinal choices have to be made by a handful of men . . . who cannot have firsthand knowledge of what those choices depend upon or what their results may be. . . .*
>
> *[And by] "cardinal choices," I mean those which determine in the crudest sense whether we live or die. For instance, the choice in England and the United States in 1940 and 1941, to go ahead with work on the fission bomb: the choice in 1945 to use that bomb when it was made.* (SNOW 1961, P. 1)

Political analyst Robert Michels believed that large formal organizations tended inevitably to become oligarchical, for the following reasons. First, democratic participation is virtually impossible in large organizations. Size alone makes it "impossible for the collectivity to undertake the direct settlement of all the controversies that may arise" (Michels 1962, p. 66). For example, Union Carbide (headquartered in Danbury, Connecticut) employs 91,459 people in 700 plants located in more than 30 countries (Derdak 1988; *International Directory of Corporate Affiliations 1988/1989*). At the time of the Bhopal crisis, Union Carbide had 14 factories, had 28 sales branches, and employed 9,000 workers in India alone (Diamond 1985a; Leprowski 1992). "It is obvious that such a gigantic number of persons belonging to a unitary organization cannot do any practical work upon a system of direct discussion" (Michels 1962, p. 65).

Second, as the world becomes more interdependent and as technology becomes increasingly complex, many organizational features become incomprehensible to workers. As a result, many employees work toward achieving organizational goals that they did not define, cannot control, may not share, or may not understand. This lack of

Michels believed that oligarchy, or rule by the few, is the inevitable tendency of large organizations. Size alone makes it impossible to get everyone's input into organizational decisions.
Brad Markel/Gamma-Liaison

knowledge prevents workers from participating in or evaluating decisions made by executives.

A danger of oligarchy is that those who make decisions may not have the necessary background to understand the full implications of the decisions. For example, Warren Anderson, the chairman of the Union Carbide Corporation, stated, "It never entered my mind that an accident such as Bhopal could happen" (Engler 1985, p. 495). In addition, decision makers may not consider the greater good and may become preoccupied with preserving their own leadership. Guarding against these effects of oligarchy requires that the average worker and the general public be interested, attentive, and informed. As more and more people hold jobs that require them to deal with science and technology, and as our daily lives become increasingly dependent on technology, people need to understand and feel responsible for understanding what is going on around them. Otherwise, technology decisions will be made by others on their behalf, and they will be forced to accept the consequences.

The reaction of Bhopal residents to the unexpected gas leak dramatizes how an uninformed and inattentive public cannot take the precautions they need to protect themselves when a crisis occurs. Although an emergency alarm sounded at the Bhopal plant, most people living near the plant did not know what it meant; they were conditioned to hearing alarms because alarms sounded at the plant an average of 20 times a week. Many assumed that the alarm signaled a change in shift or a practice drill, or that it had gone off by accident:

> *We used to hear sirens go off often. . . . We thought it was routine—a change of shift or a fire in the factory. We were never alarmed. . . . We thought they were making powder, some kind of powder. . . . We were never told anything about poison, by the company or by the government. [Sabir Kahn, Bhopal resident]* (REINHOLD 1985, P. A8)

Few residents of Bhopal knew what chemicals were manufactured at the plant, how toxic they were, how to detect escaped chemicals by smell or sight, or what to do if chemicals were released accidentally from the plant:

> *Several Bhopal residents said many of the people living near the plant thought it produced "Kheti Ki Dawai," a Hindi phrase meaning "medicine for the crops." Thus, [they thought] the plant's output was healthful. [Sabir Kahn, Bhopal resident]* (THE NEW YORK TIMES 1985A, P. A6)

The one response they knew was to run. If they had known to lie close to the ground and cover their faces with a wet cloth, most of the victims would have escaped injury or death.

In addition to this general lack of knowledge, physicians, firefighters, police, and city officials did not know the hazards that the plant presented or what precautions to take in an emergency:

> *"We had no inkling of what kind of emergency steps should be taken"* in such a situation. *[Nily Chaudhuri, chairman, Central Water and Air Pollution Board]* (STEVENS 1984, P. A10)

> *What shocked me was to find out from the mayor of Bhopal that he didn't have the vaguest idea that this could happen. [Stephen Solarz, U.S. congressman and chairman of the Subcommittee on Asian and Pacific Affairs]* (HAZARIKA 1984, P. A3)

To complicate matters even further, the physicians at the two hospitals where the victims went for treatment had no information about MIC in general or about how to treat people exposed to the chemical. It was not that the Bhopal physicians were incompetent. Rather, they could not find out this information from Union Carbide or medical journals. Physicians at one hospital—the Hemida Hospital, a 1,000-bed facility—faced 20,000 patients desperately short of medical supplies and without a firm idea of how to treat patients' symptoms (Wexler 1989).

Many newspaper reporters in the United States attributed the disorder to the reaction of a poor, illiterate peasant population whom they considered incapable of understanding sophisticated technology (Everest 1986). Residents of Institute, West Virginia, however, responded to a chemical leak at the local Union Carbide plant in the same disorganized way.[8] When they noticed a strong odor seeping into their houses, they did not know what to do, as evidenced by statements from various residents (Franklin 1985).

> *"I kept burying my head into the pillow, trying to get rid of the smell. Then I finally woke up, and I had my shoes on in no time. If it had been something really bad, it would have been too late. People would have been dead in their beds."*

> *"I didn't know what it was when I saw that white cloud go up, disperse and spread out. I just locked the building, picked up my wife, and took off."*

> *"The whistle was blowing and there was a terrible smell. We couldn't find anything on the radio or television, and my husband insisted that we leave. We didn't know what it was. There was a breakdown in communication."*

"I didn't know what to do." (*THE NEW YORK TIMES* 1985B, P. A12)

It is needless for people, even the poorest people, to be unprepared or be passive victims of the decisions of a government or a large multinational corporation. The ecological movement led by Sunderlal Bahuguna, a 60-year-old man working to save the forests of northern India, illustrates the power of individuals. Bahuguna travels from village to village teaching the people to conserve forests, replant trees when they cut them, and protect the forests from commercial enterprises. Bahuguna teaches children about the chemistry of trees and urges them to hug trees and sing a song about them: "Do not touch me with an ax, I, too, feel pain. I am your friend. I bring you fresh air. I bring you water. I always bow down before you. Why do you cut me down?" (Hutchison 1989, p. 185). The women of some villages have hugged trees in successful efforts to prevent contractors from cutting them down and have formed human chains across access roads to prevent contractors from reaching timbering sites.

Another illustration is the work of Kishan Baburao Hazare. Hazare has helped the people of Ralegan Sidhi, a village in western India, to transform the area. At one time the region was an ecological disaster due to deforestation. Now the village is able to produce enough food to be self-sufficient and export half of what it produces (Nikore and Leahy 1993). Essentially Hazare educated the people about inexpensive ways to fight soil erosion, such as terracing and planting trees along the hillside.

To this point we have discussed a number of important concepts that help us understand how some characteristics of organizations make them coordinating mechanisms with the potential for both constructive and destructive consequences. Now we turn to Karl Marx and his concept of alienation to understand how workers are dominated so strongly by the forces of production that they remain uninformed or uncritical about their role in the production process.

Alienation of Rank-and-File Workers

Human control over nature increased with the development of increasingly sophisticated instruments and tools and with the growth of bureaucracies to coordinate the efforts of humans and machines. Machines and bureaucratic organizations combined to extract raw materials from the earth more quickly and more efficiently and to increase the speed with which necessities such as food, clothing, and shelter could be produced and distributed.

Karl Marx believed that increased control over nature is accompanied by **alienation,** a state in which human life is dominated by the forces of human inventions. Chemical substances represent one such invention; they have reduced the physical demands involved in producing goods. Fertilizers, herbicides, pesticides, and chemically treated seeds give people control over nature because they eliminate the need to fight weeds with hoes, they prevent pests from destroying crops, and they help people produce unprecedented amounts of food. In the long run, however, people are dominated by the ef-

fects of this invention. Heavy reliance on chemical technologies causes the soil to erode and become less productive; it also causes insects and disease-causing agents to develop resistance to the chemicals. Chemical technologies also have altered the ways in which farmers plant crops: planting patterns have changed from many species of sustenance crops planted together to a single cash crop, planted in rows. As a result of these changes, farmers have lost knowledge of how to control insects and diseases without chemicals by interplanting a variety of flowers, herbs, and vegetables. Farmers are now economically dependent on a single crop and on the chemical industry.

Although Marx discussed alienation in general, he wrote more specifically about alienation in the workplace. He believed that alienation resulted from dividing up the production process so that a single product was assembled by many workers, each performing a specialized task. Mechanization, specialization, and bureaucratic organization have given people new control over production because

Karl Marx believed that alienation occurs when the production process is divided up so that workers are treated like parts of a machine rather than as active, creative, social beings.

Charles Harbutt/Actuality, Inc.

these technologies reduce physical effort and increase the pace of work. But, in turn, people are controlled by these same technologies, which have caused workers to lose self-direction. As a result of technology, some workers have become replaceable, or as interchangeable as machine parts. They are treated as economic components rather than as active, creative social beings (Young 1975). Marx believed that the conditions of work usually are such that they impair an individual's "capacity to become a multidimensional, authentic being with human qualities of compassion, reflection, judgment, and action" (Young 1975, p. 27).

Marx believed that workers are alienated on four levels: (1) from the process of production, (2) from the product, (3) from the family and from the community of fellow workers, and (4) from the self. Workers are alienated from the process because they produce not for themselves or for known consumers but for an abstract, impersonal market. In addition, they do not own the tools of production. Workers are alienated from the product because their roles are rote and limited; no person can claim a product as the unique result of his or her labor. Workers are alienated from their families because home and work environments are separate. In other ways, households are uprooted because large-scale enterprises take over the land or force families to move to areas where work is avail-

able. Furthermore, workers are alienated from one another because they compete for a limited number of jobs. As they compete, they fail to consider how they might unite as a force and control their working conditions. Finally, workers are alienated from themselves because "one's genius, one's skills, one's talent is used or disused at the convenience of management in the quest of private profit. If private profit requires skill, then skill is permitted. If private profit requires subdivision of labor and elimination of craftsmanship, then skill is sacrificed" (p. 28).

Alienation on these four levels is particularly evident among workers in developing countries such as India, which are trying to lure foreign business.[9] The foreign companies own the factory buildings, the tools, the machines, and the labor of workers from the host country. The work is often repetitious, mind-numbing, and tedious. In the worst case, workers handle chemicals about which "they may understand little, other than that their eyes tear, they cough harshly and they suffer recurring rashes and headaches" (Engler 1985, p. 493). Although studies on working conditions in developing countries are imprecise and sketchy, the preliminary information suggests that workers are exploited without regard for health consequences. Two studies found that one-third of Indians working in a DDT plant and one-fourth of all employees at battery plants were sick (Engler 1985). The World Health Organization found that of the thousands of industries in Delhi, many are "cramped, poorly lighted, ill-ventilated spaces with atmospheres full of dust, gas, vapors and fumes" and that they operate without health controls (Crossette 1989).

Another sign that workers in developing countries are exploited and treated as dispensable is that hazardous substances banned in the United States and other Western countries are now being produced in the developing countries. Asbestos production is a typical example.[10] The adverse effects of asbestos are well known: chronic exposure is linked to lung disease and to a fatal form of cancer (mesothelioma). In the United States asbestos products are gradually being replaced or phased out; for example, asbestos no longer is used to insulate homes and other buildings. However, for some products—most notably automobile brakes to fit cars made before 1990—asbestos is still used. EPA limits on workers' exposure to asbestos have caused many Western factories to move production offshore to Taiwan, Mexico, and India, where regulations governing workers' safety are decades behind those in the United States. Asbestos dust in one Bombay, India, plant is like the dust "behind a bus on a dirt road in the dry season" (Castleman 1986, p. 62).

When a factory is built in India, workers migrate from rural homes and gather around the factory to live in squatter settlements or slums. In Bhopal the population was 102,000 in 1961. After Union Carbide and other industries settled there in the 1960s, the population grew to 385,000 in 1971, to 670,000 in 1981, and to 800,000 by 1984.

> *It is a familiar sight in so-called underdeveloped countries to find somewhere, in the midst of great poverty . . . , a gleaming, streamlined new factory, created by foreign enterprise . . . immediately outside the gates you might find a shanty town of the most miserable kind teeming with thousands of people most of whom are unemployed and do not seem to have a chance of ever finding regular employment of any kind.*
> (SCHUMACHER 1985, P. 490)

Approximately 20 percent of Bhopal's 800,000 residents live in squatter settlements. The location of two of these squatter camps—directly across from the Union Carbide plant—explains why the deaths occurred disproportionately among the poorer residents. These people are paid poverty-level wages, which prohibits them from acquiring decent living quarters. In many parts of India, people are drawn to land surrounding industrial sites because they have been driven from their homes by large corporations. The Indian village of Kerala is a striking case. Twelve thousand families in Kerala eke out a living making mats from the bamboo and reeds that grow in the surrounding forest, but they must compete with the large paper companies that moved in to harvest the bamboo and reeds. After the forest is rapidly and fully stripped, the company will leave to find new forests, while the Indian families who remain will be left without the materials needed to make a living. As a result, many will be forced to migrate to urban areas in search of factory work (Sharma 1987).

The factory owners and shareholders make large profits, in part because they can find places where minimum wage, environmental protection, and workers' health and safety laws are not systematically enforced or are less stringent.[11] When the company loses money, the management at the corporate headquarters can neglect the plant, allowing equipment, employees' skills, and, sometimes, already inadequate safety standards to deteriorate. The workers are thankful for any job and are hesitant about organizing to improve working conditions when such an action could mean losing their jobs and source of livelihood.

This point about workers' hesitancy to speak out has led economist Robert Reich to conclude that people must evaluate corporations, particularly multinationals, in a new light. According to Reich, we cannot assume that, if corporations headquartered in a country are prosperous and profitable, the workers in that country are also prosperous. Nor can we assume that executives from foreign-owned corporations are more likely than native-born executives to make decisions that will harm the economic well-being of the host country. Why? Because "American corporations and their shareholders can now prosper by going wherever on the globe the costs of doing business are lowest—where wages, regulations, and taxes are minimal. Indeed, managers have a responsibility to their shareholders to seek out just such business climates" (Reich 1988, p. 79). Reich advises people to be leery of business executives who point to large trade deficits as evidence that they need wage concessions, trade protection, and other incentives if their corporations are to compete on a global scale. In the case of the United States, a large portion of the trade deficits with other countries is the result of American corporations' making products in those countries to be sold in the United States under their own brand name. For example, 40 percent of the trade that Mexico does with the United States originates from U.S. subsidiaries in Mexico (Clark 1993). "One of the ironies of our age is that an American who buys a Ford automobile or an RCA television is likely to get less American workmanship than if he had bought a Honda or [a] Matsushita TV" (Reich 1988, p. 78).

In view of this global context, Reich argues that people should not be overconcerned with the nationality of a company or with trade deficits. Instead, they should focus on the quality of jobs that a corporation brings to a community:

The mistake often made by local chambers of commerce and state governors when drumming up business is that they focus only on the number of jobs, not the quality of jobs. Attracting Volkswagen or Honda to set up a facility nearby employing 5,000 routine assembly-line jobs at a cost of $5 million in tax abatements may not be a great accomplishment.

Ultimately for a British, Dutch, or Japanese company to come into town isn't necessarily better or worse than a California or Massachusetts company. If they bring good jobs, build up the ability of the work force, this type of "foreign" investment is probably a big plus. But if jobs are unskilled or routine, and fail to add new skills, the investment may ultimately cause small towns to suffer in the long run. [For most towns,] another generation of low-skill, low-paid assembly-line workers is not particularly beneficial in the long run. (REICH 1990, P. 84)

Discussion

Organizations are powerful coordinating mechanisms that permit goods and services to be produced and delivered efficiently to millions of people. Although they manage people, information, goods, and services efficiently, organizations also can promote inefficient, irresponsible actions. This chapter introduces concepts that sociologists use to understand the two sides of organizations. We have focused on those aspects of organizations that promote overuse and improper handling and disposal of potentially hazardous substances, but organizations are also capable of responsible actions as well (see "Denmark's Industry").

This focus on the negative does not mean that organizations are all bad (see Table 7.5); we need organizations. For example, if we are to protect the environment and the public health, we must have organizations with well-conceived rules and procedures to monitor industrial activity on a worldwide scale. If workers are to have any bargaining power

in an economic atmosphere in which corporations shift production to those areas with the lowest wages, the best tax breaks, and the fewest worker safety and environmental regulations, strong international labor organizations must exist (Faux 1990). We have concentrated on problematic areas of organizations, however, because knowledge of such problems helps check the potentially dangerous features.

The Bhopal disaster reveals that many factors within organizations can promote ignorance, misinformation, and failure to take responsibility for known risks. These factors include (1) rational decision making that emphasizes the quickest and most cost-efficient means to achieve a goal without considering the merits of other methods and goals, (2) departures from rules and regulations, (3) excessive adherence to rules and regulations, (4) reliance on experts with compartmentalized training, (5) oligarchy, and (6) an alienated work force.

Denmark's Industry

In this chapter we have tended to emphasize the mechanisms within organizations, especially the drive for profit, that promote problems for people and the environment. However, organizations can also be agents of responsible action as well. Industrial plants in Kalundborg, Denmark, for example, manage to make profits and at the same time take actions to protect the environment from the waste they generate.

Much has been written on the environment. There have been many books and articles on what each one of us can (and should) do in our daily lives to slow down the rate of destruction—by recycling, by refusing to buy the most destructive products, by arranging our lives so as to use everything we need more efficiently and thus to need less. All of that sort of information is certainly very important, but if every one of us does all those things, it still will not be enough. It is already too late to save our planet from harm. Too much has happened already: Farms have turned into deserts, forests have been clear-cut to wasteland, the global temperature will rise, the ozone layer will continue to fray.

But this is not a hopeless cry of doom at all. We still have time to save, or restore, a large part of the gentle and benevolent environment that has made our lives possible. We can't do it easily. But, after all the things that have gone wrong, it's worth reminding ourselves that there are a few places, here and there, where they've gone *very right.*

One of those very right places is Kalundborg, and it is in Denmark. Kalundborg is a small city which contains a number of industrial plants, all of them of the kinds that we identify as serious polluters and contributors to our environmental problems.

There's the Asnaes power plant, the biggest electrical generating installation in Denmark. There's the Gyproc factory, which produces great quantities of plasterboard for construction. There's an oil refinery, and the Novo Nordisk pharmaceutical plant, as well as the usual government and retail establishments, homes, and a surrounding collar of farms.

All of them have their special needs for raw materials and other resources, and all of them produce their own wastes.

Novo Nordisk needs steam to make its enzymes and drugs. Until about 10 years ago, like everybody else in the business, it made its own steam by burning fossil fuel and boiling the water it took from the local rivers. Then Novo Nordisk made a deal with the Asnaes power plant to buy the plant's waste steam—still very hot but no longer sufficiently superhot to turn its turbines. Asnaes, like most power plants, had previously simply allowed its steam to cool and discharge into those same rivers, to their detriment.

Asnaes had been drawing its steam water from those rivers, but then it turned to the Statoil refinery (which had some warm water left over from cooling its oil-processing towers) and bought that water from them. Statoil had something else to sell, too. There are waste gases from oil refining, usually not enough of them to be economically worth collecting and selling, and generally they are flared off in permanently burning torches. Statoil sold those gases—first to Gyproc, to replace other fuel in making plasterboard, then to the power plant.

The Asnaes power plant still had surplus heat, so it began piping it to the town's domestic-heating system; this permitted some 1,500 homes in Kalundborg to close down their household furnaces. There was still more heat at the power plant, in the form of relatively low-temperature hot water, so it was used to supply a nearby fish farm.

In fermenting its pharmaceutical products, Novo Nordisk produces a lot of thick sludge. It doesn't discharge the sludge into the rivers, as many such factories do elsewhere in the world. Instead, it treats it to kill off microorganisms and then sells it to the local farmers as fertilizer. The farmers love it. They can use it without fear of heavy metals or industrial toxins; there never were any there. Meanwhile, fly-ash from the power plant makes cinderblocks, and the plant's new scrubbers will provide cheap gypsum for the Gyproc plant to make into plasterboard.

It wasn't even primarily concern for the environment that started Kalundborg on its green path. Each of those companies was looking for ways to increase profits.

Kalundborg hasn't solved all of our environmental problems. It hasn't even solved all of its own. But it shows how many of them *can* be solved in ways that actually make money and keep people employed, as well as doing essential work in the repair of Planet Earth.

SOURCE: From *Our Angry Earth: A Ticking Time Bomb,* by Isaac Asimov and Frederik Pohl, pp. 247–49. Copyright © 1991 by Isaac Asimov and Frederik Pohl. Reprinted by permission of the authors and Tom Doherty Associates.

TABLE 7.5   Organizations—What the Critics Say Versus What the Defenders Say

What the Critics Say	*What the Defenders Say*
1. Organizations force workers to do simplified, meaningless tasks, so robbing them of initiative, creativity, and independence. Look at the assembly line.	We can enrich jobs by deliberately building in autonomy and discretion. In Sweden, the assembly line has been modified to do just this. In any case, very few people actually work on assembly lines anymore. We are becoming a service society.
2. Organizations do not serve society or consumers well. Automobiles are unsafe, factories pollute the air, and the sheer size of organizations makes them dangerous. The multinationals dominate life in company towns and even overthrow governments.	We can control pollution, and make better products with quality-control circles and other advances pioneered by Japan and other countries. Large size produces economies of scale. Besides, only very large organizations can afford to do research on new products.
3. Organizations don't even work well. Cars break down before you get them home from the showroom, prisons don't rehabilitate, and schools turn out illiterates. We need alternative institutions.	You go too far. Cars give trouble because you want so much from them—trouble-free driving at high speeds with minimal maintenance. If we produced a serviceable car with minimum features, you wouldn't buy it. Prisons could rehabilitate if you would pay the price for vocational counseling and training. The schools do a great job—name another country in which over a third of all high-school students go on to higher education.
4. Organizations are out of control. An arrogant power elite of interlocking directors controls them, and an army of ever-increasing bureaucrats administers them. The client or citizen is powerless against organizations.	The evidence for any monolithic power elite is exaggerated or so biased as to be invalid. Client and citizen power, they are increasing. Look at the detail on contents now provided on labels at the supermarket. Look at the growth of consumer action groups.

SOURCE: From *Organizations in Society,* by Edward Gross and Amitai Etzioni, p. 4. Copyright © 1985 by Prentice-Hall, Inc. Reprinted by permission of Prentice-Hall, Englewood Cliffs, NJ.

We also have learned that a complex relationship exists between an organization and the environment, especially when the organization is a multinational corporation operating in a country such as India. The case of Union Carbide shows that an organization not only draws on labor and raw materials from the environment but also affects the environment through the products it manufactures, the services it provides, and the waste it disposes of.

Organizations cannot be held totally responsible for negative consequences to the environment. A nation needs an informed, attentive public, rather than a public that consumes unthinkingly. Such a public is important because it reinforces conscious decision making on the part of management and experts, who will sense that the public is watching their actions. In addition, an informed and attentive public is capable of responding intelligently if a crisis does occur. Scientist Kenneth Prewitt suggests that it would be helpful if the public simply understood that all technologies have unpredictable consequences and that complex problems do not have simple answers:

A savvy population recognizes a "fix" for the unrealistic goal that it is. It is the rare problem that is solved; more often, the problem changes

An Indian boy sells spices at a market.
The scales in the foreground are just as
much a technology—a tool for dealing
with the physical environment—as a
complex power plant.
Nicholas Devore/Tony Stone Images

*form. To control infectious diseases is not to
solve the health problem. It is to shift attention
to cancer and heart failure. And when these illnesses of late middle life are cured, the health
problem will shift to senility and related illnesses of the aged. Society copes with problems;
it doesn't solve them. If the public is to become
actively involved in debates over the introduction of major technologies—chemical fertilizers,
fluoridated water, nuclear energy, genetic engineering—then the minimum literacy necessary
is appreciation of this fact.* (PREWITT 1983,
PP. 62–63)

The absence of an informed public is not unique
to countries such as India. The United States too is
plagued by public apathy. The problem is greater in
India and other parts of the developing world,
however, because such places lack watchdog agencies and because many people are so poor they will
tolerate almost any conditions for a chance to make
a living.

The important point is that management, the experts, the workers, and the public recognize that
technology is not good or bad in itself. It carries a
variety of benefits and costs. If people do not understand these benefits and costs, they see themselves as powerless and incompetent to judge the

merits of technology. Many people believe they
can't live with technology but they can't live without it either. If they must choose, they would rather
live with technology. An Indian folk story shows
the hopelessness of such a view:

*In the [slums] of Bhopal today, an old woman is
said to visit houses unannounced. She knocks at
the door and asks for food. If she is given it, she
throws it at the house, and all the inhabitants
die. If she is refused food, she curses those who
live there, and all of them will die.* (ALVARES
1987, P. 182)

The people believe that they cannot live without
chemicals because chemicals ensure agricultural
success; yet, they believe that they cannot live with
them because they are hazardous to health and to
the environment. Actually, the problem is not the
chemicals themselves but the misuse, overuse, and
abuse of chemicals.

The issue that Bhopal represents is not technology versus no technology. The real issue is the need
for organizational mechanisms to coordinate the
conscientious production and use of potentially
hazardous materials, experts capable of understanding the full implications of technology, and
educated consumers and informed workers. A program sponsored by the multinational corporation

Ciba-Geigy in conjunction with an unnamed sub-Saharan African rice-growing country shows that these goals are not unrealistic and that good business ethics make good business sense because sustained and stable development in developing countries is more profitable to corporations in the long run. According to Klaus M. Leisinger, director of Third World relations at Ciba-Geigy Ltd., the corporation conducted a study to assess the level of need before it sold pesticides to this country. Once the study was complete, the corporation established an organizational infrastructure to coordinate the activities of all of the people involved with the pesticide, inside and outside the corporation; it trained 40,000 farmers to use and apply the pesticide; and it oversaw the safe application of the pesticide (Leisinger 1988).

FOCUS
The Problems of Value-Rational Action

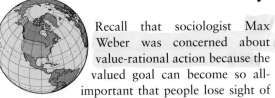

Recall that sociologist Max Weber was concerned about value-rational action because the valued goal can become so all-important that people lose sight of the negative consequences that can arise from the methods used to reach that goal. We have argued in this chapter that the failure to consider such consequences is at the heart of the "destructive side" of organizations. Many of the specific problems underlying the disaster at Bhopal such as trained incapacity and evaluating performance with narrow statistical measures represent cases in which employee flexibility, ability to respond to unusual situations, and willingness to take responsibility are sacrificed to the larger valued goal (producing chemicals at a profit). "The Case of the Hanford Nuclear Reservation" shows that the problem of value-rational action goes beyond the disaster at Bhopal.

The Case of the Hanford Nuclear Reservation

Yvonne Dupont and Karen Meyers

The disaster at Bhopal is not an incident isolated to countries such as India. In the United States we are just beginning to realize that our Bhopal is the 17 nuclear weapons plants located across the country. The Department of Energy has admitted that the procedures used to make bombs over the past 50 years have consequences that pose a serious threat to public health. Mishandling of nuclear waste has severely contaminated the air, water, and soil with cancer-causing radioactivity and toxic chemicals.

Though all 17 of the nuclear plants in the United States pose serious threats to the environment and to human life, our focus is on the Hanford Nuclear Reservation in Richland, Washington, a government-owned plant responsible for making atomic bombs. In Hanford's first decade of production, 470,000 curies of iodine 131 escaped from the plant. In contrast, an estimated 15 curies of iodine 131 were released at Three Mile Island, the site that most Americans associate with the United States' worst commercial nuclear accident (Boly 1990).

Fifty years later we ask, How could such a tragedy occur? As sociologist Max Weber warned, "Correct rational action is rare"—once people set a goal they seldom question the wisdom of the goal, consider the safest means of achieving it, or prepare for unforeseeable consequences. More often than not people choose the most efficient and expedient means of achieving valued goals. The scientists at the Hanford nuclear weapons plant had only one valued goal—to win the arms race against the Soviet Union in order to deter communism. In fact, they employed special strategies to avoid dealing with complications that might slow arms production.

At the same time, Americans, particularly those living around the plant, had only the vaguest idea about what was being produced at the plant. Because residents feared communism, they considered the production of atomic bombs necessary. Looking back at articles published at that time in

SOURCE: Yvonne Dupont and Karen Meyers, Northern Kentucky University, Class of 1990.

popular magazines, we get a glimpse of just how great the fear of communism was. In the early 1960s civil defense was the country's top priority. The Office of Civil Defense (OCD) planned to have some 250,000 public fallout shelters marked, stocked with food and water, and ready to shelter some 50 million Americans by 1962 (*Newsweek* 1961). In addition, the United States government created a stockpile of nuclear weapons large enough to destroy the world several times over (Nevin 1988). It manufactured 60,000 nuclear warheads at a cost of $750 billion at an average production rate of four warheads per day for 40 years (Norris, Cochran, and Arkin 1985).

For the most part Hanford residents had no understanding of nuclear energy (how it was produced, what part it played in the making of the bomb, who was responsible for monitoring its use, and what effects its production had on the human population). Consequently they left the matter to experts and they supported atomic bomb production. Here we can draw an analogy to the Persian Gulf War, in which the American public generally accepted government censorship on American military activities in Iraq in order to achieve the valued goal of defeating Saddam Hussein.

Because nuclear energy was such an abstract idea, people had to rely on experts to tell them if they were safe. The experts regularly sampled dairy products that came from farms near the plant to make sure the people were not being exposed to anything harmful. It did not matter that these scientists came to farms in radiation suits; the people generally did not question the experts. "When the people would ask, 'What are you looking for?' 'Oh, don't worry,' experts would say. 'If we find anything, we'll tell you'" (Boly 1990). In the atmosphere of the Cold War, people needed few explanations. Most displayed blind faith and assumed that the experts were concerned with the country's well-being.

Richland citizens were not the only people who left their fate in the hands of experts. Tens of thousands of workers moved to Richland, Washington, almost overnight to help build the Hanford plant. Since the Hanford project was classified as top secret, workers were ordered to work with hazardous material without knowing what they were handling

or building. For the most part workers were not bothered by this secrecy, a secrecy so tight that independent members of the American scientific community were not able to monitor what was occurring at the plant. The workers collected a paycheck while contributing to communism's containment (Boly 1990).

The Hanford Nuclear Reservation had one goal: the scientists and upper management at Hanford were there to produce plutonium needed to manufacture nuclear bombs quickly and cost-effectively (Boly 1990). Hanford officials viewed hazardous waste disposal as secondary to the goal of containing the Soviet threat and as something that could be later corrected. As a result of this single-minded effort, vast quantities of nuclear waste were simply dumped into ditches and unlined tanks. This waste eventually seeped into the groundwater and then into the Columbia River, a source of food, water, and recreation for millions of people (Boly 1990).

Those living downwind of the Hanford plant, or "downwinders" as they are known, have a higher than normal incidence of cancer (especially thyroid cancer), heart disease, and physical deformities in their offspring (Boly 1990). Hanford officials not only kept plant procedures and documents secret, they failed to warn citizens about health risks. We now know that Hanford officials intentionally used the downwinders as guinea pigs in experiments designed to track the effects of radiation purposely released into the air (Boly 1990). Even today some documents have not been released to the public.

Another problem of value-rational action is that governmental officials did not consider the long-term effects of nuclear bomb production. Cancer and other health consequences were examined only for the short-term. Experts did not consider that 20 years into the future thousands of people might suffer from cancer, birth defects, and other problems that would emerge as the exposed children matured and began to have families. Only in 1989 did research on long-term health effects begin.

The Hanford plant is just one example of how the quest to produce nuclear weapons damaged the environment and the lives of those in the surrounding community. Present and future generations are left to deal with a legacy requiring tens of thousands of years and billions of dollars to correct.

Key Concepts

Alienation 252

Automate 244

Bureaucracy 239

Disenchantment of the World 235

Externality Costs 236

Formal Dimension 241

Ideal Type 239

Informal Dimension 241

Informate 244

Multinational Corporation 229

Oligarchy 250

Organization 227

Professionalization 249

Rationalization 234

Trained Incapacity 243

[handwritten:] Diversity
Rank-and-FileWorker
Value-Rational

Notes

1. The Consumers Union of the United States tests and rates products on price, quality, and safety, and publishes this information in Consumer Reports and Buying Guide Issues (Lydenberg, Marlin, Strub, and the Council on Economic Priorities 1986).

2. India, with 844 million people, is the second most populous country in the world after China. This translates to 612 people per square mile (as compared to 66 in the United States). India's population increases by approximately 18 million per year (Crossette 1991).

3. Historically, the rise of multinational corporations has coincided with decolonization.

4. B.A.T Industries "perversely refuses to put a period after the T" (Moskowitz 1987, p. 37).

5. Such panic in the face of disaster might also occur in the case of nuclear war, an event for which there are civil defense evacuation plans. Would the police, for example, go out and direct traffic, or would they go home to care for their families? The civil defense plans assume an unrealistic scenario—authorities in control of a calm and orderly evacuation.

6. India's 140 colleges graduate approximately 170,000 scientists and engineers annually—the third largest number in the world (Ember 1985).

7. A number of Indian agencies and appointed officials monitored the Bhopal plant at some level. These include the Ministry of Petroleum and Chemicals, the Ministry of Agriculture, the Directorate General of Technical Development, the Ministry of Finance, the Department of Science and Technology, the Controller of Imports and Exports, the Chief Inspector of Explosives, the Plant Protection Advisor, the Bureau of Industrial Costs and Prices, the Ministry of Commerce, and the Ministry of Industry (Bleiberg 1987).

8. A Union Carbide report shows 134 leaks of phosgene, MIC, or a mixture of the two chemical substances between 1980 and 1985 at the Institute plant.

9. In the case of the Bhopal plant, Union Carbide owns 51 percent and the Indian public owns 49 percent.

10. Another example concerns arsenic, a substance used in pesticides, glass, wood preservatives, and semiconductors. The substance is considered cancer-causing and has been subjected to strict environmental regulations in the United States. In 1955 the United States imported 35 percent of its arsenic; as of 1984, 100 percent was imported. Production has shifted from the United States to Peru, Mexico, and the Philippines (Shaikh 1986).

11. In 1988 the profits of the three largest chemical companies based in the United States were $2.19 billion for DuPont (up 22.6 percent from 1987), $2.398 billion for Dow Chemical (up 93.4 percent from 1987), and $662 million for Union Carbide (up 185.3 percent from 1987) (*Fortune* 1989).

[handwritten:] Key Theorists: (PP)(main ideas)
Weber
Thorstein Veblen
Shoshana Zuboff
Robert Michels
Blau and Schoenherr
Marx

8 DEVIANCE, CONFORMITY, AND SOCIAL CONTROL

with Emphasis on the People's Republic of China

A view down busy Nanjing Road, Shanghai.
Dan Habib/Impact Visuals

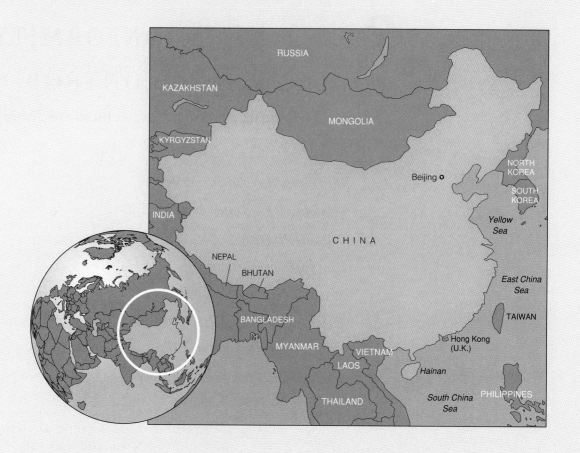

Students from China enrolled in
college in the United States for fall 1991   42,940

Chinese admitted into the United States
in fiscal year 1991 for temporary employment   13,764

People living in the United
States in 1990 who were born in China   543,000

Airline passengers flying between
the United States and China in 1991   71,641

U.S. military personnel in China in 1993   31

Applications for utility patents filed in the
United States in 1991 by inventors from China   126

Phone calls made between
the United States and China in 1991   10,156,000

When the Cultural Revolution began . . . my [family's house] was one of the first in the city to be ransacked.

I later found out that it was my mother's ignorance that started the ransacking. Both my grandfather and father were working in banks. They were well-known capitalists. At that time all the funds capitalists had in the banks were frozen. You couldn't withdraw anything. It was called "money made by exploitation." All the names of capitalists were listed just outside the banks. My mother didn't know about that. She went to get some money out. The bank clerks immediately called the Red Guards. They showed up in no time at my home and started to search and ransack our apartment.

[When I finally got there] I glanced over the rooms from the corridor. Red Guards were standing everywhere, searching for things. No sign of my family. Lots of things were in shreds, smashed and torn.

My parents moved in with my mother's family. Their home had been searched and sealed too. My parents just blocked off a tiny area with a piece of cloth in the corridor. They found some wooden boards to use as a bed. When I went to see my mother, her hair had been cut and shaved by the Red Guards.

(FENG JICAI 1991, PP. 158–60)

This event took place during the 10 years of the Cultural Revolution (1966–1976), a period in which more than a half-million Chinese were imprisoned, tortured, humiliated, and executed. During this time any person who held a position of authority, worked to earn a profit, showed the slightest leaning toward foreign ways, or had academic interests was subject to interrogation, arrest, and punishment. Included in this group were scientists, teachers, athletes, performers, artists, writers, private business owners, and people who had relatives living outside China, wore glasses, wore makeup, spoke a foreign language, owned a camera or a radio, or had traveled abroad (Mathews and Mathews 1983).

Contrast the events of the Cultural Revolution with Asian scholar Robert Oxman's description of China in the last days of 1993:

All over China people are jumping into the sea. That's the new Chinese expression for going into business, making money privately in a country that used to forbid any form of capitalism. It means saving a few dollars for the simple tools to fix bicycles at a curbside shop, or running a sidewalk shoe repair stand because new soles are still cheaper than new shoes, or cutting hair at 5:30 in the morning before going to a regular job. In the country, a wife grows produce to make extra income to buy consumer goods for her family. Unleashing the natural business instincts of the Chinese people, plus the addition of huge foreign investments have fueled one of the most explosive economic take-offs in history. Around major cities, buildings, housing projects, and roads seem to emerge almost overnight. It's even been suggested that China's national bird should be the crane.

Every urban horizon is filled with cranes. In Shanghai's frenzied harbor, ships from around the world compete with tiny river barges as they haul goods and move people. Once the people of China were caught in economic slow motion by a state which prized ideology above all else. Now the Chinese are rushing in fast forward in a national quest for prosperity.
(OXMAN 1993B, P. 9)

The topics of this chapter are deviance, conformity, and social control—some of the most complex issues in sociology because, as we will learn, almost any behavior or appearance can qualify as deviant under the right circumstances. **Deviance** is any behavior or physical appearance that is socially challenged and/or condemned because it departs from the norms and expectations of a group. **Conformity**, on the other hand, may be defined as behavior and appearances that follow and maintain the standards of a group. All groups employ **mechanisms of social control**—the methods used to teach, persuade, or force their members to conform and not to deviate from shared norms and expectations.

As shown by the vignettes at the start of this chapter, depending on the cultural circumstances a deviant act or appearance may be something as seemingly minor as wearing eyeglasses or withdrawing money from a bank.

We can illustrate the same point within U.S. culture. When sociologist J. L. Simmons asked 180 men and women in the United States from all age, educational, occupational, and religious groups to "list those things or types of persons whom you regard as deviant," 1,154 items in all were listed.

Even with a certain amount of grouping and collapsing, these included no less than 252 different acts and persons as "deviant." The sheer range of responses included such expected items as homosexuals, prostitutes, drug addicts, beatniks, and murderers; it also included liars, democrats, reckless drivers, atheists, self-pitiers, the retired, career women, divorcees, movie stars, perpetual bridge players, prudes, pacifists, psychiatrists, priests, liberals, conservatives, junior executives, girls who wear make up, and know-it-all professors. (SIMMONS 1965, PP. 223–24)

Although this study was made more than 25 years ago, Simmons's conclusions are still relevant: almost any behavior or appearance can qualify as deviant under the right circumstances. The only characteristic common to all forms of deviance is "the fact that some social audience regards them and treats them as deviant" (p. 225).

It is difficult to generate a precise list of deviant behaviors and appearances because something that

A society's definitions of deviance change over time. Considering how smokers are viewed today, it is interesting to note that cigarette smoking was not only accepted but glamorized in the United States just a few decades ago.
Cover drawing by Barry Blitt; ©1994 The New Yorker Magazine, Inc.

some people consider deviant may not be considered deviant by others. Likewise something that is considered deviant at one time and place may not be considered deviant at another. For example, wearing makeup is no longer considered a deviant behavior in China, as evidenced by the fact that Avon, a U.S. cosmetics company, has recruited 18,000 salespeople to sell its products in China (WuDunn 1993). As another example, cocaine and other now-illegal drugs once were legal substances in the United States. In fact, "We unwittingly acknowledge the previous legality of cocaine every time we ask for the world's most popular cola by brand name" (Gould 1990, p. 74). Originally Coca-Cola was marketed as a medicine that could cure various ailments. One of the ingredients used to make the drink came from the coca leaf, which is also used to produce cocaine (Henriques 1993).

Such wide variations in responses to what is deviant alert us to the fact that deviance exists only in relation to norms in effect at a particular time. The sociological contribution to understanding deviant behavior is that it goes beyond studying the individual and instead emphasizes the context under which deviant behavior occurs. Such an emphasis raises at least two general but fundamental questions about the nature of deviance. One obvious question is, How is it that almost any behavior or appearance can qualify as deviant under certain circumstances? It follows that any answers to this question must go beyond the deviant individual's personality or genetic makeup. Understanding context, for example, helps us see why it was once a crime to make a profit in China.

A second important question is, Who defines what is deviant? That is, who decides that a particular group, behavior, appearance, or person is deviant? The fact that an activity or an appearance can be deviant at one time and place and not at another suggests that it must be defined as deviant by some particular process. In the case of China, who determined that wearing eyeglasses, speaking a foreign language, and withdrawing money constituted deviant behavior and warranted such severe punishments?

In this chapter we explore the concepts and theories that sociologists use to understand deviance, and we use them to provide some answers to these fundamental but complex questions about the social nature of deviance. We pay special attention to the People's Republic of China[1] for one major reason. In a period of 12 to 15 years China has undergone a lifetime of economic reform that is startling when contrasted with its economic policies of the Cultural Revolution of the 1960s and 1970s, which made "profit making activities" criminal and which closed the country off to foreign investment and influences. During the Cultural Revolution, if a family planted small, but extra, crops to sell after the autumn harvest for money to buy food during the

The Chagan shopping center, Beijing, would be very much at home in the United States. This scene is quite remarkable when you consider that during the Cultural Revolution people were criticized for possessing foreign-made goods or appearing to have extra wealth or "bourgeoisie" status.
Adrian Bradshaw/Saba

The Importance of Ideological Commitment

In China there seem to be two groups of people: those who are ideologically sound and support the Communist Party and those who lack ideological commitment and who conspire against the party. Today the ideologically sound are those who adhere to four political principles: (1) China is a socialist country, hence the creation of not a capitalist system but a socialist market economy, (2) power and leadership resides with the Chinese Communist Party, (3) a combination of Marxism, Leninism, and Mao Zedong Thought is the guiding ideology, and (4) the state is a dictatorship by the proletariat. The principles are intentionally vague so that the party or individual leaders can claim violations against these principles when it suits their purposes (Personal Correspondence 1993). Any behavior that disrupts the progress or the smooth operation of the workplace, the neighborhood, or the country is conceived as an offense against the people or the country, which the Communist Party oversees. The most widespread formal sanction used in China is rehabilitation through re-education and labor.

Obviously, Chinese who have robbed, murdered, assaulted, or raped someone have disrupted the social fabric. In addition, those who commit adultery, disobey work assignments, lose their jobs, have no honest occupation, are vagrants, disrupt a group's discipline, are expelled or drop out of school, or engage in counterrevolutionary and countersocialist activity (behaviors and appearances that challenge communist ideology) are disruptive and in need of rehabilitation. One recent activity defined as counterrevolutionary was the series of demonstrations in Tiananmen Square by Chinese students in support of democracy,[1] eventually suppressed violently in June 1989. The students were accused of "bourgeois liberalization,"

a term that has never been clearly defined but, based on its usage by different sources in the media, it could perhaps be interpreted as wanton expression of individual freedom (individualism) that poses a threat to the stability and unity of the country. Such tendencies had to be curbed. In the government's view, students participated in the protests because they had led a sheltered life and were ignorant of the complexities of the reform process. Their youthful outburst did not take into consideration larger collective interests and concerns. (KWONG 1988, PP. 983–84)

The government undertook a number of measures to persuade the students to become more knowledgeable about the complexities of life and less responsive to subversive ideas. These measures included sending them "to rural areas to teach them to endure hardship, work hard and appreciate the daily difficulties faced by China's mostly rural population" (Kristof 1989b, p. Y1). The rationale is that proper ideological commitment can be instilled through contact with the masses and through manual labor.[2] Other measures included placing limits on the number of students

winter, they were criticized for taking the capitalist road. At that time the pressure against acquiring even small amounts of extra wealth drew such severe criticism that peasants accepted living in poverty rather than being equated with the so-called bourgeoisie (T. Bernstein 1983). In contrast, today there are an estimated 70,000 foreign-funded firms in China (Holley and Courtney 1993). In Beijing alone there are at least 100 luxury hotels, up from 9 in 1981, for foreigners and wealthy Chinese to stay in (Montalbano 1993). And in 1993, the International Monetary Fund named China the third largest economy in the world after the United States and Japan.

In spite of this vast amount of economic change and freedom, which is described as one of the largest economic experiments the world has ever known (Broadfoot 1993), China has received considerable international criticism and pressure regarding human rights. The spotlight has been on the issue of human rights in China since 1989, when the world's attention focused on what appeared to be an unprecedented and excessively harsh crackdown on student demonstrators in Tiananmen Square in Beijing.[2] The image that remains with us is that of a young Chinese man facing down a tank. Although it is estimated that hundreds were killed and thousands were wounded in

A single man risks his life to try to face down a line of tanks sent to suppress the 1989 demonstration in Tiananmen Square.
AP/Wide World Photos

entering the humanities and social sciences. In the year following the June 4 incident, almost no students were admitted to study academic areas that government officials consider to be "ideologically suspect" by nature, such as history, political science, sociology, and international studies (Goldman 1989).

1. The movement toward democracy has been suppressed many times. In 1979 it was suppressed on the grounds that it was "unstabilizing" to China. In 1983 it was labeled a case of "spiritual pollution" (Link 1989).

2. "A Chinese professor who at the end of the Cultural Revolution was sent to Anhui province, in central China . . . described what it was like to live there. Everything was made from mud, he recalled: the floors, the walls, even the beds and the stools the peasants sat on were constructed from pounded earth. The villagers had no wood for fuel—all the trees in the region had long ago been chopped down—so the women and children gathered grass and wheat stalks, depriving the earth of valuable natural nutrients. During the year the professor was there, he ate no meat, and the family he was quartered with had no matches, no soap, and most of the time no cooking oil, an essential part of the Chinese diet" (Butterfield 1982, p. 16).

this action, it was mild in comparison to other crackdowns such as the Cultural Revolution. The Tiananmen Square action, known in China as the June 4 incident, drew worldwide attention because, unlike previous crackdowns, events surrounding it were broadcast to the world (Fang 1990; Kristof 1992). The students, like the people who were struck down during the Cultural Revolution, were defined as deviant and were portrayed as presenting a threat to the authority of the Communist Party and to the smooth operation of Chinese society.

Perhaps one of the most intriguing things about China is that, since 1949, the leaders in power have adhered to the following code:

In any circumstance and at any cost, political power must be retained in its totality. This rule is absolute, it tolerates no exception and must take precedence over any other consideration. The bankruptcy of the entire country, the ruin of its credit abroad, the destruction of national prestige, the annihilation of all efforts toward overture and modernization—none of these could ever enter into consideration once the Party's authority was at stake. (LEYS 1989, P. 17)

This code still applies (see "The Importance of Ideological Commitment"). Even today, as China undergoes massive economic growth and change "its leaders have made it abundantly clear that they intend to preserve an unchallengeable Communist Party dictatorship, even as they pull back from trying to dominate every detail of economic and social life" (Holley 1993, p. H15). China illustrates clearly how those in power have the ability to define who and what is deviant and to enforce those

definitions. We will examine the Chinese system more completely in subsequent sections, comparing it to the U.S. system where appropriate. We begin with some background material on China since the Cultural Revolution before we revisit a concept that is central to the topic of deviance—norms—the violation of which constitutes deviance.

The Cultural Revolution and Beyond

It is impossible to fully understand how the Cultural Revolution disrupted Chinese lives without having experienced it (R. Bernstein 1982). The revolution, which began in 1966 and lasted approximately a decade, was a campaign inspired by Communist Party Chairman Mao Zedong to restore revolutionary spirit to China. It was also a campaign against revisionism and against entrenched authority. Mao loosely defined revisionism as "an abandonment of the goals of the [Chinese communist] revolution, and acceptance of the evils of special status and special accumulation of worldly goods" (Fairbank 1987, p. 319).[3] He loosely defined entrenched authority as the "Four Olds": old ideas, old culture, old customs, and old habits. At first Mao assigned the Red Guards (his name for the youths of China between the ages of 9 and 18) the task of finding and purging revisionists and those in entrenched positions of authority. Within three years, however, the Red Guards had disrupted and divided the country so deeply that Mao denounced them, claiming that they had failed to understand and implement his strategy. Thus, he banished 17 million youths to the countryside to live among the peasants and to be re-educated, and he reassigned the task of purging revisionists to the People's Liberation Army.

During the Cultural Revolution many artifacts from China's long history (tombstones, relics, manuscripts, art objects, books, scrolls of poetry) and any other objects that suggested the accumulation of worldly possessions were destroyed. As mentioned previously, any person in a position of authority or with the slightest leaning toward foreign ways, including farmers who planted extra crops, were suspect. If someone simply remarked that a foreign-made product such as a can opener was

better than its Chinese counterpart, or if someone wrapped some food or garbage in a piece of newspaper with Mao's picture, he or she was suspect (Mathews and Mathews 1983).

As Jung Chang writes in *Wild Swans: Three Daughters of China* (1991), such conditions reduced many people "to a state where they did not dare even to think, in case their thoughts came out involuntarily" (1992, p. 6). The slightest misstep could make one a target:

> Targets might be required to stand on a platform, heads bowed respectfully to the masses, while acknowledging and repeating their ideological crimes. Typically they had to "airplane," stretching their arms out behind them like the wings of a jet. In the audience tears of sympathy might be in a friend's eyes, but from his mouth would come only curses and derisive jeering, especially if the victim after an hour or two fell over from muscular collapse. . . .
>
> To Chinese, so sensitive to peer-group esteem, to be beaten and humiliated in public before a jeering crowd including colleagues and old friends was like having one's skin taken off. (FAIRBANK 1987, P. 336)

The reflections of a former Red Guard, summarized in Fox Butterfield's *China: Alive in the Bitter Sea* (1982) some 15 years after the event, show the intensity with which so-called revisionists and entrenched authority figures were hunted down and persecuted:

> "I was very young when the Cultural Revolution began. . . . My schoolmates and I were among the first in Peking to become Red Guards, we believed deeply in Chairman Mao. I

"We believed deeply in Chairman Mao":
The decade-long Cultural Revolution
was an attempt to rekindle revolution-
ary spirits in China and enforce con-
formity to Mao's doctrine.
Max Scheler/Black Star

could recite the entire book of the Chairman's quotations backward and forward, we spent hours just shouting the slogans at our teachers."

Hong remembered in particular a winter day, with the temperature below freezing, when she and her faction of Red Guards put on their red armbands and made three of the teachers from their high school kneel on the ground outside without coats or gloves. "We had gone to their houses to conduct an investigation, to search them, and we found some English-language books. They were probably old textbooks, but to us it was proof they were worshiping foreign things and were slaves to the foreigners. We held a bonfire and burned everything we had found."

After that, she recalled, the leader of her group, a tall, charismatic eighteen-year-old boy, the son of an army general, whose nickname was "Old Dog," ordered them to beat the teachers. He produced some wooden boards, and the students starting hitting the teachers on their bodies. "We kept on till one of the teachers start[ed] coughing blood," Hong said. . . . "We felt very proud of ourselves. It seemed very revolutionary." (BUTTERFIELD 1982, P. 183)

The forces behind the Cultural Revolution help us understand more clearly why making money

was considered a crime against the state and why such people were considered deviant. The circumstances leading to this revolution are not easy to summarize, however; one would have to go back at least 185 years to trace its origin. Even so, we can say that the Cultural Revolution was an attempt by Mao Zedong to eliminate anyone in the Communist Party and in the masses who opposed him. Mao was particularly vulnerable to attack because immediately before the Cultural Revolution an important national plan—The Great Leap Forward—had failed miserably.

The Great Leap Forward was a plan of Mao's to mobilize the masses and transform China from a country of poverty to a land of agricultural abundance in five short years. Mao mobilized hundreds of thousands of people for projects ranging from killing insects to building giant dams with shovels and wheelbarrows (Butterfield 1976). This ill-conceived, hastily planned, widespread, sweeping reorganization of Chinese society created economic disruption on a massive scale. The nature of this disruption is illustrated by the failure of a plan to increase fertilizer by cutting down trees, burning them, and spreading their ashes over the fields. This plan left peasants without fuel for cooking and heating. In one region peasants stopped harvesting crops and started to dig tunnels in their fields to find coal, which Communist Party officials believed was plentiful. No coal was found, however, and the crops rotted. As a result of the Great Leap Forward, 30 to 50 million Chinese died from human-made famine (Leys 1990; Liu Binyan 1993). In addition, the plan caused considerable environmental destruction. Desperate to grow food and obtain fuel, Chinese peasants destroyed more than half of the grasslands and one-third of the forests (Leys 1990). Mao blamed the failure of his plan on entrenched authority and on the loss of revolutionary spirit.

In view of everything that happened, why is making money acceptable in China today? Why did China bid (unsuccessfully) for the Olympic Games, an event that would attract foreigners to the country, in the year 2000? Again, these questions must be examined in the context of the Cultural Revolution. The revolution took place at the expense of China's economic, technological, scientific, cultural, and agricultural development and drained the Chinese physically and mentally.

After Mao's death in 1976, the Cultural Revolution ended, and the new leaders were faced with a great many problems that the revolution had exacerbated and generated. The Cultural Revolution created a 10-year gap in trained workers. "'I can't honestly let any of the young doctors in my hospital operate on a patient,' said a leading surgeon. 'They went to medical school. But they studied Mao's thought, planting rice or making tractor parts. They never had to take exams and a lot of them don't know basic anatomy'" (Butterfield 1980, p. 32). This situation applies to more than the medical profession; during the Cultural Revolution millions of people secured positions because they were loyal to Mao, not because they were qualified.

In order to solve these problems, the Communist Party, under the leadership of Deng Xiaoping, needed the support of those teachers, technicians, artists, and scientists who had been struck down during the Cultural Revolution. To make up for "Ten Lost Years" the Chinese leaders also sent thousands of students overseas to study in the capitalist West. They allowed farmers and factory workers to keep profits from surplus crops after meeting government quotas, and they established Special Economic Zones (SEZs—designated areas within the People's Republic of China that enjoy

A woman sells Mao memorabilia in contemporary China. Ironically, it is now acceptable to make money from objects celebrating the man who forbade any form of capitalism.

Murray White/Sipa Press

capitalist privileges). SEZs are designed to attract foreign investment and capital; foreign investors are lured there by the cheap labor force and by the potential market of more than 1 billion, 200 million Chinese consumers (Jao and Leung 1986; Pannell 1987; Ross 1984; Stepanek 1982). In 1992 foreign corporations such as Avon, McDonald's, and Procter and Gamble invested approximately $11 billion dollars into China's economy. Also in that year, China ran an $18.2 billion trade surplus with the United States creating trade tensions between the two countries (WuDunn 1993). This is truly remarkable when we remember that as recently as 1976 one could suffer imprisonment and hard labor simply for knowing some foreign words.

Chinese officials made an unsuccessful bid to host the 2000 Olympic Games with the hope of generating international recognition (Gabriel 1988). In the words of Wu Yimei, a Chinese rhythmic gymnastics coach:

Sports is just like diplomacy. . . . When we get good marks, it is a window the world can look into and tell a lot about our country. It can tell we are well fed, we are stabilized politically, we are developing quickly, for otherwise it would not be possible to train athletes and be advanced in sport. (1988, PP. 31–32)

The fact that China's ideas of what is deviant have changed so dramatically since 1966 makes it an ideal case for illustrating two major assumptions that guide sociological thinking about the nature of deviance:

1. Almost any behavior or appearance can qualify as deviant under the right circumstance.
2. Conceptions of what is deviant vary over time and place.

In Chapter 4 we examined norms, a sociological concept that underscores these two assumptions. As you recall, norms are written and unwritten rules specifying appropriate and inappropriate behaviors. The important concept of norms cannot be overlooked when discussing deviance, because it is the violation of norms that constitutes deviance.

Deviance: The Violation of Norms

Recall that in Chapter 4 we learned that some norms are considered more important than others. In that chapter we highlighted two kinds of norms—folkways and mores—distinguished by sociologist William Graham Sumner.

Folkways and Mores

Folkways are customary ways of doing things that apply to the details of life or routine matters—how one should look, eat, greet another person, express affection toward same-sex and opposite-sex persons, and even go to the bathroom.[4] When Francisco Martins Ramos, an anthropology professor who teaches in Portugal, visited the United States, he found that folkways governing how to behave in public restrooms were different from those in his country. Ramos writes that the Portuguese newcomer

will certainly be quite surprised with a form of cultural behavior never dreamt of: The American who urinates in public initiates conversation with the partner at his side, even if he does not know the latter! Themes of these occasional dialogues are the weather, football, politics, and so on. We can guess at the forced pleasure of the Portuguese, who heretofore has regarded

urination as a necessary physical function, not as a social occasion. The public restroom! Is it an extension of the bar room? (RAMOS 1993, P. 5)

Mores are norms that people define as essential to the well-being of their group or nation. Obvious examples of mores are norms that prohibit cannibalism or the unjust and deliberate taking of another person's life. People who violate mores usually are punished severely; they are ostracized, institutionalized in prisons or mental hospitals, and sometimes executed. In comparison to folkways, however, people consider mores to be "the only way" and "the truth." Consequently, people consider mores to be final and unchangeable.

The United States has a large number of mores that protect individual privacy, property, rights, and freedoms. For example, in the United States an individual has the right to marry, to have children, to choose a career and to change jobs and residences without appealing to a higher authority for permission. It does not matter that the country as a whole has a shortage of scientists, teachers, and nurses or that there are no physicians in some geographic areas and a glut of physicians in others. What matters is individual choice. It is unthinkable that Americans could *not* change jobs or residences whenever they wanted, for whatever reason.

In China, throughout its long history, its mores have reflected the traditional values of conformity, collectivism, and obedience to authority. In fact, the Communist Party, which was formed in the 1920s and eventually came to power under Mao's leadership in 1948, sought to maintain these traditional values but to shift them away from the family to the party and party-led institutions. At the same time the party wanted to get rid of many old mores that were based on a disdain for physical labor, allegiance to the family, religious beliefs, and reliance on personal networks, and to substitute mores that reflected new values including a love for physical labor, loyalty to the party and its leader, and atheism. The communists aimed to create a new style of person ("socialist man"), and they built upon some old mores and values and introduced a new set of values, mores, and sanctions to discourage deviance and enforce their own rule (Personal Correspondence 1993). In China it is unthinkable that an individual could marry, have a baby, or obtain housing without first obtaining approval from the Communist Party–controlled work unit or neighborhood committee (Oxman 1993c). Until recently, a person could not select an occupation; purchase a train ticket, a bicycle, or a television; secure a hotel room; obtain employment; or buy food without written approval from a unit or a committee.

Usually people abide by established folkways and mores because they accept them as "good and proper, appropriate and worthy" (Sumner 1907, p. 60). For the great majority of people, "the rule to do as all do suffices." Recall from Chapter 5 that socialization is the process by which most people come to learn and accept as natural the ways of their culture. Because socialization begins as soon as a person enters the world, there is little opportunity to avoid exposure to the culture's folkways and mores. If we compare the ways in which life is structured for four-year-olds in Chinese preschools and American preschools, we can see that different but important cultural lessons are incorporated into their daily activities. Even though it is impossible to generalize about preschools in countries as large and as diverse as the United States and China, there are some broad, outstanding differences. Chinese preschoolers are taught to suppress individual impulses, to play cooperatively with other children, and to attune themselves to group enterprises. In contrast, American preschoolers are taught to cultivate individual interests and to compete with other children for success and for recognition by the teacher.

Socialization as a Means of Social Control: Preschool in China and in the United States Professors Joseph Tobin and Dana Davidson and researcher David Wu filmed daily life in Chinese and American preschools to learn how teachers in each system socialize children to participate effectively in their respective societies.[5] (It is worth noting that the great majority of preschool children whom they filmed in both countries seemed happy and productive.) The researchers found that in comparison to American preschools, Chinese preschools are highly structured and social-minded: Chinese teachers discipline their four-year-old students "by stopping them from misbehaving before they even know they are about to misbehave" (Tobin, Wu, and Davidson 1989, p. 94), and they promote loyalty to the group. The bathroom scene described here is an example of the extent to which Chinese children are taught to follow instructions and attune themselves to group enterprises:

> It is now 10:00, time for children to go to the bathroom. Following Ms. Wang, the twenty-six children walk in single file across the courtyard to a small cement building toward the back of the school grounds. Inside there is only a long ditch running along three walls. Under Ms. Wang's direction and, in a few cases, with her assistance, all twenty-six children pull down their pants and squat over the ditch, boys on one side of the room, girls on the other. After five minutes Ms. Wang distributes toilet paper, and the children wipe themselves. Leaving the toilet, again in single file, the children line up in front of a pump, where two daily monitors are kept busy filling and refilling a bucket with water that the children use to wash their hands. (PP. 78–79)

When the researchers showed American parents and teachers the film portraying daily life in Chinese preschools, Americans were particularly disturbed by the bathroom scene and asked why

Forrest Anderson/Gamma-Liaison　　　　　　　　　　　　　Paul Conklin/Monkmeyer Press

Quite different mores govern behavior in American and Chinese preschools. Whereas Chinese preschool teachers emphasize structured activities and social-mindedness, teachers in the United States are more likely to teach pupils self-direction, freedom of choice, and individuality.

children were forced to go to the bathroom in this manner.[6] To this question one Chinese educator replied:

> *Why not? Why have small children go to the bathroom separately? It is much easier to have everyone go at the same time. Of course, if a child cannot wait, he is allowed to go to the bathroom when he needs to. But, as a matter of routine, it's good for children to learn to regulate their bodies and attune their rhythms to those of their classmates.* (P. 105)

In the United States preschoolers also undergo discipline, but they are more likely to be disciplined *after* they do something wrong or get out of hand. In comparison to their Chinese counterparts, American teachers encourage self-direction, freedom of choice, independence, and individuality—qualities that often depend on a supply of material items. For example, American children typically use as much paper as they want; they start a drawing, decide they don't like it, and crumple up the paper. When they play house, store, or firefighter,

they use costumes, plastic dishes, children's versions of household appliances, plastic food items, and so on. American teachers also encourage children to choose from a number of activities. A typical exchange between a preschool teacher and his or her students follows. It is difficult to imagine this exchange taking place in a Chinese classroom:

> *Who would like to paint? Michelle. Mayumi. Nicole. Okay, you three get your smocks from your cubbies and you can paint. [The teacher holds up a wooden block.] Who wants to do this? Mike? Okay, that's one. Stu, that makes two. Billy is three. . . . Here's a puzzle piece. You want to start on the puzzles? Okay? [The teacher holds up a toy frying pan.] Who wants to start in the house? Lisa, Rose, Derek. Go ahead to the housekeeping corner. Kerry, what do you want to do? The Legos? You're going to work on the radio, Carl? That's fine. Who is going to come over to the book corner to read this book? It's called* Stone Soup. *Okay, come on with me.* (P. 130)

Reaction to Socialization of Another Culture Both the Chinese and the Americans who watched the films were disturbed by the other country's system of handling preschool. Comments by Chinese viewers showed their clear preference for their own way of structuring early education. They maintained that their form of discipline expresses care and concern, and they regard the American preschools as chaotic, undisciplined, and promoting self-centeredness. As one Chinese viewer remarked, "There are so many toys in the classroom that children must get spoiled. When they have so much, children don't appreciate what they have" (p. 88). On the other side, the Americans criticized the Chinese preschools for being rigid, totalitarian, too group-oriented, and overrestrictive, "making children drab, colorless, and robot-like" (p. 138). The people in each country feel discomfort with the other's system because the lessons that are taught in the country clash with the prevailing mores or ideas about which behaviors are essential to their own country's well-being.

The point is that early socialization experiences prepare children to fit within the existing system. China is probably one of the most regulated, collectivity-oriented countries in the world, whereas the United States is one of the most (if not the most) self-managing, individually oriented countries. Each society tries to prepare its people to fit into and accept their respective environments. In the case of preschool socialization, most Chinese preschoolers begin kindergarten having learned a great deal about the need for group cooperation. Most American preschoolers graduate knowing that they will continue to be evaluated and rewarded on their individual performance. Even so, primary socialization experiences such as those that take place during preschool are uneven at best. Not all preschools are alike, and some children do not attend preschools. Even among those who attend, some children do not internalize (take as their own and accept as binding) the values, norms, and expectations that they are taught. Therefore all societies establish other mechanisms of social control to ensure conformity.

As we learned earlier, conformity may be defined as behavior and appearance that follow and maintain standards set by a group (Cooley [1902]

1964). Ideally, conformity is voluntary. That is, people are internally motivated to maintain group standards and feel guilty if they deviate from them. As we have seen, during the Cultural Revolution it was considered deviant to wear glasses, use makeup, speak a foreign language, or break or destroy items that displayed Mao Zedong's picture. Many Chinese conformed to these rules on their own and punished themselves if they violated them even if they did so by accident. The memories of one Chinese man illustrate this point:

> *As a boy, I did not know what a god looked like, but I knew that Mao was the god of our lives. When I was six, I accidentally broke a large porcelain Mao badge. Fear gripped me. In my life until that moment, the breaking of the badge seemed the worst thing I had ever done. Desperate to hide my crime, I took the pieces and threw them down a public toilet. For months I felt guilty.* (AUTHOR X, 1992, P. 22)

In this case, the author's own guilt is a sign of voluntary conformity. Oftentimes, however, if conformity cannot be achieved voluntarily, people employ various means to teach, persuade, or force others to conform.

Mechanisms of Social Control

Ideally, socialization brings about conformity and, ideally, conformity is voluntary. But when conformity cannot be achieved voluntarily, mechanisms of social control are used to convey and enforce shared norms and expectations. Such mechanisms are known as **sanctions**—reactions of approval and disapproval to behavior and appearances. Sanctions can be positive or negative, formal or informal. A **positive sanction** is an expression of approval and a reward for compliance; such a sanction may take the form of applause, a smile, or a pat on the back. A **negative sanction** is an expression of disapproval for noncompliance; the punishment may be withdrawal of affection, ridicule, ostracism, banishment, physical harm, imprisonment, solitary confinement, or even death.

Informal sanctions are spontaneous and unofficial expressions of approval or disapproval; they are not backed by the force of law. The following

Formal sanctions include written rules governing appropriate behavior, as illustrated by these signs posted in both Chinese and English at China's Shenzhen University.

James F. Hopgood, Northern Kentucky University

incident involving the son of a well-known psychologist, Sandra Bem, shows how informal sanctions usually are applied. Bem and her husband are attempting to raise their children not to be limited by those "attributes and behaviors the culture may have stereotypically defined as inappropriate" for one sex or the other (Monkerud 1990, p. 83). Their efforts to go against norms and expectations are challenged through informal sanctions at every step of the way:

> *At age four, Bem's son, Jeremy, wore barrettes to nursery school. One day a boy repeatedly told him that "only girls wear barrettes." Jeremy tried to explain that wearing barrettes didn't make one a boy or girl: only genitalia did. Finally, in frustration, he pulled down his pants to show the boy that having a penis made him a boy. The boy responded, "Everybody has a penis; only girls wear barrettes."* (P. 83)

This incident also shows how early socialization to norms works. Even at the age of four, children use sanctions to enforce the norms that they have learned from adults, teachers, and other sources such as television.

Formal sanctions are definite and systematic laws, rules, regulations, and policies that specify (usually in writing) the conditions under which people should be rewarded or punished, and that define the procedures for allocating rewards and imposing punishments. Examples of formal sanctions include medals, cash bonuses, and diplomas, on the one hand, and fines, prison sentences, and the death penalty, on the other. Sociologists use the word **crime** to refer to deviance that breaks the laws of society and is punished by formal sanctions. People in every society have different views of what constitutes crime, what causes people to commit crimes, and how to handle offenders (see "The Extent of Social Control in China").

Defining Deviance

According to sociologist Randall Collins (1982), Emile Durkheim presented one of the most sophisticated sociological theories of deviance. Durkheim

([1901] 1982) argued that, although deviance does not take the same form everywhere, it is present in all societies. He defined deviance as those acts that

The Extent of Social Control in China

In China "social control is everywhere and involves everyone" (Clark and Clark 1985, p. 109). Thus, "each person has a social duty to participate in group activities and to 'help' others in the collective living arrangement. The mandate to 'help' means assuming responsibility for others and correcting their faults" (Rojek 1985, p. 119).

Theoretically, every Chinese belongs to a work unit and to a neighborhood committee headed by Communist Party members. Until recently almost every important area of life was supervised by the unit or the committee. The work unit issued job assignments; determined salary; distributed ration coupons and other goods (light bulbs, contraceptives, bicycles, television sets); granted permission to travel, change jobs, and change residences. It still scrutinizes requests to marry and have children, and it determines whether a worker or his or her children may take college entrance examinations. The work unit maintains a confidential file[1] on every person, which contains information on education, work, class background as far back as three generations, the party's evaluation of the person (as an activist or as a counterrevolutionary), and any political charges made against him or her by informants. Members of the work unit are encouraged to report wrongdoings committed by other members. Whereas Americans are hesitant to report their suspicions of misconduct such as child abuse, spousal abuse, or drug use because they believe in a person's right to live without interference, the Chinese typically are afraid not to voice such suspicions because, by remaining silent, they are viewed as accomplices to such acts.

The neighborhood committee monitors life outside the workplace. It enforces birth control policies, monitors contacts between its members and outsiders, settles domestic problems, scrutinizes each household's activities, investigates disputes among neighbors, deals with petty crime, and educates members about new policy and law. The committee has the right to search living quarters without the occupants' consent. As in the work unit, members are encouraged to inform on other members' wrongdoings. "When society is so tightly organized, where can one run to [when one unintentionally or intentionally violates the rules]?" (Wu Han 1981, p. 39).

Among the aspects of Chinese life that are controlled by work units and neighborhood committees, perhaps the most widely publicized outside of China is procreation. Because of its huge population, China has tried to impose a limit of one child per couple in hopes of slowing population growth so that it will stabilize at 1.2 billion by the year 2000 (Tien 1990). Each province and each city is assigned a yearly quota, with regard to how many children can be born; and the neigh-

---

offend collective norms and expectations. The fact that always and everywhere there are people who offend collective sentiments led him to conclude that deviance is normal as long as it is not excessive, and that "it is completely impossible for any society entirely free of it to exist" (p. 99). According to Durkheim, deviance will be present even in a "community of saints in an exemplary and perfect monastery" (p. 100). Even in seemingly perfect societies, acts that most persons would view as minor may offend, create a sense of scandal, or be treated as crimes.

Durkheim drew an analogy to the "perfect and upright" person. Just as such a person judges his or her smallest failings with a severity that others reserve for the most serious offenses, so do societies that are supposed to contain the most exemplary people. Even among such exemplary individuals, some act or some appearance will offend simply because "it is impossible for everyone to be alike if only because each of us cannot stand in the same spot. It is also inevitable that among these deviations some assume a criminal character" (p. 100). What makes an act or appearance criminal is not so much the character or the consequences of that act or appearance but the fact that the group has defined it as something dangerous or threatening to its well-being. Wearing eyeglasses, for example, is

This billboard in Beijing is part of the Chinese effort to control population growth through a policy of one child per family.

Forrest Anderson/Gamma-Liaison

borhood committee or work unit determines which couples will be included in that quota. In essence, the work unit and the neighborhood committee decide which married couples can have a baby, determine when they can start trying, oversee contraceptive use, and even record women's menstrual cycles. Female officials check to see that IUDs are in place. If a woman misses several periods and has not been authorized to try to conceive, she is persuaded to have an abortion (Ignatius 1988). As one Chinese official told an American journalist, "We don't force her. . . . We talk to her again and again until she agrees" (Hareven 1987, p. 73). Couples who have only one child and who sign an agreement to have no more children are rewarded with positive sanctions: salary bonuses and other financial incentives, educational opportunities, housing priority, and extra living space. If couples request permission to have a second child, they are asked to wait at least four years; if they have two children, they are persuaded not to have any more. Various kinds of economic sanctions or penalties are imposed on couples who have more than one child.[2]

1. In China, access to what people living in the United States would consider in the public domain such as demographic statistics is considered classified information. Chinese students and professors interested in such information must make a formal request. Such requests are entered into their permanent file (Rorty 1982).

2. The one-child-per-couple policy has been more successful in urban areas than in rural areas. Chinese peasants have resisted government efforts because they believe that more children bring more happiness, that sons are more effective laborers in the fields than daughters, and that sons offer security to parents in old age. (When a male child marries, his wife moves in with him and his parents, and the couple supports them in their old age.)

clearly not a behavior that is harmful to others. However, as we have learned this behavior was defined as a clear indicator of other threatening behaviors, such as the crimes of revisionism and entrenched authority, that were not so easily observable.

Durkheim maintained that deviance is functional for society for at least two reasons. First, the ritual of punishment (exposing the wrongdoing, determining a punishment, and carrying it out) is an emotional experience that serves to bind together the members of a group and establish a sense of community. Durkheim argued that a group that went too long without noticing crime or doing something about it would lose its identity as a group.

Second, deviance is functional because it is useful in making necessary changes and in preparing people for change. It is the first step toward what will be. Nothing would change if someone did not step forward and introduce a new perspective or new ways of doing things. Almost every invention or behavior is rejected by some group when it first comes into existence.

Durkheim's theory offers an intriguing explanation for why almost anything can be defined as deviant. Yet, Durkheim did not address an important question: Who decides that a particular activity or

FIGURE 8.1   Categories of Deviance

	Rule Violated	Rule Not Violated
Enforced	Pure deviant	Falsely accused
Not Enforced	Secret deviant	Conformist

(Sanctions on the vertical axis: Enforced / Not Enforced)

Deviant or ministering angel? San Francisco's "Brownie Mary" has been arrested several times for giving marijuana brownies to AIDS patients to ease their discomfort. Are her acts criminal because of their harmful consequences or because society has defined them as illegal?

Fran Ortiz/San Francisco Examiner/Saba

appearance is deviant? One answer can be found from labeling theory.

Labeling Theory *Goffman*

A number of sociologists—Frank Tannenbaum (1938), Edwin Lemert (1951), John Kitsuse (1962), Kai Erikson (1966), and Howard Becker (1963)—are linked to the development of what is conventionally called labeling theory. The scholar most frequently associated with labeling theory, however, is Howard Becker. In *Outsiders: Studies in the Sociology of Deviance,* Becker states the central thesis of labeling theory: "All social groups make rules and attempt, at some times and under some circumstances, to enforce them. . . . When a rule is enforced, the person who is supposed to have broken it may be seen as a special kind of person, one who cannot be trusted to live by the rules agreed on by the group. He is regarded as an *outsider*" (1963, p. 1).

As Becker's statement suggests, labeling theorists operate under the assumption that rules are socially constructed and that rules are not enforced uniformly or consistently. These assumptions are supported by the fact that the definitions of what is deviant vary across time and place, and by the fact that some people break rules and escape detection whereas others are treated as offenders even though they have broken no rules. Labeling theorists maintain that whether an act is deviant depends on whether people notice it, and, if they do notice, on whether they label it as a violation of a rule and subsequently apply sanctions. Such contingencies

suggest that violating a rule does not make a person deviant. That is, from a sociological point of view a rule breaker is not deviant (in the strict sense of the word) unless someone notices the violation and decides to take corrective action.

Labeling theorists suggest that for every rule a social group creates, there are four categories of people: conformists, pure deviants, secret deviants, and the falsely accused. The category that one belongs to depends on whether a rule has been violated *and* on whether sanctions are applied (see Figure 8.1).

Conformists are people who have not violated the rules of a group and are treated accordingly. **Pure deviants,** on the other hand, are people who have broken the rules and are caught, punished, and labeled as outsiders. As a result, the rule breaker takes on the master status of deviant, an identification that "proves to be more important than most others. One will be identified as a deviant first, before other identifications are made" (Becker 1963, p. 33). We must remember that,

TABLE 8.1 Percent Distribution of Victimizations and Whether or Not Reported to Police		
Sector and type of crime	**Number of victimizations**	**Percent of victimizations reported to the police**
All crimes	34,730,370	38.0%
All personal crimes	18,956,060	35.3
Crimes of violence	6,423,510	48.6
Crimes of theft	12,532,550	28.5
All household crimes	15,774,310	41.2

SOURCE: U.S. Bureau of Justice Statistics (1992), p. 102.

although pure deviants undeniably violate rules, rule enforcers select the people they apprehend and punish. Consider, for example, how drug enforcement agents make a decision to stop someone in an airport because they suspect him or her of smuggling drugs. Or consider how highway patrol officers choose, from among all of the cars speeding along a highway, which drivers to pull over. A study of vehicles' and drivers' characteristics on a stretch of the New Jersey Turnpike showed that fewer than five percent of the vehicles observed were late-model cars with out-of-state license plates, driven by black males. Yet, 80 percent of the arrests made on that stretch of highway fit this profile (Belkin 1990). The drivers who are pulled over for speeding assume the status of deviant; those not stopped assume the status of secret deviant.

Secret deviants are people who have broken the rules, but whose violation goes unnoticed, or, if it is noticed, no one reacts to enforce the law. Becker maintains that "no one really knows how much of this phenomenon exists," but [he is] convinced that the "amount is very sizable, much more so than we are apt to think" (1963, p. 20). For example, a 1992 U.S. Bureau of Justice survey of crime victims documented that only 38 percent of crime victims reported the crime to police (U.S. Bureau of Justice Statistics 1992; see Table 8.1). Another area in which secret deviants seem to abound is among those who drive under the influence of alcohol. The typical drunk driver in the United States can expect to make up to 2,000 trips before he or she is stopped by police or an accident (Jacobs 1983).

The war on drugs, waged by the U.S. government since 1989 against users and sellers, is a third dramatic example of the prevalence of secret deviance. According to estimates, 25 million Americans spend $50 billion a year on drugs, including $20 billion on cocaine alone. Ninety percent of the world's supply of cocaine is grown in Peru and Bolivia, making it Bolivia's single most important export and accounting for one-third of Peru's exports. Political scientist Ethan Nadelmann (1989) estimated that for every 300 cocaine exporters there are 222,000 coca farmers, 74,000 paste processors, 7,400 paste transporters, and 1,333 refiners. In view of the large number of users and the fact that whole economies are built around cocaine, a law enforcement system can catch and punish only a very small portion of those who use, produce, and sell the drug (Gorriti 1989).

The **falsely accused** are people who have not broken the rules but who are treated as if they have done so. Like pure deviants, the falsely accused are labeled as outsiders. As with the phenomenon of secret deviants, no one knows how often people are falsely accused, but it probably happens more often than we think. In any case, the status of *accused* often lingers even if the person is cleared of all charges. For example, a sampling of records kept by drug enforcement officials at three major airports (Houston, New York, and Miami) shows that on the average, 50 percent of all people detained and taken in for X-rays were found *not* to have swallowed drug-filled pouches, as suspected. (The pouches are usually condoms sealed with tape, from which smugglers can retrieve drugs after the pouches pass through their system.) This figure, of course, varies according to the individual airport. At the Houston International Airport, for example, a sampling of records showed that 4 out of 60 persons X-rayed were found to be guilty (Belkin 1990).

The falsely accused usually are persecuted for what they *are* or because they are visible, not for what they supposedly did. A substantial number of people who are stopped by drug enforcement agents are stopped because they fit a profile of a drug courier and a drug swallower. The profile includes the following traits: having dark skin and/or a foreign appearance, being obviously in a hurry, having an exotic hairstyle or wearing brightly colored clothing, purchasing a one-way ticket, paying cash for a ticket, changing flights at the last minute, flying to or from Detroit, Miami, or another large city, and flying to or from the Caribbean, Nigeria, Jamaica, or other known drug-supplying regions (Belkin 1990).

The Stuart murder that took place in Boston in October 1989 represents another such case. Charles Stuart murdered his pregnant wife, Carole, and then deliberately and severely wounded himself. At that point he made a "frantic and heart-wrenching call" (Fox and Levin 1990, p. 66) for help to the state police. Stuart told police that he and his wife had been robbed and shot by a black man. The police then proceeded to question and round up black men in the neighborhood, one of whom Stuart identified as the killer. Later, Stuart committed suicide upon learning that investigators were on to the real story. Stuart had murdered his wife in an attempt to collect her life insurance and had inflicted injuries on himself to make his story more believable. Such events as the Stuart murder lead us to ask a larger question: Under which circumstances are people most likely to be falsely accused?

The Circumstances of the Falsely Accused

Sociologist Kai Erikson (1966) identifies a particular situation in which people are likely to be falsely accused of a crime: when the well-being of a country or a group is threatened. The threat can take the form of an economic crisis (an economic depression or recession), a moral crisis (family breakdown, for example), a health crisis (AIDS, for example), or a national security crisis (such as war). At such times people need to define a seemingly clear source of the threat. Whenever a catastrophe occurs, it is common to find someone to blame for it. Identifying the threat gives an illusion of control. The person blamed is likely to be someone who is at best indirectly responsible, someone in the wrong place at the wrong time and/or someone who is viewed as different. In China, for example, a drunken, off-duty deckhand aboard an overloaded ferry was sentenced to death for having started a brawl that attracted a crowd to his side of the boat. The uneven weight caused the boat to capsize, and 158 people died (Burns 1986).

This defining activity can take the form of a witch-hunt, a campaign against subversive elements with the purpose of investigating and correcting behavior that undermines a group or a country. In actuality, a witch-hunt may not accomplish this goal because the real cause of a problem is often complex and may lie beyond the behavior of a targeted person or group. Often the people who are defined as the problem are not in fact the cause of the threat but are used intentionally to represent the cause of a complicated situation. For example, as happened during the Cultural Revolution, sometimes the pettiest and most insignificant acts were classified as crimes against a group or country in order to divert the public's attention from the shortcomings of those in power, to unite the public behind a cause, or to take people's attention away from a disruptive event such as massive job layoffs. Another example is the timing of the 1993 debate over whether gays should be allowed to serve in the U.S. military. Sociologists might ask: Is it a coincidence that gays became the focus of media attention just when the Clinton administration and Congress were planning such large cuts in the defense budget?

The internment of more than 110,000 people of Japanese descent (80 percent of whom were American citizens) living on the West Coast of the United States during World War II is another example of a situation in which a group was targeted in conjunction with a crisis. Japanese-Americans were forced

Bart Bartholomew/Black Star

Rick Friedman/Black Star

Conflict theorists emphasize that not all rules and rule-breakers are treated equally. A sharp financial dealer like Michael Milken (shown here doing community service as part of his punishment) may inflict far more harm on more people than street thieves (right). Yet the white-collar felon often is treated far less harshly by the criminal justice system.

from their homes and taken to desert prisons surrounded by barbed wire and guarded with machine guns (Kometani 1987). There was no evidence of anti-American activity on the part of Japanese-Americans. Yet, the wartime hysteria, combined with long-standing prejudices, led to the shipping of men, women, and children to concentration camps.

Another example of false accusation took place after the war. In the late 1940s and 1950s, hundreds of Americans were accused of being communists. They were placed under investigation and, in some cases, on trial, without sufficient evidence to support the accusations.

The existence of the falsely accused underscores the fact that the study of deviance must go beyond looking at people identified or labeled as rule breakers. The roles of rule makers and rule enforcers must be examined as well.

Rule Makers and Rule Enforcers

Sociologist Howard Becker recommends that researchers pay particular attention to the rule makers and rule enforcers and to how they achieve power and then use it to define how others "will be regarded, understood, and treated" (1973, p. 204). This topic, of course, interests not only labeling theorists but conflict theorists as well. According to conflict theorists, those with the most wealth, power, and authority have the power to create laws and to create crime-stopping and monitoring institutions. Consequently we should not be surprised to learn that law enforcement efforts focus disproportionately on crimes committed by the poor and by other powerless groups rather than on those committed by the wealthy and other politically powerful groups. This uneven focus gives the widespread impression that the poor, the uneducated, and minority-group members are more prone to criminal behavior than are people in the middle and upper classes, the educated, and majority-group members. Crime exists in all social classes, but the type of crime, the extent to which the laws are enforced, access to legal aid, and the power to shape laws to one's advantage vary across class lines (Chambliss 1974). In the United States, for example, police efforts are directed at controlling crimes against individual life and property (crimes such as robbery, assault, homicide, and

The Presumption of Guilt or Innocence

In China a defendant is presumed to be guilty, not innocent until proved guilty as in the U.S. system. Before he or she appears in court, the defendant will have been scrutinized already by a work unit or neighborhood committee. If the work unit or committee members cannot correct the defendant's outlook or behavior, they will refer the case to the control committee, which can either assign a punishment or refer the case to the courts. If a case is referred to the courts, it is presumed that the offender is guilty and that some punishment is warranted. The court appearance is likely to be just a sentencing hearing, as evidenced by the fact that approximately 99 percent of persons whose cases are heard by the Supreme People's Court are convicted. Theoretically, a Chinese offender can appeal a decision, but the court system allows only one appeal (Chiu 1988).

In theory, a defendant in the United States is assumed innocent until proved guilty; the burden of proving guilt rests with the prosecution. Furthermore, sentences and punishments can be reduced through plea bargaining, shock probation (sudden and unexpected release from prison), or plea of insanity or self-defense, or simply because prisons are overcrowded. In 1990, 300,000 criminal cases were processed in New York City alone.

A man accused of counterrevolutionary activity is brought to trial before being executed. By the time his case reached the trial stage, he was already presumed guilty.
Agence Vu/JB Pictures

Only 10 percent, or 30,000 cases, were tried to completion. The remaining 270,000 cases ended in dismissals, reduction in charges, or plea bargaining (Shipp, Baquet, and Gottlieb 1991). Finally, defendants who can afford the fees are able to purchase the best legal defense and thus have the greatest chance of being found innocent or receiving lighter sentences.

The role of lawyers in each society dramatically reflects the basic differences between Chinese and American views on how defendants should be handled. In China, lawyers are paid by the state and must remit any fees to the state. Although the Chinese lawyer is trained to protect the legitimate rights of clients, those rights are secondary to the rights and interests of the state and the people. In addition, Chinese lawyers cannot use the impression management strategies that American lawyers employ to convince a jury of a client's innocence when in fact the client is guilty (Lubman 1983). If Chinese lawyers followed such techniques, they would be accused of conspiring with an enemy of the state.

---

rape) rather than against white-collar and corporate crime.

White-collar crime consists of "crimes committed by persons of respectability and high social status in the course of their occupations" (Sutherland and Cressey 1978, p. 44). **Corporate crime** is crime committed by a corporation as it competes with other companies for market share and profits. Usually white-collar and corporate crimes, such as the manufacturing and marketing of unsafe products, unlawful disposal of hazardous waste, tax evasion, and money laundering, are handled not by the police but by regulatory agencies that have little power and minimal staff. It is easier to escape punishment for white-collar crimes because offenders are part of the system: they occupy positions in the organization that permit them to carry out illegal activities discreetly. In addition, white-collar and corporate crime "is directed against impersonal—and often vaguely defined—entities such as the tax system, the physical environment, competitive conditions in the market economy, etc." (National Council for Crime Prevention in Sweden 1985, p. 13). These crimes are without victims in the usual sense because they are "seldom directed against a particular person who goes to the police and reports an offense" (p. 13).

Deviance is a consequence not of a particular behavior or appearance, but of the application of rules and sanctions to a so-called offender by others (see "The Presumption of Guilt or Innocence"). For these reasons, sociologists are concerned less with rule violators than with those persons who make the rules and those who support and enforce the rules. In view of this need to understand more about the role that rule makers play in molding deviance, we turn to the constructionist approach, which concentrates on a particular type of rule maker—the claims maker.

The Constructionist Approach

The **constructionist approach** focuses on the process by which specific groups (for example, illegal immigrants or homosexuals), activities (for example, child abuse or drug swallowing), conditions (teenage pregnancy, infertility, pollution), or artifacts (song lyrics, guns, art, eyeglasses) become defined as problems. In particular, constructionists examine the claims-making activities that underlie this process.[7] Claims-making activities include "demanding services, filling out forms, lodging complaints, filing lawsuits, calling press conferences, writing letters of protest, passing resolutions, publishing exposés, placing ads in newspapers, supporting or opposing some governmental practice or policy, setting up picket lines or boycotts" (Spector and Kitsuse 1977, p. 79). **Claims makers** are people who articulate and promote claims and who tend to gain in some way if the targeted audience accepts their claims as true.

Claims Makers

Claims makers include but are not limited to victims of discrimination, government officials, professionals (medical doctors, scientists, professors), and pressure groups (any group that exerts pressure on public opinion and government decision makers in order to advance or protect its interests). The success of a claims-making campaign depends on a number of factors, including access to the media, available resources, and the claims maker's status and skills at fund raising, promotion, and organization (Best 1989). During the Tiananmen Square incident, for example, government officials were able to control the information that the troops received about the demonstration. Essentially the soldiers were denied access to newspapers and television for about a week. When soldiers and students became friendly with one another, the government sent in replacements, some of whom came from the North and were not fluent in standard Chinese (Calhoun 1989). Thus, the replacements could not communicate with the demonstrators. Obviously the government officials, because of their positions, were better able than the students to control the soldiers' understanding of the situation.

According to sociologist Joel Best, when constructionists study the process by which a group or behavior is defined as a problem to the society, they focus on who makes the claims, whose claims are heard, and how audiences respond. Constructionists are guided by one or more of the following questions: What kinds of claims are made about the problem? Who makes the claims? Which claims are heard? Why is the claim made when it is made? What are the responses to the claim? Is there evidence that the claims maker has misrepresented or inaccurately characterized the situation? In answering this last question, constructionists examine how claims makers characterize a condition. Specifically, they pay attention to any labels that claims makers attach to a condition, the examples they use to illustrate the nature of the problem, and their orientation toward the problem (describing it as a medical, moral, genetic, educational, or character problem).

Labels, examples, and orientation are important because they tend to evoke a particular cause and a particular solution to a problem (Best 1989). For example, to call AIDS a moral problem is to locate its cause in the goodness or badness of human action and to suggest that the solution depends upon changing evil ways. To call it a medical problem is to locate its cause in the biological workings of the body or mind and to suggest that the solution rests with a drug, a vaccine, or surgery. Similarly, a claims maker who uses the example of a promiscuous homosexual male to illustrate the nature of the AIDS problem sends a much different message about AIDS than does a claims maker who uses hemophiliacs or HIV-infected children as examples. As we learned in Chapter 6, such narrow conceptions shed little light on the nature of the AIDS problem and on potential solutions.

In the section that follows, we use the constructionist approach to examine the accuracy of claims that alcohol and cigarettes are legal because they are less dangerous substances than cocaine and other now-illegal drugs. This claim is one reason that drug enforcement officials generate profiles of people likely to be drug carriers and drug swallowers but not people likely to be tobacco or alcohol vendors.

Claims About the Dangers of Legal and Illegal Substances

In his article "Taxonomy as Politics: The Harm of False Classification," biologist Stephen Jay Gould describes a war of words between two claims makers—former Surgeon General C. Everett Koop and Illinois Representative Terry Bruce—which began after Koop announced that nicotine was no less addicting than heroin or cocaine:

Representative Terry Bruce (D-IL) challenged [Koop's] assertion by arguing that smokers are not "breaking into liquor stores late at night to get money to buy a pack of cigarettes." Koop properly replied that the only difference resides in social definition as legal or illegal: "You take cigarettes off the streets and people will be breaking into liquor stores. I think one of the things that many people confuse is the behavior of cocaine and heroin addicts when they are deprived of the drug. That's the difference between a licit and an illicit drug. Tobacco is perfectly legal. You can get it whenever you want to satisfy the craving." (GOULD 1990, P. 75)

Koop's hypothesis that cigarette smokers deprived of their drug would behave like cocaine and heroin addicts is supported by events in Italy, where the government has a monopoly over the manufacture and distribution of cigarettes. In 1992 approximately 13 million smokers were prevented from purchasing cigarettes because of a strike by the workers who distributed them. The forced abstinence resulted in the following behaviors: (1) panic buying and hoarding of the cigarettes still on the shelves when the strike began, (2) a run on nicotine gum and patches, (3) a sevenfold increase in traffic into Switzerland, (4) an underground market in cigarettes, and (5) robbery of cigarette vendors at gun point (Cowell 1992).

Gould argues that illegal drugs such as cocaine should be legalized and controlled, like alcohol and tobacco, and their use strongly discouraged. He would even accept, reluctantly, a policy in which alcohol and tobacco had the same illegal status as drugs such as cocaine. Gould cannot find any rationale, however, for "an absurd dichotomy (legal versus illegal) that encourages us to view one class of substances with ultimate horror as preeminent scourges of life . . . while the two most dangerous and life-destroying substances by far, alcohol and tobacco, form a second class advertised in neon on every street corner of urban America" (Gould 1990, p. 74). Gould's claim is supported by data from the National Institute on Drug Abuse, an agency of the U.S. government: for every one cocaine-related death in 1987 there were 300 tobacco-related and 100 alcohol-related deaths (Reinarman and Levine 1989).

Gould does not deny that all such drugs mentioned are dangerous, but he finds that the illegal status for one set of substances and the legal status for another "permit a majority of Americans to live well enough with one, while forcing a minority to murder and die for the other" (p. 75). Gould argues that illegal status causes more social and medical problems than it solves and that illegal status itself causes crime. For example, the responses to the prohibition of alcohol in the United States between 1920 and 1933 are remarkably similar to those associated with cocaine's illegal status. The responses include murder, payoffs, bribes, and international intrigue. There is one argument, seldom articulated, in favor of keeping drugs like cocaine illegal: prohibition might keep the cocaine problem from reaching the tragic dimensions of the alcohol and tobacco problems.

To this point we have examined several sociological concepts—socialization, norms (folkways and mores), and mechanisms of social control—and how they relate to deviance. In addition, we have also examined Durkheim's theory of deviance for insights about how any behavior or appearance can come to be defined as deviant. We have looked at labeling theory and the constructionist approach for insights about the role that rule makers play in shaping deviance. Now we turn to the theory of structural strain, which gives us insights about how society creates deviance by virtue of its valued goals and the opportunities it offers people to achieve those goals.

Structural Strain Theory

Robert K. Merton's theory of structural strain takes into account three factors: (1) culturally valued goals defined as legitimate for all members of society, (2) norms that specify the legitimate means of achieving these goals, and (3) the actual number of legitimate opportunities available to people to achieve the culturally valued goals. According to Merton **structural strain,** or **anomie,** is any situation in which (1) the valued goals have unclear limits (that is, people are unsure whether they have achieved them), (2) people are unsure whether the legitimate means that society provides will lead to the valued goals, and (3) legitimate opportunities for meeting the goals are closed to a significant portion of the population. The rate of deviance is likely to be high under any one of these conditions. Merton uses the United States, a country where all three conditions exist, to show the relationship between structural strain and deviance.

In the United States, most people place a high value on the goal of economic affluence and social mobility. In addition, Americans tend to believe that all people, regardless of the circumstances in which they are born, can achieve monetary success. Such a viewpoint suggests that success or failure results from personal qualities and that persons who fail have only themselves to blame.

Merton does not believe that the same proportion of people in all social classes accept the cultural goal of monetary success, but he does think that a significant number of people across all classes do so. He argues that "Americans are bombarded on every side" with the message that this goal is achievable, "even in the face of repeated frustration" (1957, p. 137).

According to Merton, considerable structural strain exists in the United States because this culturally valued goal has no clear limits. He believes there is no point at which people can say they have achieved monetary success. "At each income level . . . Americans want just about twenty-five percent more (but of course this 'just a bit more' continues to operate once it is obtained)" (p. 136).

Merton also maintains that structural strain exists in the United States because the legitimate means of achieving affluence and mobility are not entirely clear; it is the individual's task to choose a path that leads to success. That path might involve education, hard work, or natural talent. The problem is that school, hard work, and talent do not guarantee success. With regard to education, for example, many Americans believe that the diploma (especially the college diploma) in itself entitles them to a high-paying job. In reality, however, the

FIGURE 8.2   Merton's Typology of Responses to Structural Strain
Merton believed that people respond to structural strain and that their response involves some combination of acceptance and rejection of valued goals and means.

Mode of Adaptation	Goals	Means
Conformity	+	+
Innovation	+	−
Ritualism	−	+
Retreatism	−	−
Rebellion	+/−	+/−

+ Acceptance/achievement of valued goals or means

− Rejection/failure to achieve valued goals or means

SOURCE: Adapted from "A Typology of Modes of Individual Adaptations" (1957). P. 140 in *Social Theory and Social Structure*, by Robert K. Merton.

diploma is only one component of many needed to achieve success.

Finally, structural strain exists in the United States because too few legitimate opportunities are available to achieve desired goals. Although all Americans are supposed to seek financial success, they cannot all expect to achieve it legitimately; the opportunities for achieving success are closed to many people, especially those in the lower classes. For example, many young African-American men living in poverty believe that one seemingly sure way to achieve success is through sports. The opportunities narrow rapidly, however, as an athlete advances: fewer than two percent of the athletes who play college basketball, for instance, even have a chance at the professional ranks.

Merton believed that people respond in identifiable ways to structural strain and that their response involves some combination of acceptance and rejection of the valued goals and means (see Figure 8.2). Merton identified the following five responses:

- **Conformity** is not a deviant response; it is the acceptance of the cultural goals and the pursuit of these goals through legitimate means.

- **Innovation** involves the acceptance of the cultural goals but the rejection of legitimate means to obtain these goals. For the innovator, success is equated with winning the game rather than playing by the rules of the game. After all, money may be used to purchase the same goods and services whether it was acquired legally or

illegally. Merton argues that when the life circumstances of the middle and upper classes are compared with those of the lower classes, the lower classes clearly are under the greatest pressure to innovate, though no evidence suggests that they do so.

- **Ritualism** involves the rejection of cultural goals but a rigid adherence to the legitimate means of those goals. This response is the opposite of innovation; the game is played according to the rules despite defeat. Merton maintains that this response can be a reaction to the status anxiety that accompanies the ceaseless competitive struggle to stay on top or to get ahead. It finds expression in the cliches "Don't aim high and you won't be disappointed" and "I'm not sticking my neck out." Ritualism can also be the response of people who have few employment opportunities open to them. If one wonders how it is possible for a ritualist to be defined as deviant, consider the case of a college graduate who can find only a job bagging groceries at $4.50 an hour. Most people react as if this person is a failure even though he or she may be working full time.

- **Retreatism** involves the rejection of both cultural goals and the means of achieving these goals. The people who respond in this way have not succeeded by either legitimate or illegitimate means and thus have resigned from society. According to Merton, retreatists are the true aliens or the socially disinherited—the outcasts, vagrants, vagabonds, tramps, drunks, and addicts. They are "in the society but not of it."

- **Rebellion** involves the full or partial rejection of both goals and means and the introduction of a new set of goals and means. When this response is confined to a small segment of society, it provides the potential for subgroups as diverse as street gangs and the Old Order Amish. When rebellion is the response of a large number of people who wish to reshape the entire structure of society, the potential for a revolution is great.

In China, one source of structural strain can be traced to the actual number of legitimate opportunities open to married couples to achieve the culturally valued goal of reproducing one child. As pointed out in "The Extent of Social Control in China," China has tried to impose a limit of one child per couple in hopes of slowing population growth. The one-child policy poses a problem because in China there is a cultural preference for boys, especially among the people living in the countryside. One reason sons are valued over daughters is that sons and their families are expected to care for parents in old age. Thus, couples who manage to produce a son can be confident that someone will care for them later in life. In China the legitimate means open to a couple include: obtain permission to have a baby, accept the sex of the child (that is, not aborting female fetuses or killing a daughter), report the birth and sex of the child to appropriate agencies, and practice birth control to avoid conceiving other children.

Merton's typology of responses to structural strain can be used to describe the reactions of couples. Those most likely to be conformists are those whose first child is a son. Conformists would also include those who have no preference as to the sex of their child, and/or those who are firmly committed to upholding the laws related to birth control because they see them as critical to China's quality of life. The majority of people in China can be classified as conformists.

Innovators accept the culturally valued goal of having one child but reject the legitimate means to obtain this goal. Upon learning she is expecting a child, the woman may undergo ultrasound to learn the sex of the fetus. If it is a girl, the couple may decide to abort the baby. Or upon the birth of a girl baby the parents may kill or arrange to have a midwife kill the infant. Such practices are blamed for the so-called missing girls problem; the 1990 census showed that for every 114 boys born there are 100 girls (Oxman 1993d).

Ritualists reject the cultural goal of having one child, but they adhere to the rules. They do not agree with the government policies, but they are afraid they will be punished if they do not follow the rules. Retreatists, on the other hand, reject the one-child goal and reject the legitimate means open to them. This category of deviants may include women and men who reject the one-child policy and are afraid to conceive out of fear that their firstborn may not be the desired sex. Retreatists might also include those couples who keep having children until they have a boy and hide the birth of the baby girls from party officials. In fact, Sterling Scruggs of the United Nations Population Fund argues that there is good evidence that many of the so-called missing girls include those who have not been reported (Oxman 1993d). In a sense, these parents are in society but, because of their secret, "not of it," as Merton would say.

Rebels reject the cultural goal of having one child and reject the legitimate means; instead they introduce new goals and new means. One could argue that this response applies to couples who are part of China's 55 ethnic minority populations that make up nine percent (91.2 million) of the total population. The majority of these ethnic groups are exempt from the one-child policy. The government permits them to have two children, and some groups are permitted to have three and in special cases four children. For these groups the cultural goal is to increase the size of the minority populations, and the means of achieving that goal is to exempt them from the one-child policy. This option has prompted many couples to claim ethnic minority status (Tien et al. 1992).

The deviant responses to structural strain leave us with the question of how people learn such responses. One answer comes from the work of Edwin H. Sutherland and Donald R. Cressey, who offer us the theory of differential association.

Differential Association Theory

Sociologists Edwin H. Sutherland and Donald R. Cressey advanced a theory of socialization called **differential association** to explain how deviant behavior, especially delinquent behavior, is learned. It refers to the idea that "when persons become criminal, they do so because of contacts with criminal patterns and also because of isolation from anticriminal patterns" (1978, p. 78). The contacts take place within **deviant subcultures**—groups that are part of the larger society but whose members adhere to norms and values that favor violation of the larger society's laws. That is, people learn techniques of committing crime from close association and interaction with people who engage in and approve of criminal behaviors.

Sutherland and Cressey maintain that impersonal forms of communication such as television, movies, and newspapers play a relatively small role in the genesis of criminal behavior. If we accept the premise that criminal behavior is learned, then criminals constitute a special type of conformist. That is, they conform to the norms of the group with which they associate. Furthermore, the theory of differential association does not explain how a person makes contact with deviant subcultures in the first place (unless, of course, the person is born into such a subculture). Once contact is made, however, the individual learns the subculture's rules for behavior in the same way that all behavior is learned.

Sociologist Terry Williams studied a group of teenagers, some as young as 14, who sold cocaine in the Washington Heights section of New York City. These youths were recruited by major drug suppliers because, as minors, they could not be sent to prison. Williams argues that the teenagers were susceptible to recruitment because they saw little chance of finding well-paying jobs and because they perceived drug dealing as a way to earn money that would enable them to pursue a new life.

To become a successful drug dealer, a youth must learn a number of skills. Williams describes some of these skills:

Each week, Max is fronted [consigned] three to five kilos (in 1985, this would have a street value of $180,000 to $350,000) to distribute to the kids. The quantity he receives varies according to the quantity he has previously sold, how much he has on hand, and how much he is committed to deliver both to the crew and directly to other customers.

The quality, variety and amount of cocaine each crew member receives, then, is determined by Max and by his suppliers. The kids might ask for more but whether or not they get more depends on how much they sold from their last consignment. Personal factors are also involved. (WILLIAMS 1989, P. 34)

Williams's findings suggest that once teenagers get involved in drug networks they learn the skills to do their jobs in the same way that everyone learns to do a job. Success in a "deviant" job is measured in much the same way that success is measured in mainstream jobs: pleasing the boss, meeting goals, and getting along with associates.

In China notions of differential association are the basis of the philosophy underlying rehabilitation: a deviant individual, whether a thief or a revisionist, becomes deviant because of "bad" education or associations with "bad" influences (see "Early Intervention and Rehabilitation in China"). The way to correct these influences is to re-educate deviant individuals politically, inspire them to support the Communist Party, and teach them a love of labor. Particular attention is paid to the five million or so university or advanced technical institute students, the selected and elite few (three percent of middle or high school students). As one Beijing University professor commented in an interview with Asian scholar and MacNeil/Lehrer Newshour special correspondent Robert Oxman (1994a), "It is very challenging job here on the campus how to learn those advanced technology and also very interesting ideas, value systems from the outside [and at] the same time to keep our own traditions" (p. 7). In other words, the Chinese government must prepare new generations of young people for global interdependence and to run the country's economy and other affairs. In doing this, students associate with people and ideas that challenge Chinese ways of doing things. To counteract such exposures, communist ideology is officially espoused

Early Intervention and Rehabilitation in China

The Chinese believe in early intervention and have a definite rehabilitation program for offenders, consisting of re-education through labor:

> The Chinese believe that it is important to intervene early, before deviance has become extreme or caused too much damage. To refrain from correcting minor expressions of deviance on the grounds that such correction interferes with the individual's civil rights strikes the Chinese as analogous to withholding early treatment of a disease on the grounds that the disease has a right to develop until it proves to be lethal. This is particularly true since society (through bad education) is at least partially to blame for the deviance; it thus has a corresponding duty to correct it. (BRACEY 1985, P. 142)

An offender's work unit, neighborhood committee, school, or parents can submit requests to local "control committees," composed of party members who evaluate the application and who can sentence a person to re-education through labor for up to four years without judicial review.

Each city and each province maintains labor camps and prisons. Offenders are placed in one of six types of correctional facilities: (1) prisons, (2) labor reform battalions located in rural areas, (3) re-education-through-labor battalions, (4) forced job placement battalions composed of ex-prisoners, (5) detention centers for the not-yet-rehabilitated, and (6) juvenile facilities (Mosher 1991). The purpose of incarceration is to rehabilitate or "rehumanize" offenders. Rehabilitation programs are designed to teach offenders a productive skill and to make their outlook social-minded, patriotic, and law-abiding. Offenders labor in mines, farms, or factories or on large infrastructure projects such as road building and irrigation. A significant number of prisoners produce goods for export. They also undergo "thought reform," which involves self-criticism and intensive study of communist principles. In sum, "Enormous effort is invested in the attempt to drive a convicted criminal into a hell of personal despair and loneliness, from which the only escape is by genuinely turning over a new 'moral leaf' and wholeheartedly accepting the political line of the Party" (Bonavia 1989, p. 164).

China's treatment of political dissidents has drawn considerable criticism. In fact, the United States House of Representatives passed a resolution asking the International Olympic Committee to deny Beijing the privilege of hosting the 2000 Olympic Games (Oxman 1993a). Many critics in the United States believe that China should be denied Most Favored Nation status (a trading status that has been given to all but about 12 countries in the world) until it improves its human rights record.

The human watch group Asia Watch maintains that 1,300 people arrested during Tiananmen Square remain in prison. Chinese officials say the figure is 70. Since China gives no monitoring group such as the International Committee of the Red Cross access to its prisons, accounts of human rights abuses such as torture, solitary confinement, and other physical and mental abuses are based on anecdotal evidence.[1] Chinese officials deny that such abuses take place and further state that such practices are against the law. Critics also claim that China uses many of its 10 million prisoners in labor camps to manufacture exports. However, the U.S.–China Business Council claims that prisoner-made exports cannot possibly exceed one percent of all exports. China has responded to the charge by opening five labor camps believed to be sites of such manufacturing activities (Oxman 1994c).

In the United States, on the other hand, no clear rationale, such as moral reform, underlies the formal sanctions. In other words, there is little agreement about the purpose of corrections: Should it be rehabilitation? Restitution? Revenge? Moreover, no coherent philosophy guides the operation of the 1,185 state and 61 federal prisons (Lilly 1991). American politicians and criminal justice personnel debate continually not only the purpose of prisons but also the effectiveness of the various methods, which include an array of educational, vocational, and therapeutic programs.

1. It is worth noting that in China, execution is one method of education. Its purpose is reflected in the Chinese saying "Kill a chicken to scare the monkeys." In a recent anticrime campaign the Chinese government executed 10,000 people between 1983 and 1986. "The victims of such instruction might be murderers; but they might also be pimps, arsonists, prostitutes, gamblers, procurers, rapists, white collar criminals, thieves, muggers or members of Chinese secret societies. In other words, disrupters of life. A suitable candidate for capital punishment is anyone who goes against the grain; and unlawful assembly, or airing allegedly seditious material, as occurred this past May and June, is going against the grain" (Theroux 1989, p. 7).

at the universities. Also, Chinese students are required to take a course in Marxism, and campus security remains tight (Oxman 1994a).

We close the chapter with a discussion of the factors that influence the Chinese and American systems of social control. They include population size, the age of the two countries, and the degree of internal turmoil.

The American System Versus the Chinese System of Social Control

The size of the Chinese population is one major reason for the rigid system of social control. In 1995 more than 1.2 billion Chinese—roughly one person out of every five people alive in the world—live in a space roughly the size of the United States. After subtracting deserts and uninhabitable mountain ranges, the habitable land area is about half that of the United States. Although almost 21.3 percent of the world's population lives in China, China has only 7 percent of the world's agriculture land (Han Xu 1989). Even though China manages to feed most of its population well, about eight percent, or 60 million, of its 800 million people who live in the countryside suffer from malnutrition (Oxman 1994b). On the average, there are 18,000 births and 7,000 deaths per day in China. If these patterns persist, the total population could increase by another 100 million by the year 2000. To complicate the situation, the population has expanded rapidly in the past 45 years, from approximately 500 million in 1949 to 1,192,000,000 in 1994 (Haub and Yanagishita 1994).

Such rapid growth strains the country's ability to house, clothe, educate, employ, and feed its people, and can overshadow industrial and agricultural advancements. When one considers that approximately one billion people live under approximately 50 independent governments in Europe, North America, and South America (Fairbank 1989), the magnitude of China's population problem appears even more overwhelming. China's large population explains in part why jobs and housing are assigned and why until recently nearly everything was rationed. Rice and other grains, cooking oil, soap, light bulbs, bicycles, and television sets all have been rationed. Rice, for example, was purchased with ration coupons. Even if a bowl of rice was purchased in a restaurant, it counted as part of a person's monthly allotted ration. In light of the rapid economic growth and the accompanying abundance of goods and food, the rationing system has become obsolete (Oxman 1993c).

The Chinese government must create work for many people, especially the large number of people awaiting employment. (Theoretically, in China there are no unemployed people—only those awaiting employment.) Currently, 580 million people (160 million in urban areas and 420 million in rural areas) constitute the work force in China. However, approximately 200 million people in the rural work force (which is growing by 20 million a year) are surplus workers. About 100 million have found work in newly created rural industries, but that leaves 120 million without work. Many of these people—the floating population—have migrated to the cities in search of work (Oxman 1993c). Consequently, there is a considerable amount of "make-work." For example, "every park employs dozens of ticket takers, some to collect the ten cents for a ticket, some to collect the ticket, and some to collect the stub" (Mathews and Mathews 1983, p. 34).

Living space is another problem. About 10 percent of housing in China is owned privately. Usually housing is allocated by the government or by work units. Space is tight. Family members sleep in the same room, and most households share bathroom and kitchen facilities with other households (Butterfield 1982).

A second reason for the difference between Chinese and American social control systems is rooted in history. Most Americans are immigrants or descendants of immigrants from many nations and cultures who have settled in the United States over the past 200 years. Geographic separation from their native lands enabled them to break with tradition and either retain or discard elements of their native cultures (Fairbank 1989). In addition, the United States is a land with abundant resources and an impressive ratio of people to resources. The material abundance, combined with geographic separation from native cultures, permitted those who immigrated to the United States to create a society in which, in theory, citizens live wherever they can afford to and where they can manage their own lives.

The Chinese, on the other hand, not only are members of the most populous country in the world but also belong to the longest-continuing civilization, which dates back some 3,700 years. This long and complex history has at least four strong and persistent traits:

1. The Confucian[8] system of ethics—which emphasizes order, justice, harmony, personal virtue and obligation, devotion to the family (including the spirits of one's ancestors), and respect for tradition, age, and authority—assigns everyone an unalterable role and place in society.

2. The system of family responsibility makes each member responsible for the conduct of other family members.

3. An imperial tradition gives rulers supreme authority over the lives of the people—historically, those in power have been above the law, and their power has not been constrained by institutionalized checks and balances.

4. No regime in China has ever relinquished its power without first resorting to bloodshed (Fairbank 1987, 1989).

A third issue that explains the rigid system of social control in China also relates to history. Since the beginning of the 20th century, the Chinese people have suffered through several wars of foreign aggression and four revolutionary civil wars.[9] The most recent civil war, the Cultural Revolution (1966–1976), ended less than 20 years ago; the last civil war in the United States took place more than 125 years ago. As a people, Americans are unfamiliar with the division, chaos, and destruction that war and revolution can bring. Often, strict control is the only way to restore order and ensure the stability needed to rebuild a society whose citizens have opposed each other violently.[10,11]

Discussion

Deviance is a complex concept because (1) definitions of deviance change over time and place, (2) not everyone who commits a deviant act is caught and not everyone who is punished actually committed the crime of which he or she is accused, and (3) rule making and rule enforcing affect how some behaviors or appearances come to be defined as deviant and others do not.

For these reasons, sociologists maintain that deviance is not the result of a particular behavior or appearance but a consequence of time, place, and the application of rules and sanctions by others to an "offender." The case of China's dramatic shift in attitude toward profit-making activities illustrates that examining a behavior or an individual personality provides little insight into the nature of deviance. As the Carole Stuart murder case suggests, some people are labeled deviant for what they are, not for what they have done. Examining the context or the historical events surrounding the behavior or appearance gives more insight into the causes of deviant behavior than does studying the individual deviants.

The existence of secret deviants and falsely accused has tremendous and important implications for the study of deviance. In the first place any person who wishes to understand the causes of deviant behavior must study more than the personal characteristics of the officially accused (pure deviants and the falsely accused). Therefore, researchers who compare prison populations with a population that has never been imprisoned in order to discover traits that distinguish the two learn little about what causes people to break rules. Such studies do not identify the characteristics that make criminals, but only those that cause a person to end up in prison. This is the case because the so-called never-imprisoned population includes a large number of people who have committed the same crimes as those who are imprisoned but who have escaped detection or conviction. To complicate matters even further the prison population can contain innocent people. The contribution of sociology to the study of deviance is that it reminds us to consider factors other than the individual defined as deviant. By considering the larger context, we will be better able to understand the true nature of deviance and design policies to change not just the individual but the larger society.

Profitmaking in the new China: street vendors in Beijing.
James F. Hopgood, Northern Kentucky University

FOCUS:
Accepting Definitions of Deviance

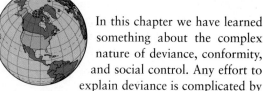

In this chapter we have learned something about the complex nature of deviance, conformity, and social control. Any effort to explain deviance is complicated by the fact that the powerful play an important role in defining what is deviant and in establishing the sanctions that should be applied to correct such behavior. This leaves us with the question of how the powerful are able to get the public to accept such definitions and apply the recommended sanctions. The work of social psychologist Stanley Milgram gives us one answer to this question.

Obedience to Authority

When Stanley Milgram conducted his research for *Obedience to Authority,* he was interested in learning about how people in positions of authority manage to get people to accept their definitions of deviance and to conform to orders about how to treat people classified as deviant. His study gives us insights about how events like the Holocaust, the Cultural Revolution, and the systematic rape of women by soldiers could have taken place. The fact

that such atrocities required the cooperation of a large number of people raises important questions about people's capacity to obey authority.

> *The person who, with inner conviction, loathes stealing, killing, and assault may find himself performing these acts with relative ease when commanded by authority. Behavior that is unthinkable in an individual who is acting on his own may be executed without hesitation when carried out under orders.* (MILGRAM 1974, P. XI)

Milgram designed an experiment to see how far people would go before they would refuse to conform to an authority's orders. The findings of his experiment have considerable relevance for understanding the conditions under which rules handed down by authorities are enforced by the masses.

The participants for this experiment were volunteers who answered an ad placed by Milgram in a local paper. When each participant arrived at the study site, he or she was greeted by a man in a laboratory jacket who explained to the participant and another apparent volunteer the purpose of the study: to test whether the use of punishment improves the ability to learn. (Unknown to the

 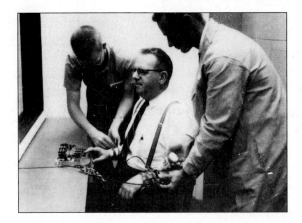

The realistic-looking "shock box" and the "learner" in Milgram's classic experiment on obedience to authority.

Stanley Milgram. From the film OBEDIENCE, distributed by The Pennsylvania State University, Audio Visual Services

subject, the other apparent volunteer was actually a **confederate**—someone who works in cooperation with the experimenter.) The participant and the confederate drew lots to determine who would be the teacher and who would be the learner. The draw was fixed, however, so that the confederate was always the learner and the real volunteer was always the teacher.

The learner was strapped to a chair, and electrodes were placed on his or her wrists. The teacher was placed where he or she could not see the learner, in front of an instrument panel containing a line of shock-generating switches. The switches ranged from 15 volts to 450 volts and were labeled accordingly from "slight shock" to "danger, severe shock." The experimenter explained that when the learner made a first mistake, the teacher was to administer a 15-volt shock and increase the voltage with each subsequent mistake. In each case, as the strength of the shock increased, the learner expressed increased discomfort. One learner even said that his heart was bothering him. After that state-

ment, there was complete silence. When the volunteers expressed concern about the learner's safety, the experimenter firmly told them to continue administering the shock. Although many of the volunteers protested against administering such severe shocks, a substantial number obeyed and continued to administer shocks, "no matter how vehement the pleading of the person being shocked, no matter how painful the shocks seemed to be, and no matter how much the victim pleaded to be let out" (1987, p. 567).

The results of Milgram's experiments are especially significant when one considers that no penalty was imposed on the participants if they refused to administer a shock. Obedience in this situation was founded simply on the firm command of a person with a status that gave minimal authority over the subject. If this level of obedience is possible under the circumstances of Milgram's experiments, one can imagine the extent to which it is possible in a situation in which disobedience brings severe penalties or negative consequences.

Key Concepts

Anomie 287	Conformity 266	Deviance 266
Claims Makers 285	Constructionist Approach 285	Deviant Subcultures 290
Confederate 295	Corporate Crime 284	Differential Association 290
Conformists 280	Crime 277	Falsely Accused 281

[handwritten annotations in margins:]

Labeling theory

Formal Sanctions 277

Informal Sanctions 276

✗ Innovation 288

Mechanisms of Social Control 266

✳ Negative Sanction 276

△ folkways/mores
△ norms

Positive Sanction 276

◑ Pure Deviants 280

✗ Rebellion 289

✗ Retreatism 288

✗ Ritualism 288

□ rule makers
□ rule enforcers

● Sanctions 276

◉ Secret Deviants 281

Structural Strain 287

White-Collar Crime 284

Witch-Hunt 282

Notes

1. Whenever the term *China* is used in this chapter, it denotes the People's Republic of China, whose capital is Beijing. The use of *China* does not include Hong Kong (currently a British colony but expected to revert to the People's Republic of China in 1997) or Macao (currently a Portuguese colony expected to revert to the People's Republic of China in 1999).

2. The Tiananmen Square incident is the most well known event in a series of demonstrations that took place in China between April 15 and June 4 of 1989. In all, 84 cities and up to three million students joined by workers, teachers, even police and soldiers were involved in demonstrations against those in power.

3. The phrase "an abandonment of the goals of the revolution" refers to the 1949 revolution that resulted in the establishment of the People's Republic of China. Before this revolution China was very poor. Between 1849 and 1949, the Chinese people were victims of every imaginable sort of exploitation by foreign countries. To survive starvation, many Chinese ate the leaves off trees and the grass off the ground, and a large portion of the population was addicted to opium. "A peasant party had come out of the hills to put an end to corruption, invasion, and humiliation; they wore straw sandals and told the truth" (Wang Ruowang 1989, p. 40). Mao led this party; he instilled a revolutionary spirit into the masses that motivated the Chinese people to stand up to these immense problems (Strebeigh 1989). Mao was a remarkable leader and hero to his followers. The Chinese intellectual Liu Zaifu noted that "in the 1950s 'we did not believe in ourselves, but only in the all-wise, all powerful Mao'" (1989, p. 40).

[handwritten in left margin:] Notes? Sheldon Milgram

4. Fox Butterfield (1982) describes several aspects of life—sleeping arrangements, living arrangements, and departures—in which Chinese folkways differ from those of Americans. In regard to sleeping arrangements, the Chinese are surprised to find that most American children have their own rooms, because Chinese are accustomed to sleeping with their entire family. According to another Chinese folkway, children live at home with their parents until they marry. Butterfield describes a response of "Oh, I'm so sorry" made by a Chinese reporter when he learned that an American woman to whom he was talking lived alone. He assumed that she lived alone because her parents had died or because she had been forced to take a job that separated her from them geographically. As to departures, when guests leave, the Chinese host walks them out to their vehicle and then stands and waves until the visitors are out of sight.

5. Actually, the authors studied preschools in three countries—China, the United States, and Japan.

6. The Chinese system seems harsh from an American point of view, but it ensures that children wipe themselves properly and that their hands are washed. This method has some practical benefits because many diseases are spread through contact with bodily substances, including saliva, mucus, and feces.

7. Sociologist Joel Best assembled 13 articles, written by various sociologists, in a single volume called *Images of Issues: Typifying Contemporary Social Problems* (1989). Articles express a constructionist viewpoint. The titles include "Horror Stories and the Construction of Child Abuse," "Dark Figures and Child Victims: Statistical

[handwritten at bottom:]
Key theorists (pp)(main ideas)
Terry Williams
Emile Durkheim
Edwin Sutherland
Donald Cressey
Robert Merton
Kai Erikson
Howard Becker
Erving Goffman

Claims about Missing Children," "AIDS and the Press: The Creation and Transformation of a Social Problem," and "The Surprising Resurgence of the Smoking Problem." In spite of the diversity of topics, all 13 authors take a constructionist approach, a method with strong roots in symbolic interaction, but one that also appeals to conflict theorists.

8. Confucius lived between 551 and 479 B.C., a time of constant war, destruction, and chaos. Therefore it is not surprising that his teachings emphasize rules that support order and harmony.

9. The four revolutionary civil wars include the Republican Revolution of 1911, the Nationalist Revolution of 1925–28, the Kuomintang-Communist Revolution of 1945–49, and the Great Proletarian Cultural Revolution of 1966–76.

10. Mao Zedong, the charismatic Chinese leader who mobilized the masses in two of these civil wars, stated: "A revolution is not a dinner party, or writing an essay, or painting a picture, or doing embroidery; it cannot be so refined, so leisurely and gentle, so temperate, kind, courteous, restrained and magnanimous. A revolution is an insurrection, an act of violence by which one class overthrows another" (Mao Tse-Tung 1965, p. 28).

11. Political purges directed by communists are not unique to the Cultural Revolution. Whenever China has faced serious economic problems or whenever the authority of Chinese officials has been questioned, campaigns have been launched to hunt down and purge those who opposed them.

9 SOCIAL STRATIFICATION

with Emphasis on South Africa

Capetown, South Africa.
Milner/Sygma

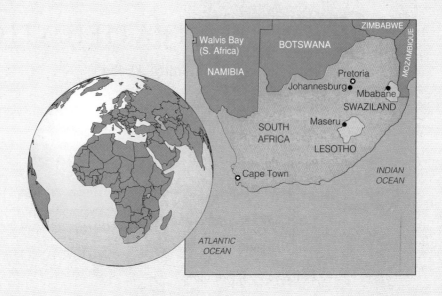

Students from South Africa enrolled in
college in the United States for fall 1991   2,101

South Africans admitted into the United
States in fiscal year 1991 for temporary employment   912

People living in the United States in
1990 who were born in South Africa   38,000

Airline passengers flying between
the United States and South Africa in 1991   8,046

U.S. military personnel in South Africa in 1993   21

People employed in 1990 by South
African affiliates in the United States   14,500

Applications for utility patents filed in the
United States in 1991 by inventors from South Africa   186

Phone calls made between the United
States and South Africa in 1991   7,422,000

We dedicate this day to all the heroes and heroines in this country and the rest of the world who sacrificed in many ways and surrendered their lives so that we could be free.

Their dreams have become reality. Freedom is their reward.

We are both humbled and elevated by the honor and privilege that you, the people of South Africa, have bestowed on us, as the first president of a united, democratic, non-racial and non-sexist South Africa, to lead our country out of the valley of darkness.

We understand it still that there is no easy road to freedom.

We know it well that none of us acting alone can achieve success.

We must therefore act together as a united people, for national reconciliation, for nation-building, for the birth of a new world.

Let there be justice for all.

Let there be peace for all.

Let there be work, bread, water and salt for all.

Let each know that for each the body, the mind and the soul have been freed to fulfill themselves.

Never, never and never again shall it be that this beautiful land will again experience the oppression of one by another and suffer the indignity of being the skunk of the world.

The sun shall never set on so glorious a human achievement! Let freedom reign. God bless Africa!

(MANDELA 1994)

On June 11, 1964, Nelson Mandela, a black South African, was convicted of treason and sentenced to life in prison. Among other things, Mandela helped establish *UmKhonto*, an organization within the African National Congress (ANC), a South African political party that was outlawed in 1960 for its agenda to abolish racial discrimination laws. *UmKhonto* was devoted to sabotage (that is, damaging property so as to disrupt the functioning of society). The acts of sabotage were directed against **apartheid,** South Africa's rigid system of racial classification. At his trial Mandela explained why sabotage was necessary:

The initial plan was based on careful analysis of the political and economic situation of our country. We believed that South Africa depended to a large extent on foreign capital and foreign trade. We felt that planned destruction of power plants, and interference with rail and telephone communications, would tend to scare away capital from the country, make it more difficult for goods from the industrial areas to reach the seaports on schedule, and would in the long run be a heavy drain on the economic life of the country, thus compelling the [white] voters of the country to reconsider their position. (MANDELA 1990, PP. 26–27)[1]

Apartheid was a system in which everyone in South Africa was put into a racial category and issued an identity card denoting his or her race. The system was one in which "you're not just born a baby, you're born a black, a colored, an Indian, or a white baby—and that label profoundly affects every aspect of your life" including access to occupation, income, property, education, food, clothing, shelter, and the vote (Mabuza 1990, p. A15).

Although South Africa is known around the world for apartheid, it is hardly the first or the only society in the world that divides its people into categories. In all societies people are categorized—whether formally, by the government and other institutions, or informally, by social status. They are categorized according to criteria such as intelligence, athletic ability, beauty, educational attainment, occupation, family background, weight, skin color, national origin, sex, age, and so on. People who occupy the highest strata (categories) have the greatest access to scarce and valued resources and enjoy a disproportionate share of those resources while those in the lowest strata have very limited access.

Sociologists pay close attention to **social stratification,** the systematic process by which people in societies are ranked on a scale of social worth. In addition, they study how the categories in which people are placed affect their **life chances,** a critical set of potential social advantages including "everything from the chance to stay alive during the first year after birth to the chance to view fine art, the chance to remain healthy and grow tall, and if sick to get well again quickly, the chance to avoid becoming a juvenile delinquent—and very crucially, the chance to complete an intermediary or higher educational grade" (Gerth and Mills 1954, p. 313).

This chapter examines several important and interrelated features of social stratification, including the symbols that designate a person's category (for example, skin color, type of car owned, occupation), types of stratification systems, reasons for stratification systems, and the effects of stratification on life chances. We look closely at apartheid, South Africa's system of social stratification, in which skin color overwhelmingly determined a person's life chances; it was a system that enabled approximately 14 to 17 percent of the population (the whites) to control the fate of the remaining 83 to 86 percent.

South Africa is the focus for several reasons. First, it is one of many countries around the world that is working to establish a democracy. In the broadest sense, a democracy is a system in which all people have an equal say in decision making. "Equal say" typically is achieved through free elections and policies of "one person, one vote." On the continent of Africa alone, South Africa is one of 52 governments that has responded to (or is responding to) internal or external pressures to hold free elections (Bratton and van de Walle 1993). The nation is moving from a system in which only whites could vote to a multiracial democracy.

Second, focusing on South Africa's attempt to dismantle its system of apartheid and the effects of that system helps us see that free elections are a first step toward achieving complete democracy. The "Freedom Charter" (1990) of the African National

African National Congress leader, anti-apartheid activist, and now President of South Africa Nelson Mandela was sent to prison as a young man in 1964 (left) and was finally released 27 years later (right).

Congress outlines at least nine characteristics of democracy in addition to free election. It includes such things as the rights to share in a country's wealth, to have a fair trial, to travel without restriction, to have decent housing, and to obtain education without distinction of skin color, racial or ethnic background, sex, or political belief. All of these characteristics are connected to a country's system of social stratification. In order to achieve a true state of democracy, a country's system of social stratification must be "democratic"; it isn't enough if the country's government institutions are set up on democratic principles. We turn now to one critical feature of stratification systems—the criteria that societies use to categorize people.

Classification Schemes

Almost any criterion can be used (and at one time or another has been used) to categorize people: hair color and texture, eye color, physical attractiveness, weight, height, occupation, age, grades in school, test scores, and many others. Two major kinds of criteria are used to categorize people: ascribed and achieved characteristics.

Ascribed characteristics are attributes that people (1) have at birth (such as skin color, gender, or hair color), (2) develop over time (such as baldness, gray hair, wrinkles, retirement, or reproductive ca-

pacity), or (3) possess through no effort or fault of their own (national origin or religious affiliation that was "inherited" from parents).

Achieved characteristics, on the other hand, are acquired through some combination of choice, effort, and ability. In other words, people must act in some way to acquire these attributes. Some examples of achieved characteristics include occupation, marital status, level of education, and income. Ascribed and achieved characteristics seem clearly distinguishable, but such is not always the case. Debate continues, for example, over whether homosexuality is achieved or ascribed. Some people argue that homosexuality is genetically based behavior over which people have no control; others insist that it is learned and ultimately a matter of choice.

Sociologists are interested in those ascribed and achieved characteristics that take on social significance and **status value:** when that occurs, persons who possess one feature of a characteristic (white skin versus brown skin, blond hair versus dark hair, physician versus garbage collector) are regarded and treated as more valuable or more worthy than persons who possess other categories (Ridgeway 1991). Sociologists are particularly interested in such situations because ascribed characteristics are attributes over which people have no control.

A classic demonstration of how ascribed characteristics can affect people's access to valued resources involves a third-grade class in Riceville, Iowa. In 1970, teacher Jane Elliot conducted an experiment in which she divided her students into two groups according to a physical attribute—eye color—and rewarded them accordingly. She did this to show her class how easy it is for people (1) to assign social worth, (2) to explain behavior in terms of an ascribed characteristic such as eye color, and (3) to build a reward system around this seemingly insignificant physical attribute. The following excerpt from the transcript of the program "A Class Divided" ("FRONTLINE" 1985) shows how Elliot established the ground rules for the classroom experiment:

ELLIOT: It might be interesting to judge people today by the color of their eyes . . . would you like to try this?

CHILDREN: Yeah!

ELLIOT: Sounds like fun, doesn't it? Since I'm the teacher and I have blue eyes, I think maybe the blue-eyed people should be on top the first day. . . . I mean the blue-eyed people are the better people in this room. . . . Oh yes they are, the blue-eyed people are smarter than brown-eyed people.

BRIAN: My dad isn't that . . . stupid.

ELLIOT: Is your dad brown-eyed?

BRIAN: Yeah.

ELLIOT: One day you came to school and you told us that he kicked you.

BRIAN: He did.

ELLIOT: Do you think a blue-eyed father would kick his son? My dad's blue-eyed, he's never kicked me. Ray's dad is blue-eyed, he's never kicked him. Rex's dad is blue-eyed, he's never kicked him. This is a fact. Blue-eyed people are better than brown-eyed people. Are you brown-eyed or blue-eyed?

BRIAN: Blue.

ELLIOT: Why are you shaking your head?

BRIAN: I don't know.

ELLIOT: Are you sure that you're right? Why? What makes you sure that you're right?

BRIAN: I don't know.

ELLIOT: The blue-eyed people get five extra minutes of recess, while the brown-eyed people have to stay in. . . . The brown-eyed people do not get to use the drinking fountain. You'll have to use the paper cups. You brown-eyed people are not to play with the blue-eyed people on the playground, because you are not as good as blue-eyed people. The brown-eyed people in this room today are going to wear collars. So that we can tell from a distance what color your eyes are. [Now], on page 127—one hundred twenty-seven. Is everyone ready? Everyone but Laurie. Ready, Laurie?

CHILD: She's a brown-eye.

ELLIOT: She's a brown-eye. You'll begin to notice today that we spend a great deal of time waiting for brown-eyed people. ("FRONTLINE" 1985, PP. 3–5)

Once Elliot set the rules, the blue-eyed children eagerly accepted and enforced them. During recess the children took to calling each other by their eye colors, some brown-eyed children got into fights with blue-eyed children who called them "brown-eye." The teacher observed that these "marvelous, cooperative, wonderful, thoughtful children" turned into "nasty, vicious, discriminating little third-graders in a space of fifteen minutes" (p. 7).

This experiment illustrates on a small scale how categories and their status value are reflected in the distribution of valued resources. Because category determines status and is related to life chances, it is essential that we examine the shortcomings of classification schemes.

One major shortcoming of any classification scheme is that not all people in a society fit neatly in the categories designated as important. For example, the third-grade teacher whose experiment we described divided her students into just two categories—the blue-eyed and the brown-eyed. Such a classification scheme, however, does not accommodate people with green eyes, hazel eyes, gray eyes, or mixed-color eyes (one blue and one brown). This shortcoming in the classification scheme leaves us unsure about what to do with someone who does not fit into any of the designated categories. Typically, when people do not fit a category, others find ways to make them fit. In the third-grade class, brown-eyed people were required to wear collars to make it absolutely clear who belonged to that category.

Such a strategy is not unique to that third-grade classroom experiment. Strategies to make people

fit into categories are a part of all classification schemes—even those for gender, age, and race. For example, there is no perfect dividing line to separate people into the categories "male" and "female." A small but significant number of babies are born hermaphrodites; that is, they have both male and female reproductive organs. In the United States, parents must choose what sex to put on a hermaphrodite child's birth certificate. Given the importance of sexual categories, physicians tell parents that it is in the child's best interest to undergo a sex-clarifying operation. Because most people accept without questioning their society's category system (see Chapter 11 on gender and sexuality), it is difficult to see clearly those categories as social constructions that everyone is more or less made to fit into and/or that the majority of people work to fit into. In order to demonstrate the many shortcomings associated with classification schemes, we will critique one major form of categorizing people—that of race.

Racial Categories

In the purely biological sense, a **race** is a group of people who possess certain distinctive and conspicuous physical characteristics. The challenge, however, lies in identifying the physical traits that distinguish one race from another. Although by convention we assign people to one of three racial categories—Caucasoid, Mongoloid, or Negroid—this classification scheme, like all schemes, has many shortcomings, which immediately become evident when we imagine using such a system to classify the more than 5.6 billion people in the world. First we would find that many people do not fit into a category because no sharp dividing lines distinguish such features as black skin from white or curly hair from wavy. This lack of clear divisions, however, does not keep people from trying to create them.

A hundred years ago in the United States there were churches that "had a pinewood slab on the outside door . . . and a fine-tooth comb hanging on

a string . . ." (Angelou 1987, p. 2). People could go into that church if they were no darker than the pinewood and if they could run the comb through their hair without it snagging. In South Africa, a state board oversaw the racial classification of everyone in the country. In the past, the board devised a number of measures designed to make decisions about an individual's race seem rational rather than arbitrary. One test was to place a pencil in a person's hair; if it fell out, the person was classified as white (Finnegan 1986). A South African woman classified as Coloured explains how racial classification worked as late as 1989:

Under South African law, I am officially considered Colored. But so is my light-skinned sister with brown hair and my brother who has kinky hair and skin even blacker than mine. The state determines what color you are. At sixteen, you

A Beauty for Our Times

And so I made a silent vow not to see things in terms of colour. Then came the Miss South Africa contest. Miss Jacqui Mofokeng was chosen as the new Miss South Africa, and all hell seemed to have broken loose. First, someone on the black radio station Metro said that this was the first "black" Miss South Africa. "Hold on," said some others, "what about the outgoing Miss South Africa?" That is Miss Amy Kleynhaas; she was supposed to have been the first black Miss South Africa last year. Well, it turns out she is a "so-called coloured."

Not my choice of words, you understand, but "coloureds" or mixed-race people always come with "so-called," to demonstrate that those who oppose apartheid do not recognise any such ethnic distinction: there are whites and everybody else is black. . . . The problem is that to the uninitiated like me, many of those classified as coloured look white, as, for example, does Miss Kleynhaas . . . on television, you understand, as I haven't met her.

All the same, there is no such ambiguity about Miss Mofokeng. She is black and a Miss Soweto. If the callers to commercial radio phone-ins are anything to judge by, a lot of white people feel that she only won the title because the organisers of the contest felt it was politically correct for South Africa to have a black beauty queen.

When Miss Mofokeng came to the radio station to be interviewed and to take calls, it made for the most difficult listening. One white woman phoned up to tell the new Miss South Africa, ". . . you are ugly, your bum is too big and you ought to get your teeth fixed!" . . .

Jacqui Mofokeng, Miss South Africa 1993.
David Sandison

Some others were not quite as direct, but the message was basically the same: "the competition had been fixed so a black woman would win."

The 21 year old Jacqui Mofokeng displayed a composure well beyond her years. "I don't think I was the most beautiful girl there," she says, "but the competition is not just about looks and I am going to work hard to win the hearts of all South Africans with a smile. . . ." And the hard work she talks about is to try and bring peace to the country. "As I wear this crown tonight," she said in her acceptance speech, "I am very much aware that people are dying in the townships. We must all try and bring an end to the violence."

An extremely well-spoken young lady. But rather difficult to accept for white South Africans who are used to milky white, preferably blond, long legged Miss South Africans. As one such exasperated white told me, "this is affirmative action at its most blatant. . . ." One black caller to the radio said, "it is time there was a black Miss South Africa and those who can't take it can emigrate. . . ."

SOURCE: From "Tales of South Africa: A Beauty for Our Times," by Elizabeth Ohene, pp. 23–24. Copyright © 1993 by *BBC Focus on Africa* Magazine. Reprinted by permission.

Bernard Gotfryd/Woodfin Camp & Associates

Rieder/Monkmeyer Press

One reason that racial classification schemes are highly problematic is that many people around the world have mixed ancestry.

have to fill out some forms, attach a photograph and send them to a state authority where your race will be decided. Differences in color are noted by official subdivisions. For example, I am a Cape Colored whereas my sister is called Indian Colored. Of course, many people categorized as Colored are of mixed Black and white ancestry. In fact, if you can prove having had a white grandparent or parent and are yourself very light skinned, you can even make an application to be reclassified from Colored to white. (CHAPKIS 1986, P. 69)

Despite the official status of classification in South Africa, the passage just cited shows that people often disagreed with and sometimes even formally appealed their racial assignments. Even if the system of formal classification completely disappears, the legacy of classification will remain for some time to come (see "A Beauty for Our Times").

A second problem with trying to classify people according to any racial scheme is that millions of people in the world have a mixed ancestry and possess the physical traits of more than one race. In addition, some people belong to one race in one society but would belong to another if they lived

elsewhere. For example, racial categories in South Africa are different from those used in the United States. In South Africa the four officially recognized racial categories are white, African, Coloured, and Asian. Whites, who constitute about 14 percent of the population, consist of two distinct groups—the Afrikaners (descendants of the early Dutch, German, and Huguenot farmers who settled in South Africa in the late 1600s and who speak Afrikaans) and the descendants of British settlers, who speak English. Afrikaners make up about 60 percent of the white population, whereas English speakers make up about 40 percent of the white population.

The other three groups are regarded by whites as "black," and include Coloureds, Asians, and Africans. Coloureds are those of mixed descent, the offspring of sexual and marital unions between white settlers and the indigenous peoples of South Africa or indentured servants from India and Malaya. The Asians, primarily Hindus of Indian descent, trace their South African residency to the late nineteenth century, when their ancestors arrived as indentured servants to work the sugar cane fields of Natal Province. Africans are divided further into 10 different ethnic categories of which the largest are the Zulu, the Xhosa, and the Sotho.

TABLE 9.1   Race and Ethnic Categories Used in Selected Decennial Censuses

Census	1860	1890	1900	1970	1990	
Race	White	White	White	White	White	Vietnamese
	Black	Black	Black (Negro	Negro or Black	Black or Negro	Japanese
	Mulatto*	Mulatto*	descent)	Japanese	American Indian	Asian Indian
		Quadroon†	Chinese	Chinese	Eskimo	Samoan
		Octoroon‡	Japanese	Filipino	Aleut	Guamanian
		Chinese	Indian	Hawaiian	Chinese	Other API§
		Japanese		Korean	Filipino	Other race
		Indian		Indian [Amer.]	Hawaiian	
				Other	Korean	
				Hispanic Origin:	Hispanic Origin:	
				Mexican	Mexican or	
				Puerto Rican	Chicano or	
				Cuban	Mexican-Am.	
				Central/So.	Puerto Rican	
				American	Cuban	
				Other Spanish	Other Spanish/	
					Hispanic	

* Three-eighths to five-eighths black.
†One-quarter black.
‡One-eighth black.
§Asian and Pacific Islander.
Note: Prior to 1970, census enumerators wrote in the race of individuals using the groups cited above. In the 1970 and subsequent censuses, respondents and enumerators filled in circles corresponding to the category with which the respondent most closely identified. Persons choosing "other race" or "Indian" were asked to write in the race or Indian tribe.

SOURCE: "American Minorities—The Demographics of Diversity," by William P. O'Hare. P. 7 in *Population Today* 47(4). Copyright © 1992 by the Population Reference Bureau Inc.

For all practical purposes, in South Africa "Coloured" is a miscellaneous category for those who do not fit easily into the other three groups (Sparks 1990). In the United States, however, there is no parallel category. According to the rule of hypodescent, descendants of a black-white union or a nonwhite-white union are not assigned to special categories but are classified arbitrarily as members of the subordinate group. At one time in the United States, people were considered black if they were known to be ⅟₃₂ black. Consequently, in the United States, "blackness is a taint . . . [and] light-skinned people have more status" (Poussaint 1987, p. 7).

A third problem with systems of racial classification is that definitions of race are often vague, contradictory, and subject to change. The U.S. Census Bureau, for example, has used several different and confusing criteria to get at a person's race (see Table 9.1). The 1980 census stated, "if persons of mixed racial parentage could not provide a single response to the race question, the race of the person's mother was used" (U.S. Bureau of the Census 1990, p. 4). In the 1970 census, if a person could not identify his or her race, the father's race was used. In South Africa, official definitions are equally contradictory, vague, and confusing. Officially, a white is "any person who in appearance obviously is or who is generally accepted as a white person other than a person who, although in appearance obviously a white person, is generally accepted as a colored person" (Lelyveld 1985, p. 85).

Finally, in trying to classify people by race, we would find a tremendous amount of variation even among people who possess all of the traits designated as belonging to a particular race. For example, people classified as Mongoloid include Chinese, Japanese, Malayans, Mongolians, Siberians,

Eskimos, and Native Americans. In other words, racial classifications ignore the great physical and cultural differences between people who are said to be of the same race.

Sociologists are particularly interested in any classification scheme that incorporates the belief that certain important abilities (such as athletic talent or intelligence), social traits (such as criminal tendencies or aggressiveness), and cultural practices (dress, language) are passed on genetically, or incorporates the belief that some categories of people are inferior to others by virtue of genetic traits and therefore should receive less wealth, income, and other socially valued items. Sociologists are interested in classification schemes because they have enormous consequences for society and the relationships among people.

For sociologists, one important dimension of stratification systems is the extent to which people "are treated as members of a category, irrespective of their individual merits" (Wirth [1945] 1985, p. 310). In this vein, they examine how "open" or "closed" a stratification system is.

"Open" and "Closed" Stratification Systems

Despite wide variations among forms of stratification systems, each falls somewhere on a continuum between two extremes: a **caste system** (or "closed" system, in which people are ranked on the basis of traits over which they have no control) and a **class system** (or "open" system, in which people are ranked on the basis of merit, talent, ability, or past performance). No form of stratification exists in any pure form. There are three characteristics that most clearly distinguish a caste from a class system of stratification: (1) the rigidity of the system (how difficult it is for people to change their category), (2) the relative importance of ascribed and achieved characteristics in determining people's life chances, and (3) the extent to which there are restrictions on social interaction between people in different categories. Caste systems are considered closed because of how rigid they are—rigid in the sense that ascribed characteristics determine life chances and social interaction among people in different categories is restricted. In comparison, class systems are open in the sense that achieved characteristics determine life chances and that no barriers exist to social interaction among people in different categories.

Class and caste systems are ideal types. We learned in Chapter 7 that an ideal type is ideal not in the sense of having desirable characteristics but as a standard against which "real" cases can be compared. Actual stratification systems depart in some ways from the ideal types.

Caste Systems

When people hear the term *caste,* what usually comes to mind is India and its caste system, especially as it existed before World War II. India's caste system, now outlawed in its constitution, was a strict division of people into four basic categories with 1,000 subdivisions. The strict nature of the caste system is shown by some of the rules that specified the amount of physical distances that were to be maintained between people of different castes: "a Nayar must keep 7 feet (2.13m) from a Nambudiri Brahmin, an Iravan must keep 32 feet (9.75m), a Cheruman 64 feet (19.5m), and a Nyadi from 74 to 124 feet (22.6 to 37.8m)" (Eiseley 1990, p. 896).

Most sociologists use the term *caste* not to refer to one specific system but to designate any scheme of social stratification in which people are ranked on the basis of physical or cultural traits over which they have no control and that they usually cannot change. Whenever people are ranked on the basis of such traits, they are part of a caste system of stratification.

The rank of any given caste is reflected in the public esteem accorded to those who belong to it, in the power wielded by members of that caste, and in the opportunities of members of that caste to acquire wealth and other valued items. People in lower castes are seen as innately inferior in intelligence, personality, morality, capability, ambition, and many other traits. Conversely, people in higher castes consider themselves to be superior in such traits. Moreover, caste distinctions are treated as if they are absolute: their significance is never doubted or questioned (especially by those who occupy higher castes), and they are viewed as unalterable and clear-cut (as if everyone is supposed to fit neatly into a category). Finally, there are heavy restrictions on the social interaction among people in higher and lower castes. For example, marriage between people of different castes is forbidden.

Apartheid: A Caste System of Stratification

Apartheid (the Afrikaans word for "apartness") was practiced in South Africa for hundreds of years. But it did not become official policy until 1948, when the conservative white Nationalist Party, led by D. F. Malan, won control of the government. Once in power, the Nationalists passed hundreds of laws and acts mandating racial separation in almost every area of life and severely restricting blacks' rights and opportunities. The ultimate result has been legalized political, social, and economic domination by whites over nonwhites.

It is virtually impossible to give an overview of the effects of apartheid, a system that has inflicted enduring misery, economic damage, and daily indignities on South Africa's nonwhite population (Wilson and Ramphele 1989). Although apartheid is dismantled on paper, the challenge lies in dismantling its effects on relationships between whites and nonwhites and on life chances for nonwhites. To fully grasp how apartheid laws determined life chances, we turn to an overview of these policies. As you read this overview, do not try to memorize names, dates, and specific purpose of the acts. Instead, think about how every aspect of life in South Africa was regulated according to skin color. And consider the challenges beckoning with the dismantling of apartheid.

Apartheid Policies

Apartheid policies centered on one aim: maintaining separate black (that is, mixed race, Asian, and African) and white areas. The Transvaal Province Law of 1922 (par. 267) stated that a nonwhite "should only be allowed to enter urban areas, which are essentially the white man's creation, when he is willing to enter and to minister

During the same period as the 1953 Separate Amenities Act, the United States had its own version—the "Jim Crow" laws—in a number of southern states that legalized racial segregation of all public facilities.

UPI//Bettmann Newsphotos

to the needs of the white man and should depart therefrom when he ceases so to minister" (Wilson and Ramphele 1989, p. 192). Other legislation—the Separate Amenities Act, the Group Areas Act, the Land Acts, and the Population Registration Act—represent additional ways in which the white-ruled government has worked to achieve this aim.

The Separate Amenities Act of 1953 authorized the creation of separate and unequal (in quantity and quality) public facilities—parks, trains, swimming pools, libraries, hotels, restaurants, hospitals, waiting rooms, pay telephones, beaches, cemeteries, and so on—for whites and nonwhites. Local governments in areas officially designated for "whites only" constructed only the most basic and necessary public facilities (such as toilets and bus stations) for blacks and prohibited them from using

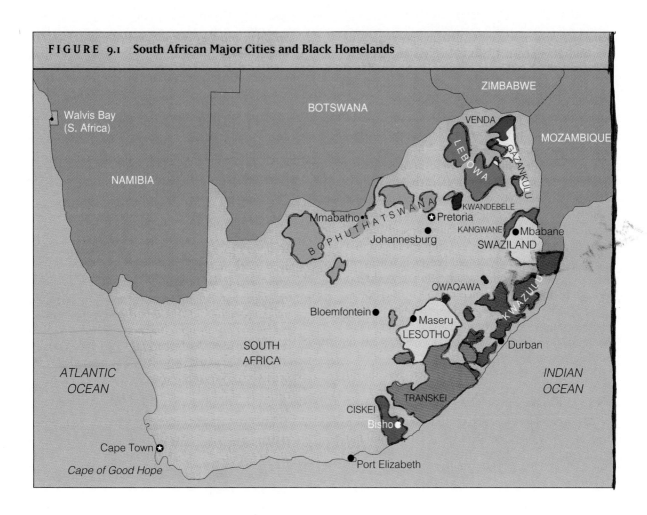

FIGURE 9.1   South African Major Cities and Black Homelands

all other public facilities (for example, parks and libraries). This act promoted the aim of apartheid by expressly forbidding blacks to enter the community except for the purpose of "ministering to the needs of whites."

The Separate Amenities Act was formally abolished on October 16, 1990. Before that date some public facilities had been integrated in the large cities, and some laws had been passed to do away with some aspects of the practice. In 1989, for example, beaches were opened to all races. Yet, until October 16, 1990, local ordinances mandating separate facilities remained in effect (Wren 1990b). Despite the 1990 legislation many in power have created ways to avoid the legal obligation to integrate public facilities. Residents of white areas, for example, can treat nonwhites as "nonresidents"

and charge them fees as high as $200 a year to use public facilities such as the library. As another example, white schools do not have to integrate unless 80 percent of the parents vote and 72 percent of those voting approve of integration (Wren 1991).[2]

The Land Acts of 1913 and 1936 reserved 85 percent of the land for five million whites (14 percent of the population) and the remaining 15 percent for the rest of the population. These acts, repealed on June 5, 1991, mandated that 10 homelands in the eastern half of the republic be set aside, one for each of the 10 major African ethnic groups (as defined by white South Africans). The homelands consisted of land that is arid, underdeveloped, and lacking in natural resources (see Figure 9.1). More than 80 percent of the people in homelands live in a state of extreme poverty (Wilson and Ramphele 1989). Needless to say, the Africans had no influence in determining the location or type of land that would constitute their homelands. For example, a homeland may be composed of as many as two dozen pieces of land, separated by large, white-owned plantations, farms, and industries (Liebenow 1986).

Regardless of where they were born or where they had lived most of their lives, each African was declared a citizen, not of South Africa, but of one of the 10 homelands. Four of these homelands were declared independent (Bophuthatswana, Transkei, Ciskei, and Venda), and the government hoped that the other six eventually would accept independent status. However, no other country in the world has recognized the independence of any of the homelands because they were established only to keep black South Africans from living in white areas while meeting the low-wage employment needs of white South Africans.

The logic underlying the Land Acts was that if all Africans had citizenship in a homeland, they had no claim to economic and political rights when they were outside their homeland. Thus the South African government could deny them a vote and could treat them as guest workers and noncitizens:

The homelands have a more sinister side. On the one hand, they are a labor pool for industry; on the other, they are a rubbish heap for the people industry has no use for—the old, the sick, the very young. By law, if you cannot work you must go to your homeland, and there you are dumped. Sometimes there are "resettlement camps" of tents or huts for groups who are moved [by the government] to a countryside that they do not know. Sometimes people are just left on the land, and told to survive off relatives. Schools are bad or nonexistent, medical facilities the same. Some people survive. Many die. The homelands policy is a policy of polite extermination. Gas chambers are not needed when people can simply starve to death. (SEIDMAN 1978, P. 111)

Many of the people sent to the homelands were born in the cities and have no strong family ties in the homelands. There are no facilities to care

for them and they are not wanted. (WEEKEND WORLD 1977)

The Group Areas Act of 1950, also repealed on June 5, 1991, designated separate living areas for whites and nonwhites. Under this act Africans were permitted to work in white areas, but they must be housed in adjacent townships or in single-sex (usually male) closed compounds called hostels. Buses brought blacks from the townships or compounds into white areas to work and then took them home after work. Most blacks spent three or more hours a day traveling to and from work:

Well, let's take a common scenario of a middle-aged black man living in a township 30 miles from his job. He lives there because that's where he's been put. About 4 a.m. he gets up so he can make the two-hour trek into the city each day. He rides a bus or train that makes our worst rush-hour scene look peaceful by comparison.

When the workday is over, the man must leave the city by a certain time because of his curfew and take his two-hour trek back home again.

Or let's say the man happens to be a miner. He'll live in a crowded dormitory and make about $5000 a year, while white miners live in rent-free homes and make around $20,000 a year for often doing the same work. It's against the law for the black miner to bring his wife and children to live with or near him or for them to visit him. Except for his two short visits home a year, he stays at the mine and works at a very hazardous job.

Now what about his wife? Well, she can't live with her husband if he works at the mines, and she can't live with her family if she's a domestic. If she works for a white family, they often will insist that she live in housing quarters behind their home. If so, her kids have to be left back in the homeland with grandma or auntie. Occasionally, she too will be permitted to travel back home to visit her family. (LAMBERT 1988, PP. 29–30)

Under the Group Areas Act, every square foot of land was officially classified as reserved for whites, Coloureds, Asians, or Africans. Anyone found living in the wrong area could be resettled or forced out. In addition, the government had the right to change the classification of an area. It is estimated that, between 1960 and 1983, the South African government "resettled" more than 3.5 million nonwhites to places where they did not choose to go (Wilson and Ramphele 1989).

South Africa's pass laws restricted the free movement of blacks. According to these laws, which were abolished in 1986, blacks were required to carry passbooks that contained information about their racial, residential, and employment status. Without their passbooks, blacks were not permitted to be present in areas designated as white (85 percent of the land). Between 1916 (when statistics on pass law violations first were recorded) and 1986, 17 million Africans were prosecuted for pass law violations. One white employer explained:

> We protested the pass laws, too. They were dreadful. If our garden man left our garden to water some plants on the other side of the hedge and if he didn't have his pass in his working shirt (not wanting to get it dirty and spoiled because without it he was a lost soul), he could have been carted off by any policeman who happened to walk by. We wouldn't know where he was. We wouldn't know what had happened. We'd ring up the police stations, and if we happened to strike somebody a bit simpatico, we might learn his whereabouts. This used to happen all the time in the suburbs. It was driving us mad. Every evening you used to see lines of these people handcuffed to one another being dragged to the nearest police station. We succeeded in getting that done away with. Of course, the passes still exist, but there is usually less harassment.
>
> When I began thinking, I was distressed to have to write a piece of paper. This was before they issued passbooks. I had an African man servant whom we absolutely adored. He had worked for us for over twenty-five years. We had built very beautiful servants' quarters for him and our other help. We all loved him. And this man was really noble. He was old enough to be my father, and yet he had to come to me if he wanted to go out. I would have to scribble,

> "Please pass Amos until six o'clock—or eight o'clock—tonight." (CRAPANZANO 1985, PP. 131–32)

The Population Registration Act of 1950 required that everyone in South Africa be classified by race and issued an identification card denoting his or her race. This act was repealed on June 18, 1991. As mentioned previously, a person's racial classification affects every aspect of a South African's life. Because so much depends on race, classification boards composed of white social workers ruled on questionable cases and heard appeals from those who wished to change their racial classification. It is not known how many people applied for such a change:

> Deliberations take months, sometimes years, with the board paying careful attention to skin tint, facial features, and hair texture. In one typical twelve month period, 150 coloreds were reclassified as white; ten whites became colored; six Indians became Malay; two Malay became Indians; two coloreds became Chinese; ten Indians became coloreds; one Indian became white; one white became Malay; four blacks became Indians; three whites became Chinese.
>
> The Chinese are officially classified as a white subgroup. The Japanese, most of whom are visiting businessmen, are given the status of "honorary whites." Absurd? Yes, if all this were part of an antiutopian novel. But in real life it is a chilling part of the most complex system of human control in the world. (LAMB 1987, PP. 320–21)

It is evident from this description of apartheid that simply dismantling the apartheid laws cannot reverse the enormous economic disparities that have been systematically created over the past 300 years (see Table 9.2). As the then African National Congress leader Nelson Mandela asked, "What is the use of repealing the Group Areas Act and Land Acts when the Government has given me no resources to take advantage of the situation?" (Mandela 1991, p. 4E).

In addition to dismantling apartheid laws, steps must be taken to correct the economic disparities. To achieve this end Mandela promised during his

T A B L E 9.2 Life Chances and Racial Classification in South Africa				
	White	*Asian*	*Coloured*	*African*
Percentage of the Total Population	15.0	3.0	9.0	73.0
Percentage of Annual Income Going to Each Caste	64.4	3.0	7.2	19.7
Infant Mortality Rate	12/1,000	18/1,000	52/1,000	110/1,000
Percentage of Babies Born Underweight				
Urban	16.0	35.0	49.0	28.0
Rural	—	—	—	43.0
Number of Pupils per Teacher	18	25	29	42
Government Expenditures per Pupil	$1,700	$1,100	$600	$220
Literacy Rate	99/100	69/100	62/100	50/100
Life Expectancy in Years	70	65	—	59

SOURCES: Wilson and Ramphele (1989); *The World Almanac and Book of Facts 1991* (1990).

campaign and again upon assuming the presidency of South Africa that he will institute plans to build more than one million low-income houses, make massive public investments in health and education, purchase and redistribute 30 percent of farmland and an unspecified amount of residential property in the first five years, take control of the mining industry so that wealth beneath the land will go to the people of South Africa, and break up white-run conglomerates (Keller 1994).

Apartheid is a classic example of a caste system of stratification: people are categorized according to physical traits such as skin color; ascribed characteristics determine life chances; and there are many barriers to social interaction among people who belong to different racial categories. Now we turn to a system of stratification that is at the opposite end on the continuum—a class system.

Class Systems

In class systems of stratification, "people rise and fall on the strength of their abilities" (Yeutter 1992, p. A13). Class systems of stratification contain economic and occupational inequality, but that inequality is not systematic. In other words, there is no connection between a person's sex, race, age, or ascribed characteristics and his or her life chances.

A class system differs from a caste system in that life chances are connected to merit, talent, ability, and past performance and not to attributes over which people have no control, such as skin color. In class systems people assume that they can achieve a desired level of education, income, and standard of living through hard work. Furthermore, people can change their class position upward during their own lifetimes, and their children's class position can be different (and ideally higher) from their own.

Movement from one class to another is termed social mobility. There are many kinds of social mobility in class systems. **Vertical mobility** exists if the change in class status corresponds to a gain or loss in rank or prestige, as when a college student becomes a physician or when a wage earner loses a job and goes on unemployment. A loss of rank is **downward mobility,** whereas a gain in rank is **upward mobility. Intragenerational mobility** is movement (upward or downward) during an individual's lifetime. **Intergenerational mobility** is a change in rank over two or more generations.[3]

In contrast to caste systems, distinctions between classes are not always clear. In principle, people can move from one class to another. It is often difficult to identify a person's class through observation alone. That is, a person may drive a BMW and wear designer clothes—symbols of solid middle-class status—but may work at a low-paying

Simply dismantling the apartheid laws will not reverse the enormous economic disparities between black and white South Africans that have developed over the last 300 years.

A. Tannenbaum/Sygma

job, live in a low-rent apartment, and have exceeded his or her line of credit. In other words, a person may have champagne taste and a beer income. Or a person may change class position through marriage, graduation, inheritance, or job promotions but may not assume the lifestyle of that class immediately.

India-born novelist Bharati Mukherjee, who now lives in the United States, was asked, "What does America mean to you as an idea?" Her answer clarifies the differences between class and caste systems of stratification:

What America offers me is romanticism and hope. I came out of a continent of cynicism and irony and despair. A traditional society where you are what you are, according to the family that you were born into, the caste, the class, the gender. Suddenly, I found myself in a country where—theoretically, anyway—merit counts, where I could choose to discard that part of my history that I want, and invent a whole new history for myself. . . .

[America represents] that capacity to dream and then try to pull it off, if you can. I think that the traditional societies in which people like me were born really do not allow the individual to dream. To dream big. (MUKHERJEE 1990, PP. 3–4)

Is Mukherjee's vision of the United States accurate? Is the United Sates a class system?

Does the United States Have a Class System?

Before we answer this question, let us recall that in a true class system, ascribed characteristics do not determine social class. Although considerable economic inequality may be present, equality of opportunity exists. On paper, the United States has a class system. The Declaration of Independence, the preamble to the Constitution, and the Bill of Rights assert that all human beings are created equal, not in innate qualities like intelligence or physical endowment, but in the human right to full equality: "the right of equitable access to justice and freedom, and opportunity, irrespective of [sex], race, or religion or ethnic origin" (Merton 1958, p. 189). If individuals differ in innate endowment, "they do so as individuals, not by virtue of their group memberships" (p. 190). The documents that support the "American creed," however, are ambiguous and subject to various legal interpretations. This feature makes them remarkable but frustrating documents of freedom—historically, many injustices have been interpreted as agreeing with them in spirit.[4]

However much Americans want to believe that they live in a true class system, evidence indicates otherwise. The chances for economic and occupa-

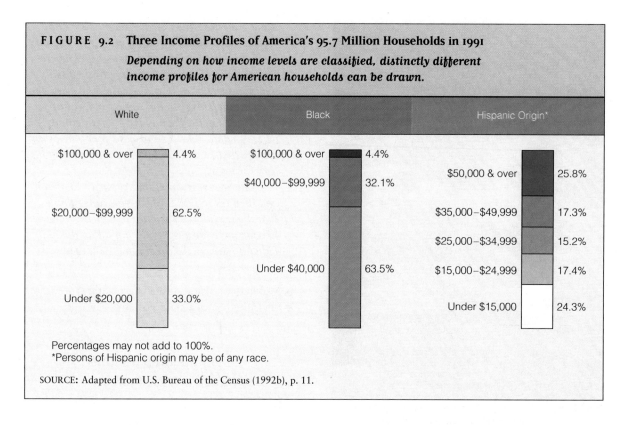

FIGURE 9.2   Three Income Profiles of America's 95.7 Million Households in 1991

Depending on how income levels are classified, distinctly different income profiles for American households can be drawn.

| White | Black | Hispanic Origin* |

$100,000 & over — 4.4%
$20,000–$99,999 — 62.5%
Under $20,000 — 33.0%

$100,000 & over — 4.4%
$40,000–$99,999 — 32.1%
Under $40,000 — 63.5%

$50,000 & over — 25.8%
$35,000–$49,999 — 17.3%
$25,000–$34,999 — 15.2%
$15,000–$24,999 — 17.4%
Under $15,000 — 24.3%

Percentages may not add to 100%.
*Persons of Hispanic origin may be of any race.

SOURCE: Adapted from U.S. Bureau of the Census (1992b), p. 11.

tional success are more often than not connected to forces over which people have little control—social background, ascribed characteristics, and massive restructuring of the economy. To put it bluntly, economic inequality follows a clear pattern in the United States, and some groups are affected more than others by major changes that occur in the economy. Although the pattern of inequality in the United States is not anywhere near as systematic as that which existed under apartheid and which exists today in South Africa, it is striking nonetheless.

Income Inequality in the United States The U.S. Bureau of the Census (1992b) classifies people according to 21 income categories (under $5,000, $5,000–9,999, and so on up to over $100,000 per year). We can generate many different income profiles for the United States, depending on how categories are combined (Hacker 1991). Figure 9.2 shows three such profiles. Each offers a distinct view of how income is distributed across the 95.7 million households in the United States. Some views are more flattering than others.

The presence of inequality alone does not refute a claim that the United States possesses a class system of stratification. But when we examine income profiles for white, black, and Hispanic households, as shown in Figure 9.3, it becomes clear that black and Hispanic households are disproportionately concentrated in lower-income categories. Such differences suggest that the United States is not a class system in the true sense of the word. If it were a true class system, the percentages of families in each income category would be the same across all three groups.

Income distribution is even more uneven when comparing types of households—couple, male-headed (no wife), female-headed (no husband). Figure 9.4 shows that female-headed households in general, and black and Hispanic female-headed households in particular, are concentrated in the lowest-income categories. In view of these findings, we should not be surprised to learn that some groups are concentrated more heavily than others in the lowest-paying and least prestigious occupational categories.

FIGURE 9.3 Income Profiles for U.S. Residents Who Declared Their Race as White, Black, or Hispanic Origin* in 1991

No matter which profile one chooses as most representative of income distribution in the United States, black and Hispanic households are concentrated in lower-income categories.

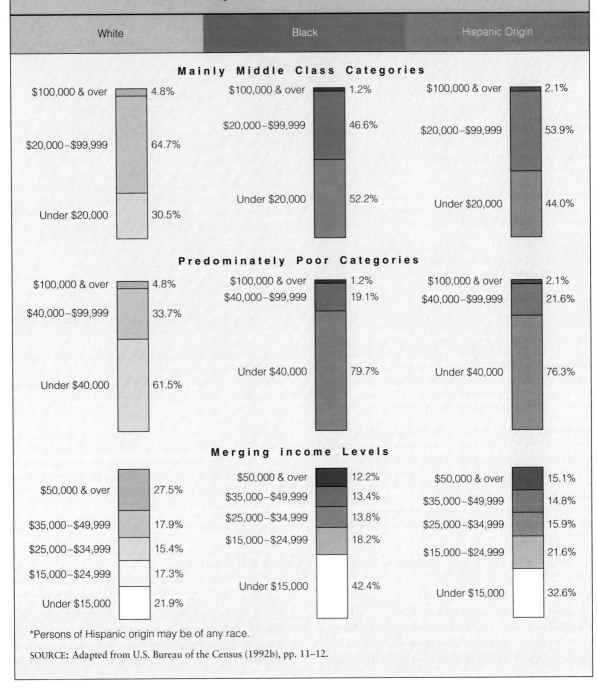

Mainly Middle Class Categories

White
- $100,000 & over — 4.8%
- $20,000–$99,999 — 64.7%
- Under $20,000 — 30.5%

Black
- $100,000 & over — 1.2%
- $20,000–$99,999 — 46.6%
- Under $20,000 — 52.2%

Hispanic Origin
- $100,000 & over — 2.1%
- $20,000–$99,999 — 53.9%
- Under $20,000 — 44.0%

Predominately Poor Categories

White
- $100,000 & over — 4.8%
- $40,000–$99,999 — 33.7%
- Under $40,000 — 61.5%

Black
- $100,000 & over — 1.2%
- $40,000–$99,999 — 19.1%
- Under $40,000 — 79.7%

Hispanic Origin
- $100,000 & over — 2.1%
- $40,000–$99,999 — 21.6%
- Under $40,000 — 76.3%

Merging income Levels

White
- $50,000 & over — 27.5%
- $35,000–$49,999 — 17.9%
- $25,000–$34,999 — 15.4%
- $15,000–$24,999 — 17.3%
- Under $15,000 — 21.9%

Black
- $50,000 & over — 12.2%
- $35,000–$49,999 — 13.4%
- $25,000–$34,999 — 13.8%
- $15,000–$24,999 — 18.2%
- Under $15,000 — 42.4%

Hispanic Origin
- $50,000 & over — 15.1%
- $35,000–$49,999 — 14.8%
- $25,000–$34,999 — 15.9%
- $15,000–$24,999 — 21.6%
- Under $15,000 — 32.6%

*Persons of Hispanic origin may be of any race.

SOURCE: Adapted from U.S. Bureau of the Census (1992b), pp. 11–12.

**FIGURE 9.4   Income Distribution, United States
Female-Headed Family Households in 1991**

Female-headed households in general are concentrated in the lowest-income categories. Households headed by Hispanic* and black females, however, are most likely to be concentrated in those categories.

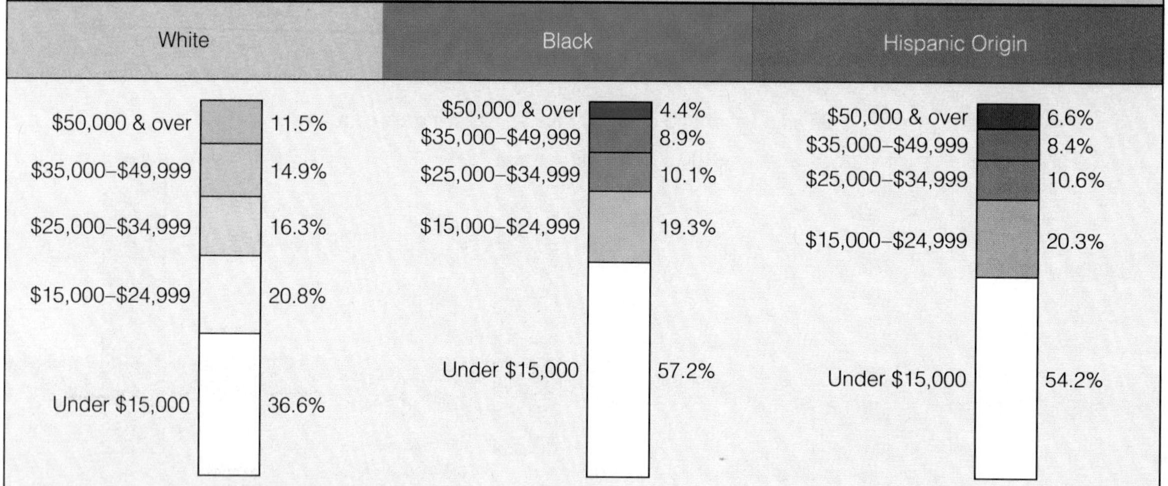

White		
$50,000 & over	11.5%
$35,000–$49,999	14.9%
$25,000–$34,999	16.3%
$15,000–$24,999	20.8%
Under $15,000	36.6%

Black		
$50,000 & over	4.4%
$35,000–$49,999	8.9%
$25,000–$34,999	10.1%
$15,000–$24,999	19.3%
Under $15,000	57.2%

Hispanic Origin		
$50,000 & over	6.6%
$35,000–$49,999	8.4%
$25,000–$34,999	10.6%
$15,000–$24,999	20.3%
Under $15,000	54.2%

Percentages may not add to 100%.
*Persons of Hispanic origin may be of any race.

SOURCE: Adapted from U.S. Bureau of the Census (1992b), pp. 11–12.

Occupational Inequality in the United States Occupations are considered to be segregated according to ascribed characteristics when some occupations (librarian, secretary, physician, lawyer, chief executive officer) are filled primarily by people of a particular race, ethnicity, sex, or age. The occupational categories shown in Table 9.3 and Table 9.4 are examples of those with the highest concentrations by race and by sex.

Occupational segregation is a problem when certain groups are concentrated in the low-paying, low-ranking jobs. When we compare the annual income of black men, black women, and white women who work year-round and full-time with those of their white male counterparts, the impact of occupational segregation becomes evident (see Table 9.5). The fact that income and occupation are connected to race (an ascribed characteristic) means that we must conclude that the United States is at best a mixture of class and caste systems of stratification.

Mixed Systems: Class and Caste

Systems of stratification are usually a combination of class and caste. In the United States, virtually every occupation contains members of different ethnic, racial, age, and sex groups. At the same time, however, some groups such as women and blacks are severely overrepresented or underrepresented in some occupations. Moreover, in the United States, a person's ascribed characteristics can overshadow his or her achievements in such a way that "no amount of class mobility will exempt

TABLE 9.3 Occupations in Which People Who Declare Themselves Black and Hispanic Are Disproportionately Underrepresented and Overrepresented*

Underrepresented (1991)

Occupation	Percent of All Employed	
	Black	Hispanic Origin
Managerial and professional	6.3	3.7
Scientists	3.3	3.6
Physicians	3.2	4.4
Dentists	1.5	2.7
Pharmacists	3.4	3.2
College professors	4.8	2.9
Lawyers and judges	2.8	1.6
Authors	1.4	1.0
Technical writers	5.3	1.8
Artists	2.7	3.2
Dental hygienists	1.1	3.3
Airplane pilots and navigators	1.5	2.9
Sales occupations	6.6	5.5
Waiters and waitresses	4.2	7.1

Overrepresented (1991)

Occupation	Percent of All Employed	
	Black	Hispanic Origin
Maids and housemen	27.2	19.7
Nurses' aides, orderlies, and attendants	31.2	6.9
Cleaners and servants (private households)	29.3	23.3
Postal clerks (except mail carriers)	27.7	6.3
Pressing machine operators	25.8	23.5
Short-order cooks	23.3	11.7
Textile sewing machine operators	20.0	22.7
Farm workers	8.6	26.8

*According to the 1990 census, blacks constitute 12.1 percent, persons of Hispanic origin constitute 9.0 percent, and whites constitute 80.3 percent of the U.S. population.

SOURCE: Adapted from U.S. Bureau of the Census (1992a), pp. 392–94.

a person from the crucial implications of . . . birth" (Berreman 1972, p. 399). For example, even though a person holds a high-ranking occupation, others may question or overlook that person's talent, merit, and accomplishments if they have dark skin or female reproductive organs. The interrelations between caste and class *can be brought readily to mind by thinking of the relative advantages and disadvantages which accrue in Western class systems to persons who occupy such occupational statuses as judge, garbage man, stenographer, airline pilot, factory worker, priest, farmer, agricultural labourer, physician, nurse, big businessman,*

TABLE 9.4 Examples of Occupations in Which Women Are Disproportionately Underrepresented and Overrepresented in 1991*

Underrepresented

Occupation	Percent of All Employed
Engineers	8.2
Dentists	10.1
Clergy	9.3
Firefighting and fire prevention	2.3
Mechanics and repairers	3.7
Construction trades	1.8
Transportation and material-moving occupations	9.0
Forestry and logging operations	5.6
Airplane pilots and navigators	3.4

Overrepresented

Occupation	Percent of All Employed
Registered nurses	94.8
Speech therapists	88.2
Prekindergarten and kindergarten teachers	98.7
Dental hygienists	99.8
Licensed practical nurses	95.0
Secretaries, stenographers, typists	98.5
Receptionists	97.1
Financial records processing (bookkeepers)	91.5
Eligibility clerks, social welfare	90.5
Data entry keyers	86.0
Teachers' aides	93.1
Child care (private household)	96.7
Cleaners and servants (private)	95.8
Dental assistants	98.2
Textile sewing	89.2

*According to the 1990 census, females constitute 51 percent of the total population in the United States.

SOURCE: Adapted from U.S. Bureau of the Census (1992a), p. 394.

beggar, etc. The distinction between class and birth-ascribed stratification can be made clear if one imagines that he encounters two Americans, for example, in each of the above-mentioned occupations, one of whom is white and one of whom is black. This quite literally changes the complexion of the matter. . . . Obviously something significant has been added to the picture of stratification in these examples which is entirely missing in the first instance— something over which the individual generally has no control, which is determined at birth, which cannot be changed, which is shared by all those of like birth, which is crucial to social identity, and which vitally affects one's opportunities, rewards, and social roles. The new

TABLE 9.5	Median Money Income in the United States of Year-Round Full-Time Workers by Declared Race and Sex	
Race/Sex	**Annual Income**	**Percent of White Men's Earnings**
White men	$30,186	100.0
Black men	21,540	71.3
Hispanic men	19,314	64.0
White women	20,840	69.0
Black women	18,518	61.3
Hispanic women	16,186	54.0

SOURCE: U.S. Bureau of the Census (1992a), p. 452.

element is race (colour), caste, ethnicity (religion, language, national origin), or sex. (BERREMAN 1972, PP. 385-86)

Filmmaker Spike Lee, historian John Hope Franklin, television broadcaster Carole Simpson, and professional basketball player Isaiah Thomas have made insightful comments on how their race affects people's definitions of what they do.

I want to be known as a talented young filmmaker. That should be first. But the reality today is that no matter how successful you are, you're black first. (LEE 1989, P. 92)

It's often assumed I'm a scholar of Afro-American history, but the fact is I haven't taught a course in Afro-American history in 30-some-odd years. They say I'm the author of 12 books on black history, when several of those books focus mainly on whites. I'm called a leading black historian, never minding the fact that I've served as president of the American Historical Association, the Organization of American Historians, the Southern Historical Association, Phi Beta Kappa, and on and on.
The tragedy . . . is that black scholars so often have their specialties forced on them. My specialty is the history of the South, and that means I teach the history of blacks and whites.
(FRANKLIN 1990, P. 13)

A radio station executive in Honolulu read my name tag and said, "Carole Simpson, ABC. What does that stand for? African Broadcasting Company?"

When [Larry] Bird makes a great play, it's due to his thinking, and his work habits. It's all planned out by him. It's not the case for blacks. All we do is run and jump. We never practice or give a thought to how we play. It's like I came dribbling out of my mother's womb.
Magic [Johnson] and Michael Jordan and me, for example, we're playing only on God-given talent, like we're animals, lions and tigers, who run around wild in a jungle, while Larry's success is due to intelligence and hard work.
(THOMAS 1987, P. D27)[5]

The experiences of Lee, Franklin, Simpson, and Thomas show how the element of race affects the ways in which people evaluate accomplishments.

Two less personal examples demonstrate how class and caste stratification can operate together in the United States. Consider how positions are filled and how players relate to one another on a coed softball team. Usually, females are assigned to the least central positions (those requiring the least amount of involvement in completing a play). Also, when a player hits the ball toward a female player, at least one male comes over to "help out." Rarely does a female charge over to aid a male teammate.

Finally, when a female player comes to the plate to bat, the four outfielders move in to the infield; when a male player bats, they stay in the outfield. (A female might be a good singles hitter, but it is difficult to hit a single when there are nine infielders.)

We could make the case that women play the least central positions and are helped by men in adjacent positions because they lack the skills and experience to field, throw, and hit the ball adequately. When talent and skill are the only criteria for assigning positions to men and women, a class system of stratification is at work. When there is a predictable relationship between an ascribed characteristic such as sex and an achievement such as shortstop, this is a clue for sociologists to investigate the source of this pattern. Often the reasons for the pattern can be traced to the fact that an undetermined, yet significant, amount of the male-female difference in talent and ability is imposed externally—that is, they are by-products of the ways in which males and females are socialized.

Starting in infancy, parents elicit more active and more physical behavior from sons than daughters. They also channel their children toward sex-appropriate sports. Girls are guided into predominantly noncontact sports, often individual sports, that require grace, flowing movements, flexibility, and aesthetic, such as gymnastics, tennis, and swimming. By contrast, boys are encouraged to participate in team sports that involve contact, lifting, throwing, catching, and running. Because boys and girls participate in different kinds of athletic activities, they develop different skills and styles. Finally, the number of organized teams for each sex (from T-ball to professional) shows that more human and monetary resources are devoted to male than to female athletic development.

In the case of coed softball, a class system of stratification is operating on one level because those who have the most ability and experience are assigned to the most central positions. Yet, on another level a caste system is also at work, because social practices contribute to differences between males' and females' talent and ability.

Caste and class systems of stratification operate together in professional sports as well. Sociologists have long noted that black athletes are concentrated in positions that require strength, speed, and agility, whereas whites are concentrated in central leadership, "thinking," and playmaking positions (Loy and Elvogue 1971; Medoff 1977). In professional baseball whites tend to play in the infield positions (including pitcher and catcher), whereas blacks tend to play in the outfield. Because no on-field position in professional baseball excludes black athletes completely, some element of class stratification must be at work. On the other hand, coaches, who are predominantly white, obviously assign blacks to noncentral positions and whites to leadership positions. This hypothesis is supported by the fact that most elementary and high school athletes play on predominantly white or black teams. As a result, there are black athletes who have experience in playing all positions. Sports sociologists believe that black athletes are removed systematically from positions of leadership and are assigned to the less central positions as they advance from high school to college to the professional ranks. This practice continues after their on-field professional career is ended: in comparison to their white counterparts, few blacks become head coaches, general managers, or executives.

From a sociological point of view, any person who explains these differences as due to biological differences between men and women and across race and ethnic groups is, in the succinct words of social psychologist E. A. Ross ([1908] 1929), "too lazy" to trace these differences to the social environment or historical conditions.

Clearly, social stratification is an important feature of a society, one with significant consequences for the life chances of the advantaged and disadvantaged alike. In the remainder of this chapter we will examine various theories that seek to explain why stratification occurs and the forms it takes. One theory—functionalism—seeks to explain why resources are distributed unequally in society. A second set of theories deals with identifying the various strata within society.

A Functionalist View of Stratification

In their classic article "Some Principles of Stratification," functionalist sociologists Kingsley Davis and Wilbert Moore (1945) ask how stratification—the unequal distribution of social rewards—contributes to maintain order and stability in society. They maintain that social inequality is the device by which societies ensure that the most functionally important occupations, particularly those that require great amounts of talent and costly and rigorous training, are filled by the best-qualified people.

The Functional Importance of Occupations

Davis and Moore concede that it is difficult to document the functional importance of an occupation, but they suggest two somewhat vague indicators: (1) the degree to which the occupation is functionally unique (that is, there are few occupations that can perform the same function adequately) and (2) the degree to which other occupations depend on the one in question. In view of these indicators, garbage collectors, although functionally important to sanitation, need not be rewarded highly because little training and talent are required to do that job. Even though we depend on garbage collectors to maintain sanitary environments, there are many people able to do the work.

Davis and Moore argue that society must offer extra incentives in order to induce the most talented individuals to undergo the long and arduous training needed to fill the most functionally important occupations. They specify that the incentives must be great enough to prevent the best-qualified and most capable people from finding less functionally important occupations as attractive as the most important occupations.

Davis and Moore concede that the efficiency of a stratification system in attracting the best-qualified people is weakened when capable individuals are overlooked or are not granted access to training, when elite groups control the avenues of training (as through admissions quotas), and when parents' influence and wealth (rather than the ability of the offspring) determine the status that their children

attain. Yet, Davis and Moore believe that the system adjusts to such inefficiencies. When there are shortages of personnel for functionally important occupations, the society must increase people's opportunities to enter those occupations; if this is not done, the society as a whole will suffer and will be unable to compete with other societies.

A functionalist might argue that such an adjustment is reflected in white South Africans' recent moves to modify the system of job reservation, which restricted each race to certain types of jobs and prohibited the advance of a nonwhite over a white in the same occupation. To operate effectively (especially after the internationally imposed antiapartheid sanctions ended in 1993), the South African economy needs an increasingly skilled work force, which cannot be maintained by the white population alone. In fact, an article in the February 1994 issue of *World Trade* named the nearly 50 percent unemployment rate, coupled with a shortage of middle managers and professionals in South Africa, as liabilities for foreign companies thinking of investing there (Jones 1994). Even though the long-standing job and mobility restrictions placed on nonwhite workers have been repealed, it will take many years to train those who were denied the educational opportunities to obtain such positions. Thus, the functionalist argument that society will adjust and that all will work out in the end receives considerable criticism because it introduces a moral question: Should the life chances of nonwhites be tied to the needs of the dominant (white) group in society? It also introduces another more general question: Is social inequality the way to ensure that the most important occupations are filled by the most qualified people?

Critique of the Functionalist Perspective

The publication of "Some Principles of Stratification" prompted a number of articles that took issue with the fundamental assumption underlying the Davis and Moore theory—that social inequality is a necessary and universal device that societies use to ensure that the most important occupations are filled by the best qualified people. Two especially insightful critiques were Melvin M. Tumin's "Some Principles of Stratification: A Critical Analysis" (1953) and Richard L. Simpson's "A Modification of the Functional Theory of Social Stratification" (1956).

Neither Tumin nor Simpson believed that a position commands great social rewards simply because it is functionally important or because the available personnel are scarce. Some positions command large salaries and bring other valued rewards even though their contribution to the society is questionable. Consider the salaries of athletes, for example. For the 1993 season, the average salary of a major league professional baseball player was $1,116,946; 40 percent of the 650 athletes who played for the 26 major league teams were paid $1 million or more (Chass 1992, 1993). Elementary and secondary teachers, on the other hand, were paid an average of $34,098 per year (*The World Almanac and Book of Facts 1994*, 1993). This difference in pay raises the question of whether professional athletes and entertainers are more essential to society than teachers—or whether there are other forces that are at least equally important in defining occupational rewards and status.

Critics of functionalism also question why a worker should receive a lower salary for the same job just because the person is of a certain race, age, sex, or national origin. After all, the workers are performing the same job, so functional importance is not the issue. This latter question is at the center of the "comparable worth" debate. Advocates of comparable pay for comparable work ask whether women who work in predominantly female occupations (registered nurse, secretary, day-care worker) should receive salaries comparable to those earned by men who work in predominantly male occupations that are judged to be of comparable worth (vocational education teacher, housepainter, carpenter, automotive mechanic). For example, assuming equivalent education and seniority, why should a female registered nurse working in the state of Minnesota get paid $1,723 a month and a male vocational education teacher working in the same state get paid $2,260 (Johnson 1989)? Why should a female day-care worker earn

less than $8,000 per year, while a male automotive mechanic earns more than $16,000?

In addition, Tumin and Simpson argued that it is very difficult to determine the functional importance of an occupation, especially in societies characterized by a complex division of labor. The specialization and interdependence that accompany a complex division of labor imply that every individual contributes to the whole operation. In light of this interdependence, one could argue that every individual makes an essential contribution. "Thus to judge that the engineers in a factory are functionally more important to the factory than the unskilled workmen involves a notion regarding the dispensability of the unskilled workmen, or their replaceability, relative to that of the engineers" (Tumin 1953, p. 388). Even if engineers, supervisors, and CEOs have the more functionally important positions, how much inequality in salary is necessary to ensure that people choose these positions over unskilled ones? In the United States, for example, the average annual salary of the CEO of a Fortune 500 corporation is $2,025,485—93 times the average salary of a factory worker. Are such high salaries really necessary to make sure that someone *chooses* the job of CEO over the job of factory worker? Probably not. But such high salaries have been justified as necessary to recruit the most able people to run a corporation in the context of a global economy. It is unclear whether such salaries accurately reflect the CEO's contribution to society relative to that of the factory worker. Even though unskilled workers might be replaced more easily than engineers or CEOs, an industrialized society depends upon motivated and qualified people in all positions.

Finally, both Tumin and Simpson argued that the functional theory of stratification implies that a system of stratification evolves as it does in order to meet the needs of the society. In evaluating such a claim, one must look at whose needs are being met by the system. In the case of apartheid, the needs of whites unquestionably were met at the expense of the needs of blacks. We can see this claim supported by the way in which electricity is still distributed to the people of South Africa. South Africa produces 60 percent of all of the electricity used throughout the African continent. But 66 percent of all South Africans (mostly nonwhites), and 80 percent of all Africans are denied access to that energy. Those without electricity face enormous difficulties in obtaining the fuel to cook food and to warm and light their houses. As a result, people without electricity must resort to gathering wood:

> *In the high grassland areas of KwaZulu [one of the 10 African homelands], for example, the average distance walked in collecting one headload [of fuel] was a little over 5 miles (8.3 km) and the average time taken in collecting the one load was 4.5 hours.*
>
> *Not only is the collecting of firewood exhausting, time-consuming, and dangerous but*

> *it has serious ecological consequences. Each household uses between three and four tons of wood a year. . . . Over the relatively brief span of the past 50 years, 200 of the 250 forests in KwaZulu have disappeared.* (WILSON AND RAMPHELE 1989, P. 44)

It is clear from this example that the long-term environmental consequences of the way in which electricity is distributed creates a situation that may benefit whites. But it does not meet the needs of society as a whole or of the planet. Whose interests should we take into account, then, in evaluating whether a system is "functional"?

Analyses of Social Class

Although sociologists use the term *class* to refer to one form of stratification, they also use *class* to denote a category that designates a person's overall status in society. Sociologists consider social class to be an important factor in determining life chances. Sociologists, however, are preoccupied with two questions: (1) How many social classes are there, and (2) what constitutes a social class? For some answers to these questions we turn to the works of Karl Marx and Max Weber.

Karl Marx and Social Class

Karl Marx viewed every historical period as characterized by a system of production that gave rise to specific types of confrontation between the exploiting and exploited classes in society. Consequently, Marx was interested in relationships between various social classes that make up a society. He gave several answers to the question, How many social classes are there? In *The Communist Manifesto* he named two: the bourgeoisie and the proletariat. In *Capital: A Critique of Political Economy* he named three social classes: wage laborers, capitalists, and landlords. In *The Class Struggles in France 1848–1850* he named at least six: the finance aristocracy, the bourgeoisie, the

petty bourgeois, the proletariat, landlords, and peasants. According to French sociologists Raymond Boudon and François Bourricaud, a careful reading of Marx's writings suggests that he believed "that the number of classes to be defined depends on the reason why we want to define them" (1989, p. 341). The fact that Marx paid so much attention to class and to the class divisions in society underscores his belief that the most important engine of change is class struggle. A brief overview of these works clarifies this point.

In *The Communist Manifesto* written with Friedrich Engels in 1848, Marx described how class conflict between two distinct classes propels society from one historical epoch to another. Over time free men and slaves, nobles and commoners, barons and serfs, and guildmasters and journeymen have confronted each other. Marx observed that the rise of factories and mechanization as a means of production created two modern classes: the bourgeoisie (the owners of the means of production) and the proletariat (those who must sell their labor to the bourgeoisie). In light of this historical theme, it is appropriate that in *The Communist Manifesto* Marx focused on the two social classes he believed would usher society out of capitalism and into another era.

In *Capital: A Critique of Political Economy* (1909), Marx named three classes: wage laborers, capitalists, and landlords. Each class is composed of people whose revenues or income "flow from the same common sources" (p. 1032). For wage laborers the source is wages; for capitalists, profit; for landowners, ground rent. Marx acknowledged that the boundaries separating landowners from capitalists are not clear-cut. In this three-category classification scheme, for example, Henry Ford, the founder of Ford Motor Company, is both a landlord (because he owned a rubber plantation in Brazil) and a capitalist (because he owned the factories and the machines and purchased the labor). Marx also acknowledged that each of the three classes can be subdivided further. The category of landowners, for example, can be divided into owners of vineyards, farms, forests, mines, fisheries, and so on. Because Marx was interested in distinguishing people according to their sources of income, a three-category social class scheme made sense.

The Class Struggles in France 1848–1850 ([1895] 1976) is a historical study of an event in progress—the 1848 revolutions against several European governments (Germany, Austria, France, Italy, and Belgium), with special emphasis on France. In this book Marx sought to describe and explain "a concrete situation in its complexity" (Boudon and Bourricaud 1989, p. 341). He described the 1848 revolution as a struggle for the necessities of life and as "a fight for the preservation or annihilation of the bourgeois order" (Marx [1895] 1976, p. 56). The latter consisted of a finance aristocracy, which lived in obvious luxury among masses of starving, low-paid, unemployed people.

Marx outlined the major factors that triggered the 1848 revolution and explained why he believed it failed. The widespread discontent was fueled by two world economic events. One was the potato blight and the bad harvests of 1845 and 1846, which raised the already high level of frustration among the people. The resulting rise in the cost of living caused bloody conflict in France as well as on the rest of the continent. The other event was a general commercial and industrial crisis which resulted in an economic depression and a collapse of international credit. The revolutions were centered in the cities, where the Industrial Revolution had created a proletariat from persons who had migrated there in search of work. Generally, the workers were paid very low wages, lived in squalor, and lacked the necessities.

Although the faces of those who ruled the French government changed as a result of the 1848 revolution, the exploitive structure remained. In the end, the workers were put down by "unheard of brutality" (Marx [1895] 1976, p. 57). Marx believed that the uprising failed because, even though the workers displayed unprecedented bravery and talent, they were "without chiefs, without a common plan, without means and for the most part, lacking weapons" (p. 56). Also, the revolution failed because the "other" classes did not support the proletariat when they moved against the finance aristocracy.

It is difficult to apply Marx's ideas about social class in a total way because, as he made clear in *The Class Struggles in France 1848–1850,* the reality of class is very complex. He left us, however, with some useful ideas with which to approach social class. First is the idea that conflict between two distinct classes propels us from one historical epoch to another. South Africa clearly contained two distinct classes: those designated as white and those designated as something else. In the United States, on the other hand, one area that has received considerable attention is the class division between skilled and unskilled workers. Unskilled workers in the United States are vulnerable because in the capitalistic economy corporations have transferred (and still are transferring) low-skilled jobs out of the country. Class conflict is an important and impending agent of social change in both South Africa and the United States.

A second important idea left to us by Marx is the concept of viewing social class in terms of the *sources* of income. Approaching social class in this way carries our understanding of social class beyond the simple notion of occupation (or relationship to the means of production). Often, however, such information on income is very hard to acquire. This information is not available for the South African population; at best, it is incomplete for the United States. In the United States the Federal Reserve sponsors the Survey of Consumer Finances, conducted every six years or so. Unfortunately, the various sources of income are available for only five income groups. Households with annual incomes of 50,000 or more are grouped into one category. Therefore we cannot determine sources of income for the wealthiest Americans.

Finally, Marx's ideas remind us that the conditions that lead to a successful revolt by an exploited class against the exploiting class are multifaceted and complex. He recognized that exploitive conditions can trigger uprisings but observed that other factors such as a well-thought-out plan, effective leadership, the support of other classes, and access to weapons determine the success or failure of the revolt.

To illustrate this point, consider that two of the best-known demonstrations against apartheid occurred in 1960 and 1976. In 1960 the residents of Sharpville, an African township, marched to the police station without their passes to protest the pass laws. The police fired into the crowd, killing 69 persons and wounding 180. Although the police response initiated riots throughout South Africa, in the end the white government won and declared all antiapartheid organizations illegal. In 1976, 20,000 Soweto schoolchildren marched in protest against the use of Afrikaans as the language of classroom instruction. The police opened fire; hundreds of children were killed and thousands were wounded. This brutal response set off riots throughout the country. The point is that nonwhite South Africans have always protested apartheid but it is only since the mid-1980s that the other class—whites—has taken steps to end the system.

What factors other than protests by the exploited class have contributed to the movement to end apartheid? First, in 1976 the United States and Europe applied new and more forceful economic sanctions while resistance from within increased. Also, the African continent and South Africa lost their strategic importance with the collapse of communism and the end of the Cold War between the United States and the Soviet Union. As a result, the United States no longer needed to support the white South African government to counter communist influences within South Africa and surrounding countries.

Max Weber and Social Class

Although Karl Marx did not consistently specify an exact number of social classes in society, he clearly stated that a person's social class was based on his or her relationship to the means of production. Max Weber, like Marx, did not specify how many social classes exist. For Weber, though, the basis for a social class was not the means of production; rather, it was the marketplace. According to Weber, class situation is ultimately market situation. It is based on the chances of acquiring goods and services, obtaining a well-paying job in the marketplace, and finding inner satisfaction.

According to Weber ([1947] 1985), people's class standing depends on their marketable abilities (work experience and qualifications), their access to consumer goods and services, their control over the means of production, and their ability to invest in property and in other sources of income (see "What Are You Worth?"). Persons completely unskilled, lacking property, and dependent on seasonal or sporadic employment constitute the very bottom of the class system. They form the "negatively privileged" property class. Those at the very top—the "positively privileged"—monopolize the purchase of the highest-priced consumer goods, have access to the most socially advantageous kinds of education, control the highest executive positions, own the means of production, and live on income from property and other investments. Between the top and the bottom of the ladder is a continuum of rungs.

Weber states that class ranking is complicated by status groups and parties, of which there are many different kinds. He defines **status group** as a plurality of persons held together by virtue of a common lifestyle, formal education, family background, or occupation and "by the level of social esteem and honor accorded to them by others" (Coser 1977, p. 229). This definition suggests that wealth, income, and position are not the only factors that determine an individual's status group. "The class position of an officer, a civil servant or a student may vary greatly according to their wealth and yet not lead to a different status since upbringing and education create a common style of life" (Weber 1982, p. 73). In South Africa, English-speaking whites form a

What Are You Worth?

According to Max Weber, a person's class standing depends on many factors, including access to various sources of income.

Personal Financial Statement of _____

Prepared as of _____

Assets   **Liabilities**

	Estimated Current Value		*Amount*
Cash in Banks and Money Market Accounts	_____	Mortgages	_____
Amounts Owed Me	_____	Broker Loans/ Margin Accounts	_____
Stocks/Bonds	_____	Bank Loans/Notes	_____
Other Investments	_____	Life Insurance Loans	_____
Life Insurance (cash surrender value)	_____	Charge Accounts	_____
IRA and Keogh Accounts	_____	Pledges to Charity	_____
Pension and Profit Sharing (vested interest)	_____	Divorce Settlement	_____
		Support Obligations	_____
Real Estate: home	_____	Alimony	_____
other	_____	Taxes Owed: income	_____
Business Interests	_____	real estate	_____
Personal Property*	_____	other	_____
_____	_____	Other Liabilities	_____
_____	_____	_____	_____
_____	_____	_____	_____
_____	_____	_____	_____

TOTAL ASSETS _____   TOTAL LIABILITIES _____

NET WORTH _____ *(Subtract liabilities from assets to find net worth)*

*Include furnishings, cars, jewelry, collections, security deposit on rent, etc.
Your personal financial statement should also include a special tax provision based on any increases in the value of your assets, so ask your CPA for help. Review your statement annually—or more often if there are major financial changes in your life.

SOURCE: American Institute of Certified Public Accountants (1991).

status group distinct from Afrikaans-speaking white Afrikaners. The two groups practice what sociologist Diana Russell calls a voluntary apartheid: "speaking different languages, attending different schools, living in different areas, and voting for different political parties" (Russell 1989, p. 4).

Political parties are organizations "oriented toward the planned acquisition of social power [and] toward influencing social action no matter what its content may be" (Weber 1982, p. 68). Parties are organized to represent class, status, and other interests; they exist at all levels (within an organization, a city, a country). The means employed to obtain power include violence, canvassing for votes, bribery, donations, the force of speech, suggestion, and fraud. In South Africa the best-known anti-apartheid organization is the African National Congress (ANC), founded in 1912. Another prominent party is the Pan-Africanist Congress, which broke from the ANC in 1959. As mentioned earlier, the Inkatha Freedom Party, a conservative Zulu organization, is led by chief Mangosuthu Buthelezi. The United Democratic Front (UDF), founded in 1983, merged more than 800 antiapartheid groups under the umbrella of one organization.[6]

Weber's conception of social class enriches that of Marx. Weber views class as a continuum of rungs on a social ladder, with the top and the bottom rungs being the positively privileged class and the negatively privileged property class. Weber argues that a "*uniform* class situation prevails only when completely unskilled and propertyless persons are dependent on irregular employment" (1982, p. 69). We cannot speak of a uniform situation with regard to the other classes because class standing is complicated by such elements as occupation, education, income, the status groups to which people belong, differences in property, consumption patterns, and so on.

Weber's idea of top and bottom rungs, with everyone else somewhere between, inspires us to compare the situation of the wealthiest person against that of the very poor. Table 9.6 shows the percentage share of household income for various population groups in what the United Nations calls high-income economies. It shows the wealthiest in society (the wealthiest 10 and 20 percent) have an-

nual household incomes that are anywhere from four to nine times that of the poorest 20 percent of the population.

Weber's ideas about social class also draw our attention to the negatively privileged classes. The proportion of negatively privileged persons tells us something important about the extremes of inequality in a society. Although some people of all racial groups in South Africa have high incomes, income distribution is clearly related to race. Table 9.7 shows the average annual per capita income by race, as well as the percentage of households in each racial group in various income categories. Fifty percent of white households have incomes of more than 8,000 rand (about $25,000 in U.S. dollars), as compared to less than 5 percent of Asian, Coloured, and African households. Inequality is even more extreme in the homelands, where 80 percent of households are in a state of dire poverty (Wilson and Ramphele 1989). The plight of the negatively privileged in South Africa can be traced directly to apartheid.

The existence of a negatively privileged property class in the United States can also be traced to structural factors—in particular, to changes in the occupational structure. This may come as a surprise to some Americans who attribute mobility, whether upward or downward, to individual effort, and do not consider changes in the occupational structure as the cause. Many Americans may not recognize that some groups are affected by changes in the occupational structure more strongly than others. In *The Truly Disadvantaged* (1987) and in other related articles and books, sociologist William Julius Wilson describes how structural changes in the U.S. economy helped create what he termed, in his 1990 presidential address to the American Sociological Association, the "ghetto poor." A number of economic transformations have taken place, including the restructuring of the American economy from a manufacturing-based economy to a service- and information-based economy; a labor surplus that began in the 1970s, marked by the entry of women and the large baby boom segment into the labor market; a massive exodus of jobs from the cities to the suburbs; and the transfer of low-skilled manufacturing jobs out of

TABLE 9.6 **Percentage Share of Household Income in High-Income Economies**

Each year the United Nations publishes the most recent statistics available for every country in the world. One rough measure of the inequality that exists in a country is the distribution of total household income in a given year. This table shows the percentage of total household income that the poorest 20 percent earn and compares it with the percentages earned by the wealthiest 20 and 10 percent of the population.

	Lowest 20 Percent of Population	Highest 20 Percent of Population	Highest 10 Percent of Population
Saudi Arabia			
Spain	6.9	40.0	24.5
Ireland			
Israel	6.0	39.6	23.5
Singapore	5.1	48.9	33.5
Hong Kong	5.4	47.0	31.3
New Zealand	5.1	44.7	28.7
Australia	4.4	42.2	25.8
United Kingdom	5.8	39.5	23.3
Italy	6.8	41.0	25.3
Kuwait			
Belgium	7.9	36.0	21.5
Netherlands	6.9	38.3	23.0
Austria			
United Arab Emirates			
France	6.3	40.8	25.5
Canada	5.7	40.2	24.1
Denmark	5.4	38.6	22.3
Germany, Fed. Rep.	6.8	38.7	23.4
Finland	6.3	37.6	21.7
Sweden	8.0	36.9	20.8
United States	4.7	41.9	25.0
Norway	6.2	36.7	21.2
Japan	8.7	37.5	22.4
Switzerland	5.2	44.6	29.8

SOURCE: Adapted from "Table 30: Income Distribution and International Comparison Program Estimates of GDP." P. 237 in *World Development Report* (1990).

the United States to offshore locations (see Chapter 2). These changes are major forces behind the emergence of the ghetto poor or **urban underclass,** a "heterogeneous grouping of families and individuals in the inner city that are outside the mainstream of the American occupational system and that consequently represent the very bottom of the economic hierarchy" (Wilson 1983, p. 80).

Wilson (in collaboration with sociologist Loïc J. D. Wacquant) looks at Chicago as a case in point. (Actually, the point applies to every large city in the United States—Los Angeles, New York, Detroit, and so on.) In 1954, Chicago was at the height of its industrial power. Between 1954 and 1982, the number of manufacturing establishments within the city limits dropped from more than 10,000 to

TABLE 9.7   Annual Per Capita Income in South Africa, By Race

		Percentage in Each Income Category				
	Average Per Capita Income	*≤ 500 Rand**	*500—1,500 Rand*	*1,501—3,000 Rand*	*3,001—8,000 Rand*	*8,000+ Rand*
White	6,242	2	2	7	39	50
Asian	2,289	12	3	34	25	4
Coloured	1,630	25	29	24	20	2
African						
–urban	1,366	31	43	21	5	0.01
–rural	388	—	—	—	—	—
Entire Population		24	32	18	14	12

*A rand is equivalent to about $3.18 U.S. dollars.

SOURCES: Adapted from Wilson and Ramphele (1989): *The World Almanac and Book of Facts 1991* (1990); *World Development Report* (1990).

5,000, and the number of jobs dropped from 616,000 to 277,000. This reduction, in conjunction with the outmigration of stably employed working class and middle class black families, fueled by new access to housing opportunities outside the inner city, had a profound impact on the daily life of people left behind. The exodus of the stably employed resulted in the closing of hundreds of local businesses, service establishments, and stores. According to Wacquant (1989), the single most significant consequence of these historical and economic events was the "disruption of the networks of occupational contacts that are so crucial in moving individuals into and up job chains . . . [because] ghetto residents lack parents, friends, and acquaintances who are stably employed and can therefore function as diverse ties to firms . . . by telling them about a possible opening and assisting them in applying [for] and retaining a job" (Wacquant 1989, pp. 515–16).

The ghetto poor are the most visible and most publicized underclass in the United States. In addition, demographers William P. O'Hare and Brenda Curry-White (1992) estimate that approximately 736,000 rural residents can be classified as an underclass. Like their urban counterparts, the rural underclass is concentrated in geographic areas with high poverty rates. They too have been affected by economic restructuring, which includes the decline of farming, mining, and timber industries and the transfer of routine manufacturing out of the United States.

The rural and urban underclass represent two distinct segments of the population that live below the poverty line, set at about $12,675 for a family of four. On the basis of this definition, almost 32 million Americans (or 13 percent of the population) live below the poverty line. Included in this 32 million are 12 million people whose income is less than half the amount officially defined as the poverty level. However, the definition of poverty encompasses diverse groups of people, some of whom might not be considered really poor (such as graduate students and retired people who have assets but who live on a low fixed income). Still, it is important to point out that two out of every three poverty-level households are headed by women. For many of these women "their only 'behavioral deviancy' is that their husbands or boyfriends left them" (Jencks 1990, p. 42); in the case of many older women, their husbands have died. For the most part, there are two reasons for women's poverty: (1) the economic burden of children and (2) women's disadvantaged position in the labor market. These issues will be explored further in Chapter 11.

Although individual effort is one important variable in upward and downward mobility, this crowded unemployment office illustrates the reality that structural changes in the economy can significantly affect people's ability to find and keep well-paying jobs.

Alon Reininger/Woodfin Camp & Associates

Discussion

In this chapter we have examined the workings of social stratification (the systematic division of people into categories). More importantly, we have learned that classification schemes have a profound impact on people's life chances. Some forms of social stratification affect life chances more strongly than others. Caste systems have a decided impact because in such systems life chances are determined by characteristics over which people have no control. Class systems, although not models of equality, permit life chances to be enhanced on the basis of individual effort. Class and caste systems are end points on a continuum.

South African society still approaches a caste system because in that country people's life chances and access to scarce and valued resources are clearly connected to race. As much as we would like to believe that the United States represents a pure class system, the evidence tells us that it is more castelike than we care to acknowledge. At the same time, the United States is classlike in the sense that every occupation contains people of different ethnic, racial, age, and sex groups. On the other hand, there is no question that some ethnic, racial, age, and sex groups are concentrated in the low-status occupations.

The most intriguing and most problematic feature of stratification systems in general involves the criteria used to rank people, especially when ascribed characteristics are the important criteria. How can there exist ranking systems in which people who belong to one category of an ascribed characteristic (such as white skin or blue eyes) are treated as more valuable or worthier than people who belong to other categories? Jane Elliot, the third-grade teacher who separated her students by eye color and rewarded them accordingly, gives one answer: "This is not something I can do alone" ("Frontline" 1985, p. 20). By this statement Elliot meant that the experiment couldn't work without the cooperation of the people on top. Her observation suggests that people cooperate in maintaining systems of stratification. But why do people cooperate? One answer comes from a blue-eyed member of Jane Elliot's class, now an adult, as he recalled the experience. "Yeah, I felt like I was—like a king, like I ruled them brown-eyes, like I was better than them, happy" (p. 13). The feeling of being better translated into an unexpected result. Jane Elliot explains:

The second year I did this exercise I gave little spelling tests, math tests, reading tests two weeks before the exercise, each day of the exercise and two weeks later and, almost without exception, the students' scores go up on the day they're on the top, down the days they're on the bottom and then maintain a higher level for the rest of the year, after they've been through the exercise. We sent some of those tests to Stanford University to the Psychology Department and they did, sort of an informal review of them, and they said that what's happening here is kids' academic ability is being changed in a 24-hour period. And it isn't possible but it's happening. Something very strange is happening to these children because suddenly they're finding out how really great they are and they are responding to what they know now they are able to do. And it's happened consistently with third graders. (P. 17)

One clear answer to the question of why the people on top cooperate to maintain a system of stratification is that they benefit (whether they know it or not) from the system of stratification and the way in which rewards are distributed. In the third-grade class, the blue-eyed children benefited from the system that distributed rewards on the basis of eye color. In South Africa, whites clearly have benefited from a system that rewards people on the basis of race. On the other hand, it is difficult for those on top in the United States to acknowledge the extent to which their own society's system of stratification benefits them. Perhaps we believe that this country is a model of equal opportunity because (1) we can find examples of people from all racial, ethnic, sex, and age groups who do achieve rewards; (2) we have no obvious laws governing ascribed characteristics and life chances; and (3) we believe that anyone can transcend his or her environment through hard work.

The case of South Africa reminds us how difficult it is for people to give up their privileges and to put into practice a new ranking and reward system. In 1990 South African President F. W. de Klerk traveled to the United States to meet with President Bush and other government leaders, and to convince the American people that the political changes in South Africa over the past year represented a sincere, irreversible effort to dismantle apartheid. During his visit he emphasized repeat-

South Africa's moves toward a racially inclusive democracy are a cause for celebration for many, but changing an entire system of social stratification will be difficult at best.

Mark Peters/Sipa Press

edly that the principles underlying the American political system (such as the Bill of Rights and the Constitution) deserved to be emulated (Wren 1990a). Earlier in this chapter we noted that these documents, remarkable as they are, also are frustrating in that they allow for many injustices to be interpreted as being in agreement with these documents in principle. Hundreds of times throughout the history of the United States, members of some group or other have been defined as less than human or as not completely human; as a result, they have been denied equitable access to justice, freedom, health care, housing, jobs, and education. Moreover, there are cases substantiating that one group's rights have been upheld at the expense of another, less powerful group. The point is that adopting a set of principles does not guarantee that they will be put into practice. Gunnar Myrdal called this situation "an American dilemma" in a book by that name (1944). The dilemma is that a substantial gulf exists between the so-called American creed and actual conduct. In addition, even if these principles were implemented immediately, the people who were penalized in the past would continue to be penalized.

The sociological perspective is valuable in that it enables us to see how social stratification systems are connected to life chances. When we know what is going on we have an obligation to work to change things. However, people usually are not so clear-sighted. The case of South Africa shows that it took pressure from outside (in the form of economic sanctions and cultural isolation) and from within (in the form of mass demonstrations, strikes, and bloodshed) to push South African whites to take the first steps toward dismantling apartheid and toward creating a multiracial democracy. The success of South Africa's attempts to dismantle apartheid will hinge on whether it can dismantle the legacy of this policy—the profound social and economic inequalities that now keep South Africans apart.

FOCUS:
Mobility in the United States

In this chapter we learned that one characteristic that distinguishes class from caste systems of stratification is mobility. In a caste system, overall status and chances for mobility are tied to ascribed characteristics. In a class system, life chances are connected theoretically to merit, talent, ability, and past performance, not to attributes over which people have no control, such as skin color, sex, or parents' social class. In "Those Born Wealthy or Poor Usually Stay So, Studies Say," Sylvia Nasar evaluates the character of mobility in the United States.

Those Born Wealthy or Poor Usually Stay So, Studies Say

Sylvia Nasar

In the continuing debate over the distribution of wealth and income—the richest 1 percent of American families control more wealth than the bottom 90 percent—one school of thought says findings like this, however extreme, do not present a true picture.

In this view, largely conservative, these data are flawed because they are mere snapshots of a single moment in time and fail to reflect the constant ebb and flow of fortunes.

Even if the raw numbers are accurate, these economists say, the portrait fails to capture the amazing fluidity and flux of American society, the hard-scrabble beginnings of many a millionaire, from Bobby Bonilla to Ross Perot. And the richest person in America these days is not a Rockefeller or a du Pont but William Gates, founder of the Microsoft Corporation, the leading software company.

But modern Horatio Algers notwithstanding, a wave of recent studies shows that rags-to-riches remains the economic exception, not the rule. Though many doors and many rewards are open to talent, being rich or poor is more likely than not to carry over from generation to generation, and certainly from year to year.

If anything, economists say, the climb out of poverty has gotten harder in the last decade or two. The United States economy has become increasingly less hospitable to the young, the unskilled and the less educated. It is highly unlikely that this year's rich man will be next year's pauper—or vice versa.

"It's certainly true that someone whose parents are poor has a lower expected income, on average, than someone whose parents are rich," said Gary Solon, a University of Michigan economist. "It's not that you inherit the same position, but there's a substantial correlation."

A child whose father is in the bottom 5 percent of earners, for instance, has only 1 chance in 20 of making it into the top 20 percent of families, according to a forthcoming article in *The American Economic Review* by Professor Solon. The same child has a 1-in-4 chance of rising above the median wealth and a 2-in-5 chance of staying poor or near poor [see Figure 9.5].

"All you have to do is look at L.A. to decide that there are lots of people who think their permanent prospects are pretty crummy," said David M. Cutler, an economist at Harvard University.

What is more, apart from changes in income and wealth that reflect the normal lifetime pattern of growing earnings and savings until retirement, followed by a gradual decline, more Americans do not move a great many rungs, up or down, in a lifetime. The changes tend to be one step forward or one step back, not from the lower half to the upper.

To be sure, Americans can still shinny up and down the economic ladder with an agility that has amazed visitors from Alexis de Tocqueville to the latest arrivals from Latin America or Asia.

"We have more inequality than other countries," said Timothy M. Smeeding, an economist at the Maxwell School at Syracuse University. "But we probably have more mobility as well."

Some conservative economists, including Bruce R. Bartlett, Deputy Assistant Secretary of the Treasury, and Lawrence B. Lindsey, a governor of the Federal Reserve Board, contend that much of the apparent increase in inequality in the last 15 years is a statistical artifact that reflects more upward mobility.

A Little Luck and Hard Work "The rising share of the top 1 percent means that there was a lot of income mobility in America during the 1980's," Mr. Lindsey said.

Similarly, Mr. Bartlett said: "Those who are on top today could easily be down and out tomorrow.

A little luck and hard work can turn today's poor into tomorrow's rich."

But recent studies that track the fortunes over time do not support the view that there is so much mobility that where people stand in any given year hardly matters.

"There's a lot of churning at the top and the bottom," said Lawrence F. Katz, an economist at Harvard University, "but the key is that the top group has just pulled away from everyone else."

Economists look at three kinds of mobility that vary mainly in whether they define income in terms of a year, the average of several years or the average over a lifetime.

"More time allows for more things—good and bad—to happen," said Paul Menchik, an economist at Michigan State University.

From one year to the next, according to Census Department data from the Survey of Income and Program Participation and from a study of taxpayers by Joel Slemrod, a University of Michigan economist, roughly one-fifth to one-fourth of the people in the top income group—whether the top 1, 10 or 20 percent—were not in that group the year before.

But much of this short-term turnover may be illusory. As Mr. Slemrod points out in a study that will be published in a National Bureau of Economic Research volume, "Tax Policy and the Economy," a large fraction of year-to-year changes, especially in the uppermost slice of income, reflects reporting error, timing of income like capital gains, episodes of illness or unemployment and other transitory effects that have little to do with true mobility.

"On average, if somebody has a high income in a given year, the odds are that they had a high income, on average, for many years," Mr. Slemrod said.

The average income of those people who earned more than $100,000 in 1983, for example, was $175,707 in that year and $153,381 on average for a seven-year period ended in 1985.

Mr. Slemrod found no evidence of more true mobility in the 1980's than in the 1960's. The relationship between the seven-year averages and a single year were also quite similar in the 1960's and the 1980's, he found.

As Mr. Slemrod's study suggests, a more sensible way to look at mobility is to consider changes

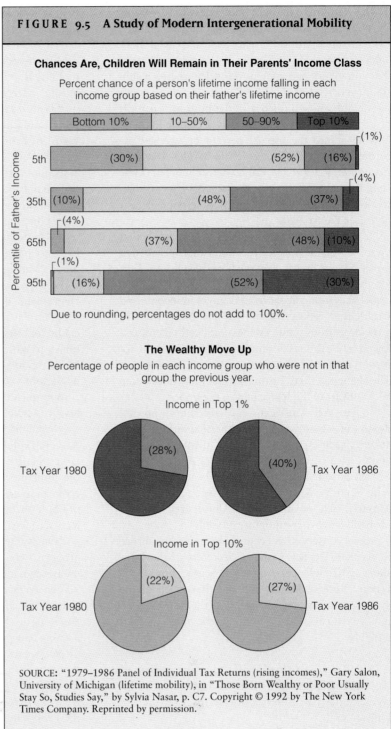

FIGURE 9.5   A Study of Modern Intergenerational Mobility

Chances Are, Children Will Remain in Their Parents' Income Class

Percent chance of a person's lifetime income falling in each income group based on their father's lifetime income

Due to rounding, percentages do not add to 100%.

The Wealthy Move Up

Percentage of people in each income group who were not in that group the previous year.

Income in Top 1%

Income in Top 10%

SOURCE: "1979–1986 Panel of Individual Tax Returns (rising incomes)," Gary Salon, University of Michigan (lifetime mobility), in "Those Born Wealthy or Poor Usually Stay So, Studies Say," by Sylvia Nasar, p. C7. Copyright © 1992 by The New York Times Company. Reprinted by permission.

Key theorists (pp) (main ideas)
Weber
Marx
Moore/Davis
Robert Merton
(Jane Elliot)

averaged over 5 or 10 years. That is what Greg J. Duncan, Timothy M. Smeeding and Willard Rogers did in a study that will be published in "Inequality at the Close of the 20th Century" by the Levy Institute.

Their main finding: During the 1980's, it became easier for those in the middle class to become rich but harder for the rich to fall out of the top 10 percent.

The explosive growth of pay for top professionals and managers in the era propelled lots of people out of the middle class into the ranks of the rich.

Harder to Escape Poverty More troubling was the finding, based on the same data, that it became harder to climb out of poverty largely because of the stagnation and outright decline of real earnings among young, less-educated men.

For example, said Sheldon Danziger, an economist at the University of Michigan, the proportion of high school graduates likely to earn more than a poverty-level income—which was $11,662 in 1989 dollars—fell to 88 percent in the 1980's from 93 percent in the 1970's among white men and to 75 percent from 84 percent among black men.

What Americans probably care about most, though, is the notion that children can do better than their parents, that being born with a plastic spoon in one's mouth does not necessarily preclude riches.

Here the evidence suggests that from one generation to the next, both rich and poor children have about a 40 percent chance of staying put and a 60 percent chance of ending up somewhere closer to the middle.

At the same time, going from rags to riches or falling from the pinnacle of wealth to poverty are rare.

"A lot of people do better or worse than their parents," said Mr. Katz at Harvard. "But the best predictor of how well you'll do is still how well your parents did."

Mr. Solon's study of mobility between generations shows that if a father's income is at the 95th percentile, his son has a 42 percent chance of being in the top 20 percent, and less than half a percent chance of ending up in the bottom 20 percent.

"There's a lot more mobility in the middle than at the bottom or the top," said David J. Zimmerman, an economist at Williams College who studied fathers' and sons' earnings. "Two-thirds of the people who start at the bottom end up in the bottom half."

No Lottery Winners Here Fortunes rise and fall more slowly than incomes, and rock stars, ball players and lottery winners who end up in the top 1 percent of income earners for a year or even several years are not likely to land on Forbes magazine's list of the 400 wealthiest, which is based on net worth.

A study by Professor Menchik and his co-author, Nancy A. Jianokotlos, an economist at Colorado State University, shows that 75 percent of those who were in the top 10 percent in wealth in 1966 were still there five years later and 62 percent were still there 15 years later.

"It's not a lottery in which people with high school degrees or poor parents have just as much chance of getting to the top as someone with college or rich parents," Professor Katz said. "People with less education are systematically falling behind."

Key Concepts

Achieved Characteristics 303

Apartheid 302

Ascribed Characteristics 303

Caste System 309

Class System 309

Downward Mobility 314

Intergenerational Mobility 314

Intragenerational Mobility 314

Life Chances 302

Race 305

Social Stratification 302

Status Group 327

Status Value 304

Upward Mobility 314

Urban Underclass 330

Vertical Mobility 314

Income

upper class / lower class / middle class

o mixed systems

Notes

1. Below are some excerpts from Mandela's speech:

   We of the ANC had always stood for a non-racial democracy, and we shrank from any action which might drive the races further apart than they already were. But the hard facts were that fifty years of non-violence had brought the African people nothing but more and more repressive legislation, and fewer and fewer rights. It may not be easy for this Court to understand, but it is a fact that for a long time the people had been talking of violence—of the day when they would fight the White man and win back their country—and we, the leaders of the ANC, had nevertheless always prevailed upon them to avoid violence and to pursue peaceful methods. When some of us discussed this in May and June of 1961, it could not be denied that our policy to achieve a non-racial State by non-violence had achieved nothing, and that our followers were beginning to lose confidence in this policy and were developing disturbing ideas of terrorism.

   At the beginning of June 1961, after a long and anxious assessment of the South African situation, I, and some colleagues, came to the conclusion that as violence in this country was inevitable, it would be unrealistic and wrong for African leaders to continue preaching peace and non-violence at a time when the Government met our peaceful demands with force.

   This conclusion was not easily arrived at. It was only when all else had failed, when all channels of peaceful protest had been barred to us, that the decision was made to embark on violent forms of political struggle, and to form Umkhonto we Sizwe. *We did so not because we desired such a course, but solely because the Government had left us with no other choice. In the Manifesto of* Umkhonto *published on 16 December 1961 . . . we said:*

   The time comes in the life of any nation when there remain only two choices—submit or fight. That time has now come to South Africa. We shall not submit and we have no choice but to hit back by all means in our power in defence of our people, our future, and our freedom. (MANDELA 1990, PP. 22–24)

2. As of May 1992, 700 of the 2,100 once exclusively white schools have admitted an estimated 6,000 nonwhite students. This is a small number (about 9 students per school) in light of the fact that there are 7 million nonwhite students in South Africa (J. Berger 1992).

3. Most sociologists who have studied mobility focus on intergenerational mobility between fathers and sons with regard to occupation and income. On the basis of this research, mobility in the United States seems to follow this pattern: there is considerable upward mobility with regard to occupation, but most of this mobility is modest and occurs between occupational categories that are adjacent in rank and prestige (a professor's son becomes a lawyer, or the son of a construction worker becomes a restaurant manager). Most upward (and downward) mobility, however, can be attributed to changes in occupational structure such as the recent loss of manufacturing jobs and the gain in information and service jobs (Berger and Berger 1983).

4. For example, the Supreme Court's Dred Scott decision (1857) proclaimed that under the Constitution blacks possessed "no rights which whites were bound to respect." The Supreme Court also declared that the Missouri Compromise (which prohibited slavery in specified areas) was unconstitutional in that it deprived people of their right to own and manage their "property" without interference.

5. About a steal that basketball player Larry Bird made in game 5 of the playoffs in the 1987 Eastern Conference finals, Thomas said:

   It was a great basketball play. I didn't put enough zip on the ball, and Laimbeer didn't step up to get it. Meanwhile, this white guy on the other team who is supposed to be very slow, with little coordination, who can't jump, all of a sudden appears out of nowhere, jumps in, grabs the ball, leaps up in the air as he's falling out of bounds, looks over the court in the space of two or three seconds, picks out a player cutting for the basket, and hits him with a dead-bullet, picture-perfect pass to win the game. You tell me this white guy—Bird—did that with no God-given talent? (THOMAS 1987, P. D27)

6. South Africa is home to the National Party and the Conservatives Party, both of which are controlled by Afrikaners. The National Party, the architect of apartheid, is currently in power and is overseeing

the dismantling of the system it created. The Conservative Party opposes the end of apartheid. Some neo-Nazi organizations also exist, the largest of which is the Afrikaner Resistance Movement. In *After Apartheid: The Future of South Africa,* Sebastian Mallaby states that the reporters love to interview "grizzled Afrikaners, who breathed fire and slaughter and wore neo-Nazi garb. . . .

"They make such marvelous footage, these wild-eyed bully boys staring into the camera, swearing that they would fight de Klerk [to the end]" (1992, pp. 96–97). Unfortunately, as Mallaby points out, this kind of media attention only serves to exaggerate the influence that such groups have on South African politics.

10 RACE AND ETHNICITY

with Emphasis on Germany

Brandenburg Gate, Berlin.
Dave Bartruff/Artistry International

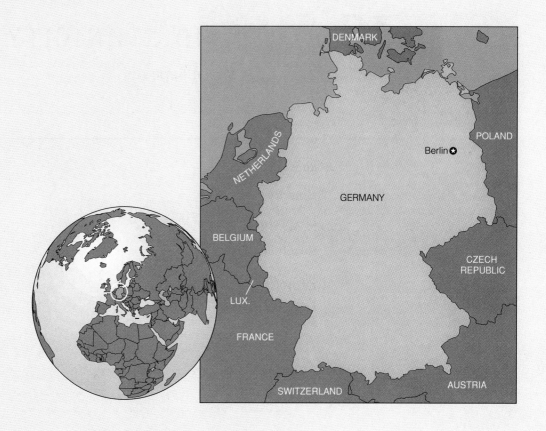

Students from Germany enrolled in college in the United States for fall 1991	7,570
Germans admitted into the United States in fiscal year 1991 for temporary employment	27,078
People living in the United States in 1990 who were born in Germany	1,163,000
Airline passengers flying between the United States and Germany in 1991	4,742,740
U.S. military personnel in Germany in 1993	109,888
People employed in 1990 by German affiliates in the United States	513,300
Applications for utility patents filed in the United States in 1991 by inventors from Germany	10,874
Phone calls made between the United States and Germany in 1991	106,091,000

Ender Bsaran is twenty-five now. He looks no different from most young Germans on his street. He has the latest German haircut—long in back, and short and spiky on top . . . with his haircut and his Pfalzisch Deutsch, and his baggy jeans and the pack of Marlboros sticking out of his shirt pocket, it's hard to think of him as "foreign." . . .

He likes to say that his family is "more of a European family," although his father goes to the mosque every day and sits with the men after prayers and listens to the gossip. . . . He wants you to know that his mother is nothing at all like the squat, ruddy Turkish women you see on the streets of Hemshof, wrapped in the bulky coats and gabardine head shawls that they brought from Turkey. He likes Turkey, but he doesn't think that either he or his mother belongs in Turkey. . . .

His father never intended to come to Germany. It was his uncle who got the job at Halberg, and the German work papers; his father "borrowed" the name and came instead. . . . His father is a small, humorous, courtly man, a family man—someone you would call a good German citizen if Turks who had lived and worked in Germany for thirty years were citizens.
(KRAMER 1993, P. 56)

On December 31, 1989, two months after the wall came down, two skinheads kicked Ender Bsaran to the sidewalk and jumped on his head with their steel-tipped [shoes] until they heard the bones crack.
(KRAMER 1993, P. 52)

I had spent more than an hour talking in my office at home with a reporter for *People* magazine. Her editor had sent her to do a story about me and how I was coping with AIDS. The reporter's questions had been probing and yet respectful of my right to privacy. Now, our interview over, I was escorting her to the door. As she slipped on her coat, she fell silent. I could see that she was groping for the right words to express her sympathy for me before she left.

"Mr. Ashe, I guess this must be the heaviest burden you have ever had to bear, isn't it?" she asked finally.

I thought for a moment, but only a moment. "No, it isn't. It's a burden, all right. But AIDS isn't the heaviest burden I have had to bear."

"Is there something worse? Your heart attack?"

I didn't want to detain her, but I let the door close with both of us still inside. "You're not going to believe this," I said to her, "but being black is the greatest burden I've had to bear."

"You can't mean that."

"No question about it. Race has always been my biggest burden. Having to live as a minority in America. Even now it continues to feel like an extra weight tied around me."
(ASHE AND RAMPERSAD 1993, P. 126)

nder Bsaran is the son of international labor migrants, people who move temporarily or permanently from one country to another for employment. Bsaran's father is one of 2.6 million migrants who came to West Germany between 1961 and 1973 as temporary workers. He came from Turkey, the homeland of many labor migrants in Germany.

American tennis champion Arthur Ashe was a descendant of international labor migrants who were forced to move from Africa to work on plantations in the United States. Ashe's ancestors were among the 50 million people imported as slaves to the Americas between 1619 and 1800 (Zinn 1980).[1]

Hundreds of books and articles have been written on the Turks who have lived in Germany for almost 40 years; thousands have been written on African-Americans, who have lived in the United States for hundreds of years. Many writers discuss why Turks or African-Americans as a group have not adapted, integrated, or assimilated into their respective host societies. In this chapter we examine this question. Specifically, we explore concepts of race, ethnicity, and assimilation and how people's ideas on these subjects have affected racial or ethnic relations on both an interpersonal and an institutional level. We give special attention to Turks in Germany and to African-Americans in the United States because they are the most visible minorities in each country. For the same reason we also give attention to Afro-Germans, a term that is "not immediately comprehended by most people who hear it. Virtually no Americans and indeed few Germans have ever heard the term at all" (Adams 1992, p. 234). This term applies to more than the "occupation babies" (babies born to white German mothers and black servicemen from the United States and from the African countries who fought in World War II). It also applies to the products of unions between men and women whose paths crossed as a result of the European colonization of the African continent. The concepts and theories in this chapter are relevant to understanding one important source of racial/ethnic tension: international labor migrants who are of a different race or ethnicity than the dominant population in the country to which they move or are forced to move (see "International Labor Migrants").

Although many Turkish people have been living in Germany for 40 years, they have not been assimilated into Germany society.
Regis Bossu/Sygma

Germany and the United States are two of the major labor-importing countries in the world.[2] Even when unemployment rates are high, both countries still admit and recruit skilled and unskilled workers. People in both countries debate constantly the role of labor migrants, especially during times of recession and high unemployment. Labor migrants are perceived as the "them" who take jobs from citizens, place strains on institutions (hospitals, schools, the welfare system, prisons), and engage in criminal activities. The debate almost always centers around labor migrants who are racially and ethnically different from the dominant population. For example, some immigration-related news stories in the United States portray immigrants as terrorists (Muslim immigrants masterminding the World Trade Center bombings), as des-

International labor migrants are very diverse. They can be contract labor migrants recruited in groups for specific projects; individual contract workers, often hired for indeterminate periods; highly skilled professional, managerial, and technical workers on short-term assignments or joint ventures; seasonal workers; au pairs; domestic servants; bar girls; entrepreneurs. Other migrants do not move primarily because of work, but subsequently enter the labor market and become part of the immigrant work force: family members rejoining a primary labor migrant, foreign students who work temporarily during their courses or who stay on permanently after qualification; asylum seekers who entered the labor market unofficially before acceptance, and who stay and work illegally despite rejection of their claims; refugees who enter the labor market legally after acceptance; and those who come as tourists, stay beyond their legal limit, and enter labor markets.

SOURCE: Adapted from "The Future of International Labor Migration," by John Salt. P. 1078 in *International Migration Review* (1992).

---

perate (Chinese immigrants who paid $30,000 each to be smuggled by boat into the United States), and as economic refugees posing as political refugees (the Haitian boat people). The media direct very little public attention toward the 300,000 skilled workers (managers, scientists, executives) who enter the United States on nonimmigrant visas and stay for indefinite periods, or toward a provision written into the Immigration Act of 1990 that allows 40,000 people from 34 countries—mostly European countries[3]—to enter the United States each year for three years (Keely 1993; Mydans 1991).

The underlying message of the news stories that make the headlines is that the United States cannot control its borders and that it is being flooded by immigrants who are racially and ethnically different from the dominant population. This concern about controlling the borders to keep out the racially/ethnically different is not unique to the United States or to the mid-1990s. Every country in the world has policies stating how many outsiders it will admit. A country that paid no attention to such matters would be overwhelmed—its culture engulfed and its citizens outnumbered by an influx of people greater than its population (M. Weiner 1990).

All labor-importing countries must determine how many foreign workers to recruit and allow into their country, but almost no country makes systematic plans to help foreigners adjust to life in their new environment once they have arrived. Adjustment almost always is a one-sided process in which labor migrants are left on their own to adapt to the society (Opitz 1992a). Perhaps by exploring the experiences of Turks and Afro-Germans in Germany we can gain insight into the intentional, unintentional, and even unconscious barriers that prevent labor migrants and their families from adapting, integrating, and assimilating into the host country. Perhaps hindsight will help us create policies that might facilitate adjustment (Hopkins 1993).

We pay special attention to Germany because it is the most economically powerful country in the European Union. Germany also is dependent on foreign labor. This demand is unlikely to decrease in the future because Germany has one of the lowest birth rates in the world and an aging population (Hoffmann 1992). Yet, despite its dependence on foreign labor, Germany is officially a nonimmigration country; its constitution "does not seek to increase its national population through naturalization of foreigners" (Holzner 1982, p. 67). The foreigners who live in Germany, even those who have lived and worked in the same town for 20 years and those who were born there and know only the German language, are considered guests (Thränhardt 1989). Under German law, people who can prove German ancestry are entitled to citizenship regardless of their country of origin. It does not matter if they cannot speak German or if they know nothing about German culture; the criterion for German citizenship is biological. This law makes life especially difficult for those who have lived in Germany all their lives but who are not biologically German or who do not look German. Afro-Germans fit into this latter category.

On the one hand, Afro-Germans qualify for citizenship because they are of German ancestry. On the other hand, they are perceived to be foreign because they have dark skin and other physical features not associated with German ancestry. Comments from some Afro-German women illustrate their marginal status:

> *I'm German, and I'm dark. But then not all that dark either. I've often looked in the mirror and asked myself what distinguishes me, what makes me so different in the eyes of others. Inside I am German because of my German environment, school, my home—just German. And yet it was always made clear to me that is exactly what I am not. But why? It's all based on externalities.* (WIEDENROTH 1992, P. 165)

> *Everywhere the same thing—in the job market, in the search for a place to live. Always I have to identify myself, prove, two times, three times, that I'm German, prove my right to exist. "Oh, we thought you were a foreigner." Foreigners are different; they are singled out; they are—as I said—"also-people."* (P. 167)

> *Given the black-white matrix in people's minds—you are placed on the nonwhite side and you are classified as an "also-person." After all, Blacks are "also" people.* (P. 166)

> *My color is black, therefore I'm perceived as a foreigner—African or American. I'm always being asked how come I speak German so well, where I come from, etc. This quizzing gets on my nerves. Most of the time I answer provocatively that I'm German. In spite of my unequivocal answer they continue: How? Why?* (ADOMAKO 1992, P. 199)

Before we examine the sociological concepts and theories that help us see the connection between labor importation and racial or ethnic relations, we will examine Germany's history as a labor-importing country since the end of World War II.

German Guest Workers and the Question of Race or Ethnicity

At the end of World War II, Germany's economy, industries, roads, cities, public facilities, and houses lay in ruins. The victorious Allies (Britain, France, the Soviet Union, and the United States) divided the country into four occupied zones. The Soviet-occupied zone became East Germany; the British-, French-, and U.S.-occupied zones became West Germany (see Figure 10.1). Germany's capital city, Berlin, lies in the east. The city was also divided into occupied zones.

Between 1945, when the war ended, and 1961, three million refugees from East Germany and nine million from former German territories migrated to West Germany.[4] These migrants provided the labor needed to offset the labor loss due to war casualties and to fuel postwar economic reconstruction. In 1961, when the East-West flow of migrants reached an average of 3,000 per day, the Soviet communists erected the Berlin Wall—103 miles of heavily guarded 10-foot-high steel-fortified concrete—to seal off East Berlin from West Berlin and to stop the flow of people from East to West (McFadden 1989). In addition, the Soviets helped construct fortified fences and walls along the 860-mile border that separated East Germany from West Germany.

When the border was sealed, West German employers lost a major labor pool. To make up for the shortage, the German government established labor recruitment offices in Turkey, Yugoslavia, Italy, Greece, Portugal, and Spain. From these offices, officials screened male and female job applicants. Those who possessed the needed occupational

FIGURE 10.1   Post—World War II Germany

Pending the reunification of Germany, its pre-war capital, situated in the German Democratic Republic (East Germany), remained under four-power Allied occupation. In August 1961 the Berlin Wall was built, dividing the Soviet from the three Western sectors.

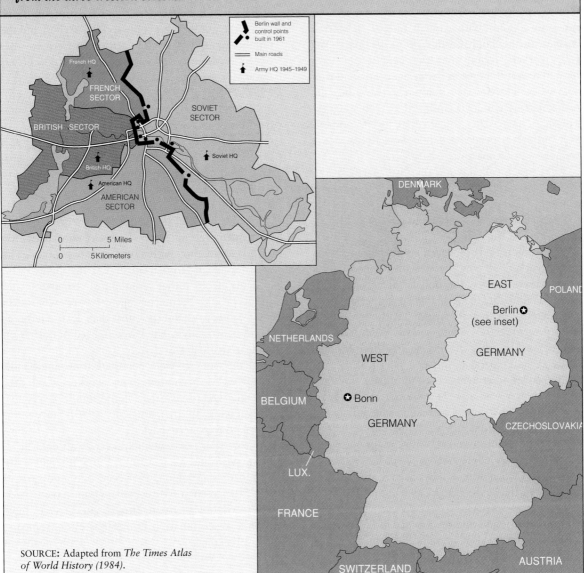

SOURCE: Adapted from *The Times Atlas of World History* (1984).

skills, had no police records, and enjoyed good health were admitted to West Germany as guest workers to live in employer-provided worksite housing. Between 1961 and 1973, some 2.6 million guest workers were admitted into West Germany on the assumption that they were temporary and would eventually return to their home countries (Castles 1986). Officials from the labor-exporting countries agreed to this arrangement because it helped lower their high unemployment rates; it created revenue because migrant workers sent home a portion of their earnings; and it gave workers an opportunity to acquire skills and training that would benefit their countries on their return (Sayari 1986). The recruits accepted guest worker status because they believed they could work in West Germany for a few years, save some money, and return home with enough money to start a business or buy land (Castles 1986). In time, however, workers found that they could not save enough to return home in better financial condition than when they had left. Temporary stays became permanent as employers lobbied to extend guest worker permits. They argued that to be competitive, businesses needed a stable work force, not a temporary and constantly rotating one.

After the so-called world oil crisis in 1973,[5] the German government stopped recruiting guest workers and initiated policies, including offers of financial incentives, to encourage guest workers to return home. The number of foreigners in West Germany increased, however, as family members joined workers already there. With the end of the guest worker system, the only legal way for outsiders to immigrate to Germany was to declare German ancestry or to seek asylum.[6] Although only a small percentage of asylum seekers eventually are granted asylum, they all receive free housing and meals at state-operated shelters, plus a small stipend to cover other living costs, until their cases are heard, often years later. In the end, the German government allows many to remain indefinitely for humanitarian reasons (for example, until a war is over in their home country), even though their applications for asylum have been rejected.

In 1988, when Mikhail Gorbachev was named president of the Soviet Union, he instituted policies of *glasnost* ("openness"), *perestroika* ("economic restructuring"), and noninterference in Eastern Europe. Gorbachev also announced plans to withdraw a half-million troops and 10,000 tanks from Eastern Europe over two years. These policies led to the overthrow of hard-line communist regimes in Poland, East Germany, Czechoslovakia, Bulgaria, and Romania. (In addition, they led to the breakup of the Soviet Union into separate and independent countries.) On November 9, 1989, the Berlin Wall was dismantled. The subsequent reunification of East Germany (with a population of 17 million)

with West Germany (with a population of 63 million) followed less than a year later, on October 3, 1990. Reunification has been costly, accounting for at least $100 billion. Projected costs are estimated to be $65 billion a year.

The political changes in the former Soviet Union and Eastern Europe since 1988 have profoundly affected the number of people seeking asylum and the number of ethnic Germans settling in Germany. Between 1988 and 1992, more than one million requests for asylum were made (Kinzer 1993; Rogers 1992). Between 1990 and 1991, over 600,000 ethnic Germans from countries such as Poland, Romania, and the former Soviet Union settled in Germany (U.S. Bureau for Refugee Programs 1992). Ethnic Germans are automatically entitled to citizenship even though most have not been exposed to German culture and do not speak German (Thränhardt 1989).

Until the fall of the Berlin Wall in 1989 and the subsequent reunification, East Germans had little or no interaction with non-German peoples. The only foreigners who lived in East Germany were students from other communist countries and about 190,000 conscript workers who worked in heavy industry and in jobs that East Germans found unattractive. The workers lived in hostels on or near worksites and were subject to strict rules: pregnant workers were deported, only single individuals were permitted to be workers, and they were allowed to visit their home country every five years (Heilig, Büttner, and Lutz 1990). Since unification, the eastern part of Germany has had to admit its share of asylum seekers awaiting a hearing. Few studies have examined how foreigners are treated in this region. However, one study found that 70 percent of foreigners report having been insulted by a German citizen, 40 percent have suffered discrimination in business establishments, and 20 percent have been attacked physically in some way (Wilpert 1991).

Today, united Germany's population is 80 million. This figure includes 5.3 million foreigners, most of whom live in large cities (Strasser 1993). The most numerous, poorest, and most visibly different foreigners are the Turks (including 400,000 Kurds), who make up one-third of the foreign-born population (Jones and Pope 1993; Martin and Miller 1990). The Turks, along with North

UPI/Bettmann

Reuters/Bettmann

In 1961, Communist authorities decided to stop the flow of people from the East to the West by erecting the Berlin Wall (left). Nearly 30 years later, demonstrators were eager to have a hand in taking down the infamous wall while East German border guards looked on placidly (right).

Africans, sub-Saharan Africans, Pakistanis, and Persians (virtually all of whom are Muslims) are considered "rejected foreigners" and are targets of prejudice and discrimination (Safran 1986; see Table 10.1).

In view of the monetary costs of reunification and the other political changes in Eastern Europe and the former Soviet Union, many German citizens (55 percent) believe that too many foreign people live in or want to enter their country (Protzman 1993; Riding 1991). Recently, German politicians tightened the asylum laws, making it more difficult for economic refugees to claim political asylum. In addition, the German government is negotiating with several governments to take back refugees, to keep ethnic Germans in their home countries, or to tighten their borders (Shearer

1993). For example, Germany is financing German ethnic communities in Russia to make life more attractive for ethnic Germans living there (Erlanger 1993). It is paying the Romanian government $21 million to take back 43,000 Romanians, 60 percent of whom are Gypsies (Green 1992). Also, the German government is paying Poland and Czechoslovakia to tighten their borders.

Germany's efforts to control immigration must be viewed in the context of the then European Community's (now called the European Union) drive to coordinate its member countries' immigration policies. In December 1992, the 12 EC governments dismantled border controls, thus permitting their 320 million citizens to move freely across the borders of member countries. It is now possible, for example, to cross from Germany into France without showing

TABLE 10.1	Attitudes Toward Foreigners of Various Ethnic Groups Who Live in Germany

Erwin Scheuch classifies the German views of foreigners into four categories and makes no mention of Austrians, Swiss, and Dutch. The people of all three countries live in Germany in large numbers. They are not highly visible and might be classified as "half foreigners" in the public mind.

Data from various fields show the same picture. Turks, for instance, have been presented most unfavorably in the media. This also includes a type of "goodwill" news which shows them as different from the Germans and not on the level of Western civilization. There are also systematic differences among migrant groups concerning occupation, housing, concentration, and educational achievement. Among migrant groups "the desire for closer contacts (with Germans) is strongest among Turks, but it meets with the strongest rejection by Germans.*

Category	Nationality of Foreigner	Attitude Toward Foreigner
1. Noble foreigners	British, French, Americans, Swedes	Positive
2. Foreigners	Spaniards, Yugoslavs, Greeks	Neutral
3. Strange foreigners	Portuguese, Italians, Vietnamese	Neutral, with tendency toward negative
4. Rejected foreigners	North Africans, Black Africans, Pakistani, Persians, Turks	Rejected by substantial parts of the population

*Ursula Neuman, *Erziehung ausländischer Kinder* (Düsseldorf, 1980), 90.

SOURCE: Adapted from "Patterns of Organization Among Different Ethnic Minorities," by Dietrich Thränhardt. P. 13 in *New German Critique* 46 (1989).

a passport. Because of free movement across borders, one country's immigration policies affect all of the countries (Marshall 1992a; M. Weiner 1992).

Citizenship and immigration status in Germany are based on race and/or ethnicity. As mentioned earlier, Turks and other non-German people, even those who have lived in Germany all their lives and are more German than Turkish in culture, are considered guests or temporary workers. People of German heritage, however, no matter where they grew up or what language they speak, are automat-ically German citizens. There are some exceptions to this rule. For example, a German Information Center publication points out: "If one parent is a German citizen, a child born in wedlock is a German citizen. A child born out of wedlock is German if the mother is German, this is true whether the child is born in Germany or in another country" (German Information Center 1993, p. 1). On the surface, the criteria for citizenship and immigration status seem clear-cut. Yet, in reality race and ethnicity are not clear-cut categories.

Race and Ethnicity

In Chapter 9 we learned that race is a biology term that refers to a group of people who possess certain distinctive and visible physical character-istics. **Ethnicity** refers to people who share a common national origin, ancestry, distinctive and visible cultural traits (religious practice, dietary

Examples of Questions to Determine Ethnicity

1. To what ethnic or cultural groups did you or your ancestors on the male and the female side belong? If parents' ethnicities are different, which of these two different ethnic or cultural backgrounds do you think is more important?

2. What language did you learn first in childhood and still are able to understand?

3. While you were growing up, was your parents' ethnic or cultural background very important to them? Was your ethnic or cultural background very important to you?

4. How would you describe the people who live in this neighborhood? Is their ethnic or cultural background mostly the same as your own?

5. Have you ever been discriminated against because of your ethnic or cultural background? Give an example.

6. How often do you do the following—frequently (1), fairly often (2), sometimes (3), very rarely (4), never (5):
   a. Attend ethnic-group dances, parties, or informal social affairs?
   b. Attend non-ethnic group dances, parties, or informal social affairs?
   c. Go to ethnic-group vacation resorts, summer camps, etc.?
   d. Eat any food that is associated with your ethnic group?
   e. Listen to ethnic-language radio broadcasts?
   f. Read ethnic-group newspapers, magazines, or other periodicals?

7. How interested are you in local issues or events that do not pertain to your ethnic group?

SOURCE: Adapted from "Appendix B, The Interview Schedule." Pp. 279–311 in *Ethnic Identity and Equality: Varieties of Experience in a Canadian City,* by Raymond Breton, et al. (1990).

---

habits, style of dress, body ornaments, or language), and/or socially important physical characteristics. A person's race is based on biological traits; a person's ethnicity is based on an almost countless number of traits including history, family name, birthplace, ancestry, language, food preferences, socialization, residence, self-image, sense of distinctiveness, physical characteristics, shared tradition, and cultural practices (Waters 1991).

When sociologist Raymond Breton and his colleagues studied ethnicity in Toronto, they asked respondents 167 questions to determine their ethnicity (Breton, Isajiw, Kalbach, and Reitz 1990). A small sample of questions from their survey shows that no single indicator can be used as a mark of ethnicity (see "Examples of Questions to Determine Ethnicity"). For example, if we know that someone comes from Turkey, it does not follow that he or she is a Turk in an ethnic, linguistic, or religious sense. Turkey's population includes 3 to 10 million Kurds, Arabs (who migrated from Egypt and North Africa in the 1930s), Armenians, Greeks, and Jews. Even the so-called Turkish population contains at least three distinct regional groups (Anatolian Turks, Rumelian Turks, and Central Asian Turks).[7]

In Chapter 9 we also learned that racial classification schemes have many shortcomings. The shortcomings become evident when we imagine using the schemes to classify the 5.6 billion people in the world. First, many people simply do not fit into any category, because no sharp dividing lines exist to distinguish such features as black skin from white or curly hair from wavy. Second, millions of people in the world have mixed ancestry and possess the physical traits of more than one race.

If racial classification schemes based on a single indicator—highly visible biological attributes—are problematic, imagine the difficulties facing ethnic classification schemes when at least 167 indicators might be considered. What answers might a person give to the 167 questions in Breton's survey to qualify as a member of a clear-cut ethnic category? If one considers how these questions might be answered by Turks such as Ender Bsaran who have lived in Germany all of their lives, and by people of German ancestry who have lived outside Germany, one must conclude that many Turks in Germany are more German than people of German heritage

East Meets West

The following excerpt from *The New York Times Magazine* points out the inevitable strains between East and West Germans that arise from a 45-year separation. It also raises the question of whether Turks who have lived in West Germany all their lives have more in common with West Germans than do East Germans.

It has been 40 years. I think it is time to stand back from this social experiment that we call the divided Germany and assess the results.

True, an experiment was the last thing the allies had in mind when they agreed to that border running through the middle of Europe. Call it an involuntary experiment, then, born of the pressures of victory, in which the allies functioned as principal researchers and the Germans as extraordinarily cooperative white mice.

Perhaps it would clarify matters if I mention a related scientific area, one in which I have some expertise as an amateur; I am speaking of research on twins.

Let us assume we are dealing with two hell-raising twins who share a criminal past. Through allied efforts, they are finally forcibly separated and sent to two extremely different boarding schools. One twin grows up in the bracing climate of Western values; first with difficulty and then with growing enthusiasm, he learns to appreciate as basic values democracy, capitalism and individual freedom; and he develops great respect for the Western principal researcher.

The other twin has quite a different fate. He is often beaten and brutalized, and finally learns just as assiduously the basic values of Eastern culture: "solidarity," "social commitment," "passion for socialism," and of course "eternal friendship" for the Eastern principal researcher.

Let us further assume that a wall is constructed between the twins and an odd system of visitation rights is established. The Western twin can move in any direction he likes, including east; he can visit his brother on the other side, chat with him and compare experiences, bring him presents, then return to the Western half to have dinner in a French restaurant.

The Eastern twin, on the other hand, has some freedom of movement north and south, and to the east has access to an almost unlimited recreational area (which, however, has only recently come to be considered a place worth visiting). But access to the West is blocked. There is the famous forbidden door, a good 1,400 kilometers wide. It still opens legally only in exceptional cases, and after a wait of at least two years; anyone who does not wish to wait that long must take his life in his hands and jump, or dig.

Let us further assume that the twin in the West, thanks to assistance such as the Marshall Plan and the Western market economy, gradually gets rich. His twin brother, meanwhile, not only has to pay war debts to the far poorer principal researcher in the East; he also has to adopt the researcher's inefficient economic system.

At least one result we can safely predict—the twin in the East will fall victim to a psychological law. Every wall in the world, whether German or Chinese, begs to be overcome. Also, because he finds himself in the awkward situation of having to wait for his Western sibling to come to him, a certain reproachful attitude forms. "That guy over there really could come to see me more often," thinks the Eastern twin. "At the very least, he could write or phone regularly. And he could be a bit more generous, because it's turned out that he has the better deal. Not that he has done anything to *deserve* it, by the way; it was pure luck that he happened to be living on the right side of the Elbe at the right time."

who did not grow up in Germany. In fact, Turks living in West Germany may have more in common with West Germans than do East Germans (see "East Meets West").

Even though racial and ethnic classification schemes are problematic, many people argue that they know a white, black, Asian, or Arabic person when they see one. If they meet someone who does not fit the image, they state, "But you don't look like someone of (African, Asian, or Arabic) descent." That's because their vision of human variety is limited by the images of people portrayed on airline posters, magazine ads, and television sitcoms (Houston 1991).[8] Sometimes people will go to ex-

"But now his success has gone to his head. Instead of sharing, he acts as though he had all the talent and claims he works harder. He can't fool me; I know him, we started out under the same roof, and he's just as industrious and just as lazy as I am. He's gotten pretty arrogant, even self-righteous. Actually, he's still living off the misery of the world's poor, whom he exploits, but he doesn't want to hear that.

"Well, if he doesn't have a conscience, at least he could show a little family feeling, a little interest in his relatives in the East, at least listen to us—is that asking too much? He's gained an awful lot of weight, by the way; even if he is rich, that doesn't look like happiness to me."

Meanwhile, the Western twin is concocting a monologue of his own. He feels pressured by what he calls his relative's "eternal posture of expectation."

"I can see that the poor guy behind his wall doesn't have an easy time of it," thinks the Western twin. "But these demands, these unspoken reproaches really cast a pall over our relationship. God knows I'm happy to give him things, but it's no fun bringing a present when the other person always expects you to. Those people over there seem to think that cars and color televisions grow on trees. But we're not *born* with a Mercedes; you have to earn it; you have debts, interest payments—concepts my brother knows only by hearsay. . . .

"I'd like to explain this, but he won't listen; he just talks and talks. Of course, it's not his fault that he always has to stand in line to buy oranges; but at least he should admit that he bet on the wrong horse, that the socialist economic system is a disaster—no one's criticizing him personally. The problem is—and it's so *German* of him—he takes any kind of criticism personally. Instead of agreeing with me, he tries to spoil my success. He claims to be an idealist—I'm glad to let him call himself that, because the poor fellow has a lot to compensate for. . . . But he accuses me of being a conspicuous consumer and a 'conformist'—a compliment I'm glad to return by pointing out that his so-called 'socialist' or even 'revolutionary' virtues are all a pose; I know him, after all.

"Sometimes I feel downright relieved when the visit is over; there's an unpleasant tension between us that we really should talk about some day. Next time."

And what happens when the Eastern twin comes west? Every immigrant is expected to declare allegiance to the Western way of life, of course, but I can hardly think of anyone to whom crossing the border has not brought culture shock. Far from softening this effect, the common language exacerbates it, because it simulates a commonality that daily experience does not bear out.

The Western twin finds it hard to understand the complaints of his Eastern brother, now safely "back home." He's never satisfied, he always finds something to criticize. What's all this whining about how "cold" it is in the West, about the lack of "real friendship" and "coziness"? Suddenly, the Western twin begins to suspect that his brother has taken on a whole raft of official socialist virtues, even when he claims to be a sworn anti-Communist; after three beers, this difficult relative begins to dream of being back behind the wall, where a word was a word and the promise "I'll give you a call" really meant something.

SOURCE: "If the Wall Came Tumbling Down," by Peter Schneider, pp. 27, 61–62 in *The New York Times Magazine* (June 25). Copyright © 1989 by The New York Times Company. Reprinted by permission.

tremes to create an ethnic and racial group that fits an ideal image.

Sociologist Paul D. Starr (1978) found that in the absence of distinctive skin color and other physical characteristics, people determine ethnicity on the basis of any number of other imprecise attributes—language, dress, type of jewelry, tattoos and other body scars, modes of expression, and residence. In determining another person's race and ethnicity, people rarely think beyond the most visible characteristics. Most people fail to consider the details of another person's life (see "Hispanic as a Social Category").

If race and ethnicity are such vague categories,

Hispanic as A Social Category

Hispanics are those individuals whose declared ancestors or who themselves were born in Spain or in the Latin American countries. Until recently, this rubric did not exist as a self-designation for most of the groups so labeled; it was essentially a term of convenience for administrative agencies and scholarly researchers. Thus, the first thing to be said about this population is that it is not a consolidated minority, but rather a group-in-formation whose boundaries and self-definitions are still in a state of flux. The emergence of a Hispanic "minority" has so far depended more on actions of government and the collective perceptions of Anglo-American society than on the initiative of the individuals so designated.

The principal reason for the increasing attention gained by this category of people is its rapid growth during the last two decades which is, in turn, a consequence of high fertility rates among some national groups and, more importantly, of accelerated immigration. In addition, the heavy concentration of this population in certain regions of the country has added to its visibility. Over [70 percent] of the [22.4] million people identified by the [1990] Census as Hispanics are concentrated in just four states—California, New York, Texas, and Florida; California alone has absorbed [over] one third (U.S. Bureau of the Census [1993]).

The absence of a firm collective self-identity among this population is an outcome of its great diversity, despite the apparent "commonness" of language and culture which figures so prominently in official writings. Under the same label, we find individuals whose ancestors lived in the country at least since the time of independence and others who arrived last year; we find substantial numbers of professionals and entrepreneurs, along with humble farm laborers and unskilled factory workers; there are whites, blacks, mulattoes, and mestizos; there are full-fledged citizens and unauthorized aliens; and finally, among the immigrants, there are those who came in search of employment and a better economic future and those who arrived escaping death squads and political persecution at home.

Aside from divisions between the foreign and the native-born, there is no difference of greater significance among the Spanish-origin population than that of national origin.

Nationality does not simply stand for different geographic places of birth; rather it serves as a code word for the very distinct history of each major immigrant flow, a history which molded, in turn, its patterns of entry and adaptation to American society. It is for this reason that the literature produced by "Hispanic" scholars until recently has tended to focus on the origins and evolution of their own national groups rather than to encompass the diverse histories of all those falling under the official rubric.

The bulk of the Spanish-origin population—at least 60%—is of Mexican origin, divided into sizable contingents of native-born Americans and immigrants. Another [12 percent] come from Puerto Rico and are U.S. citizens by birth, regardless of whether they were born in the island or the mainland. The third group in size is made up of Cubans who represent about 5% and who are, overwhelmingly, recent immigrants coming after the consolidation of a communist regime in their country. These are the major groups, but there are, in addition, sizable contingents of Dominicans, Colombians, Salvadoreans, Guatemalans, and other Central and

perhaps the most appropriate definition of a racial or ethnic group is that its members believe (or outsiders believe) that they share a common national origin, cultural traits, or distinctive physical features. It does not matter whether this belief is based on reality. The point is that membership in an ethnic or racial group is a matter of social definition, an interplay of self-definition and others' definitions.

Sociologists are interested in racial and ethnic classification schemes that incorporate the assumption that specific abilities (athletic talent, intelligence), social traits (criminal tendencies, aggressiveness), and cultural practices (dress, language) are transmitted genetically. This assumption underlies the belief that some categories of people are inferior to others because of their race and/or ethnicity and therefore can be denied equal access to scarce and valued resources such as citizenship, decent wages, wealth, income, health care, and education. In other words, sociologists are interested in the experiences of minority groups.

A Salvadoran immigrant and his daughter.

Fred Chase/Impact Visuals

South Americans, each with its own distinct history, characteristics, and patterns of adaptation (Nelson & Tienda 1985, U.S. Bureau of the Census 1983, U.S. Immigration and Naturalization Service 1984, [U.S. Bureau of the Census 1993]).

The complexity of Hispanic ethnicity is a consequence, first of all, of these diverse national origins which lead more often to differences than similarities among the various groups. Lumping them together is not too dissimilar from attempting to combine turn-of-the-century Northern Italian, Hungarian, Serbian, and Bohemian immigrants into a unit based on their "common" origin in various patches of the Austro-Hungarian empire. A second difficulty is that most Spanish-origin groups are not yet "settled," but continue expanding and changing in response to uninterrupted immigration and to close contact with events in the home countries. This dense traffic of people, news, and events between U.S.-based immigrant communities and their not-too-remote place of origin offers a far more challenging landscape than, for example, the condition of European ethnic groups whose boundaries are generally well defined and whose bonds with the original countries are increasingly remote (Glazer 1981, Alba 1985).

Alba, R. D. 1985. *Italian Americans: Into the Twilight of Ethnicity.* Englewood Cliffs, NJ: Prentice-Hall

Glazer, N. 1981. Pluralism and the new immigrants. *Society* 19(Nov–Dec):31–36

Nelson, C., Tienda, M. 1985. The structuring of Hispanic ethnicity: Historical and contemporary perspectives. *Ethnic Racial Stud.* 8(Jan): 49–74

U.S. Bureau of the Census. 1983a. *General Population Characteristics, United States Summary.* Washington, DC: USGPO

[U.S. Bureau of the Census. 1993. *Statistical Abstract of the United States: 1993,* 113th ed. Washington, DC: USGPO]

U.S. Immigration and Naturalization Service. 1984. *Annual Report.* Washington, DC: USGPO

SOURCE: Updated and adapted from "Making Sense of Diversity: Recent Research on Hispanic Minorities in the United States," by Alejandro Portes and Cynthia Truelove. Pp. 359–61 in *Annual Review of Sociology* 13. Copyright © 1987 by Annual Reviews, Inc. Reprinted by permission.

Minority Groups

Minority groups are subgroups within a society that can be distinguished from members of the dominant groups by visible and identifying characteristics, including physical and cultural attributes. Members of such subgroups are regarded and treated as inherently different from those in dominant groups. For these reasons they are systematically excluded, consciously or unconsciously, from full participation in society and are denied equal access to positions of power, prestige, and wealth. Thus, members of minority groups tend to be concentrated in inferior political and economic positions and isolated socially and spatially from members of the dominant groups.

On the basis of these characteristics, many groups can be classified as minorities, including some racial, ethnic, and religious groups, women, the very old, the very young, and the physically different (for example, visually impaired people or overweight people). Although we focus on ethnic and racial minorities in this chapter, the concepts that follow can be applied to any minority.

Sociologist Louis Wirth (1945) made a classic statement on minority groups, identifying a number of essential traits characteristic of all minority groups. First, membership is involuntary: as long as people are free to join or leave a group, no matter how unpopular the group, they do not by virtue of that membership constitute a minority. This first trait is quite controversial because the meaning of "free to join or leave" is unclear. For example, if a very light-skinned person of African and German descent can pass as German, is he or she "free" to leave the African connections in his or her life? Second, minority status is not necessarily based on numbers; that is, a minority may be the majority of the people in a society. The key to minority status, then, is not size but access to and control over valued resources. South Africa is an obvious example: roughly 85 percent of the population (Asian Indians, blacks, and people of mixed race) are controlled by the 14 percent of the population who are white. A third characteristic, and the most important, is nonparticipation by the minority group in the life of the larger society. That is, minorities do not enjoy the freedom or the privilege to move within the society in the same way as members of the dominant group. Sociologist Peggy McIntosh identifies a number of privileges that most members of the dominant group take for granted, including the following:

- *I can, if I wish, arrange to be in the company of people of my race [or ethnic group] most of the time.*

- *If I should need to move, I can be pretty sure of renting or purchasing housing in an area which I can afford and in which I would want to live.*

- *I can go shopping alone most of the time, fairly well assured that I will not be followed or harassed by store detectives.*

- *I can be late to a meeting without having the lateness reflect on my race [or ethnicity].*

- *Whether I use checks, credit cards, or cash, I can count on my skin color not to work against the appearance that I am financially reliable.* (1992, PP. 73–75)

The final and most troublesome characteristic of a minority group is that people who belong to such a group are "treated as members of a category, irrespective of their individual merits" (Wirth 1945, p. 349) and often irrespective of context. In other words, people outside the minority group focus on the visible characteristics that identify someone as belonging to a minority. This visible characteristic becomes the focus of interaction:

"It obsesses everybody," declaimed my impassioned friend, "even those who think they are not obsessed. My wife was driving down the street in a black neighborhood. The people at the corners were all gesticulating at her. She was very frightened, turned up the windows, and drove determinedly. She discovered, after several blocks, she was going the wrong way on a one-way street and they were trying to help her. Her assumption was they were blacks and were out to get her. Mind you, she's a very enlightened person. You'd never associate her with racism, yet her first reaction was that they were dangerous." (TERKEL 1992, P. 3)

The characteristics that Wirth identifies as associated with minority group status indicate that minorities stand apart from the dominant culture. Some people argue that minorities stand apart because they do not wish to assimilate into mainstream culture. To assess this claim, we turn to the work of sociologist Milton M. Gordon, who has written extensively on assimilation.

Perspectives on Assimilation

Assimilation is a process by which ethnic and racial distinctions between groups disappear. There are two main types of assimilation. One form is absorption assimilation. The other form is called melting pot assimilation.

Absorption Assimilation

In this form of assimilation, members of a minority ethnic or racial group adapt to the ways of the dominant group, which sets the standards to which they must adjust (Gordon 1978). According to Gordon, this type of assimilation has at least seven levels. That is, an ethnic or racial group is completely "absorbed" into the dominant group when it goes through all of the following levels:

1. The group abandons its culture (language, dress, food, religion, and so on) for that of the dominant group (an action known as acculturation).

2. The group enters into the dominant group's social networks and institutions (structural assimilation).

3. The group intermarries and procreates with those in the dominant group (marital assimilation).

4. The group identifies with the dominant group (identification assimilation).

5. The group encounters no widespread prejudice from members of the dominant group (attitude receptional assimilation).

6. The group encounters no widespread discrimination from members of the dominant group (behavior receptional assimilation).

7. The group has no value conflicts with members of the dominant group (civic assimilation).

Gordon advances a number of hypotheses about how the various levels of assimilation relate to one another. First, he maintains that acculturation is likely to take place before the other six levels of assimilation. Gordon also maintains, however, that even if acculturation is total (as in the cases of Ender Bsaran and some Afro-Germans), it does not always lead to the other levels of assimilation.

Gordon proposes that a clear connection exists between the structural and marital levels of assimilation. That is, if the dominant group permits people from ethnic and racial minority groups to join its social cliques, clubs, and institutions on a large enough scale, a substantial number of interracial or interethnic marriages are bound to occur:

If children of different ethnic backgrounds belong to the same play-group, later the same adolescent cliques, and at college the same fraternities and sororities; if the parents belong to the same country club and invite each other to their homes for dinner; it is completely unrealistic not to expect these children, now grown, to love and to marry each other, blithely oblivious to previous ethnic extraction. (PP. 177–78)

Of the seven levels of assimilation, Gordon believes the structural level is the most important because if it occurs, the other levels of assimilation inevitably will follow. Yet, even though the structural level is the most important, it is very difficult to achieve in practice. Members of ethnic or racial minorities, such as Turks, Afro-Germans, and African-Americans, are denied easy and comfortable access to the dominant group's networks and institutions. In fact, all of the important and meaningful primary relationships that are close to the core of personality and selfhood are confined largely to people of the same racial or ethnic group:

From the cradle in the sectarian hospital to the child's play group, the social clique in high school, the fraternity and religious center in college, the dating group within which he [or she] searches for a spouse, the marriage partner, the neighborhood of his [or her] residence, the church affiliation and the church clubs, the men's and the women's social and service organizations, the adult clique of "marrieds," the vacation resort, and then, as the age cycle nears completion, the rest home for the elderly and, finally, the sectarian cemetery—in all these activities and relationships which are close to the core of personality and selfhood—the member of the ethnic group may if he wishes follow a path which never takes him across the boundaries of his [or her] ethnic structural network. (P. 204)

This scenario especially characterizes the primary group relations of **involuntary minorities,** ethnic and racial groups that did not choose to be a part of a country. These groups were forced to become part of a country by slavery, conquest, or colonization. Native Americans, African-Americans, Mexican-Americans, and native Hawaiians are examples of involuntary minorities. Unlike **voluntary minorities,** whose members come to a country expecting to improve their way of life, members of involuntary minorities have no such expectations.

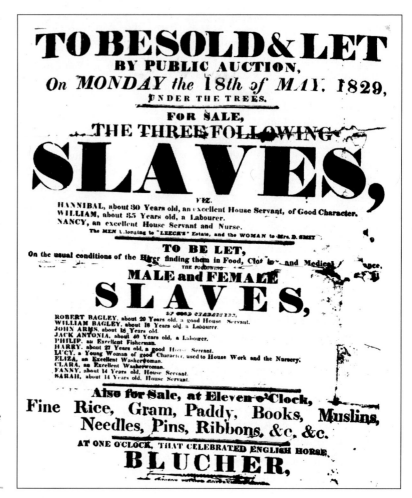

Throughout history, members of ethnic or racial groups have become involuntary minorities through conquest, annexation, or slavery.

The Bettmann Archive

Their forced incorporation involves a loss of freedom and status (Ogbu 1990).

Melting Pot Assimilation

Assimilation need not be a one-sided process in which a minority racial and ethnic group disappears, or is absorbed, into the dominant group. Ethnic and racial distinctions also can disappear in another way known as **melting pot assimilation** (Gordon 1978). In this process, the groups involved accept many new behaviors and values from one another. This exchange produces a new cultural system, which is a blend. Melting pot assimilation is total when significant numbers of people from each ethnic and racial group take on cultural pat-

terns of the other, enter each other's social network, intermarry and procreate, and identify with the blended culture.

The melting pot concept can be applied to the various African ethnic groups imported to the United States as slaves. They were "not one but many peoples" (Cornell 1990, p. 376), who spoke many languages and came from many cultures. Slave traders capitalized on this diversity: "Advertisements of new slave cargoes frequently referred to ethnic origins, while slaveowners often purchased slaves on the basis of national identities and the characteristics they supposedly indicated" (Cornell 1990, p. 376; Rawley 1981). Although slave owners and traders acknowledged ethnic differences among Africans, they treated Africans from various ethnic groups as belonging to one category of people—slaves. Because slave traders sold and slave owners purchased individual human beings, not ethnic groups, this treatment had the effect of breaking down ethnic concentrations. In addition, slave owners tended to mix together slaves of different ethnic origins in order to decrease the likelihood of the slaves' plotting a rebellion. To communicate with each other, the slaves invented pidgin and Creole languages. In addition to inventing a new language, the slaves created a common and distinctive culture based on kinship, religion, food, songs, stories, and other features. The harsh conditions of slavery, in combination with the mixing together of people from many ethnic groups, encouraged slaves to borrow aspects of each other's cultures and to create a new, blended culture.

In Germany (which calls itself a nonimmigration country) and in the United States (which calls itself a melting pot culture), assimilation has been mainly a one-sided process in that newcomers are expected to adapt to the dominant culture. Sociologist Ruth Mandel believes that the status of the labor migrant in Germany makes assimilation difficult:

> The very term by which these people are commonly referred, "Gastarbeiter," guest workers, underlies the ambiguity of their status . . . guests are by definition temporary, and are expected to return home. Guests are bound to rules and regulations of hosts. Whatever the intentions, guests rarely feel "at home" in foreign environs. The second half of the compound word, "arbeiter," worker, refers to the economic and use value of the migrant determined solely in relation to his or her labor. (1989, P. 28)

German political leaders have made some attempts to call international labor migrants something other than "guest workers." One such term is *auslaendische Mitbuerger,* or "foreign co-citizens." Yet, in the words of sociologist Czarina Wilpert: "In any other language this would be a contradiction in terms, and even in German it highlights the ambiguity of belonging to the collective of foreign settlers in Germany" (1991, p. 53).

Few people in either the United States or Germany view assimilation as a process of mutual exchange in which members of the dominant group form and identify with a blended culture (Opitz 1992d). This is not to say that the dominant culture has not been shaped and influenced by racial and ethnic minorities; rather, the dominant group more often than not fails to acknowledge the contributions of others to the society in which they live.

Stratification Theory and Assimilation

Stratification theory is one major approach to understanding forces that work against assimilation (absorption or melting pot) between dominant and minority groups. This theory is guided by two assumptions. First, ethnic and racial groups are ranked hierarchically. The group at the top is known as the **dominant group;** the groups at the bottom are called **minorities.** Second, racial and ethnic groups compete with one another for scarce and valued resources. The dominant group retains the advantage because its members are in a position to preserve the system that gives them their advantages. Stratification theorists focus on the mechanisms employed by people in the dominant group to preserve inequality (Alba 1992). These include racist ideologies, prejudice and stereotyping, discrimination, and institutional discrimination.

Racist Ideologies An **ideology** is a set of beliefs that are not challenged or subjected to scrutiny by the people who hold them. Thus, ideologies are taken to be accurate accounts and explanations of why things are as they are. On closer analysis, however, ideologies are at best half-truths, based on misleading arguments, incomplete analysis, unsupported assertions, and implausible premises. They "cast a veil over clear thinking and allow inequalities to persist" (Carver 1987, pp. 89–90). One such ideology is **racism.** People who follow this ideology believe that something in the biological makeup of an ethnic or racial group explains and justifies its subordinate or superior status.

Racist ideologies are structured around three notions: (1) people can be divided into categories on the basis of physical characteristics; (2) a close correspondence exists between physical traits and characteristics such as language, dress, personality, intelligence, and athletic capabilities; and (3) physical attributes such as skin color are so significant that

Is There Such a Thing as Black Racism?

Black racism. For some blacks it is a laughable oxymoron. ("How can the victims of racism be racist?") For some whites it is an excuse for doing little to reduce the inequalities that still plague the country. ("See, blacks are just as bad as they say *we* are.")

The recent diatribe against Catholics, Jews and homosexuals (among others) by Khalid Abdul Muhammad, a member of the hierarchy of the Nation of Islam, has focused attention on the disturbing question of how much racism permeates black America. Indeed, the public fascination with the Nation of Islam—a fringe group with negligible political and economic power—is partly explained by white fear that its views are shared by other blacks. Looking at the cleaning women, the bank tellers, partners in their law firm, some whites silently ask, "Do *you* agree with Louis Farrakhan?"

Until recently, the question of how blacks view whites has been largely unexplored territory for scholars and sociologists. Why have sociologists not probed the black psyche to see how much racial animus resides there? Perhaps because they are afraid of what they might find; perhaps because they feel that since whites are the dominant group, their views are more important; perhaps because they feel that blacks have no opinions that whites are bound to respect.

But recent work by pollsters, sociologists and political scientists has uncovered evidence of specific negative attitudes that blacks may harbor toward whites in general and Jews in particular. Seeking responses to a series of assertions (Jews tend to stick together more than other Americans; Jews wield too much power on Wall Street; Jews are more willing than others to use shady practices to get what they want), a 1992 survey by the Anti-Defamation League of B'nai B'rith found that blacks are more than twice as likely as non-Jewish whites to hold stereotypical views of Jews.

Preliminary findings in a study to be published soon by the National Conference of Christians and Jews found that blacks feel that whites think themselves superior, and that they do not want to share power and wealth with nonwhites. The poll found that black antipathy toward whites is far greater than that of other minorities, such as Latinos and Asian-Americans.

"Of course, there's black racism," said Roger Wilkins, a professor of history at George Mason University, who is black. "I think that any time that you say that a whole group of people is this or that, fill in the blanks, and the blanks are all negative adjectives, and then you go on to prove your conclusions by telling stories that support your use of those adjectives, that's racism."

But, as another black writer, Ellis Cose, points out in his 1993 book, *The Rage of a Privileged Class* (HarperCollins), the harboring of negative views by blacks does not tell the whole story. Surveys of blacks also find that many give American whites, both Jew and gentile, high marks for intelligence and for creating a democratic society. Viewed in this light, the results of the surveys may be less an affirmation of racism than an expression of envy and a plea to share power.

Racism among blacks, whatever its extent, has been like magma bubbling under black nationalism. Like other nationalist movements—Zionists, the Quebecois in Canada, Serbs seeking a "Greater Serbia"—those embraced by some blacks have used ethnic solidarity as a political organizing tool.

For some mainstream black politicians, black anger toward whites presents both opportunity

---

they explain and determine social, economic, and political inequalities. Any racial or ethnic group may use racist ideologies to explain their own or another group's behavior. (See "Is There Such a Thing as Black Racism?") One example of a racist ideology is the hypothesis offered by former Los Angeles police chief Daryl Gates to explain why so many blacks have died from restraining chokeholds: their "veins and arteries do not open up as fast as they do on normal people" (Dunne 1991, p. 26). The fact behind these statistics is that many police officers, black and white, tend to handle black suspects more harshly than they do their white counterparts.

German Nazis rationalized the extermination and persecution of Jews, Gypsies, Poles, Russians, blacks, and others "deemed unfit to breed" (including disabled, homosexual, or mentally ill persons) with the racist ideology that the Aryan race was su-

and temptation. Just as white politicians often take advantage, some discreetly and others more cynically, of racist attitudes among whites, so black politicians are aware of hostility among *their* constituents. And when they tap into it, they confront a question of conscience: When do they cross the line into demagoguery?

A call for unity is by implication a call for exclusivity. The desire to maintain that solidarity makes it difficult for many blacks to denounce black racism. But the censure of Mr. Muhammad by people like the Rev. Jesse Jackson, Benjamin Chavis, executive director of the National Association for the Advancement of Colored People, and Representative Kweisi Mfume of Maryland, chairman of the Congressional Black Caucus, shows that difficult does not mean impossible. Indeed, black officials have shown more willingness to move beyond solidarity than have members of the United States Senate who met with silence the remarks of Senator Ernest F. Hollings when he jokingly implied in December, while commenting on trade talks in Switzerland, that African heads of state are cannibals. A Hollings aide later said the Senator meant no offense.

There are those who argue that even if blacks espouse racist views, it makes little difference since they lack the power to turn those views into action. Such a rationale was articulated by Mr. Farrakhan at his recent press conference: "Really, racism has to be coupled with a sense of real power."

Mr. Farrakhan may have a point. For all the racial bombast by the Nation of Islam and other strident black nationalists, America has had no episodes of large-scale coordinated attacks by blacks against whites. But individual acts like the beating of the trucker Reginald O. Denny during the Los Angeles riots demonstrate that black bigotry can maim people and poison that atmosphere.

There is no denying that black bigotry is fed by social and economic conditions. Even discounting, as some whites do, the country's history of slavery, segregation and discrimination, there are trends that insure black resentment of whites will not diminish soon, and that black demagogues will find allies among the disaffected.

Black poverty rates remain stubbornly high. Progress toward closing the income gap between the races stalled in 1975, according to

census data, and since then the gap has widened. (For instance, male black college graduates who entered the job market in 1971 earned on average 2 percent more than their white counterparts, but by 1989 the same blacks were earning 25 percent less on average than white men who graduated in 1971.)

Blacks remain the most segregated group, according to the census. Whether it is the result of whites making blacks feel unwelcome, or of declining faith in integration among some blacks, black Americans remain huddled in their corners, afraid that any attempt to reach out will tar them with labels: bleeding heart, Uncle Tom. "We don't live next door to each other," said Lani Guinier, the law school professor whose nomination as Assistant Attorney General for civil rights was withdrawn in a controversy over her voting-rights theories. "No one talks to each other. Racial stereotypes fester in isolation."

SOURCE: "Behind a Dark Mirror: Traditional Victims Give Vent to Racism," by Steven A. Holmes. Copyright © 1994 by The New York Times Company. Reprinted by permission.

perior to other races. According to this ideology, the Jews were responsible for Germany's defeat in World War I and did not even belong to a race of people; they were "nonhuman; blood suckers, lice, parasites, fleas . . . organisms to be squashed or exterminated by chemical means" (Fein 1978, p. 283).

The premise of racial superiority is also at the heart of other rationalizations used by one group to

dominate another. Sociologist Larry T. Reynolds (1992) observes that race as a concept for classifying humans is a product of the 1700s, a time of widespread European exploration, conquest, and colonization that did not begin to subside until the end of World War II.[9] Racist ideology also supported Japanese annexation and domination of Korea, Taiwan, Karafuto (the southern half of the former Soviet island Sakhalin), and the Pacific Is-

Among the victims of the Nazi ideology of a "pure race" were Gypsies like those being held at this concentration camp in Belzec, Poland. Historians estimate that between 20 and 50 percent of European Gypsies died in the "Gypsy Holocaust."

Archives of Mechanical Documentation, Warsaw, Poland, Courtesy of the United States Holocaust Memorial Museum

lands prior to World War II. Both the Japanese and Europeans used racial schemes to classify the people they encountered; the idea of racial differences became the "cornerstone of self-righteous ideology," justifying their right by virtue of racial superiority to exploit, dominate, and even annihilate conquered peoples and their cultures (Lieberman 1968).

In *The Mismeasure of Man* Stephen Jay Gould describes instances in which so-called scientists have used the scientific method to verify the accuracy of racist ideologies (Gould 1981b). His efforts in this area reveal how millions of lives have been ruined by spurious correlations, by the mystique of quantification, or simply by sloppy science (see Chapter 3). Gould's article, "The Politics of Census," a particularly troubling account, explains how the U.S. Census figures for 1840 were used to justify slavery. In 1840 the census takers counted the mentally ill for the first time in census history. When Dr. Edward Jarvis, an authority on medical statistics, examined census tables that showed the number of mentally ill persons by race and by state, he found that "one in 162 blacks was insane in free states, but only one in 1,558 in slave states. Moreover, insanity among blacks seemed to decrease in even gradation" as one traveled from the northernmost part of the country to the deep south. Jarvis

also found that location had no bearing on rates of insanity among whites. On the basis of these findings, he "drew the conclusion that so many other whites would advance": slavery must be beneficial to blacks. A slave actually gains from not having "hopes and responsibilities which the free, self-thinking and self-acting enjoy and sustain, for bondage saves him from some of the liabilities and dangers of active self-direction" (Gould 1981a, p. 20).

Dr. Jarvis, however, was troubled by these conclusions and wondered whether slavery could possibly exert such a significant effect on mental stability. He decided to investigate, and found a great many errors in the gathering and tabulation of the data. For example, the census showed that in Worcester, Massachusetts, 133 blacks out of a total black population of 151 were insane. Jarvis discovered, however, that the 133 were actually white patients in the state mental hospital at Worcester. He attempted to have the census declared invalid, but his efforts met with considerable opposition, led by John C. Calhoun, head of the U.S. State Department. Consequently, the data stood as fact, despite the errors. This so-called scientific data existed to support the racist ideology that blacks benefit from slavery because, unlike whites, blacks cannot handle the pressures and responsibilities that come with freedom.

Prejudice and Stereotyping A **prejudice** is a rigid and usually unfavorable judgment about an outgroup (see Chapter 5) that does not change in the face of contradictory evidence and that applies to anyone who shares the distinguishing characteristics of that group. Prejudices are based on **stereotypes**—exaggerated and inaccurate generalizations about people who are members of an outgroup. Many Germans stereotype Turks as backward people who come from villages where "their houses are huts, the streets are unpaved, . . . and where there are goats, sheep, and chickens, and fields of corn, wheat, and melons" (Teraoka 1989, p. 111).

Stereotypes "give the illusion that one [group] knows the other" and "confirms the picture one has of oneself" (Crapanzano 1985, pp. 271–72). Germans who claim that Turks are backward are saying that they, by definition, are not. In addition, Germans who hold such stereotypes about Turks fail to see Turks in Germany as a highly heterogeneous group with regard to social origin, cultural norms, and length of residence in Germany (Teraoka 1989).

Stereotypes are supported and reinforced in a number of ways. One way is through **selective perception,** whereby prejudiced persons notice only those behaviors or events that support their stereotypes about an outgroup. In other words, people experience "these beliefs, not as prejudices, not as prejudgments, but as irresistible products of their own observations. The facts of the case permit them no other conclusion" (Merton 1957, p. 424).

In a recent editorial, columnist Clarence Page reflected on the fact that stereotypes about black teenagers and violence can cause people to fear contact with any black teenager:

> As much as everyone tells us how cute and handsome our little boy is now (and they're right!), I also know that someday he will be a teen-ager and many adults have an outright fear of black male teen-agers.
>
> It's getting to be an old story. See a group of small black children coming toward you on the street and you want to scoop the little darlings up and hug them. See the same kids approaching a few years later as teen-agers in sneakers and fashionably sculpted haircuts and, depend-

> ing on who you are, you very well might want to cross the street, thinking to yourself about how you can't be too careful.
>
> I dread the experiences of a family friend who is trying to help his seventh-grade son cope with his transformation from the cuddly cuteness of childhood to the imposing presence of a full-sized African American teen-ager. Formerly greeted with delight, he now is sometimes greeted with apprehension.
>
> Positive thinker that I am, I cannot help but think negatively about the damage done by the negative signals of fear we send young black males, simply because some have sinned.
> (PAGE 1990, P. A10)

Stereotypes also persist in another way: when prejudiced people encounter a minority person who contradicts the stereotype, they see that person as an exception. The fact that they have encountered a minority group member who is "different" only serves to reinforce their stereotypes.

In addition, prejudiced people use facts to support their stereotypes. A prejudiced person can point to the small number of black quarterbacks, pitchers, and baseball managers as evidence that blacks possess inferior qualities. Former Los Angeles Dodgers general manager Al Campanis used that fact to support his view that blacks do not have what it takes to be coaches and managers. As the following exchange between Campanis and journalist Ted Koppel reveals, it did not occur to Campanis that prejudice and discrimination by white coaches and management could explain why few blacks hold leadership positions in baseball:

KOPPEL: Is there still that much prejudice in baseball today?

CAMPANIS: No, I don't believe it's prejudice. I truly believe that they may not have some of the necessities to be, let's say, a field manager or, perhaps, a general manager.

KOPPEL: Do you really believe that?

CAMPANIS: Well, I don't say that all of them, but they certainly are short. How many [black] quarterbacks do you have? How many pitchers do you have that are black? ("NIGHTLINE" 1987, P. 22)

In a similar vein, the stereotype that young blacks are more violent than their white counterparts can be supported by FBI statistics showing that blacks are three times as likely as whites to be arrested for aggravated assault. National Crime Survey data, however, based on interviews with victims, show that blacks and whites are equally likely to commit aggravated assaults (32 per 1,000 blacks and 31 per 1,000 whites). The discrepancy between FBI figures and National Crime Survey data suggests that blacks are more likely than whites to be caught, arrested, and convicted, not that they are more likely to engage in violent acts (Stark 1990).

Finally, prejudiced individuals keep stereotypes alive when they evaluate the same behavior differently at different times, depending on the person who exhibits that behavior (Merton 1957). For example, incompetent behavior on the part of racial and ethnic minority members often is attributed to innate flaws in their biological makeup, whereas incompetence exhibited by someone from the dominant group is almost always treated as an individual issue. Similarly, prejudiced people treat certain ideas, when expressed by members of a minority group, as more threatening than when those ideas are expressed by members of a dominant group.[10]

In *Immigrant Workers and Class Structure in Western Europe,* Stephen Castles and Godula Kosack (1985) argue that the roots of racist ideologies and prejudice are not "based on real dislike for the physical appearance of the groups concerned. Rather, the physical appearance and the inferiority which is alleged to go with it is [sic] used as an excuse" to justify the subordinate position of racial and ethnic minorities in society (p. 456). "The exploited group is stigmatized as inherently inferior in order to justify [that position]" (p. 458). The subordinate position then is pointed out as proof of the group's inferiority.

Prejudice and racism directed toward international labor migrants serve another function: they direct attention away from the deficiencies of the profit-driven capitalist system, which lowers production costs by hiring persons who will work for lower wages, by introducing labor-saving devices, and by moving production from high-wage to low-wage zones; all of these measures create unemployment. The blame for unemployment is placed on labor migrants rather than on the economic system, which plays off workers against one another and creates the unemployment (Bustamante 1993). Because prejudice and racism against labor migrants direct attention away from the capitalist system, which treats labor as a commodity, they undermine the chances that workers will unify and make real improvements in their wages, work conditions, and job security.

Discrimination In contrast to prejudice, **discrimination** is not an attitude but a behavior. It is intentional or unintentional unequal treatment of individuals or groups on the basis of attributes unrelated to merit, ability, or past performance. Discrimination is behavior aimed at denying members of minority groups equal opportunities to achieve valued social goals (education, health care, long life) and/or blocking their access to valued goods and services. The U.S. Commission on Civil Rights describes some examples of common discriminatory actions:

- *Personnel officers whose stereotyped beliefs about women and minorities justify hiring them for low level and low paying jobs exclusively, regardless of their potential experience or qualifications for higher level jobs*

- *Teachers who interpret linguistic and cultural differences as indications of low potential or lack of academic interest on the part of minority students*

- *Guidance counselors and teachers whose low expectations lead them to steer female and ethnic minority students away from "hard" subjects, such as mathematics and science, toward subjects that do not prepare them for higher-paying jobs*

- *Parole boards that assume minority offenders to be more dangerous or more unreliable than white offenders and consequently more frequently deny parole to members of minorities than to whites convicted of equally serious crimes* (1981, P. 10)

Robert K. Merton explores the relationship between prejudice (the attitude) and discrimination (the behavior). He distinguishes between two types

---

FIGURE 10.2 A Typology of Ethnic Prejudice and Discrimination*

	Attitude Dimension: Prejudice and Nonprejudice	Behavior Dimension: Discrimination and Nondiscrimination
Type I: Unprejudiced nondiscriminator	+	+
Type II: Unprejudiced discriminator	+	−
Type III: Prejudiced nondiscriminator	−	+
Type IV: Prejudiced discriminator	−	−

* + Attitude/behavior supports equal opportunity

− Attitude/behavior rejects equal opportunity

SOURCE: Adapted from "Discrimination and the American Creed," by Robert K. Merton. P. 192 in *Sociological Ambivalence and Other Essays*. (1976).

---

of individuals: the nonprejudiced (those who believe in equal opportunity) and the prejudiced (those who do not). Merton asserts that people's beliefs about equal opportunity are not necessarily related to their conduct, as he makes clear in his four-part typology (see Figure 10.2).

Nonprejudiced nondiscriminators (all-weather liberals) accept the creed of equal opportunity, and their conduct conforms to that creed. They represent a "reservoir of culturally legitimized goodwill" (Merton 1976, p. 193). In Germany, some religious, industrial, and political groups have formed a coalition and have launched Action Courage, a national campaign encouraging Germans to challenge antiforeign and anti-ethnic remarks, to intervene and protect foreigners from discrimination, and to wear a button bearing the word *courage* to symbolize their commitment to these goals (Marshall 1992b). People who act to realize these goals would qualify as nonprejudiced nondiscriminators in Merton's typology.

Unprejudiced discriminators (fair-weather liberals) believe in equal opportunity but engage in discriminatory behaviors because it is to their advantage to do so or because they fail to consider the discriminatory consequences of some of their ac-

tions. In one example, unprejudiced persons decide to move out of their neighborhood after a black family moves in because they are afraid that property values might start to decrease. In another instance, a white personnel officer tells friends and neighbors (who also are likely to be white) about a job opening. This word-of-mouth method of recruiting reduces the chances that a minority candidate will learn of the opening.

Examples of unprejudiced discriminators specific to Germany include those who fail to challenge young fans for yelling "Jew to Auschwitz" at referees who make "bad" calls, or those who join in singing songs whose rhyming words note that the Christmas season is the time "to burn a Turk" (Marshall 1992b, A13). Fair-weather liberals also might include teachers who fail to incorporate information about the contributions of different racial and ethnic groups to a country's way of life because no appropriate classroom materials are readily available. In Germany, attempts to help minority students see the connections between classroom material and their own lives is the exception rather than the rule (Opitz 1992a). The same can be said about the United States: typically, teachers introduce material related to racial or ethnic groups

Bob Daemmrich/Sygma

Adenis/Sipa Press

Ku Klux Klan members in the United States (left) and neo-Nazis in Germany (right) exemplify active bigotry based on a racist ideology.

on special days or months of the year (for example, during a week of Black History Month, Foreigners' Day in Germany), or the materials are relegated to a special topic within a discipline. For example, textbook authors who include a chapter on race and ethnicity and confine all references to race and ethnicity to that chapter as if the topic had no relevance to the general subject qualify as unprejudiced discriminators.

Prejudiced nondiscriminators (timid bigots) do not accept the creed of equal opportunity but refrain from discriminatory actions primarily because they fear the sanctions they may encounter if they are caught. Timid bigots do not often express their true opinions about racial and ethnic groups; instead, they use code words. In *Race: How Blacks and Whites Think and Feel About the American Obsession*, Studs Terkel cites several examples of speaking in code:

- *In Chicago, during the black mayor's first campaign, his white opponent's slogan was "Before It's Too Late."*

- *It was a slow day. The waitress and I were engaged in idle talk. . . . She was large, genial, motherly. The neighborhood was an admixture of middle-class and blue-collar. On its periphery was an enclave of black families.*

As I was leaving, she said, "Of course, we're moving. You know why." It was an offhand remark, as casual as "See ya around." (1992, P. 3)

Prejudiced discriminators (active bigots) reject the notion of equal opportunity and profess a right, even a duty, to discriminate. They derive "large social and psychological gains from [the] conviction that any white man (including the village idiot) excels anyone from the hated group" (Merton 1976, p. 198). The members of this group are most likely to believe that they "have the moral right" to destroy the people whom they see as threatening their values and way of life. Of the four categories in Merton's typology, prejudiced discriminators are the most likely to commit **hate crimes,** actions aimed at humiliating minority-group people and destroying their property or lives.

Sociologists Jack Levin and Jack McDevitt (1993) studied more than 450 hate crimes committed in Boston between 1983 and 1987. They found that almost 60 percent were turf-related; that is, they were directed against people who were walking in, driving through, moving into, or working in areas where their attackers did not believe they belonged. Levin and McDevitt found that two-thirds of the attackers were under age 29 and that 108 of the crimes involved whites attacking blacks; 95 involved blacks attacking whites. Almost 30 percent of the victims were women. Sociologist Ivan Light believes that, although hate crimes often are motivated by deep prejudices against a race or ethnic group, "the roots of intergroup conflict are as much in economic competition as it is in negative stereotypes" (1990, p. B7). During tight economic times, for example, some whites may blame their unemployment on affirmative action policies that they believe favor minorities, and blacks may blame white discrimination and prejudice as the source of their economic problems.

The government of Saxony (one of 15 states in Germany) published a profile of 1,244 persons suspected of committing hate crimes between 1991 and 1992. Ninety-six percent were males; two-thirds were age 18 or younger; 50 percent had been drinking before the incident; 90 percent of the hate crimes occurred near the suspects' homes (within 12 miles); more than 98 percent of the suspects had less than a 10th-grade education. Surprisingly, only 20 percent of the suspects were unemployed or school dropouts. Eighty percent were first-time offenders. Approximately 30 percent justified their violent action with firm right-wing ideological views (Marshall 1993). This last figure contradicts the idea that the person accused of a hate crime is motivated solely by ideology. Rather, a complex set of factors is involved.

Institutionalized Discrimination Sociologists distinguish between individual and institutionalized discrimination. **Individual discrimination** is any overt action on the part of an individual that depreciates minority persons, denies members of minorities opportunities to participate, or does violence to minority members' lives and property. **Institutionalized discrimination,** on the other hand, is the established and customary way of doing things in society—the unchallenged rules, policies, and day-to-day practices that impede or limit minority members' achievements and keep them in a subordinate and disadvantaged position. It is the "systematic discrimination through the regular operations of societal institutions" (Davis 1978, p. 30).

Two examples of institutionalized discrimination in Germany are laws that limit the proportion of foreign children in German classrooms to 20 to 30 percent and laws that limit the percentage of foreigners living in an area to 9 percent (Wilpert 1991). Another example of institutionalized discrimination, both in the United States and in Germany, is the selective law enforcement practice in which one racial or ethnic group is more likely than another to be arrested or to receive stiffer penalties for breaking the law.

In the United States, surveys consistently show that the rate of drug use is the same among blacks and whites. Blacks, however, are four times as likely as whites to be arrested on drug-related charges because police are more likely to stop and search blacks than whites. As a result, official crime statistics reflect a connection between race and drug arrests (Meddis 1993). This is an example of institutionalized discrimination because it involves the questions, "How men of color are brought to the system in the first place and how the system makes the decision about who is charged or what they're charged with" (De Witt 1993, p. E9).

Of the two levels of discrimination, institutionalized discrimination is the more difficult to identify, condemn, hold in check, and punish, because it can exist in society even if most members are not prejudiced. Institutionalized discrimination cannot be traced to the motives and actions of particular individuals; discriminatory actions result from simply following established practices that seem on the surface to be impersonal and fair. After all, the same rules supposedly apply to everyone. Consider the United Nations's definition of a **refugee:** "someone who has a well-founded fear of persecution on the basis of his or her race, religion, nationality, political opinion or membership in a particular social group" (Walsh 1993, p. A6). This definition excludes gender-based persecution as a reason to declare asylum. Violence against females in the form of rape, infanticide, female circumcision, and forced marriage, abortion, or sterilization typically is treated as a private matter or as lawful practice in a woman's home country.

Another instance of institutionalized discrimination is a provision written into the U.S. Immigration Act of 1990 allowing people from 34 countries, most of which are European, to take part in an annual "lottery." The first 40,000 applications received after midnight on October 14 of each year

James Kamp/Sipa Press   Tony Savino/Sygma

The very different treatment given by the U.S. Coast Guard to refugees from Cuba (left)
and Haiti (right) is a good example of institutionalized discrimination.

for three years will win green cards for the applicants, entitling them to legal permanent residency in the United States. Currently, applicants from the Republic of Ireland (not Northern Ireland) are allotted 40 percent of the slots, or 16,000 of the 40,000 green cards. The goal of the lottery, though, is to diversify immigration to the United States. The rationale behind the plan is that, since 1965, when national origin was discarded as a criterion for entry into the United States, immigrants from Europe have been outnumbered by those from Latin America, Asia, and the Caribbean. In fact, the proportion of European immigrants has fallen from 50 percent to 8 percent of the total admitted. In 1990 immigrants from Central and South America, the Caribbean, and Asia constituted 80 percent of the total (Keely 1993; Lacayo 1991; Mydans 1991).

As a final example of institutionalized discrimination, consider how asylums have been awarded since the end of World War II. People who have escaped from communist and communist-backed countries are granted asylum automatically; people fleeing death squads from U.S.-backed governments of noncommunist totalitarian governments are jailed and deported (T. Weiner 1993). Consider, for example, the different treatment that U.S. Coast

Guard authorities regularly give Cuban and Haitian refugees whom they intercept at sea. Cuban boat people are taken to Miami, where they are free to live and work; but Haitians are returned to their country (Feen 1993). The rationale underlying this difference in treatment is that Cubans are fleeing a communist government, whereas the Haitians are considered economic refugees.[11]

Institutionalized discrimination can also be due to laws or practices that penalize minorities for the consequences of past prejudice and discrimination. Such discrimination can be overt or subtle. It is overt when laws and practices are designed with the clear intention of keeping minorities in subordinate positions as in the 1831 Act Prohibiting the Teaching of Slaves to Read.[12]

Institutionalized discrimination can also be subtle, as when the discriminatory consequences of a practice are neither planned nor intended. One might even argue that the way in which the dominant group implement affirmative action programs is a subtle form of institutionalized discrimination. As essayist and English professor Shelby Steele observes, affirmative action programs, as they typically have been carried out, have some troubling side effects because *the quality that earns minorities preferential treatment is an implied inferiority. However this inferiority is explained—and it is easily enough explained by the myriad deprivations that grew out of our oppression—it is still inferiority.*

The effect of preferential treatment—the lowering of normal standards to increase black representation—puts blacks at war with an expanded realm of debilitating doubt. (1990, PP. 48–49)

One wonders whether this side effect would exist if blacks were recruited for jobs and college scholarships in the same way they are recruited for college sports. Few people seem to be disturbed when black athletes are recruited to play at predominantly white colleges, possibly because school officials, coaches, recruiters, and students believe that the black athlete's presence on campus benefits the school. If the same energy and positive attitude were applied to recruiting black college students or job applicants, some of the troublesome features of the affirmative action programs might be alleviated. In many ways the on-the-field integration of major league baseball is a model (although not a perfect one) for affirmative action programs. Syndicated columnist William Raspberry pointed this out on the "MacNeil/Lehrer Newshour":

The Major Leagues [until 1948 had] been an exclusive preserve of white men, skilled white men, but white men. And it was also very clear then that there were a lot of black baseball players as good as many of those whites who were in the big leagues. But the rules said you couldn't be a major league ballplayer if you were black. Branch Rickey made the breakthrough in hiring Jackie Robinson to say we're going to bust this up. . . . But once the decision was made to bust it up, a couple of things happened. One had to change . . . the ways baseball looked for and recruited ballplayers. You couldn't just look to the minor leagues anymore because blacks weren't there either. They had to look in some places they weren't accustomed to looking and had to use some techniques they weren't accustomed to using to find prospects. That's affirmative action in one sense and I buy it absolutely. But there was another piece of it. Nobody supposed that pretty good black baseball players who because they had been denied opportunity to play in the big leagues before then should have been given special breaks and brought into the big leagues with lesser skills than their white counterparts. The assumption was, and we bought it, black and white, that if we were given the opportunity to use our skills, and to develop our skills, you didn't have to cut us any breaks. (RASPBERRY 1991)

Raspberry's point is that people lose sight of the real purpose of affirmative action policies: to "bust up" a flawed hiring system in which not everyone who was qualified had a chance to apply or be considered. Instead the attention focuses on whether the person who is hired is qualified. No one seems to remember that the hiring system that affirmative action replaced was a program that benefited the friends and acquaintances of those in power.

To this point we have looked at barriers that exist in society to prevent racial and ethnic minorities (particularly labor migrants from a racial or ethnic group different from the dominant group) from adapting and/or assimilating. These barriers include racist ideology, dependence on foreign labor, prejudice, and discrimination (individual and institutional). Although we have viewed these barriers in a general way, we have not examined how they operate in everyday interaction. In *Stigma: Notes on the Management of Spoiled Identity*, sociologist Erving Goffman gives us such a framework.

Social Identity and Stigma

When people encounter a stranger, they make many assumptions about what that person ought to be. Judging by an array of clues such as physical appearance, mannerisms, posture, behavior, and accent, people anticipate the stranger's **social identity**—the category (for example, male or female, black or white, professional or blue-collar, under 40 or over 40, American-born or foreign-born) to which he or she belongs and the qualities that they believe, rightly or wrongly, to be "ordinary and natural" (Goffman 1963, p. 2) for a member of that category.

Goffman was particularly interested in social encounters in which one of the parties possesses a stigma. We learned in Chapter 6 that a stigma is an attribute that is defined as deeply discrediting in that person's society. A stigma is considered discrediting because it overshadows all other attributes that a person might possess. According to Goffman, stigmas can be of three broad varieties: (1) physical deformities; (2) character blemishes attributed to factors such as sexual orientation, mental hospitalization, or imprisonment; and (3) stigmas of race, ethnicity, nationality, or religion as reflected in skin color, dress, language, occupation, and so on.

The important quality underlying all stigmas is that those who possess them are not seen by others as multidimensional, complex persons but as one-dimensional beings. When many Germans encounter a Turkish woman wearing a head scarf, for example, they do not see the scarf in all its complexities—as "signifying the relation between the wearer and any number of things, such as her male relatives, her personal religio-political views, her financial means, or her region of origin" (Mandel 1989, p. 30). Instead, they see only a foreigner and they point to the scarf as proof that Turks cannot integrate into German society, and they use it as evidence to prove why Turks should be sent home. In other words, this single attribute overshadows any other attribute the woman might possess.

The same dynamic applies to how most Germans view people of Afro-German ancestry. Many Germans see only one attribute—dark skin—and fail to acknowledge other attributes such as German ancestry and ability to speak German. Two Afro-German women explain how their dark skin dominates the course of interaction:

I have dark skin, too, but I am a German. No one believes that, without some further explanation. I used to say that I was from the Ivory Coast, in order to avoid further questions. I don't know that country, but to me it sounded so nice and far away. And after this answer, I *didn't get any more questions either. Germans are that ignorant. I could tell people any story I wanted to, the main thing was that it sounded foreign and exotic. But no one ever believes that I am German. When I respond to the remark, "Oh, you speak German so well" by saying "So do you," people's mouths drop open.* (EMDE 1992, PP. 109–10)

It often happens with me that people have their own expectations and ignore what I say. When I tell them that I grew up here [in Germany] and have spent my entire life here, the question still might come afterward: "Yes, and when are you going back?" Crazy. Now and then I have the feeling of not belonging anywhere; on the other hand I've grown up here, speak this language, actually feel secure here, and can express myself as I want. I share a background with these people here even if they don't accept me. "Yes, I'm German," I say, perhaps out of spite, to shake up their black-and-white thinking. (OPITZ 1992B, P. 150)

According to Goffman, it is important to remember that the discrediting trait itself is not the stigma. Rather, the stigma consists of the set of beliefs about the trait. Furthermore, from a sociological point of view, the person who possesses the trait is not the problem; the problem is how others react to the trait. Thus, Goffman maintained that sociologists should not focus on the attribute that is defined as a stigma. Instead, the focus should be on interaction—specifically, interaction between the stigmatized and **normals.** Goffman did not use the term *normal* in the literal sense of "well-adjusted" or "healthy." Instead, he used it to refer to those people who are in the majority or those who possess no discrediting attributes. Goffman's choice of this word is unfortunate because some readers forget how Goffman intended it to be used.

Mixed Contact Between the Stigmatized and the Majority Population

In keeping with this focus, Goffman wrote about **mixed contacts,** "the moments when stigmatized and normals are in the same 'social situation,' that is, in one another's immediate physical presence, whether in a conversation-like encounter or in the mere co-presence of an unfocused gathering" (Goffman 1963, p. 12). According to Goffman, when normals and the stigmatized interact, the stigma comes to dominate the course of interaction. First, the very anticipation of contact can cause normals and stigmatized individuals to try to avoid one another. One reason is that interaction threatens their sense of racial and ethnic "purity" or loyalty.

Sometimes the stigmatized and the normals avoid each other in order to escape the other's scrutiny. Persons of the same race may prefer to interact with each other so as to avoid the discomfort, rejections, and suspicions they encounter from people of another racial or ethnic group. ("I have a safe space. I don't have to defend myself or hide anything, and I'm not judged on my physical appearance" [Atkins 1991, B8].)

Another reason that stigmatized persons and normals make conscious efforts to avoid one another is that they believe widespread social disapproval will undermine any relationship. With regard to interracial relationships, some people believe that racist attitudes will destroy even the "most perfect and loving interracial relationships [because] racism waits like a cancer, ready to wake and consume the relationship at any, even the most innocuous, time" (Walton 1989, p. 77).

An article in *Time* magazine profiled the friendship of two six-year-old girls, one black (Jennifer) and one white (Phoebe), to demonstrate how societal pressures can interfere with the development of an interracial relationship. Whenever Jennifer accompanies Phoebe and her parents on outings, she worries that strangers will think she is adopted (Buckley 1991). Consequently, Jennifer cannot relax and enjoy Phoebe's company because she feels pressured to explain why she's with Phoebe. The lesson of this example is that for this six-year-old, interracial relationships are so rare that she cannot imagine people thinking she is Phoebe's friend.

Goffman observed a second pattern that characterizes mixed contacts: upon meeting each other, each party is unsure how the other views him or her or will act toward him or her. Thus, the two parties are self-conscious about what they say and about their behavior:

I must battle, like all humans, to see myself. I must also battle, because I am black, to see myself as others see me; increasingly my life, literally, depends upon it. I might meet Bernhard Goetz on the subway; my car might break down in Howard Beach; the armed security guard might mistake me for a burglar in the lobby of my building. And they won't see a mild-mannered English major trying to get home. They will see Willie Horton. (WALTON 1989, P. 77)

How was I to know which whites were good and which were bad?

What litmus test could I devise? I distanced myself from everyone white, watching, listening, for hints of latent prejudice. But there were no formulas to follow. (MCCLAIN 1986, P. 36)

When you're black, you cannot help but second-guess everything. Does this person like me because I'm black or because I'm me? Does this person hate me because I'm black or because I'm a jerk? You're always second-guessing, unless it's another black person. (KING 1992, P. 401)

For the stigmatized, the source of the uncertainty is not that everyone they meet will view them in a negative light and treat them accordingly. Rather, the chance that they might encounter prejudice and discrimination is great enough to give them reason to be cautious about all encounters. According to the Kolts Report (a report written in response to the Rodney King beating, issued by Special Counsel James C. Kolts, a retired superior court judge, and his staff), there were 62 "problem" deputies out of 8,000 in the Los Angeles County Sheriff's Department. The investigators concluded that "nearly all deputies treat nearly all individuals, most of the time, with at least minimally acceptable levels of courtesy and dignity" (*Los Angeles Times* 1992, p. A18). Although only a small proportion of deputies (less than one percent) were identified as "problems," the cases of mistreatment were "outrageous enough and frequent enough to poison the well in some communities" (p. A18).

Similarly, the majority of German people do not discriminate against Turks and other dark-skinned

Journal Entry July 10, 1993

The following journal entry was written by a college student. It highlights the fact that many students, regardless of race, often lack a historical understanding of race relations, and this lack of knowledge affects the way they think about race.

Over the past two weeks, we've watched 3 films dealing with the issue of race: *Master Harold and the Boys, The Sky Is Grey,* and *Black Hair Blues.* The first film focuses on the relationship between Master Harold, a white South African youth and two Black South African men who work in his mother's cafe. *The Sky Is Grey* focuses on a day in the life of an African-American adolescent male during the Jim Crow period. *Black Hair Blues* focuses on dilemmas that African-American women face with regard to straightening their hair.

Although these films were very interesting and informative, what I found even more interesting and intriguing was the way my classmates (six white females, one white male, and one African-American male) reacted to these films. Specifically, four or five almost always remained silent throughout the discussions so it was difficult to know how they felt about each film. Those who did talk seemed to have a very simplistic view of race. I think that the silence and simple views can be explained by the fact that almost no one in the class has background knowledge about race or has had experience interacting with people of different races. I will try to explain how my classmates reacted to the films.

With regard to the film *Master Harold and the Boys* my classmates seemed to operate on the assumption that the larger social context has no bearing on how blacks and whites relate to each other in South Africa. For example, the students who spoke to the question, "Where do you see each of the characters (Master Harold, Willie, and Sam) in ten years?" answered that the characters would remain friends and Master Harold would continue to seek Sam's advice about important matters. No one could give such an "uncomplex" answer if he or she had a basic understanding of apartheid. Even when the professor asked the students to think about how apartheid might impact on the way the characters interacted with each other, everyone was silent because no one seemed to know what apartheid is.

After the class watched *The Sky Is Grey,* the professor asked us what question we would ask the director

people in conscious or overt ways. Yet, there are about 6,000 skinheads in Germany who identify with Nazi ideology. The government believes that they are responsible for as many as five extremely brutal attacks per day against foreigners (Kramer 1993), a number large enough to make life dangerous for minorities in Germany.

A third pattern characteristic of mixed contacts is that normals often define accomplishments by the stigmatized, even minor accomplishments, "as signs of remarkable and noteworthy capacities" (Goffman 1963, p. 14) or as evidence that they have met someone from the minority group who is an exception to the rule. In *Two Nations: Black and White, Separate, Hostile, Unequal,* Andrew Hacker wrote that whites attending professional meetings or panel discussions tended to applaud longer and more strenuously at the introduction of black participants and at the end of their remarks. Although Hacker acknowledges that in some instances the applause is well deserved, he also maintains that many whites do this to let the black speakers know that they are "in the company of friendly whites" (Hacker 1992, p. 56).

As well as defining the accomplishments of the stigmatized as something unusual, normals also tend to interpret the stigmatized person's failings, major and even minor (such as being late for a meeting, or cashing a bad check), as related to the stigma.

A fourth pattern characteristic of mixed contacts is that the stigmatized are likely to experience invasion of privacy, especially when people stare:

if we had the chance. Some questions mentioned were: (1) Why did the mother hit the child? (2) Why was the father absent? and (3) Who's the "guy" hanging around the aunt? Such questions suggest that people have difficulty understanding the larger context and cannot see how segregation affected the lives of people in the film. (It is as if history doesn't matter.)

Black Hair Blues was a short documentary film about black women and their reactions to getting their hair straightened. This time what surprised me most about my classmates' reactions was that no one seemed curious about this practice. (I was disappointed by this because I came to class ready to answer questions.) I expected my colleagues to ask: "What's the difference between [white people's] curly perm and [black people's] straightening perm?" I would explain: the difference is that hair is

curled using the chemical ammonium thioglycolate and is straightened using the chemical lye. Upon learning this I expected someone to ask: "How does it feel?" I would answer, "It burns, especially if the person has scratched her scalp in the past 24 hours." In that case, the lye "eats" at the scalp and can sometimes leave sores. In addition lye, improperly applied, can damage the hair. One person in the class did say she could relate to the film because she used to iron her hair. Her comment made me think of the lines to the song "The Right Life" by Rashida Oji:

You don't know what it's like
Being black and feeling like
   white only is right
(Trying to fit in the great white
   life)
You can't possibly know
How could you know
Unless you've tried it?

I'm almost there when I straighten my hair. . . .

Please don't misunderstand the point I am trying to make. I am simply trying to point out that open and constructive dialogue between people of different races is difficult because people have had so little background knowledge and experience. For instance, during one of the class discussions, one of my classmates had a very difficult time just getting the words "black people" out. What worries me most about this is that the people in this class are social work, psychology, and human service majors, people who are supposed to be more aware of larger social issues. It makes me wonder how they will deal with race issues in their work lives when their reactions really matter.

SOURCE: Renée D. Johnson, Northern Kentucky University, Class of 1994.

I am tired of walking into restaurants, especially with a group of African-Americans, and having the patrons and proprietors act as if they were being visited by Martians.

I'm tired of being out with my 8-year-old nephew and his best friend and never being completely at ease because I know they will act like little boys. Their curiosity will be perceived as criminal behavior. (SMOKES 1992, P. 14A)

If the stigmatized show their displeasure at such treatment, normals often treat such complaints as exaggerated, unreasonable, or much ado about nothing. Their argument is that everyone suffers discrimination in some way and that the stigmatized do not have the monopoly on oppression. They announce that they are tired of the complain-

ing and that perhaps the stigmatized are not doing enough to help themselves (Smokes 1992). Normals are apt to respond:

I'm tired of them. (SMOKES 1992, 14A)

Why can't they put past discrimination behind them?

I think they've been whining too long, and I'm sick of it. (O'CONNOR 1992, P. 12Y)

My grandparents, when they came here from Italy, didn't ask anyone to speak Italian for them. They became American, and they never asked for anything. (BARRINS 1992, P. 12Y)

Such responses suggest that normals do not understand the large social context and historical forces that shape interaction (see "Journal Entry"). The discussion thus far may lead you to believe that members of racial and ethnic minorities are passive victims who are at the mercy of the dominant group. This is not the case, however; minorities respond in a variety of ways to being treated as members of a category.

Responses to Stigmatization

In *Stigma: Notes on the Management of Spoiled Identity*, Goffman describes five ways in which the stigmatized respond to people who fail to accord them respect or who treat them as members of a category. One way is to attempt to correct that which is defined as the failing, as when people change the visible cultural characteristics that they believe represent barriers to status and belonging. A person may undergo plastic surgery or do other things to alter the shape of nose, eyes, or lips, or may enroll in a school to change an accent.

> *When I was about thirteen I started to straighten my "horse hair" so that it would be like white people's hair that I admired so much. I was convinced that with straight hair I would be less conspicuous. I would squeeze my lips together so that they appeared less "puffy." Everything, to make myself beautiful and less conspicuous.* (EMDE 1992, P. 103)

> *They came here [to American Accent Training, a Berkeley, California, school for correcting accents] for "corrective surgery.". . . Correcting an accent problem requires that students be willing to separate their identities from the way they talk. They have to be in effect willing to talk white middle class.* (NATIONAL PUBLIC RADIO 1990)

Turks who change their religion from Islam to Christianity, who Germanize their names, or who give up wearing head scarves in order to fit into German society represent direct attempts to correct the stigma. Often these kinds of responses are not easy because such persons may be considered traitors to their racial or ethnic group (Safran 1986). People may experience conflicting reactions to the visible changes they make, too:

> *My wife is 25 years old, she is beginning to dress German now, she is becoming more and more beautiful in her German clothes. . . . [M]y friends grumble because my wife dresses so modern, they want her, if she's going to wear German clothes, to wear at least a headscarf and long skirts; why does everyone have to see from far away that my wife is a Turk[?]* (TERAOKA 1989, P. 110)

The stigmatized also may respond in a second way. Instead of taking direct action and changing the "visible" attributes that normals define as failings, they may attempt an indirect response. That is, they may devote a great deal of time and effort to trying to overcome the stereotypes or appear as if they are in full control of everything around them. They may try to be perfect—to always be in a good mood, to outperform everyone else, or to master an activity ordinarily thought to be beyond the reach of or closed to people with such traits. The first two statements below were made by African-Americans, the third and fourth by Afro-Germans:

> *[You have to be on guard all the time;] there is no way to get away from it. Because if you do something like close the door to your room, people will start saying, "Is she being angry? Is she being militant?" You can't even afford to be moody. A white girl can look spacey and people will say, "Oh, she's being creative." But if you walk around campus with anything but a big smile on your face, they'll wonder, "Why is she being hostile?"* (ANSON 1987, P. 92)

> *Don't ever do anything bad, because people are always looking for you to do something bad. You not only have to be good, you have to be perfect. If you do something bad, it's not a mark against yourself but a mark against the entire race.* (P. 129)

> *As far as that's concerned, I have a mask, too. I give the appearance of going through life in full control. Sometimes I can actually feel myself, right when I'm stepping out of the house, pull my shoulders back and take on a perfectly erect bearing. I can't walk relaxed at all. Walking*

Protesting stigmatization: This Turkish immigrant to Germany is wearing a yellow star identifying herself as a Turk. The star recalls the Nazi period, when German Jews were forced to wear the Star of David.

Coopet/Sipa Press

Parks, a black seamstress from Montgomery, Alabama, refused to give her seat on the bus to a white person. Her actions, which challenged a law requiring her to do so, triggered a boycott of Montgomery buses and was revolutionary in sparking the civil rights movement.

Another example is the declared goal of the African American Marketing and Media Association. As the organization's president announced, one goal of the organization is to combat stereotypes in advertising (*The New York Times* 1990). Although such actions to change the way normals respond are in one way very direct, they fall into Goffman's category of indirect responses because the stigmatized person does not try to change his or her traits but rather attempts to change the way that normals respond to those traits.

Sometimes the stigmatized respond in a third way: they use their subordinate status for secondary gains, including personal profit or "an excuse for ill success that has come [their] way for other reasons" (Goffman 1963, p. 10). If a black person, for example, levels a charge of racism and threatens to file a lawsuit in a situation in which he or she is justly sanctioned for poor work, academic, or other performance, that person is using his or her status for secondary gains. A fourth response is to view discrimination as a blessing in disguise, especially for its ability to build character or for what it teaches a person about life and humanity. Finally, the stigmatized can condemn all of the normals and view them negatively:

> *You build up these perceptions of whites, that whites are mean and vile, never trust a white person.* (ANSON 1987, P. 127)

> *This coldness here in Germany makes me sick; I am still homesick today, sometimes even more strongly than five years ago; homesickness is an illness, and the illness can only be cured in Turkey; but in Anatolia there is no work for me, no gain, no possibility to move up in life, to live like a human being, with a house and a steady income: I must stay in Germany for now, I must live with this illness; in this cold; the coldness here in Germany, that is its people.* (TERAOKA 1989, P. 107)

> *through the street like that, I'm unapproachable.* (WIEDENROTH 1992, P. 176)

> *For a while I actually practiced walking erect. That was during my school days, when I felt I wasn't accepted or taken seriously on several different sides. I built up a facade of "Nobody can do anything to me."* (OPITZ 1992C, PP. 176–77)

In another type of indirect response, the stigmatized take issue with the way that normals define a particular situation, or they take action to change the way normals respond to them. In 1955 Rosa

Discussion

In August 1993, National Public Radio broad-casted a series of programs on the subject of immigration into the United States and some European countries. One excerpt from that series stands out:

> *The most heart-rending road sign in all America can be seen on the freeway speeding south along the California coast about an hour before one reaches Mexico. It's a caution sign, an orange rectangle with a set of black silhouettes in the middle where you might expect to see a picture of a deer. Instead the silhouette on this sign shows an entire family dashing heads down, hair flying—a man, behind him a woman, then a child hanging wildly off his mother's arms.*
> (NATIONAL PUBLIC RADIO 1993)

This sign is intended to warn automobile drivers to watch out for illegal immigrants dashing across the highway to avoid border patrol agents (see photo on page 39 in Chapter 2). The sign is only one example of the images that come to mind for most Americans when they think about immigration. Germany has comparable images of labor migrants. In a recent issue of the magazine *Der Spiegel*, a graphic shows the country surrounded by big black arrows, each representing a country from which refugees are coming. "It looks like a battle plan. As if Germany is about to be invaded" (National Public Radio 1993). The mental images of labor migrants that these graphics conjure up take into account only a small number of actual labor migrants in these countries.

It goes almost unnoticed that both of these countries recruit skilled and unskilled workers even during times of high unemployment. Recall that the United States issues 300,000 so-called nonimmigration visas to highly skilled workers, who stay for indefinite periods (Keely 1993). Moreover, it goes unnoticed that German and American corporations and government offices export jobs to other countries. (Even the United States Post Office has stamps printed overseas.)

Why the relentless focus on dark-skinned and other highly visible immigrants? One of the most convincing answers to this question was proposed by Castles and Kosack (1985), who believe that the focus on the most visible international labor migrants as the cause of a country's economic and social problems diverts attention from the deficiencies of the profit-driven capitalist system. This system lowers production costs by hiring people who will work for lower wages, by introducing labor-saving devices, and by moving production from high-wage to low-wage zones; all of these measures create unemployment. People blame labor migrants rather than the economic system, which plays off workers against one another and creates the unemployment (Bustamante 1993). The focus on the most visible labor migrants also undermines the chances that workers will unify and make real improvements in their wages, work conditions, and job security.

This argument makes sense when we consider that race as a concept for classifying humans is a product of European exploitation, conquest, and colonization. The Europeans used racial schemes to classify the people they encountered; the idea of racial differences justified their right, by virtue of self-proclaimed racial superiority, to exploit, dominate, enslave, and even annihilate conquered peoples and their cultures (Lieberman 1968).

Castles and Kosack's argument also helps explain why some ethnic and racial minorities—involuntary minorities—in the United States and Germany have not assimilated into the mainstream society. These groups have been exploited most severely by the capitalist economic system. Members of the dominant group are able to maintain the system that has given them the advantages (even if the advantages are slight in some cases). They exercise a variety of mechanisms—racist ideologies, prejudice, individual discrimination, and institutional discrimination—to preserve the inequality created by the capitalist system.

What can we learn about immigration policies by studying Afro-Germans and Turks in Germany and African-Americans in the United States? First, we must place our images of the most visible migrants in the context of international labor migration as a whole. We live in a time when corporations recruit labor on a global scale. Seeing the larger context will discourage people from blaming the most heavily exploited workers as the cause of

The hopeful symbolism of candles graces a demonstration by young Germans against xenophobia, or fear of foreigners.
Rainer Unkel/Saba

the economic problems in a country. Second, we must design more enlightened immigration policies. Agency official W. R. Böhning (1992) argues that the policies must be based on three principles: nondiscrimination (especially at the institutional level), respect for the immigrants' culture and language, and demarginalization. The last principle has to do with eliminating discriminatory practices that push immigrants into the margins of society (into ethnic ghettos, into the underclass, and into underachievement in schools).

FOCUS
Race as a Social Construction

Earlier in this chapter we pointed out that race and ethnic classification schemes both have major shortcomings, so much so that it is appropriate to question whether race and/or ethnicity have any biological validity. As sociologist Prince Brown, Jr., argues in the focus section below, if we simply consider the fact that "from the perspective of evolutionary biology we are all Africans—sharing common ancestors who evolved first in Africa"—how can anyone claim that he or she is a member of one racial category? The case of Africans and Native Americans is a powerful example of the idea that race is a social construction. As you read this section, consider that by social convention the people we call Native Americans are classified as belonging to the Mongoloid race and that the people we call Africans are classified as belonging to the Negroid race. In light of the material that Brown presents, does such a classification scheme make sense?

Why "Race" Makes No Scientific Sense: The Case of Africans and Native Americans
Prince Brown, Jr.

In the futile efforts to create "racial" classification schemes, people born to the combination of African and Native American parents have been overlooked. Examining this particular case of the fusion of biology and culture assists understanding of human variation and makes clear that fixed, distinct, and exclusive categories (races), never have existed and do not now exist.

SOURCE: Prince Brown, Jr., Northern Kentucky University

Everywhere ships anchored in the Americas, Asia, Africa, and the various islands there was immediate exchange and sharing of human genes. In the case of Africans and Native Americans, this process was set in motion more than 500 years ago and continues unabated. Hence, many persons labeled African American have Native American ancestry, and the reverse is true. According to the African American historian Carter G. Woodson, the history of the relationship between Africans and Native Americans is long, deeply intertwined, and largely unwritten (Woodson 1920, p. 45).

In this essay the term *Native American* shall be used to refer to the native inhabitants of North, Central, and South America and the Caribbean Islands. *African* shall be used to refer to people descendant from the African continent; *European* shall, in the same manner, be used to reference people from Europe.

Conventional history teaches that Native American–African contact commenced with Columbus's arrival in the Americas in 1492. We may start then by calling attention to the physical characteristics of the Spanish and Portuguese sailors making up the crews of the first voyages of exploration (1492–1520). Many of them can be labeled "mulatto," which, according to Forbes (1993, p. 140), may have derived from the Arabic, translated as half-caste, mestizo, hybrid, half-breed, or half-blood (that is, any person with one Arabic and one non-Arabic parent). The European inhabitants of Spain and Portugal had known many dark-complected North African Moors (CKSSG 1989, p. 403) as well as Africans from south of the Sahara as conquerors for more than 800 years. Thus, the first contact between Native Americans and part-Africans (Portuguese and Spanish sailors) occurred in the Caribbean with the arrival of Columbus's ships. It is safe to assume that some of the sailors may have been unmixed Africans as well since many were brought to Spain and Portugal as slaves after about 1440.

Immediately upon the arrival of Spanish ships in the Caribbean the enslavement of Native Americans began. Columbus's disdain of the natives, his disregard for their humanity, and his perception that their mere existence signaled their availability as objects to be used in furthering his ambitions earned him the title "Father of Native American Slavery." It was he who suggested to the Spanish monarchs that the docile Native Americans who welcomed his crew would make ideal slaves (Forbes 1993, pp. 21–25). It is not well known that thousands of Native Americans were shipped to Europe as slaves and in subsequent years served as soldiers for the Portuguese and Spanish in their wars for control of the West African slave trade (Forbes 1993). Thus, African-Native American interbreeding occurred in Europe and Africa as well as in the Americas.

It was, on the other hand, Columbus's contemporary, Bartolome de la Casas, whose concern over the high rate of death among enslaved Native Americans led to his suggestion that Africans be used in that capacity instead. La Casas was a priest who observed firsthand the appalling treatment of Native Americans.

We know that almost as soon as they arrived Africans started to run away and join Native Americans (Forbes 1993), establishing joint communities (Katz 1986, Forbes 1993) and independent ones (Campbell 1990). They clearly preferred to face the unknown rather than live as slaves. These runaways were called Cimarrons in South America and Maroons in the Caribbean. Women of Native American ancestry were the first mothers of children born to European and African men in the initial phase of settlement (Forbes 1993), because there were no European women and few African women present. As the slave trade grew in scope, more and more African women were brought in (Rogers 1984). It was not until Europeans were able to create secure zones containing permanent housing that they were joined by sizable numbers of women from Europe.

Forbes argues that, given the initial absence of European women and the fact that the number of imported slaves continued to grow, people of African and Native American descent made up the majority of the population in the colonial territories (Forbes 1993). The situation was only slightly different in the United States. Relative to the rest of the Americas, it was settled much later and European women were among the early colonists.

Keep in mind that many of the first African slaves brought in to the colonies that later became the United States had been slaves in the Caribbean. Therefore, many were already mixed biologically with Native Americans. As the number of African slaves continued to grow, population ratios changed quickly. In fact, by 1765 a Massachusetts census showed more Africans than Native Americans in some counties. The increase in the slave population led to more runaways and, consequently, increased absorption of Native Americans into the African population.

Africans took advantage of the presence of nearby Native American settlements as primary destinations when they decided to run away from their owners. This process started very early and continued until the end of slavery. It does not appear that Native Americans made any strong effort to return slaves, and Europeans were apparently too busy establishing the colony to invest much time in slave retrieval. Massachusetts passed the first laws enslaving Africans (initially they were free persons) between 1639 and 1661. This followed the failed effort to enslave Native Americans, who were held in joint bondage with Africans for a while. Prior to the passage of laws enslaving them, Africans (the majority of whom were males) could, of their own volition, choose to join Native American communities. It is logical to assume that they would have taken wives primarily from the Native American community, because European women would not have been available to them.

Native Americans and Africans shared a number of social traits and values that facilitated their absorption into each other's culture. The recognition of kinship traced matrilineally (on the female side) and the practice of polygyny (multiple wives) constituted common forms of social organization. Other shared views centered on a spirituality that celebrated the mutual dependency of humans and animals. This belief was manifested through the social structure known as the clan and represented by an animal icon. Africans and Native Americans possessed a reverence for the earth, which they considered sacred. These shared traits made the social acceptance of one group by the other easier than it otherwise might have been. Also helpful was the fact that the subtropical setting, as found in the southeastern United States, was one with which both groups were familiar.

Africans brought to Native American communities an intimate knowledge of Europeans, skilled farming techniques acquired in their former slave roles, and a spirit of resistance to domination, evidenced by their status as runaways. They emerged as important political and military leaders in the communities in which they settled (Katz 1986). Europeans frequently expressed fear of African-inspired revolts with Native American support (Porter 1932, Forbes 1993, Mullin 1992). These hybrid communities were the centers of the most successful defenses against European domination. The Seminoles of Florida provide an outstanding example of successful autonomy. This group of Africans and Native Americans engaged in protracted military struggle with the United States army. Though outnumbered and less well armed, they defended their community with tenacity to avoid being resettled in Indian Territory (what later became Oklahoma). After the Civil War, many Seminoles served as guards on the Mexican border and as scouts for the same army that they had fought in Florida.

The widespread interbreeding between Native Americans and Africans led to a number of efforts by writers and legislators to develop terms that specify different biological combinations. But, in fact, terms like *mulatto*, *mustee*, and *colored* (terms that were later adopted by social scientists) were not used consistently within or between societies. They carried no more specificity than the expression "people of color," which has been applied to Native American–African, European-African, European–Native American, and Asian peoples. *Mustee* evolved in the United States to refer to persons of African–Native American ancestry (Johnston 1929).

In South America the terms *mulatto*, *zambo*, and *zambaigo* have been used for the same purpose at different times (Forbes 1993). Likewise, the term *negro* has not always been used to refer only to African ancestry. Therefore, it cannot be assumed that the literature on the subject specifying ancestry carries the definition assigned by the reader. Paul Cuffee, a well-known "free person of color," is often presented as being of African extraction when, in fact, he is of African–Native American ancestry. Off the reservation, persons of African–Native American ancestry are commonly perceived and spoken of as African. On the reservation, those same persons are Native Americans.

Classification laws determining the legal status of people of color changed with each event that Europeans defined as a threat. Depending on location and

Note the mixed facial features in the members of this Seminole family.
Courtesy of the Oklahoma Historical Society

time, a person could be legally free if born to a Native American mother, or a slave if born to an African mother. Thus, it was an advantage for persons so labeled to be born to a Native American mother.

On the several occasions between 1869 and 1902 when the state of Massachusetts re-allocated lands assigned to Native Americans, many persons who were compensated with small pensions were more African than Native American in appearance (Woodson 1920, p. 57). Katz reports the existence of "Black Indian" societies (1986, p. 129) in almost every seaboard state between the American Revolution and the Civil War.

Europeans in King William County, Virginia, petitioned the legislature to take away the land deeded to the Pamunkies because "these people had become Negro and hence had no claim to the rights of the Indians" (Johnston 1929, p. 29). The charge against another group of Native Americans, the Gingaskin, in 1784 was that their "land is at present an asylum for free Negroes and other disorderly persons, who build huts thereon and pillage and destroy the timber without restraint to the great inconvenience of the honest inhabitants of the vicinity, who have ever considered it a den of thieves and a nuisance to the neighborhood" (p. 32). The Gingaskin were successfully removed from their lands by a court order in 1812.

Legislative actions demonstrate but one of the ways in which "racial" classification schemes are used to subordinate less powerful groups. This pattern of removing Native Americans from land formally ceded to them by treaty was one that was repeated many times in the United States.

The slave-based colonial culture literally pushed Africans and Native Americans together. In fact, some Native American groups adopted the European practice of enslaving Africans. But if benign slavery were ever the case in the United States, it occurred among the Five Civilized Tribes (so labeled because of their adoption of European customs) of the Southeast. The Cherokee, Chickasaw, Choctaw, Creek, and Seminole were centered in Florida, Georgia, Alabama, and Mississippi. The first two did display some color-prejudice toward Africans. For the most part, however, researchers are agreed that relations between these masters and slaves involved minimum supervision and often approached equality as they struggled against a common enemy.

The emulation of the European practice of slavery did not prevent the removal of these groups, with their slaves, to Indian Territory (which later became Oklahoma) after the passage of the Indian Removal Act of 1830. With the implementation of this act (during which more than 60 different Native American ethnic groups were forced to move), Oklahoma became known as an "Indian" state (Strickland, 1980). Perhaps a more accurate description may be that it became an "Indian and African" state. Further north and west (Minnesota) some of the first pioneers to live with and marry Native Americans were of African descent (Katz 1987).

The extensive biological mixing of Africans and Native Americans clearly demonstrates the compatibility of human genes deriving from groups and individuals where some aspects of their physical appearance are different. The process of absorption of Native Americans into the African population largely accounts for the fact that there is such a small population of people labeled Native Americans on the East Coast of the United States. Jack D. Forbes contends that the view that indigenous peoples in the Americas were replaced [killed off or died out] and the area repopulated by Europeans and Africans is erroneous. Rather, he suggests that

> *[Native] American survivors and African survivors (because huge numbers of Africans also died in the process) have merged together to create the basic modern populations of much of the Greater Caribbean and adjacent mainland regions.* (1993, P. 270)

The case of African and Native Americans makes clear that the label *race*, when assigned to a particular set of observable human features, is socially derived. The categories widely used to denote "race" are inappropriate since they do not capture "real" biological distinctions (Levin, 1991) but rather reflect social and cultural conventions. The point is that it is not possible to identify biological ancestry simply by referencing physical features. Indeed, from the perspective of evolutionary biology we are all Africans—sharing common ancestors who evolved first in Africa.

Key Concepts

Notes

1. The figure 50 million includes an estimated 35 million slaves who died en route to the Americas because of the inhumane conditions aboard ship. According to ships' logbooks:

   The height, sometimes, between decks, was only eighteen inches; so that the unfortunate human beings could not turn around, or even on their sides. . . . [H]ere they are usually chained to the decks by the neck and legs. In such a place the sense of misery and suffocation is so great, that the Negroes . . . are driven to frenzy. (ZINN 1980, P. 28)

   When they had the opportunity, many captives jumped into the ocean and held themselves under water until they drowned to escape these conditions.

2. Other such countries include Argentina, Australia, Austria, Bahrain, Belgium, Canada, France, Iraq, Ivory Coast, Kuwait, Libya, Malaysia, New Zealand, Netherlands, Nigeria, Oman, Peru, Qatar, Saudi Arabia, South Africa, Sweden, Switzerland, United Arab Emirates, United Kingdom, and Venezuela. Labor migrants make up 10 percent of the labor force in Germany and 9 percent in the United States (Miller and Ostrow 1993). In some countries the percentage is much greater. In Switzerland, for example, labor migrants account for as much as 25 percent of the work force. Two-thirds of the labor force in the Persian Gulf countries (Kuwait, Qatar, Bahrain, Oman, and United Arab Emirates) are migrants.

3. The 34 countries are Albania, Algeria, Argentina, Austria, Belgium, Bermuda, Czechoslovakia, Denmark, Estonia, Finland, France, Germany, Gibraltar, Guadeloupe, Hungary, Iceland, Indonesia, Ireland, Italy, Japan, Latvia, Liechtenstein, Lithuania, Luxembourg, Monaco, New Caledonia, Netherlands, Norway, Poland, San Marino, Sweden, Switzerland, Tunisia, and United Kingdom.

4. The Potsdam Conference was a post–World War II meeting in which the Allies planned the occupation of Germany and transferred German territory to Poland, the Soviet Union, Czechoslovakia, and Hungary. After the Potsdam Conference, millions of Germans in these territories were relocated back to Germany.

5. The so-called oil crisis in 1973 was triggered by an armed conflict initiated by Egypt against Israel. Israel struck back and defeated Egypt for the third time since the new nation's inception. The Arabs blamed Israel's military superiority on U.S. sponsorship and initiated an oil embargo against Western countries. As a result the price of oil jumped from $3 to $12 a barrel (Halberstam 1986).

6. The liberal German asylum laws were adopted as partial atonement for the systematic persecution and genocide of Jews and other so-called foreigners during World War II.

7. Although the Turkish government claims officially that Turkey is a homogeneous society, its

constitution reads: "[T]he Turkish state, with its territory and nation, is an indivisible entity. Its language is Turkish" (Pitman 1988).

8. Those who design ads and posters and cast actors for movies and television sitcoms choose people who "look like" members of the racial and ethnic group they are supposed to represent. Some very light-skinned African-Americans thus are overlooked as actors who can play black characters (Proffitt 1993). In reality, there is no such thing as an "ethnic look." For example, "from a historical perspective, the 'Asian look' is one that ranges across a wide spectrum; from the paler-than-Caucasian look . . . to the chocolate brown Afro-Asian face of a Micronesian" (Houston 1991, p. 55).

9. Social scientists, including anthropologists and sociologists, unwittingly supported racist ideology when they classified societies around the world on the basis of their technological development, using white European technology as the standard for judging other cultures. By this definition most non-European peoples were "people of nature" and, by extension, "in comparison to Europeans were 'underdeveloped,' 'primitive,' 'uncivilized,' or 'backward'" (Opitz 1992d, p. 229).

10. Spike Lee drew considerable criticism for advocating violence as a means for solving problems when he ended his film *Do the Right Thing* with two quotes—one by Martin Luther King, rejecting violence as evil and ultimately ineffectual, and the other by Malcolm X: "I don't even call it violence; when it's self-defense, I call it intelligence." From Lee's point of view, such criticism meant only that "we're not allowed to do what everyone else can. The idea of self-defense is sup

posed to be what America is based on. But when black people talk about self-defense, they're militant. When whites talk about it, they're freedom fighters" (McDowell 1989, p. 92).

11. The Haitians are fleeing a country in which 90 percent of the people cannot read or write and 90 percent of the working-age population is unemployed. On the political side, the ruling elite historically has persecuted all of those who oppose its policies. Haiti is ruled by military dictators who overthrew a democratically elected leader (Aristide) in a bloody coup. In 1994 the flood of refugees became so large and controversial that the Clinton administration set up refugee and processing centers at Guantánamo Bay.

12. The law forbidding slaves to learn to read was written as follows:

Whereas the teaching of slaves to read and write, has a tendency to excite dissatisfaction in their minds, and to produce insurrection and rebellion, to the manifest injury of the citizens of this State:

Therefore: Be it enacted by the General Assembly of the State of North Carolina . . . That any free person, who shall hereafter teach, or attempt to teach any slave within the State to read or write, the use of figures excepted, or shall give or sell to such slave or slaves any books or pamphlets, shall be liable to indictment . . . and upon conviction, shall, at the discretion of the court, if a white man or woman, be fined not less than one hundred dollars, nor more than two hundred dollars, or imprisoned; and if a free person of color, shall be fined, imprisoned, or whipped, at the discretion of the court, not exceeding thirty-nine lashes, nor less than twenty lashes. (GENERAL ASSEMBLY OF THE STATE OF NORTH CAROLINA 1831)

II GENDER

with Emphasis on the Former Yugoslavia

Sarajevo before the Bosnian war.
Fridmar Damm/Leo de Wys, Inc.

Note: This map shows the traditional boundaries before the breakup of Yugoslavia.

Students from Yugoslavia enrolled
in college in the United States for fall 1991   1,430

Airline passengers flying between
the United States and Yugoslavia in 1991   97,188

Applications for utility patents filed in the
United States in 1991 by inventors from Yugoslavia   30

Phone calls between the
United States and Yugoslavia in 1990   4,363,042

Those from the former Yugoslavia admitted into the
United States in fiscal year 1991 for temporary employment   18,980

In *The Balkan Express: Fragments from the Other Side of War,* Croatian writer Slavenka Drakulić includes the essay "A Letter to My Daughter." In the letter Drakulić tells her daughter about an article that appeared in a local newspaper. The letter appeared soon after the republic of Croatia declared its independence from Yugoslavia and the Serb-dominated Yugoslavian army invaded Croatia:

Entitled 'Will You Come to My Funeral?' it was about the younger generation and how they feel about the war. I remember an answer by Pero M., a student from Zagreb.

"Perhaps I don't understand half of what is going on, but I know that all this is happening because of the fifty or so fools who, instead of having their sick heads seen to, are getting big money and flying around in helicopters. I'm seventeen and I want a real life, I want to go to the cinema, to the beach . . . to travel freely, to work. I want to telephone my friend in Serbia and ask how he is, but I can't because all the telephone lines are cut off. I might be young and pathetic-sounding, but I don't want to get drunk like my older brother who is totally hysterical or to swallow tranquillizers like my sister. It doesn't lead anywhere. I would like to create something, but now I can't."

And then he said to the reporter interviewing him something that struck me:

"Lucky you, you are a woman, you'll only have to help the wounded. I will have to fight. Will you come to my funeral?"

This is what he said in the early autumn of 1991. I could almost picture him, the street-wise kid from a Zagreb suburb, articulate, smart, probably with an earring and a T-shirt with some funny nonsense on it, hanging out in a bar with a single Coke the entire evening, talking about this or that rock group. The boy bright enough to understand that he might die and that there is nothing that he could do about it. But we—me, you, that woman reporter—we are women and women don't get drafted. They get killed, but they are not expected to fight. After all someone has to bury the dead, to mourn and to carry on life and it puts us in a different position in the war.
(1993A, PP. 133–34)

This chapter is about **gender**, which sociologists define as social distinctions based on culturally conceived and learned ideas about appropriate behavior and appearance for males and females. "Appropriate" male and female behavior varies according to time and place. In the 1950s, for example, men were not expected to witness their children's births. Today, however, it is taken for granted that the father will be present. As another example, women in the United States typically remove hair on their faces (even very small amounts of hair), on their legs, and under their arms. In contrast, people in many countries in Europe and elsewhere do not expect women to be hairless in these areas.

If we simply think about the men and women we encounter every day, we quickly realize that people of the same sex vary in the extent to which they meet their society's gender expectations. Some people conform to gender expectations; others do not. This variability, however, does not stop most people from using their society's gender expectations to evaluate their own and others' behavior and appearances in "virtually every other aspect of human experience, including modes of dress, social roles, and even ways of expressing emotion and experiencing sexual desire" (Bem 1993, p. 192).

Sociologists find gender a useful concept, not because all people of the same sex look and behave in uniform ways, but because a society's gender expectations are central to people's lives whether they conform rigidly or resist. For many people, failure to conform to gender expectations, even if they fail deliberately or conform only reluctantly, is a source of intense confusion, pain, and/or pleasure (Segal 1990). Pero M. is confused and pained because, as a male, he is expected (required) to serve his country. From his point of view, women are lucky because their sex puts them in a different, safer position than his. On the other hand, we might speculate that Ila Borders, the first female to pitch in a men's NAIA collegiate baseball game, experienced considerable pleasure from accomplishing a feat that goes against gender expectations. She threw a three-hit, seven-inning shutout to help her team (Southern California College) defeat Claremont College. Yet, at the same time, Borders finds it annoying that she must continually remind peo-

ple that she is "not a bruiser" and likes being a girl, using lipstick, and wearing feminine clothes (Stevenson 1994, p. B10).

In exploring the concept of gender, we focus on the former Yugoslavia, which at the time of this writing has split into five separate countries. Four of the six republics of the former Yugoslavia—Croatia, Slovenia, Bosnia-Herzegovina, and Macedonia—began to declare independence about two years after the fall of the Berlin Wall. The two remaining republics—Serbia and Montenegro—call themselves the new Federal Republic of Yugoslavia (see "The Breakup of Yugoslavia"). Although the former Yugoslavia is now five countries, it is impossible to speak about one of these countries without considering the other four. For all practical purposes, the fate of one is intertwined with that of the others.

We pay special attention to the former Yugoslavia because it offers an interesting contrast with the United States. In the United States the question of gender differences, specifically the position of women in society relative to that of men, is subject to constant analysis and debate. In view of the ongoing, severe economic upheaval that has taken place in the former Yugoslavia over the past decade and in light of the wars that have been fought since 1991, gender is not a pressing issue there. In fact, gender was never an issue in Yugoslavia in the same way as in the United States. Officially, at least, the question of inequality between women and men was resolved under communism. This does not mean, however, that there is no connection between gender and life chances in the former Yugoslavia.

In this chapter we explore the concepts used by sociologists to analyze the connection between gender and life chances. In outlining this connection, sociologists distinguish between sex (a biologically based classification scheme) and gender (a socially constructed phenomenon). They also focus on the extent to which society is gender-polarized—that is, organized around the male-female distinction. In addition, sociologists seek to explain gender stratification and the mechanisms by which people learn and perpetuate their society's expectations about appropriate behavior and appearances for males and females. Finally, sociologists explore the

Ila Borders with her teammates.
Jan Sonnenmair/NYT Pictures

interactions between gender and such variables as race and ethnicity. Before discussing these concepts, we examine why gender is not considered a pressing issue in the former Yugoslavia.

The Question of Gender in the Former Yugoslavia

One reason that gender is not an issue in the former Yugoslavia in the same sense as in the United States may be that there is virtually no feminist voice there. In the broadest sense of the word, a **feminist** is a man or woman who actively opposes gender scripts (learned patterns of behavior expected of males and females) and believes that men's and women's self-image, aspirations, and life chances should not be constrained by those scripts (Bem 1993). For example, a man should be free to choose to stay home and take care of the children rather than pursuing a full-time career; a female athlete should be able to develop her physique beyond what is considered feminine. Unfortunately for many people, the term *feminist* evokes extremely negative images and stereotypes of mannish-looking women who hate men and who find vocations such as mother and wife oppressive and unrewarding. In the former Yugoslavia, even more than in the United States, very few men or women will declare themselves feminists, even if they live according to the basic feminist principle (Drakulić 1993b).

In the former Yugoslavia (and throughout Eastern Europe, for that matter), only a handful of feminists talk and write about gender issues. The unofficial feminist organizations that exist are concentrated in three cities—Belgrade, Zagreb, and Ljubljana—and they are "small and without money or institutional support" (Drakulić 1993b, p. 127; Ramet 1991). Colleges and universities in the former Yugoslavia offer no women's studies or gender studies programs. It was not until 1979 that Slavenka Drakulić (the best-known feminist of her country) and about 30 other persons, including some male journalists and intellectuals, founded Yugoslavia's first feminist group (Drakulić 1993b;

The Breakup of Yugoslavia

One can argue that the breakup of Yugoslavia in the early 1990s actually began with the death of Marshal Josip Tito, Yugoslavia's "President-for-Life," in 1980. Tito established a communist government in 1945 after uniting various ethnic factions (Serbs, Croats, Muslims, and others) who were at war with each other to defeat the German army occupying the country at that time.*

While Tito was in power he used a careful, if often brutal, balance of reward and repression to realize his ideal of ethnic tolerance in a multicultural society (C. Williams 1993). People who claimed to be Serb, Croat, or other ethnic category first and Yugoslav second risked arrest for being a nationalist (Ignatieff 1993). In addition, Tito distributed political power according to ethnicity so that each ethnic group dominated a specific republic or autonomous region.

Under this arrangement, the Serbs, Croats, Slovenes, Macedonians, Montenegrins, and Muslims were considered nations. It is worth noting that Tito gave Muslims nation status in the late 1970s in order to put to rest the Serb-Croat conflict over whether the Muslims were Serbian or Croatian. The answer would have allowed either Serbs or Croats to achieve majority status in Bosnia (Curtis 1992).

Tito's death created a leadership vacuum that proved particularly problematic in 1989, the year in which the Berlin Wall was dismantled. By that time Yugoslavia's economy had deteriorated to the point of collapse.‡ With the end of the Cold War and the overthrow of communist governments in Eastern Europe, Yugoslavia was no longer able to play the Soviet and American blocs against each other to gain economic aid (Borden 1992). This economic and political atmosphere intensified the forces of fragmentation within Yugoslavia.

In 1990 Slovenia, Bosnia, Macedonia, and Croatia elected noncommunist leaders who favored market reforms. Serbia and Montenegro, on the other hand, elected communist leaders by decisive margins. Particularly interesting was Serbia's election of Slobodan Milosevic, a fervent nationalist who claimed that the Yugoslav government had shortchanged Serbian interests ever since Tito came to power (Borden 1992). He used the government-controlled media to advocate a policy of uniting all of the Serbs of Yugoslavia into one state. Before and after his 1990 election, Milosevic focused his attention on Kosovo, an autonomous region within Serbia in which 92 percent of the population were Albanian and 8 to 10 percent were Serbian. Milosevic sent the national army into the region to prevent Kosovo from declaring its independence from Serbia and subsequently abolished its autonomous status.

Soon after this event, Slovenia and Croatia, the two wealthiest republics, declared their independence. Because the Serbian members dominated the votes in the collective federal presidency, Milosevic was able to use the Yugoslavian People's Army—minus the Croats and Slovenes, who had been dismissed—to forcibly preserve Yugoslavia's territorial integrity. The ensuing war with Slovenia lasted about 10 days. The war with Croatia lasted approximately six months, until a U.N.–monitored cease-fire was declared. By that time the Serbian-controlled armed forces had managed to take control of 25 percent of Croatia's territory.

In Bosnia, the multiethnic parliament, not wishing to be part of the Serbian-dominated "new" Yugo-

Ethnic Composition of the Three Largest Republics of the Former Yugoslavia, 1991[†]

		%
Bosnia-Herzegovina	4,365,639	
Muslims	1,900,000	44
Serbs	1,450,000	33
Croats	750,000	17
Yugoslavs	250,000	6
Others	15,639	0.4
Croatia	4,703,941	
Croats	3,500,000	74
Serbs	700,000	15
Yugoslavs	400,000	9
Others	103,941	2
Serbia	9,721,177	
Serbia Proper	5,753,825	
Serbs	5,500,000	96
Muslims	125,000	2.2
Gypsies	50,000	0.9
Croats	40,000	0.7
Others	38,825	0.6
Kosovo	1,954,747	
Albanians	1,630,000	83
Serbs	250,000	12.8
Muslims	40,000	2.0
Gypsies	30,000	1.5
Others	4,747	0.2
Vojvodina	2,012,605	
Serbs	1,400,000	70
Hungarians	450,000	22
Croats	100,000	5
Romanians	50,000	2.5
Others	12,605	0.6

A woman in Sarajevo visits the grave of a loved one during a cease-fire.
Chris Rainier/JB Pictures

slavia, declared its sovereignty. A referendum was placed before the Bosnian people and passed with 99 percent of the vote in favor. A significant number of Bosnian Serbs and Croats, however, boycotted the election. The Serbs threatened to secede in order to become part of Serbia if the Bosnian government declared independence, while the Croats (17 percent of the population) warned that they would secede to become part of Croatia if Bosnia did *not* declare independence.

After the referendum passed, the European Community and the United States recognized Bosnia as a new state. Serbs opposed to independence took control of roads leading into Sarajevo, Bosnia's capital and a 600-year-old city with an ethnically and culturally mixed population of 600,000, making it impossible for anyone to enter or leave. The so-called Yugoslav national army entered Bosnia to take control of the cities and towns that border Serbia.

At first Croatians and Bosnians (including some Bosnian Serbs) fought as allies against the Serbian-dominated armies. The Bosnia-Croatia alliance collapsed, however, as nationalistic Croats from both Croatia and Bosnia sought to claim their share of Bosnian territory.

Nationalistic Bosnian Serbs and Croats declared Serbian and Croatian states within Bosnia. Each boasted military units supported with arms, money, and recruits from Serbia and Croatia, respectively. Between them, the Serbs and Croats seized or surrounded so much of Bosnia's territory that the Bosnian Muslims were rendered practically stateless. Bosnia, a new country with no army, weapons, or military traditions, was no match for the Croatian and Serbian forces. The Bosnian army—which is predominantly Muslim but which contains significant numbers of Croats, Serbs, and other ethnic groups opposed to the formation of ethnic states—was particularly affected by

a U.N.–mandated arms embargo to the region.

As of mid-1994, Serbian armed forces surrounding Sarajevo were slowly strangling the city with the goal of making Sarajevo part of Greater Serbia. Despite the attack by Serbian forces, approximately 90,000 Serbs remained in Sarajevo with Muslims, "Yugoslavs,"§ Croatians, and other ethnic groups who refused to be a part of the nationalist policies (Glenny 1992). "The Serb leadership, for its part, has made a concerted effort to see that as many educated Muslims as possible are killed, so that even if some sort of independent Bosnia and Herzegovina remains when the war ends, what will exist will be a state bereft of people who can make it work" (Rieff 1992, p. 84).

---

*In 1974 Tito instituted a government structure to succeed him after his death. Executive power was to be held by a committee with one representative from each of the six republics and the two autonomous regions within Serbia. The chairmanship was to rotate among the members. This created a power vacuum, however, because no one person was in charge and no one was able to enforce Tito's ideal of ethnic tolerance in a multicultural society.
‡At that time at least one million people were unemployed and one-third of the work force lived "below poverty." Inflation was at four figures and the standard of living had declined to the level of 30 years earlier.
†From *Yugoslav Survey, XXXII* (March, 1990–91), 5. The numbers for republics and autonomous provinces are actual; the numbers for the population groups are estimates.
§"Yugoslavs" are people who claim no national (ethnic) identification. That is, instead of identifying themselves as Croat, Muslim, Serb, and so on, they identify themselves with the country.

Kinzer 1993), and feminism still has not gained strength as a movement. In fact, Drakulić's book *The Deadly Sins of Feminism,* the first feminist book produced in Yugoslavia, was not published until 1984 (Drakulić 1990).

By contrast, feminist literature in the United States has a relatively long history. The feminist scholarship that has been published by men and women, especially since 1960, is vast and wide-ranging (Komarovsky 1991). In the United States Drakulić is quite popular; she has appeared on talk shows, and her two books of essays, *How We Survived Communism and Even Laughed* and *The Balkan Express,* are selling quite well. They are not available in Croatia, where the most vocal critics consider her a "cheap and defective" writer who took up feminism as a way to "rape Croatia" (Kinzer 1993, p. 4Y). In some newspaper editorials, Drakulić and other feminists have been labeled "witches" (Kirka 1993).

Another reason that organized feminism fails to thrive in Yugoslavia, and in Eastern Europe in general, is that the communist system as implemented has failed to meet people's basic needs and has made life hard for everyone. Thus, it is difficult to view men as advantaged or as able to achieve their aspirations: "It's hard to see them as an opposite force, men as a gender . . . perhaps because everyone's identity is denied" (Drakulić 1992, p. 109). In comparison with the former Yugoslavia, the United States has an impressive ratio of people to resources, abundant job opportunities, and a wide choice of services and products. Because opportunities for mobility are relatively plentiful in the United States, mobility and the factors that enhance or prohibit mobility are an important question here.

Finally, there is almost no feminist voice in the former Yugoslavia because, as mentioned earlier, the official position is that "the woman question" ceased to exist after World War II. The end of that war also marked the end of the Yugoslavian National Liberation War against Fascist Germany. During this war the Yugoslav Communist Party formed and coordinated the activities of the Anti-Fascist Women's Front of Yugoslavia, which supported male-dominated resistance groups. This organization "brought together some 2,000,000

Croatian writer Slavenka Drakulić helped to found Yugoslavia's first feminist group in 1979.
Filip Horvat/Saba

women during the war; approximately 100,000 took part in regular partisan military units, and of that number 25,000 were killed and 40,000 badly wounded" (Milic 1993, p. 111). Many other women, through their veterans' and political activities, attained leading positions in the party and in local administrative bodies, or became national heroes.

Many women who participated in the national struggle for freedom found it a liberating experience. It gave them opportunities to experience something beyond the closed world of the traditional patriarchal family, which limited their life choices to marriage or domestic service. After the war they could not return to this world.

When a new communist government was formed after the war, its leaders made women's rights a part of official doctrine, but with a twist. Government leaders took the position that, because women's liberation was achieved through the pursuit of a higher goal—national liberation—their equality would come with the pursuit of other more important goals, such as class struggle and economic stability. This assumption underlay all efforts to address "the woman question." The communists passed legislation that gave women the right to vote, equal pay for equal work, access to the salaried labor force, publicly funded child care, the right to abortion, and paid maternity leave, among other things. The Communist Party set a 30 percent quota to encourage women to participate in the political arena.

In spite of this kind of legislation, however, "women remained subordinated and segregated in all walks of life" (Milic 1993, p. 111). Drakulić argues, for example, that, despite the quota for political participation, women never achieved an independent, secure voice in the political sphere (Drakulić 1993b). As evidence of this claim, she points to the fact that after the 1990 elections, the proportion of women in government declined from 30 to somewhere between 5 and 10 percent (depending on the republic).

Critics argue that the overall economic and social gains made under communism have eroded steadily since Tito's death in 1980. The gains allowed Yugoslavia to be the most liberal and economically well-off communist country. They gave people the feeling that the communist plan was working. The gains were possible because post–World War II Yugoslavia was unique among Eastern European countries. It resisted alignment with the Soviet Union and maintained friendly, if distant, relations with the United States. Yugoslavia was unique in that the communist government permitted its people to travel and work outside the country.[1] As a result, Yugoslavia benefited from money that its people earned as guest workers (in countries such as West Germany and Austria) and sent home. Yugoslavia's economy also depended heavily on tourism, a service industry sensitive to seasonal fluctuations and political and economic events (Curtis 1992).[2]

Yugoslavia's prosperity—built on remittances from Yugoslav guest workers working abroad, tourism, and loans from the International Monetary Fund—began its decline in 1979 when the world recession forced many guest workers to return home. The accompanying declines in living standards, together with inflation and unemployment, left some groups more vulnerable than others, making clear who benefited most under the communist system. The gains made by women under Tito were especially vulnerable, as evidenced by the disappearance of women from political life. In addition, newly elected leaders in all of the former republics are conservative. In the atmosphere of rapid change and severe economical upheaval connected with the fall of communism, with severe economic hardship, and with the wars, gender issues still are considered irrelevant to the larger national concerns that must be solved first.

In light of this history, there is very little literature on gender issues in Yugoslavia or in Eastern Europe for that matter. The first anthology of essays devoted to the status of women in Eastern Europe was just published in the United States in 1993 (Katzarova 1993).

Distinguishing Sex and Gender

Although many people use the terms *sex* and *gender* interchangeably, the two terms do not have the exact same meaning. Sex is a biological concept, whereas gender is a social construct. In the following section we will pursue this distinction further because it will help illustrate how the social differences between males and females develop.

Sex as a Biological Concept

A person's sex is determined on the basis of **primary sex characteristics**, the anatomical traits essential to reproduction. Most cultures divide the population into two categories—male and female—largely on the basis of what most people consider to be clear anatomical distinctions. Like race (see Chapter 10), however, even biological sex is not a clear-cut category, if only because a significant (but unknown) number of babies are born **intersexed**. This is a broad term used by the medical profession to classify people with some mixture of male and female biological characteristics. The intersexed group includes three very broad categories: *true hermaphrodites*, persons who possess one ovary and one testis; *male pseudohermaphrodites,* persons who possess testes and no ovaries, but some elements of female genitalia; and *female pseudohermaphrodites,* persons who have ovaries and no testes but some elements of male genitalia.

Why, then, is there no intersexed category? Instead, parents of intersexed children collaborate with physicians to assign their offspring to one of the two recognized sexes. Intersexed infants are treated with surgery and/or hormonal therapy. The rationale underlying medical intervention is the belief that the condition "is a tragic event which immediately conjures up visions of a hopeless psychological misfit doomed to live always as a sexual freak in loneliness and frustration" (Dewhurst and Gordon 1993, p. A15).

Even adding a third category would not do justice to the complexities of biological sex. French endocrinologist Paul Guinet and Jacques Descourt estimate that on the basis of variations in the appearance of external genitalia alone, the category "true hemaphrodite" may contain as many as 98 subcategories (Fausto-Sterling 1993). This variation within only one intersex category suggests that "No classification scheme could [do] more than suggest the variety of sexual anatomy encountered in clinical practice" (p. A15).

The picture becomes even more complicated when we consider that a person's primary sex characteristics may not match the sex chromosomes. Theoretically one's sex is determined by two chromosomes: X (female) and Y (male). Each parent contributes a sex chromosome: the mother contributes an X chromosome and the father an X or a Y depending on which one is carried by the sperm that fertilizes the egg. If this chromosome is a Y, then the baby will be a male. Although we cannot possibly know how many people's sex chromosomes do not match their anatomy, the results of mandatory "sex tests" of female athletes over the past 25 years have shown us that such cases exist and that a few women are disqualified from each Olympic competition and from other major international competitions because they "fail" the tests (Grady 1992). Perhaps the most highly publicized after-the-fact case is that of Spanish hurdler Maria José Martinez Patino, who although "clearly a female anatomically, is, at a genetic level, just as

clearly a man" (Lemonick 1992, p. 65). Upon giving her the test results, track officials advised her to warm up for the race but to fake an injury so as not to draw the media's attention to her situation (Grady 1992). Patino lost her right to compete in amateur and Olympic events but subsequently spent three years challenging the decision. The IAAF (International Amateur Athletic Federation) restored her status after deciding that her X and Y chromosomes gave her no advantage over female competitors with two X chromosomes (Kolata 1992; Lemonick 1992).[3,4]

In addition to primary sex characteristics and chromosomal sex, **secondary sex characteristics** are used to distinguish one sex from another. These are physical traits not essential to reproduction (breast development, quality of voice, distribution of facial and body hair, and skeletal form) that result from the action of so-called male (androgen) and female (estrogen) hormones. We use the phrase "so-called" because, although testes produce androgen and ovaries estrogen, the adrenal cortex produces androgen and estrogen in both sexes (Garb 1991). Like primary sex characteristics, none of these physical traits has any clear dividing lines to separate males from females. For example, biological females have the potential for the same hair distribution as biological males—follicles for a mustache, a beard, and body hair. Moreover, females produce not only estrogen but also androgen, a steroid hormone that triggers hair growth.

Given this information, we must ask, Why is it that women seem to have different patterns of facial and body hair growth than men? Before we answer this question, we must consider the interrelationship between sex in the physical and reproductive sense and the concept of gender.

Gender as a Social Construct

Whereas sex is a biological distinction, gender is a social distinction based on culturally conceived and learned ideas about appropriate appearance, behavior, and mental or emotional characteristics for males and females (Tierney 1991). The terms **masculinity** and **femininity** signify the physical, behavioral, and mental or emotional traits believed to be characteristic of males and females, respectively (Morawski 1991).

To grasp the distinction between sex and gender, we must note that no fixed line separates maleness from femaleness.[5] The painter Paul Gauguin pointed out this ambiguity in his observations about Maori men and women, which he recorded in a journal that he kept while painting in Tahiti in 1891. These observations are influenced by the norms regarding femininity around the turn of the century:

> *Among peoples that go naked, as among animals, the difference between the sexes is less accentuated than in our climates. Thanks to our cinctures and corsets we have succeeded in making an artificial being out of woman. . . . We carefully keep her in a state of nervous weakness and muscular inferiority, and in guarding her from fatigue, we take away from her possibilities of development. Thus modeled on a bizarre ideal of slenderness . . . our women have nothing in common with us [men], and this, perhaps, may not be without grave moral and social disadvantages.*
>
> *On Tahiti, the breezes from forest and sea strengthen the lungs, they broaden the shoulders and hips. Neither men nor women are sheltered from the rays of the sun nor the pebbles of the sea-shore. Together they engage in the same tasks with the same activity. . . . There is something virile in the women and something feminine in the men.* (GAUGUIN [1919] 1985, PP. 19–20)

Often we attribute differences between males and females to biology, when in fact they are more likely to be socially created. In the United States, for example, norms specify the amount and distribution of facial and body hair appropriate for females: it is acceptable for women to have eyelashes, well-shaped eyebrows, and a well-defined triangle of pubic hair, but not to have hair above their lips, under their arms, on their inner thighs (outside the bikini line), or on their chin, shoulders, back, chest, breasts, abdomen, legs, or toes. Most men, and even women, do not realize that women work to achieve these cultural standards and that their compliance makes males and females appear more physically distinct on this trait than they are in

reality. We lose sight of the fact that significant but perfectly normal biological events—puberty, pregnancy, menopause, stress—contribute to the balance between two hormones, androgen and estrogen. Changes in the proportions of these hormones trigger hair growth that departs from societal norms about the appropriate amount and texture of hair for females. When women grow hair as a result of these events, they tend to think something is wrong with them instead of seeing it as natural. A "female balance" between androgen and estrogen is one in which a woman's hair is consistent with these norms.[6]

The extreme measures taken by some women to eliminate facial and body hair are reflected in reports of physicians who intervened after women were harmed by commercial treatments. These treatments caused adverse side effects including wrinkling, scarring, discoloration, and cancerous growths due to X-ray treatments,[7] as well as paralysis caused by depilatories containing thallium acetate, a highly toxic substance (Ferrante 1988).

Just as women strive to meet norms for facial and body hair, they work to achieve the ideal standards of feminine beauty as portrayed in such places as magazines and television. In *How We Survived Communism and Even Laughed*, Drakulić (1992) powerfully describes the profound effect of magazines such as *Vogue* on the thinking of people who read them, although these magazines were rarely available in most Eastern European countries before the end of the Cold War. Just holding the magazine was "almost like holding a pebble from Mars or a piece of a meteor that accidentally fell into your yard" (p. 27). One Budapest woman, an editor of a scientific journal, told Drakulić in an interview that *Vogue* magazine "makes me feel so miserable I could almost cry. Just look at this paper—glossy, shiny, like silk. You can't find anything like this around here. Once you've seen it, it immediately sets not only new standards, but a visible boundary. Sometimes I think that the real Iron Curtain[8] is made of silky, shiny images of pretty women dressed in wonderful clothes, of pictures from women's magazines" (p. 27).

Drakulić observes that even though women in the West are bombarded on all sides with these images, they still notice them. She maintains that the average Western woman "still feels a slight mixture of envy, frustration, jealousy, and desire while watching this world of images" (p. 28) and proceeds to buy the products that promise that image. For the average Eastern European woman, who studied "every detail with the interest of those who had no other source of information about the outside world" (p. 28), the images created hatred for the reality that surrounded them.

To this point we have made a distinction between sex and gender. Although sociologists ac-

knowledge that there are no clear biological markers to distinguish males from females, they would not argue that biological differences do not exist. Sociologists are interested in the extent to which differences are socially induced. To put it another way, they are interested in the actions men and women take to accentuate differences between them. As we see in the next section, these actions lead to gender polarization.

Gender Polarization

In *The Lenses of Gender*, Sandra Lipsitz Bem defines **gender polarization** as "the organizing of social life around the male-female distinction," so that people's sex is connected to "virtually every other aspect of human experience, including modes of dress, social roles, and even ways of expressing emotion and experiencing sexual desire" (1993, p. 192). To understand how just about every aspect of life is organized around this distinction, we consider research by Alice Baumgartner-Papageorgiou.

In a paper published by the Institute for Equality in Education, Baumgartner-Papageorgiou (1982) summarizes the results of a study of elementary and high school students, in which she asked the students how their lives would be different if they were members of the opposite sex. Their responses reflect culturally conceived and learned ideas about sex-appropriate behaviors and appearances. The boys generally believed that as girls their lives would change in negative ways. Among other things, they would become less active and more restricted in what they could do. In addition, they would become more conscious about their appearance, about finding a husband, and about being alone and unprotected in the face of a violent attack:

- "I would start to look for a husband as soon as I got into high school."
- "I would play girl games and not have many things to do during the day."

- "I'd use a lot of make-up and look good and beautiful. . . . I'd have to shave my whole body."
- "I'd have to know how to handle drunk guys and rapists."
- "I couldn't have a pocket knife."
- "I would not be able to help my dad fix the car and truck and his two motorcycles." (PP. 2–9)

The girls, on the other hand, believed that if they were boys they would be less emotional, their lives would be more active and less restrictive, they would be closer to their fathers, and they would be treated as more than sex objects:

- "I would have to stay calm and cool whenever something happened."
- "[I could sleep later in the mornings] since it would not take [me] very long to get ready for school."
- "My father would be closer because I'd be the son he always wanted."
- "I would not have to worry about being raped."
- "People would take my decisions and beliefs more seriously." (PP. 5–13)

These beliefs about how the character of one's life depends on one's sex seem to hold even among the college students enrolled in my introductory sociology classes.[9] In the fall 1993 semester I asked students to take a few minutes to write about how their lives would change as members of the other sex. The men in the class believed they would be more emotional and more conscious of their physical appearance and that their career options would narrow considerably. Here are some of their responses:

- "I would be much more sensitive to others' needs and what I'm expected to do."

- "I wouldn't always have to appear like I am in control of every situation. I would be comforted instead of always being the comforter."

- "People would put me down for the way I look."

- "I would be more emotional."

- "I would worry more about losing weight instead of trying to gain weight."

- "I probably wouldn't really feel any different, but people would see me as a female and respond accordingly. If I stayed in the construction program I would have to fight the belief that men are the only real construction workers."

- "My career options would narrow. Now I have many career paths to choose from, but as a woman I would have fewer."

- "I would have to be conscious of the way I sit."

Notice that the first two responses suggest that the "feminine" traits would in some ways be a plus (being "more sensitive to others' needs" and being "comforted instead of always being the comforter"). It is important to realize that men as well as women can feel constrained by their gender roles.

The women in the class believed that as men they would have to worry about asking women out and about whether their major was appropriate. They also believed, however, that they would make more money, be less emotional, and be taken more seriously. Here are some of their responses:

- "I would worry about whether a woman would say 'yes' if I asked her out."

- "I would earn more money than my female counterpart in my chosen profession."

- "People would take me more seriously and not attribute my emotions to PMS."

- "My dad would expect me to be an athlete."

- "I'd have to remain cool when under stress and not show my emotions."

- "I think that I would change my major from 'undecided' to a major in construction technology."

These comments by high school and college students show the extent to which life is organized around male-female distinctions. They also show that students' decisions about how early to get up in the morning, what subjects to study, whether to show emotion, how to sit, and whether to encourage a child's athletic development are **gender-schematic decisions**. Decisions and viewpoints about any aspect of life are gender-schematic if they are influenced by a society's polarized definitions of masculinity and femininity rather than on the basis of other criteria such as self-fulfillment, interest, ability, or personal comfort. For example, college students make gender-schematic decisions about possible majors if they ask, even subconsciously, what is the "sex" of the major, and, if it matches their own sex, consider it a viable option, or, if it does not match, reject it outright (Bem 1993). In *Paths to Power*, Natasha Josefowitz (1980) gives an example of gender-schematic evaluation when she describes how the same workplace incidents and behavior are viewed differently, depending on whether a man or a woman is involved.

Even sexual desire between men and women is organized around male-female characteristics unrelated to reproduction. Bem argues that "neither women nor men in American society tend to like heterosexual relationships in which the woman is bigger, taller, stronger, older, smarter, higher in status, more experienced, more educated, more talented, more confident, or more highly paid than the man, they do tend to like heterosexual relationships in which the man is bigger, taller, stronger, and so forth, than the woman" (1993, p. 163).

The negative consequences of channeling sexual desire according to age differences is evident when

TABLE 11.1	Selected Gender Statistics for U.S. and Yugoslavia	
	United States	**Yugoslavia**
Ratio of females to males 60 years and older	138/100	141/100
Life expectancy (female to male)	+7.1 years	+5.9 years
Average age at first marriage		
males	25.2	26.2
females	23.3	22.2
Ratio of females to males in selected occupational groups		
administrative/managerial	61/100	15/100
clerical/sales/service	183/100	138/100
production/transportation	23/100	23/100
agricultural	19/100	88/100
Percentage of adults who smoke		
males	30	57
females	24	10

SOURCE: Adapted from United Nations (1991), pp. 22, 26, 67, 104.

we consider that the average woman outlives her spouse by about nine years in the United States and by about seven years in the former Yugoslavia. In the United States, the average life expectancy for women is seven years longer than that of men; in the former Yugoslavia, the average life expectancy for women before the war was six years longer than that of men. (The difference in life expectancy can be explained in part by the fact that men tend to do the most hazardous jobs in society.) In both countries, men tend to marry younger women (see Table 11.1). This practice, in combination with differences in life expectancy, means that women can expect to live a significant portion of their lives as widows.

Not only is sexual desire between men and women influenced strongly by gender-polarized ideas, but emotions toward persons of the same sex are also influenced. In Chapter 4 we learned that **social emotions** are internal bodily sensations that we experience in relationships with other people and that **feeling rules** are norms specifying appropriate ways to express those sensations. When I asked students in my class to comment on social emotions or "internal bodily sensations" that they had felt and expressed toward someone of the same sex, most

indicated that other people made them feel uncomfortable and defensive about such feelings (see "Expressing Affection Toward Same-Sex Friends").

A society's feeling rules are so powerful that they even affect how people solve problems. For example, when human evolutionists discovered petrified footprints believed to be 3.5 million years old, they inferred that the prints belonged to a man and a woman, not to two women, two men, or an adult and a child. Ian Tattersall, curator of the American Museum of Natural History, explains the logic underlying this conclusion:

We know that [the people who left the footprints] were walking side by side because even though the individuals are of different size, because their footprint sizes are different, their stride lengths are matched. They must have been walking together. And if they were walking together, the footprints are so close that they must have been in some kind of bodily contact with each other. What the nature of that contact was we don't know. We have chosen to put the arm of the, of the male around the shoulder of the female. It's a bit anthropomorphic, but we couldn't think of a less, a more non-committal, if you want, kind of a gesture. (1993, P. 13)

Expressing Affection Toward Same-Sex Friends

The comments of Introduction to Sociology students show that their relationships with same-sex friends, specifically expressing affection, are influenced by norms specifying appropriate ways to express such sensations.

I have noticed that some people struggle with how to show affection. I coached a sixth-grade boys' volleyball team. On this team was a little boy who happened to be a very tactile kid. When he talked to me he would always grab my hand or arm and shake or swing it according to how intense he felt about what he was telling me. When people on the team did something good I think he wanted to give them a big hug, but instead he ended up slapping them around.

My best friend and I are really close. Sometimes when we haven't seen each other for a long time we run to each other and hug and kiss. Our boyfriends look at us like we're crazy. I explain to my boyfriend that she is like my sister and that is the way I love her, not the same way I love him. It's sad that two friends can't be close without others looking at them as if they are weird.

I don't know of any guys that I am good friends with that I would consider touching. For the most part, I think my friends would think I was gay if I did. It is all right for two guys to be friends but I can't picture two guys holding hands and not having some sexual feelings for each other. Because I am an athlete, people always ask me why do guys pat each other on the butt and my response is—"I don't know."

This discussion about "feeling rules" has opened up ideas I've held most of my life. I have few friends, but feel very close to the ones I do

have. I never hesitate to show affection, albeit sometimes in a subdued manner. Sometimes I can feel tension if I hug someone or pat them on their back, but that's usually for the first time. I feel affection is the easiest way to transfer positive energy. I enjoy giving and receiving affection.

Unfortunately, in our society people think that if two guys touch and they are not playing sports, then they're gay. I noticed some people's reactions in the class when you asked us about the way we express positive feelings toward someone of the same sex and (mostly guys) acted as if you had said something gross and immoral. In other cultures, men do hold hands, even in public: it's accepted and almost expected to hug and touch each other. A lot of people my age are so homophobic that they are afraid to express their true selves around same-sex friends. It's sick.

I don't know who I love more, my mom, dad, sister or my husband. I know my sister Marcia is my favorite person to be around, because she's most like me. In ideas, mannerisms, thinking, we just click. I have such strong feelings for her, I just want to show her how much I love her every time I see her. I hug and kiss her on the cheek when we part. But since my brother David asked if we were lesbians, she shies away from my affection.

There is an unwritten rule my friends and I follow: men shouldn't

Feeling rules in the United States tend to restrict physical contact between members of the same sex, particularly men, to specific situations such as sports. Even though it is common to see male athletes hugging, many people would regard this gay couple's casually affectionate behavior as inappropriate.

Les Stone/Impact Visuals

touch each other. We say, "If you're going to touch me—make it hurt."

SOURCE: Introduction to Sociology Class, fall 1993, Northern Kentucky University.

The point is that Tattersall could not imagine someone putting his or her arm on the shoulder of a same-sex person. In the United States physical contact between same-sex persons is reserved for specific situations. Men can give full body hugs during a sports contest but cannot hold hands or put their hands on each other's shoulders while walking down the street. As of 1993 servicemen and service women seen holding hands with someone of the same sex are subject to investigation (Lewin 1993). These norms against touching someone of the same sex are so powerful that some museum curators assume they existed 3.5 million years ago and construct exhibits that reflect such norms.

The information on gender and gender polarization suggests that one's sex has a profound effect on life chances—determining how long one can expect to live, the major subject one chooses to study in college, and whether one dates a shorter or taller person. A person's sex then is an important variable in determining his or her position in a society's system of social stratification.

Gender Stratification

Recall that in Chapter 9 we learned that social stratification is the system societies use to place people in categories. When sociologists study stratification they study how the category people are placed in effects their perceived social worth and their life chances. They are particularly interested in how persons who possess one category of a characteristic (male reproductive organs versus female reproductive organs) are regarded and treated as more valuable than persons who possess the other category. Sociologist Randall Collins (1971) offers a theory of sexual stratification to analyze this phenomenon.

Economic Arrangements

Sociologist Randall Collins (1971) offers a theory of sexual stratification based on three assumptions: (1) people use their economic, political, physical, and other resources to dominate others; (2) any change in the way that resources are distributed in a society changes the structure of domination; and (3) ideology is used to justify one group's domination over another. In the case of males and females, males in general are physically stronger than females. Collins argues that, because of differences in strength between men and women, the potential for coercion by males exists in every encounter of males with females. He maintains that the ideology of **sexual property**—which he defines as the "relatively permanent claim to exclusive sexual rights over a particular person" (p. 7)—is at the heart of sexual stratification and that for the most part women historically have been viewed and treated as men's sexual property. (See "Khasi Society of India" for an example of one exception.)

Collins believes that the extent to which women are viewed as sexual property and are subordinate to men historically has depended, and still depends, on two important and interdependent factors: (1) women's access to agents of violence control, such as the police, and (2) women's position relative to men in the labor market. On the basis of these factors, Collins identifies four historical economic arrangements: low-technology tribal societies, fortified households, private households, and advanced market economies.

The four economic arrangements that Collins identifies are ideal types; the reality is usually a mixture of two or more types. (Note that his theory does not account for the type of communistic self-management that characterizes Yugoslavia or communist economies in general.) Each arrangement is characterized by distinct relationships between men and women. We begin with the first type, characteristic of low-technology tribal societies.

Low-technology tribal societies include hunting-and-gathering societies with technologies that do not permit the creation of surplus wealth, or wealth beyond what is needed to meet the basic needs (food and shelter). In such societies, sex-based division of labor is minimal because all members must contribute if the group is to survive. Some evidence, however, shows that in hunting-and-gathering societies women perform more menial tasks and work longer hours than men. Men hunt large animals,

Khasi Society of India

Although for the most part women historically have been viewed and treated as men's sexual property, there are exceptions to this rule. *The Times of India* reporter Syed Zubair Ahmed reports on a matrilineal society in India where the reverse seems to be true.

The matrilineal Khasi society in northeastern India, one of the few surviving female bastions in the world, is making a fervent effort to keep men in their place.

Though an all-male organization that is battling the centuries-old matrilineal system has yet to make any significant dent, the rebels claim to have enlisted the support of some prominent Khasi women. Their struggle to break free, they say, has resulted in small victories; some have begun to have a say in family affairs and are even inheriting property. But they constitute an insignificant minority in the 800,000-member Khasi society.

The men say the Khasi women are overbearing and dominating. "We are sick of playing the roles of breeding bulls and baby sitters," complains Mr. A. Swer, who heads the organization of maverick males. Another member laments: "We have no lines of succession. We have no land, no business. Our generation ends with us."

The demand for restructuring Khasi society in the patriarchal mold is a fallout from the growing number of women who are marrying outsiders. That, according to male opinion, has resulted in the bastardization of Khasi society.

Following custom, the youngest daughter inherits the property and after marriage her husband moves into the family house. Outsiders are said to marry the Khasi women for their property, while the women say they prefer to marry outsiders because their own tribesmen tend to be irresponsible in family matters.

In rebuttal, many Khasi men say the outsiders take advantage of the immaturity, youth and vulnerability of the youngest daughters and devour all their property and business. As a result, many Khasi men become paupers. And if the young men are often lazy and have no sense of a family, the rebels argue, it's the matrilineal system that is to blame.

Another problem caused by these marriages is the disintegration of families. About 27,000 Khasi women were divorced by their non-Khasi husbands in recent years, the highest number among India's northeastern tribes.

The identity crisis has led the Khasi Student Union to issue a stern warning to young Khasi women against marrying "non-tribals," saying they may be ostracized if they do. The Student Union is against switching over to the patrilineal system, however. So is a prominent Khasi scholar, H. W. Sten, who cautions that a patrilineal shift "would result in cross-marriages between clans, which is taboo in Khasi society," and adds, "Ultimately, it would lead to genetic defects in the offspring."

He points out that a Khasi son or daughter takes the surname of the mother. Therefore, if two sisters marry two men of different clans, in a patriarchal system the surnames of their children would be different and the marriages between cousins would be valid. "This goes against the basic principle of Khasi custom," he said.

At the same time, Mr. Sten condemns those who are opposed to Khasi women's marrying outside the tribe. "Khasi culture is very flexible," he said. "No problem if a non-tribal wants to marry a Khasi girl as long as he is prepared to live with her and follow the Khasi custom. It will only add to the variety in Khasi society."

But Mr. Swer, president of the male group of social reformers, says such liberalism is the root cause of bastardization of his tribe. "Today, we have over 2,000 clans, but very few of them are pure Khasis," he observed. His demand for change, he adds, would stop outsiders from chasing Khasi young women, since under the patrilineal system their wives could not inherit property. But what about men marrying outside their tribe? "The girls will be taken into the Khasi fold," he replied. "The children from the wedlock will automatically be Khasis."

While some men would like to end female domination, they do not support Mr. Swer's movement to abandon the deeply held tradition. "We Khasis underestimate the contributions of our fathers to the family," said Mr. H. T. Wells, a cousin of Mr. Sten. "Our fathers do a lot, but the credit goes to the mothers. I would love to have the patriarchal system but for the respect of our custom."

Mr. Swer admits that the Khasi men's demand for a patrilineal society is still a distant hope. But people like Mr. Wells, half converts to his idea, sustain his dream.

SOURCE: "What Do Men Want?" by Syed Zubair Ahmed. Copyright © 1994 by *The Times of India.*

for example, while women gather most of the food and hunt smaller animals. Because there is almost no surplus wealth, marriage between men and women from different families does little to increase a family's wealth or political power. Consequently, daughters are not treated as sexual property in the sense that they are not used as bargaining chips to achieve such aims.

Fortified households include preindustrial arrangements in which there is no police force, militia, national guard, or other peacekeeping organization. Therefore, the household is an armed unit, and the head of the household is its military commander. Fortified households "may vary considerably in size, wealth, and power, from the court of a king or great lord . . . down to households of minor artisans and peasants" (Collins 1971, p. 11). All fortified households, however, have a common characteristic: the presence of a **nonhouseholder class** consisting of propertyless laborers and servants. In the fortified household, "the honored male is he who is dominant over others, who protects and controls his own property, and who can conquer others' property" (p. 12). Men treat women as sexual property in every sense: daughters are bargaining chips for establishing economic and political alliances with other households; male heads of household have sexual access to female servants; and women (especially in poorer households) bear many children, who eventually become an important source of labor. In this system, women's power depends on their relationship to the dominant men. Because of their position, the servants in the household have few opportunities to form stable marriages or family lives.

Private households emerge with a market economy, a centralized, bureaucratic state, and the establishment of agencies of social control that alleviate the need for citizens to take the law into their own hands. Thus, private households exist when the workplace is separate from the home, where men are still heads of households but assume the role of breadwinner (as opposed to military commander), and where women remain responsible for housekeeping and childrearing. Men, as heads of households, control the property; it is a relatively new practice in the United States, for example, to put a house or credit in the names of both husband

and wife. Moreover, men monopolize the most desirable and most important economic and political positions. Collins states that a decline in the number of fortified households, the separation of work from home, smaller family size, and the existence of a police force to which women can appeal in cases of domestic violence give rise to the notion of romantic love as an important ingredient in a marriage. In the marriage market, men offer women economic security because they dominate the important, high-paying positions. Women offer men companionship and emotional support, and strive to be attractive—that is, to achieve the ideals of femininity, which may include possessing an 18-inch waist or removing most facial and body hair. At the same time, they try to act as sexually inaccessible as possible because sexual access is something they offer men in exchange for economic security.

Advanced market economies offer widespread employment opportunities for women. Although women are far from being men's economic equals, some women now can enter relationships with men with more than an attractive appearance; now they can offer an income and other personal achievements. Having more to offer, women can demand that men be physically attractive and meet the standards of masculinity. This situation may explain why more commercial attention has been given to males' appearance—body building, hair styles, male skin and cosmetic products—in the past decade.

Collins's classic theory of sexual stratification suggests that men and women cannot be truly equal in a relationship until women are men's economic equals. To reach this goal—still unrealized—fathers must share equally in household and childrearing responsibilities.

Household and Childrearing Responsibilities

In the United States, research shows consistently that even among couples in which the spouses earn the same income and share egalitarian ideals, the husbands spend considerably less time than the wives in preparing meals, taking care of children, shopping, and other household tasks (Almquist 1992). The same is true of the division of household labor in pre-war Yugoslavia and other Eastern European countries (United Nations 1991). On the other hand, in 1990, the last year for which data is available, Yugoslavia allowed men and women 105–210 paid days for parental leave (Drakulić 1990). In the United States, it was not until August 1993 that men and women who had worked for an employer for at least one year could take up to 60 days of unpaid leave without the fear of losing their job and other benefits.[10] In addition to having equality in the labor market, women must also have access to agents of violence control.

Access to Agents of Violence Control

Collins (1971) argues that women must have access to agents of violence control if they are to be men's equals. Even today in the United States, when women are raped or otherwise physically abused, they must prove to police, intake personnel, court officers, prosecutors, judges, and sometimes family members that they did not do something to provoke the attackers. In the United States, rape laws vary according to state; the strictest laws require that the defense prove force on the part of the man, and the most liberal laws make it a crime for a man to have sexual intercourse without the woman's consent (Burns 1992). Although legal reforms over the past 20 years have made it easier to prosecute rape, the jurors who decide the case "still tend to blame the victim, particularly if she used alcohol or drugs, kept late hours, frequented bars, had an active sex life, or—in short—was in their eyes [of] 'questionable moral character'" (Jones 1994, p. 14). Furthermore, jurors are swayed by the appearance of the accused: the man has to "look like a rapist" (Jones 1994). Women's access to violence control is particularly an issue in wartime. To illustrate, we consider the case of Bosnia.

In all countries, wars typically have been fought primarily by men. In most of these conflicts, raping of women has been a more than incidental occurrence—whether as a by-product of war's savagery or as a deliberate instrument of warfare. These observations lead to questions about the relationship between war rape and gender roles. The case of Bosnia provides an illustration of these questions.

Although soldiers on all sides of the war in Bosnia raped women, the Serbian soldiers were under orders to rape non-Serbian women and even Serbian women who opposed the establishment of an ethnically pure Greater Serbia. This systematic mass rape[11] is part of the Serbian policy of ethnic cleansing, a euphemism for forcing people of ethnic heritage different from the dominant group out of a territory. According to Jeri Laber (1993), who investigated war crimes against women in Bosnia and Croatia, the rapes seemed to follow several different patterns. One pattern occurred when Serbian military units invaded a village or town populated primarily by Muslims or Croatians. Rape was used as a weapon to terrorize non-Serbs so they would flee their homes or sign documents indicating that they were leaving voluntarily. Other rapes occurred in Serbian-operated detention centers or camps, where hundreds or thousands of women were held. Still other rapes took place in Serbian-controlled houses, schools, or hotels.

The Serbian soldiers' rape of women resembled other war-related rapes. First, the rape was often a public event: it occurred in front of witnesses. The witnesses may have been other villagers, neighbors, or the woman's family. It was public so that the enemy, especially the woman's male family members, were humiliated (Swiss and Giller 1993). The rape was used to symbolize ultimate control and was carried out with extreme cruelty. "Rape by a conquering soldier destroys all remaining illusions of power and property for men of the defeated side" (Brownmiller 1975, p. 38). Second, more than one soldier and as many as 20 participated; all women were vulnerable, regardless of age. Third, the soldiers who raped often killed their victims because they were disgusting to them. One Serbian soldier commented, "I only remember that I was twentieth, that her hair was a mess, that she was disgusting and full of sperm, and that I killed her in the end" (Mladjenovic 1993, p. 14). In the Bosnian war, some women reported knowing the men who raped them.

The Serbian soldiers also had been ordered to forceably impregnate non-Serbian women to produce *chetnicks* (babies).[12] Some women were held

Among the casualties of war are the orphans produced by rape of "enemy" women. This baby was conceived after her 17-year-old Bosnian Muslim mother was repeatedly raped in a Serb-run detention center. The mother refused to see her after her birth.
AP/Wide World Photos

in camps until they were far enough along in their pregnancies to make abortion dangerous (Mladjenovic 1993). One woman interviewed by United Nations human rights workers described the horror of the camps:

> *It is a nightmare that cannot be talked about, or described, or understood.*
>
> *Sometimes I think I will go crazy and that the nightmare will never end. Every night in my dreams I see the face of Stojan, the camp guard. He was the most ruthless among them. He even raped 10-year-old girls, as a delicacy. Most of those girls didn't survive. They murdered many girls, slaughtered them like cattle.* (MIRSADA 1993, P. 32)

It is difficult to determine how many women have been raped in this war. Official statistics never have been collected on the number of women killed or injured as a result of wartime rape; these wars in the former Yugoslavia are no exception.

Women of all ethnic groups in the former Yugoslavia who were raped during the war lack access to agents of violence. It was not until 1993, in the wake of the massive and systematic rapes in Bosnia, that the United Nations passed a resolution identifying rape as a war crime and called for an international tribunal to prosecute those who ordered, committed, or did nothing to stop rape (Swiss and Giller 1993). This tribunal has yet to be held. The problem is that it is difficult to have rape recog-

nized as a war crime unless there is clear proof that high-ranking military officials ordered the rapes. Otherwise rape is considered an individual act (Brew 1994). In a 1993 article in the *Journal of the American Medical Association,* Shana Swiss and Joan Giller specified medical technologies that might be used to establish rapists' identities, given that many of the raped women did not know their attackers. These include collecting and storing sperm collected from the genital tract, taking placental tissue (from abortions or delivery), and drawing blood samples from mother and child. DNA from these specimens could be matched later against DNA obtained from the alleged perpetrator's blood and hair follicles. The point is that this medical technology can be used for violence control if a potential rapist realizes that a rape can be traced to him.

We have seen that there are a number of sources and expressions of inequality based on gender. Even where there is a physical basis for these inequalities (as in the case of physical strength), the inequalities themselves are social rather than biological realities. Even in the case of war rape, soldiers are not acting out a biological imperative. Their behavior reflects, among other things, the way their social reality has been redefined so that women are considered suitable targets of aggression and rape is considered acceptable under the circumstances.

Benazir Bhutto, prime minister of Pakistan, inspects the first women's police group in that country. Sociologist Randall Collins's theory implies that giving women access to agents of violence control is an important step toward removing inequality in the status of men and women.
Saeed Khan/AFP

Mechanisms of Perpetuating Gender Expectations

As mentioned at the onset of this chapter, people vary in the extent to which they conform to their society's gender expectations. This fact, however, does not prevent us from using gender expectations to evaluate our own and other behavior. For many people, failure to conform (whether that failure is deliberate or reluctant) is a source of intense confusion, pain, and pleasure. This leads sociologists to ask what mechanisms explain how we learn and perpetuate a society's gender expectations. To answer this question, we examine three important factors: socialization, situational constraints, and ideologies.

Socialization

In Chapter 5 we learned that **socialization** is a learning process that begins immediately after birth and continues throughout life. By this process, "newcomers" develop their human capacities, acquire a unique personality and identity, and internalize, or take as their own and accept as binding, the norms, values, beliefs, and language they need to participate in the larger community. Socialization theorists argue that an undetermined but significant portion of male-female differences are products of the ways in which males and females are treated.

Child development specialist Beverly Fagot and her colleagues observed how toddlers in a play group interacted and communicated with each other and how teachers responded to the children's attempts to communicate with them at age 12 months and at age 24 months (Fagot et al. 1985). Fagot found no real sex differences in the interaction styles of 12-month-old boys and girls: all of the children communicated by gestures, gentle touches, whining, crying, and screaming. The teachers, however, interacted with the toddlers in gender-polarized ways. They were more likely to respond to girls when the girls communicated in gentle, "feminine" ways and to boys when the boys communicated in assertive, "masculine" ways. That is, the teachers tended to ignore assertive acts by girls and to respond to assertive acts by boys. Thus, by the time these toddlers reached two years of age, the differences in their communication styles were quite dramatic.

Fagot's findings may help explain the differing norms governing body language for males and females. According to women's studies professor Janet Lee Mills, norms governing male body language suggests power, dominance, and high status, whereas norms governing female body language suggest submissiveness, subordination, vulnerability, and low status. Mills argues that these norms are learned and that people give them little thought until someone breaks them, at which point everyone focuses on the rule breaker. Such norms can prevent women from conveying a sense of security and control when they are in positions that demand these qualities, such as lawyer, politician, or physician. Mills suggests that women face a dilemma: "To be successful in terms of femininity, a woman needs to be passive, accommodating, affiliative, subordinate, submissive, and vulnerable. To be successful in terms of the managerial or professional role, she needs to be active, dominant, aggressive, confident, competent, and tough" (Mills 1985, p. 9; we return to this topic in the Focus section at the end of this chapter).

Children's toys figure prominently in the socialization process along with the ways in which adults treat children. Barbie dolls, for example, have been on the market for more than 30 years and currently are available in 67 countries. Executives at Mattel, the company that created Barbie, are studying Eastern Europe as a potential new market. The company considers Barbie to be an aspirational doll—that is, the doll is a role model for the child. Barbie accounts for approximately half of all toy sales by Mattel (Boroughs 1990; Cordes 1992; Morgenson 1991; Pion 1993). Ninety-five percent of girls between ages 3 and 11 have Barbie dolls, which come in several different skin colors. Market analysts attribute Mattel's success to the fact that "they generally have correctly assessed what it means to a little girl to be grown-up" (Morgenson 1991, p. 66).[13]

In *How We Survived Communism and Even Laughed*, Slavenka Drakulić (1992) describes how the standard of beauty in Yugoslavia, at least since World War II, has come from Hollywood. She tells how the faces and bodies of Hollywood women such as Rita Hayworth, Ava Gardner, and Brigitte Bardot, who appeared in a fashion magazine that their parents forbade them to read, were the models that she and her friends used for paper dolls, along with paper outfits and accessories. Drakulić recalled that she and her friends "painted [the dolls'] little lips and nails bright red, and dressed them in tight sexy dresses" (p. 61). Reflecting on this childhood activity, Drakulić remarked: "Sometimes I think that at that early age I learned everything

about my sex from these paper dolls. . . . Later on, it took me—and our whole generation of women—years and years of hard work to unglue ourselves from those paper idols; to break through into another dimension, away from the dolls of our childhood, to which we were constantly reduced" (pp. 61–62).

In an effort to spare her daughter, Rujana, this struggle to overcome Hollywood ideas of beauty, Drakulić gave her stuffed animals. A few days before Rujana's twenty-second birthday when Drakulić asked her what she wanted, her daughter asked for a Barbie doll, which Drakulić believed was "the symbol of bimbodom" (p. 63). At first, Drakulić thought she could not give Rujana the doll because it ran counter to her feminist principles. She changed her mind after Rujana reminded her of an incident she had forgotten: when Rujana "was about ten years old, she had stolen a Barbie doll from her friend and hid it in her room, under the bed. For two weeks she played with her every afternoon" (p. 63), until her grandmother found out and told Drakulić. Rujana asked her mother, "Didn't you remember your own dolls, your own longing?" (p. 64). Drakulić then bought her daughter a Barbie doll, concluding that "dolls are important—the dolls that we had, that we didn't have, that we longed for, that we betrayed and left behind" (p. 64).

Structural or Situational Constraints

Situational theorists agree with socialization theorists that the social and economic differences between men and women are not explained by something in their biological makeup. In their view, the causes of these differences are structural and situational. For example, a structural constraint is that occupations are segregated by sex, so that women tend, more often than men, to be concentrated in low-paying, low-ranking, dead-end jobs. (Recall the discussion in Chapter 9 of occupations in which women are disproportionately underrepresented and overrepresented, and of the inequality in men's and women's earnings. Also see Table 11.2.) Even when women are in professional and management positions, they are concentrated in specialties and fields that handle children and young adults, that involve supervising other women, and/or that are

TABLE 11.2 Characteristics of Selected Nations and Nationalities*

Characteristic	Albanians	Croats	Hungarians	Macedonians
Sex ratio	109	93	90	102
Birthrate (%)	36.4	16.2	12.4	18.5
Percent highly educated (total population)	1.2	3.3	1.7	3.7
Percent highly educated men	1.8	4.4	2.2	4.9
Percent highly educated women	0.5	2.3	1.2	2.5
Women's share of highly educated population (percent)	21.4	36.6	38.7	33.7
Labor force participation (percent)				
Women	15.3	40.7	35.0	48.5
Men	66.8	66.6	37.9	70.2
Directors (percent)				
Women	0.3	0.5	0.4	0.4
Men	10.9	2.5	1.6	2.8
Women's share of directorial positions	4.2	12.6	13.0	9.5

Characteristic	Montenegrins	Muslims	Serbs	Slovenes
Sex ratio	102	101	98	92
Birthrate (%)	17.6	23.1	15.4	15.7
Percent highly educated (total population)	5.0	1.6	3.1	3.4
Percent highly educated men	8.5	2.3	4.2	4.7
Percent highly educated women	2.1	0.9	2.0	2.3
Women's share of highly educated population (percent)	23.4	29.0	33.5	35.3
Labor force participation (percent)				
Women	35.8	25.9	47.7	58.8
Men	65.7	69.3	73.2	69.6
Directors (percent)				
Women	0.7	0.4	0.5	1.0
Men	4.5	1.6	2.2	3.5
Women's share of directorial positions	7.9	9.1	11.6	20.7

*Data based on 1981 census, the last year for which census data is available.

SOURCE: Adapted from Darville and Reeves (1992), p. 281.

otherwise considered feminine (a professor of social work versus a professor of mathematics and computer sciences).

These structural differences affect expectations about gender. The different social and physical demands and skills required to perform the jobs held by men and women function toward "channeling their motivations and their abilities into either a stereotypically male or a stereotypically female direction" (Bem 1993, p. 135). This point does not preclude the fact that men and women may limit their job search to positions that are considered "sex-appropriate." On the other hand, considerable evidence supports the hypothesis that once women are hired, management steers male and female employees into different assignments and of-

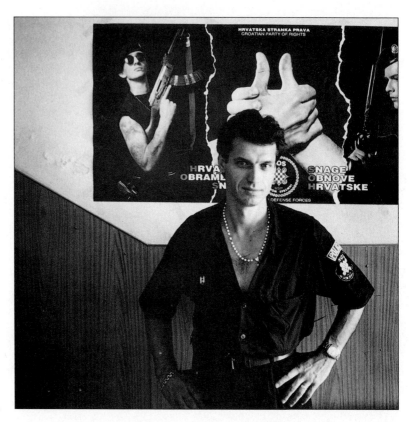

The poster seen in this photograph, used to recruit men into the Croatian armed forces, revolves around symbols that depict "masculine" ideals and stereotypes.

Tuen Voeten/Impact Visuals

fers them different training opportunities and chances to move into better-paying positions.

A case in point is Lucky Stores, Inc., which operates 188 stores in northern California. The company lost a class-action sex discrimination suit because (among other things) it failed to post job openings but filled them instead at management's discretion. Lawyers for Lucky Stores argued that women were steered into part-time, low-paying, dead-end jobs in the delicatessen and bakery rather than jobs in stocking and receiving because women preferred such work. Once the store began to post job openings, however, the percentage of women in entry-level management positions jumped from 12 to 58 percent (Gross 1993).

The case of Lucky Stores shows that women are no less highly motivated than men to seek advancement in the workplace. Many other cases also serve to support this point. The implication is clear: if we remove structural barriers to advancement, women will seek to improve their position.

Sociologist Renee R. Anspach's research illustrates vividly how one's position in a social structure can channel behavior in stereotypically male or female directions. Anspach spent 16 months conducting field research (observing and holding interviews) in two neonatal intensive care units (NICUs). Among other things, she found that nurses (almost all of whom were female) and physicians (usually male) used different criteria to answer the question, "How can you tell if an infant is doing well or poorly?" Physicians tended to draw on so-called objective (technical or measurable) information and immediate perceptual cues (skin color, activity level) obtained during routine examination:

Well, we have our numbers. If the electrolyte balance is OK and if the baby is able to move one respirator setting a day, then you can say he's probably doing well. If the baby looks gray and isn't gaining weight and isn't moving, then you can say he probably isn't doing well.

The most important thing is the gestalt. In the NICU, you have central venous pressure, left atrial saturations, temperature stability, TC (transcutaneous) oxymeters, perfusions (oxygenation of the tissues)—all of this adds in. You get an idea, when the baby looks bad, of the baby's perfusion. The amount of activity is also important—a baby who is limp is doing worse than one who's active. (ANSPACH 1987, PP. 219–20)

Although immediate perceptual and measurable signs were important to the nurses as well, Anspach found that the nurses also considered interactional clues such as the baby's level of alertness, ability to make eye contact, and responsiveness to touch:

I think if they're doing well they just respond to being human or being a baby. . . . Basically emotionally if you pick them up, the baby should cuddle to you rather than being stiff and withdrawing. Do they quiet when held or do they continue to cry when you hold them? Do they lay in bed or cry continuously or do they quiet after they've been picked up and held and fed . . . Do they have a normal sleep pattern? Do they just lay awake all the time really interacting with nothing or do they interact with toys you put out, the mobile or things like that, do they interact with the voice when you speak? (P. 222)

Anspach concluded that the differences between nurses' and physicians' responses to the question "How can you tell if an infant is doing well or poorly?" could be traced to their daily work experiences. In the division of hospital labor, nurses interact more with patients than do physicians. Also, doctors and nurses have access to different types of knowledge about infants' condition, which correspond to our stereotypes of how females and males manage and view the world. Because physicians have only limited amounts of daily interaction and contact with infants, they tend to rely on perceptual and technological (measurable) cues. Nurses, on the other hand, are in close contact with infants throughout the day; consequently, they are more likely to consider interactional cues as well as perceptual and technological ones.

Anspach (1987) suggests that one's position in the division of labor "serves as a sort of interpretive lens through which its members perceive their patients and predict their futures" (p. 217). Her findings suggest that when physicians make life-and-death decisions about whether to withdraw or continue medical care, they should collaborate with NICU nurses so that they can consider interactional as well as technological and immediate perceptual cues. Anspach's findings also suggest that if nurses' experiences and opinions counted more in medical diagnosis, we might see a corresponding increase in the prestige and salary associated with this largely female position.

Anspach's research on the relative weight assigned to physicians' and nurses' experiences points to another problem that affects men and women's life chances: institutionalized discrimination. In Chapters 9 and 10 we learned that **institutionalized discrimination** is the established and customary way of doing things in society—the collection of unchallenged rules, policies, and day-to-day practices that impede or limit people's achievements and keep them in subordinate and disadvantaged positions on the basis of ascribed characteristics. This situation is "systematic discrimination through the regular operations of social institutions" (Davis 1979, p. 30). A case in point is the typical way of determining how much child support fathers should contribute toward the care of their children. Currently the father's needs are regarded as more important than the children's: judges and lawyers generally assume that these men need 80 percent of their earnings to support themselves and thus have only 20 percent to give toward supporting their children. This assumption, coupled with the fact that most men (as many as 75 percent) do not meet their child support obligations, may be the most important reason why, on average, a man's standard of living increases by 42 percent after divorce, while a woman's decreases by 73 percent (Weitzman 1985; see Figure 11.1).

Sexist Ideology

In Chapters 2 and 10 we learned that ideologies are ideas that support the interests of the dominant group but that do not stand up to scientific investigation. They are taken to be accurate accounts and explanations of why things are as they are. On

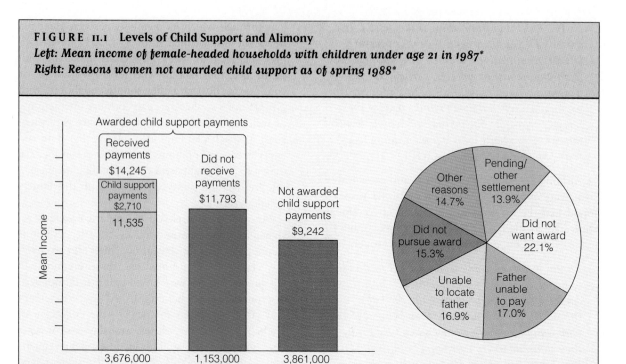

FIGURE 11.1   Levels of Child Support and Alimony
Left: Mean income of female-headed households with children under age 21 in 1987*
Right: Reasons women not awarded child support as of spring 1988*

*Most recent data available.
SOURCE: U.S. Bureau of the Census (1990).

closer analysis, however, we find that ideologies are at best half-truths, based on misleading arguments, incomplete analysis, unsupported assertions, and implausible premises.

Sexist ideologies are structured around three notions:

1. People can be classified into two categories, male and female.

2. There is a close correspondence between a person's primary sex characteristics and characteristics such as emotional activity, body language, personality, intelligence, the expression of sexual desire, and athletic capability.

3. Primary sex characteristics are so significant that they explain and determine behavior and social, economic, and political inequalities that exist between the sexes.

Sexist ideologies are so powerful that "almost everyone has difficulty believing that behavior they have always associated with 'human nature' is not human nature at all but learned behavior of a particularly complex variety" (Hall 1959, p. 67). One example of a sexist ideology is the belief that men are prisoners of their hormones, making them powerless in the face of female nudity or sexually suggestive dress or behavior. Another example is the belief that men are not capable of forming relationships with other men that are as meaningful as those that women form. Since the 1980s, dozens of books were written by men in response to these stereotypes (Shweder 1994).

We might also add a fourth characteristic about sexist ideology: people who behave in ways that depart from concepts of masculinity or femininity are considered deviant, in need of fixing, and subject to negative sanctions ranging from ridicule to physical violence.

Ideologies are reflected in social institutions. A good example is the military. One ideology that has

The issue of military personnel's sexual orientation involves ideology as much as policy.

© Jim Borgman. Reprinted with special permission of King Features Syndicate

dominated U.S. military policy since World War II is stated in a U.S. Department of Defense directive:

> *Homosexuality is incompatible with military service. The presence of such members adversely affects the ability of the Armed Forces to maintain discipline, good order, and morale; to foster mutual trust and confidence among the members; to ensure the integrity of the system of rank and command; to facilitate assignment and worldwide deployment of members who frequently must live and work under close conditions affording minimal privacy; to recruit and retain members of the military services; to maintain the public acceptability of military services; and, in certain circumstances, to prevent breaches of security.* (1990, P. 25)

There is no scientific evidence, however, to support this directive. In fact, it seems that whenever Pentagon researchers (with no links to the gay and lesbian communities and with no ax to grind) found evidence that ran contrary to this directive, high-ranking military officials refused to release the information or found the information unacceptable and directed researchers to rewrite the reports. For example, when researchers found that sexual orientation is unrelated to military performance and that men and women known to be gay or lesbian displayed military suitability that is as good as or better than that of men and women believed to be het-

erosexual, the U.S. Deputy Undersecretary of Defense, Craig Alderman, Jr., wrote the researchers that the "basic work is fundamentally misdirected" (Alderman 1990, p. 108). He explained that the researchers were to determine whether there was a connection between being a homosexual and being a security risk, not to determine whether homosexuals were suitable for military service. Although the researchers found no data to support a connection between sexual orientation and security risks, Alderman maintained that the findings were not relevant, useful, or timely.[14] (See "Ideology to Support Military Policy.") The research that Alderman dismissed would have gone unnoticed if Congressman Gerry Studds and House Arms Subcommittee Chairwoman Patricia Schroeder had not insisted it be released.

This example shows the role that ideologies play in setting policy. In this case the ideologies are that homosexuality is incompatible with military service and that being homosexual represents a security risk to the United States. The case of the military also alerts us to the fact that other variables, such as a person's sexual orientation, race, ethnicity, and social class, interact with gender to make the experience of being male and female different. To illustrate this interaction, we turn to the work of sociologists Floya Anthias and Nira Yuval-Davis, who have written about the interconnection among gender, race and ethnicity, and country (the state).

Ideology to Support Military Policy

People who oppose the presence of gays and lesbians in the military stereotype homosexuals as sexual predators who are waiting to pounce on a heterosexual person while he or she is showering, undressing, or sleeping. Opponents seem to believe that any same-sex person is attractive to a gay or lesbian person. But as one gay ex-midshipman notes, "Heterosexual men have an annoying habit of overestimating their own attractiveness" (Schmalz 1993, p. B1). The excerpt from Pentagon research included below shows that there is no evidence to support such a stereotype.

Those who resist changing the traditional policies support their position with statements of the negative effects on discipline, morale, and other abstract values of military life. Buried deep in the supporting conceptual structure is the fearful imagery of homosexuals polluting the social environment with unrestrained and wanton expressions of deviant sexuality. It is as if persons with nonconforming sexual orientations were always indiscriminately and aggressively seeking sexual outlets. All the studies conducted on the psychological adjustment of homosexuals that we have seen lead to contrary inferences. The amount of time devoted to erotic fantasy or to overt sexual activity varies greatly from person to person and is unrelated to gender preference. In one carefully conducted study, homosexuals actually demonstrated a lower level of sexual interest than heterosexuals.

Homosexuals are like heterosexuals in being selective in their choice of partners, in observing rules of privacy, in considering appropriateness of time and place, in connecting sexuality with the tender sentiments, and so on. To be sure, some homosexuals are like some heterosexuals in not observing privacy and propriety rules. In fact, the manifold criteria that govern sexual interest are identical for homosexuals and heterosexuals, save for only one criterion: the gender of the sexual partner.

Age, gender, kinship, class membership, marital status, size and shape, social role, posture, manners, speech, clothing, interest/indifference signalling, and other physical and behavioral criteria are all differentiating cues. They serve as filters to screen out undesirable or unsuitable potential sex partners. With such an array of cues, many (in some cases, all) potential objects of interest are rejected. For most people, only a small number of potential partners meet the manifold criteria. Whether in an Army platoon or in a brokerage office, people are generally selective in their choice of intimate partners and in their expression of sexual behavior. Heterosexuals and homosexuals alike employ all these variables in selecting partners, the only difference being that the latter include same-gender as a defining criterion, the former include opposite-gender.

SOURCE: Sarbin and Karols (1990), p. 37.

Gender, Ethnicity, Race, and the State

Ethgender refers to people who share (or are believed by themselves or others to share) the same sex, race, and ethnicity. This concept acknowledges the combined (but not additive) effects of gender, race, and ethnicity on life chances. Ethgender merges two ascribed statuses into a single social category. In other words, a person is not a Croat *and* a woman but a Croatian woman; a person is not an African-American and a man but an African-American man (Geschwender 1992). To complicate matters, the country or state that people of a particular ethgender inhabit (and their legal relationship to the state—as citizen, refugee, or temporary worker) has a significant effect on their life chances. We use the term **state** here to mean a governing body organized to manage and control specified activities of people living in a given territory.

Everyone has some legal relationship to the state, whether as a citizen by birth or naturalization, a refugee, a temporary worker, an immigrant, a permanent resident, or an illegal alien. Sociologists Floya Anthias and Nira Yuval-Davis (1989) give special attention to women, their ethnicity, and the state. They argue that "Women's link to the state is complex" and that women "are a special focus of state concerns as a social category with a specific role (particularly human reproduction)" (p. 6). In the broadest sense, reproduction includes biological reproduction, especially in relation to the birth of children who become the state's citizens and future labor force. Anthias and Yuval-Davis maintain that the state's policies and discourse reflect its concerns about the kinds of babies (that is, their ethnicity) to which women give birth and about the ways in which the babies are socialized. They identify five areas of women's lives over which the state may choose to exercise control. One should not conclude, however, that women accept the policies and programs without resistance that the state directs at them. In fact, women often work to modify these policies.

1. Women as Biological Reproducers of Babies of a Particular Ethnicity or Race

As factors that can underlie a state's population control policies, Anthias and Yuval-Davis name "fear of being 'swamped' by different racial and ethnic groups" or fear of a "demographic holocaust" (that is, a particular racial or ethnic group will die out or become too small to hold its own against other ethnic groups). Such policies can range from physically limiting numbers of a particular racial or ethnic group deemed undesirable to actively encouraging the "right kind" of women to produce more children. Policies that limit numbers include immigration control (limiting or excluding members of certain ethnic groups from entering a country and subsequently producing children), physical expulsion (which includes ethnic cleansing), extermination, forced sterilization, and massive birth control campaigns. Policies that encourage the "right kind" of biological reproduction include ideological mobilization (appeals to a woman's duty to her country), tax incentives, maternal leave, and other benefits.

Even before the wars in Yugoslavia and its eventual breakup, various leaders depending on their ethnicity and political aims pointed to the "fear of being swamped" by other ethnic groups as a reason for their women to bear children. Most notably, Serbian leaders such as Slobodan Milosevic pointed to the high birthrate among Albanians and Muslims as a serious threat to Serbian autonomy and quality of life. Similarly, some Croatian leaders of parties such as the Croatian Democratic Union have asked that the right to an abortion be made illegal and have asked Croatian women to bear at least three, but ideally five, children (Drakulić 1990; Enloe 1993).

2. Women as Reproducers of the Boundaries of Ethnic or National Groups

In addition to implementing policies intended to encourage or discourage women in "having children who will become members of the various ethnic groups within the state" (Anthias and Yuval-Davis 1989, p. 9), states also implement policies that define the "proper ways" to reproduce offspring. Examples include laws prohibiting sexual relationships with men or women of another race or ethnicity, laws specifying legal marriage if the child is to be recognized as legitimate, and laws connecting the child's ethnic and legal status to the ethnicity of the mother and/or father.

Although these laws apply to both men and women, the woman generally pays a heavier social price when the law is broken. For example, in *Wake Up Little Susie: Single Pregnancy and Race Before Roe v. Wade,* Rickie Solinger (1992) documents the options open to unmarried girls and women who faced pregnancy between 1945 and 1965. They included "futilely appealing to a hospital abortion committee, [which at that time were not concerned about the question of when human life begins but about punishing single mothers]; being diagnosed as neurotic, even psychotic by a mental health professional; [being] expelled from school (by law until 1972); [becoming] unemployed; [enrolling] in a Salvation Army or some other maternities home; [and being] poor, alone, ashamed, threatened by the law" (p. 4). Solinger argues that the policies and programs encouraged white women to give up their babies for adoption but encouraged black women to keep their babies and to prevent them from having more.

Today, by contrast, many school districts across the United States offer on-site day care, private tutoring, and special classes to pregnant girls. Still, considerable political debate revolves around the questions of whether single mothers should receive welfare, especially after they have a second child and whether it might be to society's benefit to bring the stigma of the past back (L. Williams 1993). For the most part, these debates rarely focus on stigmatizing the fathers of these children. Although there are federal and state programs in place to collect child support from so-called dead-beat dads, only about 25 percent of the women receive the full amount (Brownstein 1993).

3. Women as Transmitters of Social Values and Culture

The state can institute policies that either encourage women to be the main socializers of their offspring or that leave socialization in the hands of the state. Examples include tying welfare payments to nonemployment so that mothers are forced to stay home with the children, instituting liberal or restrictive maternity leave policies, and subsidizing day-care centers, providing opportunities to enroll children in preschools. Sometimes state leaders become concerned that children of particular ethnic or racial groups are not learning the cultural values and/or language they need to succeed in the dominant culture. This concern motivates them to fund programs that expose children to the necessary personal, social, and learning skills.

4. Women as Signifiers of Ethnic and Racial Differences

Political leaders often use various images of women to symbolize the most urgent issues they believe the state faces. In wartime the state is represented as "a loved woman in danger or as a mother who lost her sons in battle" (Anthias and Yuval-Davis 1989, pp. 9–10). Men are called to battle to fight and protect the women and children. Often the leaders present the image of a woman who meets the culture's ideal

of femininity and who belongs to the dominant ethnic group. Sometimes political leaders use veiled language to evoke images of women of a certain ethnic or racial group as the source of a country's problems (for example, Albanian women who produce many children; African-American welfare mothers with no economic incentives to practice birth control). Often with a check of the facts such images are unfounded or there is no evidence to support such generalizations. In "Fertility Among Women on Welfare: Incidence and Determinants," sociologist Mark R. Rank (1989) maintains that "it is impossible to calculate with any precision the fertility rate of women on public assistance" (p. 296) because the data available has serious flaws. "There is no way of judging whether the fertility rate of women on welfare is high or low" (p. 296) relative to the fertility rate of other women.

5. Women as Participants in National, Economic, and Military Struggles

States implement policies governing the roles that women and men can assume in crises, notably in war. Historically, women have played supportive and nurturing roles, even in situations in which they have been exposed to great risks. In most countries, women are not drafted; they volunteer to serve. If they are drafted, the state defines acceptable military roles. If women do fight, they often do so as special units or in an unofficial capacity. In February 1994, Bosnian-Serb leaders announced that "The entire able-bodied population will be mobilized, either into military or labor units, and special women's units will be formed" (Kifner 1994, p. A4).

Regardless of their official roles in the war, women are affected by war. It is estimated that since World War II, civilians have suffered 80 percent of the deaths and casualties in war (Schaller and Nightingale 1992). Women are killed, taken prisoner, tortured, and raped. Even so, they often are not trained formally and systematically to fight. As a result, women occupy a different position than men during war.

The fact that women's combat roles are limited does not mean that women are incapable of combat, however. A small number of Serbian female battalions are fighting in Bosnia, and we hear occasional accounts of women "warriors," such as the profile of a Sarajevan female sniper that aired on the Canadian Broadcasting Corporation (CBC) in 1993:

> They say nobody loves their city the way the people of Sarajevo do, and when I saw it all, the way they are destroying the city, destroying people in it, I knew it had to be like this. The first person I shot at was a soldier; he was also a sniper. I shot at the very last second because he was aiming at me, so if I had waited another second, we would not be sitting here talking now. There is no time to think about it. . . .
>
> If I counted all of my victims I doubt I could preserve my sanity. It's not easy to pull the trigger but when I do I save at least ten other lives. The others shoot children at play, men and women standing in the bread line, or civilians walking down the street. I don't think I kill; I try to save as many lives as possible, the lives of civilians, the lives of innocent people, lives in general. (CANADIAN BROADCASTING CORPORATION 1994)

Through its military institutions, the state even establishes policies that govern male soldiers' sexual access to women outside military bases, both in general and in times of war. The Serbs, for example, captured some women and sent them to places resembling concentration camps in which many were raped, but they also kept other women in brothel-like houses and hotels. We also know that during World War II Japanese military authorities forcibly recruited[15] between 60,000 and 200,000 women, mostly Korean but also Chinese, Tai-wanese, Filipina, and Indonesian to work as sex slaves in army brothels in the war zone (Doherty 1993; Hoon 1992). They referred to them as "comfort women."

In *Let the Good Times Roll: Prostitution and the U.S. Military in Asia*, Saundra Pollock Sturdevant and Brenda Stoltzfus (1992) examine "the sale of women's sexual labor outside U.S. military bases" (p. vii). They present evidence that the U.S. military helps regulate prostitution; that retired military officers own some of the clubs, massage parlors, brothels, discotheques, and hotels; and that the military provides the women with medical care so as to prevent the spread of sexually transmitted diseases between the soldiers and the women. In 1993, thousands of Filipina women who live near the Subic Bay Naval Base filed a class action suit against the United States, arguing that the United States has moral and legal responsibilities to support the estimated 8,600 children fathered by U.S. servicemen stationed at Subic Bay.[16] Their suit is further evidence of the U.S. military's involvement in the lives of women abroad. It is also evidence of the military's disregard for and neglect of civilian populations, especially women and children.

In identifying areas over which the state controls men's, but especially women's lives, Anthias and Yuval-Davis highlight the effects of race and ethnicity in combination with gender on people's life chances. In this chapter we've explored the origins, manifestations, and consequences of the social differences between men and women. A key theme has been that these differences are social constructions rather than biological facts. We now apply these ideas to an issue of great concern to many college-age men and women in the United States: date rape.

Gender and Date Rape

In 1987 Mary P. Koss, Christine A. Gidycz, and Nadine Wisniewski published their findings from a nationwide survey[17] about college students' sexual experiences. The researchers found that 57 percent of female college students said they had experienced some form of sexual victimization (from fondling to forced anal or oral intercourse) during the past academic year. Fifteen percent of female college students said they had been victims of rape, and 12 percent said they had been victims of attempted rape.[18] The researchers also found that most of the women knew the perpetrators. Because slightly more than one-fourth of the female student population reported that they had experienced a rape or an attempted rape, the researchers concluded that rape takes place more frequently than we are led to believe by official statistics such as Uniform Crime Reports or the National Crime Survey.

Koss and her colleagues used the term **hidden rape** to refer to rape that goes unreported. Media accounts of this research coined the catchier if less precise phrases, *date rape, acquaintance rape,* and *campus rape.* The last of these terms is especially imprecise because Koss did not ask whether the rapists were also college students or attended the same institution as the victim. A number of articles appearing in popular magazines such as *Time* and *Vogue* suggest that an epidemic of "date rape" is occurring, especially on college campuses.

Sociologists G. David Johnson, Gloria J. Palileo, and Norma B. Gray (1992) observed that between 1987, when Koss's study was published, and 1991, the media changed its focus from discovering and labeling a social problem on campus to debating whether so-called date rape is not actually something women claim after a "bad sexual experience" or when they don't hear again from their partner. In January 1991 these researchers replicated the Koss study. To do so they surveyed a sample of 1,177 male and female college students attending a southern university. They hoped (among other things) to clarify a number of unanswered questions clouding this debate, including the following:

1. Is there an epidemic of date rape on U.S. college campuses?

2. Were the *female respondents* defining "forcible" sexual experiences as rape or were the *researchers* defining such experiences as rape? In Koss's study, for example, female students were asked, "Have you ever had sexual intercourse when you didn't want to because a man threatened you or used some degree of physical force (twisting your arm, holding you down, etc.) to make you?" Critics point out that women who say *yes* to this question are not asked whether they considered that sexual experience to be rape, even though the experience, as worded, constitutes the legal definition of rape.

3. Is so-called date rape something we might attribute to miscommunication rather than sexual aggression? Critics of the Koss findings often focused on this question. "This criticism emphasizes the ambiguity of sexual speech and nonverbal communication: When does 'no' really mean no, and when does it mean maybe, or even yes?" (Johnson, Palileo, and Gray 1992, p. 38).

Johnson and his colleagues found that the prevalence of rape at the campus they studied was strikingly similar to the prevalence recorded in Koss's national study in 1987. This fact suggests that date rape is not an epidemic, in the strict sense of the word, on the campus they studied.

With regard to the second question, approximately half (51 percent) of the 149 women in this study who said they had experienced forced sexual relations said the experience was not rape; 12 percent said it was rape; 37 percent said that "some people would think it was close to rape" or that "many would call it rape."

The researchers also found considerable miscommunication between men and women. Figure 11.2 shows the percentage of female respondents who said "no" to sex when they meant "yes." Slightly more than one-third of the women reported that they never say "no" when they mean "yes," and 66 percent reported that they have said "no" when they meant "yes." This finding shows

a significant number of female university students miscommunicate their sexual intentions. We expected this outcome, but were surprised by the magnitude of the miscommunication. That one in six females at a college campus today always says "no" when she means "yes" indicates the presence of a significant problem. Clearly the ambiguity of sexual communication is a significant barrier for the achievement of nonexploitive sexual relationships between men and women. (JOHNSON, PALILEO, AND GRAY 1992, P. 41)

In interpreting the data we must be careful about the conclusions we draw. For one thing, Johnson and his colleagues did not ask women who said they had experienced force whether they had

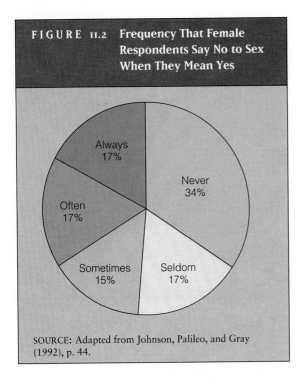

FIGURE 11.2 Frequency That Female Respondents Say No to Sex When They Mean Yes

Always 17%

Never 34%

Often 17%

Sometimes 15%

Seldom 17%

SOURCE: Adapted from Johnson, Palileo, and Gray (1992), p. 44.

said "no" when they meant "yes" in those instances. Also, we cannot conclude that the women were defining as rape those instances when they reported that they had said "no" but meant "yes." Finally, on the basis of questions asked in the Koss and Johnson studies, we cannot conclude that women are crying "rape" the morning after having "bad sex."

Some social commentators such as Katie Roiphe (1993) argue that the media hype and so-called feminist focus on date rape has created a mass hysteria on college campuses and that women are confused about what is rape. Such conclusions go far beyond any data that we have on this subject. Linda A. Fairstein (1993) takes an entirely different perspective than Roiphe and argues that she does not see hysteria on college campuses and does not think young women are confused about what is rape and what is not rape. Fairstein believes that unsubstantiated conclusions such as Roiphe's trivialize the experience of rape.

This debate aside, we are still left with the question of why 66 percent of the female respondents at this university at one time or another miscommunicated their intentions to their dates. For one answer to this question we turn to a classic ethnographic study of a U.S. high school and its students, conducted by Jules Henry (1963). Although the study is more than 30 years old, Henry's observations are still relevant today. Henry argues that sending mixed signals is one strategy women can use to communicate their sexual interests while maintaining their "reputation" and not appearing "easy." The woman gives double messages (saying "no" while indicating in other subtle ways that she means "yes"), because the feeling rules are such that she can't communicate her desires directly without being considered "loose" or a "slut."

Discussion

In this chapter we have emphasized various societal and cultural forces that make gender expectations central to people's lives regardless of the extent to which they meet those expectations. What can we learn from gender nonconformity? **Gender nonconformists** include:

1. Persons whose primary sex characteristics are not clear-cut (the intersexed)

2. Persons whose secondary sex characteristics depart from the ideal conceptions of masculinity and femininity

3. Persons whose interests, feelings, sexual orientation, choice of occupation, or academic major do not match gender-polarized scripts

4. Persons "who actively oppose the gender scripts of the culture" (Bem 1993, p. 167).

Bem argues that the existence of gender nonconformists challenges us to critically examine the connections we make between primary sex characteristics and the expression of almost every other trait. Specifically, it challenges us to ask whether it is reasonable to expect that one variable, a person's

These Bosnian boys, dressed like fighters and brandishing toy machine guns, are learning through socialization that fighting in war is part of being a male.
AP/Wide World Photos

primary sex characteristics, can channel sexual desires, ambitions, emotional makeup, interests, and so on in gender-polarized directions. Bem maintains that there are too many "mismatches" to support the argument that the variety existing within a sexual category is an aberration that we should try to correct.

In this chapter we have critically examined not only the connections we make between primary sex characteristics and appearance but the conse-

quences of making such connections on people's life chances, experiences, and opportunities. Sociologists do not deny that there are biological differences between males and females. Nor do they argue that bearing and raising children is something that women should not want to do. However, they question the ways in which socially and culturally conceived ideas about appropriate behavior and appearances for males and females channel behaviors to affect life chances.

FOCUS
Gender-Specific Communication Styles

We close this chapter by returning to this issue of gender polarization, which we defined as "the organizing of social life around the male-female distinction" so that people's sex is connected to "virtually every other aspect of human experience" (Bem 1993, p. 192). In this section we turn to

the work of communication specialist Janet Mills, who writes about the rules governing male and female body language.

Body Language Speaks Louder Than Words
Janet Lee Mills

I am a professional body watcher, and I love to turn others into body watchers, too. And that is pre-

SOURCE: "Body Language Speaks Louder Than Words," Janet Lee Mills. Copyright © 1985 by Janet Lee Mills. Reprinted by permission.

cisely what I do in my university classes and in executive training seminars.

As a specialist in male–female communications, I propose that males and females in our culture speak different body languages. To illustrate this point I recruited Richard Friedman, Assistant to the President at the University of Cincinnati, to model with me for the accompanying photographs. The photos in which Friedman and I posed in our usual male and female roles, respectively, were easy. Then came the hard part, the part that proves my point. I posed us in postures typical of the opposite sex. The results in the photos illustrate the old maxim, "One picture is worth a thousand words!"

The photos contain two basic sets of behavioral cues—affiliative cues and power cues. Male nonverbal behavior typically includes very few affiliative displays, such as smiles and head cants, and many power cues, such as expanded limb positions and serious facial expressions. Female nonverbal behavior, however, is ordinarily just the opposite, containing many affiliative displays and few power cues. The overall impression males create is one of power, dominance, high status, and activity, particularly in contrast to the overall impression females create, which is one of submissiveness, subordination, low status, and passivity.

Sensitizing students and professional groups to these sex-role differences and the functions they serve in social and business contexts is my business. In my seminars, I illustrate how men spread out their upper and lower limbs, expanding to take up space; how men sit and stand in loose, relaxed postures; how they gesture widely, speak in loud, deep tones, and engage in either direct or detached eye-contact patterns. All these behaviors communicate power and high status, especially when men are communicating with subordinates. I also illustrate how women constrict their arms and legs; sit in attentive, upright postures; gesture diminutively; speak in soft, breathy voices; and lower their eyes frequently. These behaviors give away power and announce low status.

In addition, I point out to audiences that women smile often, cant their heads, nod their heads, open their eyes in wide-eyed wonder, and posture themselves in positions of unstable balance. Men return smiles or not, at will, engage in far less head cant-

ing and nodding, keep their eyes relaxed, and posture themselves in stable balance.

These sex differences are socially learned and publicly performed, but are relatively unconscious in both the sender and receiver—until someone breaks the rules or norms. But when the rules are broken all attention is focused on the person breaking them.

Women in management are of particular concern to me. Managerial and professional women simultaneously play two roles, that of "woman" and that of "manager" (or professor, doctor, accountant, and so forth). And the role of woman and the role of manager or professional each has a different set of rules—contradictory rules. From the depths of sex-role socialization comes the demand "be feminine," but from the context of the managerial work place comes the demand "be powerful." The story is told in body language, through which relatively unconscious messages are continually expressed.

To be successful in terms of femininity, a woman needs to be passive, accommodating, affiliative, subordinate, submissive, and vulnerable. To be successful in terms of the managerial or professional role, she needs to be active, dominant, aggressive, confident, competent, and tough. In their everyday communications with others, women face a dilemma. Behaviors indicative of traditional femininity are taken by others as evidence of the inability of women to cope with the demands of the professional role; behaviors appropriate to the professional role are taken as evidence of women's lack of femininity. This no-win situation constitutes a double bind: She's damned if she does, damned if she doesn't.

After several years of research and consulting with women and men in management, I developed a model describing how a woman's nonverbal behavior evolves as she responds to the contradictory injunctions: "be feminine," "be powerful." In each stage, nonverbal behavior is assumed to be an externally observable expression of her internal responses to the role conflicts and double binds she experiences. (Though it isn't addressed in this article, the model I developed also describes the developmental stages men go through as they encounter professional women.)

Mills is famous for this pose, which she captions, "Could you say no to this woman?" In it she violates many traditional female behaviors with relaxed posture; arms, legs positioned away from body; direct, confrontive eye contact; no affiliative smile.

All photos courtesy of Janet Lee Mills

Mills demonstrates the "power spread," another typical high-status male pose, with hands behind head, elbows thrust out, legs in a "broken-four" position, and an unaffiliative facial expression. Women dressed for success appear shocking in this pose.

Friedman does not appear shocking in the same pose, because many male executives conduct business from a similar position.

Ah, but doesn't Mills look feminine in this typically feminine pose with canted head, affiliative smile, ankles crossed, and hands folded demurely?

And doesn't Friedman look ridiculous in the same pose?

Power is often wielded in postural alignment, gesture, and use of objects. Mills holds power, along with papers on which attention is focused; while Friedman signals deference with lowered gaze, and a constricted body.

Now Friedman holds power, along with papers, with lower limbs spread away from his body; while Mills signals deference with an attentive pose, submissively bowed head, and a hand-to-mouth gesture of uncertainty.

In a typical office scene, Friedman holds power with an authoritative stance, hand in pocket and at mid-chest, straight posture, and head high; while Mills is submissive, subordinate with canted head, smile, arms, hands close to her body. Note Friedman's wide, stable stance; Mills's unstable stance. Many women tend to slip into a similar posture when talking to a shorter male authority figure.

Now the tables are turned. Friedman defers to authority in a feminine, subordinate posture with scrunched up spine, constricted placement of arms, legs, canted head, and smiling attentiveness.

In Stage I: "Sex-Role Enactment," a woman enters the professional work place enacting her well-learned, traditional sex-role behaviors. Nonverbally she signals subordination, submissiveness, affiliation, and low status. She presents a social-self that others respond to as weak, ineffective, and unimportant. The smiling, head canting, attentive looking, gaze aversion, head nodding, and attentive postures elicit quizzical stares or incredulous replies in the work place. The same behaviors that won her popularity, dates, and admiration in her social life now cost her her credibility in the work place. The basic problem is that her communications are affiliative, but not powerful.

In Stage II: "Sex-Role Reversal," a woman begins to sense her powerlessness in the interpersonal domain of the work place. After all, the negative feedback abounds. She decides, consciously or unconsciously, "When in Rome, do as the Romans do," and will likely adopt a role model or several role models to emulate. These are, of course, most likely to be men. The nonverbal messages she sends begin to be assertive and aggressive—masculine, if you please. She begins to adopt relaxed postures, take up more space, gesture big, talk louder and longer, interrupt others, smile infrequently, use direct eye contact, and square her shoulders while engaged in a conflict. Others respond to her in various ways. Some find her amusing, an imitation man, a plastic man; some find her vulgar, a competitive bitch, an emasculating female. Many try to tease her back into her feminine role. The problem with her communications is the opposite of that in the sex-role enactment stage; here, she communicates power, but little or no affiliation, warmth, or friendliness.

A third stage, "Shuttle," relies on behavior that actually alternates between Stages I and II. Although this may seem, at first, a desirable resolution of the "be powerful/be feminine" paradox, it is not. Very simply, the woman in Stage III responds to situations in either the "feminine" mode or the "power" mode. While she views herself as situationally flexible and able to meet different demands, others tend to view her as unstable, whimsical, "off the wall," undependable, and uncertain. Her flexibility gains her the reputation of being flighty or inconsistent. The crux of the ultimate failure of Stage III behavior is that she is still responding to only one role demand at a time. Be feminine here, be powerful there.

But there is a way for women to extricate themselves from the double bind and rise to the top of the corporate ladder with their femininity intact. To succeed ultimately requires the simultaneous expression of both femininity and power.

Stage IV, "Integration," involves a collapsing of the two inconsistent roles into a unified whole. The nonverbal behaviors of the woman in this stage include both messages of affiliation and power, delivered simultaneously. Neither is missing, so no one can fault her for being a corporate cheerleader or mascot, nor can they fault her for being cold, callous, and too stern. She smiles as she looks you straight in the eye. She spreads out her arms, cants her head, knits her brows in thought, and speaks with clear diction, resonant voice tones, and good volume. She may be dynamic; she may be regal and stately—but her nonverbal presentation is an expression of the integrated personality she has developed. In short, she is what *Fortune 500* companies ask for in women executives: powerful and feminine.

Fortunately, communication experts can help clients shorten considerably the time involved in such personal/professional evolution. In workshops for women and men, both sexes not only see the sex-role differences, but *practice* them as well, which provides the advantage of allowing them to feel kinesthetically within their own body the differences in nonverbal sex-role behaviors. I've found that watching a model dramatize gender-identifying nonverbal behavior generates a very powerful learning experience. Getting into the postures of feminine constriction sensitizes men to a new understanding of their female counterparts. For women, expressing power nonverbally often stirs images and feelings of new competence and confidence, as well as a new understanding of men.

The best part of every seminar for me is the moment when many members of the audience recognize that our sex-role socialization is real, consequential, and undeniable in its behavioral forms. At that moment, everyone steps out of the game able to view it from a new vantage point.

And now a challenge: If you are a women, assume the masculine poses pictured here; if you are a man, assume the female poses on these pages. By actually experiencing poses typical of those of the opposite sex, you may gain new insight into your own sex-role training—and learn to know a thing by its opposite.

Key Concepts

Advanced Market Economies 402

Ethgender 412

Feeling Rules 398

Femininity (or feminine characteristics) 394

Feminist 389

Fortified Households 402

Gender 388

Gender Nonconformists 417

Gender Polarization 396

Gender-Schematic Decisions 397

Hidden Rape 416

Institutionalized Discrimination 409

Intersexed 393

Low-Technology Tribal Societies 400

Masculinity (or masculine characteristics) 394

Nonhouseholder Class 402

Primary Sex Characteristics 393

Private Households 402

Secondary Sex Characteristics 394

Sexist Ideologies 410

Sexual Property 400

Social Emotions 398

Socialization 405

State 412

Notes

1. Slavenka Drakulić describes how the people of Yugoslavia were caught between Eastern and Western Europe:

   People in the West always tend to forget one key thing about Yugoslavia, that we had something that made us different from the citizens of the Eastern bloc: we had a passport, the possibility to travel. And we had enough surplus money with no opportunity to invest in the economy (which was why everyone who could invested in building weekend houses in the mid-sixties) and no outlet but to exchange it on the black market for hard currency and then go shopping. Yes, shopping to the nearest cities in Austria or Italy. We bought everything—clothes, shoes, cosmetics, sweets, coffee, even fruit and toilet paper. I remember times when my mother who lives in a city only a short drive from Trieste would go there every week to get in stores what she couldn't get here. Millions and millions of people crossed the border every year just to savour the West and to buy something, perhaps as a mere gesture. But this freedom, a feeling that you are free to go if you want to, was very important to us. It seems to me now to have been a kind of a contract with the regime: we realize you are here forever, we don't like you at all but we'll compromise if you let us be, if you don't press too hard. (1993A, P. 135)

2. In 1988, nine million tourists visited Yugoslavia. Two republics—Slovenia and Croatia—benefited the most from guest worker programs and tourism. The government transferred about 25 percent of the two republics' earnings to other less prosperous republics.

3. Medical researchers hypothesize that a mutation in the Y chromosome in the case of anatomical females (who are genetically male), or a mutation in the X chromosome in the case of anatomical males (who are genetically female), respectively suppresses or fails to suppress the production of excess testosterone.

4. In Olympic history there has been only one case in which a man competed as a women. In 1936, Nazi officials forced a German male athlete to enter the woman's high jump competition. Three women jumped higher than he did (Grady 1992).

5. For more on this topic see Denise Grady's article, "Sex Test of Champions" in the June 1992 issue of *Discover*, pp. 78–82.

6. Although "excess" hair sometimes warns a doctor to check for the presence of an underlying endocrine disorder such as an ovarian tumor, 99 percent of the time such hair is not associated with any pathology.

7. The severity of damage caused by X-ray treatments is illustrated by the case of Mrs. A. E. C., one of many reported in the *Journal of the American Medical Association*:

   Mrs. A. E. C., aged 32, with a history of treatment five years ago by the Tricho [X-ray] system in Boston. She had 15 treatments on her chin for [superfluous hair]. She was treated every two weeks, and three areas on the underside of her chin were treated each time. Twice she was treated by the bookkeeper when the nurse was not in. Three years ago red blotches began to appear and they have continued. She shows definite signs of telangiectasis [a chronic dilation of blood vessels causing dark red blotches] and atrophy over an area of about three inches in diameter just below the edges of the chin. (AMERICAN MEDICAL ASSOCIATION BUREAU OF INVESTIGATION 1929, P. 286)

8. The "Iron Curtain" is a term coined by British Prime Minister Winston Churchill to describe the division between communist Eastern Europe and noncommunist Western Europe after World War II.

9. Northern Kentucky University is primarily a commuter campus, largely serving students who cannot afford to attend college away from home or who are committed to staying in the area because of job and family responsibilities.

10. The Family Leave Act is not as straightforward as it might appear. For example, it excludes persons who are employed at worksites with fewer than 50 employees. And employers may deny leave to salaried employees who are in the top 10 percent of pay categories.

11. In an April 1993 editorial in *Ms.*, Robin Morgan writes that mass systematic rape is nothing new to history:

    An aberration? "Rape and take spoil" is an ancient motto of war. An isolated incident? Then what happened to the Sabine women? What did Alexander's armies do? Caesar's legions? The conquistadors and all the other colonizers? What of Chaka's Zulu army? The rape of Belgian and French women by German troops during World War I? What did their sons do to Russian women in World War II—and what did Russian troops do to German women "in response"? What about the institutionalized concentration-camp brothels where Jewish women were forced to supply "Enjoyment Duty" to their Nazi captors? The thousands of Chinese, Korean, and Filipina women conscripted into sexual slavery as "comfort women" for the Japanese army? The Chinese city where that army's actions were so extreme historians termed it "the Rape of Nanking"? The virtually ignored mass rape of Bengali women (estimated as high as 400,000) during the 1971 Bangladesh-Pakistan war? What about Vietnam? The Iraqi rapes of Kuwaiti women—and their Asian servants? (P. 1)

12. *Chetnik* is the term for Serbian guerrilla fighters during World War II.

13. There is some debate over whether Barbie is a "bimbo" or a "feminist." Some critics say that in real life Barbie's figure would measure 36-18-33 and that she is quite materialistic. Accessories range from "toe nail polish to a pink RV camper to haul around her never-quite-big-enough wardrobe" (Cordes 1992, p. 46). Others argue that Barbie's résumé is quite impressive and that she allows little girls to dream of what they can be. Over the years Barbie has been marketed as a fashion model, ballerina, stewardess, teacher, fashion editor, medical doctor, Olympic athlete, TV news reporter, corporate executive, and animal rights volunteer (*Harper's* 1990).

14. Alderman (1990) wrote to the researchers: "It is as if *Consumers' Reports* commissioned research on the handling characteristics of the Suzuki Sammurai, and received instead a report arguing that informal import quotas for Japanese automobiles were not justified" (p. 108).

15. *Recruit* is the word the Japanese used to describe how they managed to bring Korean women to war zones. Documents show that the Japanese acquired the women through civilian agents, village raids, and applying pressure on Korean school administrations to provide girls (Hoon 1992).

16. Under current U.S. immigration law, Filipino children of servicemen are not permitted to migrate to the United States with sponsorship from any American, as can the children of servicemen stationed in South Korea, Thailand, Cambodia, and Laos (Lambert 1993).

17. The researchers surveyed 6,159 male and female college students enrolled at 32 institutions.

18. In the Koss, Gidycz, and Wisniewski study, the section "Measurement of Sexual Oppression or Victimization" explains how to read the survey data table.

12 POPULATION AND FAMILY LIFE

with Emphasis on Brazil

Street scene in São Paolo, Brazil.
Paulo Fridman/Sygma

Students from Brazil enrolled in
college in the United States for fall 1991 — 4,260

Brazilians admitted into the U.S. in
fiscal year 1991 for temporary employment — 7,957

People living in the United
States in 1990 who were born in Brazil — 94,000

Airline passengers flying between
the United States and Brazil in 1991 — 1,226,704

U.S. military personnel in Brazil in 1993 — 52

People employed in 1990 by
Brazilian affiliates in the United States — 2,300

Applications for utility patents filed in the
United States in 1991 by inventors from Brazil — 124

Phone calls made between the
United States and Brazil in 1991 — 30,946,000

Brasil 82 17,00

CENTENARIO DO NASCIMENTO DE MONTEIRO LOBATO   JO OLIVEIRA

As you read the following excerpts, think of what they say about family life. The first two excerpts, by Christy Brown and Robert Sayre, respectively, give a personal point of view. The remaining three portray scenes that provide insight into some aspects of family life. What do these families have in common?

I was born in the Rotunda Hospital in Dublin, Ireland, on June 5th, 1932. There were nine children before me and twelve after me. . . . Out of this total of twenty-two, seventeen lived, but four died in infancy, leaving thirteen still to hold the family fort.

Mine was a difficult birth, I am told. Both mother and son almost died. A whole army of relations queued up outside the hospital until the small hours of the morning, waiting for news and praying furiously that it would be good.

After my birth Mother was sent to recuperate for some weeks and I was kept in the hospital while she was away. I remained there for some time, without name, for I wasn't baptized until my mother was well enough to bring me to church.
(C. BROWN 1992, P. 85)

One of the common experiences of people in their early forties, which they seem to need to talk about, is having parents who are in their sixties and seventies. The situation itself is only a biological and statistical inevitability, at least in contemporary American culture. But I have been surprised, just the same, at the frequency with which discussion of parents now comes up among people I know. Ten years ago, when we were in our early thirties, we talked about our children—about pregnancies and births, bottles versus breast-feeding, how to get "them" to sleep through the night, and then about toilet training and schools. We were primarily parents, and our own parents were secondary subjects. They were just grandparents and in-laws who were or were not helpful or demanding, visiting, vacationing, or whatnot. Only in the last four or five years, I realize, have I been telling and hearing stories about *parents,* usually with friends, but sometimes with people I have just met, if they are my age.
(SAYRE 1983, P. 124)

Perhaps half of rural households in Kenya are, in practice, headed by women. As the population grows and landholdings shrink, an increasing proportion of men are migrating to the cities for work, leaving the wife [like Rachel Mwangene] behind to look after farm and family.

Rachel's day is long. She rises "when the sky is beginning to lighten," cooks breakfast, gets the children off to school and cleans the house. Then she sets off for the holding, two miles away up and downhill, where she tethers the animals to graze and gets down to planting, digging or weeding her corn, cassava and cowpeas, eating a snack in the field. On her way home she gathers whatever firewood she finds. Then she fetches water, half an hour's walk away, with the return journey uphill. As the sun begins to set, she cooks the evening meal of *ugali* (maize porridge) in a pot balanced on three stones, until it is as stiff as bread dough: "You are stirring solidly for an hour," she complains, "and it gets harder and harder until at the end the sweat is pouring off you." Getting the maize ground is another chore: twice a week she must trek two miles to the nearest neighbour who possesses a handmill.

(HARRISON 1987, PP. 438–439)

The Aratanha family of Rio de Janeiro lives in a fenced-in, highly guarded highrise with many amenities (soccer field, tennis courts, a gym). The family consists of a mother and father (both of whom are physicians), and two children (ages 18 months and 3 years). They have a maid who helps care for the children. "[The father] gets out of bed quickly without waking his wife . . . he leaves very early to avoid the traffic jams . . . [the mother is] grateful for her new car. It saves her half an hour every morning."

(TREMBLAY 1988, P. 31)

[In the United States] the ideal [that] fifties women were to strive for was articulated by *McCall's* in 1954: togetherness. A family was as one, its ambitions were twined. The husband was designated leader and hero, out there every day braving the treacherous corporate world to win a better life for his family; the wife was his mainstay on the domestic side, duly appreciative of the immense sacrifices being made for her and her children. There was no divergence within. A family was a single perfect universe—instead of a complicated, fragile mechanism of conflicting political and emotional pulls. Families portrayed in women's magazines exhibited no conflicts or contradictions or unfilled ambitions. Thanks, probably, to the drive for togetherness, the new homes all seemed to have what was called a family room.

(HALBERSTAM 1993, P. 591)

In the broadest sense of the word, a **family** consists of two or more people related to one another by blood, marriage, adoption, or some other socially recognized criteria. On reading this assortment of scenes of family life, one might wonder if there are any sociological frameworks for thinking about the family that would be relevant to such a wide range of experiences and issues. What do the families of Christy Brown, Robert Sayre, and Rachel Mwangene have in common with the Aratanha family and the ideal family of the 1950s? As diverse as these scenes are, we can say that each

is affected by key episodes common to all family life. No matter what kind of family people are born into, live with, or form later, their lives are shaped by the following key episodes:

1. Birth (even if it is only their own birth), which includes the number of children born and the spacing (amount of time) between births

2. Death, which includes how and when (infancy through old age) a family member dies

3. How much and what kind of work each family member must do, and where family members

must travel to find work to sustain the family's standard of living.

The sociologists who focus most on key episodes are those who study human **populations**. The study of population includes an interest in the number of people in and composition (the various ages, the ratio of males to females, the percentage of people in various racial and ethnic groups, and so on) of social groupings (those who live within the boundaries of country, state, province, or other geographical area) and the factors that lead to changes in that social grouping's size and composition (Pullman 1992). Obviously the number of births and deaths and the number of people moving into and out of a geographic region (migration) affects the region's population size and composition.

Population and family life are paired together in this chapter because the factors that affect population size and composition represent the sum of the key episodes common to all family life. In this chapter we look at the concepts and theories that sociologists draw upon to explain childbearing experiences (number and spacing of children), the type and timing of death, and migration, especially movement related to the search for work.

One important event that has shaped the key episodes of family life is the Industrial Revolution, which is still in progress. Traditionally we think of the Industrial Revolution as an event that began in England in the late eighteenth century and spread to other countries in western Europe and to the United States. Actually, this event forced people from even the most remote regions of the planet into a worldwide division of labor. The extent to which the Industrial Revolution has been realized in other parts of the world varies according to country and, in most cases, according to regions within countries.

We give special emphasis to Brazil, which has the fifth largest population in the world and the second highest standard of living in Latin America. The World Bank classifies Brazil as a country with a lower-middle- to middle-income economy. Yet, as many as 60 percent of Brazil's people live in extreme poverty. The unequal distribution of wealth is reflected by the fact that the percentage share of household income for the most affluent 20 percent

of the population is 63 percent, while the percentage share for the least affluent 40 percent of the population is only 8 percent (The World Bank 1990). In Brazil, as we will learn, the effects of industrialization vary across and within its five major regions:

- The North (Amazon Basin covering half the country with vast reaches of largely uninhabited tropical forest)
- The Northeast (semiarid scrubland prone to periodic drought and massive flooding, heavily settled and poor)
- The South (rich farmland and pasture lands and large modern cities with a large and relatively prosperous population)
- The Southeast (huge, densely populated urban centers, including the city and state of São Paulo)[1]
- The Central West (home of one of the earth's major ecological frontiers and some of the largest cattle ranches in South America, sparsely populated)[2]

In view of this variety, a focus on Brazil allows us to consider childbearing experiences, types and timing of death, and migration as it occurs in a variety of contexts: environments touched directly by industrialization, on the fringes of industrialization, and bypassed, exploited, and abandoned by industrialization.

We begin by discussing the problem of defining what constitutes a family and why a focus on the key episodes of family life is more useful than a focus on defining families. We explore why the Industrial Revolution (in all its forms) is such an important factor in shaping the episodes of family life, and then we consider how the Industrial Revolution has affected Brazil. Next we examine the theory of the demographic transition, a model that outlines historical changes in births and deaths in western Europe and the United States, and we discuss how well the theory applies to non-Western countries such as Brazil. Finally, we explore how large-scale changes in births, deaths, and the nature of work present families with new and different challenges.

What Kinds of People
Constitute a Family?

On August 20 and September 5, 1977, the United States launched the *Voyager 1* and *Voyager 2* probes into outer space to explore and photograph Jupiter, Saturn, Uranus, and Neptune. In 1990 the spacecraft left the earth's solar system. Attached to the outside of each was a gold-coated copper phonograph that contained 118 photographs of the planet and its inhabitants, 90 minutes of music from countries around the world, a collection of the sounds of the earth, and greetings in 60 different languages. This "portfolio" of the planet was made to "send to any possible extraterrestrial auditors information about the Earth and its inhabitants" (Sagan 1978, p. 33). Among the 118 photographs was one entitled "Family Portrait" (see Figure 12.1).

Imagine for a moment the photograph you would have selected to represent all of the various family arrangements on earth. Choosing a single photograph to represent family life in the United States—not to mention the entire planet—would be an overwhelming challenge. The difficulty is rooted in the fact that, even though every person is a member of a family (if only in the biological sense), "there is no concrete group which can be universally identified as 'the family'" (Zelditch 1964, p. 681). There is an amazing variety of family arrangements worldwide—a variety reflected in the numerous norms that specify how two or more people can become a family. These include norms that govern the number of spouses a person can have, the way a person should select a spouse, the ideal number and spacing of children, the circumstances under which offspring are considered legitimate (or illegitimate), the ways in which people trace their descent, and the nature of the parent-child relationship over the child's and parent's lives (see Table 12.1). In light of this variability, we should not be surprised to learn that it is difficult to construct a definition of family and that "no general theory or universal model of the family can be formulated" (Behnam 1990, p. 549).

Most official definitions of family emphasize blood ties, adoption, or marriage as criteria for membership, and the function of procreation and socialization of offspring. The U.S. Bureau of the Census (1993) uses the term *family* to mean "a group of two or more persons related by birth, marriage, or adoption and residing together in a household" (p. 5). This definition has been in effect since 1950. Between 1930 and 1950, however, the definition of family revolved around a head of the household and reflected living arrangements in a more rural America:

> *a private family comprises a family head and all other persons in the house who are related to the head by blood, marriage, or adoption, and who live together and share common house- keeping arrangements. The term "private household" is used to include the related family members (who constitute the private family) and the lodgers, servants, or hired hands, if any who regularly live in the home.* (U.S. BUREAU OF THE CENSUS 1947, P. 2)

Until the mid-1970s the official definition of family in Brazil included the following criteria:

> *(1) [that] family is synonymous with legal mar- riage, (2) marriage lasts until a spouse dies; (3) the husband is the breadwinner and the sole earner; (4) the wife is a full-time home-maker and her work has no [formal] economic value; (5) the husband is the legal head of the family.* (GOLDANI 1990, P. 525)

The Brazilian government's official definition of what constitutes a family has changed several times in the past 25 years such that this definition of family had been abandoned constitutionally. For an example, the 1988 constitution recognizes a "stable union" between a man and a woman as a family and children living with one parent as a family. Such changes mean that the official definition recognizes that "family-like" arrangements can exist outside legal marriage (Goldani 1990).

Changes in the official definition reflect dramatic changes in the size and composition of families that have occurred in both Brazil and the United States

FIGURE 12.1   Family Portraits

Can a single image represent all that we mean by "family"? What idea of "family" is conveyed by these pictures? The examples shown here are just a few of the many types of family arrangements.

Among the 118 photographs on board the Voyager 1 and 2 spacecraft was this one entitled "Family Portrait."
Nina Leen/Life Picture Service

A Mexican family celebrating a birthday.
Chip and Rosa María de la Cueva Peterson

A father with his adopted children.
J. Patrick Forden/Sygma

An African mother and her children (absent the father, who lives and works in the city).
Betty Press/Woodfin Camp & Associates

Lesbian parents with their child.
Chris Maynard/Gamma Liaison

T A B L E 12.1   Various Norms That Govern Family Life

Marriage Systems

Monogamy (one husband, one wife)
Polygamy (multiple spouses)
    Polygyny (one husband, multiple wives)
    Polyandry (one wife, multiple husbands)

Choice of Spouse

Arranged (parents select their children's marriage partners)
Romantic (a person selects a marriage partner on the basis of love)
Endogamy (marriage within one's social group)
Exogamy (marriage outside one's social group)
Homogamy (marriage to people who share similar social characteristics such as similar social class, religion, and level of education)

System of Authority

Patriarchal (male-dominated)
Matriarchal (female-dominated)
Egalitarian (equal authority)

System of Descent

Patrilineal (traced through father's lineage)
Matrilineal (traced through mother's lineage)
Bilateral (traced through both mother's and father's lineage)

Family Type

Nuclear (husband, wife, and immediate children)
Extended (three or more generations)
Single-parent (mother or father)
Household (people who share the same residence)

Family Residence

Patrilocal (wife lives with the husband's family or in the husband's village)
Matrilocal (husband lives with the wife's family or in the wife's village)
Neolocal (residence separate from parents)

---

in the past 20 to 30 years (see Tables 12.2 and 12.3). The changing definitions also show that focusing on membership criteria and specific functions excludes many primary groups that have family-like qualities, including child-free couples, couples whose children no longer live at home, elderly people who live with their adult offspring, and unmarried heterosexual and homosexual couples with children.

One problem with definitions of families is that they can be used to deny family-related benefits to people who don't fit the definition of a family. Such

Total households	1960	1992
	53,021,000	95,669,000
I. Family household		
A. Nuclear		
Married couple with children	52.18	26.46
Married couple with no children	22.00	21.64
Father–child(ren)	2.33	3.16
Mother–child(ren)	8.48	12.22
B. Extended[†]		
Grandparent(s), grandchildren, both parents	N/A	.52
Grandparent(s), grandchildren, mother only	N/A	1.82
Grandparent(s), grandchildren, father only	N/A	.15
Grandparent(s), grandchildren, no parents	N/A	.91
II. Non-family household		
One person living alone	14.88	25.06
Institutions[‡]	3.61	3.48
III. Other living arrangements[§]	6.71	3.46

TABLE 12.2 Distribution (Percent) of Family and Non-family Households in the United States

*Percentages do not total 100% because of inconsistencies and overlaps in the categories defined by the U.S. Bureau of the Census.
[†]Only those extended family categories used by the U.S. Bureau of the Census.
[‡]Institutions include those living in correctional institutions, nursing homes, and juvenile centers.
[§]Other living arrangements include college dormitories, military quarters, emergency shelters, and other unrelated persons living together.
SOURCES: U.S. Bureau of the Census, p. 40 (1961); pp. 64–65 (1993).

benefits include insurance coverage, housing, time off from work to care for another family member, inheritances (especially in the absence of a will), and custody of children. In this vein, a number of recent court decisions in the United States have ruled that criteria other than the socially recognized bonds of blood, marriage, or adoption must be considered in defining whether a person belongs to a family and thus is entitled to family-related benefits. Other factors that influence family relationships include exclusivity, longevity, emotional support, and financial commitment (Gutis 1989a, 1989b).[3,4]

Given all of this, the operative definition of family may be the broad one given earlier: two or more people related to one another by blood, marriage, adoption, or some other socially recognized criteria. Changing definitions of family in Brazil and the United States suggests that it is not especially useful to think about the family in terms of specific memberships or according to a single function (such as the procreation or socialization function). Instead, we would do better to take a long view and consider key episodes that affect family size and composition and an event that transformed the character of family life in very general but fundamental ways. This event is the Industrial Revolution.

Family and non-family household	1960*	1984†
I. Family household	92.8	93.2
A. Nuclear	68.9	70.3
Married couple with children	54.1	46.5
Married couple with no children	8.5	12.8
Father or mother child-family	6.3	11.0
B. Extended	22.6	14.1
Married couple with children, unmarried and relatives	13.1	6.6
Married couple, no children and other relatives	2.9	1.6
Other extended family (no married couple, only father or mother child-family and relatives)	6.6	5.9
C. Complex	1.5	8.7
Married couple with children unmarried and non-relatives	1.5	8.7
II. Non-family household	7.2	6.8
Men or women living alone	5.3	5.5
Other non-family	1.9	1.3
Total percent	100	100
Absolute values in thousands	13,532	31,075

TABLE 12.3 Distribution (Percent) of Family and Non-family Arrangements in Brazilian Households

*For the 1960 census the family-household classification did not include maids and lodgers. At other times only maids are not included.
†1984 is the last year for which data is available. The rural population of the Central West region was not considered in 1976, nor was the North region considered in 1984.

SOURCE: Goldani (1990), p. 527. Original sources: Brazilian Population Censuses, tapes of 1% in 1970 and 75% in 1980; and Household Survey, 1976 and 1984 tapes.

The Industrial Revolution and the Family

The Industrial Revolution cut people off from the family and from the clan-oriented ways of life in which they had existed for most of history (Riesman 1977). The phrase "family and clan-oriented" refers to an environment in which change is slow (although not completely absent), in which family influences almost all facets of life (where one works, whom one marries, what one does), and in which people follow patterns that have endured for centuries. In clan-oriented societies life expectancy at birth is low and the relationship to the food supply is precarious. In addition, the division of labor is simple: the majority of people perform labor-intensive and subsistence-oriented[5] tasks that are allocated on the basis of age, gender, and physical condition. Under these circumstances adult roles change very little from one generation to the next; parents can assume that their children's daily lives will be much like their own and that their children will face virtually the same challenges and problems. Consequently, parents pass on to their children a set of roles and time-tested rules and rituals that promote the group's survival (see the discussion in Chapter 6 on organic solidarity). Because life expectancy is short, children assume these adult roles very early (Riesman 1977).

As we learned earlier in this chapter, the Industrial Revolution was not something unique to Western Europe and the United States. It was an event that forced people from even the most remote regions of the planet into a worldwide division of labor, and its effect is not uniform but varies according to country and according to regions within countries. In general, industrialization in Western societies is associated with dramatic changes in women's childbearing experiences, in the timing and causes of death, and in the kinds of work the majority of people do. Specifically, the following changes occur:

Birth

- The number of children that women bear and the proportion of their reproductive life that women give to childbearing decreases substantially.

Death

- Degenerative diseases and diseases often caused by lifestyle choices (such as cancer and heart disease) displace infectious and parasitic diseases (smallpox, yellow fever, polio, influenza, typhoid, scarlet fever, and measles) as the primary causes of death. The decline in infectious and parasitic diseases results in a dramatic increase in the number of infants who survive the first year of life and the number of women who survive childbirth.

Work

- Natural sources of power (human muscle, animal strength, wind, and water) are replaced with created power (fueled by oil, natural gas, coal, nuclear power, and hydroelectric power) and programmed machines (Bell 1989). Consequently, the proportion of the population engaged in manual labor (the extraction and transformation of raw materials) declines and a greater proportion is engaged in human and professional services (social work, education, health care, entertainment, publishing, sales, marketing, and so on).

At first glance this list of changes may mean very little, but try to imagine the profound effects of such changes on family life. To appreciate what it means for women to have control over the number and spacing of children, consider the reproductive experiences of two women of comparable social status, who lived at different times—Alice Wandesworth Thorton, who lived in the seventeenth century, and Jane Metzroth, a twentieth-century woman and a vice president in investment banking:

At last [Alice] was persuaded to accept the proposal of one William Thorton, Esq., a gentleman less well-to-do than various other suitors,

but more pious. Alice became pregnant almost immediately, and over the next fifteen years bore nine children, three of whom lived—a typical survival rate for her time. She remembers her pregnancies in terms of fevers, sweatings, nosebleeds, faintings, and agues; her deliveries were inevitably perilous, and her recoveries of many months duration. The babies who lived were subject to diarrhea, convulsions, or being rolled upon by their wet nurses. (SOREL 1984, PP. 318–19)

There were a couple of years' worth of thinking that went into the decision to have another child. We already had a school-age son, and for a long time I thought we'd just have one child, because working and raising a family and trying to have a normal kind of life, too, was kind of hard work. But after Erik was in school, and I saw him developing into a really neat little person, I kept trying to decide in my own mind whether to do it again. . . . I felt if we were going to add to the family, this would be the time. There were also financial considerations, and child care arrangements; we've had the same lady for almost four years now, and she's really terrific. And that helped me make the decision. . . . So with all those things combined, we said, "Now's the time." Plus, I'm in my thirties now, so I didn't want to put it off and then find out it wouldn't be possible. (SOREL 1984, PP. 60–61)

These kinds of changes, as reflected in the lives of these two women, made industrialization a revolution: "Within a few decades a social order which had existed for centuries vanished, and a new one, familiar in its outline to us in the late twentieth century, appeared" (Lengermann 1974, p. 28).

We can classify the countries of the world into two broad categories with regard to industrialization. These categories are the mechanized rich and the labor-intensive poor. Comparable but misleading dichotomies include developed and developing, industrialized and industrializing, and First World and Third World. They are misleading names because they suggest that a country is either industrialized or is not. The dichotomy implies that a failure to industrialize is what makes a country poor,

FIGURE 12.2   The Mechanized Rich and the Labor-Intensive Poor Countries
The labor-intensive poor countries and areas include sub-Saharan Africa, the Near East and North Africa, Latin America and the Caribbean, Asia (except Japan), and Oceania (except Australia and New Zealand). The mechanized rich nations and areas include North America and Europe, Russia and some of the former Soviet states, Japan, Australia, and New Zealand.

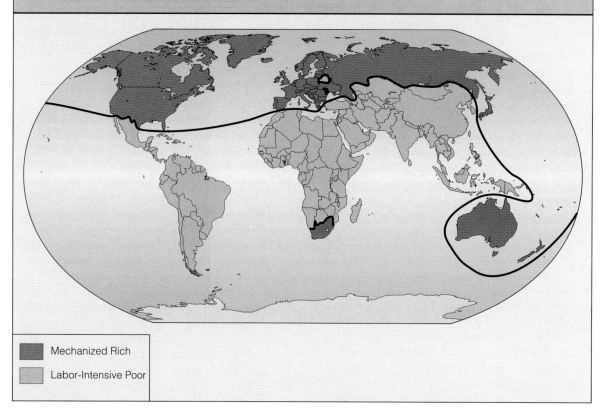

Mechanized Rich
Labor-Intensive Poor

and it camouflages the fact that as Europe and North America plunged into industrialization they took possession of Asia, Africa, and South America and then established economies oriented to their industrial needs. The point is that labor-intensive poor countries were part of the Industrial Revolution from the beginning.

The World Bank, the United Nations, and other international organizations use a number of indicators to distinguish between mechanized rich and labor-intensive poor countries. These indicators include **doubling time** (the estimated number of years required for a country's population to double in size), **infant mortality** (the number of deaths in the

first year of life for every 1,000 live births), **total fertility** (the average number of children women bear over their lifetime), **per capita income** (the average income that each person in a country would receive if the country's gross national product[6] were divided evenly), the percentage of the population engaged in agriculture, and the **annual per capita consumption of energy** (the average amount of energy each person consumes over a year). When per capita energy consumption is low, it suggests that, for the vast majority of people, work is labor-intensive rather than machine-intensive. In labor-intensive work, considerable physical exertion is required to produce food and goods. When the term

TABLE 12.4 **Differences Between Labor-Intensive Poor and Mechanized Rich Countries**

Labor-intensive poor countries are very different from mechanized rich countries
with regard to indicators used by international organizations to classify countries.
Data is presented for the 10 most populous countries in the world.

	Population Doubling Time (years), 1992	Infant Mortality (per 1,000 births), 1992	Total Fertility (children born per woman), 1992	Per Capita GNP Income (in U.S. dollars), 1990	Annual Per Capita Consumption of Energy (in kilograms of oil equivalent), 1990
Labor-Intensive Poor					
Nigeria	23	114	6.5	370	138
Bangladesh	29	120	4.9	200	57
Pakistan	23	109	6.1	380	233
Brazil	37	69	3.1	2,680	286
India	34	91	3.9	350	231
Indonesia	40	70	3.0	560	272
China (mainland)	53	34	2.2	370	598
Mechanized Rich					
United States	89	9	2.0	21,700	7,822
Germany	(—)	7.5	1.4	16,200 (1989)	3,491
Japan	217	4.6	1.5	25,430	3,563

SOURCES: Adapted from *1991 World Population Data Sheet* (Haub, Kent, and Yanagishita 1991);
1992 World Population Data Sheet (Haub and Yanagishita 1992);
Human Development Report 1993, pp. 184–85, 209 (1993).

labor-intensive poor (or an equivalent term) is used to characterize a country, it means that country differs markedly on these and other indicators from countries considered to be industrialized (Stockwell and Laidlaw 1981). See Figure 12.2 and Table 12.4. According to these measures, 163 countries are labor-intensive poor and 45 are mechanized rich (U.S. Bureau of the Census 1989).

The official statistics used to identify labor-intensive poor and mechanized rich countries are counts of how many people have experienced an event. For example, infant mortality is an annual count of how many infants die for every 1,000 born. A high infant mortality rate tells us that for many families, the death of an infant is a common experience.

Industrialization and Brazil: The Context Shaping Key Episodes in Family Life

Brazil has a powerful economy that is likely to surpass the economies of Canada, Italy, and Great Britain in the 1990s. Moreover, after the United

States, it is the world's largest exporter of agricultural products. Its major exports include soybeans, coffee, transport equipment, footwear, orange

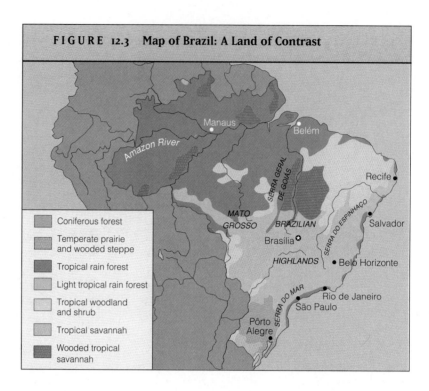

FIGURE 12.3   Map of Brazil: A Land of Contrast

juice, iron ore, and steel products. The benefits we associate with industrialization, however, have bypassed the majority of Brazilians, who live at or below subsistence level. Brazil is a country in which poverty is widespread and chronic (not the result of some temporary misfortune). Sixty percent of the population can be classified as poor (Calsing 1985).

Approximately 25 percent of Brazil's people live in the southeast region, dominated by the metropolises of São Paulo and Rio de Janeiro—2 of the world's 20 largest cities and centers of multinational commercial, industrial, and agricultural activity (see Figure 12.3). As many as 30 percent of the people in these two cities live in urban squatter settlements called *favelas*, in dwellings constructed from cardboard, metal, or wood, with inadequate sewers, running water, and electricity. Many of the people who live in these squatter settlements have been pushed off the surrounding plantation land because of mechanization and have come to the cities in search of a better life. Latin American Studies scholar Thomas G. Sanders maintains that outsiders stereotype *favelas* as centers of crime, prostitution, and extreme family disorganization

but that most studies show the opposite. The vast majority of *favelas* dwellers are "honest, employed, hardworking, and have high aspirations for their children. They live in *favelas* because of low cost, location near their workplaces, and often lack of viable alternative housing solutions" (Sanders 1988, p. 5).

About one-third of Brazil's people, most of whom are the descendants of slaves who worked the Portuguese sugarcane plantations between the sixteenth and the nineteenth centuries, live on the drought-stricken and exhausted land in the northeastern part of the country. Even after slavery was abolished in 1888, the plantation agricultural system, which was oriented toward export, continued. Today the most productive land in the Northeast is still used to grow sugarcane, soybeans, and other export crops, leaving the peasants with no land on which to grow subsistence crops.

Over the years landless peasants from the southeast and northeast regions have migrated to overcrowded cities in search of work or have sought land in the sparsely populated interior of the country. The interior houses the Amazon forest, the largest tropical jungle and rain forest in the world. The rate of migration to the Amazon region accelerated when the government began constructing a network of highways and roads in the 1960s to connect the Amazon region with the rest of the country. The network of roads opened the land to foreign and Brazilian investors, to those living on the fringes of the cities, and to the landless and unemployed from the Northeast. To convert forest to pasture and farmland, settlers cut the trees and other vegetation, allowed them to dry, and then set them afire during the dry season, which runs from June to October. (American and Brazilian scientists, monitoring this practice via satellite, counted 170,000 fires in 1987.) The ash from the burning trees and other foliage acts as a fertilizer for a few years, but after that time the land no longer can support crops. The land then is abandoned to cattle ranchers, whose herds graze it for a few years until it is totally exhausted (Simons 1988).[7]

Although isolated geographically from the rest of the country, the Amazon region was not uninhabited; it was occupied by indigenous people, rubber tappers, nut gatherers, and others whose forest-centered livelihood was disrupted by the highway construction and subsequent human migration. The forest dwellers, especially the indigenous people, suffered (and still suffer) cultural and physical extinction at the hands of mining companies that extract mineral wealth buried beneath the rain forest, lumber companies in search of rare jungle trees, cattle ranchers grazing their herds on deforested land, government development projects, and land-hungry peasants. Although the government has set aside land for these peoples, the forest dwellers are under pressure to abandon their language and culture and to enter the dominant society. Unfortunately, if they enter society they do so as landless peasants, low-paid laborers, or beggars (Caufield 1985).

The background information on Brazil sets the context for thinking about how various industrialization scenarios affect key episodes in family life. As we will learn later in this chapter, the place in which a family resides (*favelas*, suburbs, plantations, Amazon town, forest, and so on), the kind of work that family members do, and where they must travel to find work affect such things as childbearing experiences and life expectancy.

We turn now to a model that outlines historical changes in birth and death rates in Western Europe and the United States, especially as these changes are affected by the Industrial Revolution. This model is the theory of the demographic transition. After we describe the model, we discuss how well it applies to non-Western countries such as Brazil.

The Theory of Demographic Transition

One way to study the effects of social forces on the key episodes of family life is through demography. **Demography,** the study of population trends, is an area of specialization within sociology. Demographers study births, deaths, and migration, and their contribution to changes in population size (see "How Demographers Measure Change"). In the 1920s and early 1930s, demographers observed birth and death rates in various countries and noticed a pattern: both birth and death rates were high in the countries of Africa, Asia, and South America; death rates were declining while birthrates remained high in Eastern and Southern Europe; and birthrates were declining and death rates were low in Western Europe and North America.

Demographers observed that the countries of Western Europe and North America had experienced all three of the conditions mentioned previously according to the following sequence:

1. Birth and death rates were high until about the middle of the eighteenth century, at which time death rates began to decline.

2. As the death rates decreased, the population grew rapidly because there were more births than deaths. The birthrates began to decline around 1800.

3. By 1920 both birth and death rates had dropped below 20 per 1,000 (see Figure 12.4).

How Demographers Measure Change

In order to discuss change, demographers must specify a time period (a year, a decade, a century) over which they keep track of how many times an event (a birth, a death, a move) occurs. The accompanying table shows the number of births and deaths that occurred in Brazil and the United States between July 1, 1991, and June 30, 1992.

The simplest way to express change is in absolute terms—that is, to state the number of times an event occurred. (In 1991 there were an estimated 3,955,000 births in Brazil and 3,553,000 births in the United States.) Expressing change in absolute terms is not very useful, however, for making comparisons

between countries that have different population sizes. Consequently, for comparative purposes demographers calculate *rates* of births, deaths, and migrations, usually per 1,000 people in the population. Rates are calculated by dividing the number of times an event occurs by the size of the population at the onset of the year and then multiplying that figure by 1,000. The 1991 death rate for Brazil is calculated as follows:

$$\frac{1,107,000}{158,202,000} \times 1,000 = 7$$

When the total number of people in the population is used as the denominator, the rates are called crude

rates. Sometimes demographers wish to know how many births, deaths, or moves occur within a specific segment of the population (among males or among females ages 15–44). In these cases the denominator is the number of people in that segment of the population. For example, there are 35,356,000 women between the ages of 15 and 44 in Brazil. The age-specific birth rate for these women thus is calculated as follows:

$$\frac{3,955,000}{35,356,000} \times 1,000 = 112$$

	Mid-1992 Population	Births	Deaths	Births per 1,000 Population	Deaths per 1,000 Population
Brazil	158,202,000	3,955,000	1,107,000	25	70
United States	254,521,000	3,553,000	2,291,000	14	90

SOURCES: Text by Joan Ferrante, Northern Kentucky University (1991). Other data adapted from *The World Factbook, 1992,* pp. 47 and 358 (1992).

On the basis of these observations, demographers put forth the **theory of the demographic transition:** a country's birth and death rates are linked to its level of industrial or economic development.

This model documents the general situation and should not be construed as a detailed description of the experiences of any single country. Even so, we can say that all countries have followed or are following the essential pattern of the demographic transition, although they differ with regard to the timing of the declines and with respect to the rate at which their populations increase after death rates begin to decline. The theory of the demographic transition includes more than this pattern; it is also an explanation of the events that caused birth and death rates to drop in the mechanized rich coun-

tries (with the exception of Japan). The factors that underlie changes in birth and death rates in the labor-intensive poor countries, however, are fundamentally different from those that caused such changes in developed countries. After we examine each stage of the demographic transition in greater detail, we will consider the ways in which the demographic transition in labor-intensive poor countries such as Brazil differs from the experience of mechanized rich countries.

Stage 1: High Birth and Death Rates

For most of human history—the first two to five million years—populations grew very slowly (if at all). World population remained below 1 billion

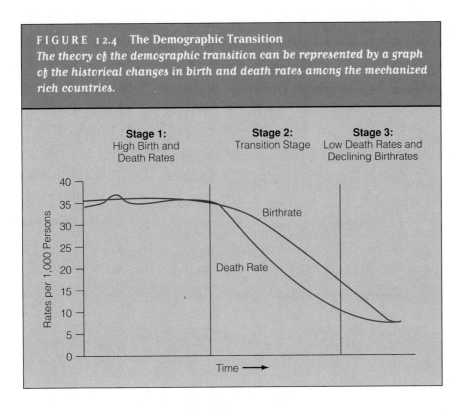

FIGURE 12.4   The Demographic Transition
The theory of the demographic transition can be represented by a graph of the historical changes in birth and death rates among the mechanized rich countries.

Stage 1: High Birth and Death Rates

Stage 2: Transition Stage

Stage 3: Low Death Rates and Declining Birthrates

Rates per 1,000 Persons

Birthrate

Death Rate

Time ⟶

until A.D. 1800, at which point it began to grow explosively. By 1930 the world's population had increased to 2 billion; during the next 50 years it increased by another 2.5 billion. During the 1980s the world's population increased by about 750 million to its present size of 5.4 billion. Demographers speculate that growth until 1800 was slow because **mortality crises**—frequent and violent fluctuations in the death rate caused by war, famine, and epidemics—were a regular feature of life (Watkins and Menken 1985).

Stage 1 is often referred to as the stage of high potential growth: if something happened to cause the death rate to decline—for example, improvements in agriculture, sanitation, or medical care—population would increase dramatically. In this stage life is short and brutal; the death rate is almost always above 50 per 1,000. When mortality crises occur, the death rate seems to have no limit. Sometimes half of the population is affected, as when the Black Plague struck Europe, the Middle East, and Asia in the middle of the fourteenth century and recurred for approximately 300 years. It

is estimated that within 20 years of its onset, the plague killed up to three-fourths of the people in the affected populations. The medieval Italian writer Giovanni Boccaccio recorded his impressions of the event:

> *It began in both men and women with certain swellings either in the groin or under the armpits, some of which grew to the size of a normal apple and others to the size of an egg (more or less), and the people called them* gavoccioli . . . *within a brief space of time [it] . . . began to spread indiscriminately over every part of the body; and after this, the symptoms of the illness changed to black or livid spots appearing on the arms and thighs, and on every part of the body, some large ones and sometimes many little ones scattered all around. [F]ew of the sick were ever cured, and almost all died after the third day of the appearance of the . . . symptoms.* (BOCCACCIO [1353] 1984, P. 728)

Another mortality crisis, which has not received as much attention as the Black Plague, affected the indigenous populations of North America when the Europeans arrived in the fifteenth century. A large proportion of the native population died because they had no resistance to diseases such as smallpox, measles, tuberculosis, and influenza, which the colonists brought with them. Others simply were killed by colonists because they refused to work as slaves on plantations. Historians debate what proportion of the native population died as a result of this contact; estimates range between 50 and 90 percent.

The point is that in stage 1 average life expectancy at birth remained short—perhaps between 20 and 35 years—with the most vulnerable groups being women of reproductive age, infants, and children under age 5. (The high infant and child mortality rate pulled down the average life expectancy at birth; many people managed to live well beyond age 30.) It is believed that women gave birth to large numbers of children, often as many as 10 or 12. Families remained small, however, because one infant in three died before reaching age 1 and another died before reaching adulthood. If the birthrate had not remained high, the society would have become extinct. Demographer Abdel R. Omran (1971) estimates that in societies in which life expectancy at birth is 30 years, each woman must have an average of seven live births in order to ensure that two children survive into adulthood. She must bear six sons to ensure that at least one survives until the father reaches age 65 (if the father lives that long).

In Western Europe, before 1650 high mortality rates were associated closely with food shortages. Even when people did not die directly from starvation, they died from diseases that preyed on their weakened physical state. Thus Thomas Malthus, a British economist and an ordained Anglican minister, concluded that "the power of population is so superior to the power in the earth to produce subsistence for man, that premature death must in some shape or other visit the human race" (Malthus [1798] 1965, p. 140). According to Malthus, positive checks served to keep population size in line with the food supply. He defined **positive checks** as events that increase mortality, including epidemics of infectious and parasitic disease, war, and famine. Malthus believed that the only

moral way to prevent populations from growing beyond a size that could not be supported by the food supply was delayed marriage and celibacy. He regarded any other method—such as infanticide, homosexuality, or sterility caused by sexually transmitted diseases—as immoral.

In most cases, however, famines do not occur simply because there are too many people and not enough food. Human affairs also play a major role. In the drought-stricken northeastern region of Brazil, many people are chronically hungry and malnourished. The plantation economy, controlled by a relatively small number of landholders and worked by massive numbers of landless peasants, is oriented toward producing export crops, such as sugarcane, rather than subsistence crops for the peasants (see "A Body Feeding On Itself").

Stage 2: The Transition Stage

Around 1650, mortality crises became less frequent in Western Europe, and by 1750 the death rate began to decline slowly in that region. The decline was triggered by a complex array of factors associated with the onset of the Industrial Revolution. The two most important factors were (1) increases in the food supply, which improved the nutritional status of the population and increased its ability to resist diseases, and (2) public health and sanitation measures, including the use of cotton to make clothing and new ways of preparing food. The following excerpt elaborates:

> The development of winter fodder for cattle was important; fodder allowed the farmer to keep his cattle alive during the winter, thereby reducing the necessity of living on salted meats during half of the year. . . . [C]anning was discovered in the early nineteenth century. This method of food preservation laid the basis for new and improved diets throughout the industrialized world. Finally, the manufacture of cheap cotton cloth became a reality after midcentury. Before then, much of the clothes were seldom if ever washed, especially among the poor. A journeyman's or tradesman's wife might wear leather stays and a quilted petticoat until they virtually rotted away. The new cheap cot-

A Body Feeding On Itself

In many parts of the world, hunger and starvation are facts of everyday life. People live with the condition of inadequate food supplies and other factors that lead to chronic hunger. What happens to the human body as it undergoes starvation? The following article explains.

1. When food is first denied, the body feeds on a starchy substance called glycogen, which is stored in the liver.

2. As less food is taken in, energy drops. Heartbeat, pulse and blood pressure fall.

3. The body begins to lose water and to feed on fat and muscle. Weight lost is composed of 50% water, 25% muscle and 25% fat. As water is depleted from the cells, the body shrinks. Victims take on a gaunt, hollow-cheeked look.

4. Loss of water causes an imbalance of fluids. The distended, swollen bellies typical of starving people are caused by edema, fluid collecting in the cells.

5. The human body can withstand the loss of 25% of its normal weight and survive. Beyond that, life expectancy is 30–50 days. The body's organs begin to waste away. For example, the heart of a healthy adult weighs 11 to 14 ounces; in late starvation, the heart may shrink to 5 to 6 ounces. Intestines wither, making it impossible for the body to digest food.

6. Disease hastens starvation. Intestinal parasites—rampant in the unsanitary conditions of Somalia's refugee camps—cause diarrhea, which in turn prevents the body from absorbing what few nutrients it takes in. The immune system begins to break down, leaving the body defenseless against other diseases.

7. Starvation has a devastating effect on mental health. Hope dwindles, and personalities slip away. A person becomes confused, disoriented and irritable. Self-preservation often leads starving mothers to snatch food from their children as panic sets in.

SOURCE: From "Mission to Somalia: The Anatomy of Starvation," by Michael Hall. Copyright © 1992 by the *Los Angeles Times*. Reprinted by permission.

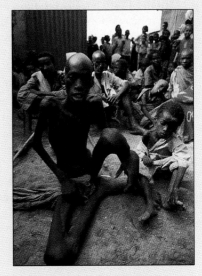

Starvation has taken a terrible toll in the war-torn country of Somalia, Africa.
Bernard Bisson/Sygma

ton garments could easily be washed, which increased cleanliness and fostered better health.
(STUB 1982, P. 33)

Contrary to popular belief, advances in medical technology had little influence on death rates until the turn of the twentieth century, well after improvements in nutrition and sanitation had caused dramatic decreases in deaths due to infectious diseases.

Over a 100-year period, the death rate fell from 50 per 1,000 to below 20 per 1,000, and life expectancy at birth increased to about 50 years of age. As death rates declined, fertility remained high. It may even have increased temporarily, because improvements in sanitation and nutrition enabled women to carry more babies to term. With the decrease in the death rate, the **demographic gap**—the difference between birthrates and death rates—widened and population size increased substantially. **Urbanization,** an increase in the number of cities and in the proportion of the population living in cities, accompanied the unprecedented increases in population size. (As recently as 1850, only two percent of the world's population lived in cities with populations of 100,000 or more.)

Around 1880, fertility began to decline. The factors that caused birthrates to drop are unclear and are subject to debate among demographers. But one thing is certain: the decline was not caused by innovations in contraceptive technology, because the methods available in 1880 had been available throughout history. Instead, the decline in fertility seems to be associated with several other factors. First, the economic value of children declined in industrial and urban settings as children were no longer a source of cheap labor but became an economic liability to their parents. Second, with the decline in infant and childhood mortality, women no longer had to bear a large number of children in order to ensure that a few survived. Third, a change in the status of women gave them greater control over their reproductive lives and made childbearing less central to women's lives. Scholars disagree, however, about the specific conditions under which women are able to control their reproductive lives.

Stage 3: Low Death
Rates and Declining Birthrates

In the industrialized West, around 1930, both birth and death rates fell below 20 per 1,000 and the rate of population growth slowed considerably. Life expectancy at birth surpassed 70 years, an unprecedented age. The remarkable successes in reducing

infant, childhood, and maternal mortality rates were such that accidents, homicides, and suicide have become the leading causes of death among young people. Because the risk of dying from infectious diseases is reduced, people who would have died of infectious diseases in an earlier era survive into middle age and beyond, when they face the elevated risk of dying from degenerative and environmental diseases (heart disease, cancer, strokes, and so on). For the first time in history, persons 50 years of age and over account for more than 70 percent of the annual deaths. Before this stage, infants, children, and young women accounted for the largest share of deaths (Olshansky and Ault 1986).

As death rates decline, disease prevention becomes an important issue. The goal is to live not only a long life but a quality life (Olshansky and Ault 1986; Omran 1971). As a result, people become conscious of the link between their health and their lifestyle (sleep, nutrition, exercise, and drinking and smoking habits). In addition to low birth and death rates, stage 3 is distinguished by an unprecedented emphasis on consumption (made possible by advances in manufacturing and food production technologies).[8]

Theoretically, all of the mechanized rich countries are in stage 3 of the demographic transition.[9] At one time some sociologists and demographers maintained that the so-called Third World countries would follow this model of development as they industrialized. However, the nature of industrialization in labor-intensive poor countries is so fundamentally different from the version that occurred in the mechanized rich countries that these countries are unlikely to follow the same path.

The Demographic Transition in
Labor-Intensive Poor Countries

One reason that the nature of industrialization in labor-intensive poor countries was so fundamentally different from that of mechanized rich countries is that most labor-intensive poor countries were once colonies of mechanized rich countries.

The mechanized rich countries established economies oriented to their own industrial needs, not the needs of the countries they colonized. In the case of Brazil, the Portuguese forced the indigenous peoples to grow crops and mine metals and materials for export to the mother country, but they did not allow the people to develop and establish their own native industries. For more than a century following its independence from Portugal in 1822, Brazil possessed a one-crop, export-oriented economy that was dominated first by sugarcane, then by rubber, and then by coffee. As is usually the case with one-product export economies, Brazil's economy experienced cycles of boom and bust. The rubber boom is illustrative. Between 1900 and 1925 the Brazilian city of Manaus in the Amazon was the rubber capital of the world, supplying 90 percent of the world's demand. The profits were reaped by a handful of Amazon rubber barons, who controlled huge plantations on which Indians and peasants from northeastern Brazil worked under conditions resembling slavery. The rubber boom came to an end when Henry Wickham, a British businessman, smuggled 70,000 rubber plant seedlings out of Brazil to plant in Singapore. Eventually, Asian rubber priced Brazilian rubber out of the world market (Nolty 1990; U.S. Department of the Army 1983; Revkin 1990). Even today, multinational corporations employ workers in developing countries to do low-skill, low-paying, labor-intensive work whose products are exported to people in the developed countries.

The fact of colonization helps explain why the model of the demographic transition does not apply to labor-intensive poor countries. When compared to mechanized rich countries, labor-intensive poor countries differ on several characteristics: they have a faster decline in death rates, relatively high birthrates despite declines in the death rate, a more rapid increase in population size, and greater (unprecedented) levels of rural-to-urban and rural-to-rural migration.

Death Rates

The decline in death rates in the labor-intensive poor countries occurred much faster than in the mechanized rich countries; only 20 to 25 years (rather than 100 years) elapsed before the death rate fell from 50 per 1,000 to less than 10 per 1,000. Demographers attribute the relatively rapid decline to cultural diffusion (see Chapter 4). That is, the labor-intensive poor countries imported some Western technology—pesticides, fertilizers, immunizations, antibiotics, sanitation practices, and higher-yield crops—which caused an almost immediate decline in the death rates. Brazil, for example, imported DDT to eradicate the mosquitoes that carried diseases such as yellow fever and malaria. In addition, the Brazilian government has developed antidotes for snake, spider, and scorpion bites and has distributed them to clinics around the country. Among other things, the government is also working to stamp out Chagas' disease and schistosomiasis, serious disorders caused by parasites in the blood (Nolty 1990).

The swift decline in death rates has caused the populations in developing countries to grow very rapidly. Some demographers believe that developing countries may be caught in a **demographic trap**—the point at which population growth overwhelms the environment's carrying capacity:

> *Once populations expand to the point where their demands begin to exceed the sustainable yield of local forests, grasslands, croplands, or aquifers, they begin directly or indirectly to consume the resource base itself. Forests and grasslands disappear, soils erode, land productivity declines, water tables fall, or wells go dry. This in turn reduces food production and incomes, triggering a downward spiral.* (L. BROWN 1987, P. 28)

The countries of Nepal (between China and India) and Costa Rica (in Central America) represent two cases of the demographic trap. Because of population pressures, the peasants of Nepal have been forced to cultivate steep and forested hillsides, and the women have no choice but to collect feed for livestock and wood for cooking and heating from these sites. As the forests recede, the daily journey for fodder and fuel grows longer and cultivation becomes more difficult. As a result, family incomes have dropped and diets have deteriorated. In fact, the malnutrition rates in the Nepal villages are correlated strongly with deforestation rates (Durning 1990).

In Costa Rica, a country once covered by tropical forest, the demographic trap is fueled by population growth and by land policies of the past two decades, which favor a small number of cattle

As the example of rural Brazil demonstrates, the issue of population growth
is not simply a question of sheer numbers, but also involves the carrying
capacity of the land.
Alain Keeler/Sygma

ranch owners (perhaps 2,000) and disregard the needs of peasants who have become landless under these policies. About half the nation's cultivatable land is used to raise cattle, an industry that requires little labor and hence provides few employment opportunities for landless peasants. "The rising tide of landlessness has spilled over into expanding cities, onto the fragile slopes, and into the forests, where families left with little choice accelerate the treadmill of deforestation" (Durning 1990, p. 146).

Keep in mind that the point at which population growth overwhelms the environment's carrying capacity is usually not caused by population pressures alone. In 1980 the Brazilian government built highway BR364 to link São Paulo with the Amazon state of Rondônia. Significant numbers of landless peasants were offered 100 acres of free land in the Amazon forest provided they clear the land and construct a house. Unfortunately, much of the land turned out to be sandy and unfit to sustain crops; millions of acres of forest land were destroyed without easing the land problem. Technically, one could argue that the population growth in the Amazon overwhelmed its capacity to feed people. On the other hand, critics of the plan argue that government officials sent peasants blindly into the jungle, unprepared for how difficult it would be to make a living. In addition, the 100-acre plots were divided without reference to terrain (many plots consisted of rocky hills and many had no source of water). Most importantly, the chronic landlessness

is caused by the system of land distribution, not by a lack of land: 50 percent of the cultivatable land in Brazil is owned by less than 1 percent of all landowners, and much of that land remains idle. Thus, in the final analysis, it is land distribution policies, not population pressures per se, that send peasants into the hillsides, forests, and grasslands (Sanders 1986).

Birthrates

Birthrates, although still high, are beginning to show signs of decline in most developing countries. The demographic gap remains wide, however (see Figure 12.5). It is not clear exactly which factors have caused total fertility to decline in developing countries. Sociologist Bernard Berelson (1978) identified some important "thresholds" associated with industrialization and with declines in fertility. These include the following conditions:

1. Less than 50 percent of the labor force is employed in agriculture.

2. At least 50 percent of persons between the ages of 5 and 19 are enrolled in school.

3. Life expectancy is at least 60.

4. Infant mortality is less than 65 per 1,000 live births.

5. Eighty percent of the females between the ages of 15 and 19 are unmarried.

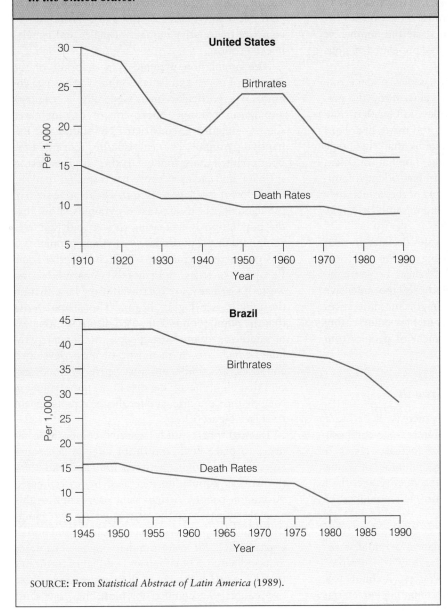

FIGURE 12.5   Crude Birth and Death Rates: United States Versus Brazil
The graphs below show that as the death rate in each country declined, the demographic gap was much wider and more persistent in Brazil than in the United States.

SOURCE: From *Statistical Abstract of Latin America* (1989).

Most of these conditions have been met in Brazil. Although total fertility is still high, it has declined over the past two decades from 5.75 children to 3.0 children per woman (Brooke 1989; U.S. Bureau of the Census 1991). The factors identified by Berelson may indeed be responsible for this decline, along with other factors such as changes in government family policy and the near-universal access to television. During the 1960s, for example, the Brazilian government established pronatalist policies designed to increase the size of the population, with particular emphasis on increasing the population in its Amazon states. Those who had children received tax breaks and maternity bonuses (U.S. Department of the Army 1983). Finally, however, the government abandoned these policies because they only served to increase the number of people living in already densely populated regions of the country.

Wider access to television has played some role in declining fertility in Brazil. Since 1960 the percentage of households with television sets has risen from 5 percent to 72 percent, and many Brazilians now are able to watch programs that encourage new norms. For example, the popular Brazilian soap operas feature small, consumer-oriented families. When large families are part of the drama, they usually are depicted as poor and miserable (Brooke 1989).

The speed at which death rates declined in relation to birthrates in developing countries is only one of several important differences between the mechanized rich countries and the labor-intensive poor countries. Two other important differences are (1) the rate at which population is increasing and (2) **migration,** the movement of people from one area to another.

Population Growth

The rate at which populations increase is tied to a number of factors including the sex-age composition. Generally, population size increases faster in countries with a disproportionate number of young adults (men and women of reproductive age) than in countries with a disproportionate number of middle-aged and older people.

A population's age and sex composition is commonly represented by a **population pyramid,** a series of horizontal bar graphs each of which represents a different five-year age cohort. (A **cohort** is a group of people who share a common characteristic or life event; in this case it includes everybody born in a specific five-year period.) Two bar graphs are constructed for each cohort, one for males and another for females; the bars are placed end to end, separated by a line representing zero. Usually, the left-hand side of the pyramid depicts the number or percentage of males that make up each age cohort

and the right-hand side depicts the number or percentage of females. The graphs are stacked according to age; the age 0–4 cohort forms the base of the pyramid and the 80+ age cohort is at the apex (see Figure 12.6). The population pyramid allows us to view the relative sizes of the age cohorts and to compare the relative number of males and females in each cohort.

The population pyramid is a snapshot of the number of males and females in the various age cohorts at a particular time. Generally, a country's population pyramid approximates one of three shapes—expansive, constrictive, or stationary. **Expansive pyramids,** characteristic of labor-intensive poor countries, are triangular; they are broadest at the base, and each successive bar is smaller than the one below it (see Figure 12.6a). The relative sizes of the age cohorts in expansive pyramids show that the population is increasing in size and that it is composed disproportionately of young people.

Constrictive pyramids, which characterize some European societies, most notably Switzerland and western Germany, are narrower at the base than in the middle (see Figure 12.6b). This shape shows that the population is composed disproportionately of middle-aged and older people. **Stationary pyramids,** which are characteristic of most developed countries, are similar to constrictive pyramids except that all of the age cohorts in the population are roughly the same size, and fertility is at replacement level (see Figure 12.6c).

Demographers study age-sex composition in order to place birth and death rates in a broader context. For example, between 1990 and mid-1991 the world's population increased by an estimated 90 million people. A significant proportion of this annual increase can be attributed to two countries—the People's Republic of China and India. Even though 90 countries had higher birthrates than India and 134 countries had higher birthrates than China (U.S. Bureau of the Census 1991), these two countries accounted for more than one-third (33 million) of the increase; China's population grew by 16 million, and India's grew by 17 million. Such a large share can be attributed to the fact that both countries have large populations to begin with (1.15 billion people live in China and 870 million live in India) and to the fact that more than 50 percent of the Indian and the Chinese populations are

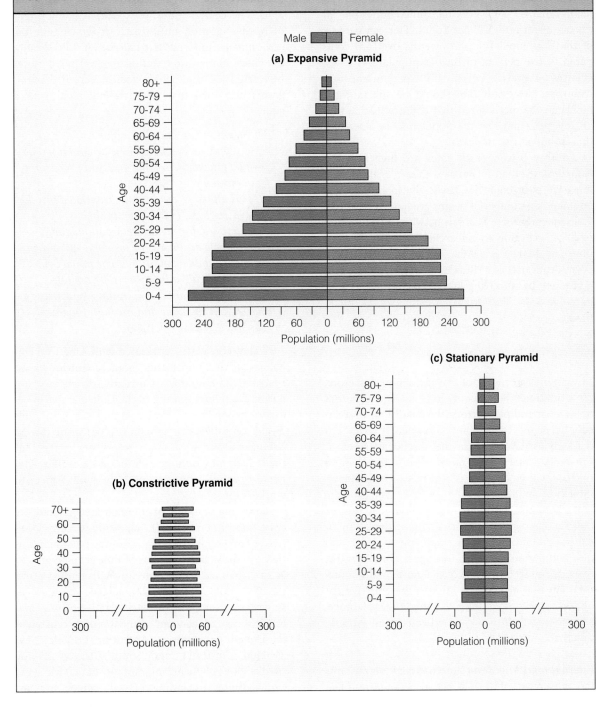

FIGURE 12.6 Population Pyramids
The three graphs show (a) the expansive pyramid characteristic of labor-intensive poor countries, (b) the constrictive pyramid characteristic of some European countries such as Switzerland, and (c) the stationary pyramid characteristic of most developed countries.

of childbearing age—between 15 and 49 (U.S. Bureau of the Census 1991).

Age-sex composition helps us understand in part why death rates in labor-intensive poor countries are the same as those in mechanized rich countries (and sometimes even lower). Brazil has an official death rate of 7 per 1,000; the United States has an official death rate of 9 per 1,000. Thirteen percent of the U.S. population, however, is over 65, compared to five percent of Brazil's population. If the chances of survival were truly equal in the two countries, the death rate should be substantially higher in the mechanized rich countries because a greater percentage of the population is older and thus at higher risk of death.

Another reason death rates are higher in the United States than in Brazil is that in many labor-intensive poor countries, especially in rural areas, an unknown number of deaths are never registered. Consequently there is a gap between statistics and reality. According to researchers Marilyn K. Nations and Mara Lucia Amaral, "for all official purposes death occurs in Brazil only when the event is registered by surviving family members" (1991, p. 207). See "The Official Meaning of Death in Brazil."

Migration

Migration is a product of two factors: **push factors,** the conditions that encourage people to move out of an area, and **pull factors,** the conditions that encourage people to move into a particular area. "On the simplest level [and in the absence of force] it can be said that people move because they believe that life will be better for them in a different area" (Stockwell and Groat 1984, p. 291). Some of the most common push factors include religious or political persecution, discrimination, depletion of natural resources, lack of employment opportunities, and natural disasters (droughts, floods, earthquakes, and so on). Some of the most common factors that pull people into an area are employment opportunities, favorable climate, and tolerance. Migration falls into two broad categories: international and internal.

International Migration **International migration** is the movement of people between countries. De-

mographers use the term **emigration** to denote the departure of individuals from one country and the term **immigration** to denote the entrance of individuals into a new country. Unless their countries are severely underpopulated, most governments restrict the numbers of foreign people who are allowed to enter. The United States, for example, had a policy of open migration until 1822, at which time the government imposed restrictions. At Ellis Island, the chief immigration station in the United States between 1822 and 1943, doctors screened immigrants and then marked each individual with a code:

> Those who were branded with symbols such as E, H, Pg, X or X with a circle around it would then be plucked from line and held for further examination.
>
> This alphabet of obstacles translated into E for eye diseases, H for heart problems, Pg for pregnancy, X for mental retardation, and X with a circle around it for insanity—all potentially excludable conditions. The initial "six-second medical" administered by doctors watching applicants walk upstairs had already weeded out the lame and the halt. Judgment was swift! (BASS 1990, P. 89)

Today, consulates associated with American embassies in the immigrants' home countries screen potential immigrants. Currently, foreigners are barred from immigrating to the United States by 38 different conditions. These include criminal, immoral, or subversive activities; physical or mental handicaps; and economic factors suggesting that the individual would be a drain on the welfare system or would compete unfairly with American workers for jobs.

Three major flows of intercontinental (and, by definition, international) migration occurred between 1600 and the early part of the twentieth century: (1) the massive exodus of European peoples to North America, South America, Asia, and Africa to establish colonies and commercial ventures, in some cases eventually displacing native peoples and establishing independent countries (this occurred in the United States, Brazil, Argentina, Canada, New Zealand, Australia, and South Africa); (2) the smaller flow of Asian migrants to East Africa, the United States (including Hawaii, which did not

The Official Meaning of Death in Brazil

The demographics of labor-intensive poor countries often depend on unreliable statistics. In Brazil, for example, the official registration of births is more common than that of deaths. The following article tells why and explains how poverty and social practices contribute to unreliable population statistics.

For all official purposes death occurs in Brazil only when the event is registered by surviving family members at a *cartório,* a government-authorized registry. Since 1973 Public Registry Law 6015 has guaranteed (in principal) all Brazilian citizens the right to register the births and deaths of their children, irrespective of their financial means (this translates into free of charge for poor families).

In theory, the new Federal Constitution of Brazil, adopted by the Republic in 1988, has facilitated the registration process by creating a seven-day paternity leave specifically to allow the father to register the baby and aid his wife after the birth. However, in practice only contracted workers (which implies their paying taxes along with receiving social benefits) can exercise this right, and most poorer people work *empreitada*—i.e., in temporary, unstable jobs. In addition, the bureaucratic red tape of completing registration often inhibits grieving families from tackling or completing the chore.

Filing a death requires first that the child's birth has been registered at an official *cartório.* Impoverished parents who want to take advantage of the free government-sponsored registration process must first seek a financial authorization from the Brazilian Assistance League (LBA) office, located in Fortaleza. Families must travel great distances, navigating the unfamiliar and bewildering maze of high-rise office buildings and bustling streets to do this. Single mothers are often homebound caring for the newborn or observing the traditional 40-day resting period (*resguardo*). Once in the LBA central office, required data about the child must be provided to the staff, who complete the review process in 2–5 working days, a delay that demands a costly return trip or overnight stay. With LBA's authorization in hand, the family now locates an affiliated official *cartório,* where the birth or death can be registered and documentation provided. If the 15-day postpartum filing period has passed, one must also first submit a petition justifying to lawyers of the *Juizado de Menores* (The Minors' Court) why the newborn was not registered within the time period stipulated by the law. With this legal exemption in hand, one returns to the LBA office, where a new financial authorization is prepared (after a waiting period, of course), and this can then be filed at the official *cartório.* Death registration is very similar: first, the family must contact the *Informador Popular,* a representative of the LBA, who solicits a doctor to write the death testimony. After acquiring the medical testimony (which may take a few days, since the doctor is not paid for the service and runs several liability risks when he signs a testimony), the parent must take it to a *cartório* to receive the actual death certificate. True, the poor pay no direct fees for registry of births or deaths, but the indirect costs to turn a private birth or loss into an official reality are considerable: lost time, energy, wages, and out-of-pocket expenses. When one considers that some parents cannot afford the time and money to go to the hospital and claim the body of their dead child, it is no wonder that few trek through the entire death registration process.

Yet parents do complete the arduous birth registration process in order to gain access to national social benefits. Most basic services—receiving food supplements, enrolling in school, voting, and (in some instances) receiving health care—depend on registering a child's birth. In contrast, the social benefits of making a child's death official are nil for the family. Motivation to do so is therefore very weak. Although Brazilian law states that only corpses with death certificates can be buried in an official cemetery, this is not a steadfast rule. Receiving a negative sanction for registering a child's death is more often the case. Families lose benefits (milk and food supplements) when death is known officially.

This lopsided registration of births over deaths further skews mortality rates. The greater insensitivity of the official registry system to death results in the numerator of the mortality ratio being artificially low in comparison with the denominator of "total live births." The resulting official mortality rate is thus extremely biased toward a low quotient.

SOURCE: From "Flesh, Blood, Souls, and Households," by Marilyn K. Nations and Mara Lucia Amaral, *Medical Anthropology Quarterly* 5:3, pp. 207–08. Copyright © 1991 by American Anthropological Association. Reprinted by permission.

become a state until 1959), and Brazil, where they provided cheap labor for major transportation and agricultural projects; and (3) the forced migration of some 11 million Africans by Spanish, Portuguese, French, Dutch, and British slave traders to the United States, South America, the Caribbean, and the West Indies. In all the Americas Brazil imported the greatest number of African slaves (Skidmore 1993).

These three migration flows are responsible for the ethnic composition of Brazil: "There are few other [countries] on earth where such a wide spread of skin tones, from whitest white to yellow to tan to deepest black, are grouped under one nationality" (Nolty 1990, p. 6). The mixing process began when the Portuguese sailors who colonized Brazil intermingled (often forceably) with native women and African slave women. After slavery was abolished in 1888, the Brazilian government encouraged Europeans and Asians to immigrate into the country to replace slave labor. Between 1884 and 1914, a large number of Japanese and Italians immigrated to Brazil as indentured servants; they worked on large plantations until they had earned enough to buy their way out of servitude.

Internal Migration In contrast to international migration, **internal migration** is movement within the boundaries of a single country—from one state, region, or city to another. Demographers use the term **in-migration** to denote the movement of people into a designated area and the term **out-migration** to denote movement out of a designated area.

One major type of internal migration is the rural-to-urban movement (urbanization) that accompanies industrialization.[10] However, urbanization in labor-intensive poor countries is substantially different from that which occurred in mechanized rich societies:

> *The world has never seen such extremely rapid urban growth. It presents the cities, especially in the developing countries, with problems new to human experience, as well as old problems— urban infrastructure, food, housing, employment, health, education—in new and accentuated forms. Many developing countries will have to plan for cities of sizes never conceived*

of in currently developed countries. High population growth in developing countries, whatever other factors enter the process, is inseparable from this phenomenon. (RUSINOW 1986, P. 9)

The unprecedented growth of urban areas is caused by a number of factors. First, the cities in the mechanized rich countries grew in closer proportion to the number of jobs created during their industrialization process a century and a half ago. Many Europeans who were pushed off the land were able to emigrate to sparsely populated places like North America, South America, South Africa, New Zealand, and Australia. If those who fled to other countries in the eighteenth and nineteenth centuries had been forced to make their livings in European cities, the conditions would have been much worse than they actually were:

> *Ireland provides the most extreme example. The potato famine of 1846–1849 deprived millions of peasants of their staple crop. Ireland's population was reduced by 30 percent in the period 1845–1851 as a joint result of starvation and emigration. The immigrants fled to industrial cities of Britain, but Britain did not absorb all the hungry Irish. North America and Australia also received Irish immigrants. Harsh as life was for these impoverished immigrants, the new continents nonetheless offered them a subsistence that Britain was unable to provide.*
>
> *. . . there are no longer any new worlds to siphon off population growth from the less industrialized countries.* (LIGHT 1983, PP. 130–31)

In Brazil, the problem of urbanization is compounded by the fact that many migrants who come to the cities are from some of the most economically precarious segments of Brazil. In fact, most rural-to-urban migrants are not pulled into the cities by employment opportunities but are forced to move there because they have no alternatives. In the northeast region of Brazil, peasants are pushed out because of droughts and floods, because land is deforested and exhausted, and because the land is concentrated in the hands of a few people. When the migrants come to the cities, they face not only unemployment but also a shortage of housing and a lack of services (electricity, running water, waste disposal). One of the most distinguishing characteristics of cities in labor-intensive poor countries is the prevalence of slums and squatter settlements, poorer and larger than even the worst slums in the mechanized rich countries. Urban squatter settlements are not unique to Brazil; they are almost universal in developing countries.

Another type of massive internal migration—rural-to-rural—is taking place in many labor-intensive poor countries. We know very little about the extent of this migration (except that it may involve as many as 370 million people worldwide) or about its impact on the communities that the migrants enter and leave. We do know, however, that in this kind of migration peasants move to increasingly more marginal or fragile lands. Generally, the move does not improve their social and economic status. In other words, peasants do not move in order to find better jobs than those they have; they move to find any kind of work or to search for land (Feder 1971).

In Brazil there are 10.5 million workers without land who "migrate all over the country . . . invading any empty patch which may seem unclaimed" (Cowell 1990, p. 137). Some of the most desperate migrate into regions of the Amazon states. These peasants do not move into the wilderness, clear the ground of trees and rocks, and become prosperous farmers, however. More often than not, they barely grow enough food to survive. This situation is not unique to Brazil. Sociologist Alfredo Molano estimates that in Colombia settlers—"fleeing violence, political persecution, or economic deprivation—go to the forests looking for another way of finding subsistence" (Molano 1993, p. 43). The desperate activities of migrants in search of wood for cooking and land to grow food and set up households results in a level of deforestation that surpasses the number of acres taken by the lumber industry (Semana 1993).

Industrialization, whatever its form, has been a major influence on family life. In this section we have identified various trends such as the relatively rapid decline in death rates coupled with relatively slow decline in birthrates that affect family size and composition. Obviously an increase in life expectancy affects such things as childbearing experiences (a decrease in infant and maternal mortality, for example) and the likelihood that parents will live to see their children into adulthood. Likewise, population pressures on land and other resources resulting from the demographic gap, problematic resource distribution policies, large numbers of people of childbearing ages, and the high rates of rural-to-urban and rural-to-rural migration affect a family's standard of living, household income, and opportunities to earn a livelihood.

Industrialization and Family Life

Sociologists have been more successful in identifying how the Industrial Revolution affected the key episodes of family life in mechanized rich countries (see Table 12.5), but they have failed to identify how it has affected family life in labor-intensive poor countries. We've learned that industrialization is not the same experience for everyone affected by it. For example, the landless peasants of Brazil, the *maquilas* workers of Mexico, the Mbuti of Zaire, the laid-off factory workers in the United States, white-collar workers, and countless other groups have all been affected in different ways by industrialization.

The sheer diversity of experiences makes it impossible to make blanket statements about how industrialization affects family life. Consider the differences among three Brazilian families that sociologist Hélène Tremblay stayed with as part of her 60-country, 118-family research circuit to learn about how families around the world live. Excerpts from her diaries about how each family begins the day give some insights into the impossibility of generalizing about family life in a specific country:

The Yanomami family of Amazonia lives in a shabono, a communal hut that houses 15 fire

TABLE 12.5   Effects of Industrialization on Key Episodes in Family Life

Key Episode	Before Industrialization	Mature Phase of Industrialization
Childbearing Experiences		
Birthrates (annual)	30–50 births per 1,000 people	Fewer than 20 births per 1,000 people
Total Fertility	9	2
Spacing	Entire reproductive life (15–49)	Short segment of reproductive life
Maternal Mortality (annual)	Over 600 deaths per 100,000 births	30 deaths per 100,000 births
Estimated Lifetime Chance of Dying from Pregnancy-related Causes	1 in 21	1 in 10,000
Chances of Survival		
Death Rate (annual)	50+ deaths per 1,000 people	Fewer than 10 deaths per 1,000 people
Life Expectancy	20–30	70+
Major Causes of Death	Epidemic sources Deficiency diseases Malnutrition Parasite diseases	Degenerative and man-made
Nature of Labor		
Livelihood	Labor-intensive agriculture	Information and services
Source of Power	Human muscles and other sources of power	Machines and created power
Economy	Subsistence-oriented	Consumption-oriented
Residence	Rural	Urban

SOURCE: Joan Ferrante, Northern Kentucky University (1991).

sites, one for each family unit. The family consists of a mother (one of three wives, two of whom live elsewhere), a father, and three children (ages 18 months, 4 years, and 7 years). "Mother plucks a few bananas hanging over her head and throws them on the coals along with some palm leaves stuffed with nutmeats. Breakfast will soon be ready.

   After feeding her children [she] joins the other women on their way to fish. [Father] is off hunting with the other men, taking advantage of the final days of the dry season." (TREMBLAY 1988, P. 3)

The Mariano de Souza Caldas family of Guriri lives in a mud hut on a government-issued land plot. The smell is appalling because they live just 50 meters from the city garbage dump. The family consists of a pregnant mother, a father, and 10 children who range in age from 11 months to 17 years old. The father "gently pushes his sleeping son and gets out of bed. He moves the baby's crib, which is blocking the doorway . . . and heads for the well at the bottom of the yard. . . . He showers and rinses his mouth . . . will milk the cows . . . and will spend all morning combing the dumps in search of items that he can recycle for money. (P. 23)

The Aratanha family of Rio de Janeiro lives in a fenced-in, highly guarded high rise with many amenities (soccer field, tennis courts, a gym).

Stephanie Maze/Woodfin Camp & Associates

Paulo Fridman/Sygma

The effects of industrialization on family life are too complex to be captured by sweeping generalizations. Brazil, for example, includes regions where children remain sources of cheap labor in a non-mechanized economy. Yet it also includes cities where middle-class children, like their counterparts in the United States, play the role of consumers in a mechanized-rich economy.

The family consists of a mother and father (both of whom are physicians) and two children (ages 18 months and 3 years). They have a maid who helps care for the children. The father "gets out of bed quickly without waking his wife . . . [and] leaves very early to avoid the traffic jams. . . . [The mother is] grateful for her new car. It saves her half an hour every morning." (P. 31)

Another reason that it is difficult to generalize about how industrialization affects family life is that industrialization and the changes that are thought to accompany it may run up against some political, cultural, or historical stumbling blocks. For example, birthrates are expected to decline with industrialization and urbanization. Yet, sociologists and demographers are still trying to understand the forces underlying decisions by American couples to have relatively large families after World War II (between 1946 and 1963—a period of dramatic industrialization and the years that produced the baby boomers).

In addition, government policies may have a strong influence on birthrates. For example, after the Romanian government legalized abortion in 1957, there were approximately 4,000 abortions for every 1,000 live births (about 80 percent of pregnancies ended in abortion), and the birthrate was about 15 per 1,000. In 1967 President Nicolai Ceausescu, in an effort to increase the population and the number of future workers, drastically restricted the sale of contraceptives and placed tight restrictions on abortions. Under Ceausescu's policies, Romanian women were required to undergo regular pregnancy tests in order to ensure that all pregnancies were carried to term, and they were given access to contraceptives and abortions only if they were over 45 years of age or already had four living children. Physicians who administered illegal abortions faced 25 years in prison and even the death penalty. Under these policies the birthrate climbed to 27.4 per 1,000 (Burke 1989; van de Kaa 1987). By 1983, however, the birthrate had tumbled to 14 per 1,000, and Ceausescu announced more restrictions and closer monitoring.

A final reason that it is difficult to generalize about industrialization's effects on family life is that every family is unique and is composed of diverse personalities. Uniqueness makes it difficult to predict the effects of migration (and other events associated with industrialization) on family life. For example, increased migration unquestionably accompanied industrialization, but any statement about the effects of migration must consider the kind of migration (rural-to-rural, urban-to-urban, international, and so on) and the reason people migrate. Even if we narrow our analysis to a specific kind of migration for a particular reason, we must acknowledge that families do not respond in uniform ways.

These mixed findings are characteristic of family research in general. Consequently, in the sections that follow we will consider how the effects of geographic mobility and the changes in life expectancy, residence, and the nature of work put new pressures on family life and may lead to changes in function, composition, and the status of children. We will also consider how these factors may lead to changes in the division of labor between men and women and to changes in the reasons people marry.

The Effects of Geographic Mobility on Family Life

Industrialization has brought increased geographic mobility, which in turn affects family life. Some sociologists have found that geographic separation virtually guarantees that family members will meet less frequently than if they live close to one another and that over time this spatial distance results in emotional distance as well (Bernardo 1967; Parsons 1966). Other sociologists found that, although geographic separation hinders face-to-face interaction, it does not necessarily disrupt kinship relations because family members can keep in touch by phone or mail. In fact, this latter group of sociologists found substantial interaction despite distance and concluded that separation can even enhance relationships. That is, separation teaches family members not to take one another for granted but to enjoy and appreciate the limited time available for interaction (Adams 1968; Allan 1977; Leigh 1982; Litwak 1960).

Some researchers, however, found that substantial interaction is more likely among persons recently separated than among those who have been separated for an extended period. Other researchers maintain that it is important to consider the variables that may confound the effects of distance—in particular, the reasons that people move away from family members and the extent of geographic dispersion of family members. In some cases lack of contact cannot be totally attributed to distance if the move liberates a family member from an already problematic family situation. As for geographic dispersion, if the majority of family members live in one geographic area and one or two members live outside that area, the modest degree of dispersion makes it easier for those who are visiting to see everyone in the family. On the other hand, it is difficult for one person to visit all family members if each one lives in a different city. By definition, some family members will not be contacted as often as others (Allan 1977).

The Consequences of Long Life

Since the turn of the century, the average life expectancy at birth has increased by 28 years in the mechanized rich countries and by 20 years (or more) in the labor-intensive poor countries. In *The Social Consequences of Long Life* sociologist Holger R. Stub (1982) describes at least four ways in which increases in life expectancy have altered the composition of the family since 1900. First, the chances that children will lose one or both parents before they reach 16 years of age has decreased sharply. In 1900 there was a 24-percent chance of such an occurrence; today the probability is less than 1 percent. At the same time, parents can expect that their children will survive infancy and early childhood. In 1900, 250 of every 1,000 children born in the United States died before age 1; 33 percent did not live to age 18. Today only 10 of every 1,000 children born die before they reach age 1; fewer than 5 percent die before reaching age 18.

Second, the potential length of marriages has increased. Given the mortality patterns in 1900, newly married couples could expect their marriage to last an average of 23 years before one partner died (if we assume they did not divorce). Today, assuming that they do not divorce, newly married couples can expect to be married for 53 years before one partner dies. This structural change may be one of several factors underlying the high divorce rates today. At the turn of the century,

> *death nearly always intervened before a typical marriage had run its natural course. Now, many marriages run out of steam with decades of life remaining for each spouse. When people could expect to live only a few more months or years in an unsatisfying relationship, they would usually resign themselves to it. But the thought of 20, 30, or even 50 more years in an unsatisfying relationship can cause decisive action at any age.* (DYCHTWALD AND FLOWER 1989, P. 213)

According to Holger Stub, divorce dissolves today's marriages at the same rate that death did 100 years ago. In view of this information, we might question whether the recent surge in the number of divorces among couples in the People's Republic of China is connected to improvements in life expectancy over the past 10 years. We might also question whether an improved standard of living changes the reasons people marry.

Third, people now have more time to choose and get to know a partner, settle on an occupation, attend school, and decide whether they want children. Moreover, an initial decision made in any one of these areas is not final. The amount of additional living time enables individuals to change partners, careers, or educational and family plans, a luxury not shared by their turn-of-the-century counterparts. Stub argues that the mid-life crisis is related to long life because many people "perceive that there yet may be time to make changes and accordingly plan second careers or other changes in lifestyle" (Stub 1982, p. 12).

Finally, the number of people surviving to old age has increased. (In countries where fertility is low or declining, the proportion of old people in the population—not merely the number—is also increasing.) In 1970, about 25 percent of people in their late fifties had at least one surviving parent; in 1980, 40 percent had a surviving parent. Even more astonishing, in 1990 20 percent of people in their early sixties and 3 percent of people in their early seventies had at least one surviving parent

(Lewin 1990). Although it has always been the case that a small number of people have lived to age 80 or 90, "There is no historical precedent for the aging of our population. We are in the midst of a new phenomenon" (Soldo and Agree 1988, p. 5).

Much has been written about the growing numbers of people over age 65 in the world, especially about issues of caring for the disabled and the frail elderly. Yet, the emphasis on this segment of the older population should not obscure the fact that the elderly are a rapidly changing and heterogeneous group. They differ according to gender, age (a 30-year difference separates persons age 65 from those in their nineties), social class, and health status. Most older people today are in relatively good health; in the United States only five percent of persons age 65 and older live in nursing homes (Eckholm 1990).

In most countries, including the United States, most disabled and frail elderly persons are cared for by female relatives (Stone, Cafferata, and Sangl 1987; Targ 1989). In the United States, approximately 72 percent of the caregivers are women. Of this 72 percent, 22.7 percent are spouses, 28.9 percent are daughters, and 19.9 percent are daughters-in-law, sisters, grandmothers, or some other female relative or nonrelative (Stone, Cafferata, and Sangl 1987). The major issue faced by even the closest of families is how to meet the needs of the disabled and frail elderly so as not to "constrain investments in children, impair the health and nutrition of younger generations, impede mobility," or impose too great a psychological, physical, and/or time-demand stress on those who care for them. This problem may be even more intense in developing countries where the presence of older dependents "may alter the level of subsistence of other members in the family" (United Nations 1983, p. 570). In subsistence cultures, even in those that display considerable respect toward the elderly, "the decrepit elderly may be seen as too much of a burden to be supported during periods of deprivation and may be killed, abandoned, or forsaken" (Glascock 1982, p. 55). Although many programs exist and are being developed to support caregivers and to address the issues associated with caring for the disabled and frail elderly, most of these programs are still in the pilot stages. "It is going to take as many

The Economic Role of Children in a Labor-Intensive Poor Country

Children often are an important source of cheap labor, especially for nonmechanized industries. The childhood experiences of Chico Mendes, the son of an Amazon rubber tapper, illustrate the economic role of children in the kind of labor-intensive work that takes place in rubber plantations.

Chico Mendes was an important Brazilian grass-roots environmentalist and union activist dedicated to preserving the Amazon and to improving rubber tappers' economic standard of living as compared with cattle ranchers and rubber barons. He was murdered in 1988 by Amazon cattle ranchers. Indians and rubber tappers are now working together against those who misuse or otherwise destroy the resources of the forests.

Childhood for Chico Mendes was mostly heavy work. . . . If all his siblings had lived to adulthood, Chico would have had seventeen brothers and sisters. As it was, conditions were so difficult that by the time he was grown, he was the oldest of six children—four brothers and two sisters.

When he was five, Chico began to collect firewood and haul water. A principal daytime occupation of young children on the *seringal* has always been lugging cooking pots full of water from the nearest river. . . . Another chore was pounding freshly harvested rice to remove the hulls. A double-ended wooden club was plunged into a hollowed section of tree trunk filled with rice grains, like an oversize mortar and pestle. Often two children would pound the rice simultaneously, synchronizing their strokes so that one club was rising as the other descended.

By the time he was nine, Chico was following his father into the forest to learn how to tap. . . . Well before dawn, Chico and his family would rise. . . . [T]hey would grab the tools of their trade—a shotgun, . . . a machete, and a pouch to collect any useful fruits or herbs found along the trail. They left before dawn, because that was when the latex was said to run most freely. . . .

As they reached each tree, the elder Mendes grasped the short wooden handle of his rubber knife in two hands, with a grip somewhat like that of a golfer about to putt. . . . It was important to get the depth [of the cut] just right. . . . A cut that is too deep strikes the cambium [generative tissue of the tree] and imperils the tree; a cut that is too shallow misses the latex-producing layer and is thus a waste. . . .

Chico learned how to position a tin cup, or sometimes an empty Brazilian nut pod, just beneath the low point of the fresh cut on a crutch made out of a small branch. . . . The white latex immediately began to dribble down the slash and into the cup. . . .

Chico quickly adopted the distinctive, fast stride of the rubber tapper. . . . The pace has evolved from the nature of the tapper's day.

as 15 or 20 years to sort out what works" (Lewin 1990, p. A11).

The Status of Children

The technological advances associated with the Industrial Revolution decreased the amount of physical exertion and time needed to produce food and other commodities. As human muscle and time became less important to the production process, children lost their economic value. In nonmechanized, extractive economies children are an important source of cheap and unskilled labor for the family. This fact may explain in part why the states in Amazonian Brazil have some of the highest total fertility rates found anywhere in the world (Butts and Bogue 1989). The total fertility rate in rural Acre is 8.03, a level that demographers have defined as approximating the human capacity to reproduce. (See "The Economic Role of Children in a Labor-Intensive Poor Country.")

Middle- and upper-class children in mechanized rich economies are likely to live in settings that strip them of whatever potential economic contribution they might make to the family economy (Johansson 1987). In these economies the family's energies shift away from production of food and other necessities and toward the consumption of goods and services (see "Tooning In to a Tiny Market").

The 150 and 200 rubber trees along an *estrada* are exasperatingly spread out. A simple, minimal bit of work is required at each tree, but there is often a 100-yard gap between trees. Thus, a tapper's morning circuit can be an 8- to 11-mile hike. And that is just the morning. Many tappers retrace their steps in the afternoon to collect the latex that has accumulated from the morning cuts.

By the time Chico and his father came full circle on an *estrada* and returned to the house, it would be close to midday, time for a lunch. . . . Then, they would retrace their steps on the same trail, to gather the latex that had flowed from the trees during the morning. Only rarely would they return home before five o'clock. By then they would be carrying several gallons of raw latex in a metal jar or sometimes in a homemade, rubber-coated sack.

The very best rubber is produced when this pure latex is immediately cured over a smoky fire.

The smoking of the latex was done in an open shed that was always filled with fumes from the fire. The smokier the fire, the better. Usually, palm nuts were added to the flames to make the smoke extra thick. Chico would ladle the latex onto a wooden rod or paddle suspended over a conical oven in which the nuts and wood were burned. As the layers built up on the rod, the rubber took on the shape of an oversize rugby football. The smoking process would continue into the evening. After a day's labor of fifteen hours or more, only 6 or 8 pounds of rubber were produced. The tappers often developed chronic lung diseases from exposure to the dense, noxious smoke.

Toward December, with the return of the rainy season, the tappers stopped harvesting latex and began collecting Brazil nuts.

During the rainy season, the Mendeses often crouched on their haunches around a pile of Brazil nut pods, hacking off the top of each one with a sharp machete blow, then tossing the loose nuts onto a growing pile. For tappers in regions with Brazil nut trees, the nuts can provide up to half of a family's income. One tree can produce 250 to 500 pounds of nuts in a good season and some tappers' trails pass enough trees to allow them to collect more than 3 tons of nuts each season. While that might sound like a potential windfall, even as late as 1989, tappers received only 3 or 4 cents a pound for the harvest—which later sold for more than $1 a pound at the export docks.

When the Mendeses were not harvesting latex or nuts, they tended small fields of corn, beans, and manioc.

SOURCE: From *The Burning Season* by Andrew Revkin, pp. 69–76. Copyright © 1990 by Andrew Revkin. Reprinted by permission of Houghton Mifflin Company.

The U.S. Department of Agriculture estimates that the average family earning an annual income of at least $50,000 will spend about $270,000 to house ($84,000), feed ($61,000), clothe ($19,000), transport ($40,000), and supply medical care ($18,000 not covered by insurance) to a child until he or she is 22 years old (Rock 1990). Moreover, the department estimates that economies of scale are minimal—it estimates that two children cost $419,000, three children $569,000, and four children $759,000. These figures represent a basic sum; the costs grow even higher when extras (summer camps, private schools, sports, music lessons) are included. Even if children go to work when they reach their teens, usually they use that income to purchase items for themselves and do not contribute to household expenses. Demographer S. Ryan Johansson argues that in the developed economies, couples who choose to have children bring them into the world to provide intangible, "emotional" services—love, companionship, an outlet for nurturing feelings, enhancement of dimensions of adult identity (Johansson 1987).

Urbanization and Family Life

Urbanization encompasses two phenomena: (1) the migration of people from rural areas to cities so that an increasing proportion of the population comes to live in cities and (2) a change in the ties

Tooning In to a Tiny Market

In mechanized rich economies the family's energies shift away from the production of food and other necessities and toward the consumption of goods and services. In fact, marketers view even very young children in these countries as potential customers who exercise considerable influence over their parents' decisions to purchase various brands of cereals, toys, and clothes. The following essay, which appeared in a 1992 Cincinnati marketing and advertising newsletter, supports this point.

Welcome to 'Toontown, a magical territory where "anything can happen, the natural laws don't apply and imagination is the only currency that counts," according to *Newsweek*. With all these attractions, you can bet 'Toontown's no ghost town. Far from it. Especially with the help of cartoons and cartoon characters like The Real Ghostbusters, Teenage Mutant Ninja Turtles and many more. 'Toontown's alive and well and aiming for kids.

Ready . . .

The changing family structure has made children a more visible target for product promoters. With more parents in the work force, a greater number of children are being left alone at home at a younger age. As a result, these children are making more decisions on their own and becoming more independent. Many kids now play a more decisive role not only in the home, but in the marketplace—they're referred to by *Marketing and Media Decisions* as "brand managers" of the home, and are "making decisions—spending their own money and . . . spending the money and influencing the decisions of the entire family." Which means that they may have a say in anything from the brand of cereal the family eats to the type of car the family drives.

Aim . . .

Marketers are catering to this young independence and influence by advertising to the younger audience, often with animated cartoon—'Toontown—characters. Fun and simplicity are the two main ingredients in this recipe for product advertising. Marketers promote anything from toe-tappin' tubes of toothpaste to "food products designed to just pop in the microwave so kids can totally fend for themselves," such as Tyson's Looney Tunes microwave meals featuring Bugs Bunny, Daffy Duck, Tweety Bird and the rest of the bunch.

'Toontown natives keep a high profile wherever youngsters may happen to be. 'Toontowners mainly appear on TV—either as cartoons or commercials; in grocery stores, department stores and toy stores. They've even made their way into movie theaters and fast-food restaurants.

Saturday morning TV may contain the highest concentration of animated characters whose job is not only to entertain kids, but to sell

that bind people to one another. (See the discussion of mechanical and organic solidarity in Chapter 6.) In mechanized rich countries people who live in urban settings usually are not part of self-sufficient economic units composed of family members. They have opportunities to make contact with people outside the family network, and they come to depend on people and institutions other than the family (the workplace, hospitals and clinics, counseling services, schools, day-care centers, the media) to meet various needs. People's relationships with most of the individuals they meet in the course of a day are transitory, limited, and impersonal. This does not mean that the family ceases to be important to people's lives, however. Instead, it is important in a different way: it becomes less an economic unit and more a source of personal support. Most sociological research on the American family shows that the great majority of people have some family members living near them, that they interact with them regularly, and that they define these interactions as meaningful and important (Goldenberg 1987).

Urbanization in itself, however, does not necessarily reduce the economic function of the family. Janice Perlman (1967), in her studies of the urban poor in Brazil, found that the majority of migrants to Rio de Janeiro (67 percent) came to the city in

various products to them. On TV, the Nestle Quik bunny paints a mouth-watering picture of "Quikland," home of his "chocolatey-dream-come-true" drink mix. Rockin' and reelin' fruit drink cartons celebrate "wacky, wild Kool-Aid style," while Tony the Tiger still tells kids that his cereal is "Grrreat!"

'Toontowners promote from supermarket shelves—in all species, colors and sizes—enticing children and wallet keepers alike from the snack foods aisle to the school supplies section. At the grocery, a shopper can set out on a Cheetos cheetah chase or follow his nose to the Froot Loops shelf. He can snatch a pack of Flintstones Franks and nab a few "Nintendo Power" notebooks.

But that's not all, folks. Toy stores are stacked to the ceiling with cartoon character action figures, board games, stuffed dolls, accessories—almost everything imaginable for the ultimate 'Toontown buff. Offering anything from classic Mickey Mouse watches to Ghostbuster water zappers, a shopper is sure to find just about any toy featuring a tyke's favorite character. Some toys have met with such success that toy makers have had trouble keeping up with consumer demand for certain products, such as Masters of the Universe.

'Toontown marketers also sell their wares in department stores and restaurants, often as a tie-in with a latest motion picture release. In 1988, J.C. Penney had teamed up with Pizza Hut to promote *Land Before Time* with items from kids' clothing to food for the family. Similarly, *An American Tail* became a hit with the support of successful merchandising by Sears and McDonalds. And the marketing potential of 'Toontown characters has been proven with over $600 million in Teenage Mutant Ninja Turtle paraphernalia, from Turtle t-shirts to shoe lace holders.

Movie stars and mutant reptiles aren't the only 'Toons making it big in the stores. As the new family on the 'Toontown block, The Simpsons' fame has shot up higher than Marge Simpson's huge, blue hair-do. Simpsonites all across town may be found sporting the popular apparel touting Bart's words of wisdom.

Fire!

With all the magic of 'Toontown, what better way is there to reach a young audience spellbound by bright, bold colors with imaginations that run rampant when movin', groovin' tunes and witty words are added to action that nearly jumps out of the set? Or off the shelf and out of the package? And who could ever doubt anything that their green "heroes in a half shell" claim as being "totally awesome"? 'Toontown can be a sure-fire hit with kids—and with product promoters. Ready . . . aim . . .

Bull's-Eye!!!

SOURCE: Lori Ling. This essay about the youngest consumers in the United States appeared in a marketing and advertising newsletter that circulates in Cincinnati, Ohio (1992).

---

the company of relatives or went to the homes of relatives upon arrival, and that many make every effort to live as close to one another as possible. Thus Perlman found that some neighborhoods were dominated by large kin networks. The research also shows that family networks are extremely useful to people living in precarious financial conditions. Perlman found that 63 percent of the people she studied had gotten their first job through friends or kin.

Anthropologist Claudia Fonseca (1986) also learned the significance of family relationships in Brazil. She studied 68 mothers from a shantytown of 750 squatters living on an empty lot in a middle-class neighborhood in southern Brazil. She found that 50 percent of the mothers had sent a child away to live under the care of another person while they recovered from illness or financial setbacks. Most often, this caretaker was a relative.

Judith Goode (1987) has written extensively about Colombian street children, and Thomas Sanders (1987a, 1987b) has written about Brazilian street children. Both groups of children are frequently misperceived by government officials and middle-class people as abandoned by their parents, as delinquents, and as a reflection of widespread family pathology among squatters. Both writers conclude, however, that street children (usually

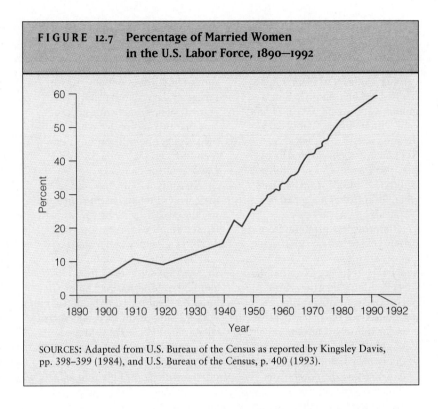

FIGURE 12.7   Percentage of Married Women in the U.S. Labor Force, 1890–1992

SOURCES: Adapted from U.S. Bureau of the Census as reported by Kingsley Davis, pp. 398–399 (1984), and U.S. Bureau of the Census, p. 400 (1993).

between the ages of 10 and 15) are from poverty-stricken families "who are trying to maximize their household incomes by utilizing the money-earning capacities of their children in such activities as shining shoes, carrying packages, watching cars, and begging" (Sanders 1987b, p. 1). Goode writes:

> *Attempts to classify types of income for street children often distinguish between those activities which are legal, such as performing services as shoeshine boys, street vendors, and car watchers; those which are more marginal, such as begging and singing on buses; and those which are illegal and anti-social, such as purse-snatching, picking pockets, and other forms of theft.* (GOODE 1987, P. 6)

Women and Work

In 1984 sociologist Kingsley Davis published "Wives and Work: The Sex Role Revolution and Its Consequences." In this article, Davis identified "as clear and definite a social change as one can find"

(p. 401): between 1890 (the first year for which reliable data exists) and 1980, the proportion of married women in the labor force rose from less than 5 percent to more than 50 percent. Davis found that this pattern holds true for virtually every industrialized country except Japan. He attributed this change to the Industrial Revolution and its effect on the division of labor between men and women (see Figure 12.7). It is important to keep in mind that Davis's theory is intended to explain how industrialization in the mechanized rich countries affected the division of labor between males and females. It does not systematically address how industrialization is affecting the division of labor between males and females in labor-intensive poor countries.

The Division of Labor and the Breadwinner System Before industrialization—that is, for most of human history—the workplace was the home and the surrounding land. The division of labor was based on sex. In nonindustrial societies (which includes two major types, the hunting-and-gathering and the agrarian),[11] men provided raw materials through hunting or agriculture, and women processed these materials. Women also worked in agriculture and provided some raw materials by gathering food and hunting small animals, and they took care of the young.

The Industrial Revolution separated the workplace from the home and altered the division of labor between men and women. It destroyed the household economy by removing economic production from the home and taking it out of the women's hands:

> *The man's work, instead of being directly integrated with that of wife and children in the home or on the surrounding land, was integrated with that of non-kin in factories, shops, and firms. The man's economic role became in one sense more important to the family, for he was the link between the family and the wider market economy, but at the same time his personal participation in the household diminished. His wife, relegated to the home as her sphere, still performed the parental and domestic duties that women had always performed. She bore and reared children, cooked meals, washed clothes, and cared for her husband's personal needs, but to an unprecedented degree her economic role became restricted. She could not produce what the family consumed, because production had been removed from the home.*
> (DAVIS 1984, P. 403)

Davis called this new economic arrangement the "breadwinner system." From a historical point of view, this system is not typical. Rather, it is peculiar to the middle and upper classes and is associated with a particular phase of industrialization—from the point at which agriculture loses its dominance to the point where only 25 percent of the population still works in agriculture. In the United States, "the heyday of the breadwinner system was from about 1860 to 1920" (p. 404). This system has been in decline for some time in the United States and other industrialized countries, but it tends to recur in countries undergoing a particular phase of development.

Davis asks, "Why did the separation of home and workplace lead to this system [in most mechanized rich countries]?" The major reason, he believes, is that women had too many children to engage in work outside the home. This answer is supported by the fact that family size, in the sense of the number of living members, reached its peak from the mid-1800s to the early 1900s. This occurred because infant and childhood mortality declined while the old norms favoring large families persisted.

The Decline of the Breadwinner System The breadwinner system did not last long because it placed too much strain on husbands and wives and because a number of demographic changes associated with industrialization worked to undermine it. The strains stemmed from several sources. Never before had the roles of husband and wife been so distinct. Never before had women played less than a direct, important role in producing what the family consumed. Never before had men been separated from the family for most of their waking hours. Never before had men had to bear the sole responsibility of supporting the entire family. Davis regards these events as structural weaknesses in the breadwinner system. In view of these weaknesses the system needed strong normative controls to survive: "The husband's obligation to support his family, even after his death, had to be enforced by law and public opinion; illegitimate sexual relations and reproduction had to be condemned, divorce had to be punished, and marriage had to be encouraged by making the lot of the 'spinster' a pitiful one" (p. 406).

Davis maintains that the normative controls collapsed because of the strains inherent in the breadwinner system and because of demographic and social changes that accompanied industrialization. These changes included decreases in total fertility, increases in life expectancy, increases in the divorce rate, and increases in opportunities for employment perceived as suitable for women.

Declines in Total Fertility The decline in total fertility began before married women entered the labor force. This fact led Davis to conclude that the decline itself changed women's lives in such a way that they had the time to work outside the home, especially after the children entered school. During the 1880s total fertility among white women was approximately 5.0; in the 1930s it averaged 2.4; during the 1970s it averaged 1.8. Not only did the number of children decrease but reproduction ended earlier in women's lives and the births were spaced closer together than in earlier years. (The median age of mothers at the last birth was 40 in 1850; by 1940 it had fallen to 27.3.) Davis attributes the changes in childbearing patterns to the forces of industrialization, which changed children from an economic asset to an economic liability, and to the "desire to retain or advance one's own and one's children's status in a rapidly evolving industrial society" (p. 408).

Increased Life Expectancy In view of the relatively short life expectancy in 1860 and the age at which women had their last child (age 40), the average woman was dead by the time her last child left home. By 1980, given the changes in family size, spacing of children, and age of last pregnancy, the average woman could expect to live 33 years after her last child left home. As a result, child care came to occupy a smaller proportion of a woman's life. In addition, although life expectancy has increased for both men and women, on the average women can expect to live longer than men. In 1900 women outlived men by about 1.6 years on the average; in 1980 they outlived men by approximately 7 years. Yet, because brides tend to be three years younger on the average than their husbands, married women can expect to live past their husbands' death for an average of 10 years. In addition, the distorted sex-ratio caused by males' earlier death decreases the probability of remarriage. Although few women think directly about their husbands' impending death as a reason for working, the difference in mortality remains a background consideration.

Increased Divorce Davis traces the rise in the divorce rate to the breadwinner system, specifically to the shift of economic production outside the home:

> *With this shift, husband and wife, parents and children, were no longer bound together in a close face-to-face division of labor in a com-*
> *mon enterprise. They were bound, rather, by a weaker and less direct mutuality—the husband's ability to draw income from the wider economy and the wife's willingness to make a home and raise children. The husband's work not only took him out of the home but also frequently put him into contact with other people, including young unmarried women [who have always worked] who were strangers to his family. Extramarital relationships inevitably flourished. Freed from rural and small-town social controls, many husbands either sought divorce or, by their behavior, caused their wives to do so.*
> (PP. 410–11)

Davis notes that an increase in the divorce rate preceded married women's entry into the labor market by several decades. He argues that once the divorce rate reaches a certain threshold (above a 20 percent chance of divorce), more married women seriously consider seeking employment to protect themselves in case of divorce. When both husband and wife are in the labor force, the chances of divorce increase even more. Now both partners interact with people who are strangers to the family. Moreover, they live in three different worlds, only one of which they share.

Increased Employment Opportunities for Women Davis believes married women are motivated to seek work by changes in childbearing experiences, increases in life expectancy, the rising divorce rate, and the inherent weakness of the breadwinner system. This motivation was realized as opportunities to work increased for women. With improvements in machine technology, productivity increased, the physical labor required to produce goods and services decreased, and wages rose in industrialized societies. (Chapter 2 explains how industrialization proceeds in labor-intensive poor countries.) As industrialization matures, there is a corresponding increase in the kinds of jobs perceived as suitable for women (traditionally, nursing, clerical and secretarial work, and teaching).

The two-income system is not free of problems, however. First, the system at present "lacks normative guidelines. It is not clear what husbands and wives should expect of each other. It is not clear what ex-wives and ex-husbands should expect, or

As reflected in the Ms. Foundation for Women's "Take Our Daughters to Work" campaign, the growing importance of employment and careers to women in industrialized societies represents a profound social change as well as an economic one. Here a nine-year-old girl is spending a day at work with her mother, a state supreme court justice.
Marty Lederhandler/AP Photos

children, cohabitants, friends, and neighbors. Each couple has to work out its own arrangement, which means in practice a great deal of experimentation and failure" (Davis 1984, p. 413). Second, although the two-income system gives wives a direct role in economic production, it also requires that they work away from the home, a situation that makes child care difficult. Third, even in the two-income system the woman is still primarily responsible for domestic matters. Davis maintains that women bear this responsibility because men and women are unequal in the labor force. (On the average women earn 66 cents for every dollar earned by males.) As long as women make an unequal con-tribution to the household income, they will do more work around the house.

The problems of the new system are stressed fur-ther by the fact that a large percentage of married women (almost 40 percent) do not work outside the home. Davis believes that this lack of participa-tion reflects the psychosocial costs of employment to married women. The dilemma women face can be summarized as follows: if married women work and have no children, they are selfish; if they stay home and raise a family, they are considered under-employed at best; if they work and raise a family, people wonder how they can possibly do either job right.

Discussion

We opened this chapter with five scenes from fam-ily life, in part to show the difficulty of constructing a definition of family. As diverse as these scenes are, we can say that no matter what kind of family peo-ple are born into, live with, or form later in life, their lives are affected by key episodes: births, deaths, and the kind of work (including migration to find work) they must do to sustain a standard of living. The Industrial Revolution is one important event that has shaped the key episodes of family life on a global scale. We focused on Brazil because that country allows us to consider family life in a variety of environments that have been affected in different ways by industrialization.

The theory of the demographic transition is a model that outlines historical changes in birth and death rates in Western Europe and the United States. This theory suggests that a country's birth and death rates are linked to its level of industrial or economic development. However, it gives us only a partial picture of how the industrialization process characteristic of colonizing countries has shaped the key episodes of family life. From a global perspective industrialization is an uneven process; its effects vary according to country and according to regions within countries. Perhaps the most important lesson of this chapter is that industrialization, whatever its form, has had powerful effects on the character of family life but that those effects are far from uniform.

Recall the variety of family-related situations we have covered in this chapter: we began with five scenes from family life; we compared the reproductive histories of two women (one from the seventeenth century, the other from the late twentieth century); we read about how people in rural Brazil register births and deaths; we glimpsed the way in which three families—the Yanomami family of Amazonia, the Mariano de Souza Caldas family of Guriri, and the Aratanha family of Rio de Janeiro—begin their days; we explored the childhood of Chico Mendes; and we looked at the breadwinner system. The common factor in all of these scenes is the family's relationship to the means of production.

If relationship to the means of production is so important, why the intent focus on the composition of the family? In the United States, for example, the media and various policymakers advance the idea that families' problems in meeting the challenges of life are tied to the composition of the family or what the family looks like. (If only the family looked as it did in the 1950s!) The media offer us glimpses of family life in poor countries, showing female-headed households, high rates of total fertility, fathers who are absent because they have migrated to find work. We are left with the simple conclusion that things would improve if only the families were more stable and couples had fewer

Finding work in a changing economy can mean learning new skills, such as how to read—but it can also mean disrupting traditional patterns of family life as workers go where the jobs are.
Stephanie Maze/Woodfin Camp & Associates

children. The problem is that people fail to focus on the source of the instability, which in many (but not all) cases is the family's relationship to the means of production.

FOCUS
A Broader Definition of Family

At the beginning of the chapter we defined family as two or more people related to one another by blood, marriage, adoption, or some other socially recognized criteria. According to this definition, a primary group qualifies as a family if some official institution such as a government or church recognizes it as such. Recall that one problem with official definitions of families is that they can be used to deny family-related benefits to people who don't fit the definition of family. In the article that follows, sociologists Teresa Marciano and Marvin B. Sussman maintain that we need a wider definition of family and suggest that a commitment to care for one another be included along with blood, marriage, and adoption as a criteria of family membership.

The Definition of Family Is Expanding

Teresa Marciano and Marvin B. Sussman

No adequate definition of "the family" exists, though anthropologists have spent decades elaborating family types. Any text or monograph will offer its own definition of family, or borrow a definition from someone else according to the author's preferences or topical concerns. At best the student of families describes the characteristics that families share with one another . . . .

Characteristics include most or all of the following: the idea that children must or were or will be present, thus fulfilling the procreative function; legal marriage will have occurred resulting in rules for responsibility and orderly descent; kinship bonds will be the foundation of family; that bonding and inheritance transfers will effect intergenerational continuity: implied lifetime commitment provides time and sanctuary for expressions of intimacy, nurturance, caring, and love; and nuclear units in the family will share a lifestyle based on the parental or ancestral socioeconomic status. Whatever lifestyle is lived will have its genesis in the family.

Defining the Wider Family The notion of wider families does not ignore some of these family realities, but maintains that families exist as social constructions, coexisting with the traditionally defined families and frequently overlapping with them in expressing family-like behaviors. Wider families' characteristics include:

- Wider families form without being limited to age distributions found in traditionally defined

SOURCE: Adapted from "Wider Families: An Overview," by Teresa Marciano and Marvin B. Sussman. Pp. 1-8 in *Wider Families: New Traditional Family Forms*, edited by T. Marciano and M. B. Sussman. Copyright © 1991 by Haworth Press. Reprinted by permission.

families. Relationships between generational cohorts are unnecessary. The actual mixture by age is a result of the reason the family got formed. . . .

- Legal marriage need not ever bind any of the members of a wider family to each other. It is a voluntary association. Individuals can leave and are under no legal obligation to remain. If contracts exist such as those regarding property ownership, settlements are made according to contract law.

- Wider families may exist which include kin-bonded people, but kinship bonds are incidental to the wider family.

- Permanence varies, and is not objectively normative for families, i.e., there are no legal, contractual bases for the continuation of wider families; there is no necessary degree of intergenerational permanence, though it may occur. Wider families have varying degrees of permanence and fluidity, and may disappear altogether or fragment and reconstitute in different forms, with sub-groups of the original wider family and with new members. . . .

- Wider families arise spontaneously in or apart from planned activities. They are logically or emotionally the "next step" for people connected by needs, activities, and interests.

- Wider families are autonomous, free, uncoerced types of families. If anything, coercion cancels the wider family notion, which springs from free choices made to enhance rather than limit one's life.

- Wider families can become primary groups for many members. The intensive interaction found in many of them results in in-group formation where members create a "we" or "in" group and view all others as outsiders. . . .

What Families Do Wider families come into existence for economic and emotional concerns. Services, exchanges, help, caring, advice, and emotional support can be found in wider families when traditional family resources are unavailable or inimical to the person's lifestyle. . . . Education, health, job, children, become major ingredients requiring services and supports. Longevity extends those needs, which often create burdens for traditional families. The wider family is the frequent and possible—though by no means inevitable—outcome, not only here and today but also in other times and places in the world.

The communards whose numbers grew so visibly and significantly in the Sixties and Seventies were forming wider families out of wider arrays of economic and social options. . . . The enlargement of possibilities, the assertion of rights of oppressed or stigmatized groups, found their dynamic in collective action, the civil rights and liberation movements for people of color, women, gay people. These assertions of freedom to live nonviolently in one's own pursuit of happiness, have had outcomes such as . . . lesbian mothers forming supportive communities. Redefinitions of "family," when those new definitions are lived, become powerful forces for changing contemporary ideas of acceptable and legal precepts and institutions.

The wider family has become more visible, is evolving rapidly and is being more enthusiastically embraced. . . .

The social dynamic of wider families, whether or not recognized as families by the legal or larger social systems, draws much power from the bonds of affection, remembrance, nurturance, support and "there-ness" of members for one another. When major changes have occurred, whether divorce, marriage, or on a more far-reaching scale when successive generations of a biological family live different lifestyles in different class and cultural arenas, the wider family may be the only group which meets deep personal needs. . . .

Continuing Change The concept of the wider family is a sensitizing rather than a definitive concept. It points to the path of inquiry rather than telling us what will or must be found. Sensitizing concepts are more open to dramatic or subtle empirical changes. They enable quicker naming, and therefore understanding of the processes we create and which influence us.

This is ever more important as we look to what becomes of aging adults who may never have married, or who have had no children, or who live without contact with biological kin. Those who respond to dependent members of wider families,

providing needed care, nurturance and emotional support, are substituting for traditional family members and emerge into the patterns of serial reciprocity. Thus, families are "made" out of other affectionate and service-providing groupings—wider families—where generational transfers of equity, status transfers, whole new lines of monetary and class inheritance, become possible. Children and grandchildren who are that because they act like that, friends' children who are "shared" children, who inherit from nonbiological "aunts" and "uncles," are such examples.

Certainly there is more freedom to be creative in wider families, since they are not constrained by the structures, normative patterns, and legal obligations of traditional families. There is the possibility of concentrating on fewer goals, or concentrating on goals obscured or blocked by the very memories and involvements that form so many of the bonds of biological families. Time is also available to do things in wider families that traditional family requirements may prevent. Holidays, birthdays, days of mourning and days of rejoicing, can be remade in voluntaristic ways, free of traditions and their sanctions. Wider families are arenas of creation, so that levels of expectation emerge, rather than being pre-existent to action and character development.

Still, habitual action does become patterned, and long-lived wider families may give rise to new kinds of traditions. These remain to be studied, as well as the question of whether over time, the spontaneous, transformed into the expected, may cause the perpetuation or the breakup of wider families. Wider families may be temporary, complementary, or latent groups. Temporary in the sense that through spontaneous formation they meet a current need or desire or solve a vexing problem. Then they cease to exist. Complementarity is functioning in tandem with traditional groups such as biological families, enhancing and supporting the life ways of its members. Latent wider families maintain their informal networks, experience times of quiescence and become active when issues stimulate members to activity.

Wider families are rescues both from isolation and independence; in all cases, the freedom to enter and the freedom to leave the wider family may be matters of optimal time and mass, of the right circumstances and the right people, at the right time. . . .

Define It, and Knowledge Follows Today, we hear so much about the minimal impact sociology seems to have on larger social understandings. Sociological, and psychological, and economic, and scientific terminology do get into everyday discourse, but the disparity between what has been learned from studies and the public awareness of it, signifies insufficient contribution on sociology's part to enriching that discourse. People want to make sense out of their lives. Religion traditionally has helped to do so, but it is not the only place that the quest for understanding can be served. In fact, to make sense out of the world we must be able to recognize what is happening, to name the phenomena of which we are part and which we create; things that have names are far less frightening than those which are nameless.

The notion of "wider family" emerges from observation, elaborating a grounded theoretical for-

mulation. If it is observed, if it can be explored and explained, it should be done in order to serve the needs of those who seek sense and definitions in their lives. For example, narrow social definitions may enable narrow legal definitions to persist; both cause much pain and deprivation. This is experienced by same-sex couples who say they are married, act lovingly and caringly, but must fight in court for access to each other if illness claims one and separates the couple, and for survivors' rights on the death of one. "Domestic partnership" laws which protect such survivors still encounter questions of duration or permanence, and conditions of dissolution. Perhaps just legitimating, through more empirical research, the existence of families that do not, or only partly, meet legal definitions of family, laws will be less calcified than they currently are in comparison to the dynamic flow of relationships in our society.

Key Concepts

Notes

1. According to Latin American Studies scholar Warren Dean, "São Paulo has to be explained. The rest of Brazil and indeed the rest of Latin America has not experienced that city's and state's breakthrough of sustained economic development based on diversified and technologically complex industrial production" (1991, p. 649).

2. The major city in this region (the Central West of Brazil) is Cuiaba, which is the site of a satellite tracking station maintained by NASA (Beresky 1991).

3. Of the 91 million households in the United States, 73 percent do not fit the so-called traditional definition of family—that is, two parents living with children (Gutis 1989a).

4. A legal case in Berkeley, CA, involves Michele G. and Nancy S., a lesbian couple who lived together for 11 years and were the parents of a daughter (the product of artificial insemination between Ms. S. and a sperm donor). Ms. G's attorney argued that "parental status can be conferred by exposure and experience as well as biology and adoption" (Margolick 1990, p. Y10).

   Sexual status aside, this couple's situation raises important questions about the definition of a family. Although biology and adoption are concrete criteria, they tell us little about the quality of the relationship between people. As Ms. G. stated, "To sit there and say with a straight face that someone who has stayed up all night nursing a child, swabbing her chicken pox, taking joy in her every advancement, picking her up every time she's skinned her knee, or singing her to sleep is not a 'mother' is an absurdity" (quoted in Margolick 1990, p. Y10). On the other hand, "quality of the relationship" cannot be used as the only criterion for a family relationship. Otherwise, babysitters, teachers, and friends would qualify as family.

5. Subsistence-oriented means that the family works to provide enough food to meet its needs but does not grow enough food to sell.

6. The gross national product is the monetary value of all goods and services produced within a specified time period (usually a year).

7. The Amazon forest receives considerable media attention in the United States and other industrialized countries, and there is public outcry about the burning of the Amazon vegetation. From the Brazilian point of view, this campaign detracts from the environmental damage that acid rain and nuclear waste of industrialized countries have caused. They argue that critics must consider the whole industrial model, not just the Amazon (Simons 1989).

8. This emphasis on consumption is illustrated most clearly by the typical American supermarket, which is stocked with 18,000 different packaged food items, 80 percent more in 1990 than in 1980. As the shoppers move down each aisle, they glance at an average of 10 items per second in contrast to 5.5 items in 1980 (Ramirez 1990). Because of the great number of items from which to select, product name, package design, and shelf location become as important as the quality of product. As a result, companies devote as much human energy (or more) to marketing the product as to producing it.

9. Japan followed an accelerated model: it underwent a rapid decline in mortality followed by an equally rapid decline in fertility. The rapid decline in fertility was due to high rates of abortion.

10. There are 94 cities in the world with populations exceeding two million. Of these 94 cities, 39 are in mechanized rich nations and 55 are in labor-intensive poor countries. By the year 2000, 34 additional cities will be added to this list; all but 4 of these new cities will be in developing countries. The proportion of the population that lives in the largest cities (two million or more) varies considerably by country. Almost 25 percent of Mexico's population lives in Mexico City; about 20 percent of Brazil's population lives in São Paulo and Rio de Janeiro. In contrast, only 9 percent of the United States's population lives in its largest city, New York City.

11. Until about 10,000 years ago, before plants and animals were domesticated, all humans belonged to hunting-and-gathering societies. These small, nomadic, egalitarian societies relied primarily on hunting wild animals, fishing, and gathering wild fruits, nuts, and vegetables. Some hunting-and-gathering societies exist today in remote parts of the world. In these societies, division of labor is based primarily on sex and age: men hunt large

animals and women take care of children, hunt small animals, and gather fruits, nuts, and berries.

Agrarian societies emerged with the development of agriculture, the conscious and planned cultivation of soil to produce food, and the domestication of animals for food. This system generated greater amounts of food, allowing people to remain in one place rather than hunt for food. Men, women, and children often worked as an economic unit. Moreover the increased food supply caused populations to expand in size and triggered principles of organization:

Stable communities were possible, setting into motion a more urban way of life. Political leaders often emerged as one kin group came to dominate others. Markets where goods, labor, and services were traded expanded and became a major mechanism of distribution. Money came into use as a means for determining the value of goods, labor, and services. Inequality and stratification became more pronounced as some hoarded more resources than others. Warfare became a more frequent event as stable communities fought each other for territory, resources, and power. (TURNER 1978, P. 453)

13 EDUCATION

with Emphasis on the United States

College graduation ceremony.
William Strode/Woodfin Camp & Associates

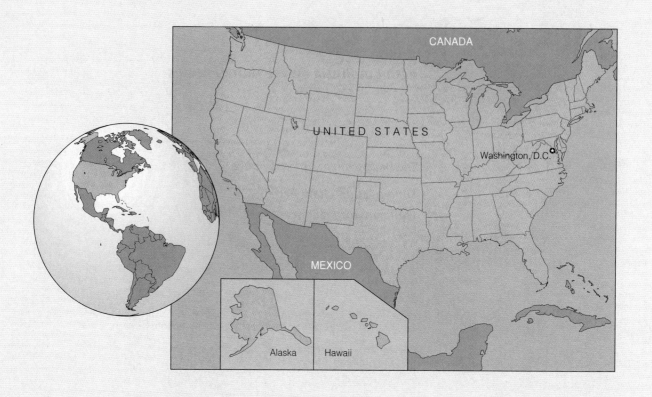

Foreign students enrolled in U.S. colleges for fall 1991   420,000

Percentage of college degrees conferred
in ethnic studies in 1991   0.42%

Percentage of total degrees conferred
in foreign language in 1990   1.11%

Percentage of doctoral degrees conferred
to foreign students in 1991   28.4%

Percentage of college students
enrolled in a foreign language course for fall 1991   8.7%

Number of U.S. colleges offering
study abroad programs in 1992   1,289

To the Editor [of *The New York Times*]:

As a student who attended a high school that offered a European education and now attends an American university, I can see the difference between the knowledge acquired by American and by European students.

American students, regardless of whether they graduate from public or private high schools, seem to have been bombarded with facts and figures that they were forced to memorize. European students, on the other hand, are taught the same subjects, but instead of memorizing them, they are forced to understand them. This may seem a small detail, but as a result of this difference in teaching, European students have a better understanding of the subjects taught. They are more likely to remember the facts than American students because what is understood lasts longer than what is memorized.

European schools also tend to teach useful skills that schools in the United States do not. Critical thinking, analysis of subjects in depth and research techniques are skills that I and other European-educated students learned in our high schools but have to learn again at our American universities.

MARIA A. DE ICAZA (1991)

In 1981 the Reagan administration formed The National Commission on Education to study the state of American education. Two years later the commission reported its conclusions in *A Nation at Risk*. The most famous lines in this report are:

> *The educational foundations of our society are presently being eroded by a rising tide of mediocrity that threatens our very future as a Nation and a people. . . .*
>
> *If an unfriendly foreign power had attempted to impose on America the mediocre educational performance that exists today, we might well have viewed it as an act of war. . . . We have, in effect, been committing an act of unthinking, unilateral educational disarmament.* (THE NATIONAL COMMISSION ON EXCELLENCE IN EDUCATION 1983, P. 5)

The dramatic charges in this report, exemplified by the excerpt above, set off a tidal wave of reforms across the 50 states and the more than 1,500 school districts. These included lengthening the school year, installing no pass/no play rules for sports, fining parents who failed to comply with attendance laws, extending the school day, enforcing competency testing of teachers and students, and increasing the number of required mathematics, English, foreign language, and science courses (Gisi 1985; White 1989).

Yet, in spite of the many reforms implemented, one phrase captures their basic intent: more of the same. The reforms stipulated more tests, more class time, more homework, and more courses, but they failed to address the content and the quality of the educational experience. Critics argue that there was a gross disparity between the urgent tone of *A Nation at Risk* and the superficial nature of the reforms that were enacted (Danner 1986).

Apparently the reforms proved ineffective, because in September 1989 at the Governors Conference President Bush announced new educational goals to be achieved by the year 2000. In April 1994 President Clinton signed the Goals 2000 Act which sets aside $700 million in federal funds for states and school districts that implement programs aimed at meeting guidelines related to these goals which include a 90 percent high school graduation rate and 100 percent literacy rate. Other goals included:

- Every American child must start school prepared to learn, sound in body and sound in mind.

- All students in grades four, eight, and twelve will be tested for progress in critical subjects.

- American students must rank first in the world in achievements in mathematics and science.

- Every adult must be a skilled, literate worker and citizen, able to compete in a global economy.

- Every school must be drug free and offer a disciplined environment conducive to learning (*Time* 1990, p. 54).

The belief that the inadequacies of the schools threaten the country's well-being is not unique to the 1980s and 1990s. Neither is the call for a restructuring of the American educational system. Indeed, the recommendation that the educational system be more inclusive dates back to the mid-1800s when educational leaders debated whether to establish universal public education. Although the problems and recommendations have remained essentially unchanged over the past 200 years, various events throughout this period have placed them in a different context and have given them a seemingly new sense of urgency. Between 1880 and 1920, for example, the public was concerned about whether the schools were producing qualified workers for the growing number of factories and businesses and whether they were instilling a sense of national identity (that is, patriotism) in the ethnically diverse student population. When the United States was involved in major wars—World War I, World War II, Korea, and Vietnam—the public was concerned about whether the schools were turning out recruits physically and mentally capable of defending the American way of life. When the Soviet *Sputnik* satellites were launched in the mid-1950s, Americans were forced to consider the possibility that the public schools were not educating their students as well as the Soviets were in mathematics and the sciences. In the 1960s civil rights events forced Americans to question whether the public schools were offering children from less advantaged ethnic groups and social classes the knowledge and skills to compete economically with children in more advantaged groups. During the

High school photography students in Indianapolis look on as their lockers are searched for drugs. Scenes like this one reflect the fact that schools are "a stage on which a lot of cultural crises get played out."

Kelly Wilkinson/The Indianapolis Star/Sipa Press

late 1970s, throughout the 1980s, and now into the 1990s, as multinational corporations draw from a worldwide labor pool, Americans confront the possibility that their schools are educating an inferior work force, one that will be unable to compete in a global labor market.

The ongoing nature of the so-called education crisis in the United States and the corresponding criticisms aimed at the American system of education suggest that the schools are very visible and highly vulnerable, that "they are the stage on which a lot of cultural crises get played out" (Lightfoot 1988, p. 3). Inequality, poverty, chronic boredom, family breakdown, unemployment, illiteracy, drug abuse, child abuse, and ethnocentrism are crises that transcend the school environment. Yet, we confront them whenever we go into the schools (Lightfoot 1988). Consequently, the schools seem to be, in some way, both a source of problems and a solution for our problems.

In this chapter we give special emphasis to public education, the system in which 89 percent of students are enrolled in the United States, for two reasons. First, many critics, at home and abroad, maintain that the U.S. system of education is not adequate for meeting the challenges associated with global interdependence. Many employees claim that they are unable to find enough workers with a level of reading, writing, mathematical, and critical thinking skills needed to function adequately in the workplace. Such a human capital deficit weakens U.S. competitiveness in the global marketplace (U.S. Department of Education 1993a). Second, the research that compares the performance of American students with that of students in other countries (particularly in Asian and European countries) points to two clear trends. "First, compared with their peers in Asian and European countries, American students stand out for how little they work . . . [and] for how poorly they do" (Barrett 1990, p. 80).

What Is Education?

In the broadest sense, education includes those experiences that stimulate thought and interpretation or that train, discipline, and develop the mental and

physical potentials of the maturing person. An experience that educates may be as commonplace as reading a sweater label and noticing that it was made in Taiwan or as intentional as performing a scientific experiment to learn how genetic makeup can be altered deliberately through the use of viruses. In view of this definition and the wide range of experiences it encompasses, we can say that education begins when people are born and ends when they die.

Sociologists make a distinction, however, between formal and informal education. **Informal education** occurs in a spontaneous, unplanned way. Experiences that educate informally occur naturally: they are not designed by someone to stimulate specific thoughts or interpretations or to impart specific skills. Informal education takes place when a child puts her hand inside a puppet, then works to perfect the timing between the words she speaks for the puppet and the movement of the puppet's mouth. **Formal education** is a purposeful, planned effort intended to impart specific skills and modes of thought. Formal education, then, is a systematic process (for example, military boot camp, on-the-job training, programs to stop smoking, classes to overcome fear of flying) in which someone designs the educating experiences. We tend to think of formal education as consisting of enriching, liberating, or positive experiences, but it can include impoverishing and narrowing occurrences (such as indoctrination or brainwashing) as well. In any case, formal education is considered a success when the people instructed internalize (or take as their own) the skills and modes of thought that those who design the experiences seek to impart. This chapter is concerned with a specific kind of formal education—schooling.

Schooling is a program of formal and systematic instruction that takes place primarily in classrooms but also includes extracurricular activities and out-of-classroom assignments. In its ideal sense: "Education must make the child cover in a few years the enormous distance traveled by mankind in many centuries" (Durkheim 1961, p. 862). More realistically, schooling is the means by which those who design and implement programs of instruction seek to pass on the values, knowledge, and skills they define as important for success in the world. This latter conception implies that what is taught in schools is only a part of the knowledge accumulated and stored throughout human history. This

point, of course, raises questions about who has the power to select, from the vast amounts of material available, what students should study. Moreover, what constitutes an ideal education—the goals that should be achieved, the material that should be covered, the best techniques of instruction—is elusive and debatable. Conceptions vary according to time and place; they differ according to whether schools are viewed primarily as mechanisms by which the needs of a society are met (see Chapter 8 for a comparison of Chinese and American preschools) or the means by which students learn to think independently, thus becoming free from the constraints on thought imposed by family, culture, and nation.

Social Functions of Education

Sociologist Emile Durkheim believed that education functions to serve the needs of society. In particular, schools function to teach children the things they need to adapt to their environment. To ensure this end, the state (or other collectivity) reminds teachers "constantly of the ideas, the sentiments that must be impressed" upon children if they are to adjust to the milieu in which they must live. Otherwise, "the whole nation would be divided and would break down into an incoherent multitude of little fragments in conflict with one another." Educators must achieve a sufficient "community of ideas and sentiments without which there is no society" (Durkheim 1968, pp. 79, 81). Such logic underscores efforts to use the schools as mechanisms for meeting the needs of society, whether they be to strip away identities of ethnicity and social origin in order to implant a common national identity, to transmit values, to train a labor force, to take care of children while their parents work, or to teach young people to drive.

According to another quite different conception, education is a liberating experience that releases students from the blinders imposed by the accident of birth into a particular family, culture, religion, society, and time in history. Schools therefore should be designed to broaden students' horizons so they will become aware of the conditioning influences around them and will learn to think independently of any authority. When schools are de-

Schooling can serve different purposes. Training in the use of computers, for example, may be designed primarily to equip students with the skills they need to adapt to their environment. But it can also be used to broaden students' intellectual horizons and encourage creativity and independent thinking.

James Wilson/Woodfin Camp & Associates

icti n as

ry if
e i divid-
ual. For example,   e market system require an informed public that is capable of independent thought. Most sociological research, however, suggests that schools are more likely to be designed to meet the perceived needs of society rather than to liberate minds. (This point raises a question: Who defines the needs of society?) In spite of this intention, however, a significant percentage of the population in every country seems to be functionally illiterate—that is, they do not possess the level of reading, writing, and calculating skills needed to adapt to the society in which they live. In fact, to many critics of the U.S. educational system, illiteracy in America has reached crisis proportions. In order to evaluate the "literacy crisis," we need to define illiteracy, to distinguish among kinds of literacy, and to ask how the United States compares to other countries when it comes to preparing its students to be educated workers and citizens.

Illiteracy in the United States

In the most general and basic sense, **illiteracy** is the inability to understand and use a symbol system, whether it is based on sounds, letters, numbers, pictographs, or some other type of character. Although the term *illiteracy* is used traditionally in reference to the inability to understand letters and their use in reading and writing, there are as many kinds of illiteracy as there are symbol systems— computer illiteracy, mathematical illiteracy (or innumeracy), scientific illiteracy, cultural illiteracy, and so on.

If we confine our attention merely to languages, of which there are thought to be between 3,500 and 9,000 (including dialects), we can see that the potential number of literacies is overwhelming (Ouane 1990) and that people cannot possibly be literate in every symbol system. If a person speaks, writes, and reads in only one language, by definition he or she is illiterate in perhaps as many as 8,999 languages. Yet, such a profound level of illiteracy rarely presents a problem because usually people need to know and understand only the language of the environment in which they live.

This point suggests that illiteracy is a product of one's environment—that is, people are considered illiterate when they cannot understand or use the symbol system of the surrounding environment in which they wish to function. Examples include not being able to use a computer, to access information, to read a map in order to find a destination, to make change for a customer, to read traffic signs, to follow the instructions to assemble an appliance, to fill out a job application, or to comprehend the language of those around them.

The contextual nature of illiteracy suggests that it is not "some sort of disease . . . like a viral infection that debilitates an otherwise healthy organism. . . . Illiteracy is a social phenomenon, not a natural one" in that it changes form whenever an environment changes to the point at which old literacy skills are no longer sufficient (Csikszentmihalyi 1990, p. 119).

In the United States (and in all countries, for that matter) some degree of illiteracy has always existed, but conceptions of what people needed to know to be considered literate have varied over time. At one time, people were considered literate if they could sign their names and read the Bible. At other times, a person who had completed the fourth grade was considered literate. The National Literacy Act of 1991 defines literacy as "an individual's ability to read, write, and speak English and compute and solve problems at levels of proficiency necessary to function on the job and in society, to achieve one's goals, and to develop one's knowledge and potential" (U.S. Department of Education 1993b, p. 3). Today there are various estimates of the number of functionally illiterate adults in the United States. The U.S. Bureau of the Census (1982) estimates that 13 percent of the adult population (or 26 million adults) are illiterate. The National Alliance of Business estimates that 30 percent of high school students cannot write a letter seeking employment or information and that one 17-year-old in eight cannot read beyond a fifth-grade level (Remlinger and Vance 1989).

As the number of low-skill but relatively high paying manufacturing jobs moved off-shore, starting in the 1970s and continuing into the present (see Chapter 2), ideas of what constituted literacy have become more complex. During this time people who previously had been gainfully employed suddenly came to be defined as functionally illiterate: they lacked the reading, writing, and calculating skills needed to find re-employment in the emerging service and information economy (Limage 1990). One of the striking characteristics of this group of functional illiterates is that they are

adults who have attended school and have had some contact with reading and writing. They may know the alphabet and even be just capable of deciphering a few words. They may be able to write a little. They may recognize figures and be able to do a few sums, but their knowledge in these fields is rudimentary and insufficient for them to cope easily with everyday life. (VELIS 1990, P. 31)

In 1988 the U.S. Congress requested that the Department of Education define literacy in the context of the new economic order and attempt to estimate how many Americans are illiterate. The project involved a representative sample of 26,091 adults (13,600 were interviewed and 1,000 were surveyed in each of 11 states.[1]) In addition, 1,100 federal and state prison inmates were interviewed. The project was completed in 1993. The researchers found that 21 to 23 percent of those contacted "demonstrated skills in the lowest level of prose, document, and qualitative proficiencies (level 1)" (U.S. Department of Education 1993b, p. xiv). Approximately 25 to 28 percent of those contacted performed at the next higher level of literacy proficiency.[2] (See "Definitions and Levels of Literacy.")

Illiteracy and Schools

The fact that almost a quarter of the adult population could function at the lowest level in a society with mandatory school attendance policies leads social critics, most notably American government officials and business leaders, to point to the schools as one source of the problem. Upon close analysis, we can see that the literacy problem cannot be solved by schools alone. Most Americans seem to believe that if "schools just did their jobs more skillfully and resolutely, the literacy problem would be solved" (L. Resnick 1990, p. 169). Not so, says Lauren B. Resnick, director of the Learning Research and Development Center at the University of Pittsburgh. She argues that policies designed to end illiteracy must consider situations in which people use or value written materials.

Resnick identifies six literacy situations (Resnick and Resnick 1989). One of these she terms "the

Definitions and Levels of Literacy

A society's definition of literacy changes whenever an environment changes to the point at which old definitions of literacy are no longer sufficient to meet the level of reading, writing, critical thinking, and quantitative skills that the new environment demands. In the United States, the most recent definition of literacy identifies three kinds of literacy: prose, document, and quantitative. Each kind of illiteracy is further divided into at least six levels.

Prose Literacy	Document Literacy	Quantitative Literacy
The knowledge and skills needed to understand and use information from texts that include editorials, news stories, poems, and fiction—for example, finding a piece of information in a newspaper article, interpreting instructions from a warranty, inferring a theme from a poem, or contrasting views expressed in an editorial.	The knowledge and skills required to locate and use information contained in materials that include job applications, payroll forms, transportation schedules, maps, tables, and graphs—for example, locating a particular intersection on a street map, using a schedule to choose the appropriate bus, or entering information on an application form.	The knowledge and skills required to apply arithmetic operations, either alone or sequentially, using numbers embedded in printed materials—for example, balancing a checkbook, figuring out a tip, completing an order form, or determining the amount of interest from a loan advertisement.
1. Identify country in short article	Sign your name	Total a bank deposit entry
2. Underline meaning of a term given in government brochure on supplemental security income	Locate intersection on a street map	Calculate postage and fees for certified mail
3. Write a brief letter explaining error made on a credit card bill	Identify information from bar graph depicting source of energy and year	Using calculator, calculate difference between regular and sale price from an advertisement
4. State in writing an argument made in lengthy newspaper article	Identify the correct percentage meeting specified conditions from a table of such information	Determine correct change using information in a menu
5. Compare approaches stated in narrative on growing up	Use information in table to complete a graph including labeling axes	Determine shipping and total costs on an order form for items in a catalog
6. Compare two metaphors used in poem	Use table of information to determine pattern in oil exports across years	Using eligibility pamphlet, calculate the yearly amount a couple would receive for basic supplemental security income

SOURCES: U.S. Department of Education, pp. 3–4, 10 (1993b); U.S. Department of Education (1992a).

useful situation," which applies to circumstances in which people use printed materials to perform practical activities such as following instructions to install equipment, consulting bus schedules, filling out job applications or other forms, reading receipts, and using computer software packages. Resnick argues that this "read-do" literacy is acquired primarily outside the classroom. Even before children can read, they observe how adults use texts to carry out tasks. Thus, through observation, they learn general patterns of interacting with texts. Read-do literacy is an important ingredient of workplace literacy. If in fact this literacy is acquired primarily outside the classroom, policies aimed at improving workplace literacy must be designed with this fact in mind.

Despite the complexity of solving the literacy problem, most people are still bothered by the fact that a high school diploma is no guarantee that one has acquired the skills needed to function in today's economy. The high estimates of illiteracy rates (one person in three), even among those who completed high school, lead people to ask how so many students could attend school for at least 12 years without acquiring enough reading, writing, and problem-solving skills to deal effectively with the work-related problems encountered in the new kinds of entry-level jobs. This question is complicated further by findings that American school-children lag behind their Asian and European counterparts in nearly every subject, especially science and mathematics (Lapointe, Mead, and Phillips 1989). This finding holds true even for American students who score in the top 5 to 15 percent on achievement tests compared with their foreign counterparts (Cetron 1988; Thomson 1989). "Generally the 'best students' in the United States do less well on the international surveys when compared with the 'best students' from other countries" (U.S. Department of Education 1992b, p. viii). These findings have prompted many critics to examine the ways in which schools in European and Pacific Rim countries differ from schools in the United States.

Insights from Foreign Education Systems

At the beginning of the chapter, we noted that in comparison with their counterparts in Pacific Rim and European countries, American students—even the most able ones—do little school-related work and perform poorly in academic subjects.

> *For example, only 13 percent of a select group of American 17-year-old students achieved algebra scores equal to [those of] 50 percent of 17-year-old Hungarians . . . 25 percent of Canadian 18-year-old students knew as much chemistry as a very select 1 percent of American*

> *high school seniors who had taken an advanced, second-year chemistry course . . . [and] 30 percent of South Korea's 13-year-old students were able to apply "advanced scientific knowledge" compared to 10 percent of American students of the same age.* (THOMSON 1989, PP. 52–53)

A 1993 Department of Education study on gifted children (those who score in the top 3 to 5 percent of achievement and IQ tests) provides insights about why even the best American students perform poorly in the context of the international arena: most report that they study less than one hour a day and that they are bored in class because much of what is taught is a rehash of what they already know (*Los Angeles Times* 1993). Such results are not meant to imply that Pacific Rim and European schools are operated perfectly or that American schools should emulate their systems. They do suggest, however, that it is important to learn why Americans, as a group, perform so poorly in the international arena.

Amount of Time Spent on Schooling In "The Case for More School Days," Michael Barrett describes how "American children receive hundreds of hours less schooling than many of their European or Asian mates and the resulting harm promises to be cumulative and lasting" (1990, p. 87). This generalization seems to hold no matter how time spent on schooling is measured: by the length of the school year, week, or day; by the amount of time spent doing homework; by the amount of time that parents spend helping their children with homework; by the number of minutes that teachers spend on instruction (as opposed to disciplining or otherwise managing students); by the rate of absenteeism; or by the dropout rate. Barrett maintains that most Americans dismiss suggestions that the amount of time devoted to school-related learning, especially the number of school days, be increased. Instead, most Americans argue that we should learn to use more efficiently the time already allotted. Although Barrett recognizes that increases in time alone cannot improve our ability to compete internationally, he does suggest that Americans need to equalize the time they commit to learning.

Barrett also notes the arrogance of thinking that we can accomplish in 180 days what the Europeans and Asians are accomplishing in 200 to 235 days, especially when that 180-day school year includes field trips, schoolwide assemblies, snow days, and teacher in-service days (in-service days count as official school days but only teachers are required to attend school).

One of the most systematic and well-designed studies comparing the time spent on academic activities in three countries was done by Harold W. Stevenson, Shin-ying Lee, and James W. Stigler (1986). They compared mathematics achievement as well as the classroom and home environments of kindergartners, first-graders, and fifth-graders in three cities: Minneapolis (United States), Sendai (Japan), and Taipei (Taiwan). They found that, across all three grades, the Taiwanese, but especially the Japanese, consistently outperformed their American counterparts. At the fifth-grade level, however, the differences in test scores were the most striking:

> The highest average score of an American fifth-grade classroom was below that of the Japanese fifth-grade classroom with the lowest average score. In addition, only one Chinese [Taiwanese] classroom showed an average score lower than the American classroom with the highest average score. Equally remarkable is the fact that the lowest average score for a fifth-grade American classroom was only slightly higher than the average score for the best first-grade Chinese classroom. (P. 694)

After thousands of hours of classroom observation and after interviews with both mothers and teachers, Stevenson and his colleagues concluded that Americans devote significantly less time to academic activities either in school or at home and that American parents help their children less with homework. American parents, however, are more likely than their Taiwanese and Japanese counterparts to rate the quality of education at the schools their children attend as good or excellent: 91 percent of American parents, 42 percent of Chinese parents, and 39 percent of Japanese parents rate the quality as good or excellent and are satisfied with the qualities of education (see Table 13.1). Steven-

son (1992) maintains that one reason U.S. parents rate quality high is that their school system gives them no clear guidelines about what academic skills children in various grades should possess. Thus they have no baseline by which to judge their child's performance.

Cultural and Economic Incentives Scott Thomson (1989), executive director of the National Association of Secondary School Principals, argues that the poor international showing by Americans reflects a lack of cultural and economic incentives to do well. He offers the cases of South Korea and Germany as examples. With regard to cultural incentives, American students are less likely than South Korean and German students to receive parental assistance with homework and to come from homes where the family value structure is supportive of education. Moreover, in the United States fewer television programs are aimed at educating youth, and the content of the programs and accompanying advertisements aired on American networks encourages students to consume, not to develop the intellect. When school is the backdrop to a television program or commercial, "more often than not, the school principals are portrayed as grumbling misfits, the teachers are blithering incompetents. Students who show the slightest interest in their studies are invariably depicted as wimps. The heroes are those who can best foul up the system" (O'Connor 1990, p. B1).

Economic incentives are another factor in the quality of education. Compared to South Koreans and Germans, Americans devote a smaller percentage of their gross national product to education, they pay their teachers lower salaries, and they make less of a direct connection between academic achievement in high school and the quality of future employment opportunities. Thomson believes that "this tendency to ignore classroom achievement in high school appears to be a peculiarly American phenomenon" (1989, p. 56) and it may help to explain U.S. Secretary of Labor Robert Reich's claim that "America may have the worst school-to-work transition system of any advanced industrial country. Short of a college degree, there is no way someone can signal to an employer that he or she possesses world-class skills" (1993, E1).

TABLE 13.1	Comparison of Chinese (Taiwanese), Japanese, and American Kindergartners, First-Graders, and Fifth-Graders		
Homework	**Americans**	**Taiwanese**	**Japanese**
Minutes per Day Doing Homework*			
first-graders	14 min.	77 min.	37 min.
fifth-graders	46 min.	114 min.	57 min.
Minutes Doing Homework (Saturday)			
first-graders	7 min.	83 min.	37 min.
fifth-graders	11 min.	73 min.	29 min.
Time Parent Spent Helping with Homework			
fifth-graders	14 min.	27 min.	19 min.
Percentage Who Possess a Desk at Home			
fifth-graders	63%	95%	98%
Percentage of Parents Who Purchase Workbooks for Children to Get Extra Practice			
fifth-graders (math)	28%	56%	58%
fifth-graders (science)	1%	51%	29%
Classroom			
Percentage of Time Devoted to Academic Activities			
first-graders	69.8%	85.1%	79.2%
fifth-graders	64.5%	91.5%	87.4%
Hours per Week Devoted to Academic Interests			
first-graders	19.6 hrs.	40.4 hrs.	32.6 hrs.
fifth-graders	19.6 hrs.	40.4 hrs.	32.6 hrs.
Total Hours per Week Spent in School			
first-graders	30.4 hrs.	44.1 hrs.	37.3 hrs.
fifth-graders	30.4 hrs.	44.1 hrs.	37.3 hrs.
Percentage of Time a Child Known to Be in School Was Not in Classroom			
fifth-graders	18.4%	<0.2%	<0.2%
Proportion of Time Teacher Spends Imparting Information (all grades)	21% or 6 hrs.	58% or 26 hrs.	33% or 12 hrs.
Mother's Perceptions			
Evaluation of Child's Achievement in Math on a Scale of 1 to 9[†]	5.9	5.2	5.8
Evaluation of Child's Intellectual Ability[†]	6.3	6.1	5.5
Percentage Who Believe School Is Doing a Good or Excellent Job Educating Children	91%	42%	39%
Percentage Very Satisfied with Child's Academic Performance	40%	<6%	<6%
Evaluation of What Is More Important to Success—Ability or Effort?	ability	effort	effort
Percentage Who Gave Child Assistance in Math	8%	2%	7%

*Mothers' estimates.
[†]With 1 being much below average and 9 being much above average.

SOURCE: Data from "Mathematics Achievement of Chinese, Japanese, and American Children," by Harold W. Stevenson, Shin-ying Lee, and James W. Stigler in *Science.* Copyright © 1986 by the *AAAS.* Reprinted by permission of Harold W. Stevenson and AAAS.

European and Pacific Rim employers, even when hiring clerical and blue-collar workers, express significantly more interest in job applicants' academic achievements. In fact, in many European countries grades and test scores are part of employment résumés (U.S. Department of Education 1992c). In South Korea, "school is considered too important for students to work . . . students simply do not hold jobs while attending school; their time and energies are directed toward learning" (Thomson 1989, p. 57). In Japan, "educational credentials and educated skills are central to employment, to promotion, and to social status in general" (Rohlen 1986, pp. 29–30).

As discouraging as these cross-national findings are for the country as a whole, it is important to acknowledge that students in states such as Iowa, North Dakota, and Minnesota score as high in mathematics, for example, as do students in Japan and Switzerland, countries with the best math scores (Sanchez 1993). Such individual state accomplishments aside, things look bleak when we consider that within the United States the average elementary and secondary student misses approximately 20 days of school per year (Schlack 1992). Moreover, the less advantaged social classes and minority groups have higher dropout rates,[3] higher rates of absenteeism, and lower scores on standardized tests than do those from more advantaged groups (Horn 1987). Consider the following facts related to dropout rates:

- Approximately 4,231,000 American students drop out of high school each year.

- The public high school graduation rate is 71.2 percent. The figure varies by state with a high of 89.5 percent in Minnesota and a low of 54.3 percent in Mississippi.

- The high school dropout rate for white Americans is 17 percent. The rates for African-Americans, Hispanics, and Native Americans are 28 percent, 35 percent, and 45 percent, respectively.[4]

- Young people who live in households with annual incomes of less than $15,000 constitute 60 percent of the high school dropouts.[5]

The fact that American students spend little time on academic activities compared to their foreign counterparts, in conjunction with the high dropout rate especially among the more disadvantaged groups, suggests that something about the American system of education turns many people away from academic pursuits. In the next sections we first examine the historical background of contemporary American education and then consider some of the most general and distinguishing characteristics that may contribute to Americans' ambivalence about education and learning.

The Development of Mass Education in the United States

The United States was the first country in the world to embrace the concept of mass education. In doing so, it broke with the European view that education should be limited to an elite few (for example, the top five percent or those who could afford it). In 1852 Massachusetts legislators passed a law making elementary school mandatory for all children, and within sixty years all of the states had passed compulsory attendance laws. A number of factors other than the legal mandate encouraged parents to comply with attendance laws, however. First, as the

pace of industrialization increased, jobs moved away from the home and out of neighborhoods into factories and office buildings. As mechanization increased, apprenticeship opportunities gradually disappeared. As family farms and businesses disappeared, parents could no longer train their children because familiar skills were becoming obsolete. Thus, with the home and work environments independent of each other, parents were no longer available to oversee their children. Second, a tremendous influx of immigrants to the United

States between 1880 and 1920 created a large labor pool—a surplus—from which factory owners could draw workers, thereby eliminating the need for child laborers. This combination of events created an environment in which there was no place for children to go except to school.

At least two prominent features of early American education have endured to the present: (1) textbooks modeled after catechisms and (2) single-language instruction.

Textbooks

The most vocal early educational reformers such as Benjamin Rush, Thomas Jefferson, and Noah Webster believed that schools were an important mechanism by which a diverse population could acquire a common culture. They believed that the new "perfectly homogeneous" American was one who studied at home (not abroad) and who used American textbooks. To use Old World textbooks "would be to stamp the wrinkles of decrepit age upon the bloom of youth" (Webster 1966, p. 32).

The first textbooks in the United States were modeled after **catechisms,** short books covering religious principles written in question-and-answer format. Each question had one answer only, and in repeating the answer the question's wording was adhered to strictly (Potter and Sheard 1918). This format discouraged readers from behaving as active learners "who could frame questions, interpret materials, and reflect on the significance of what was presented. . . . No premium was placed on generating and inventing ideas or arguing about the truth or value of what others had written (D. Resnick 1990, p. 18). The reader's job was to memorize the "right" answers for the questions. With this as the model, not surprisingly, textbooks tend to be written in such a way that the primary reason to read them is to find the "right" answers to the accompanying questions.

The influence of the catechisms on learning today is evident whenever students are assigned to read a chapter and answer the list of questions at the end. Many students discover that they do not need to read the material in order to answer the questions; they can simply skim the text until they find key words that correspond to those in the

question and then they copy the surrounding sentences.

In the past few years educators have questioned the value of textbooks as a learning tool. Although we know that some schools have abandoned textbooks, especially at the elementary level, we do not know the scope of this trend. It seems that educators in the field of elementary reading are leading the way as they increasingly abandon traditional textbooks modeled after catechisms in favor of children's literature books and daily writing projects (Richardson 1994). Historical novels and edited volumes that include essays, various perspectives, and so-called "real" or authentic literature are among the alternatives to textbooks.

Single-Language Instruction

The "peopling of America is one of the great dramas in all of human history" (Sowell 1981, p. 3). It involved the conquest of the native peoples, the annexation of Mexican territory along with many of its inhabitants (who lived in what is now New Mexico, Utah, Nevada, Arizona, California, and parts of Colorado and Texas), and an influx of millions of people from practically every country in the world. School reformers, primarily people of Protestant and British background, saw public education as the vehicle for "Americanizing" a culturally and linguistically diverse population, for instilling a sense of national unity and purpose, and for training a competent workforce. As Benjamin Rush argued, "Let our pupil . . . be taught to love his family but let him be taught, at the same time, that he must forsake and even forget them, when the welfare of his country requires it" (1966, p. 34). In order to meet these nation-building objectives, people had to learn to speak a common language. Consequently, students were taught in English, the language of the established elite. Early reformers believed that the welfare of the country depended on a common culture that required people to forget their families' language.

Although the United States is hardly unique in pressuring its people to abandon their native tongues and learn to speak a common language, it is probably the only country in the world that places so little emphasis on learning at least one other language. Education critic Daniel Resnick believes that the absence of serious foreign language instruction contributes to the parochial nature of American schooling. The almost exclusive attention to a single language has deprived students of the opportunity to appreciate the connection between language and culture and to see that language is a thinking tool that enables them to think about the world. Resnick states that the focus on a single language "has cut students off from the pluralism of world culture and denied them a sense of powerfulness in approaching societies very different from their own" (1990, p. 25). It also denies those students from non-English-speaking heritages the means of reflecting on and fully appreciating their ancestors' lives.

In the United States, English became the language people needed to learn to speak if they were to enter the mainstream of society and have a chance at upward mobility. In this respect English was positioned against the language and even regional accents of parents and grandparents. English was the language "through which one's ethnic self was converted into one's acquired American personality . . . [and] as an instrument through which one could hide and mask ethnic (and regional) origins" (Botstein 1990, p. 63). Noah Webster, an early influential education reformer, wrote in the preface of his spelling book that the United States must promote a "uniformity and purity of language" and "[demolish] those odious distinctions of provincial dialects which are subject to reciprocal ridicule in different states" (Webster 1966, p. 33). The national memory of learning English, of breaking with the past, remains with subsequent generations in their ambivalent attitudes not just toward language acquisition but toward the meaning of learning (Botstein 1990).

Fundamental Characteristics of Contemporary American Education

There are a number of characteristics that distinguish American education from other systems of education. They include the availability of college, the lack of a uniform curriculum, funding that varies by state and community, the belief that schools can be the vehicle for solving a variety of social problems, and ambiguity of purpose and value. This section discusses these characteristics.

The Availability of College

One of the most distinctive features about the United States is that in theory anyone can attend a college if he or she has graduated from high school or has received a GED. As a result the United States has the world's highest post-secondary enrollment ratio. Sixty-three percent of 1992 high school graduates enrolled in college in fall of 1992 (*The World Almanac and Book of Facts 1994* 1993, p. 195). In fact, 20 percent of American four-year colleges and universities accept students regardless of what courses they took, what grades they earned in high school, or what scores they received on ACT or SAT tests. Seventy-four percent of colleges and universities offer remedial courses in reading, writing, and mathematics for those students who lack skills necessary to do college-level work. Thirty percent of all freshman students who entered college in 1989 took one or more remedial courses (U.S. Bureau of the Census 1993).

In most other countries, education beyond high school is available only to a small minority. In England and Wales, 16-year-olds who make low scores on the General Certificate of Secondary Education Test cannot take the advanced courses required for college. Similarly, scores on the French baccalaureate examination determine who can attend college (Chira 1991). From this perspective, the educational opportunities in the United States are admirable. A college education is open to everyone (who can find money to pay the tuition) regardless of previous educational failures; it is not reserved for a privileged few. This policy reflects the American belief in equal opportunity to compete at whatever point in life one wishes to enter the competition: "We may not all hit home runs, the saying goes, but everyone should have a chance at bat" (Gardner 1984, p. 28).

At the same time, the unrestricted right to a college education seems to be connected with a decline in the value of a high school education and the effort put into achieving that level of education. As high school enrollments increased over time to include the middle class and eventually the poor and minority groups, the perceived value of the high school diploma declined and the perceived need for a college degree increased. As college enrollments increased and as a college degree came to be defined as important for success, the high school was no longer the last stop before young people entered the mainstream labor force. Consequently, high schools were subject to less pressure to make sure that their graduates had acquired the skills necessary for literacy in the workplace. As a result, the high school curriculum became less important and less rigorous (Cohen and Neufeld 1981).

Defenders of the American public schools argue that a less rigorous curriculum is a necessary consequence of trying to educate *all* citizens: if everyone is to pass through, the standards must be reduced. Even if we accept this logic, the consequences of such a policy are clear: the high school diploma loses its value, and the goal of achieving equality through compulsory and free education is undermined. Mass education accomplished through **social promotion** (passing students from one grade to another on the basis of age rather than academic competency), through awarding high school diplomas to people with fifth-grade reading abilities, and through issuing certificates of achievement to those who fail minimum competency tests, is not the same as educating everyone. True equality in education is achieved only if everyone has an opportunity to earn a degree that is valued.

Differences in Curriculum

In the United States there is no uniform curriculum. Each of the 50 states sets broad curriculum requirements for kindergarten through high school; each school interprets and implements these requirements. Consequently, even with state guidelines, the textbooks, the assignments, the instructional methods, the staff qualifications, and the material covered vary across the schools within each state.

In addition to curriculum differences across states, students in the same school are usually grouped or "tracked" on the basis of tests or past performance. For example, students enrolled in standard diploma (versus college preparatory, honors, or advanced) tracks often take fewer mathematics courses and different kinds of mathematics (general math instead of algebra) to meet a state's mathematics requirement. And they take English composition rather than creative writing to meet the state English requirement (*The Book of the States* 1992).

Although most countries also track students at some point in their careers, all students are exposed to a core curriculum, even if not everyone can assimilate the material. The Japanese, for example, believe that a standardized curriculum is the only way to ensure that everyone has an equal chance at the rewards that education brings (Lynn 1988). "They put nearly their entire population through twelve tough years of basic training" (Rohlen 1986, p. 38). Although the Japanese have three tracks at the secondary level—academic, specialized vocational, and comprehensive—all Japanese students take classes in foreign language, social studies, mathematics, science, health and physical education, and fine arts. The Japanese schools do not have gifted or self-paced learning programs (Rohlen 1986). They do not teach some students general mathematics and others algebra; everyone is taught algebra. In elementary school, all Japanese children learn to read music and to play a wind instrument and a keyboard instrument (Rohlen 1986). Likewise, in West Germany all students take a foreign language, German, physics, chemistry, biology, mathematics, and physical education.

Another source of differences in curriculum in the United States is that some minority group members are more likely to take or be assigned to less demanding classes. For example, for 46 percent of white 17-year-olds algebra II is the highest level mathematics course they have taken. By contrast, algebra II is the highest level mathematics class taken by 41 percent of African-American and 32 percent of Hispanic 17-year-olds (National Science Board 1991).

E. D. Hirsch, Jr., author of a best-selling and controversial book, *Cultural Literacy,* maintains that the nonuniform and diverse nature of American curriculum has created *one of the most unjust and inegalitarian school systems in the developed world.*

It happens that the most egalitarian elementary-school systems are also the best. . . .

The countries that achieve these results tend to teach a standardized curriculum in early grades. Hungarian, Japanese, and Swedish children have a systematic grounding in shared knowledge. Until third graders learn what third graders are supposed to know, they do not pass on to fourth grade. No pupil is allowed to escape the knowledge net. (HIRSCH 1989, P. 32)

Surprisingly, there is little evidence that tracking students into remedial or basic courses contributes to intellectual growth, corrects academic deficiencies, prepares students for success in higher tracks, or increases interest in learning. Instead, the special curricula exaggerate and widen differences among students and perpetuate beliefs that intellectual ability varies according to social class and ethnic group (Oakes 1986a, 1986b).

Differences in Funding

A 1993 report on education in 24 countries sponsored by the Organization for Economic Cooperation and Development (OECD) based in Paris concluded that no country has greater educational disparities between rich and poor than the United States (Sanchez 1993). In the United States schools differ not only with regard to curriculum requirements but also with regard to funding. Elementary and secondary schools receive approximately 6 percent of their funding from the federal government, 48 percent from the state government, and the rest from the local sources, primarily property taxes (U.S. Bureau of the Census 1993). Heavy reliance on state revenue is problematic because the less wealthy states generate less tax revenue than do the wealthier states. Of the 24 countries studied, the United States is third in spending per student at the secondary level and first at the primary level (Celis 1993a). On the other hand, the size of school funding disparities between the poorest and wealthiest states is considerably greater in the United States than in other industrialized countries. For example, as in the United States the 12 Cana-

dian provinces spend different amounts of money per student; expenditures range from a high of $6,500 per year per student to a low of $3,000. Thus a difference of about $3,500 separates the richest province from the poorest province. In the United States, state-level expenditures range from a high of $8,645 to a low of $2,960; differences of $5,685 separate the richest and poorest states (*The World Almanac and Book of Facts 1994* 1993).

Likewise an even heavier reliance on local revenue is also problematic because it causes funding disparities among schools within states. In this regard, the courts in at least 28 states are in the process of evaluating claims that methods of financing have helped create unequal school systems within the state or have ruled that the methods of school financing are unconstitutional (Celis 1992, 1993b). A dramatic case in point is Kentucky. In response to a lawsuit filed by 66, mostly rural, school districts in 1986 who challenged the state's system of funding education, the Kentucky Supreme Court declared on June 8, 1989, that it

found the entire state system of public education deficient and unconstitutional. The court declared that every aspect of the public school system should be reconsidered and a new system created no later than April 15, 1990.

The court concluded that a school system in which a significant number of children receive an inadequate education or ultimately fail is inherently inequitable and unconstitutional. (FOSTER 1991, P. 34)

To remedy this inequity, the court made the academic success of all students a constitutional obligation and required the state legislature to devise a system to ensure that every student was "learning at the highest level of which he or she is capable" (Foster 1991, p. 36). In 1990 the legislature passed the Kentucky Education and Reform Act (KERA), which set into motion the restructuring of education's rules, roles, and relationships. The sociological significance of KERA is that the state recognized that inequality could only be corrected if there was a complete rethinking and overhaul of the way education is delivered to students.

Despite gross inequalities between the poorest and richest schools, in general the amount of money spent per pupil is not the only factor that contributes to a school's overall academic performance. The logic of this conclusion can be supported by examining the profiles of two school districts in the city of Detroit (see "Profile of Two School Districts"). Each school district spends about $3,400 per pupil; yet, there is a substantial difference between the two districts with regard to dropout rate, average daily attendance, percentage of students who pass state mathematics and reading tests, and number of students suspended and expelled. The point is that although money is important in that the more money a school has, the more resources it can provide, money alone cannot solve dropout, attendance, behavior, and performance problems.

Education-Based Programs to Solve Social Problems

The United States uses education-based programs to address a variety of social problems including parents' absence from the home, racial inequality, drug and alcohol addictions, malnutrition, teenage pregnancy, sexually transmitted diseases, and illiteracy. Although all countries have education-based programs that address social problems, the United States is unique in that education is viewed as the primary solution to many of its problems. In the United States

> *The process became familiar: discover a social problem, give it a name, and teach a course designed to remedy it. Alcoholism? Teach about temperance in every school. Venereal disease? Develop courses in social hygiene. Youth unemployment? Improve vocational training and guidance. Carnage on the highways? Give driver education classes to youth. Too many rejects in the World War I draft? Set up programs in health and physical education. . . . In practice, turning real problems—death behind the wheel, the syphilitic body, the frustrated job seeker— into classroom issues [gives] . . . concerned citizens the reassuring feeling that something [is] being done—however symbolically—about real problems.* (TYACK AND HANSOT 1981, P. 13)

The importance of these programs notwithstanding, the schools alone cannot solve such com-

plex problems. Other programs must be implemented in concert with education-based programs. For example, an estimated 350,000 infants each year are born with drug- and alcohol-related problems. A disproportionate number of these babies are born to poor women who have had virtually no access to prenatal care (Blakeslee 1989; Dorfman 1989). When these children reach school age, they will present special problems to the school system. In *The Broken Cord*, Michael Dorris summarizes the concerns of teachers who work with children affected by alcohol before birth, including "difficulty staying on task, distracting other children, poor use of language, inability to structure their work time, and a constant need for monitoring and attention" (1989, p. 241).[6]

The root causes of this totally preventable problem are connected, in part, to a health-care system that uses heroic measures[7] to save low–birth-weight babies, many of whom are born to drug- and alcohol-addicted mothers, but that does not provide adequate prenatal care. Other countries, such as France, have more adequate policies directed at preventing low birth weights and at taking care of children in general (Hechinger 1990; see "Why France Outstrips the United States in Nurturing Its Children"). In the long run these policies result in less strain on the schools.

Ambiguity of Purpose and Value

As we noted earlier, in comparison to the people of other countries, Americans tend to be ambivalent about the purpose and value of an education. In general, the American public supports without question mass education and the right to a college education despite past academic history. Yet, for many Americans elementary school, high school, and college are merely something to be endured; students count the days until they are "out." Ernest Boyer interviewed hundreds of students from public schools around the United States and found no one who could articulate why he or she was in school:

> *The most frequent response was, "I have to be here." They know it's the law. Or, "If I finish this, I have a better chance at a job." The "this" remains a blank. Or, "I need this in order to*

Profile of Two School Districts

Statistics for two school districts in the city of Detroit show that each district spends about the same amount of money per year educating each student.

Still, there is a substantial difference in the dropout rate, daily attendance, and many other variables associated with effective academic environments.

Wayne-Westland, Detroit

Statistical Profile	Data
Grade organization: K–12*	
Percentage of schools built before 1955	20%
Total enrollment	17,682
Average Daily Attendance (ADA)	95%
Current expense per pupil in ADA	$3,350
Dropout rate	15%
Percentage of eligible students who took the ACT	80%
Average combined score on ACT	20.5
Percentage of graduates who enrolled in 2- or 4-year colleges	32%
Percentage of students who passed the state reading test	85%
Percentage of students who passed the state math test	80%
Teacher–student ratio in elementary grades	1:28
Teacher–student ratio in secondary grades	1:29
Counselor–student ratio in secondary grades	1:300
Number of students per music specialist in elementary grades	800
Number of students per art specialist in elementary grades	800
Number of volumes in senior high school libraries (average)	13,345
Beginning teacher's salary	$18,967
Maximum teacher's salary	$38,073
Average years of experience among teachers	15
Number of students suspended and expelled	NK
Subjects in which Advanced Placement courses are offered	None

Highland Park, Detroit

Statistical Profile	Data
Grade organization: K–12*	
Percentage of schools built before 1955	56%
Total enrollment	6,854
Average Daily Attendance (ADA)	88%
Current expense per pupil in ADA	$3,402
Dropout rate	40%
Percentage of eligible students who took the SAT	35%
Average combined score on SAT	695
Percentage of graduates who enrolled in 2- or 4-year colleges	66%
Percentage of students who passed the state reading test	66%
Percentage of students who passed the state math test	46%
Teacher–student ratio in elementary grades	1:28
Teacher–student ratio in secondary grades	1:24
Counselor–student ratio in secondary grades	1:320
Number of students per music specialist in elementary grades	None
Number of students per art specialist in elementary grades	None
Number of volumes in senior high school libraries (average)	19,500
Beginning teacher's salary	$15,656[†]
Maximum teacher's salary	$34,666[†]
Average years of experience among teachers	12
Number of students suspended and expelled	1,638
Subjects in which Advanced Placement courses are offered	None

*The district misinterpreted this statistic.
[†]1986–87 statistics.

SOURCE: From *Public Schools USA: A Comparative Guide to School Districts*, by Charles Harrison, pp. 141 and 148. Copyright © 1988 by Charles Harrison. Reprinted by permission of Peterson's Guides, Inc., Princeton, NJ.

Why France Outstrips the United States in Nurturing Its Children

In a contest between child care in France and the United States, American children are the losers.

"Why don't you change it? It is incredible for a large rich country to accept the terrible infant-mortality rate. It's crazy."

That was how a woman who helps oversee child-care and pre-school education in France reacted to the way America treats its infants and children. The official, Solange Passaris, special counselor on early childhood and motherhood to the Ministry of Health and Social Protection, said in an interview that because Europe was no longer in economic crisis it could now deal with children's welfare as part of a country's economic well-being.

She said she did not understand how the United States could accept ranking 19th in the world in the prevention of infant deaths while France ranked fourth.

National commitment may be at the heart of the matter. Even discounting the French passion for dramatic style, there is a political message in President François Mitterrand's statement that "France will blossom in its children." Or, as the president of the French-American Foundation, Edward H. Tuck, said more prosaically in a recent report by the foundation, "A Welcome for Every Child," France's success is at-

tributable to "national policies focused on children, highly trained staffs, clearly defined responsibilities among agencies and a committed leadership at all levels of government."

Child care in France enjoys strong support from business and industry. "We try to create a partnership between business, the unions and government," Mrs. Passaris said.

Maternal leaves with pay begin six weeks before birth and continue for 10 weeks after delivery. Corporate leaders do not appear to be concerned over any possible negative effect on productivity and profits.

By contrast, American business interests lobbied so successfully against legislation that would have provided 10 weeks of *unpaid* maternity leaves that President Bush vetoed it.

To strengthen the family, French parents may take off two years without pay after a child's birth, knowing that their jobs remain protected.

On major issues, this is truly a tale of two countries.

Pre- and postnatal health care, lacking for so many poor women and their babies in the United States, is firmly anchored in French national policy. So is immunization, which in the United States has become increasingly erratic, leading

to the revival of children's diseases that had been considered virtually extinct.

French educators, health experts and politicians appreciate the close link between preventive health care and children's future success in school and life.

In the United States, millions of children are exposed to unlicensed day care. In high-quality centers, the cost is high. That creates great disparities in the lives of rich and poor children.

In France, the cost of care for children up to the age of 3 ranges from $195 a year for the poorest families to a maximum of $4,700 for the wealthiest.

Preschool for 3-to-5-year-olds is free, and 98 percent of the children attend it, more than three times the American percentage. Safe and nurturing places for young children of working parents are available from early morning to late afternoon.

In the United States, low pay and low status severely hamper the staffing of child-care centers. According to a recent report by the General Accounting Office, 40 percent of the workers in child-care centers in the United States leave each year.

French preschool teachers hold the equivalent of a master's degree in early childhood and elementary

---

go to college." Or, "This is where I meet my friends." Not once in all our conversations did students mention what they were learning or why they should learn it. (1986, P. 43)

Most Americans tend to equate education with increased job opportunities even though, since the

late 1960s and early 1970s, the country has produced college graduates faster than the economy could absorb them, at least into the kinds of jobs the college-educated expect (Guzzardi 1976). In 1990 (the last year for which figures are available) approximately 122.6 million workers were in the labor force. Of these workers, 99.3 million held

Public kindergarten in rural France.
Stephanie Maze/Woodfin Camp & Associates

education. The directors of child-care centers are pediatric nurses, with additional training in public health and child development. Staff members have the equivalent of two years of college plus a two-year course in child development and early childhood education.

As an incentive, France offers students of preschool education free college tuition plus a stipend in return for a pledge to work in the field for at least five years after graduation.

Home care, which allows families to take in three children in addition to their own, is also licensed. In addition to daily compensation, family day-care providers are covered by Social Security, disability and unemployment insurance.

The home-care system is augmented with small-group centers to allow the children to mingle with others once a week.

A look at child care in industrial societies suggests that under regulated capitalism, as in France, or in social democracies, as in Scandinavia, children's welfare is protected because it is viewed as crucial to the children's and the nations' futures.

By contrast, leaving child care exposed to the uncertainties of a largely unregulated free market, as in the United States and Britain, has created conditions that Mrs. Passaris calls crazy. The practice leaves many children with inadequate care and permanently damaged, a costly liability to society.

SOURCE: "About Education: Why France Outstrips the United States in Nurturing Its Children," by Fred M. Hechinger. Copyright © 1990 by The New York Times Company. Reprinted by permission.

jobs that did not require a college degree (see Figures 13.1 and 13.2). Slightly less than 20 percent of American workers were employed in jobs that required a college degree (Shelley 1992). Because approximately 29 million workers have had four or more years of college and only 23.2 million jobs require a college degree, about one college graduate in five is underemployed. This gap between the number of college-educated workers and the number of jobs requiring a college education is reflected in a survey of college degree recipients (1989–90): 44 percent reported that they did not believe a degree was required for the job they had obtained in 1991. This figure is up from 37 percent in 1985

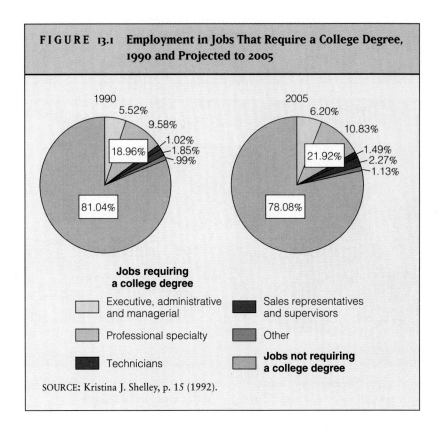

FIGURE 13.1   Employment in Jobs That Require a College Degree, 1990 and Projected to 2005

1990

5.52%
9.58%
1.02%
1.85%
.99%
18.96%
81.04%

2005

6.20%
10.83%
1.49%
2.27%
1.13%
21.92%
78.08%

Jobs requiring a college degree

☐ Executive, administrative and managerial

☐ Professional specialty

☐ Technicians

■ Sales representatives and supervisors

▨ Other

Jobs not requiring a college degree

SOURCE: Kristina J. Shelley, p. 15 (1992).

(U.S. Department of Education 1993a; see Figure 13.3).

These findings do not mean that level of education is unrelated to occupation or income. Rather, they indicate that a large proportion of college graduates are *under*employed, if only because there are not enough high-skill jobs available to absorb the increasing number of graduates. In view of this trend, "job" seems a narrow criterion by which to evaluate an education. Yet, high school and college students commonly evaluate their courses, especially general requirements, as useless because "I will never use it in the real world"—in particular, on the job.

The American tendency to associate education almost exclusively with job advancement means that other benefits of education are underemphasized, two of which include personal empowerment and civil engagement:

Personal empowerment requires that people be able to think analytically and examine informa-

tion critically; that they be able to think creatively—[to] go beyond the analysis and challenge assumptions, leap out of the present and imagine beyond where they are; and that they be able to act with a clear sense of integrity. Civic engagement requires that people learn how to use these skills while taking full part in the life of the larger community. (BOYER 1986, P. 43)

Open college enrollments, diverse and special curricula, unequal funding, the problem-solving burden, and a national ambivalence toward education explain in part the dropout rate, the high number of functional illiterates, and the United States's poor academic showing relative to its European and Pacific Rim counterparts. In the next section we look more closely at the role of the classroom environment.

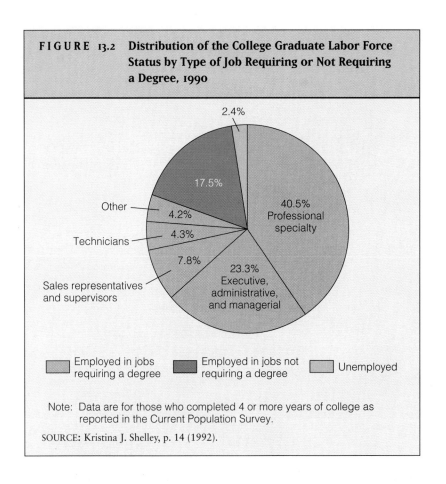

FIGURE 13.2 **Distribution of the College Graduate Labor Force Status by Type of Job Requiring or Not Requiring a Degree, 1990**

2.4%

17.5%

40.5%
Professional
specialty

Other — 4.2%

Technicians — 4.3%

7.8%

23.3%
Executive,
administrative,
and managerial

Sales representatives
and supervisors

Employed in jobs requiring a degree   Employed in jobs not requiring a degree   Unemployed

Note: Data are for those who completed 4 or more years of college as reported in the Current Population Survey.

SOURCE: Kristina J. Shelley, p. 14 (1992).

A Close-Up View:
The Classroom Environment

The classroom is where schooling takes place. In this section we examine what goes on in the classroom: the curriculum to which many students are exposed, the practice of tracking, how students are tested, and the problems that teachers face. The section focuses on practices that lead to boredom and failure, and that undermine the time and energy devoted to academic pursuits. Certainly there are schools in the United States with stimulating classroom environments. The problem is that there are not enough of them.

The Curriculum

Teachers everywhere in the United States teach two curricula simultaneously—a formal curriculum and a hidden curriculum. The various academic subjects—mathematics, science, English, reading, physical education, and so on—make up the **formal curriculum.** Students do not learn in a vacuum, however. As teachers instruct students and as students complete their assignments, other activities are going on around them. Social anthropologist Jules Henry (1965) maintains that these other activities are important and that they are the **hidden curriculum.** The hidden curriculum, then, is all the things that students learn along with the subject matter. The teaching method, the types of assignments and tests, the tone of the teacher's voice, the attitudes of classmates, the number of students absent, the frequency of the teacher's absences, and the number of interruptions during a lesson are

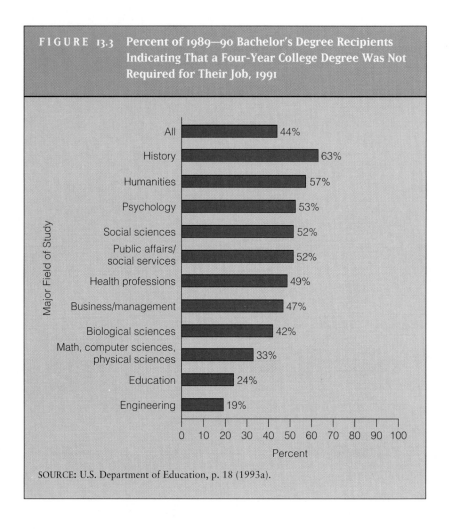

FIGURE 13.3   Percent of 1989—90 Bachelor's Degree Recipients Indicating That a Four-Year College Degree Was Not Required for Their Job, 1991

All — 44%
History — 63%
Humanities — 57%
Psychology — 53%
Social sciences — 52%
Public affairs/social services — 52%
Health professions — 49%
Business/management — 47%
Biological sciences — 42%
Math, computer sciences, physical sciences — 33%
Education — 24%
Engineering — 19%

Major Field of Study

Percent

SOURCE: U.S. Department of Education, p. 18 (1993a).

examples of things going on as students learn the formal curriculum. These so-called extraneous events function to convey messages to students not only about the value of the subject but about the values of society, the place of learning in their lives, and their role in society.

Societal Messages Conveyed Through Hidden Curriculum: The Case of Spelling Baseball

Henry uses typical classroom scenes (acquired from thousands of hours of participant observation) such as a session of "spelling baseball" to demonstrate the seemingly ordinary process by which a hidden curriculum is transmitted and to show how

students are exposed simultaneously to the two curricula. Although Henry observed this scene in 1963, his observations hold more than 30 years later:

> The children form a line along the back of the room. They are to play "spelling baseball," and they have lined up to be chosen for the two teams. There is much noise, but the teacher quiets it. She has selected a boy and a girl and sent them to the front of the room as team captains to choose their teams. As the boy and girl pick the children to form their teams, each child chosen takes a seat in orderly succession around the room. Apparently they know the game well. Now Tom, who has not yet been chosen, tries to call attention to himself in order to be cho-

Jules Henry maintains that when American children are sent to the blackboard, especially to do math problems, they learn to "fear failure" and to "envy success." If they get the problem wrong, a classmate will be called upon to correct the mistake. Hence success is achieved at the expense of another's failure.

John Ficara/Woodfin Camp & Associates

sen. *Dick shifts his position to be more in the direct line of vision of the choosers, so that he may not be overlooked. He seems quite anxious. Jane, Tom, Dick, and one girl whose name the observer does not know, are the last to be chosen. The teacher even has to remind the choosers that Dick and Jane have not been chosen.*

The teacher now gives out words for the children to spell, and they write them on the board. Each word is a pitched ball, and each correctly spelled word is a base hit. The children move around the room from base to base as their teammates spell the words correctly.

The outs seem to increase in frequency as each side gets near the children chosen last. The children have great difficulty spelling "August." As they make mistakes, those in the seats say, "No!" The teacher says, "Man on third." As a child at the board stops and thinks, the teacher says, "There's a time limit; you can't take too long, honey." At last, after many children fail on "August," one child gets it right and returns, grinning with pleasure, to her seat. . . . The motivation level in this game seems terrific. All the children seem to watch the board, to know what's right and wrong, and seem quite keyed up. There is no lagging in moving from base to base. The child who is now writing "Thursday" stops to think after the first letter, and the children snicker. He stops after another letter. More

snickers. He gets the word wrong. There are frequent signs of joy from the children when their side is right. (HENRY 1963, PP. 297–98)

According to Henry, learning to spell is not the most important lesson that students learn from this exercise. They are also learning important cultural values from the way in which spelling is being taught: they are learning to fear failure and to envy success. In exercises like spelling baseball, "failure is paraded before the class minute upon minute" (p. 300) and success is achieved after others fail. And, "since all but the brightest children have the constant experience that others succeed at their expense they cannot but develop an inherent tendency to hate—to hate the success of others" (p. 296).

In an exercise such as spelling baseball, students also learn to be absurd. *To be absurd*, as Henry defines it, means to make connections between unrelated things or events and not to care whether the connections are appropriate or inappropriate. From Henry's point of view, spelling baseball teaches students to be absurd because there is no logical connection between learning to spell and baseball. "If we reflect that one could not settle a baseball game by converting it into a spelling lesson, we see that baseball is bizarrely *irrelevant* to spelling" (p. 300). Yet, most students participate in classroom exercises like spelling baseball without questioning their purpose. Although some children may ask, "Why are we doing this? What is the point?" and may be told, "So you can learn to spell" or "To make spelling fun," few children challenge further the purpose of this activity. Students go along with the teacher's request and play the game as if spelling is related to baseball because, according to Henry, they are terrified of failure and because they want so badly to succeed.

Henry argues further that classroom activities such as spelling baseball prepare students to fit into a competitive and consumption-oriented culture. Because the American economy depends on consumption, the country benefits if its citizens purchase nonessential goods and services. The assignments that children do in school don't prepare them to question false or ambiguous statements made by advertisers; schools do not properly prepare demanding individuals to "insist that the world stand up and prove that it is real" (p. 49). Henry argues that this sort of training—this hidden curriculum—makes possible an enormous amount of selling that otherwise could not take place:

> [I]n order for our economy to continue in its present form people must learn to be fuzzy-minded and impulsive, for if they were clear-headed and deliberate, they would rarely put their hands in their pockets. . . . If we were all logicians the economy [as we know it] could not survive, and herein lies a terrifying paradox, for in order to exist economically as we are we must . . . remain stupid. (P. 48)

Reading Assignments The way in which many children are taught to read provides another example of one way the hidden curriculum can function to convey more than the subject matter. Teachers, parents, and other adults tell children that reading is useful and important if they want to participate in society and to succeed in life. Yet, children typically are assigned dull stories such as this one:

> *Raccoon and Groundhog wanted to play a trick on Rabbit. They dashed into an old building to hide.*
>
> *"He'll never find us here," said Raccoon.*
>
> *"What is that rope doing here?" Groundhog asked. Raccoon started to climb up the rope. Suddenly a bell rang. The bell rang and rang, until Raccoon jumped down. Rabbit poked his*

head through the door. "Did you ring for me?" he asked. (EARLY 1987, P. 97)

After reading this story, students are expected to answer painfully detailed questions such as the following:

"Did you ring for me?" he _____.

☐ cried   ☐ asked   ☐ shouted

"He will never find us here," said _____.

☐ Rabbit   ☐ Groundhog   ☐ Raccoon

Students are asked to accept the idea that books are an important source of information and knowledge, even though the assignments do not support this idea. Nobody in their right mind would want to learn to read just so they could answer questions like these (Bettelheim and Zeland 1982). In this case the hidden curriculum—the content of stories—conveys the message that reading is not a meaningful experience and that it adds little to life. In the long run a steady diet of these kinds of reading assignments and exercises teaches children to hate to read. It is no wonder that many students (even at the college level) report the following symptoms:

- Inability to feel pleasure while reading.

- Inability to read at a normal tempo for 20 minutes or more.

- Reading the same sentence over and over again, and it doesn't make sense.

- Inability to read without constantly wondering if you have attention-deficit syndrome. (Adapted from Clements 1992, p. A11)

The implications of dull reading assignments are far-reaching because "The ability to read is of such singular importance to a child's life in school that his [or her] experience in learning it more often than not seals the fate, once and for all, of his [or her] academic career" (Bettelheim and Zeland 1981, p. 5).

When we compare the content of school-related reading that students do in the United States with that in other countries, as Bruno Bettelheim and Karen Zeland did in *On Learning to Read: The Child's Fascination with Meaning*, we understand more fully why students learn to hate to read (see

Two First-Grade Stories

The American story, "Around the City," represents the kinds of stories the typical American first-grader reads. "Mami, Please" represents the kinds of stories the typical Austrian first-grader reads. The American story leaves the reader with no clue as to why boys and girls run up and down the street. The Austrian story, on the other hand, gives young readers insights about why their mothers are not always able to give them undivided attention.

Around the City

All around the city,
All around the town,
Boys and girls run up the street,
Boys and girls run down.
Boys come out into the sun.
Boys come out to play and run.
Girls come out to run and play,
Around the city, all the day.

All around the city,
All around the town,
Boys and girls run up the street,
Boys and girls run down.

Mami, Please

"Mami, please, a piece of bread!"
  says the child.

"Yes," says the mother and cuts a
  piece of bread for the child.
"Mami, please, read me a story!"
  says the child.
"Later," says the mother.
"Why later?" asks the child.

"Listen!" says the mother. "Don't
  you hear anything?"
At that the child is very quiet and
  listens.
"Mami, please wash us!" call the
  dishes.
"Mami, please polish us!" call the
  shoes.
"Mami, please mend us!" call the
  stockings.
"Mami, please sweep me!" calls the
  floor.

"Mami, please fetch the milk!" calls
  the milk jug.
"Mami, please iron me!" calls the
  laundry in the basket.

Oh, what an awful noise!
The child covers his ears.
At that the mother says:
"That's how it goes all day."

Now the child says:
"Jug, come, we'll help Mother. The
  two of us will go and fetch
  milk."

SOURCE: From *On Learning to Read* by Bruno Bettelheim and Karen Zeland, pp. 250, 283–84 (1981).

"Two First-Grade Stories"). Bettelheim and Zeland argue that Austrian children learn to read faster and better than American children, which they attribute to the content of the stories. The Austrian children read stories that address children's concerns (for example, why their mothers aren't always available for them). Therefore, they are more likely to view reading as something that can add to their lives. In general, the foreign primers reviewed by Bettelheim and Zeland "treat the beginning reader with respect for his [or her] intelligence, for his [or her] interest in the more serious aspects of life, and with the recognition that from the earliest age on he [or she] will respond positively to writings of true literary merit" (p. 303).[8, 9]

Jules Henry believes that students who have the intellectual strength to see through absurd assignments such as spelling baseball and who find it impossible to learn to accept such assignments as important may rebel against the system, refuse to do the work, drop out, or come to think of themselves as stupid. We cannot know how many students do poorly in school because they cannot accept the manner in which subjects are taught. The United States has diverse public school systems; different schools teach reading in different ways. Yet, the vast majority of Americans, especially middle- and lower-class Americans, typically learn to read and to carry out these kinds of assignments in the manner described here. In addition, students, especially young students, cannot articulate what they don't like about school; many come to believe that school is not for them and think they have failed rather than believing that school has failed them.

Classroom Morale A steady diet of absurd assignments and questions also can affect classroom morale. Bracha Alpert, an Israeli social scientist, observed three high school classrooms of upper middle-class, primarily white students in the United States and was struck by the way students responded to the teachers' questions. Alpert gave the following description of a typical exchange between teacher and students in one classroom:

> Students in this class did not respond to the teacher's questions that attempted to stimulate discussions. On occasions where a student did react, the reaction was short and uttered in a quiet tone of voice that may be described as "mumbling." Since the response rarely reached the whole class, the teacher often had to repeat it aloud. A combination of silence and mumbling is exemplified in the following excerpt of a typical transaction in this classroom.
>
> TEACHER: . . . now, ah, the first four stanzas certainly do create a mood for the reader. Now, what adjectives would you use to describe that mood?
>
> STUDENTS: (Silence)
>
> TEACHER: Think about it a little bit and then try to run it through your mind. How do we describe moods? Cheerful? Light headed? Sympathetic?

STUDENTS: (Silence)

TEACHER: What adjectives would you use to describe this one?

STUDENTS: (Silence, then a student mumbles) Mellow.

TEACHER: Mellow? OK I can buy that to a certain extent, but what [else]?

A STUDENT: (Mumbles) Solemn.

TEACHER: OK. Sarah suggests the word "solemn." Does that sound good?

SOME STUDENTS: (Mumble) No. (ALPERT 1991, P. 354)

Alpert maintains that this lack of response from students is likely to occur when teachers emphasize facts and clear-cut answers. When Alpert questioned students about their "reluctant participation," they said that it was a "mode of behavior purposely chosen" as a reaction to the teacher's instructional style.

Tracking

Most schools in the United States arrange students in instructional groups according to similarities in past academic performance and/or on standardized test scores. In elementary school the practice is often known as **ability grouping**; in middle school and high school it is known as **streaming** or **track-**

ing. Under this sorting and allocation system, students may be assigned to separate instructional groups within a single classroom; they may be sorted with regard to selected subjects such as mathematics, science, and English; or they may be separated across the entire array of subjects.

The rationales that underlie ability grouping, streaming, or tracking (hereafter referred to as "tracking") are that:

1. Students learn better when they are grouped with those who learn at the same rate: the brighter students are not held back by the slower learners and the slower learners receive the extra time and special attention needed to correct academic deficiencies.

2. Slow learners develop more positive attitudes when they do not have to compete with the more academically capable.

3. Groups of students with similar abilities are easier to teach.

In spite of these rationales, the evidence suggests that tracking does not lead to these benefits.

The Effects of Tracking Sociologist Jeannie Oakes investigated how tracking affected the academic experiences of 13,719 middle school and high school students in 297 classrooms and 25 schools across the United States.

The schools themselves were different: some were large, some very small; some in the middle of cities; some in nearly uninhabited farm country; some in the far West, the South, the urban North, and the Midwest. But the differences in what students experienced each day in these schools stemmed not so much from where they happened to live and which of the schools they happened to attend but, rather, from differences within each of the schools. (OAKES 1985, P. 2)

Oakes's findings were consistent with the findings of hundreds of other studies of tracking in terms of how students were assigned to groups, how they were treated, how they viewed themselves, and how well they did.

- *Placement:* Poor and minority students are placed disproportionately in the lower tracks.

- *Treatment:* The different tracks are not treated as equally valued instructional groups. There are clear differences with regard to the quality, content, and quantity of instruction and to classroom climate as reflected in the teachers' attitude and in student-student and teacher-student relationships. Low-track students consistently are exposed to inferior instruction—watered-down curriculum and endless repetition—and to a more rigid, more emotionally strained classroom climate.

- *Self-image:* Low-track students do not develop positive images of themselves because they are identified publicly and are treated as educational discards, damaged merchandise, or unteachable. Overall, among the average and the low-track groups, tracking seems to foster lower self-esteem and to promote misbehavior, higher dropout rates, and lower academic aspirations. Placement in a college preparatory track has positive effects on academic achievement, grades, standardized test scores, motivation, educational aspirations, and attainment—"And this positive relationship persists even after family background and ability differences are controlled" (Hallinan 1988, p. 260).

- *Achievement:* The brighter students tend to do well regardless of the academic achievements of the students with whom they learn.

These findings are reflected in the written answers that teachers and students gave to various questions asked by Oakes and her colleagues. For example, when teachers were asked about the classroom climate, high-track teachers tended to reply in positive terms:

There is a tremendous rapport between myself and the students. The class is designed to help the students in college freshman English composition. This makes them receptive. It's a very warm atmosphere. I think they have confidence in my ability to teach them well, yet because of the class size—32—there are times they feel they are not getting enough individualized attention. (OAKES 1985, P. 122)

Low-track teachers replied in less positive terms:

This is my worst class. Kids [are] very slow—underachievers and they don't care. I have no discipline cases because I'm very strict with them and they are scared to cross me. They couldn't be called enthusiastic about math—or anything, for that matter. (P. 123)

There also were clear differences in the high-track and the low-track students' responses to the question, What is the most important thing you have learned or done so far in this class? The replies of high-track students centered around themes of critical thinking, self-direction, and independent thought:

The most important thing I have learned in this [English] class is to loosen up my mind when it comes to writing. I have learned to be more imaginative. (P. 87)

The most important thing I have learned in this [math] class is the benefit of logical and organized thinking, learning is made much easier when the simple processes of organizing thoughts have been grasped. (P. 88)

Low-track students were more likely to give answers that centered around themes of boredom and conformity:

I think the most important is coming into [math] class and getting out folders and going to work. (P. 89)

To be honest, nothing. (P. 71)

Nothing I'd use in my later life; it will take a better man than I to comprehend our world. (P. 71)

In addition to these effects, tracking can create self-fulfilling prophecies by affecting teachers' expectations of the academic potential and abilities of students placed in each track.

Teachers' Expectations and Self-Fulfilling Prophecies Tracking can become a **self-fulfilling prophecy**, a deceptively simple yet powerful concept that originated from an insight by William I. and Dorothy Swain Thomas: "If [people] define situations as real, they are real in their consequences" ([1928] 1970, p. 572). A self-fulfilling prophecy begins with a false definition of a situation. The false definition,

however, is assumed to be accurate, and people behave as if the definition were true. In the end the misguided behavior produces responses that confirm the false definition (Merton 1957).

A self-fulfilling prophecy can occur if teachers and administrators assume that some children are "fast," "average," or "slow" and expose them to "fast," "average," and "slow" learning environments. Over time, real differences in quantity, quality, and content of instruction cause many students to actually become (and believe that they are) "slow," "average," or "fast." In other words, the prediction or prophecy of academic ability becomes an important factor in determining academic achievement:

The tragic, often vicious, cycle of self-fulfilling prophecies can be broken. The initial definition of the situation which has set the circle in motion must be abandoned. Only when the original assumption is questioned and a new definition of the situation is introduced, does the consequent flow of events [show the original assumption to be false]. (MERTON 1957, P. 424)

In the tradition of symbolic interactionism, Robert Rosenthal and Lenore Jacobson designed an experiment to test the hypothesis that teachers' positive expectations about students' intellectual growth can become a self-fulfilling prophecy and lead to increases in students' intellectual competence. Rosenthal and Jacobson were influenced by animal experiments in which trainers' beliefs about the genetic quality of the animals affected the animals' performances. When trainers were told that an animal was genetically inferior, the animal performed poorly; when trainers were told that an animal was genetically superior, the animal's performance was superior. This happened despite the fact that there were no such genetic differences between the animals defined as dull or bright.

Rosenthal and Jacobson's experiment took place in an elementary school called Oak School, a name given the school to protect its identity. The student body was largely from lower income families and predominantly white (84 percent); 16 percent of the students were Mexican-Americans. Oak School sorted students into ability groups based on teachers' judgments and on reading achievement.

At the end of a school year, Rosenthal and Jacobson gave a test, purported to be a predictor of academic "blooming," to those students who were expected to return in the fall. Just before classes began in the fall, all full-time teachers were given the names of the white and Hispanic students from all three ability groups who had supposedly scored in the top 20 percent. The teachers were told that these students "*will* show a more significant inflection or spurt in their learning within the next year or less than will the remaining 80 percent of the children" (Rosenthal and Jacobson 1968, p. 66). Teachers were also told not to discuss the scores with the students or the students' parents. Actually, the names given to teachers were chosen randomly: the differences between the children earmarked for intellectual growth and the other children were in the teachers' minds. The students were retested after one semester, at the end of the academic year, and after a second academic year.

Overall, intellectual gains, as measured by the difference between successive test scores, were greater for those students who had been identified as "bloomers" than they were for those not identified. Although "bloomers" benefited in general, some bloomers benefited more than others: first- and second-graders, Hispanic children, and children in the middle track showed the largest increases in test scores. It is important to note that the "bloomers" received no special instruction or extra attention from teachers—the only difference between them and the unidentified students was the belief that the "bloomers" bore watching. Rosenthal and Jacobson speculated that this belief was communicated to "bloomers" in very subtle and complex ways, which they could not readily identify:

> To summarize our speculations, we may say that by what she said, by how and when she said it, by her facial expressions, postures, and perhaps by her touch, the teacher may have communicated to the ["bloomers"] that she expected improved intellectual performance.
>
> It is self-evident that further research is needed to narrow down the range of possible mechanisms whereby a teacher's expectations become translated into a pupil's intellectual growth. (P. 180)

The research on tracking and teachers' expectations shows that the learning environment affects academic achievement and that tracking and expectations are two mechanisms that contribute to the unequal distribution of knowledge to American citizens. Because teachers draw on test results to form their expectations about students' academic potential and because academic personnel place students in different ability groups on the basis of test results (along with teacher evaluations), tests represent another mechanism that contributes to the unequal distribution of knowledge and skills.

Tests

Tests are the primary tools used by teachers to measure academic achievement. The United States and Japan are the only two mechanized rich countries (see Chapter 12) that regularly use multiple-choice examinations. Most other countries use essay tests, oral presentations, or demonstrations of skills. The United States is the only country in which commercial test publishers develop assessment instruments (SAT, ACT, and various achievement tests) and define test content (National Endowment for the Humanities 1991).

Almost everyone agrees that tests (especially the multiple-choice or true-false variety), when used as the primary measure of students' performance, reduce the motivations for learning and encourage rote memorization. For these reasons alone it is important to consider what tests actually measure. Frederick Erickson (1984) argues that tests, as currently designed and administered, do not measure overall cognitive competencies. Instead, tests measure the test-taker's ability to determine what it is the test-maker is looking for. Moreover, test-makers devise questions that require a single correct answer and that do not permit complex answers. Because tests are usually timed, they measure the student's ability not to get bogged down with the meaning of the questions. Erickson argues that cultural differences in interpreting test questions can lead to answers that make the child appear unintelligent. Erickson notes that teachers often take questions from test banks supplied by a textbook's publisher and that even the teachers have difficulty determining what the test-maker is asking. Erickson was struck by "their frustration at not being able to explain simple confusions their students may have about particular items . . . the child's reasoning is on the right track, it's just that the child is having trouble 'reading' the task cues of the item" (p. 534). In such cases we can say that the test measures a student's ability to answer confusing questions.

An example provided by A. R. Luria (1979) illustrates the kinds of problems that arise if students cannot "see" what the test-maker wants or if what the test-maker wants does not make sense. This example involves a seemingly simple exercise that most American first-graders are exposed to: determine which one of four objects does not belong. When Luria gave such a test to adult Russian peasants with no formal schooling, he could not get them to understand the exercise. From the peasants' perspective the answers he was looking for did not make sense.

Rakmat, a thirty-year-old illiterate peasant from an outlying district, was shown drawings of a hammer, a saw, a log, and a hatchet. "They're all alike," he said. "I think all of them have to be here. See, if you're going to saw, you need a saw, and if you have to split something, you need a hatchet. So they're all needed here."

We tried to explain the task by saying, "Look, here you have three adults and one child. Now clearly the child doesn't belong in this group."

Rakmat replied, "Oh, but the boy must stay with the others! All three of them are working, you see, and if they have to keep running out to fetch things, they'll never get the job done, but the boy can do the running for them. . . . The boy will learn; that'll be better, then they'll all be able to work well together."

"Look," we said, "here you have three wheels and a pair of pliers. Surely, the pliers and the wheels aren't alike in any way, are they?"

"No, they all fit together. I know the pliers don't look like the wheels, but you'll need them if you have to tighten something in the wheels."

"But you can use one word for the wheels that you can't for the pliers—isn't that so?"

"Yes, I know that, but you've got to have the pliers. You can lift iron with them and it's heavy, you know."

"Still, isn't it true that you can't use the same word for both the wheels and the pliers?"

"Of course you can't."

We returned to the original group, including hammer, saw, and hatchet. "Which of these could you call by one word?"

"How's that? If you call all three of them a 'hammer,' that won't be right either."

"But one fellow picked three things—the hammer, saw, and hatchet—and said they were alike."

"A saw, a hammer, and a hatchet all have to work together. But the log has to be here, too!"

"Why do you think [the fellow] picked these three things and not the log?"

"Probably he's got a lot of firewood, but if we'll be left without firewood, we won't be able to do anything."

"True, but a hammer, a saw, and a hatchet are all tools?"

"Yes, but even if we have tools, we still need wood. Otherwise we can't build anything."
(LURIA 1979, PP. 69–70)

Imagine a student taking a timed test with words, directions, or a task that does not make sense to him or her. Because students rarely have a chance to explain to teachers the logic underlying their answers, teachers are unlikely to learn why some students miss seemingly simple questions. Consequently, teachers are likely to label students who perform poorly on tests as "slow learners."

Joy Hakim, author of the 10-volume series *A History for Us* (1993) for fifth-graders, found that adult authors are unable to anticipate which words 10-year-olds might become confused over. She recalled an incident in which a child, upon reading the sentence "Ulysses S. Grant sends a wire to President Lincoln asking him to join him in his ship on the James River," asked, "Why would Grant send Lincoln a piece of wire?" When students repeatedly fail tests and assignments because the directions do not make sense or because the students have not acquired an ability to see what the test-maker wants, they are likely to give up. These findings do not suggest that we should stop testing students. They do, however, suggest that teachers should pay more attention than they do to the kinds of questions they ask and the directions they give.

We have examined various practices within the schools that contribute to the unequal distribution of knowledge and skills. We cannot blame teachers and other school personnel entirely for this outcome, however. Teachers do not have exclusive control over the classroom environment and they cannot single-handedly create students who are interested in learning. For many teachers the environment in which they work can make teaching problematic.

The Problems That Teachers Face

Teachers' jobs are complex; teachers are expected to undo learning disadvantages generated by larger inequalities in the society and to handle an array of discipline problems. Over 50 percent of elementary and high school teachers surveyed in a recent Gallup Poll responded that discipline is a very serious or fairly serious problem in their school, as are uncompleted homework assignments, cheating, stealing, drugs and alcohol, truancy, and absenteeism. Such widespread problems explain why 40 percent of beginning teachers drop out of teaching within five years (Elam 1989).

In addition to facing discipline problems, American teachers work in environments that discourage systematic learning outside the classroom and collaboration with other teachers. Psychologist Harold Stevenson (1992) finds that Asian schools plan extracurricular activities after school hours. During this time they teach students computer skills and do not have to use classroom time to do so. In addition Stevenson finds that Asian teachers work together very closely in preparing lesson plans. The level of collaboration is equivalent to that needed to produce a theatrical production. In contrast, teachers in the United States prepare lesson plans on their own. Asian teachers have more time to collaborate because they teach about 60 percent of the school day; in the remaining time, they discuss ideas with other teachers. American teachers, on the other hand, are in the classroom at least 85 percent of the time.

The job of teaching is further complicated by the social context of education. Teachers in the United States must deal with students from diverse family and ethnic backgrounds and with a student subculture that values and rewards athletic achievement, popularity, social activities, jobs, cars, and appearance at the expense of academic achievement. We examine these aspects of social context in the next section.

The Social Context of Education

We turn to the work of sociologist James S. Coleman (the 1991 president of the American Sociological Association), who has studied both family background and the adolescent student subculture, two factors that affect the classroom atmosphere and the learning experience.

Family Background

James S. Coleman (1966) was the principal investigator of *Equality of Educational Opportunity*, popularly known as the Coleman Report. The project was supported by the U.S. government under the directive of the 1964 Civil Rights Act, which prohibited discrimination for reasons of color, race, religion, or national origin in public places (restaurants, hotels, motels, and theaters); mandated that the desegregation of public schools be addressed; and forbade discrimination in employment (racial segregation in public schools was ruled unconstitutional by the Supreme Court in 1954). Coleman's intent was to examine the degree to which public education is segregated and to explore inequalities of educational opportunity in the United States. Coleman and his six colleagues surveyed 570,000 students and 60,000 teachers, principals, and school superintendents in 4,000 schools across the United States. Students filled out questionnaires about their home background and educational aspirations and took standardized achievement tests for verbal ability, nonverbal ability, reading comprehension, mathematical ability, and general knowledge. Teachers, principals, and superintendents answered questionnaires about their backgrounds, training, attitudes, school facilities, and curricula.

Coleman found that a decade after the Supreme Court's famous desegregation decision in 1954—*Brown* v. *Board of Education*—the schools were still largely segregated: 80 percent of white children attended schools that were 90 to 100 percent white, and 65 percent of African-American students attended schools that were more than 90 percent African-American. Almost all students in the South and the Southwest attended schools that were 100 percent segregated. Although Mexican-Americans, Native Americans, Puerto Ricans, and Asian-Americans also attended primarily segregated schools, they were not segregated from whites to the same degree as African-Americans. The Coleman Report also found that white teachers taught African-American children but that African-American teachers did not teach whites: approximately 60 percent of the teachers who taught African-American students were African-American, whereas 97 percent of the teachers who taught white students were white. When the characteristics of teachers of the average white student were compared with those of teachers of the average African-American student, the study found no significant differences in professional qualifications (as measured by degree, major, and teaching experience).

Coleman found sharp differences among ethnic groups with regard to verbal ability, nonverbal ability, reading comprehension, mathematical achievement, and general information as measured by the standardized tests. The white students scored highest, followed by Asian-Americans, Native Americans, Mexican-Americans, Puerto Ricans, and African-Americans.

Contrary to what Coleman expected to find, there were on the average no significant differences in quality between schools attended predominantly by the various ethnic groups and schools attended by whites. (Quality was measured by age of buildings, library facilities, laboratory facilities, number of books, class size, expenditures per pupil, extracurricular programs, and the characteristics of teachers, principals, and superintendents.) Surprisingly, variations in the quality of a school did not have much effect on the students' test scores.

Test scores were affected, however, by family background and by the attributes of other students. The average minority group member was likely to come from an economically and educationally disadvantaged household and was likely to attend school with students from similar backgrounds. Fewer of his or her classmates would complete high school, maintain high grade point averages, enroll in college preparatory curricula, or be optimistic about their future. Coleman also found some support for the idea that "The higher achievement of all racial and ethnic groups in schools with greater proportions of white students is largely, perhaps wholly, related to effects associated with the student body's educational background and aspirations" (1966, pp. 307, 310). This finding does not mean that there is something magical about a white environment. The Coleman Report examined the progress of African-Americans who had participated in school integration programs and found that their scores were higher than those of their counterparts who attended schools with members of the same social class. The important variable is the social class of one's classmates and not ethnicity:

> *Taking all these results together, one implication stands out above all: That schools bring little influence to bear on a child's achievement that is independent of his background and general social context; and that this very lack of an independent effect means that the inequalities imposed on children by their home, neighborhood, and peer environment are carried along to become the inequalities with which they confront adult life at the end of school. For equality of educational opportunity through the schools must imply a strong effect of schools that is independent of the child's immediate social environment, and that strong independent effect is not present in American schools.* (1966, P. 325)

Coleman's findings that school expenditures are not an accurate predictor of educational achievement (as measured by standardized tests) was used to support arguments against allocating additional funds to the public school system. Yet, the finding

A number of studies indicate that the home environment is an important variable—though not the only one—in explaining academic success.

George Goodwin/Monkmeyer Press

schools tied to residence, the end result is the demise of the common school attended by children from all economic levels. In its place is the elite suburban school . . . the middle-income suburban school, the low-income suburban school, and the central-city schools of several types—low-income white schools, middle-income white schools, and low or middle-income black schools. (1977, PP. 3–4)

The findings of this Coleman study do not imply that a person is trapped by family background. Coleman never claimed that family background explains all of the variation in test scores. He did claim, however, that it was the single most important factor in his study.

Coleman's findings about school segregation have changed very little over the past three decades. In 1968 the federal government reported that 76 percent of black students and 55 percent of Hispanic students attended predominantly minority schools (schools in which 50 percent or more of the students are black, Asian, Native American, and/or Hispanic). In 1991 the Harvard Project on School Desegregation found that figure to be 66 percent for blacks and 74.3 percent for Hispanics. In some states such as Illinois, Michigan, New York, and New Jersey more than 50 percent of the schools are 90–100 percent minority. As in the 1960s black and other minority students are significantly more likely to find themselves in schools where overall academic achievement is undervalued and low. The Harvard Project recommended that busing, the most widely used strategy for integrating schools, be supplemented by other strategies such as finding ways to integrate neighborhoods and enforcing desegregation laws (Celis 1993b).

that schools *do not* make a difference does not mean that schools *cannot* make a difference. A more accurate interpretation of this finding is that schools, as currently structured, have no significant effect on test scores; this conclusion implies that the educational system needs restructuring.

Coleman's findings about the composition of the student body and about the higher test scores earned by economically disadvantaged African-Americans in predominantly middle-class schools were used to support busing as a means of achieving educational equality. Although Coleman initially supported this policy, he later retracted his endorsement because busing hastened "white flight," or the migration of middle-class white Americans from the cities to the suburbs. This migration only intensified the racial segregation in city and suburban schools. As the ratio of white to black students dropped sharply, the positive effects of desegregation proved to be shortlived. As a result, economically and educationally disadvantaged blacks were sent from their deficient schools into equally deficient lower-class and lower middle-class white neighborhoods. Coleman adamantly maintained that court-ordered busing alone could not achieve integration:

With families sorting themselves out residentially along economic and racial lines, and with

Many subsequent studies support the importance of family background (Hallinan 1988). For example, the International Association for the Evaluation of Educational Achievement tested students in 22 countries on six subjects. The association found that the "home environment is a most powerful factor in determining the level of school achievement of students, student interest in school learning, and the number of years of schooling the children will receive" (Bloom 1981, p. 89; Ramirez and Meyer 1980). Yet, in this international study, in the Coleman study, and in other studies, home background (as measured by parents' ethnicity, income, education, and occupation) explains only about 30 percent of the variation in students' achievement. This finding suggests that factors other than socioeconomic status affect academic performance:

> In most if not all societies, children and youth learn more of the behavior important for constructive participation in the society outside of school than within. This fact does not diminish the importance of school but underlines the nation's dependence on the home, the working place, the community institutions, the peer group and other informal experiences to furnish a major part of the education required for a child to be successfully inducted into society. Only by clear recognition of the school's special responsibilities can it be highly effective in educating its students. (TYLER 1974, P. 74C)

In view of these findings, the special responsibility of the schools is to not duplicate the inequalities outside the school. Over the past three decades researchers have found that "schools exert some influence on an individual's chances of success, depending on the extent to which they provide equal access to learning" (Hallinan 1988, pp. 257–58). Unfortunately, as we have learned, several characteristics of American education and practices within the schools—hidden curriculum, test biases, self-fulfilling prophecies, and tracking—work to perpetuate social and economic inequalities. Now we turn to another problematic phenomenon that teachers confront daily—a student value system that de-emphasizes academic achievements.

Adolescent Subcultures

Around the turn of the century—the early decades of late industrialization—less than 10 percent of teenagers 14 to 18 years of age attended high school in the United States. Young people attended elementary school to learn the three Rs, after which they learned from their parents or from their neighbors the skills needed to make a living. As the pace of industrialization increased, jobs moved away from the home and out of the neighborhood into factories and office buildings. Parents no longer trained their children because the skills they knew

were becoming outdated and obsolete, so children came to expect that they would not make their livings as their parents did. In short, as the economic focus in the United States shifted from predominantly farm and small-town work environments to the factory and office, the family became less involved in the training of its children and, by extension, less involved in children's lives. The transfer of work away from the home and neighborhood removed opportunities for parents and children to work together. Under this new arrangement family occasions became events that were consciously arranged to fit everyone's work schedule.

Coleman argued that this shift in training from the family to the school cut adolescents off from the rest of society and forced them to spend most of the day with their own age group. Adolescents came "to constitute a small society, one that has most of its important interactions *within* itself, and maintains only a few threads of connection with the outside adult society" (Coleman, Johnstone, and Jonassohn 1961, p. 3).

Coleman surveyed students from 10 high schools in the Midwest to learn about adolescent society. He selected schools representative of a wide range of environments: five schools were located in small towns, one in a working-class suburb, one in a well-to-do suburb, and three in cities of varying sizes; one of the schools was an all-male Catholic school. Coleman was interested in the adolescent **status system,** a classification of achievements resulting in popularity, respect, acceptance into the crowd, praise, awe, and support, as opposed to isolation, ridicule, exclusion from the crowd, disdain, discouragement, and disrespect. To learn about this system, Coleman asked students questions similar to the following:

> How would you like to be remembered—as an athlete, as a brilliant student, as a leader in extracurricular activities, or as most popular?
>
> Who is the best athlete? The best student? The most popular? The boy the girls go for most? The girl the boys most go for?
>
> What person in the school would you like most to date?
>
> To have as a friend?
>
> What does it take to get in with the leading crowd in this school?

Based on the answers to these and other questions, Coleman was able to identify a clear pattern common to all 10 schools. "Athletics was extremely important for the boys, and social success with boys [accomplished through being a cheerleader or being good-looking] was extremely important for girls" (1961, p. 314). Coleman found that girls in particular do not want to be considered as good students, "for the girl in each grade in each of the schools who was most often named as best student has fewer friends and is less often in the leading crowd than is the boy most often named as best student" (Coleman 1960, p. 338). A boy most often could be a good student or dress well or have enough money to meet social expenses, but to really be admired he must also be a good athlete. Coleman also found that the peer group had more influence over and exerted more pressure on adolescents than did their teachers, and he found that a significant number of adolescents were influenced more by the peer group than by their parents.

In comparison to athletic and other achievements, why does the adolescent society penalize academic achievement? Coleman maintained that the manner in which students are taught contributes to their lack of academic interest: "They are prescribed 'exercises,' 'assignments,' 'tests,' to be done and handed in at a teacher's command" (1961, p. 315). The academic work they do does not require creativity but conformity. Students show their discontent by choosing to become involved in and acquiring things they can call their own—athletics, dating, clothes, cars, and extracurricular activities. Coleman, Johnstone, and Jonassohn noted that this reaction is inevitable given the passive roles that students are asked to play in the classroom.[10]

[One] consequence of the passive, reactive role into which adolescents are cast is its encouragement of irresponsibility. If a group is given no authority to make decisions and take action on its own, the leaders need show no responsibility to the larger institution. Lack of authority carries with it lack of responsibility; demands for obedience generate disobedience as well. But when a person or group carries the authority for his own action, he carries responsibility for it. In politics, splinter parties which are never in power often show little responsibility to the political system; a party in power cannot show such irresponsibility. . . . An adolescent society is no different from these. (P. 316)

Athletics is one of the major avenues open to adolescents, especially males, in which they can act "as a representative of others who surround [them]" (1961, p. 319). Others support this effort, identify with the athletes' successes, and console athletes when they fail. Athletic competition between schools generates an internal cohesion among students that no other event can. "It is as a consequence of this that the athlete gains so much status: he is doing something for the school and the community" (p. 260).

Coleman argues that, because athletic achievement is widely admired, everyone with some ability will try to develop this talent. With regard to the relatively unrewarded arena of academic life, "those who have most ability may not be motivated to compete" (p. 260). This reward structure may explain why top students in the United States have difficulty competing with top students in many other countries: the United States does not draw into the competition everyone who has academic potential.

Coleman's findings should deliver the message that the peer group is a powerful influence on learning, but they should not leave the impression that the peer group's world does not overlap with the family or the classroom. In fact, it seems more appropriate to consider how the multiple contexts of students' lives are interrelated. From data collected during interviews and observations with 54 ethnically and academically diverse youth in four urban desegregated high schools in California, educators Patricia Phelan, Ann Locke Davidson, and Hanh Cao Yu (1991, 1993, 1994) generated a model of the interrelationships between students' family, peer, and school worlds. The Students' Multiple Worlds Model describes the ways in which sociocultural aspects of students' worlds (for example, norms, values, beliefs, expectations, actions) combine to affect their thoughts and actions with respect to school and learning. These researchers are particularly concerned with understanding students' perceptions of boundaries and borders between worlds and adaptation strategies that

Students' Multiple Worlds Model and Typology

There are many factors that affect students' academic performances, including peer groups, the classroom environment, and family background. Education researchers Patricia Phelan, Ann Locke Davidson, and Hanh Cao Yu generate an important attempt to understand the interrelatedness of students' worlds.

Rather than compartmentalizing aspects of students' lives, it gives us a way of looking more holistically at the processes young people use to manage, more or less successfully, the transitions between their various contexts.

Congruent Worlds/Smooth Transitions

These students describe values, beliefs, expectations, and normative ways of behaving as similar across their worlds. Moving from one setting to another is harmonious and uncomplicated and boundaries are easily managed. This does not mean that students act exactly the same way or discuss the same things with teachers, friends, and family members, but rather that commonalities among worlds override differences. Students who exhibit this pattern say that their worlds are merged by their common sociocultural components rather than bounded by conspicuous differences. While many of these youths are white, upper middle-class, and high achieving,

this is not always the case. Some minority students describe little difference across their worlds and experience transitions as smooth. Likewise, academically average students can also exhibit patterns that fit this type.

Different Worlds/Border Crossings Managed

For some adolescents, differences in family, peer, and/or school worlds (with respect to culture, ethnicity, socioeconomic status and/or religion) require students to adjust and reorient as movement among contexts occurs. For example, a student's family world may be dominated by an all-encompassing religious doctrine in which values and beliefs are contrary to those

found in school and peer worlds. For other students, home and neighborhood are viewed as starkly different than school—particularly for students of color who are transported. And for still other students, differences between peers and family are dominant themes. However, regardless of differences, students in this category are able to utilize strategies that enable them to manage crossings successfully (in terms of what is valued in each setting). However, this does not mean that crossings are always easy, that they are made in the same way, or that they cannot result in personal and psychic consequences. It is not uncommon for high-achieving minority youth to exhibit patterns common to this type.

students employ as they move from one context to another. While they report that they found a good deal of variety in students' descriptions of their worlds and in their perceptions of boundaries they also uncovered four distinctive patterns that students utilize as they move between and adapt to different contexts and settings (see "Students' Multiple Worlds Model and Typology").

As Phelan and her colleagues point out, the patterns that students describe "are not necessarily stable for individual students but rather can be affected by external conditions such as classroom or school climate conditions, family circumstances, or

changes in peer group affiliation" (1991, p. 228). Also, the typology does not divide students along ethnic, achievement, or gender lines but rather focuses on the congruency of students' worlds and the borders that they face. In other words, youth of the same ethnicity or students who achieve at the same level can be found within any of the four types.

To this point we have examined a number of factors that help explain why American students, even the most able, learn so little relative to their Pacific Rim and European counterparts. These factors also help explain how skills and knowledge are transmitted unevenly across different social classes

Different Worlds/Border Crossings Difficult

In this category, like the former, students define their family, peer, and/or school worlds as distinct. They say they must adjust and reorient as they move across worlds and among contexts. However, unlike students who manage to make adjustments successfully, these students have either not learned, mastered, or have not been willing to adapt all of the strategies necessary for successful transitions. For example, a student may do poorly in a class where the teacher's interaction style, the student's role, or the learning activity are oppositional to what takes place within the student's peer or family worlds. Likewise, some students in this category describe their comfort and ease at school and with peers, but are essentially estranged from their parents. In these cases, parents' values and beliefs are frequently more traditional, more religious, or more constrained than those of their children, making adaptation to their home world difficult and conflictual. Other youth say that the socioeconomic circumstances of their families work against their full engagement in school. For youth who exhibit patterns common to this type, border crossing involves friction and discomfort and in some cases is possible only under particular conditions. This pattern often includes adolescents on the brink between success and failure, involvement and disengagement, commitment and apathy. These are some of the students for whom classroom and school climate conditions can mean the difference between staying in school or dropping out.

Different Worlds/Border Crossings Resisted

In this type, the values, beliefs, and expectations across students' worlds are so discordant that students perceive borders as insurmountable and actively or passively resist transitions. When border crossing is attempted, it is frequently so painful that, over time, these students develop reasons and rationales to protect themselves against further distress. In such cases, borders are viewed as insurmountable and students actively or passively resist attempts to embrace other worlds. For example, some students say that school is irrelevant to their lives while others immerse themselves fully in the world of their peers. Rather than moving from one setting to another, blending elements of all, these students remain constrained by borders they perceive as rigid and impenetrable. While low-achieving students (seemingly unable to profit from school and classroom settings) are typical of this type, high-achieving students who do not connect with peers or family also exhibit Type IV patterns.

SOURCE: Patricia Phelan, Ann Locke Davidson, and Hanh Cao Yu (1991, 1993, 1994). Synthesized by Patricia Phelan, University of Washington, Bothell.

---

and ethnic groups. The educational problems in the United States, however, seem to go beyond the uneven transmission of skills and knowledge to certain disadvantaged groups. A large segment of the student population simply tunes out. It is not that they cannot learn, but that they do not want to learn (Csikszentmihalyi 1990). Education critic Mihaly Csikszentmihalyi argues that few students, even good ones, pay attention while being taught:

In a series of studies teachers were given electronic pagers, and both they and their students were asked to fill out a short questionnaire whenever the pagers signaled (the signal was set to beep at random moments during the fifty-minute periods). In a typical high school history class, the pager went off as the teacher was describing how Genghis Khan had invaded China in 1234. At the same moment, of the twenty-seven students only two were thinking about something even remotely related to China. One of these two students was remembering a dinner she had had recently with her family at a Chinese restaurant; the other was wondering why Chinese men used to wear their hair in ponytails. (P. 134)

Discussion

Now that you have read the material in this chapter, what similarities or differences can you find in your own educational experiences? Do you believe the sociological perspective captures the realities of public education in the United States? If you would like to share your reactions to any part of this chapter please write me c/o Northern Kentucky University, Sociology Program, Highland Heights, Kentucky 41099.

The material in this chapter—particularly the fact that American students perform poorly compared with their European and Pacific Rim counterparts—leads us to ask, If Americans study the successes of public education in foreign countries, can they borrow those strategies to improve their own system of education? It depends. First, Americans must define concisely the nature and origins of the problems they wish to solve. If the problem is defined superficially as low test scores in relation to the scores of foreign students, we will have few clues as to what it is about the American system that is in need of reform. Consequently, we will not focus on the parts of the foreign system that might serve as a model of reform. For example, Japanese and Taiwanese students outperform American students in science and math. Some researchers believe that these performance differences are related to differences in the amount of time American, Japanese, and Taiwanese students spend in school— American students attend school 178 days of the year while Japanese and Taiwanese students attend school 240 days a year. Although American students spend 30.4 hours a week in the classroom, they spent only 64.5 percent of this time working on academic activities. The remainder of the time is spent maintaining discipline, attending assemblies, going to the bathroom, and so on. In contrast, Taiwanese students spend 40.4 hours in the classroom, 90 percent of which is devoted to academic activities; Japanese students spend 37.3 hours in the classroom, 80 percent of which is devoted to academic activities. On the basis of these findings, Americans could lengthen the school year to 240 days and increase the number of classroom hours to 44 a week. However, such changes would not address the real problem—the content and quality of the curriculum and the national ambivalence toward education.

The point is, Americans cannot expect that increasing the length of the school year will improve test scores. If we agree that the way the curriculum is designed and delivered is a major cause of the high dropout rate, illiteracy, and level of passivity, then we need to examine how the curriculum is designed and delivered in other countries. If we decide that a foreign technique might be useful in solving American problems, we also have to ask what it is about the country (its history, its family structure, economic incentives, cultural values, the way in which teenagers spend free time) that supports the practice and permits it to work well.

In "Japanese Education: If They Can Do It, Should We?" Japanese studies professor Thomas Rohlen (1986) offers some suggestions for how to use other countries' models of education to improve our model. In the case of Japan, Rohlen argues:

> *We would be foolish to see Japanese education as a model for our own efforts; but as a mirror showing us our own weaknesses and as a yardstick against which to measure our efforts, it has great values for us. We cannot allow ourselves to ignore or to imitate its approach. Rather, it is possible to look periodically into the "Japanese mirror" while we quite independently set out to strengthen our schools and our system within our own cultural and social context.* (ROHLEN 1986, PP. 42–43)

On the other hand, it is important not to overemphasize differences between educational systems and come to believe that no feature can be borrowed without major adjustments. Such a perspective would lead us to exaggerate differences and reject all ideas as inappropriate (Cummings 1989). But as Secretary of Education William Bennett concluded in a U.S. Department of Education study, "much of what seems to work well for Japan in the field of education closely resembles what works best in the United States—and most likely elsewhere. Good education is good education" (Bennett 1987, p. 71; see "Some Highlights of Japanese Education").

Some Highlights of Japanese Education

Below are some highlights from a study sponsored by the U.S. Department of Education, *Japanese Education Today* (1987). These highlights cover the strengths and weaknesses of the Japanese education system, the reforms underway in Japan, and implications for education in the United States.

Strengths and Weaknesses

- Japanese society is education-minded to an extraordinary degree: success in formal education is considered largely synonymous with success in life and is, for most students, almost the only path to social and economic status.

- Formidable education accomplishments result from the collective efforts of parents, students, and teachers; these efforts are undergirded by the historical and cultural heritage, a close relationship between employers and education, and much informal and supplementary education at preschool, elementary, and secondary levels.

- During 9 years of compulsory schooling, all children receive a high quality, well-balanced basic education in the 3 Rs, science, music, and art.

- Both the average level of student achievement and the student retention rate through high school graduation are very high.

- Japanese education also distinguishes itself in motivating students to succeed in school, teaching effective study habits, using instructional time productively, maintaining an effective learning environment, sustaining serious attention to character development, and providing effective employment services for secondary school graduates.

- Japanese education is not perfect. Problems include rigidity, excessive uniformity, and lack of choice; individual needs and differences that receive little attention in school; and signs of student alienation. A related problem in employment is overemphasis on the formal education background of individuals.

Education Reform in Japan

- Japan is concerned about its education system and is making extensive efforts to improve it. The reform movement includes vigorous public debate between proponents of change and defenders of the status quo.

- Reformers take a farsighted view of the national interest: they are coming to terms with societal needs in the 21st century while grappling with such complex issues as finding a new balance between group harmony and individual creativity in Japanese education.

Implications for American education

Some American education ideals may be better realized in Japan than in the United States. A close look at Japanese education provides a potent stimulus for Americans to reexamine the standards, performance, and potential of their own system. Some lessons worth considering:

- The value of parental involvement from the preschool years on;

Schooling in the arts is a key component of Japanese education.

J.P. Laffont/Sygma

- The necessity of clear purpose, strong motivation, and high standards, and of focusing resources on education priorities;

- The importance of maximizing learning time and making effective use thereof;

- The value of a competent and committed professional teaching force; and

- The centrality of holding high expectations for all children and a firm commitment to developing a strong work ethic and good study habits—recognizing that hard work and perseverance are essential elements in a good education.

SOURCE: U.S. Department of Education, p. vi (1987).

In this chapter we have gained insights into the American system by comparing it with foreign systems. Clear insights are the first step toward meaningful change. We also have learned that the problems of American education are not confined to the classroom: education is the stage on which a whole host of cultural crises become evident. Clearly, the content and quality of the curriculum is one critical area in need of immediate reform. The good news is that reforming the content of the curriculum is manageable. There are many examples of quality education in many foreign countries as well as in a few American classrooms from which we can draw. The following model describes one such example:

There are books in a corner, some written, published, and bound by the children. . . . There is a map of birthplaces—children's, their parents', and grandparents'—marked on a large world map. Two children are working on a graph of the numbers and their divisors. Two, bent on painting a potted tomato with the right leaf color, are carefully measuring and mixing pigments. The walls are full of children's reports of work completed. There is a computer in the corner for mathematical games and puzzles and other forms of self-help instruction. A traffic survey of children's routes to school is in progress, with a map of the school and its neighborhood, a surveyor's trundle wheel for the children to borrow for mapping and finding distances from home. In the hallway a school play is being rehearsed. (HAWKINS 1990, P. 7)

The future of education is a critical issue for our diverse and rapidly changing society.
Alon Reininger/Woodfin Camp & Associates

FOCUS
The "Americanizing" Function of Education

Recall earlier in this chapter we noted that the most vocal early education reformers such as Benjamin Rush, Thomas Jefferson, and Noah Webster believed that schools were an important mechanism by which a diverse population could acquire a common culture. They believed that the welfare of the country depended on a common culture that re-quired people to forget their families' language and culture. In "On Becoming a Chicano" Richard Rodriguez describes the often subtle processes by which he came to abandon the language and culture of his family and the ambivalence he felt about that loss.

On Becoming a Chicano

Richard Rodriguez

Today I am only technically the person I once felt myself to be—a Mexican-American, a Chicano. Partly because I had no way of comprehending my racial identity except in this technical sense, I gave up long ago the cultural consequences of being a Chicano.

The change came gradually but early. When I was beginning grade school, I noted to myself the fact that the classroom environment was so different in its styles and assumptions from my own family environment that survival would essentially entail a choice between both worlds. When I became a student, I was literally "remade"; neither I nor my teachers considered anything I had known before as relevant. I had to forget most of what my culture had provided, because to remember it was a disadvantage. The past and its cultural values became detachable, like a piece of clothing grown heavy on a warm day and finally put away.

Strangely, the discovery that I have been inattentive to my cultural past has arisen because others—students, colleagues and faculty members—have started to assume that I am a Chicano. The ease with which the assumption is made forces me to suspect that the label is not meant to suggest cultural, but racial, identity. Nonetheless, as a graduate student and a prospective university faculty member, I am routinely expected to assume intellectual leadership *as a member of a racial minority.* Recently, for example, I heard the moderator of a panel discussion introduce me as "Richard Rodriguez, a Chicano intellectual." I wanted to correct the speaker—because I felt guilty representing a nonacademic cultural tradition that I had willingly abandoned. So I can only guess what it would have meant to have retained my culture as I entered the classroom, what it would mean for me to be today a *Chicano intellectual.* (The two words juxtaposed excite me; for years I thought a Chicano had to decide between being one or the other.)

Does the fact that I barely spoke any English until I was nine, or that as a child I felt a surge of

self-hatred whenever a passing teenager would yell a racial slur, or that I saw my skin darken each summer—do any of these facts shape the ideas which I have or am capable of having? Today, I suspect they do—in ways I doubt the moderator who referred to me as a "Chicano intellectual" intended. The peculiar status of being a "Chicano intellectual" makes me grow restless at the thought that I have lost at least as much as I have gained through education.

I remember when, 20 years ago, two grammar-school nuns visited my childhood home. They had come to suggest—with more tact than was necessary, because my parents accepted without question the church's authority—that we make a greater effort to speak as much English around the house as possible. The nuns realized that my brothers and I led solitary lives largely because we were barely able to comprehend English in a school where we were the only Spanish-speaking students. My mother and father complied as best they could. Heroically, they gave up speaking to us in Spanish—the language that formed so much of the family's sense of intimacy in an alien world—and began to speak a broken English. Instead of Spanish sounds, I began hearing sounds that were new, harder, less friendly. More important, I was encouraged to respond in English.

The change in language was the most dramatic and obvious indication that I would become very much like the "gringo"—a term which was used descriptively rather than pejoratively in my home—and unlike the Spanish-speaking relatives who largely constituted my preschool world. Gradually, Spanish became a sound freighted with only a kind of sentimental significance, like the sound of the bedroom clock I listened to in my aunt's house when I spent the night. Just as gradually, English became the language I came not to *hear* because it was the language I used every day, as I gained access to a new, larger society. But the memory of Spanish persisted as a reminder of the society I had left. I can remember occasions when I entered a room and my parents were speaking to one another in Spanish, seeing me, they shifted into their more formalized English. Hearing them speak to me in English troubled me. The bonds their voices once secured were loosened by the new tongue.

This is not to suggest that I was being *forced* to give up my Chicano past. After the initial awkwardness of transition, I committed myself, fully and freely, to the culture of the classroom. Soon what I was learning in school was so antithetical to what my parents knew and did that I was careful about the way I talked about myself at the evening dinner table. Occasionally, there were moments of childish cruelty: a son's condescending to instruct either one of his parents about a "simple" point of English pronunciation or grammar.

Social scientists often remark, about situations such as mine, that children feel a sense of loss as they move away from their working-class identifications and models. Certainly, what I experienced, others have also—whatever their race. Like other generations of, say, Polish-American or Irish-American children coming home from college, I was to know the silence that ensues so quickly after the quick exchange of news and the dwindling of common interests.

In addition, however, education seemed to mean not only a gradual dissolving of familial and class ties but also a change of racial identity. The new language I spoke was only the most obvious reason for my associating the classroom with "gringo" society. The society I knew as Chicano was barely literate—in English *or* Spanish—and so impatient with either prolonged reflection or abstraction that I found the academic environment a sharp contrast. Sharpening the contrast was the stereotype of the Mexican as a mental inferior. (The fear of this stereotype has been so deep that only recently have I been willing to listen to those, like D. H. Lawrence, who celebrate the "noncerebral" Mexican as an alternative to the rational and scientific European man.) Because I did not know how to distinguish the healthy nonrationality of Chicano culture from the mental incompetency of which Chicanos were unjustly accused, I was willing to abandon my nonmental skills in order to disprove the racist's stereotype.

I was wise enough not to feel proud of the person education had helped me to become. I knew that education had led me to repudiate my race. I was frequently labeled a *pocho*, a Mexican with gringo pretensions, not only because I could not speak Spanish but also because I would respond in English with precise and careful sentences. Uncles

would laugh good-naturedly, but I detected scorn in their voices. For my grandmother, the least assimilated of my relations, the changes in her grandson since entering school were especially troubling. She remains today a dark and silently critical figure in my memory, a reminder of the Mexican-Indian ancestry that somehow my educational success has violated.

Nonetheless, I became more comfortable reading or writing careful prose than talking to a kitchen filled with listeners, withdrawing from situations to reflect on their significance rather than grasping for meaning at the scene. I remember, one August evening, slipping away from a gathering of aunts and uncles in the backyard, going into a bedroom tenderly lighted by a late sun, and opening a novel about life in nineteenth-century England. There, by an open window, reading, I was barely conscious of the sounds of laughter outside.

With so few fellow Chicanos in the university, I had no chance to develop an alternative consciousness. When I spent occasional weekends tutoring lower-class Chicano teenagers or when I talked with Mexican-American janitors and maids around the campus, there was a kind of sympathy—a sense, however privately held—that we knew something about one another. But I regarded them all primarily as people from my past. The maids reminded me of my aunts (similarly employed); the students I tutored reminded me of my cousins (who also spoke English with barrio accents).

When I was young, I was taught to refer to my ancestry as Mexican-American. *Chicano* was a word used among friends or relatives. It implied a familiarity based on shared experience. Spoken casually, the term easily became an insult. In 1968 the word *Chicano* was about to become a political term. I heard it shouted into microphones as Third World groups agitated for increased student and faculty representation in higher education. It was not long before I *became* a Chicano in the eyes of students and faculty members. My racial identity was assumed for only the simplest reasons: my skin color and last name.

On occasion I was asked to account for my interests in Renaissance English literature. When I explained them, declaring a need for cultural assimilation on the campus, my listener would disagree. I sensed suspicion on the part of a number of my fellow minority students. When I could not imitate Spanish pronunciations of the dialect of the barrio, when I was plainly uninterested in wearing ethnic costumes and could not master a special handshake the minority students often used with one another, they knew I was different. And I was. I was assimilated into the culture of a graduate department of English. As a result, I watched how in less than five years nearly every minority graduate student I knew dropped out of school, largely for cultural reasons. Often they didn't understand the value of analyzing literature in professional jargon, which others around them readily adopted. Nor did they move as readily to lofty heights of abstraction. They became easily depressed by the seeming uselessness of the talk they heard around them. "It's not for real," I still hear a minority student murmur to herself and perhaps to me, shaking her head slowly, as we sat together in a class listening to a discussion on punctuation in a Renaissance epic.

I survived—thanks to the accommodation I had made long before. In fact, I prospered, partly as a result of the political movement designed to increase the enrollment of minority students less assimilated than I in higher education. Suddenly grants, fellowships, and teaching offers became abundant.

In 1972 I went to England on a Fulbright scholarship. I hoped the months of brooding about racial identity were behind me. I wanted to concentrate on my dissertation, which the distractions of an American campus had not permitted. But the freedom I anticipated did not last for long. Barely a month after I had begun working regularly in the reading room of the British Museum, I was surprised, and even frightened, to have to acknowledge that I was not at ease living the rarefied life of the academic. With my pile of research file cards growing taller, the mass of secondary materials and opinions was making it harder for me to say anything original about my subject. Every sentence I wrote, every thought I had, became so loaded with qualifications and footnotes, that it said very little. My scholarship became little more than an exercise in caution. I had an accompanying suspicion that whatever I did manage to write and call my dissertation would be of little use. Opening books so dusty that they must not have been used in decades, I began to doubt the value of writing what only a few people would read.

Obviously, I was going through the fairly typical crisis of the American graduate student. But with one difference: After four years of involvement with questions of racial identity, I now saw my problems as a scholar in the context of the cultural issues that had been raised by my racial situation. So much of what my work in the British Museum lacked, my parents' culture possessed. They were people not afraid to generalize or to find insights in their generalities. More important, they had the capacity to make passionate statements, something I was beginning to doubt my dissertation would ever allow me to do. I needed to learn how to trust the use of "I" in my writing the way they trusted its use in their speech. Thus developed a persistent yearning for the very Chicano culture that I had abandoned as useless.

Feelings of depression came occasionally but forcefully. Some days I found my work so oppressive that I had to leave the reading room and stroll through the museum. One afternoon, appropriately enough, I found myself in an upstairs gallery containing Mayan and Aztec sculptures. Even there the sudden yearning for a Chicano past seemed available to me only as nostalgia. One morning, as I was reading a book about Puritan autobiography, I overhead two Spaniards whispering to one another. I did not hear what they said, but I did hear the sound of their Spanish—and it embraced me, filling my mind with swirling images of a past long abandoned.

I returned from England, disheartened, a few months later. My dissertation was coming along well, but I did not know whether I wanted to submit it. Worse, I did not know whether I wanted a career in higher education. I detested the prospect of spending the rest of my life in libraries and classrooms, in touch with my past only through the binoculars nostalgia makes available. I knew that I could not simply re-create a version of what I would have been like had I not become an academic. There was no possibility of going back. But if the culture of my birth was to survive, it would have to animate my academic work. That was the lesson of the British Museum.

I frankly do not know how my academic autobiography will end. Sometimes I think I will have to leave the campus, in order to reconcile my past and present. Other times, more optimistically, I think that a kind of negative reconciliation is already in progress, that I can make creative use of my sense of loss. For instance, with my sense of the cleavage between past and present, I can, as a literary critic, identify issues in Renaissance pastoral—a literature which records the feelings of the courtly when confronted by the alternatives of rural and rustic life. And perhaps I can speak with unusual feeling about

the price we must pay, or have paid, as a rational society for confessing seventeenth-century Cartesian faiths. Likewise, because of my sense of cultural loss, I may be able to identify more readily than another the ways in which language has meaning simply as sound and what the printed word can and cannot give us. At the very least, I can point up the academy's tendency to ignore the cultures beyond its own horizons.

Key Concepts

Ability Grouping 502	Illiteracy 481	Status System 510
Catechisms 488	Informal Education 480	Streaming 502
Formal Curriculum 497	Schooling 480	Tracking 502
Formal Education 480	Self-Fulfilling Prophecy 504	
Hidden Curriculum 497	Social Promotion 490	

Notes

1. The states involved in the Department of Education study project were California, Illinois, Indiana, Iowa, Louisiana, New Jersey, New York, Ohio, Pennsylvania, Texas, and Washington (U.S. Department of Education 1993b, p. 6).

2. The findings in this study indicate:

   The approximately 90 million adults who performed in Levels 1 and 2 did not necessarily perceive themselves as being "at risk." Across the literacy scales, 66 to 75 percent of the adults in the lowest level and 93 to 97 percent in the second lowest level described themselves as being able to read or write English "well" or "very well." Moreover, only 14 to 25 percent of the adults in Level 1 and 4 to 12 percent in Level 2 said they get a lot of help from family members or friends with everyday prose, document, and quantitative literacy tasks. It is therefore possible that their skills, while limited, allow them to meet some or most of their personal and occupational literacy needs. (U.S. DEPARTMENT OF EDUCATION 1993B, P. XV)

3. In this chapter high school dropout rates reflect the number of people who leave high school be

fore they complete 12 grades. Many dropouts, however, obtain a GED at a later point. Whether or not a dropout earns a GED, it is significant that a substantial number of students choose to leave school or are faced with circumstances that force them to drop out of the formal educational process.

4. Daunting as these statistics are, they may actually underestimate the number of young people annually who drop out of high school (Cooke, Ginsburg, and Smith 1985). There is no central authority to audit the accuracy of dropout counts tabulated by school officials. Consequently, each school uses its own system of keeping track of dropouts. For example, the Chicago Board of Education reported a dropout rate of 10 percent through the 1970s and early 1980s. A later state investigation, however, revealed that the rate was 50.7 percent—38 percent for whites, 56 percent for African-Americans, and 57 percent for Hispanics (Hahn 1987).

   The higher high school dropout rate is not unique to Chicago. When figured by this formula, the dropout rate is 50 percent for St. Louis, 49 percent for New York City, 45 percent for Baltimore, and 40 percent for Cleveland (Horton, Leslie, and Larson 1988). One reason for the discrepancy was that every student who leaves school before graduation is placed into 1 of 19 categories, only one of which is labeled "dropout." Other categories include "lost—not coming to school," "needed at home," "married," and "cannot adjust" (Hahn 1987; Lefkowitz 1987). Another reason that the dropout rate is possibly underestimated is that typically it is tabulated on an annual basis; that is, it is calculated by comparing the number of high school students enrolled at the beginning of a school year with the number of students enrolled at the year's end. Each year this figure is determined without considering the number of high school age persons who should be enrolled at the beginning of the school year. Calculated in this way, an annual dropout rate is less accurate than a rate based on following a freshman class through the twelfth grade to determine how many members dropped out before graduation. (The 50.7 percent rate for Chicago determined by state officials was based on tracking freshman classes.)

5. Members of some ethnic groups are more likely to be poor than members of other ethnic groups. Overall, the poverty rate for the nation is 12 percent. The poverty rate for Native Americans living on reservations is 41 percent, and 22 percent for those living off the reservations (U.S. Bureau of Indian Affairs 1988). Thirty-one percent of African-Americans and 29 percent of Hispanics live in poverty (Folbre 1987). With regard to children (18 and under), 12 percent of white children live below the poverty level, compared with 40 percent of African-American and Hispanic children. In other words, two of every five African-American and Hispanic children are poor (Currie and Skolnick 1988).

6. Dorris maintains that the most striking characteristic of children with fetal alcohol syndrome is their lack of imagination:

   not so much the ability to make up a fictional story, but the basic act of foreseeing a possible consequence in one's own life. Most of us do it all the time: if I do x today, then y will occur tomorrow. If I work at a job, I will earn money. If I earn money, I can purchase more options for myself. If I go to sleep early, I will be rested when I wake up. If I eat my lunch at nine in the morning, I will be hungry at two. (DORRIS 1989, P. 245)

7. The emphasis on heroic measures rather than on simpler and more practical remedies may be rooted in the American belief in the self-made person—the person who transcends the environment despite all odds. Few themes have captured Americans' imaginations as intensely as the discovery of talent in unexpected places—the boy who couldn't make his high school baseball team but who later leads the National League in hitting, the poor boy or girl who starts a business from scratch, or a premature, low-weight baby who defies all odds and grows up to be president.

8. The criticisms that Bettelheim and Zeland make are pervasive, but especially for readings and assignments aimed at the slow reader. The following excerpt from a review of a series of four paperback texts illustrates the nature of these criticisms. The series begins with Columbus and ends with the present. It is, in the words of one of its editors, "aimed at low-achieving readers who have had difficulty with textbooks." The age group aimed at ranges from sixth grade through high school. Each textbook doubles as a workbook. Exercises precede and follow every bit of historical writing in the book.

   These four textbooks are not history; they are not even books. A book of history tells its story without interruption. A book assumes that the attention span of its readers is longer than two minutes. A book aims at making its readers think and reflect by virtue of its densely attractive subject matter. A real book does not interrupt itself every two hundred words to prod and cajole the student. (SWEETLAND 1989, P. 4)

9. What group benefits if a large segment of the population cannot read or hates to read? One answer is advertisers. When people do not read, most of their information about what is going on in the world and about products comes from the television and the radio, two media that depend heavily on advertising revenue.

10. When critics maintain that students should become more involved in shaping academic life, it is tragically taken to mean that they should be left on their own and should tell teachers what they want to learn. Such a policy misses the point, which is that students need to be taught how to organize meaningful academic projects and should be guided to carry out meaningful projects.

14 RELIGION

with Emphasis on Lebanon

A street in Beirut, Lebanon.
Bill Foley/Woodfin Camp & Associates

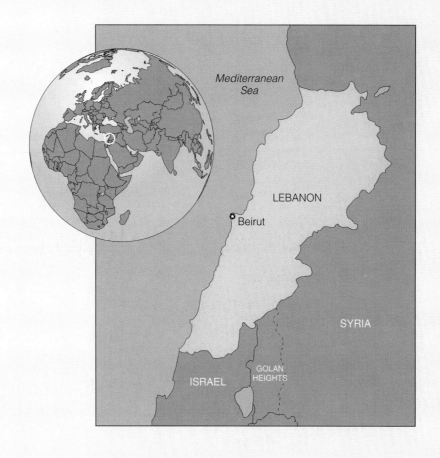

Mediterranean
Sea

LEBANON

⭐ Beirut

SYRIA

ISRAEL   GOLAN
         HEIGHTS

Students from Lebanon enrolled in
college in the United States for fall 1991   3,080

People living in the United States
in 1990 who were born in Lebanon   91,000

Airline passengers flying between
the United States and Lebanon in 1991   130

U.S. military personnel in Lebanon in 1993   2

People employed in 1990 by
Lebanese affiliates in the United States   3,500

Phone calls made from
the United States to Lebanon in 1991   270,000

Lebanese admitted into the United States
in fiscal year 1991 for temporary employment   400

From the mid-1980s until 1992, Lebanese-based kidnappers from various Islamic groups held American, British, French, and Soviet foreign nationals hostage. Among those held was U.S. journalist Terry Anderson. In an interview published in *Time* magazine, Anderson describes his captors' view of religion and the views he formed about two books he read while captive—the Koran and the Bible.

Q: During your years as a captive [in Lebanon], you were constantly exposed to the beliefs of your kidnappers about themselves and the rest of the world. What were they saying?

A: They were radicals within the fundamentalist movement. The way they interpret their religion allows them to do things or to justify to themselves doing things that any normal reading of the Koran would find insane or evil. I've read the Koran; I'm not an Islamic scholar, but the words and the concepts seem to me fairly plain, and they're not all that different from Christianity at base.

They are paranoid in the way they look at the world. They see America as the Great Satan that does everything wrong, and yet it is all-powerful, and therefore all American acts must be deliberate; they can't be the result of accident or misunderstanding, or simply stupid policy.

Q: Do you think Westerners understand this mentality?

A: No, not at all. Even many of the hostages after some years of it could not understand it, could not grasp it.

Q: What did they allow you to read in captivity?

A: At various times we did have a lot of books. The book I got first was the Bible, and I kept that almost throughout my captivity, though not the same copy. I read that over and over and over and over and over again and thought about it. That book was by far the most important to me and remains the most important to me.

(AIKMAN 1992, P. 58. INTERVIEW WITH FORMER HOSTAGE TERRY ANDERSON)

ostage taking is only one of the images that many Americans associate with the Muslim religion, or Islam. Islam also evokes images of the 1993 World Trade Center bombing in New York City; death threats against author Salman Rushdie; rousing speeches of Iran's religious leader, the late Ayatollah Khomeini; the enforced detention of 62 American diplomats in Teheran, Iran for 444 days, which ended finally on January 20, 1981; the May 1983 explosion at the U.S. Embassy in Beirut, Lebanon, which killed 50 people, followed by the October 1983 suicide attacks that killed 241 U.S. soldiers and 58 French soldiers; the dropping of the body of a dead U.S. Navy diver from a hijacked TWA plane in 1985; the 1988 bombing of Pan Am Flight 103 over Lockerbee, Scotland; the 1989 hanging of U.S. Marine Lieutenant Colonel William Higgins in Lebanon; and stepped-up airport security in the wake of the Gulf War with Iraq in 1991.

In spite of the wide media coverage given to each of these events, one idea dominates: some Islamic group—the Movement of the Islamic Amal, the Islamic Jihad Organization, the Hezbollah (Party of God), or Salvation from Hell—was responsible. Each of these events generally was reduced to the actions of a religious fanatic acting solely from primitive and irrational religious convictions. For many Americans the simple fact that the terrorists' religion was Islam explained their destructive actions.

In a 1991 PBS television program, "Images of God in the Arab World," journalist Bill Moyers asked Yvonne Haddad, a Syrian-born college professor who teaches history at the University of Massachusetts, to comment on the narrow view of Muslims held by most Americans. Moyers also asked Haddad to comment on the statement that Islam has given religion a bad name. Professor Haddad replied:

I think the West is giving Islam a bad name because basically we have decided to demonize Islam. Vice President Quayle made a statement in May [1990] in which he said this century we have had three evils ... Nazism, Communism, and Islamic fundamentalism, which I think is very unfortunate. The images that television can bring are very selective; they do not cover all Muslims. There are Muslim terrorists. There are terrorists who happen to be Muslim. There are Christian terrorists. There are terrorists who happen to be Christian. ... The KKK does not stand for all Christianity, nor does the Jewish Defense League [stand for Judaism], nor do Muslim terrorists stand for Islam. (HADDAD 1991)

In this chapter we examine religion from a sociological perspective. Such a perspective is useful because it allows us to step back and view in a detached way a subject that is often charged with emotion. Detachment and objectivity are necessary if we wish to avoid making sweeping generalizations about the nature of religions that are unfamiliar to us, as Islam is to many Americans.

When sociologists study religion, they do not investigate whether God or some other supernatural force exists, whether certain religious beliefs are valid, or whether one religion is better than another. Sociologists cannot study such questions because they adhere to the scientific method, which requires them to study only observable and verifiable phenomena. Rather, sociologists investigate the social aspects of religion, focusing on the characteristics common to all religions, the types of religious organizations, the functions and dysfunctions of religion, the conflicts within and between religious groups, how religion shapes people's behavior and their understanding of the world, and how religion is intertwined with social, economic, and political issues.

We give special emphasis to Lebanon for several reasons. First, since the mid-1970s Lebanon has suffered through at least three civil wars. One of these wars is being waged between Christians and Muslims for control of the largest share of power in the Lebanese government. The other wars are between Muslims and Muslims (for example, Sunni Muslim versus Shia Muslim) and between Christians and Christians (for example, Christian Phalangist militia versus the Christian-led Lebanese army); those wars concern the share of power to be held by each of these groups once the larger Christian-Muslim civil war is resolved (Friedman 1989).

Second, in Lebanon, as in the rest of the Middle East, a person's religious affiliation is very important; one's political position with regard to the civil

American Muslims at prayer. How does this picture compare with the images of Islam that you are likely to see in the mass media?

John Van Hasselt/Sygma

wars is influenced profoundly by religion. Yet, it would be wrong to assert that the Lebanese civil wars are being fought for strictly religious reasons. Christians are not killing and being killed because they profess a belief in the divinity of Jesus; nor are Muslims killing and being killed because they do not recognize the Christian Trinity (Genet 1984). It would be equally mistaken to suggest that the various battle lines are drawn strictly on the basis of religion. A more appropriate method of classifying the divisions in Lebanese society is to distinguish between those who seek basic and comprehensive change—the leftists—and those who would maintain the existing order—the rightists—(Barakat 1979; Wenger 1990).

A number of Muslim leaders, factions, or countries support the rightist camp, which is dominated by Maronite Christians; a significant number of Lebanese Christians support the leftist camp. As we will learn, the actual state of affairs that exists in Lebanon is conflict between groups of people who identify each other in religious terms. This focus on religious affiliations masks more important factors such as economic and political inequalities between conflicting parties. On the other hand, because religious differences are used to distinguish the opposing groups and to mobilize people to fight each other, we also must investigate the qualities of religion that make it a factor in such violent conflicts (Stavenhagen 1991).

Third, Lebanon is a microcosm of the Middle East. The issues that have kept Lebanon in continuous turmoil for nearly a generation are the same problems that must be resolved if peace and stability are to come to the whole region (Lamb 1987). These issues revolve around the Israeli-Palestinian conflict; conflict among the various religious, ethnic, and family factions; rapid modernization; grossly inequitable distributions of wealth; ruthless and autocratic rulers; and foreign interference in internal affairs.

With regard to foreign involvement, the various Christian and Muslim factions within Lebanon have been supported with arms from the former Soviet Union, the United States, Libya, Saudi Arabia, France, Britain, Iran, Iraq, Israel, and Syria. Moreover, at one time or another, many foreign armies— the Palestinian Liberation Organization (P.L.O.) and Syrian, Israeli, American, British, and Italian forces—have occupied Lebanon and have aligned themselves with one faction or another in order to stabilize the country or to install leaders friendly to their respective governments (Fisk 1990). If anything, foreign interference has helped harden the boundaries between various factions. A close look at Lebanon can help us understand better the focus at work in this volatile and vital region. We begin with a closer look at the roots and the nature of the civil wars in Lebanon.

A Society in Turmoil:
The Civil Wars in Lebanon

Although Lebanon has a population of three million people, it is physically smaller than the state of Connecticut. Bounded on the west by the Mediterranean Sea, on the east by Syria, and on the south by Israel, Lebanon is a gateway to the Middle East and Asia and a center of trade and transportation. Between 1516 and 1916, Lebanon was part of a territory within the Ottoman Empire known as Greater Syria (now Lebanon, Syria, Jordan, and Israel). After the Allies defeated the Ottoman Empire in World War I, they carved Greater Syria into a number of nation-states. Britain took control of the area now known as Jordan and Israel; France took charge of the area now known as Syria and Lebanon.

In 1920 the French created the Republic of Lebanon by combining various sectarian communities living in the northern mountains (predominantly Maronite Christian communities[1]), the southern mountains (predominantly Druze communities), the port cities of Tripoli, Sidon, and Tyre (predominantly Sunni areas), and the Akkar and the Bekaa valleys (predominantly Shia areas). A **sectarian community** is a geographically distinct group of people who profess allegiance to a particular religion, such as Islam or Christianity, and who also have strong ties to a powerful family, clan, or ethnic group (U.S. Department of the Army 1989). By joining together this mosaic of communities, the French hoped to create a country in the Middle East in which Christians were the majority and at the same time to ensure that Lebanon was economically viable. To meet this second objective, the French had to include port cities and agricultural lands that were populated by Sunni Muslims, Shia Muslims, and Druze (members of an offshoot Muslim sect).

The French also wrote a constitution that allocated government power on the basis of religious affiliation and ensured that Maronite Christians held the most powerful positions (president and commander-in-chief of the army). The French, however, still retained ultimate control over government policies and decisions. Except for the Maronites, most of the religious communities within

in Lebanon and Syria opposed the creation of Lebanon. Many, for example, wished to remain connected to Syria. The French formed coalitions with various communities to put down rebellions by other communities. In 1925, for example, they brutally suppressed a Druze rebellion with the help of Armenians who had emigrated recently to the mountains of Lebanon after millions of their people were killed by the Turks.

During World War II the French, British, Turks, Germans, Austrians, and Hungarians formed coalitions with the various sectarian communities, further factionalizing Lebanon. In 1943, after the Vichy French had been ousted by a British-led force, Lebanon was granted independence. (The French, however, did not fully relinquish their control until 1946.) The Maronite president worked with Sunni leaders to create the National Pact, an unwritten agreement that outlined power sharing and other issues. Essentially the pact called for a Maronite president and commander-in-chief of the Lebanese army, a Sunni prime minister, a Shia speaker of the parliament, and a Druze chief of staff. The legislature, the civil service, and the army were to be staffed according to a ratio of six Christians to every five Muslims. In addition, the pact stipulated that Lebanon would be completely independent; Christians were not to form alliances with the West and Muslims were not to form alliances with Arab states. Arabic would replace French as the official language of the country (although the Lebanese acknowledged spiritual and intellectual ties with France). Finally, Lebanon would belong to the family of Arab states but would remain neutral in conflicts between other Arab states.

At the time of Lebanon's independence, there were approximately 18 religions in the country. The four largest groups were composed of Maronite Christians, Sunni Muslims, Shia Muslims, and Druze. Religious loyalties were complicated by loyalties to powerful families within Lebanon (Norton 1986; U.S. Department of the Army 1989).

Shortly after attaining independence, Lebanon opened its border to 120,000 Palestinian refugees who fled the newly created state of Israel. Most of

Palestinian refugees at Shatila Refugee Camp, Beirut.
J.P. Laffont/Sygma

these refugees settled in southern Lebanon and in United Nations refugee camps on the fringes of Tyre, Beirut, and Tripoli. Because the majority of refugees were Sunnis, the Maronites distrusted them, regarding them as potential allies of Lebanon's Muslim groups.

Then events elsewhere in the Middle East complicated foreign relations for Lebanon. In the mid-1950s, under the leadership of Jamal Abdul Nasser, Egypt nationalized the British-owned Suez Canal and merged with Syria to form the United Arab Republic. At that time Lebanon faced pressures, from both within and outside, to join the larger pan-Arab movement for self-determination. The Lebanese prime minister, fearing that he might be overthrown by the movement, invoked the Eisenhower Doctrine, which guaranteed that the signing countries would receive military assistance from the United States if they were threatened by a communist or communist-backed country. The Lebanese president argued that the Soviet-armed Syrians were arming Lebanese Muslims. U.S. troops accordingly were sent to Lebanon to keep the peace.

Between 1958 and 1964, when Fuad Shihab was president, Lebanon enjoyed prosperity, and relationships between sectarian communities were more stable than ever. Shihab asked the United States to withdraw its troops from Lebanon, and he pursued friendly but neutral relations with both Arab and Western countries:

> *[Shihab] was determined to observe the terms of the National Pact and to have the government serve Christian and Muslim groups equally. [He] also concentrated on improving Lebanon's infrastructure, developing an extensive road system, and providing running water and electricity to remote villages. Hospitals and dispensaries were built in many rural areas . . . .*
> (U.S. DEPARTMENT OF THE ARMY 1989, P. 24)

In 1967 a second wave of Palestinian refugees fled to Lebanon after Israel took control of the West Bank of the Jordan River, East Jerusalem, and Gaza in the Six-Day War (see Chapter 5). A third wave of Palestinian refugees emigrated to Lebanon in the wake of the Jordanian-Palestinian civil war, bringing the total number of Palestinians in Lebanon to 350,000 (Fisk 1990). The P.L.O. created a state within a state, using southern Lebanon to train its military force and to launch military operations against Israel. Consequently, southern Lebanon became a war zone. The Lebanese army was not strong enough to neutralize the fighting between the P.L.O. and Israel. Maronite leaders also feared that the army might become polarized—that the Christian soldiers would favor Israel and the Muslim soldiers would favor the Palestinians. The

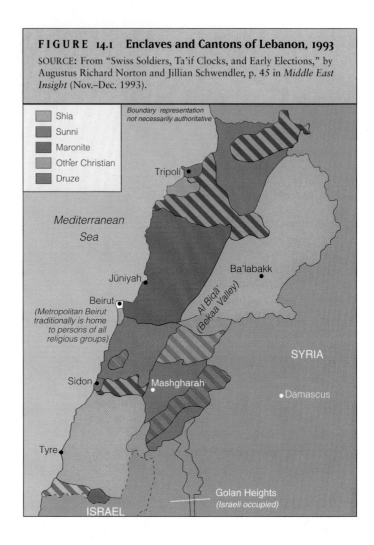

FIGURE 14.1   Enclaves and Cantons of Lebanon, 1993

SOURCE: From "Swiss Soldiers, Ta'if Clocks, and Early Elections," by Augustus Richard Norton and Jillian Schwendler, p. 45 in *Middle East Insight* (Nov.–Dec. 1993).

war in southern Lebanon triggered a massive migration of Shia Muslims to Beirut, where they were forced to live in the "Belt of Misery. They lived at the gates of Beirut, but the city never really admitted them—not socially, not politically, and not economically. By the early 1980s, the [Shia] of Lebanon were the largest single religious community in the country, making up close to half the total population . . . and were looked down on by the Sunni aristocracy as much as by the Maronites" (Friedman 1989, p. 226).

Many of the Palestinians formed coalitions with Lebanese Muslim factions, who were increasingly dissatisfied with the power-sharing arrangement. The latter argued that Christians were no longer in the majority and that power must be reallocated to reflect the religious composition of the society. The Christian leaders refused to allow a census to be conducted in order to assess the merits of the Muslims' claim. The details of the events at this point are quite complicated but the themes are clear: in subsequent years the balance of power was adjusted in a most violent and bloody manner, and the various rivals in Lebanon (up to 20 identifiable political groups) formed private militias and coalitions "in order to contain whoever among them might be emerging as a predominant power" (Whetten 1979, p. 75).[2]

The civil wars divided the city of Beirut into east (the Christian side) and west (the Muslim side), and the country of Lebanon for the most part into autonomous cantons and enclaves (see Figure 14.1).

Although religiously integrated areas such as the Hamra neighborhood in Beirut still exist, many people who had lived in integrated areas before 1975 moved to areas where their own group predominated. The frequent interaction with and dependence on people in an enclave, the geographic isolation, the hardships of war, and the shared religious beliefs promoted a strong sense of solidarity with and loyalty to the group. Intrusion into these areas by outsiders was frequently interpreted as a "surrender of their identity, an outright attack upon their religious or sectarian status" (Friedman 1989).

As of 1993, all the militias in Lebanon were disarmed, with the exception of the Hezbollah (or Party of God). The U.S. government refuses to lift a ban on citizens' travel to Lebanon, however, until the Iranian-, Syrian-, and Lebanese-backed Hezbollah is disarmed. Although the Prime Minister of Lebanon declared the civil war over in 1993, 40,000 Syrian troops were stationed in Lebanon to maintain control over certain strategic areas in southern Lebanon and Bekaa. In addition, conflict continues unabated between Israeli forces in Israel's self-proclaimed security zone (which includes 10 percent of Lebanon's territory) and various anti-Israeli guerrilla forces located in southern Lebanon. Both sides launch attacks against the other and then retaliate when the other responds. Usually the fighting is low-grade but at various times it has become intense, as in July 1993 when Israel responded to guerrilla attacks and systematically bombed and launched an estimated 13,000 shells into southern Lebanon in an effort to destroy some 70 towns and villages just north of the security zone, making the area uninhabitable for the various guerrilla militias. The military action caused massive refugee flows (150,000 to 200,000) from the area (Hedges 1993; Murphy 1993a, 1993b).

In view of this information, we ask, What role does religion play in conflicts such as those in Lebanon? To identify religion's contribution, we must first define religion. This is a surprisingly difficult task and one with which sociologists have been greatly preoccupied. The discussion that follows is a general overview of the nature of religion. When we examine the functions of religion and related concepts, we begin to understand how religion is intertwined with the strife in Lebanon. We will learn that the civil wars are based not on religion itself, but on the ways in which people use religion.

What Is Religion?
Weber and Durkheim's View

In the opening sentences of *The Sociology of Religion*, Max Weber stated: "To define 'religion,' to say what it is, is not possible at the start of a presentation such as this. Definition can be attempted, if at all, only at the conclusion of the study" (1922, p. 1). Despite Weber's keen interest and his extensive writings about religious activity, he could offer only the broadest of definitions: religion encompasses those human responses that give meaning to the ultimate and inescapable problems of existence—birth, death, illness, aging, injustice, tragedy, and suffering (Abercrombie and Turner 1978). To Weber, the hundreds of thousands of religions, past and present, represented a rich and seemingly endless variety of responses to these problems. In view of this variety, he believed it was virtually impossible to capture the essence of religion in a single definition.

Like Max Weber, Emile Durkheim believed that there was nothing as vague and diffuse as religion. In the first chapter of his book *The Elementary Forms of the Religious Life*, Durkheim cautioned that when studying religions, sociologists must assume that "there are no religions which are false" ([1915] 1964, p. 3). Like Weber, Durkheim believed that all religions are true in their own fashion—all address in different ways the problems of human existence. Consequently, Durkheim said, those who study religion must first rid themselves of all preconceived notions of what religion should be. We cannot attribute to religion the characteristics that reflect only our own personal experiences and preferences.

In *The Spiritual Life of Children*, psychiatrist Robert Coles recounts his conversation with a 10-year-old Hopi girl, which illustrates Durkheim's point. The conversation reminds us that if we approach the study of religion with preconceived notions, we will lose many insights about the nature of religion in general:

> *"The sky watches us and listens to us. It talks to us, and it hopes we are ready to talk back. The sky is where the God of the Anglos lives, a teacher told us. She [the teacher] asked where our God lives. I said, 'I don't know.' I was telling the truth! Our God is the sky, and lives wherever the sky is. Our God is the sun and the moon, too; and our God is our [the Hopi] people, if we remember to stay here [on the consecrated land]. This is where we're supposed to be, and if we leave, we lose God."*
> *[The interviewer then asked the child if she had explained all of this to the teacher.]*
> *"No."*
> *"Why?"*
> *"Because—she thinks God is a person. If I'd told her, she'd give us that smile."*
> *"What smile?"*
> *"The smile that says to us, 'You kids are cute, but you're dumb; you're different—and you're all wrong!'"*
> *"Perhaps you could have explained to her what you've just tried to explain to me."*
> *"We tried that a long time ago; our people spoke to the Anglos and told them what we think, but they don't listen to hear us; they listen to hear themselves."* (COLES 1990, P. 25)

The conversation between Coles and the Hopi child shows that the teacher's preconceived notions of what constitutes religion closed her off to other kinds of religious beliefs and experiences.

Indeed, the nature of religion is elusive. And there are many varieties of religious experiences (see "The World's Major Non-Christian Religions"). In spite of these obstacles, Durkheim identified three essential features that he believed were common to all religions, past and present: (1) beliefs about the sacred and the profane, (2) rituals, and (3) a community of worshipers. Thus Durkheim defined **religion** as a system of shared rituals and beliefs about the sacred that bind together a community of worshipers.

Beliefs About the Sacred and the Profane

At the heart of all religious belief and activity stands a distinction between two separate and opposing domains—the sacred and the profane. The **sacred** includes everything that is regarded as extraordinary and that inspires in believers deep and absorbing sentiments of awe, respect, mystery, and reverence. The deep and absorbing sentiments motivate people to safeguard what is sacred from contamination or defilement. In order to find, preserve, or guard that which they consider sacred, people have gone to war, sacrificed their lives, traveled thousands of miles, and performed other life-endangering acts (J. Turner 1978).

Definitions of what is sacred vary according to time and place. Sacred things include objects (chalices, sacred documents, and books), living creatures (cows, ants, birds), elements of nature (rocks, mountains, trees, the sea, sun, moon, or sky), places (churches, mosques, synagogues, birthplaces of religious founders), days that commemorate holy events, abstract forces (spirits, good, evil), persons (Christ, Buddha, Moses, Muhammad, Zarathustra, Nanak), states of consciousness (wisdom, oneness with nature), past events (the crucifixion, the resurrection, the escape of the Jews from Egypt, the birth of Buddha), ceremonies (baptism, marriage, burial), and other activities (holy wars, just wars, confession, fasting, pilgrimages). Durkheim ([1915] 1964) maintained that the sacredness springs not from the item, ritual, or event itself, but from its symbolic power and from the emotions that people experience when they think about the sacred thing or when they are in its presence. The emotions are so strong that believers feel part of something larger than themselves and are outraged when others behave inappropriately in the presence of the sacred.

Ideas about what is sacred are such an important element of religious activity that many researchers classify religions according to the type of phenomenon that their followers consider sacred. One such typology consists of three categories: sacramental, prophetic, and mystical religions (Alston 1972). These categories are not mutually exclusive, however, because most religions identify phenomena from each category as sacred. Consequently, religions can be placed in a single category, although one of these categories usually predominates.

In **sacramental religions** the sacred is sought in places, objects, and actions believed to house a god or a spirit. These may include inanimate objects (relics, statues, crosses), animals, trees, plants, foods, drink (wine, water), places, and certain processes (such as the way in which people prepare for a hunt or perform a dance). Most primitive religions can be classified as sacramental.

In **prophetic religions** the sacred revolves around items that symbolize significant historical events or around the lives, teachings, and writings of great people. Sacred books such as the Christian Bible, the Muslim Koran, and the Jewish Torah hold the records of these events and revelations. In the case of historical events, God or some other higher being is believed to be directly involved in the course and the outcome of the event (a flood, the parting of the Red Sea, the rise and fall of an empire). In the case of great people, the lives and inspired words of prophets or messengers reveal a higher state of being, "the way," a set of ethical principles, or a code of conduct. Followers seek to imitate this life. Some of the best-known prophetic religions include Judaism (as revealed to Abraham in Canaan and to Moses at Mt. Sinai), Confucianism (founded by Confucius), Christianity (founded by Jesus Christ), and Islam (founded by Muhammad).

In **mystical religions** the sacred is sought in states of being that, at their peak, can exclude all awareness of one's existence, sensations, thoughts, and surroundings. In such states the mystic is caught up so fully in the transcendental experience that all earthly concerns seem to vanish. Direct union with the divine forces of the universe is of utmost importance. Not surprisingly, mystics tend to become involved in practices such as fasting or celibacy in order to separate themselves from worldly attachments. In addition, mystics meditate to clear their minds of worldly concerns, "leaving the soul empty and receptive to influences from the divine" (Alston 1972, p. 144). Buddhism and philosophical Hinduism are two examples of religions that emphasize physical and spiritual discipline as a means of transcending the self and earthly concerns.

According to Durkheim, the sacred encompasses more than the forces of good: "There are gods [that is, satans] of theft and trickery, of lust and war, of sickness and of death" ([1915] 1964, p. 420). Evil and its various representations, however, are almost always portrayed as inferior and subordinate to the forces of good: "in the majority of cases we see the good victorious over evil, life over death, the powers of light over the powers of darkness" (p. 421). Even so, Durkheim considers evil phenomena to be sacred because they are endowed with special powers and are the object of rituals (confessions, baptisms, penance, fasting, exorcism) designed to overcome or resist their negative influences.

One such ritual—an exorcism—was aired in April 1991 on ABC's television program "20/20" with the consent of a Roman Catholic bishop (although not without controversy). The ritual, which has been part of Catholicism since medieval times, is performed to cleanse the bodies and souls of those possessed by satanic influences (Goodman 1991). The consenting bishop supported his decision to permit the broadcast, in spite of the risk of being accused of sensationalism, with the argument that showing the exorcism is a way to prove that the devil "is powerful and actively at work in the world" (Steinfels 1991, p. B1).

Religious belief, doctrines, legends, and myths detail the origins, virtues, and powers of sacred things and describe the consequences of mixing the sacred with the profane. The **profane** is everything that is not sacred, including things opposed to the sacred (the unholy, the irreverent, the contemptuous, the blasphemous) and things that stand apart from the sacred, although not in opposition to it (the ordinary, the commonplace, the unconsecrated, the temporal, the bodily) (Ebersole 1967). Believers often view contact between the sacred and the profane as dangerous and sacrilegious, as threatening the very existence of the sacred, and as endangering the fate of the person who allowed the contact. Consequently, people take action to safeguard those things they regard as sacred by separating them from the profane. For example, some believers refrain from speaking the name of God in frustration; and they believe that a woman must cover her hair or her face during worship and that a man must remove his hat during worship.

The World's Major Non-Christian Religions

Buddhism

Buddhism has 307 million followers. It was founded by Siddhartha Gautama, known as the Buddha (Enlightened One), in southern Nepal in the sixth and fifth centuries B.C. The Buddha achieved enlightenment through meditation and gathered a community of monks to carry on his teachings. Buddhism teaches that meditation and the practice of good religious and moral behavior can lead to Nirvana, the state of enlightenment, although before achieving Nirvana one is subject to repeated lifetimes that are good or bad depending on one's actions (*karma*). The doctrines of the Buddha describe temporal life as featuring "four noble truths": existence is a realm of suffering; desire, along with the belief in the importance of one's self, causes suffering; achievement of Nirvana ends suffering; and Nirvana is attained only by meditation and by following the path of righteousness in action, thought, and attitude.

Confucianism

A faith with 5.6 million followers, Confucianism was founded by Confucius, a Chinese philosopher, in the sixth and fifth centuries B.C. Confucius's sayings and dialogues, known collectively as the *Analects,* were written down by his followers. Confucianism, which grew out of a strife-ridden time in Chinese history, stresses the relationship between individuals, their families, and society, based on *li* (proper behavior) and *jen* (sympathetic attitude). Its practical, socially oriented philosophy was challenged by the more mystical precepts of Taoism and Buddhism, which were partially incorporated to create neo-Confucianism during the Sung dynasty (A.D. 960–1279). The overthrow of the Chinese monarchy and the Communist revolution during the twentieth century have severely lessened the influence of Confucianism on modern Chinese culture.

Hinduism

A religion with 648 million followers, Hinduism developed from indigenous religions of India in combination with Aryan religions brought to India c. 1500 B.C. and codified in the Veda and the Upanishads, the sacred scriptures of Hinduism. Hinduism is a term used to broadly describe a vast array of sects to which most Indians belong. Hindu beliefs include the acceptance of the caste system, which ranks people from birth based on religious practice, employment, locale, and tribal affiliation, among other categories, and classifies society at large into four groups: the Brahmins or priests, the rulers and warriors, the farmers and merchants, and the peasants and laborers. The goals of Hinduism are release from repeated reincarnation through the practice of yoga, adherence to Vedic scriptures, and devotion to a personal guru. Various deities are worshiped at shrines; the divine trinity, representing the cyclical nature of the universe, are Brahma the creator, Vishnu the preserver, and Shiva the destroyer.

Islam

Islam has 840 million followers. It was founded by the prophet Muhammad, who received the holy scriptures of Islam, the Koran, from Allah (God) c. A.D. 610. Islam (Arabic for "submission to God") maintains that Muhammad is the last in a long line of holy prophets, preceded by Adam, Abraham, Moses, and Jesus. In addition to being devoted to the Koran, followers of Islam (Muslims) are devoted to the worship of Allah through the Five Pillars: the statement "There is no god but God, and Muhammad is his prophet"; prayer, conducted five times a day while facing Mecca; the giving of alms; the keeping of the fast of Ramadan during the ninth month of the Muslim year; and the making of a pilgrimage at least once to Mecca, if possible. Consumption of pork and alcohol, as well as usury, slander, and fraud, are prohibited. The two main divisions of Islam are the Sunni and the Shiite [or Shia]; the Wahabis are the most important Sunni sect, while the Shiite sects include the Assassins, the Druses [or Druze], and the Fatimids, among countless others.

Judaism

Stemming from the descendants of Judah in Judea, Judaism was founded c. 2000 B.C. by Abraham, Isaac, and Jacob and has 18 million followers. Judaism espouses belief in a monotheistic God, who is creator of the universe and who leads His people, the Jews, by speaking through prophets. His word is revealed in the Hebrew Bible (or Old Testament), especially in that part known as the Torah. Jews believe that the human condition can be improved, that the letter and the spirit of the Torah must be followed, and that a Messiah will eventually bring

The famous Daibutsu in Kamakura, Japan.
Frilet/Sipa Press

the world to a state of paradise. Judaism promotes community among all people of Jewish faith, dedication to a synagogue or temple (the basic social unit of a group of Jews, led by a rabbi), and the importance of family life. Religious observance takes place both at home and in temple. Judaism is divided into three main groups who vary in their interpretation of those parts of the Torah that deal with personal, communal, international, and religious activities: the Orthodox community, which views the Torah as derived from God, and therefore absolutely binding; the Reform movement, which follows primarily its ethical content; and the Conservative Jews, who follow most of the observances set out in the Torah but allow for change in the face of modern life.

Shinto

Shinto, with 3.5 million followers, is the ancient native religion of Japan, established long before the introduction of writing to Japan in the fifth century A.D. The origins of its beliefs and rituals are unknown. Shinto stresses belief in a great many spiritual beings and gods, known as *kami*, who are paid tribute at shrines and honored by festivals, and reverence for ancestors. While there is no overall dogma, adherents of Shinto are expected to remember and celebrate the *kami*, support the societies of which the *kami* are patrons, remain pure and sincere, and enjoy life.

Taoism

Both a philosophy and a religion, Taoism was founded in China by Lao-tzu, who is traditionally said to have been born in 604 B.C. Its number of followers is uncertain. It derives primarily from the *Tao-te-ching*, which claims that an ever-changing universe follows the Tao, or path. The Tao can be known only by emulating its quietude and effortless simplicity; Taoism prescribes that people live simply, spontaneously, and in close touch with nature and that they meditate to achieve contact with the Tao. Temples and monasteries, maintained by Taoist priests, are important in some Taoist sects. Since the Communist revolution, Taoism has been actively discouraged in the People's Republic of China, although it continues to flourish in Taiwan.

SOURCE: Excerpts from *The New York Public Library Desk Reference* (pp. 189–91). A Stonesong Press Book. Copyright © 1989 by The New York Public Library and The Stonesong Press, Inc. Used by permission of the publisher, Webster's New World/A division of Simon & Schuster, New York.

The distinctions between the sacred and the profane do not mean that a person, object, or idea cannot pass from one domain to another or that something profane cannot ever come into contact with the sacred. Such transformations and contacts are authorized through rituals—the active and most observable side of religion. For example, people whom Christianity defines as "profane" or pagan can be saved through baptism.

Rituals

In the religious sense, **rituals** are rules that govern how people must behave in the presence of the sacred. These rules may take the form of instructions detailing the appropriate context, the roles of various participants, acceptable attire, and the precise wording of chants, songs, and prayers. Participants must follow instructions closely if they want to achieve a specific goal, whether the goal is to purify the participant's body or soul (confession, immersion, fasting, seclusion), to commemorate an important person or event (pilgrimage to Mecca, the Passover, the Last Supper), or to transform profane items into sacred items (water to holy water, bones to sacred relics). During rituals, behavior is "coordinated to an inner intention to make contact with, or to participate in, the invisible world" or to achieve a desired state (Smart 1976, p. 6).

Rituals can be as simple as closing the eyes to pray or having the forehead marked with ashes; they can be an elaborate combination and sequence of activities such as fasting for three days before entering a sacred place to chant, with head bowed, a particular prayer for forgiveness. Although rituals usually are enacted in sacred places, some rituals are codes of conduct aimed at governing the performance of everyday activities—sleeping, waking, eating, defecating, washing, dealing with the opposite sex. Durkheim maintains that the nature of the ritual is relatively unimportant. The important element is that the ritual be shared by a community of worshipers and evoke certain ideas and sentiments that help individuals feel part of something bigger than themselves.

Community of Worshipers

Durkheim used the word **church** to designate a group whose members hold the same beliefs with regard to the sacred and the profane, who behave in the same way in the presence of the sacred, and who gather in body or spirit at agreed-upon times to reaffirm their commitment to those beliefs and practices. Obviously, religious beliefs and practices cannot be unique to an individual; they must be shared by a group of people. If this were not the case, the beliefs and practices would cease to exist when the individual who held them died or if he or she chose to abandon them. In the social sense, religion is inseparable from the idea of church. The gathering and the sharing create a moral community and give worshipers a common identity (see "Spiritual Journey"). The gathering, however, need not take place in a common setting. When people perform a ritual on a given day or at given times of day, the gathering is spiritual rather than physical.

Durkheim used the term *church* loosely, acknowledging that it could assume many forms: "Sometimes it embraces an entire people . . . sometimes it embraces only a part of them . . . sometimes it is directed by a corps of priests, sometimes it is almost completely devoid of any official directing body" ([1915] 1964, p. 44). Sociologists have identified at least five broad types of religious organizations (communities of worshipers): ecclesiae, denominations, sects, established sects, and cults. As is the case with most classification schemes, the categories overlap on some characteristics because the criteria by which religions are classified are not always clear.

Ecclesiae An **ecclesia** is a professionally trained religious organization governed by a hierarchy of leaders, which claims as its members everyone in a society. Membership is not voluntary; it is the law. Consequently, considerable political alignment exists between church and state officials, which makes the ecclesiae the official church of the state. Ecclesiae formerly existed in England (the Anglican church), France (the Roman Catholic church), and Sweden (the Lutheran church). Today Islam is the official religion of Bangladesh and Malaysia; Iran has been an Islamic republic since the Ayatollah Khomeini took power in 1979; and Saudi Arabia is a monarchy based on Islamic law (*The World Almanac and Book of Facts 1991* 1990).

Individuals are born into ecclesiae; newcomers to the society are converted; dissenters often are persecuted. Those who do not accept the official view emigrate or occupy the most marginal status in the society. The ecclesia claims to be the one true faith and often does not recognize other religions as valid. In its most extreme form it directly controls all facets of life. Such is the case in Saudi Arabia, where social, political, and economic policies are framed according to the Koran. Saudi women's lives are particularly restricted. For example, they are not allowed to drive and must wear veils in public; education is strictly segregated; and women's employment opportunities are limited severely because they cannot work in settings where they mix with men. "This is a nation where Kentucky Fried Chicken shops, Baskin-Robbins ice cream parlors and Pizza Huts are liable to periodic patrol by Islamic vice squads, knows as Muttawa, who zealously enforce closing for prayer hours, restrictions on mingling between the sexes and dress codes to cover women's bodies" (LeMoyne 1990, p. A7).

Denominations A **denomination** is a hierarchical organization in a society in which church and state are usually separate; it is led by a professionally trained clergy. In contrast to an ecclesia, a denomination is one of many religious organizations in the society. For the most part, denominations are tolerant of other religious organizations; they may even collaborate to address and solve some problems in the society. Membership is considered to be voluntary, but most people who belong to denominations did not choose to do so. Rather, they were born to parents who are members. Denominational leaders generally make few demands on the laity, and most members participate in limited and specialized ways: they choose to send their children to church-operated schools, attend church on Sundays and religious holidays, donate money, or attend church-sponsored functions. The leaders of a denomination do not oversee all aspects of the members' lives. Yet, even though laypeople vary widely in lifestyle, denominations frequently attract people of particular races and social classes: that is, their members are drawn disproportionately from specific social and ethnic groups.

There are eight major religious denominations in the world—Buddhism, Christianity, Confucianism, Hinduism, Islam, Judaism, Taoism, and Shinto—each dominant in different areas of the globe. For example, Christianity predominates in Europe, North and South America, New Zealand, Australia, and the Pacific Islands; Islam in the Middle East and North Africa; Hinduism in India.

Sects A **sect** is a small community of believers led by a lay ministry, with no formal hierarchy or official governing body to oversee the various religious gatherings and activities. Sects typically are composed of people who broke away from a denomination because they came to view it as corrupt. Therefore they created the offshoot to reform the religion from which they separated.

All of the major world religions encompass splinter groups that have sought, at one time or another, to preserve the integrity of their religion. In Islam, for example, the most pronounced split occurred 1,300 years ago, approximately 30 years after the death of the Prophet Muhammad, over the issue of Muhammad's successor. The Shia maintained that the successor should be a blood relative of Muhammad; the Sunni believed that the successor should be selected by the community of believers and need not be related by blood. When Muhammad died, the Sunni (the great majority of Muslims) accepted Abu-Bakr as the caliph (successor). The Shia supported Ali, Muhammad's first cousin and son-in-law, and they called for the overthrow of the existing order and a return to the pure form of Islam. The Druze are another religious group that began as a sect of Islam; over time, its members developed such distinct rules and rituals that they are no longer regarded as Muslim.

Similarly, several splits have occurred within the Christian churches. In 1054, for example, the Eastern Orthodox churches rejected the pope as the earthly deputy of Christ and questioned the papal claim of authority over all Catholic churches in the world. The Protestant religions owe their origins to Martin Luther, who also challenged the papal authority and protested many of the practices of the Roman Catholic church. This protest is known as the Reformation because it involved efforts to reform the Catholic church and to cleanse it of corruption, especially with regard to paying for indulgences (the forgiveness of sins upon saying specific prayers or performing specific good deeds at the order of a priest). Luther believed that a person is

Spiritual Journey

The Hajj, or the annual Muslim pilgrimage to Mecca and Medina, embodies the characteristics of a church (as defined by Durkheim): the participants hold the same beliefs about the sacred and profane, behave in the same way in the presence of the sacred, and gather at an agreed-upon time to reaffirm commitment to beliefs and practices.

Ehsan Khan's prayers were finally answered this year.

He, along with 90 other Muslims from Los Angeles and 3 million Muslims from across the world, recently visited Saudi Arabia to complete the Hajj—the pilgrimage to the holy cities of Mecca and Medina.

"It's not like a vacation," said Khan, a Reseda resident who grew up in Canada but whose parents are Pakistani. "I've been asking God for years to be able to go, and this time I was accepted."

Every physically and financially able Muslim is required to complete the Hajj during his or her life. A pilgrim must be an adult to go on the Hajj.

Many people in countries distant from Arabia look upon the completion of the Hajj as the climax of a lifetime. But Khan,who is 32, is one of a growing number of people embarking on the pilgrimage earlier in life.

Hassan Khan, 29, who also recently returned from Mecca, said he is encouraging people to go on the Hajj when they are young.

"To get to the height of spirituality, why wait until you are old to mend or correct your ways?" asked Hassan Khan (no relation to Ehsan Khan), a Sun Valley salesman who came to the United States from Pakistan when he was 18. "You might as well shape up your behavior at an earlier age."

The pilgrimage begins on the eighth day of Zul-Hijjah, the last month of the Islamic year, and ends on the 12th day. The Islamic year runs on a lunar calendar, which is about 10 days shorter than the solar calendar. This year the Hajj started on June 9.

The Saudi government usually allows only 2 million people to visit Mecca during the Hajj, and each country has its own pilgrim quota. Visits to Mecca declined last year because of the Gulf War, so the Saudi government increased the quota to 3 million this year.

Visiting the Kaaba, a cube-shaped brick structure about 40 feet tall, is one of the most important rituals of a pilgrim's stay in Mecca. The Kaaba is considered the House of Allah, and signifies the oneness of God. Every Muslim in the world must pray in the direction of the Kaaba.

"There is no other particular place in the world where you can see people of every color, dress, language and culture, going and doing the same thing," said Ehsan Khan, a certified public accountant. "It's beautiful, and spiritually uplifting. There is nothing like it."

In the old days, people visited the holy city on foot, by camel or by whatever means they had available. Today's pilgrims come by air and sea, and once they arrive, they travel by bus.

In the United States, travel agencies offer Hajj packages. From Los Angeles, the cost of a month in Saudi Arabia, including travel and lodging, is about $3,500 per person. The Saudi Hajj department also assigns each group a guide, and that cost is included in the package.

The pilgrimage is extremely taxing. Visitors are confronted by tremendous heat and crowding, as millions attempt to complete the necessary rituals. The Saudi government provides tents and water, as well as medical facilities, to make the process easier and safer. But historically, it has not been uncommon for older people to die during the Hajj. It is considered a blessing to die and be buried in Mecca.

Despite the heat and crowding, those who made the pilgrimage from the United States say the hardships they endured were worth it.

"Before I went, I didn't know what to expect," said Artis Terrell, 41, a South Gate chiropractor who

---

saved not by the intercession of priests or bishops, but by private and individual faith (Van Doren 1991). Divisions also exist within various Protestant sects and between Catholic sects or denominations. Lebanon, for example, contains several offshoots of the Roman Catholic church, including Maronites, Greek Catholics, Greek Orthodox, Jacobites, and Gregorians.

Pilgrims at Mecca.
Mohamed Dunes/Gamma-Liaison

converted to Islam in 1973. "But once I was there, in Mecca, at the Kaaba, the place where we prostrate ourselves five times a day, it's an overwhelming feeling. Going there is like a personal invitation God extends to the believer."

Muslims pray five times a day facing Mecca. Those prayers represent the second of the five pillars of Islam. The Hajj represents the fifth.

Before the pilgrims enter Mecca, the men must put on the *ihram* garments, two white seamless sheets wrapped around the body to symbolize the equality of every man in the eyes of God. Women retain their usual Muslim dress, and are completely covered except for their faces and hands.

Jane Farra, a Shadow Hills resident who converted to Islam 11 years ago, is accustomed to Western-style dress. But she said she had no problem dressing as a Muslim woman while completing the Hajj.

"It felt fine, it felt perfect," said Farra, 53, whose husband of 32 years is Muslim. "You are doing a religious act, and this is the way you are told to dress at the time. It was no problem.

When you are there, you don't think of men and women, you are just a Muslim among all those millions of Muslim people," she said. "For myself, it was the most wonderful thing. I felt close to my religion and in the presence of God."

Once the pilgrims enter Mecca, they walk seven times around the Kaaba inside the Great Mosque, and kiss or gesture a kiss to the Black Stone, the oldest part of the Kaaba. The Muslims believe that the Black Stone came from the heavens in the time of Abraham.

Next they pray in the direction of the Kaaba, drink the holy water from the well of Zam Zam, and walk briskly seven times between the small hills of Mt. Safa and Mt. Marwah. Each action symbolizes the search for water by Hagar, the wife of Abraham, according to the Koran.

At the second stage of the ritual, the pilgrims visit the holy places outside Mecca, including Arafat and Mina. The Hajj is considered incomplete if the pilgrim is not in Arafat on the ninth day. In Muslim tradition, Arafat is the place where Adam and Eve first met.

At Mina, the pilgrim throws seven stones on three successive days at three pillars representing Satan. Later, an animal is sacrificed in commemoration of Abraham's potential sacrifice of his son Ishmael. Afterward, the male pilgrim cuts his hair or shaves his head.

Before he leaves Mecca, he again circles the Kaaba seven times. Either before or after the Hajj, pilgrims usually visit the Prophet's Mosque in Medina, where Mohammed, the founder of Islam, is buried. Only Muslims are allowed inside Mecca and Medina.

"As a Muslim, once you go and see the Kaaba for the first time, the feeling is not explainable," Ehsan Khan said. "That's the meaning of life right there, the closest you can get to God. You feel swept away, there are no words to describe it."

SOURCE: "Spiritual Journey," by Andrea Heiman. Copyright © 1992 by Andrea Heiman. Reprinted by permission of the author.

People are not born into sects, as they are with denominations; theoretically, they convert. Consequently, newborns are not baptized; they choose membership later in life, when they are considered able to decide for themselves. Sects vary on many levels, including the degree to which they view society as religiously bankrupt or corrupt and the extent to which they take action to change people in society.

Established Sects In some ways, **established sects** resemble both denominations and sects. They are renegades from denominations or ecclesiae but have existed long enough to acquire a significantly large membership and to achieve respectability. At this stage in their history it is appropriate to refer to groups such as the Druze, the Shia, and the Baptists as established sects.

Cults Generally, **cults** are very small, loosely organized groups, usually founded by a charismatic leader who attracts people by virtue of his or her personal qualities. Because the charismatic leader plays such a central role in attracting members, cults often dissolve after the leader dies. For this reason few cults last long enough to become established religions. Even so, a few manage to survive, as evidenced by the fact that the major world religions began as cults. Because cults are formed around new and unconventional religious practices, outsiders view them with considerable suspicion.

These groups vary according to purpose and to the level of commitment that the cult leaders demand of converts. Cults may draw members on the basis of highly specific but eccentric interests such as astrology, UFOs, or transcendental meditation. Members may be attracted by the promise of companionship, of a cure from illness, of relief from suffering, or of enlightenment. A cult may meet infrequently and strictly voluntarily (such as at conventions or monthly meetings), or the cult leaders may require members to break all ties with family, friends, and jobs, and thus come to rely exclusively on the cult to meet all their needs.

A Critique of Durkheim's Definition of Religion

Durkheim's definition of religion revolves around the most outward, most visible characteristics of religion. Critics argue, however, that the three essential characteristics—beliefs about the sacred and the profane, rituals, and a community of worshipers—are not unique to religious activity. This combination of characteristics, they say, can be found at many gatherings—sporting events, graduation ceremonies, reunions, and political rallies—and in many political systems (for example, Marx-

ism, Maoism, and fascism). On the basis of these characteristics alone, it is difficult to distinguish between an assembly of Christians celebrating Christmas, a patriotic group supporting the initiation of a war against another country, and a group of fans eulogizing James Dean (see the Focus article at the end of this chapter).

In other words, religion is not the only unifying force in society to make use of these three elements. **Civil religion,** another such force, is "any set of beliefs and rituals, related to the past, present and/or future of a people (nation), which are understood in some transcendental fashion" (Hammond 1976, p. 171). A nation's beliefs (such as individual freedom or equal opportunity) and rituals (parades, fireworks, singing of the national anthem, 21-gun salutes, and so on) often assume a sacred quality. Even in the face of internal divisions based on race, ethnicity, region, religion, or gender, national beliefs and rituals can inspire awe, respect, and reverence for the country. These sentiments are most notable on national holidays that celebrate important events or people (such as Washington's Birthday, Martin Luther King, Jr. Day, or Independence Day), in the presence of national monuments or symbols (the flag, the Capitol, the Lincoln Memorial, the Vietnam Memorial), and at times of war or other national crises.

Often political leaders appeal to these sentiments in order to win an election, to legitimate their policies, to rally a nation around a cause that requires sacrifice, or to motivate a people to defend their country, as President George Bush attempted to do in his State of the Union Address on January 7, 1991:

> *I come to this house of the people to speak to you and all Americans, certain that we stand at a defining hour.*
>
> *Halfway around the world, we are engaged in a great struggle in the skies and on the seas and sands. We know why we're there. We are Americans—part of something larger than ourselves.*
>
> *For two centuries, we've done the hard work of freedom. And tonight we lead the world in facing down a threat to decency and humanity.*
>
> *For two centuries, America has served the world as an inspiring example of freedom and*

Many civil celebrations, such as honoring Martin Luther King Day, share some of the characteristics and unifying function of religious observances.
J.P. Laffont/Sygma

democracy. For generations, America has led the struggle to preserve and extend the blessings of liberty. And today, in a rapidly changing world, American leadership is indispensable. Americans know that leadership brings burdens, and requires sacrifice.

But we also know why the hopes of humanity turn to us. We are Americans; we have a unique responsibility to do the hard work of freedom. And when we do, freedom works.
(BUSH 1991)[3]

Unifying forces other than civil religion can be blood ties, the village, or a powerful leader. In Lebanon and in most Middle Eastern countries, the people feel a strong loyalty toward their families (see "The Family in Lebanon"), which in turn are connected to other powerful family households. In addition, each community—Roman Catholic, Druze, Shia, and so on—is essentially a nation within a country and is linked to the ruling power through its leader:

The Lebanese individual traditionally derived his social identity and psychological support from . . . family, neighborhood, or religious community, but rarely from the [country] as a whole. He was always a Druse, a Maronite, or a Sunni before he was a Lebanese; and he was always a member of the Arslan or Jumblat Druse clans before he was a Druse, or a member of the Gemayel or Franjieh Maronite clans before he was a Maronite. The civil war and the Israeli invasion only reinforced this trend, dividing Lebanese into tighter-knit micro-families, or village and religious communities, but pulling them further apart as a [country]. (FREIDMAN 1989, P. 46)

The traits that Durkheim cites as characteristics of religion apply to other events, relationships, and forces within society.

The shortcomings of Durkheim's definition of religion are also reflected in other definitions of religion. Sociologist Herbert Spencer suggested that religion is the recognition that all things are derived from or are dependent on a "Power" that transcends our knowledge (Spencer 1972). This definition encounters similar problems as Durkheim's definition. That is, the "Power" need not be a deity; it could refer just as well to nature, magic, astrology, or science.

In *Boundaries: Psychological Man in Revolution*, psychiatrist Robert Jay Lifton described how people who witnessed nuclear explosions viewed the bombs as if they were technological deities "capable of both apocalyptic destruction and unlimited creation" (1969, p. 64). Brigadier General Thomas Farrell wrote the following account after he witnessed the first atomic bomb test at Alamogordo:

The effects could well be called unprecedented, magnificent, beautiful, stupendous and terrifying. No man-made phenomenon of such tremendous power had ever occurred before. . . . The whole country was lighted by a searing

The Family in Lebanon

The primacy of the family manifests itself in all phases of Lebanese life, including political, financial, and personal relationships. In the political sphere, families compete with each other for power and prestige, and kin combine forces to support family members in their quest for leadership. In business, employers give preference to hiring relatives, and brothers and cousins often consolidate their resources in operating a family enterprise. Wealthy family members are expected to share with less prosperous relatives, a responsibility that commonly falls to expatriate and urban relatives who help support their village kin.

In the personal sphere, the family has an equally pervasive role. To a great extent, family status determines an individual's access to education and chances of achieving prominence and wealth. The family also seeks to ensure an individual's conformity with accepted standards of behavior so that family honor will be maintained. An individual's ambitions are molded by the family in accordance with the long-term interests of the group as a whole. Just as the family gives protection, support, and opportunity to its members, the individual member offers loyalty and service to the family.

The traditional form of the family is the three-generation patrilineal extended family, consisting of a man, his wife or wives, their unmarried children of both sexes, and their married sons, together with the sons' wives and children. Some of these groups live under one roof as a single household, which was the norm in earlier generations, but most do not.

The family commands primary loyalty in Lebanese society. In a

A traditional Lebanese family is patrilineal (the family line is traced through the male members and their families) and includes three generations.
L. Delahaye/Sipa Press

study conducted by a team of sociologists at the American University of Beirut in 1959, loyalty to the family ranked first among both Christians and Muslims, among both males and females, and among both politically active and noncommitted students. Next to the family in order of importance were religion, nationality or citizenship, ethnic group, and finally the political party. The results of this study probably reflected the attitudes of the Lebanese in 1987. If anything, primordial ties appear to have increased during the 1975 Civil War. The rise of Islamic and Christian fundamentalism encouraged the development of ethnic and familial consciousness. Among Maronites, there has always been an emphasis on the family; for example, the motto of the Phalange Party is "God, the Fatherland, and the Family."

The family in Lebanon has been a means through which political leadership is distributed and perpetuated. In the Chamber of Deputies of 1960, for example, almost a quarter of the deputies "inherited" their seats. In the 1972 Chamber of Deputies, Amin Jumayyil (who became president in 1982) served with his father, Pierre Jumayyil, after inheriting the seat of his uncle Maurice Jumayyil. Because "political families" have monopolized the representation of certain sects for over a century, it has been argued that family loyalty hinders the development of a modern polity.

SOURCE: From *Lebanon: A Country Study*, edited by Thomas Collelo. Copyright © 1989 by the U.S. Government as represented by the Secretary of the Army. Reprinted with permission.

light with the intensity many times that of the midday sun. It was golden, purple, violet, gray and blue. It lighted every peak, crevasse and mountain range with a clarity and beauty that cannot be described but must be seen to be imagined. It was the beauty the great poets dream about but describe most poorly and inadequately. Thirty seconds after the explosion came, first the air blast pressing hard against people and things, to be followed almost immediately by the strong, sustained awesome roar which warned of doomsday and made us feel that we puny things were blasphemous to dare tamper with the forces heretofore reserved to The Almighty. Words are inadequate tools for the job of acquainting those not present with the physical, mental and psychological effects. It had to be witnessed to be realized. (LIFTON AND HUMPHREY 1984, P. 65)

Other reactions at Alamogordo were summed up, "The Sun can't hold a candle to it," or, better

known, "I am become death, the shatterer of worlds," and "One felt as though he had been privileged to witness the birth of the world" (Lifton 1969, pp. 27–28).

Narrower definitions of religion are also problematic. Suppose that religion were defined as the belief in an ever-living god. This definition would exclude polytheistic religions such as Hinduism, which has more than 640 million adherents. It would also exclude religions in which a deity plays little or no role, such as Buddhism, which has more than 300 million adherents. So, narrow definitions of religion are no improvement over broad ones.

Despite its shortcomings, Durkheim's definition of religion is one of the best and most widely used definitions. No sociologist with any standing in the discipline can study religion without encountering and addressing Durkheim's definition. Besides proposing a definition of religion, Durkheim also wrote extensively about the functions of religion. This work laid the foundation for the functionalist perspective of religion.

Functionalist and Conflict Perspectives on Religion

In addition to identifying the characteristics common to all religions, sociologists also examine the various social functions that religion serves for the individual and the group. Among other things, religion functions to comfort believers in the face of uncertainty, to promote group unity and solidarity, to bind the individual to a group, and to regulate society in the face of severe disturbances and abrupt change. History shows, however, that religion is not always a constructive force. Conflict theorists point out that religion can also have repressive, constraining, and exploitative qualities.

The Functionalist Perspective

As far as we know, some form of religion has existed as long as humans have lived (at least two million years). In view of this fact, functionalists maintain that religion must serve some vital social functions for the individual and for the group. On

the individual level, people embrace religion in the face of uncertainty: they draw on religious doctrine and ritual in order to comprehend the meaning of life and death and to cope with misfortunes and injustices (war, drought, illness). Life would be intolerable without reasons for existing or without a higher purpose to justify the trials of existence (Durkheim 1951). In the wake of the destruction of Kuwait by invading Iraqi soldiers, one Kuwaiti Muslim commented to a reporter for *The New York Times:*

A crisis like this makes a Muslim search within himself for values to counter the bestiality of man. We look around at what they have done, and the only thing you can say is: Thank God, who is, in a situation like this, the only refuge for those who cannot understand why man would do this to his fellow man. (IBRAHIM 1991, P. A7)

People also turn to religious beliefs and rituals to help them achieve a successful outcome (the birth of a healthy child, a job promotion) and to gain answers to questions of meaning: How did we get here? Why are we here? What happens to us when we die?

According to Durkheim, people who have communicated with their god or with other supernatural forces (however conceived) report that they gain the inner strength and the physical strength to endure and to conquer the trials of existence. "It is as though [they] were raised above the miseries of the world. . . . Whoever has really practised a religion knows very well . . . these impressions of joy, of interior peace, of serenity, of enthusiasm which are, for the believer, an experimental proof of his beliefs" (Durkheim [1915] 1964, pp. 416–17).

Religion functions in several ways to promote group unity and solidarity. First, the shared doctrine and rituals help create emotional bonds among those who believe. Second, all religions strive to raise individuals above themselves—to help them live a life better than they would lead if they were left to their own impulses. In this sense, religion offers ideas of proper conduct that carry over into everyday life. When believers violate this code of conduct, they feel guilt and remorse. Such feelings, in turn, motivate them to make amends.

Third, although many religious rituals function to alleviate individual anxieties, uncertainties, and fears, they also establish, reinforce, or renew social relationships, binding individuals to a group. Death, for example, strikes more than the person who dies and the people who had been closest to him or her. Death also disrupts the total network of social relationships in which the deceased once participated. Consequently, this network must be rebuilt. Religious services not only help cushion the blow of death; they also mark the point at which the survivors must meet without the deceased and readjust their relationships to one another for the first time (Chinoy 1961). Sometimes a gathering to honor the dead reinforces the group's identity and sense of purpose. In *Pity the Nation: The Abduction of Lebanon*, Robert Fisk states that in Lebanon "death is important because it provides a mandate for the living." Fisk gives accounts of two Lebanese men whose deaths have only added to their purpose:

More than two years after he was torn apart by a bomb on his helicopter, Rashid Karami still speaks to the Sunnis of Lebanon; on the cassettes sold by street vendors of Hamra Street, Karami still lectures his people on Lebanon's Arab identity, the importance of national unity and the struggle against Zionism.

Bashir [Gemayel] can be heard regularly on the Voice of Lebanon, broadcasting on behalf of the Phalange from transmitters. . . . It is seven years since he was crushed to death in the bombing of his party headquarters in Beirut, but on the air he still shouts to his supporters. You can hear them cheering him on the recordings. A Lebanon free from all foreign interference, a strong Lebanon, a respected Lebanon. (1990, P. 93)

Finally, religion functions as a stabilizing force in times of severe social disturbances and abrupt change. During such times, many regulative forces in society break down. For example, the ongoing civil wars have left Lebanon unable to enforce laws that protect the environment. Fishermen use illegal methods such as dynamite and poison bait to catch large quantities of fish in the Mediterranean Sea. Lebanese industries dump solid waste and chemicals into the sea. The Lebanese have cut down many of the cedar forests for firewood and to prevent camouflage; the resulting deforestation has led to soil erosion and a reduction in supplies of fresh water. (Without the forests, rainwater and topsoil flow unimpeded into the sea.) The bird population has been virtually wiped out because so many people have conducted target practice on birds flying overhead. The forests also are affected by these killings because, with few birds to eat them, tree-eating insects multiply unchecked (Boling 1985).

When such regulative forces are absent, people are more likely to turn to religion in search of a force that will bind them to a group. This tie helps people to think less about themselves and more about some common goal (Durkheim 1951), whether the goal is to work for peace or to participate more fervently in armed conflicts. The fact that religion functions to meet individual and societal needs, in combination with the fact that people create sacred objects and rituals, led Durkheim to reach a controversial but thought-provoking conclusion: the "something out there" that people worship is actually society.

Society as the Object of Worship If we operate under the assumption that "all religions are true in their own fashion" and that the variety of religious responses is virtually endless, we find support for Durkheim's conclusion that everything encompassed by religion—gods, rites, sacred objects—is created by people. That is, people play a fundamental role in determining what is sacred and how people should act in the presence of the sacred. Consequently, at some level, people worship what they (or those before them) have created. This point led Durkheim to conclude that the true object of worship is society itself—a conclusion that many critics cannot accept (Nottingham 1971).

Let us give Durkheim the benefit of the doubt, however, and ask, Is there anything about the nature of society that makes it deserving of such worship? In reply to this question, Durkheim gave what sociologist W. S. F. Pickering called a "virtual hymn to society, a social Gloria in Excelsis" (Pickering 1984, p. 252). Durkheim maintained that society transcends the individual life because it frees us from the bondage of nature (as in nature and nurture). How is this accomplished? Chapter 5 presents cases that show the consequences of extreme isolation, neglect, and limited social contact. Such cases make it clear that "it is impossible for a person to develop without social interaction" (Mead 1940, p. 135). In addition, studies of mature and even otherwise psychologically and socially sound persons who experience profound isolation—astronauts orbiting alone in space, prisoners of war placed in solitary confinement, individuals who volunteer to participate in scientific experiments in which they are placed in deprivation tanks—show that when people are deprived of contact with others, they lose a sense of reality and personal identity (Zangwill 1987). The fact that we depend so strongly on society supports Durkheim's view that for the individual "it is a reality from which everything that matters to us flows" (Durkheim 1984, p. 252).

Durkheim, however, did not claim that society provides us with perfect social experiences: "Society has its pettiness and it has its grandeur. In order for us to love and respect it, it is not necessary to present it *other than it is.* If we were only able to love and respect that which is *ideally perfect,* . . . God Himself could not be the object of such a feeling, since the world derives from Him and the world is full of imperfection and ugliness" (p. 253). Durkheim observed that whenever a group of people have a strong conviction (no matter what kind of group it is), that conviction almost always takes on a religious character. Religious gatherings and affiliations become ways of affirming convictions and mobilizing the group to uphold them, especially when they are threatened. In this sense, religion is used for many purposes. For example, Islam has been used

> to legitimate monarchies (Saudi Arabia and Morocco), military regimes (Pakistan, Libya, and Sudan), and a theocracy (Iran). These self-styled Islamic regimes span the ideological spectrum, from Libya's radical socialist "state of the masses" to the conservative monarchy of Saudi Arabia. Islamic actors display a similar diversity: clerical and lay, traditionalist and modernist, highly educated and illiterate, moderate and terrorist. (ESPOSITO 1986, P. 57)

A Critique of the Functionalist View of Religion To claim that religion functions as a strictly integrative force is to ignore the long history of wars between different religious groups and the many internal struggles among factions within the same religious group. Therefore, in speaking of the integrative function of religion, it is important to specify the segments of society to which this idea applies.

Sociologist Robert K. Merton (1957) argued that if religion were truly integrative in every sense of the word, there would be no conflict or tensions among religious groups within the same society. For example, all of the Lebanese Shia would hold the same views about the future of Lebanon. This is not the case, however. The Shia are divided into at least three camps: (1) the Groups of the Lebanese Resistance (AMAL), who seek an independent Lebanon in which they have a fair share of the power; (2) the Islamic AMAL, a breakaway faction of AMAL allied with Iran; and (3) the Hezbollah (Party of God, code named "Islamic Jihad"), who seek to establish an Islamic republic modeled after Iran (Doerner 1985). Furthermore, the Shia are divided with regard to their support of the Palestinians. Some feel loyalty to the Palestinian cause; others, especially those driven out of southern Lebanon by the fighting, resent the Palestinian presence. The Shia is not the only community with internal factions. On paper there are at least 13 different militias—perhaps as many as seven different organizations. There are also Christian-dominated militias and various Sunni and Druze organizations.

Moreover, if religion were entirely an integrative force, religious beliefs and sacred symbols would never be important to ingroup-outgroup distinctions. That is, religious symbols that unite a community of worshipers would not unite them so strongly that they would be willing to destroy persons who did not belong to their religion. For example, between 1975 and 1992 tens of thousands of Lebanese Christians and Muslims were kidnapped, 14,000 of whom disappeared and are now presumed dead. Many were kidnapped because they were in the wrong parts of town. Most of the kidnappings were retaliatory; that is, a Christian militia would kidnap 10 Muslims and a Muslim militia would respond by taking 10 Christians hostage (Murphy 1993a). The two excerpts that follow show the destructive tendencies that religion can be used to fuel:

> *Being a Christian in Lebanon means much more than what church you go to. I live in East Beirut, the center of the Christian part. We are on one side and the Palestinians, Syrians and Moslems are on the other. We are fighting them so we can free Lebanon from the Palestinians and Syrians.* (SHAMES 1980, P. 6)

> *The city itself is divided along the infamous "Green Line" separating East (Christian) from West (Muslim). In both sectors, rival militias control various suburbs, the boundaries changing with each round of fighting. Imagine traveling from the south side of Minneapolis to its northern suburbs, having to stop and show your passport at each suburb's border. At some border crossings, there are nothing but warning signs, machine gun pillboxes, and a hundred feet of "no man's land." Travel across such areas amounts to providing bored soldiers with target practice, which explains why Lebanese*

> *have learned to take the longer, more circuitous mountain routes.* (CRYDERMAN 1990, P. 44)

The functionalist perspective, then, tends to overemphasize the constructive consequences associated with religion's unifying, bonding, and comforting functions. Strict functionalists who focus only on the consequences that lead to order and stability tend to overlook the fact that religion can also unify, bond, and comfort believers in such a way that it supports war and other forms of conflict between ingroups and outgroups. The conflict perspective, on the other hand, acknowledges the unifying, comforting functions of religion but, as we shall see, views such functions as ultimately problematic.

The Conflict Perspective

Scholars who view religion from the conflict perspective focus on how religion turns people's attention away from social and economic inequality. This view stems from the work of Karl Marx, who believed that religion was the most humane feature of an inhumane world and that it arose from the tragedies and injustices of human experience. He described religion as the "sigh of the oppressed creature, the sentiment of a heartless world, and the soul of soulless conditions. It is the *opium* of the people" (*The World Treasury of Modern Religious Thought* 1990, p. 80). People need the comfort of religion in order to make the world bearable and to justify their existence. In this sense, Marx said, religion is analogous to a sedative.

Even while Marx acknowledged the comforting role of religion, he focused on its repressive, constraining, and exploitative qualities. In particular, he conceptualized religion as an ideology that justifies the status quo, rationalizing existing inequities or downplaying their importance. This aspect of religion is especially relevant with regard to the politically and economically disadvantaged. For them, said Marx, religion is a source of "false consciousness." That is, religious teachings encourage the oppressed to accept the economic, political, and social arrangements that constrain their chances in this life because their suffering will be compensated in the next.

Consider the view that some Christian groups have of heaven. Material published in 62 languages by the Watchtower Bible and Tract Society and distributed worldwide describes life in God's Kingdom:

God's Kingdom will bring earthly benefits beyond compare, accomplishing everything good that God originally purposed for his people to enjoy on earth. Hatreds and prejudices will cease to exist. . . . The whole earth will eventually be brought to a gardenlike [paradise]. . . . No longer will people be crammed into huge apartment buildings or run-down slums. . . . People will have productive, satisfying work. Life will not be boring. (WATCHTOWER BIBLE AND TRACT SOCIETY 1987, PP. 3–4)

This kind of ideology led Marx to conclude that religion justifies social and economic inequities and that religious teachings inhibit protest and revolutionary change. He went so far as to claim that religion would not be needed in a **classless society**—a propertyless society providing equal access to the means of production. In the absence of material inequality, there would be no exploitation and no injustice—experiences that cause people to turn to religion.

In sum, Marx believed that religious doctrines turn people's attention away from unjust political and economic arrangements and that they rationalize and defend the political and economic interests of the dominant social classes. For some contemporary scholars, this legitimating function is reflected by the fact that most religions allow only a specific category of people—men—to be leaders and to handle sacred items. A letter written to the editor of *Christianity Today* in reaction to the article "Women in the Seminary: Preparing for What?" shows how the "facts" of the Bible can be used to explain and justify such inequalities:

If the Lord meant for a woman to lead the church in such roles as preacher, elder, pastor, minister, prophet, priest, et cetera, why didn't he provide early Christians with a scriptural prototype? Where in Scripture can a woman priest be found? A woman (literary) prophet? A woman apostle? A woman elder or pastor?

Could it be the Lord didn't intend for a woman to serve in any of these positions? It makes me wonder. (CHRISTIANITY TODAY 1986, P. 6)

Sometimes religion can be twisted in ways that serve the interests of dominant groups. During the days of slavery, for example, some Christians prepared special catechisms for slaves to study. The following questions and answers were included in such catechisms:

Q: What did God make you for?

A: To make a crop.

Q: What is the meaning of "Thou shalt not commit adultery"?

A: To serve our heavenly Father, and our earthly Master, obey our overseer, and not steal anything. (WILMORE 1972, P. 34)

Sometimes political leaders use religion to unite their country in war against another. Both Iraqi President Saddham Hussein and U.S. President George Bush invoked the name of God to rally people behind their causes in the Gulf War of 1991:

In the name of God, the merciful, the compassionate: Our armed forces have performed their holy war duty of refusing to comply with the logic of evil, imposition and aggression. They have been engaged in an epic, valiant battle that will be recorded by history in letters of light. (SPOKESMAN FOR THE IRAQI GOVERNMENT 1991, P. Y1)

This we do know: Our cause is just. Our cause is moral. Our cause is right.

May God bless the United States of America. (BUSH 1991, P. A8)

A Critique of the Conflict Perspective of Religion

The major criticism of Marx and the conflict perspective on religion is that religion is not always the "sigh of the oppressed creature." On the contrary, the oppressed have often used religion as a vehicle for protesting or working to change social and economic inequities. **Liberation theology** represents such an approach. Liberation theologians maintain that they have a responsibility to demand social justice for the marginalized peoples of the world, especially landless peasants and the urban poor, and to take an active role at the grass-roots level in order to bring about political and economic justice. Ironically, this doctrine is inspired by Marxist thought in that it advocates raising the consciousness of the poor and teaching them to work together to obtain land and employment and to preserve their cultural identity.

Sociologist J. Milton Yinger (1971) identified at least two conditions in which religion can become a vehicle of protest or change. In the first condition a government or other organization fails to achieve clearly articulated ideals (such as equal opportunity, justice for all, or the right to bear arms). In such cases, disadvantaged groups may form sects or cults and may use seemingly eccentric features of the new religion to symbolize "their sense of separation" (p. 111) and to rally their followers to fight against the establishment. The United States, for example, has not achieved the ideal of "equal opportunity irrespective of race, sex, religion, or ethnic group." One religion that emerged in reaction was the Nation of Islam or the Black Muslims.

In the 1930s black nationalist W. D. Fard, who went by a variety of names including Farad Muhammad, founded the Nation of Islam and began preaching in the Temple of Islam in Detroit. (When Farad disappeared in 1934, his chosen successor, Elijah Muhammad, took his place.) Farad taught that the white man was the personification of evil and that black people were Muslim but that their religion had been taken from them after they came to America as slaves. Farad also taught that the way out was not through gaining the "devil's" (the white man's) approval but through self-help, discipline, and education. Members received an X to replace their "slave names" (hence Malcolm X). In the social context of the 1930s, this message was very attractive:

You're talking about Negroes. You're talking about "niggers," who are the rejected and the despised, meeting in some little, filthy, dingy little [room] upstairs over some beer hall or something, some joint that nobody cares about. Nobody cares about these people. . . . You can pass them on the street and in 1930, if they don't get off the sidewalk, you could have them arrested. That's the level of what was going on. (NATIONAL PUBLIC RADIO 1984A)

The Nation of Islam is only one example of a religious organization working to improve life for African-Americans. Historically the African-American churches have reached out to millions who have felt excluded from the system (Lincoln and Mamiya 1990). "They have demonstrated their capability for empowerment, emancipating black people from the ravages of generations of slavery and equipping them to recover a sense of identity, to forge ties of communal loyalty, to help themselves and others and to create cultural expressions that embody their hopes and aspirations" (Forbes 1990, p. 2). For example, African-American churches were instrumental in the overall successes of the civil rights movement. In fact, some observers argue that the movement would have been impossible if the churches had not been involved (Lincoln and Mamiya 1990).

Religion in a Polarized Society In the second condition in which a religion can become a vehicle of protest or change, a society is polarized along class, ethnic, or sectarian lines. In 1975, when the civil wars began in Lebanon, a tremendous gap existed between the rich (predominantly Christian Maronites) and the poor (predominantly Sunni and Druze, but especially Shia and the dispossessed Palestinians). Although such a gap always had existed in Lebanon, it became especially visible as large numbers of poverty-stricken and neglected people settled in slums around the city of Beirut.

The poor came to Beirut for two reasons. One group, predominantly Shia, had been driven out of southern Lebanon by the fighting between P.L.O. and Israeli forces. The other group consisted of migrants from the rural Akkar region in the north and the Bekaa valley in the east. These regions had been severely neglected by the Maronite-dominated government: the roads, schools, and health care facilities were inadequate; few industrial and commercial enterprises existed; and few funds were funneled into agriculture. On paper, Lebanon was prosperous: "Lebanon, as the gate to the Arab East, served as a headquarters for business, financial, and administrative activities prompted by the discovery of oil in the Middle East" (Barakat 1979, p. 2). The wealth generated from these activities, however, went to a small segment of the population (about

18 percent total, and predominantly Christian). The dramatic contrast between the Muslim poor and the Christian rich, combined with the fact that the Christian-led government made few credible attempts to correct the inequities, created a situation in which the disadvantaged came to believe they had no stake in the system and had nothing to lose if the system collapsed. Any improvement in their lives seemed to be tied to the dismantling of the Christian-dominated political structure.

These situations show that the larger social, economic, and political context must be incorporated into any discussion of the role of religion in shaping human affairs. It is especially important to examine how religion influences people's behavior in a specific context and how it gives legitimacy to people's actions. This line of investigation was particularly interesting to Max Weber, who examined how religious beliefs direct and legitimate economic activity.

Max Weber: The Interplay Between Economics and Religion

Max Weber was interested in understanding the role of religious beliefs in the origins and development of **modern capitalism,** "a form of economic life which involved the careful calculation of costs and profits, the borrowing and lending of money, the accumulation of capital in the form of money and material assets, investment, private property, and the employment of laborers and employees in a more or less unrestricted labor market" (Robertson 1987, p. 6).

In his book *The Protestant Ethic and the Spirit of Capitalism* (1958), Weber asked why modern capitalism emerged and flourished in Europe rather than in China or India (the two dominant world civilizations at the end of the sixteenth century). He also asked why business leaders and capitalists in Europe and the United States were overwhelmingly Protestant. To answer these questions, Weber studied the major world religions and some of the societies in which these religions were practiced. He focused on understanding how norms generated by different religious traditions influenced the adherents' economic orientations and motivations.

On the basis of his comparisons, Weber concluded that a branch of Protestant tradition—Calvinism—supplied a "spirit" or ethic that supported the motivations and orientations that capitalism required. Unlike other religions that Weber studied, Calvinism emphasized **this-worldly asceticism**—a belief that people are instruments of divine will and that their activities are determined and directed by God. Consequently, people glorify God

when they accept the tasks assigned to them and carry them out in exemplary and disciplined fashion, and when they do not indulge in the fruits of their labor (that is, when they do not use money to eat or drink, or otherwise relax to excess). In contrast, Buddhism, a religion that Weber defined as the Eastern parallel and opposite of Calvinism, "emphasized the basically illusory character of worldly life and regarded release from the contingencies of the everyday world as the highest religious aspiration" (Robertson 1987, p. 7).

The Calvinists conceptualized God as all-powerful and all-knowing; also they emphasized **predestination,** the belief that God has foreordained all things including the salvation or damnation of individual souls. According to this doctrine, people could do nothing to change their fate. To compound matters, only relatively few people were destined to attain salvation.

Weber maintained that such beliefs created a crisis of meaning among adherents as they tried to determine how they were to behave in the face of their predetermined fate. Such pressures led them to look for concrete signs that they were among God's chosen people, destined for salvation. Consequently, accumulated wealth became an important indicator of whether one was among the chosen. At the same time, this-worldly asceticism "acted powerfully against the spontaneous enjoyment of possessions; it restricted consumption, especially of luxuries" (Weber 1958, p. 171). Frugal behavior encouraged people to accumulate wealth and make investments, important actions for the success of capitalism.

This calculating orientation was not part of Calvinist doctrine per se. Rather, it grew out of this-worldly asceticism and the doctrine of predestination. In view of this distinction, it is important that we do not misread the role that Weber attributed to the Protestant ethic in supporting the rise of a capitalistic economy. According to Weber, the ethic was a significant ideological force; it was not the sole cause of capitalism but "*one* of the causes of *certain aspects* of capitalism" (Aron 1969, p. 204). Unfortunately, many people who encounter Weber's ideas overestimate the importance he assigned to the Protestant ethic for achieving economic success, and they draw a conclusion that Weber himself never reached: the reason that some groups and societies were disadvantaged was simply that they lacked this ethic.

Finally, let us remember that Weber was writing about the origins of industrial capitalism, not about the form of capitalism that exists today, which heavily emphasizes consumption and self-indulgence. Weber maintained that once capitalism was established, it would generate its own norms and would become a self-sustaining force. In fact, Weber argued that "[c]apitalism produces a society run along machine-like, rational procedures without inner meaning or value and in which men operate almost as mindless cogs" (B. Turner 1974, p. 155). In such circumstances religion becomes increasingly insignificant in maintaining the capitalist system. Some sociologists believe that industrialization and scientific advances cause society to undergo unrelenting secularization, a process in which religious influences become increasingly irrelevant not only to economic life but also to most aspects of social life. On the other hand there are those who argue that as religion becomes less relevant to economic and social life in general, a significant number of people become fundamentalist— that is, they seek to re-examine their religious principles in an effort to identify and return to the most fundamental and basic principles (from which believers have departed) and to hold those principles up as the definitive and guiding blueprint for life.

Two Opposing Trends: Secularization and Fundamentalism

Secularization and fundamentalism are processes that have become increasingly popular in the recent past. Each has grown in spite of the other's growth, or possibly in opposition to it. This section examines these two opposing trends.

Secularization

In the most general sense, **secularization** is a process by which religious influences on thought and behavior are reduced. It is difficult to generalize about the causes and consequences of secularization because they vary across contexts. Americans and Europeans associate secularization with an increase in scientific understanding and in technological solutions to everyday problems of living. In effect, science and technology assume roles once filled by religious belief and practice. Muslims, on the other hand, do not attribute secularization to science or to modernization; there are many devout Muslims who are physical scientists. In a National Public Radio interview, a Muslim geologist explained that Islam was compatible with science:

> *I realized that there is something—there's a glue, there's a creator—behind all those marvelous facts of life in the field of geology, of chemistry. [I had] to think and rethink about my attitude towards life, and that there should be a creator and a very wise creator who has already given all of us laws, and produced all these facets of life.* (NATIONAL PUBLIC RADIO 1984C)

From a Muslim perspective, secularization is a Western-imposed phenomenon—specifically, a result of exposure to what many people in the Middle

East consider the most negative of Western values. This point is illustrated by the observations of a Muslim who attended college in Britain and by a Muslim artist:

> *If I did not watch out [while I was in college] I knew that I would be washed away in that culture. In one particular area, of course, was exposure to a society where free sexual relations prevailed. There you are not subject to any control, and you are faced with a very serious challenge, and you have to rely upon your own strength, spiritual strength to stabilize your character and hold fast to your beliefs.* (NATIONAL PUBLIC RADIO 1984B)

> *You can't be against foreign influence because art and culture and science are things that you learn from people and [that] we exchange. It depends on what kind of foreign influence is being exerted. For example, if we speak of the United States, the U.S. has a tradition of science and of art and of culture . . . which can be extremely useful and can enrich our own culture if there is an exchange. But the type of culture that we are getting from the U.S. at the moment is the "Dallas" series, the cowboy serials, crimes, commodity values, big cars, luxury, things like that. This kind of culture can't help.* (NATIONAL PUBLIC RADIO 1984B)

We can speak of two kinds of secularization: subjective and objective (Berger 1967). **Subjective secularization** is a decrease in the number of people who view the world and their place in it from a religious perspective. They shift from a religious understanding of the world, grounded in faith, to an understanding grounded in observable evidence and the scientific method (see Chapter 3). In the face of uncertainty, people often turn away from religion. They come to perceive the supernatural as a distant, impersonal, and even inactive phenomenon. Consequently, they come to believe less in direct intervention by the supernatural (forces that can defy the laws of nature and can affect the outcome of an event) and to rely more strongly on human intervention or scientific explanation:

> *Consider the case of the lightning rod. For centuries, the Christian church held that lightning was the palpable manifestation of divine wrath and that safety against lightning could be gained only by conforming to divine will. Because the bell towers of churches and cathedrals tended to be the only tall structures, they were the most common targets of lightning. Following damage or destruction of a bell tower by lightning, campaigns were launched to stamp out local wickedness and to raise funds to repair the tower.*

> *Ben Franklin's invention of the lightning rod caused a crisis for the church. The rod demonstrably worked. The laity began to demand its installation on church towers—backing their demands with a threat to withhold funds to restore the tower should lightning strike it. The church had to admit either that Ben Franklin had the power to thwart divine retribution or that lightning was merely a natural phenomenon.* (STARK AND BAINBRIDGE 1985, PP. 432–33)

There is considerable debate over the extent to which subjective secularization is taking place. Data collected by the Gallup Organization over the past 20 years show little change in the degree of importance that Americans say they assign to religion. More than 90 percent have a religious preference; almost 70 percent are members of a church, mosque, temple, or synagogue; 40 percent attend church weekly; and almost 60 percent state that religion is very important in their lives (Gallup and Castelli 1989). Recent polls also show that almost 80 percent of Americans are "sometimes very conscious of God's presence" and that more than 80 percent agreed that "even today, miracles are performed by the power of God" (p. 58).

These findings, however, do not necessarily prove that religion influences behavior and thought as strongly today as in the past. To reach such a conclusion, we would have to determine whether people are as willing now as they were in the past to leave matters in God's hands. We might speculate, for example, that people were more willing to put their trust solely in God before the advent of sophisticated medical technology. Today, people who take such a position may be viewed as ignorant, stubborn, or uncaring. Consider the publicity surrounding cases of children who have died of diabetes, meningitis, tumors, and obstructed bowels because their parents believed that physical ailments could be cured by spiritual means and did not seek medical treatment for the children. The publicity and the accompanying outrage suggest that although an overwhelming number of Americans believe in miracles, their belief is not strong enough to leave medical matters in the hands of supernatural forces.

Objective secularization is the decline in the control of religion over education, medicine, law, and politics, and the emergence of an environment in which people are free to choose from many equally valid religions the one to which they wish to belong. For several reasons, it is difficult to make generalizations about this type of secularization and the extent to which it exists in a society. First, objective secularization is not an even, inevitable process that eventually results in the dominance of science and rationality and in the end of religion. No matter how strongly science comes to dominate human life, it cannot provide solace for the inescapable problems of existence—birth, death, illness, aging, injustice, tragedy, and suffering—nor can it "formulate a coherent plan for life" (Stark and Bainbridge 1985, p. 431).

Another reason that it is difficult to generalize about objective secularization is that no aspect of political life seems to be fully secularized. For example, the American currency includes the slogan "In God We Trust"; it does not say "In the Federal Reserve We Trust" (Haddad 1991). In fact, religion often is at the center of American politics, in spite of the division between church and state that most Americans assume exists. In the 1988 presidential campaign, for example, "the candidates included two ministers [Pat Robertson and Jesse Jackson] and a senator accused of sin [Gary Hart]; where the Pledge of Allegiance, describing 'one nation under God,' became a major campaign issue [for Michael Dukakis and George Bush]; [and] the wife of the victorious vice-presidential candidate [Marilyn Quayle] professed her belief in the literal truth of Noah's Ark" (Wills 1990, jacket cover).

Finally, it is difficult to generalize about objective secularization because as religion becomes less relevant to social affairs, a significant number of people in a society may become more fundamentalist. The secularization process actually provides the conditions that support the rise of fundamentalism.

In other words, believers react to the declining influence of religion and seek to find ways to restore that influence and make it more central in human affairs.

Fundamentalism

In "Popular Conceptions of Fundamentalism," anthropologist Lionel Caplan offers his readers one of the clearest overviews of a complex religious phenomenon—**fundamentalism,** a belief in the timeless nature of sacred writings and a belief that such writings apply to all kinds of environments. This term is applied popularly to a wide array of religious groups in the United States and around the world, including the Moral Majority in the United States, Orthodox Jews in Israel, and various Arab groups in the Middle East.

Religious groups labeled as fundamentalist are usually portrayed as "fossilized relics . . . living perpetually in a bygone age" (Caplan 1987, p. 5). Americans frequently employ this simplistic analysis to explain events in the Middle East, especially the causes of the political turmoil that threatens the interests of the United States (including its need for oil). But such oversimplification misrepresents fundamentalism and cannot explain the widespread appeal of contemporary fundamentalist movements within several of the world's religions.

The Complexity of Fundamentalism Fundamentalism is a more complex phenomenon than popular conceptions lead us to believe. First, it is impossible to define a fundamentalist in terms of age, ethnicity, social class, or political ideology because fundamentalism appeals to a wide range of people. Moreover, fundamentalist groups do not always position themselves against those in power; they are just as likely to be neutral or to fervently support existing regimes. Perhaps the most important characteristic of a fundamentalist is the belief that a relationship with God, Allah, or some other supernatural force provides answers to personal and social problems. In addition, fundamentalists often wish to "bring the wider culture back to its religious roots" (Lechner 1989, p. 51).

Caplan suggests a number of other traits that seem to characterize fundamentalists. First, fundamentalists emphasize the authority, infallibility, and timeless truth of sacred writings as a "definitive blueprint" for life (Caplan 1987, p. 19). This characteristic should not be taken to mean that a definitive interpretation of sacred writing actually exists. Any sacred text has as many interpretations as there are groups that claim it as their blueprint. For example, members of fundamentalist sects in both Islam and Christianity disagree about the true meaning of the texts they follow.

Second, fundamentalists usually conceive of history as a "process of decline from an original ideal state, [and] hardly more than a catalog of the betrayal of fundamental principles" (p. 18). They conceptualize human history as a "cosmic struggle between good and evil": the good is a result of dedication to principles outlined in sacred scriptures, and the evil is an outcome of countless digressions from sacred principles. To fundamentalists, truth is not a relative phenomenon; it does not vary across time and place. Truth is unchanging, and is knowable through the sacred texts.

Third, fundamentalists do not distinguish between the sacred and the profane in their day-to-day lives. All areas of life, including family, business, and leisure, are governed by religious principles. Religious behavior, in their view, does not take place only in a church, a mosque, or a temple.

Fourth, fundamentalist religious groups emerge for a reason, usually in reaction to a perceived threat or crisis, real or imagined. Consequently, any discussion of a particular fundamentalist group must include some reference to an adversary.

Fifth, one obvious concern of fundamentalists is to reverse the trend toward gender equality, which they believe is symptomatic of a declining moral order. In fundamentalist religions, women's rights often are subordinated to ideals that the group considers more important to the well-being of the society, such as the traditional family or the right to life. Such a priority of ideals is regarded as the correct order of things.

Islamic Fundamentalism In *The Islamic Threat: Myth or Reality?*, professor of religious studies John L. Esposito maintains that most Americans' understanding of fundamentalism does not apply very well to contemporary Islam. The term *fundamen-*

talism has its roots in American Protestantism and the 20th-century movement that emphasizes the literal interpretation of the Bible. Fundamentalists are portrayed as static, literalist, retrogressive, and extremist. Just as we cannot apply the term *fundamentalism* to all Protestants in the United States, we cannot apply it to the entire Muslim world, especially when we consider that Muslims make up the majority of the population in at least 45 countries. Esposito believes that a more fitting term is *Islamic revitalism* or *Islamic activism*. The form of Islamic revitalism varies from one country to another but seems to be characterized by the following themes:

> *A sense that existing political, economic, and social systems had failed; a disenchantment with, and at times a rejection of the West; a quest for identity and greater authenticity; and the conviction that Islam provides a self-sufficient ideology for state and society, a valid alternative to secular nationalism, socialism, and capitalism.* (ESPOSITO 1992, P. 14)

In "Islam in the Politics of the Middle East" Esposito asks, "Why has religion [specifically Islam] become such a visible force in Middle East politics?" He believes that Islamic revitalism is a "response to the failures and crises of authority and legitimacy that have plagued most modern Muslim states" (1986, p. 53). Recall that after World War I, France and Britain carved up the Middle East into nation-states with the boundaries drawn to meet the economic and political needs of Western powers. Lebanon, for example, was created in part to establish a Christian tie to the West; Israel was created as a refuge for persecuted Jews when no country seemed to want them; the Kurds received no state; Iraq was virtually landlocked; and resource-rich territories were incorporated into states with very sparse populations (Kuwait, Saudi Arabia, the Emirates). Many of the leaders who took control of these foreign creations were viewed by the citizens "as autocratic heads of corrupt, authoritarian regimes that [were] propped up by Western governments and multinational corporations" (p. 54).

When Arab armies from six states lost "so quickly, completely, and publicly" to Israel in 1967 (see Chapter 5), Arabs were forced to question the political and moral structure of their societies (Hourani 1991, p. 442). Had the leaders and the people abandoned Islamic principles or deviated too far? Could a return to an Islamic way of life restore confidence to the Middle East and give it an identity independent of the West? Questions of social justice also arose. Oil wealth and modernization policies had led to rapid increases in population and urbanization, and to a "vast chasm between rich and poor" (Esposito 1992, p. 15) within countries and between countries—that is, between the oil-rich countries such as Kuwait and Saudi Arabia and the poor, densely populated countries such as Egypt, Pakistan, and Bangladesh. Western capitalism, which was believed to be one of the primary forces behind these trends, seemed to be blind to social justice, promoting unbridled consumption and widespread poverty. Marxist socialism (a godless alternative) likewise had failed to produce social justice. In 1975, on the eve of the civil war, Muslims in Lebanon were asking these kinds of questions.

For many people Islam offers an alternate vision for society. According to Esposito (1986), Islamic activists (who are of many political persuasions, from conservative to militant) are guided by five beliefs:

1. Islam is a comprehensive way of life relevant to politics, state, law, and society.

2. Muslim societies fail when they depart from Islamic ways and follow the secular and materialistic ways of the West.

3. An Islamic social and political revolution is necessary for renewal.

4. Islamic law must replace Western-inspired or Western-imposed laws.

5. Science and technology must be used in ways that reflect Islamic values in order to guard against the infiltration of Western values.

Muslim groups differ dramatically as to how quickly and by what methods these principles should be implemented. Most Muslims, however, are willing to work within existing political arrangements; they condemn violence as a method of bringing about political and social change.

The information presented in this section points to the interplay of religion with political, economic, historical, and other social forces. Fundamentalism cannot be viewed in simple terms. Any analysis must consider the broader context. A focus on context allows us to see that fundamentalism can be a reaction to many events and processes including secularization, foreign influence, failure or crisis in authority, the loss of a homeland, or rapid change.

Discussion

The sociological perspective on religion helps us step back and view in a detached way a subject that is often charged with emotion. Such an approach encourages us not to apply preconceived notions to the study of religion. Many Americans, for example, view the *hijab*, or modest dress of Muslim women, as evidence that the women are severely oppressed. Although women in the Middle East certainly do not have the same rights as men, American observers should not be so quick to apply their own standards as the model. Consider the view that some Muslim women hold toward American dress customs:

If women living in western societies took an honest look at themselves, such a question [as to why Muslim women are covered] would not arise. They are the slaves of appearance and the puppets of male chauvinistic society. Every magazine and news medium (such as television and radio) tells them how they should look and behave. They should wear glamorous clothes and make themselves beautiful for strange men to gaze and gloat over them.

So the question is not why Muslim women wear hijab, *but why the women in the West, who think they are so liberated, do not wear*

Carrying on with life in the midst of civil war: friends visit in a ruined Beirut hotel.
Gary Matoso/Contact Press Images

hijab? (*MAHJUBAH: THE MAGAZINE FOR MOSLEM WOMEN* 1984)

In this chapter, sociological concepts and perspectives have helped us see that the civil wars in Lebanon cannot be reduced to the actions of fanatics acting from primitive and irrational religious convictions. Religion is important in understanding the civil wars, but its explanatory importance depends on understanding how religion is used rather than on knowing the religious group to which the majority in each faction belongs.

Durkheim's ideas about the nature of religion, despite their shortcomings, allow us to consider the functions of religion for the individual and for the group. Most important, religion is a rich and seem-ingly endless variety of responses to problems of human existence. People embrace religion in the face of uncertainty in order to cope with misfortunes and injustices, when they have a great emotional investment in securing a successful outcome and to find answers to agonizing questions about the meaning of life and death.

In addition, religion contributes to group unity and solidarity. Whenever a group of people has a strong conviction (no matter what kind of group and no matter what the conviction), the conviction almost always takes on a religious character. Religious gatherings and affiliations become ways of affirming belief and mobilizing the group members to uphold their beliefs, especially if they are threatened. When religion is used in this way, it can unite a community of worshipers so strongly that they would be willing to destroy those who do not share their views. The convictions that religion is called on to legitimate may be honorable or unprincipled.

Karl Marx was concerned about religion's repressive, constraining, and exploitative qualities. He believed that religious doctrines can be used to turn people's attention away from unjust political and economic arrangements and that they rationalize and defend the political and economic interests of the dominant classes. For example, those in political power may use religion to unite the society in war against another society or to dominate in some other way. Karl Marx ignored the possibility that the oppressed have used religion as a vehicle by which to protest or to work to change the existing social and economic inequities.

Max Weber alerts us to yet another way in which religion affects economic life by outlining the role that religious beliefs played in the origins and development of modern capitalism. Yet, at the same time Weber argued that once capitalism was established it would generate its own value-rational logic, which in turn would make religion less relevant to economic activity. (See Chapter 7 for a discussion of value-rational action.) In fact, the advances in science and technology that accompany modern capitalism cause society to become secularized. Ironically, the same forces that support secularization processes support the rise of fundamentalism.

The uses to which religion are turned show that religion is a multifaceted and complex phenomenon that cannot be discussed in isolation. To understand the civil wars in Lebanon, we must understand the social, economic, and political context. With this information in hand we can examine how religion shapes people's behavior and how it legitimates their actions.

FOCUS:
The Question Revisited—What Is Religion?

In this chapter we have learned that sociologists focus on the most outward and visible characteristics of religion. Such an approach allows us to step back and view religion (a subject that is often charged with emotion) in a detached way. This is important if we wish to avoid making sweeping generalizations about the nature of unfamiliar religions. On the other hand, it is difficult to identify outward and visible characteristics that are unique to religious activity. Many of these characteristics can be found in other gatherings such as sports events, graduation ceremonies, reunions, political rallies. Therefore it is difficult for sociologists to make a distinction between people on a spiritual journey such as the Youth Day that took place in Denver, Colorado, in 1993 and a group of fans eulogizing James Dean. The observations of James F. Hopgood, which follow, illustrate.

What Would Durkheim Say About James Dean?

James F. Hopgood

In March 1989 I began a research project that—little did I know—would become increasingly complex and multifaceted. The occasion, in fact, was the return trip to Cincinnati by car from the Central States Anthropological Society Annual Meeting at Notre Dame. For several years I had read of a "strange" and "exotic" happening in a small Indiana town each September, when thousands of people would visit the town and its special sites on "Museum Days." More recently, I had read of a

SOURCE: James F. Hopgood, Northern Kentucky University (1991).

new gallery being opened there by a man from New York. The town was only an hour or so out of my way, so I was determined to check it out. I have to admit, however, that this was not *entirely* new to me. In 1955, as an adolescent of 12 growing up in Missouri, I first learned of the town and its most famous native. The town is Fairmount, and the reason for all the activity and attention is the late actor James Dean (1931–1955).

I picked up an expressly prepared map of Fairmount at the James Dean Gallery. The map located the major places associated with Dean when he lived in Fairmount: his uncle and aunt's farm where he grew up following his mother's death, the high school he attended, the motorcycle shop he frequented, the church he attended, his grave, a memorial erected for him in the cemetery (with its bust of Dean stolen long ago), the local historical museum with the "James Dean Room," and the new gallery dedicated to him. It was like a mapping of sacred places and sites for all "Deaners" to visit. The more dedicated, I learned later, know of other special places not on the public map.

I have returned to Fairmount on several occasions for more participant observation and interviewing. The following excerpt from my field notes, written during Fairmount's "Museum Days" on September 28, 1990, indicates the "on-the-ground" problem in distinguishing Durkheim's classic "sacred" and "profane."

Our first stop this morning was the Park Cemetery. People were coming and going at a steady rate. Some came alone, others in small groups, and there were even a few van loads. All ages were represented. . . .

We noticed license plates from IN, NY, VT, MD, Ontario, IL, MO, KY, OH, and MI. Flow-

The tokens of devotion left by "pilgrims" to the grave of James Dean illustrate the difficulty of making absolute distinctions between the sacred and the profane.

James F. Hopgood

ers at the grave were placed there by people from Iowa, Michigan, France, England, and Canada. Some of the flowers didn't have a place of origin, only the names of those who sent them. An assortment of the usual sorts of objects such as coins and cigarettes had been left at the stone. One unusual object was a pair of clip-on sunglasses of the style worn by Dean. There were also some poems and messages. One message with flowers read: "Dearest Jimmy [:] Your star continues to shine. The light has touched many a heart . . . Especially mine. Love & memories." Someone had left a typed message of considerable length (maybe 1000 words) titled: "The Key to Eternal Life." It contained a testimonial [along with complimentary references to Dean] of how its author had found God and Jesus Christ, and how the reader can find the same. . . .

Many of the pilgrims to the grave made a point of touching the grave stone; some with only a brief brush, others with longer sustained contact. Some held only a finger, others an entire hand, to the stone for several minutes. Some people drove by the grave very slowly with a very brief pause in front of the stone. Others parked at various places in or near the cemetery and walked to the grave. Some had walked from town. One man rode a bike to the grave.

Some took photos of the grave, with or without companions. . . . Dean's current stone, by the way, is suffering from the same fate as the two or three preceding ones—it is being slowly chipped and carried away, fragment by fragment.

I noticed two girls who were there for a long time. They sat or stood across from the grave at various locations. At each spot maybe five minutes or so. In any case, they were there a longer time than anyone else. They didn't talk much to each other or anyone else and appeared to be quietly reflecting on something of great importance.

Two young men, one from NY and the other from CT, asked me to take a photo of "the three of them" at Dean's grave stone, using one of their cameras. We later saw them out front of the Winslow farm taking photos of each other.

A black car with very heavily tinted windows drove up to the grave. A man in the back seat rolled his window three-quarters of the way down and peered out at the grave for a minute or so. He never got out. He only leaned forward to intently view the grave. Then, with the window back up, they drove on. I couldn't see the driver and I didn't get that one on film—too bad!!

Visiting Dean's grave is only one of numerous activities in his honor during "Museum Days." Many of the devoted prefer, in fact, not to visit the grave with so many people coming and going. It's better to be there when it's tranquil and they can pause to reflect on their love and devotion to "Jimmy" and his very special, personal meaning to them. Another, more solemn event is the memorial service held each September 30, the anniversary of his death, at the little Back Creek Friends Church. The church, located between the farm and the cemetery, takes on an added sacred quality for the more devoted because of its connection to Dean.

As I dig deeper I find the terms *secular* and *profane* increasingly inappropriate in thinking about the "Deaners." Does that mean this is a "religious" matter? If not, then what is to be made of the "Deaners"? Overall, the Dean phenomenon seems somewhere between the religious and the secular, between the sacred and the profane.

While grappling with the inadequacy of Durkheim's dichotomy, I have moved closer to the position that the devotion of many "Deaners" is, after all, simply religious. Of course, there is no accepted, recognized godhead of the movement or any claim of a special relationship to Adonai, Yahweh, or other recognized deity by Dean. Dean, in fact, is never depicted in life as being an especially religious person. I am tempted to argue that it is Dean himself who is the deity. Based on my observations, I have to say many Deaners behave *as if* he is godlike. Of course, other followers do not. In either case, it is not yet clear to me where they place Dean in their personal pantheons of the immortals.

Key Concepts

Church 536

Civil Religion 540

Classless Society 547

Cults 540

Denomination 537

Ecclesia 536

Established Sects 540

Fundamentalism 552

Liberation Theology 547

Modern Capitalism 549

Mystical Religions 533

Objective Secularization 552

Predestination 549

Profane 533

Prophetic Religions 533

Religion 532

Rituals 536

Sacramental Religions 533

Sacred 532

Sect 537

Sectarian Community 528

Secularization 550

Subjective Secularization 551

This-Worldly Asceticism 549

Notes

1. The Maronite religion has been a distinct Christian community since the seventh century, at which time it separated from the Papal church over the doctrine of monotheism (a belief that there is only one God). In the nineteenth century Maronites reestablished communication with the Pope. Maronite priests are permitted to marry. The head of the Maronites under the Pope is the Patriarch of Antioch. Antioch, located near the Mediterranean Sea, was a significant city in religious history because it was an early center of Christianity and a place where Peter and Paul, disciples of Jesus Christ, preached. Antioch is the archaeological site of the Great Chalice of Antioch, which some believe to be the Holy Grail, or the Last Supper chalice (*The New Columbia Encyclopedia* 1975).

2. Events since the late 1970s reflect the complex relationships of groups and countries of the Middle East: The P.L.O. mount military raids against Israeli citizens. . . . The Israeli army attacks P.L.O.

guerrilla camps and villages. . . . Israeli troops occupy southern Lebanon. . . . Civil war breaks out (the P.L.O. and leftist Muslim militias battle Maronite and other Christian militias). . . . The war takes some 60,000 lives and causes billions of dollars worth of damage. . . . Syrian troops intervene and battle Palestinian groups. . . . A cease-fire is reached. . . . Fighting breaks out between Syrians and Christians. . . . Other groups join the fighting in a war of each against all. . . . The Israeli army attacks Palestinian positions at Tyre, Tulin, and Beirut. . . . Israeli forces invade Lebanon in coordinated land, sea, and air attacks to crush the P.L.O. . . . Israeli and Syrian forces battle in the Bekaa valley. . . . The Israelis engage in massive bombings of Beirut. . . . The P.L.O. withdraws from Lebanon under supervision by UN troops. . . . Israeli troops withdraw but re-enter following the assassination of the Christian prime minister they had helped to install. . . . With the tacit approval of the Israelis, Lebanese Christian troops enter two refugee camps and massacre hundreds of Palestinian refugees. . . . Terrorist bombings and kidnappings follow. . . . U.S. peacekeeping troops arrive. . . . Explosion at the U.S. Embassy kills 50 people. A few months later Muslim suicide attacks kill 241 U.S. soldiers and 58 French soldiers. . . .
Civil war erupts between Christian forces and Druze and Shia Muslims. . . . Full-scale fighting takes place between the Shia and the P.L.O. and among various groups of Christians, Druze, Sunni, and Shia. . . . Israel withdraws its forces but continues to defend strategic parts of southern Lebanon. (Adapted from THE WORLD ALMANAC AND BOOK OF FACTS 1991, 1990, p. 727)

3. The astonishing level of support for the Persian Gulf War—an estimated 90 percent of Americans—prompted some media critics to predict that war coverage in the future will be modeled after the coverage in this war, that it will be more concerned with sustaining patriotism than with offering accounts of events (Leo 1991). In fact, the patriotic fervor generated by the themes and the initial success of Operation Desert Storm resulted in the silencing of those who opposed the war. Perhaps the most visible silencing was that of Marco Lokar, an athlete from Italy who attended Seton Hall University on a four-year basketball scholarship. The crowd booed Lokar whenever he touched the ball because they knew he refused to wear the American flag on his uniform during the war. (Lokar's refusal was grounded in his religious conviction that all wars are wrong.) In addition, Lokar and his pregnant wife received many threatening phone calls. In the end the couple returned to Italy (Shulman 1991; Vecsey 1991).

15 SOCIAL CHANGE

with Emphasis on the Aftermath of the Cold War

A statue of Lenin being taken down in post–Cold War Germany.
Regis Bossu/Sygma

Since the mid 1970s the number of:

- Foreign students enrolled
  in U.S. colleges and universities   increased 90 percent

- Doctoral degrees conferred by
  U.S. colleges to non–U.S. citizens   increased 100 percent

- People living in the United States
  who were born in other countries   increased 75 percent

- Passengers on flights into and out of the United States   increased 330 percent

- People employed by foreign affiliates in the United States   increased 286 percent

- Applications for utility patents filed in the
  United States by inventors from other countries   increased 109 percent

- Phone calls into and out of the United States   increased 356 percent

- U.S. military personnel stationed in other countries   decreased 29 percent

In 1991 a 13-member panel made up of an archaeologist, three anthropologists, two astronomers, a geologist, two materials scientists, a linguist, an artist, an architect, and a cognitive psychologist met to discuss the best design for a marker that would warn people to stay away from the Waste Isolation Pilot Plant (WIPP). WIPP is a vast nuclear bedded-salt waste depository 2,150 feet below the earth's surface about a 45-minute drive from Carlsbad, New Mexico. Congress authorized the plant's construction in 1979 and the Environmental Protection Agency stipulated that it be constructed in such a way that the estimated 900,000 drums of plutonium-tainted waste from nuclear weapons facilities to be stored there would not leak out for 10,000 years (Burdick 1992; Idelson 1991, 1992; Seltzer 1991). In addition, the agency mandated that a warning marker be constructed that can remain intact for 10,000 years and be understandable to anyone who might approach it. The panel divided into two teams—A and B—which met separately and then reconvened two months later to present their ideas. Representatives from five other countries (England, Belgium, Sweden, France, and Spain) attended to learn how they might mark their own sites. Some of the suggestions they heard are listed below:

> [A]rchaeologist Maureen Kaplan, leader of Team A, flashed transparencies on the overhead projector. To reach all those who might approach WIPP, she explained, the message must come in multiple forms, from the simple ("Danger: Poisonous Radioactive Waste Buried Here. Do Not Dig Until A.D. 12,000") to the complex (a detailed listing of the repository's contents). And in multiple languages: English, Spanish, Chinese, Russian, Arabic, and French; and perhaps, too, in the tongue of the local Mescalero Apache. Of course, as the panelists readily conceded, no modern language will be widely understood in 10,000 years. Who speaks Etruscan anymore? . . .
>
> Team B opted for a series of pictographic panels, a sort of stick-figure version of "See Dick Run (From Radioactive Death)." Team A has disavowed all symbols but two: a Munch-style expression of facial horror; and the radioactive trefoil, which, though conceived only four decades ago, probably will be all-too-well understood as the centuries progress.
>
> The conversation turned next to what the marker itself should be made of—or, more precisely, the marker system, since it was becoming evident that any monument agreed upon would have to be composed of numerous elements. . . .
>
> And so there was proposed by Team A the Landscape of Thorns: one square mile of randomly spaced basalt spikes, 80 feet high, erupting from the ground at all angles. Or its kin, the Spike Field, with thorns evenly spaced and erectile. Or the Black Hole: a boundless pad of black concrete that would absorb so much heat it would be impossible to approach. . . . Or the team favorite, Menacing Earthworks: an expansive empty square surrounded by 50-foot-high earthen berms jolting outward like jagged bolts of lightning. And at

the center of the square, a 2,000-foot-long, walk-on global map displaying all the world's nuclear dumps, including this one. Rudimentary message kiosks would dot the periphery. A sealed room just below-ground would harbor the most intimate details of the repository's contents.

At one point the Spanish representative sheepishly raised her hand. In halting English she recognized that the discussion of the last two days had addressed the fate of the repository's contents in the coming ten millennia. But, she pointed out, won't the waste be radioactive for some twenty times longer, closer to 240,000 years? Why, please, this interest only in 10,000?

Rip Anderson handled the question confidently. "Because the regulation says so."

(BURDICK 1992, PP. 63–65)

This chapter is about **social change,** which sociologists define as any significant alteration, modification, or transformation in the organization and operation of social life. Social change is an important topic in the discipline of sociology. In fact, it is fair to say that sociology emerged as a discipline in an attempt to understand social change. Recall that the early sociologists were obsessed with understanding the nature and consequences of the Industrial Revolution—an event that triggered dramatic and seemingly endless changes in every area of social life.

When sociologists study change, they first must identify the aspect of social life they wish to study, which has changed or is undergoing change. The list of possible topics is virtually endless. Some examples include changes in the resources people use; changes in what, when, with whom, and where people learn; changes in how people communicate with each other; changes in the amount of goods and services that people produce, sell, or buy from others; changes in the world population and the average life span; and changes in how and where people work (Martel 1986). Often change is something that can be measured: that is, sociologists strive to identify the amount, range (how widespread), duration, pace (speed), and/or direction of change over a specified period (see Table 15.1). Upon identifying a topic, sociologists ask at least two key questions: (1) What factor(s) are causing that change? and (2) What are the consequences of that change for social life?

In this chapter we look at the concepts and theories that sociologists use to answer these complex questions. We give special attention to a major change: the end of the Cold War and the rise of what some people have called the New World Order. The Cold War is the name given to the political tension and military rivalry that have existed between the United States and the Soviet Union from approximately 1945 until its symbolic end on November 9, 1989. On that date the Berlin Wall—the "great divide" of the Cold War—was dismantled and transformed into a "dance floor" and then into "nothing but a heap of stone" (Darnton 1990, p. 12).[1,2]

The Cold War included an arms race, an intense and ongoing military buildup by the Soviet Union and the United States in which each side competed to match or surpass any advances made by the other in the quantity or technological quality of weapons. Both countries created a vast, permanent armaments industry; both made weapons research and development (especially of nuclear weapons) their highest priorities. This 40-year arms race increased the world's inventory of nuclear weapons from zero to some 50,000.

The Cold War is giving way to what some call the New World Order, a term that hints at the emergence of new political and economic relationships between the countries of the former Soviet Union and the United States and among the nations of the world, many of which had aligned themselves on the side of the Soviet Union or the United

TABLE 15.1 Change in the Amount of Interactions Between U.S. Residents and Foreign Residents

We can use a number of indicators to measure changes in the extent to which people in the United States are interacting with people from other countries. In some five-year periods the rate of change has increased dramatically over the past 25 years.

	Amount of Change				Rate of Change		
	1975	1980	1985	1990	1975–80	1980–85	1985–90
Foreign students enrolled in U.S. colleges and universities	154,600	305,000	344,700	391,000	97.3%	13.0%	13.4%
Doctoral degrees conferred by U.S. colleges to non–U.S. citizens	5,313*	4,203	5,317	8,875	−21.0%	26.5%	66.9%
People living in the United States who were born in other countries	N/A	14,080,000	N/A	21,632,000	N/A	N/A	N/A
Passengers on flights into and out of the United States	16,000,000	24,000,000	25,000,000	42,000,000	50.0%	4.2%	68.0%
People employed by foreign affiliates in the United States	1,218,711†	2,033,932	3,228,896	4,705,300	66.9%	58.8%	45.7%
Applications for utility patents filed in the United States by inventors from other countries	36,569	42,231	53,132	76,351	15.5%	25.8%	43.7%
Phone calls into and out of the United States	62,211,237	N/A	433,556,000	2,279,150,000‡	N/A	N/A	425.7%
U.S. military personnel stationed in other countries	485,000	488,000	516,000	609,000	0.6%	5.7%	18.0%

* 1973
† 1977
‡ 1991
SOURCES: Compiled by Joan Ferrante (1994). See Appendix for references.

States during the Cold War. Although the outlines of the new order are still unclear, what has emerged is dramatically different from the largely bipolar order that emerged from the ashes of World War II.

As one concrete indicator that the Cold War has ended, the United States, Russia, and the New States of Eurasia (the former Soviet republics) have shifted their energies from nuclear arms proliferation to nuclear arms reduction and from nuclear weapons production to cleanup, storage, and disposal of radioactive waste. The Strategic Arms Reduction Treaty between the United States and Russia, signed in January 1993, called for the two countries to eliminate 75 percent of their nuclear warheads. At the time this treaty was signed, each side employed more than one million people to maintain and build their nuclear arsenals. Ironically, as the number of nuclear weapons decreases, the stockpiles of nuclear waste increase. In addition to "deciphering four decades of constantly evolving technology" (Jerome 1994, p. 46), the countries involved must dismantle the work forces involved in producing and maintaining nuclear weapons.

The sociological perspective offers some concepts and theories to help us think about the meaning of this historical change. Before we discuss the Cold War, however, let us consider how sociologists approach the study of change.

Social Change: Causes and Consequences

When we think about the factors that cause a specific social change, we usually cannot name a single factor as the cause of an identified change. More often than not, change results from a sequence of events. An analogy may help clarify this point. Suppose that a wide receiver, after catching the football and running 50 yards, is tackled at the five-yard line by a cornerback. One could argue that the cornerback caused the receiver to fall. Such an account, however, would not fully explain what actually happened. For one thing, a tackle is not the act of one person; "there is present a simultaneous conflict of forces" (Mandelbaum 1977, p. 54) between the tackler (who is attempting to seize and throw his weight onto the person with the ball) and the wide receiver (who is doing everything in his power to elude the tackler's grasp). To complicate matters, the wide receiver and the tackler are each members of a team, and their teammates' actions help determine how the play develops and ends. Similarly we can argue that the end of the Cold War caused a shift from proliferation to reduction of nuclear weapons, but we cannot name that event as the sole cause without considering the factors that brought about an end to the Cold War or, for that matter, without considering the factors that led initially to the Cold War in the first place.

Yet even though change is caused by a seemingly endless sequence of events, sociologists can identify some key events or factors that trigger changes in social life. These agents of change include the following:

1. **Innovation:** the development of something new, whether an idea, a practice, or a tool.

2. The actions of leaders—people in positions of authority, including charismatic leaders (those who possess the power, by virtue of their compelling personal style, to make people want to change their way of life) and the power elite (those in positions that give them the power to make and enforce decisions that change the lives of millions of people).

3. Conflict, which Karl Marx believed was the greatest agent of change throughout history. In its most basic form, conflict involves clashes between groups over their shares of wealth, power, prestige, and other valued resources.

4. **Capitalism:** an economic system in which natural resources and the means of production and distribution are privately owned. Capitalism is a system motivated by competition, the pursuit of profit, and a free market, and has been evolving for at least five centuries. This economic system is responsible for the enormous and unprecedented rate of increase in the amount of goods and services produced since the Industrial Revolution.

An analysis of a social change is not complete until sociologists assess its consequences for the life of the society. As in identifying a cause, it is difficult to predict exactly how a specific change will affect society if only because we cannot pinpoint a time in the future when it will cease to have an impact. The effects of nuclear bomb proliferation are a particularly dramatic example. Some materials will remain radioactive for at least 240,000 years—a time "so long that from the human standpoint, they might as well last forever" (Wald 1989, p. Y19).

Another reason why sociologists find it difficult to predict the effects of a specific change is that people react to the change, and their reactions play a role in shaping the consequences. This does not mean that change is random or that it has no clear pattern. Rather, we should view change as a "complex, unrepeatable, unpredictable historical result" that "is explainable after it happens." At the same time, we must acknowledge that "if we wind the tape of life back and start over it may not happen again" (Gould 1990).

The unpredictable element of change—how people choose to react to it—should not be viewed as problematic for the study of change and its consequences. In fact, this is a positive characteristic because it means that people are not passive agents; they create conditions that lead to change, and they react to those conditions as well. Before we examine the four main agents of change let us survey some of the causes and consequences of the Cold War, the event that set the context for the proliferation of nuclear arms and the accompanying problem of radioactive waste.

The New World Order and the Legacy of the Cold War

The Waste Isolation Pilot Plant site was chosen after a long selection process that began in the 1950s. If it opens, it will change the way in which the U.S. Department of Energy stores radioactively contaminated materials which have been produced since the early 1940s, when the United States began its nuclear defense program. Currently, a variety of above-ground and shallow-land burial methods are used to store both civilian and defense-related radioactive waste temporarily until the WIPP is declared a suitable permanent repository and is ready to open (U.S. Department of Energy 1988). A large portion of the radioactive material that will be transported to and stored at WIPP

> *consists of everyday items used by people working at national defense facilities. Examples of these items are rags, rubber gloves, shoe covers, cloth lab coats, plastic bags, and discarded laboratory glass and metalware. Pumps, valves, motors, hand tools, and some machine tools such as lathes and milling machinery must occasionally be scrapped, too, because they become contaminated during routine operations.* (U.S. DEPARTMENT OF ENERGY 1988, P. 3)

The world's radioactive waste problem began in June 1942 in connection with the Manhattan Project, a U.S. government–sponsored effort to develop an atomic bomb. The idea for the project began when Albert Einstein, a German Jew who immigrated to the United States when Hitler came to power in 1934, sent a letter to President Franklin Roosevelt advocating that the United States undertake research on an atomic bomb. Einstein, along with other refugee scientists, suspected that German scientists were working on such a bomb. Debate continues to this day; but, "[g]iven the context of total war and the certainty that every potential weapon would be used with utter ruthlessness, it is scarcely surprising that most scientists everywhere placed their services at the disposal of the country of their birth (or their choice)" (Dyer 1985, p. 95). On July 16, 1945, the United States exploded the world's first atomic bomb in a New Mexico desert. On August 6, 1945, the U.S. military dropped an atomic bomb on Hiroshima and three days later dropped one on Nagasaki, Japan.[3]

Until 1944, the Soviet Union had employed only about 100 researchers to work on the A-bomb.

After the defeat of Germany, however, and the subsequent destruction of Hiroshima in 1945, "Stalin ordered an all-out effort to create a Soviet atomic bomb, and a Scientific-Technical Council was established to set policy for this undertaking. A rapid expansion of institutes, factories, and technical schools ensued" (Parrott 1983, p. 118).

After World War II, the U.S. government built an industrial complex with plants in 12 states to produce the plutonium, uranium, and tritium needed to manufacture nuclear weapons. In 1949, after the Soviet Union detonated a nuclear weapon, President Truman authorized the production of a hydrogen bomb.[4] Britain, France, and China detonated their own bombs in 1952, 1960, and 1964, respectively.

As the number of nuclear weapons in the world grew from zero to 50,000, the volume of waste increased even more dramatically. For example, for every pound of plutonium produced at the Hanford, Washington, nuclear production site, 170 gallons of high-level radioactive liquid waste and 27,500 gallons of low-level radioactive waste also were produced (Steele 1993).

During the Cold War, relations between the United States and the Soviet Union fell short of direct, full-scale military engagement; even so, as many as 120 wars were fought in so-called Third World countries. In many of these wars the United States and the Soviet Union supported opposing factions with weapons and other military equipment, combat training, medical supplies, economic aid, and food. Three of the best-known proxy wars were fought in Korea, Vietnam, and Afghanistan.

Soviet and American leaders justified their direct or indirect intervention in these proxy wars on the grounds that it was necessary to contain the spread of the other side's economic and political system, to protect national and global security, and to prevent the other side from shifting the balance of power in its favor.[5]

In the mid-1960s, President Lyndon Johnson justified U.S. involvement in Vietnam in this way:

If we allow the Communists to win in Vietnam, it will become easier and more appetizing for them to take over other countries in other parts of the world. We will have to fight again some place else—at what cost no one knows. That is why it is vitally important to every American family that we stop the Communists in South Vietnam. (JOHNSON 1987, P. 907)

The Soviets justified their involvement on the grounds that they were blocking "the export of 'counterrevolution,' or actions by Western powers that would hinder the historic progress of socialism" (Zickel 1991, p. 681). The Soviets defined it as their "internationalist duty" to defend their socialist allies and to give military and economic support to wars of national liberation in Third World countries. They rationalized such actions according to "the Marxist belief that workers around the world are linked by a bond that transcends nationalism" (pp. 999–1000).

On the Soviet side, the end of the Cold War can be measured by cutbacks in military spending, production, and manpower, and by the withdrawal of Soviet troops from Central Europe and Afghanistan. It can be measured as well by an unprecedented increase in the amount of information now available for public consumption: "the former USSR has been transformed from a secretive regime about which useful data were sometimes impossible to acquire into a group of countries releasing masses of information too large for scholars to evaluate in a timely fashion" (Dawisha and Parrott 1994, pp. 3–4).

On the American side, the end of the Cold War can be measured by the U.S. government's release in 1990 of some of the health data it had collected on people who had worked at nuclear plants since World War II (Schneider 1990b), by reductions in American military spending, and by the downsizing of U.S. troops in Europe. Perhaps one of the most significant signs that the Cold War had ended came with the December 7, 1993, announcement by the U.S. secretary of energy Hazel O'Leary that the Department of Energy wanted to come clean with all of the nuclear secrets it had been keeping from the American public and the world. "We've got to expose the impact of the Cold War, both in terms of its environmental, health, safety impacts and also impacts, if you will, on the psyche of the nation" (O'Leary 1993). O'Leary initiated a project to declassify millions of Department of Energy records. She acknowledged that such a policy opens the department up to lawsuits but also acknowledged that "there is a great need to assess what happened and to own up to what happened" (Schneider 1994, p. E3).

In spite of these momentous changes, however, one must question whether the Cold War is completely finished, if only because of the large number of nuclear weapons still existing and the massive amounts of nuclear waste that must be cleaned up after decades of bomb building. "Recent arms treaties have spelled out how delivery systems such as rockets and cruise missiles should be crushed, burned, cut up or otherwise reduced to rubble. But none has addressed the most basic challenge of all: destroying the nuclear warheads that have kept the world on edge for so long" (Griffin 1992, p. 496).

This brief account of the causes of the Cold War, its consequences, and the shift to the New World Order barely begins to capture the complexity of this decisive period in world history. Author Richard Rhodes attempts to trace the details in just one event of this period in his book *The Making of the Atomic Bomb* (winner of the Pulitzer Prize, the National Book Award, and the National Book Critics Circle Award). Rhodes takes 886 pages to tell how the bomb was developed, beginning with Leo Szilard, a Hungarian-born Jewish immigrant and theoretical physicist. On a Tuesday morning (September 12, 1933), eight months after Hitler had been appointed Chancellor of Germany, Szilard was standing on a curb in London waiting for a stoplight to change.

"As the light changed to green and I crossed the street," Szilard recalls, "it . . . suddenly oc-curred to me that if we could find an element which is split by neutrons and which would emit two neutrons when it absorbs one neutron such an element, if assembled in sufficiently large mass, could sustain a nuclear chain reaction.

"I didn't see at the moment just how one would go about finding such an element, or what experiments would be needed, but the idea never left me. In certain circumstances it might be possible to set up a nuclear chain reaction, liberate energy on an industrial scale, and construct atomic bombs." (RHODES 1986, PP. 27–28)

Thousands of events like this must be included in any discussion of the factors that led to the making and dropping of the first atomic bombs on Hiroshima and Nagasaki in August 1945, the point at which Rhodes's book ends. On the surface it seems that no conceptual framework is powerful enough to include the seemingly infinite number of interactions and reactions that brought us to where we are today. Sociologists, however, make sense of all these events by emphasizing "the more *general* properties of social experience, the everyday patterns of activity that appear again and again in the life of the social order" (Erikson 1971, p. 64). From a sociological perspective, "Every event has properties that can be subsumed under a more general heading," such as the creation of and reactions to innovation, the actions of leaders, conflict over scarce and valued resources, or the behaviors and policies shaped by capitalist principles. The remainder of this chapter discusses each of these agents of change, beginning with innovations.

Innovations

Innovation is the invention of something new—an idea, a process, a practice, a device, or a tool. Although all innovations build on existing knowledge and materials, they also go beyond that which already exists. They are syntheses, refinements, new applications, and reworkings of existing inventions.

Innovations also include discoveries, the uncovering of something that had existed before but had remained hidden, unnoticed, or undescribed. Inno-vations are sociologically significant because they change the ways in which people think and relate to one another, as illustrated by the automobile, the microwave oven, and the atomic bomb. The atomic bomb, for example, sharply reduced the amount of human power needed to destroy a city. One atomic bomb surpassed the destructive power of 3,000 tons of bombs dropped on the island of Honshu on July 10, 1945, by 700 planes of the U.S. Third Fleet

Bettmann Archive

A. Persstein/Sygma

Material innovations often lead to dramatic and unforeseen social changes. For example, computers started out as room-sized machines invented to perform super-fast mathematical calculations. Today the same processing power can be packed into a portable, battery-operated notebook computer that serves any number of purposes. In what ways does the businessman pictured on the right illustrate White's point that "invention is the mother of necessity"?

and 550 B-29s (*Life* 1945). A single aircraft carrying one atomic bomb killed an estimated 70,000 people in five minutes and turned the city of Hiroshima into a "boiling black mass" (Tibbett 1985, p. 96).

Innovations can be classified broadly in one of two categories—basic or improving. The distinction between these is not always clear-cut. **Basic innovations** are revolutionary, unprecedented, or ground-breaking inventions that are the cornerstones for a wide range of applications. Examples of basic innovations include the invention of mass production by Henry Ford in 1904, the accidental discovery of radioactivity by French physicist Antoine Henri Becquerel around 1898, Pierre and Marie Curie's discovery of radium and polonium in 1898, and the creation of the first sustained nuclear chain reaction by scientists working at the University of Chicago in 1942.[6]

Improving innovations, on the other hand, represent modifications of basic inventions in order to improve on them—that is, to make them smaller, faster, less complicated, or more efficient, attractive, durable, or profitable. Every modification or change in nuclear weapons after the first explosion

in the New Mexico desert in 1945 represents an improving innovation. Examples include the hydrogen bomb (700 times more powerful than the atomic bomb), which was tested successfully in 1952, the "dry" hydrogen bomb (a lighter version of the 65-ton 1952 model), the minuteman intercontinental ballistic missile with a thermonuclear warhead (Broad 1992, 1994), and the "multi-warhead land-based missiles . . . which can each launch many separately aimed nuclear warheads" (Schmemann 1993, p. A1). Anthropologist Leslie White maintains that once a basic or an improving innovation has been invented, it becomes part of the cultural base, the size of which determines the rate of change.

Innovations and the Rate of Change

Anthropologist Leslie White (1949) maintains that the rate of change is tied to the size of the **cultural base,** the number of existing inventions. White defines an invention as the synthesis of existing inventions. For example, the first airplane was a synthesis of many preexisting inventions, including the gasoline engine, the rudder, the glider, and the wheel. Likewise, many commercial inventions are a synthesis of nuclear-related technology and medical, engineering, energy, and other technologies. Some examples are nuclear medicine (using radioisotopes to track the progress of medicine through the body, or radiation therapy for cancer); nuclear energy (turning water to steam for electrical power); irradiation (sterilizing medical products and foodstuffs by killing harmful bacteria); and radiocarbon dating—using radioactive decay rates to date archaeological and geologic events (Cobb 1989).

White maintains that the number of inventions in the cultural base increases geometrically—1, 2, 4, 8, 16, 32, 64, and so on. (Geometric growth is equivalent to a state of runaway expansion.) He argues that if a new invention is to come into being, the cultural base must be large enough to support it. If the Wright brothers had lived in the fourteenth century, for example, they never could have invented the airplane because the cultural base did not contain the ideas, materials, and inventions to support its creation. Physicist Robert Oppenheimer, who led the research and development of the atomic bomb in the 1940s but who opposed the building of the hydrogen bomb in the 1950s, would have agreed with White on this idea. Oppenheimer said, "It is a profound and necessary truth that the deep things in science are not found because they are useful; they are found because it was possible to find them" (1986, p. 11).

The seemingly runaway expansion, or increases in the volume, of new inventions prompted White to ask: Are people in control of their inventions, or do our inventions control us? For all practical purposes, he believed that inventions control us. He supported this conclusion with two arguments. First, he suggested that the old adage "Necessity is the mother of invention" is naive: in too many cases, the opposite idea—that invention is the mother of necessity—is true.

> *We invent the automobile to get us between two points faster, and suddenly we find we have to build new roads. And that means we have to invent traffic regulations and put in stop lights [and build garages]. And then we have to create a whole new organization called the Highway*

> *Patrol—and all we thought we were doing was inventing cars.* (NORMAN 1988, P. 483)

Second, White argued, when the cultural base is capable of supporting an invention, that invention will come into being whether people want it or not. White supported this conclusion by pointing to **simultaneous-independent inventions,** situations in which the same invention is created by two or more persons working independently of one another at about the same time (sometimes within a few days or months). He cited some 148 such inventions—including the telegraph, the electric motor, the microphone, the telephone, the microscope, the steamboat, and the airplane—as proof that someone will come along to make the necessary synthesis if the cultural base is ready to support a particular invention. In other words, inventions such as the light bulb and the airplane would have come into being whether or not Thomas Edison and the Wright brothers (the people we traditionally associate with these inventions) had ever been born. According to White's conception, inventors may be geniuses, but they also have to be born at the right place and the right time. In other words, they must live in a society with a cultural base sufficiently developed to support their invention.

White's theory suggests that if the parts are present, someone eventually will come along and put them together. The implications are that people have little control over whether an invention should come into being and that they adapt to inventions after the fact. The case of nuclear weapons is a clear example of an invention that scientists chose to create and one whose consequences (nuclear waste) scientists chose to address later (see Table 15.2). As one critic of the nuclear weapons program commented as early as 1950: "Nuclear waste is like getting on a plane, and in mid-air you ask the pilot, how are we going to land? He says, we don't know—but we'll figure it out by the time we get there" (Bauerlein 1992, p. 34). Sociologist William F. Ogburn called this failure to adapt to a new invention **cultural lag.**

Cultural Lag

In his theory of cultural lag, William Ogburn (1968) distinguished between material and nonma-

TABLE 15.2 High-Level Waste Burial Programs

The United States is not the only country that is investigating permanent storage sites for materials contaminated by radioactivity. Fifteen other countries in the world have programs planned.

Country	Earliest Planned Year	Status of Program
Belgium	2020	Underground laboratory in clay at Mol.
Canada	2025	Independent commission conducting four-year study of government plan to bury irradiated fuel in granite at yet-to-be-identified site.
China	none announced	Irradiated fuel to be reprocessed; Gobi desert sites under investigation.
Finland	2020	Field studies being conducted; final site selection due in 2000.
France	2010	Two sites to be selected and studied; final site not to be selected until 2006.
Germany	2008	Gorleben salt dome sole site to be studied.
India	2010	Irradiated fuel to be reprocessed, waste stored for twenty years, then buried in yet-to-be-identified granite site.
Italy	2040	Irradiated fuel to be reprocessed, and waste stored for 50–60 years before burial in clay or granite.
Japan	2020	Limited site studies. Cooperative program with China to build underground research facility.
Netherlands	2040	Interim storage of reprocessing waste for 50–100 years before eventual burial, possibly sub-seabed or in another country.
Russia	none announced	Current Russian program uncertain.
Spain	2020	Burial in unidentified clay, granite, or salt formation.
Sweden	2020	Granite site to be selected in 1997; evaluation studies under way at Aspo site near Oskarshamn nuclear complex.
Switzerland	2020	Burial in granite or sedimentary formation at yet-to-be-identified site.
United States	2010	Yucca Mountain, Nevada, site to be studied and, if approved, to receive 70,000 tons of waste.
United Kingdom	2030	Fifty-year storage approved in 1982; explore options including sub-seabed burial.

SOURCE: Adapted from "Nuclear Waste: The Problem that Won't Go Away," by Nicholas Lenssen, *Worldwatch Institute*, pp. 24–25, 36 (December 1991).

terial culture. Recall that in Chapter 4 you learned that material culture includes tangible creations or objects—including resources (oil, trees, land), inventions (paper, guns), and systems (factories, sanitation facilities)—that people have created or, in the case of resources such as oil, have identified as having the properties to serve a particular purpose. Nonmaterial culture, on the other hand, includes intangible creations such as beliefs, norms, values, roles, and language.

Although Ogburn maintained that both components were important agents of social change, his theory of cultural lag emphasized the material component, which he suggested was the more important of the two. Ogburn believed that one of the most urgent challenges facing people today is adapting to material innovations in thoughtful and constructive ways. He used the term **adaptive culture** for the portion of the nonmaterial culture (norms, values, and beliefs) that adjusts to material innovations. Such adjustments are not always immediate. Sometimes they take decades; sometimes they are never made. Ogburn used the term *cultural lag* to refer to a situation in which adaptive culture fails to adjust in necessary ways to a material innovation. One can argue that the United States and the Soviet Union failed to adjust in constructive and responsible ways to the existence of nuclear weapons because both put weapons production and testing ahead of environmental, health, and safety concerns.

Ogburn, however, was not a **technological determinist,** someone who believes that human beings have no free will and are controlled entirely by their material innovations. For one thing, he noted people do not adjust to new material innovations in predictable and unthinking ways; rather, they choose to create them and after they create them they choose how to use them. For example, according to Nobel Prize physicist Isidor I. Rabi considerable U.S. and Soviet leadership, ingenuity, and planning went into the creation of the atomic bomb.

I would also like to say a few things about the war period and what happened in American physics. This was a most interesting period which deserves some very profound study as a piece of intellectual history apart from the weaponry, because we did two things. We killed all pure research in the universities. We were told that this is a total war and that we were fighting two enemies who believed in total war. So, in good old American competitive spirit, we went in for total war. It turned out that we became infinitely more total than the Germans or the Japanese. With very few exceptions, we went around to the various laboratories and took every productive scientist from his laboratory;

we took scientists from their teaching laboratories, and we put them into large laboratories for the war effort. The Radiation Laboratory at Cambridge was one of them; another was at Los Alamos, and so on. There they were, in very large numbers, devoted to making weapons, to the problems of war. This was to my knowledge the first great attempt at marrying the military and the scientific. . . .

This combination of nonmilitary thinkers and fighting men to apply their innovations, as history will show, turned out to be a tremendously potent force—so much so that in the short time from 1940 to this day, which is twenty-seven years, we have become deeply frightened about the future, not only of the country but of the whole race. This is a new discovery, this application of the most forward-looking science to the military. . . . Something has happened which is irreversible. (RABI 1969, PP. 37–38)

If people have the power to create material innovations, they also have the power to destroy them and to ban or modify their use. This point can be illustrated by many policies and initiatives aimed at controlling the testing and production of nuclear weapons. In July 1991, the United States and what was then the U.S.S.R. signed the Strategic Arms Reduction Treaty, the first treaty ever to reduce the number of long-range nuclear weapons in both countries. In September 1991, President Bush announced that the United States would eliminate its tactical nuclear weapons in Europe and Asia; Soviet President Mikhail Gorbachev responded with a call for even deeper cuts in nuclear arsenals in his country. When the Soviet Union disbanded in December 1991 to become the Commonwealth of Independent States, the U.S. Congress passed the Soviet Nuclear Threat Reduction Act, authorizing 400 million dollars to help Russia dismantle and store nuclear weapons located in Belarus, Ukraine, and Kazakhstan. (The U.S. General Accounting Office estimates the cost of dismantling weapons and cleaning up the production sites in the United States to be between 100 billion and 400 billion dollars. In light of this fact, the amount authorized for Russia is actually just a drop in the bucket [Griffin 1992]).

The human ingenuity, effort, and planning that went into the construction of the atomic bomb and the many policies and initiatives aimed at controlling testing and production point out that people can exercise control over their inventions. The challenge lies, however, in convincing people that they need to address an invention's potential disruptive consequences (which are usually known in advance because someone points them out but the warning is often ignored) before they have a chance to materialize.

In our discussion about innovations as a trigger of social change we have emphasized material inventions (devices, tools, and equipment). Innovations can also be nonmaterial inventions such as a revolutionary idea.

Revolutionary Ideas

In *The Structure of Scientific Revolutions* Thomas Kuhn (1975) maintains that most people perceive science as an evolutionary enterprise: that is, over time scientists move closer to finding the solution to problems by building on their predecessors' achievements. Physicist Leo Szilard's idea about how to create atomic energy, which came to him while waiting for the traffic light to change, was an idea that built on achievements of many who came before. In fact, "the atom as an idea—as an invisible layer of eternal, elemental substance below the world of appearances where things combine, teem, dissolve and rot—is ancient" (Rhodes 1986, p. 29). In this regard, "Szilard was not the first to realize that the neutron might slip past the positive electrical barrier of the nucleus; that realization had come to other physicists as well. But he was the first to imagine a mechanism whereby more energy might be released in the neutron's bombardment of the nucleus than the neutron itself supplied" (p. 28). The point is that Szilard built upon the paradigms of physics but, more importantly, he departed from them as well.

Kuhn takes issue with this evolutionary view; he argues that some of the most significant scientific advances have been made when someone breaks away from or challenges the prevailing paradigms. According to Kuhn, **paradigms** are the dominant and widely accepted theories and concepts in a particular field of study. Paradigms gain their status not because they explain everything, but because they offer the best way of looking at the world for the time being. The dominant paradigms of an academic discipline generally are recorded in its textbooks, which summarize the body of accepted theory and illustrate it with the most successful applications, the most exemplary observations, and the most supportive experiments. The dominant paradigms of a government are recorded in its constitution and other official documents. On the one hand, paradigms are important thinking tools; they bind a group of people with common interests into a scientific or national community. Such a community could not exist without agreed-upon paradigms. On the other hand, paradigms can be blinders, limiting the kinds of questions that people ask and the observations they make.

The explanatory value, and hence the status, of a paradigm is threatened by **anomaly,** an observation or observations that it cannot explain. The existence of an anomaly alone, however, is usually not enough to cause people to abandon a particular paradigm. According to Kuhn, before people abandon old paradigms, someone must articulate an alternative paradigm that accounts convincingly for the anomaly. Kuhn hypothesized that the people most likely to put forth new paradigms are those who are least committed to the old paradigms—the young and those new to a field of study.

A **scientific revolution** occurs when enough people in the community break with the old paradigm and change the nature of their research or thinking in favor of the incompatible new paradigm. Kuhn considers a new paradigm incompatible with the one it replaces because it "changes some of the field's most elementary theoretical generalizations" (Kuhn 1975, p. 85). The new paradigm causes converts to see the world in an entirely new light and to wonder how they could possibly have taken the old paradigm seriously. "[W]hen paradigms change, the world itself changes with them. Led by a new paradigm, scientists adopt new instruments and look in new places" (p. 111).

The Actions of Leaders

The actions of leaders represent a second major trigger of social change. In the most general sense, a leader is someone who has the power to influence others or who is in charge or in command of a social situation. Recall from Chapter 9 that Max Weber defined power as the probability that an individual can realize his or her will even against the resistance of others (Weber 1947). The probability increases if that individual can force people to obey his or her commands or if the individual has authority over others. **Authority** is legitimate power in which people believe that the differences in power are just and proper—that is, people see a leader as entitled to give orders. Max Weber identified two types of authority—charismatic and legal-rational—that have important implications with regard to social change.[7]

Charismatic Leaders as Agents of Change

Charismatic authority rests on the exceptional and exemplary qualities of the person issuing the commands. Charismatic leaders are obeyed because their followers believe in and are attracted irresistibly to the vision that they articulate. Because the source of charismatic authority resides in the leader's exceptional qualities and not in tradition or established rules, the charismatic leader's actions and visions are not bound by rules or traditions. Consequently, these leaders, by virtue of their special qualities, have the ability to unleash revolutionary changes; they can ask their followers to behave in ways that depart from rules and traditions.

Charismatic leaders often appear during times of profound crisis (such as economic depressions or wars), when people are most likely to be drawn to someone with exceptional personal qualities who offers them a vision of a new order different from the current, seriously flawed situation. A charismatic leader is more than popular, attractive, likable, or pleasant; a merely popular person, "even one who is continually in our thoughts" (Boudon and Bourricaud 1989, p. 70), is not someone for whom we would break all previous ties and give up our possessions. Charismatic leaders are so demanding as to insist that their followers make ex-

traordinary personal sacrifices, cut themselves off from ordinary worldly connections, or devote their lives to achieving a vision that leaders have outlined.

The source of the charismatic leader's authority, however, does not rest with the ethical quality of the command or vision. Adolf Hitler, Franklin D. Roosevelt, Mao Zedong, and Winston Churchill all were charismatic leaders. All assumed leadership of a country during turbulent times. All conveyed a powerful vision (right or wrong) of their countries' destinies. A description such as the following is typical of charismatic leaders:

> *He had a powerful sense of both his nation's destiny and his own . . . he saw no difference between the two. Difficult, egocentric, vainglorious, he demanded complete loyalty from those beneath him but did not always bestow comparable loyalty on those above him. . . . His belief in his own vision and fate was so strong that few other men dared challenge it.* (HALBERSTAM 1986, P. 111)

Charismatic authority is a product of the intense relationships between leaders and followers. From a relational point of view, then, charisma is a "highly asymmetric power-relationship between an inspired guide and a cohort of followers" (Boudon and Bourricaud 1989, p. 70) who believe in the promises and visions offered by the person with special qualities. Many Soviets, for example, believed that Joseph Stalin, the general secretary of the Soviet Union from 1922 to 1953 and the Soviet premier from 1941 to 1953, was laying the foundation for a new kind of society in which people would eventually live in harmony and experience absolute economic security. Similarly, many Germans believed that Adolf Hitler's vision could help them recover from the humiliation and the massive destruction they suffered at the hands of the Allies (Britain, France, Russia, Italy, and the United States) in World War I.

Charismatic leaders and their followers come to constitute an "emotional community" devoted to achieving a goal and sustained by a belief in the leader's special qualities. Weber argues, however,

Channel 9 Australia/Gamma Liaison

UPI Bettmann

Cult leader David Koresh (left) and President Franklin D. Roosevelt illustrate Weber's theory of charismatic leaders whose authority depends upon the emotional bonds they forge with their followers. Koresh's influence evaporated when he and most of his followers died in a shootout with law enforcement officers in Waco, Texas. In contrast, Roosevelt's leadership became "routinized" and gave rise to such institutional legacies as Social Security that have endured long after his death in office.

that at some point the followers must be able to return to a normal life and to develop relationships with one another on a basis other than their connections to the leader. Attraction and devotion cannot sustain a community indefinitely, if only because the object of these emotions—the charismatic leader—is mortal. Unless the charisma that bonds a community is routinized, the community may disintegrate from exhaustion or from a void in leadership. **Routinized charisma** develops as the community establishes procedures, rules, and traditions to regulate the members' conduct, to recruit new members, and to ensure the orderly transfer of power. Authority must come to rest on legal-rational grounds. That is, it must be grounded in the position, not in personal qualities.

The Power Elite: Legal-Rational Authority and Change

Legal-rational authority rests on a system of impersonal rules that formally specifies the qualifications for occupying a powerful position. The rules also regulate the scope of power and the conduct appro-

priate to someone holding a particular position. In cases of legal-rational authority, people comply with commands, decisions, and directives because they believe that those who have issued them have earned the right to rule.

Sociologist C. Wright Mills (1959, 1963, 1973) argues that the causes of **great changes**—events whose causes lie outside ordinary people's characters or their immediate environments but profoundly affect their life chances—can be traced to the decisions made by the **power elite,** those few people positioned so high in the social structure of leading institutions that their decisions have consequences affecting millions of people worldwide. For the most part, the source of this power is legal-rational and resides not in the personal qualities of those in power but in the positions that the power elite have come to occupy. "Were the person occupying the position the most important factor, the stock market would pay close attention to retirements, deaths, and replacements in the executive ranks" (Galbraith 1958, p. 146).

The amount of power wielded by the elite over the lives of others is related to the nature and the quality of instruments that they can use, by virtue of their position, to rule, control, and influence others. Instruments might include weapons, surveillance equipment, and specialized modes of communication. Mills argued that since World War II rapid advances in technology have allowed power to become concentrated in the hands of a few; those with access to such power can exercise an extraordinary influence not only over their immediate environment but over millions of people, tens of thousands of communities, entire countries, and the globe.

Joseph Stalin represents one case of the extraordinary influence one person can wield over an entire country when a leader has obtained a position that gives him or her such complete power over mechanisms of social control. In the spirit of his predecessors, Stalin believed that if the Soviet Union did not industrialize, it could not compete with the more highly industrialized European nations, which were expanding their territories and fighting to gain influence. Stalin maintained that the Soviet Union was 50 years behind these countries and that the gap must be closed within 10 years. To accomplish this goal, he established a reign of terror. He forced millions of peasants to work in factories, and seized millions of private land holdings and created massive state-owned agricultural collectives. In addition, Stalin established a brutal system of repression (secret police, forced labor camps, mass deportations, guilt by association, and purges) as means of controlling or eliminating anyone who opposed him, or even thought of opposing him. His system was so relentless and so brutal that more than 20 million people died as a result of it. Because Stalin controlled and censored all information, the Soviet people learned only about his vision of socialism and heard only good things about his policies. Finally, Stalin strengthened control over the economy so that state bureaucrats decided what should be produced, how much should be produced, how much the products should cost, and where they should be distributed.[8]

Individual factory managers not only were not free to produce what they wished but could not ship their output to whomsoever they chose, could not fire unwanted or inefficient workers, could not refuse to accept output shipped to them under the plan directive, and could not even spend their factory's money to buy unauthorized electric light bulbs. . . .

From what we know now about how anarchic the planning process was, the wonder is not that the Soviet economy has given out but that it went on as long as it did. . . . The remarkable achievements in space exploration and the high level of war matériel were obtained by giving them absolute priority and close supervision, unlike the lack of attention accorded to "unimportant" goods like buttons, toilet paper, needles, thread, and diapers, which therefore regularly disappeared from the market. (HEILBRONER 1990, P. 94)

By virtue of his position, Joseph Stalin had a virtual monopoly over the mechanisms of social control that, in turn, allowed him to realize his goals even as others resisted them. In writing about the power elite, Mills did not focus on any one individual but focused almost exclusively on the power elite who occupied the highest positions in the leading institutions in the United States.

According to C. Wright Mills, the leading institutions are the military, corporations (especially the 200 or so largest American corporations), and government. "The power to make decisions of national and international consequence is now so clearly seated in political, military, and economic institutions that other areas of society seem off to the side and, on occasion, readily subordinated to these" (Mills 1963, p. 27).

The origins of these institutions' power can be traced to World War II, when the political elite mobilized corporations to produce the supplies, weapons, and equipment needed to fight the war. For one measure of the extent to which government, military, and corporations worked together, see advertisements published in *Life* magazine during the war years. As one example, consider this Bell Telephone System advertisement in the July 30, 1945, issue of *Life*:

In the last five years the Bell System has furnished millions of telephones for war, including 1,325,000 head sets for air and ground forces and more than 1,500,000 microphones. . . . Also more than 1,000,000 airplane radio transmitters and receivers . . . 4,000,000 miles of

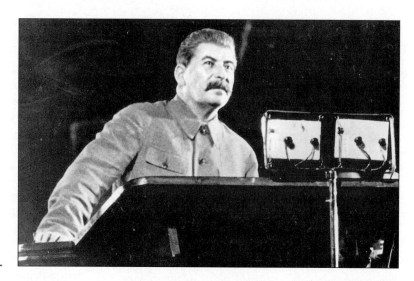

Joseph Stalin's reign of terror is an extreme case of concentration of power in a single person. But a society does not have to be an autocratic dictatorship for power to be distributed unevenly. According to Mills, in our own society power elites make decisions that have enormous influence on the lives of ordinary citizens.
UPI/Bettmann

telephone wire in cables . . . a vast quantity of switchboards, gun directors and secret combat equipment. That helps to explain why we are short of all kinds of telephone facilities here at home. (BELL TELEPHONE SYSTEM 1945)

After the war, as Stalin moved to consolidate his power in Eastern Europe, Japan and the countries of Western Europe—their populations demoralized, their economies in ruins, their infrastructures devastated—had little choice but to accept help from the United States in the form of the Marshall Plan. American corporations, unscathed by the war, were virtually the only companies able to offer the services and products that the war-torn countries needed for rebuilding. The interests of the government, the military, and corporations became further entangled when the political elite decided that a permanent war industry was needed in order to contain the spread of communism.

Thus, over the past 45 or 50 years, these three institutions have become deeply and intricately interrelated in hundreds of ways, as in the following examples:

- On May 24, 1993, a U.S. federal judge ruled that the Worker Adjustment and Retraining Notification Act, which requires employers to give workers 60-day advance notice if a massive layoff is to occur, does not apply to defense contractors. As a result, General Dynamics Corpo-

ration did *not* have to give backpay to 3,000 former employees who were laid off in 1991 with less than 60 days' notice "after the U.S. Navy had scrapped a $57 billion A-12 Stealth attack-plane project" with the company (*Facts on File* 1993, p. 470).

- Of the 435 congressional districts, 138 depend heavily on military contracts (Barnet 1990). In addition, military bases are important to the economic well-being of hundreds of communities around the country. When a seven-member federal commission announced 31 base closings on June 31, 1991, representatives from the affected districts protested loudly. Landlords, real estate brokers, and cinema operators and restaurant owners who had come to rely on the patronage of military personnel and their families now had to worry about the future of their businesses (Barron 1991; Ifill 1991).

- Now that the Cold War is over, Pentagon planners project the need for rapid-deployment forces that can respond quickly to instability in resource-rich or strategically located developing countries (Barnet 1991).

- The U.S. Department of Defense awarded approximately $137 billion in defense contracts in the fiscal year 1991, the last year for which data are available (U.S. Department of Defense 1991). As the amount allocated to defense de-

TABLE 15.3 Top Defense Contractors

These are the top 21 defense contractors, ranked according to the amount of their contracts awarded in 1991.

| | Total Contracts | |
Company	Amount ($ million)	Defense Business (percentage)
McDonnell Douglas Corp.	8,057	55.6
General Dynamics Corp.	7,848	59.3
General Electric Co.	4,866	8.0
General Motors Corp.	4,427	3.3
Raytheon Company	4,059	44.1
Northrop Corp.	3,319	65.5
United Technologies Corp.	2,825	13.6
Martin Marietta Corp.	2,689	28.5
Lockheed Corp.	2,666	20.3
Grumman Corp.	2,363	73.2
Westinghouse Electric Corp.	1,811	15.6
Rockwell International Corp.	1,707	15.7
Litton Industries Inc.	1,600	29.1
FMC Corp.	1,466	39.0
UNISYS Corp.	1,378	17.7
Loral Corp.	1,282	38.4
LTV Corp.	1,254	30.1
Boeing Co.	1,166	4.6
TRW Inc.	1,092	13.7
Textron Inc.	996	11.4
Texas Instruments Inc.	982	11.5

SOURCES: U.S. Department of Defense, pp. 9–12 (1991) and *Fortune*, pp. 220–26 (1994). Compiled by Renée D. Johnson (1994).

creases, many large procurement, research, development, and construction contracts are signed each year between corporations and government (see Table 15.3). Even if the defense budget is decreased to $200 billion a year, opportunities still exist for contractors. As one management specialist put it, "$50 billion for hardware is still a good-sized market" (Lambert 1992, p. 61). Now in light of the new emphasis on cleanup, corporations that formerly depended on weapons contracts are establishing divisions to bid for government contracts related to the cleanup of toxic wastes at 11,000 domestic sites and hundreds of overseas military installations (Schneider 1991).

- Contracts between the U.S. Department of Defense and companies that operated the nuclear production facilities specify that the government must pay any legal fees incurred by the companies—fees related to radioactive exposure, or any claims awarded to plaintiffs who win such settlements (D'Antonio 1994).

Because the three realms of institutions (military, government, and corporations) are interdependent and because decisions made by the elite of one realm affect the other two, Mills believes that it is in everyone's interest to cooperate. Shared interests cause those who occupy the highest positions in each realm to interact with one another. Out of necessity, then, a triangle of power has emerged. This is not to say that the alliance among the three is untroubled, that the powerful in each realm are of one mind, that they know the consequences of their decisions, or that they are joined in a conspiracy to shape the fate of a country or the globe:

> *At the same time it is clear that they know what is on one another's minds. Whether they come together casually at their clubs or hunting lodges, or slightly more formally at the Business Advisory Council or the Committee for Economic Development or the Foreign Policy Association, they are definitely not isolated from each other. Informal conversation elicits plans, hopes, and expectations. There is a community of interest and sentiment among the elite.*
> (HACKER 1971, P. 141)

Mills gives no detailed examples of the actual decision-making process at the power elite level. He is more concerned about understanding the consequences of this alliance than about understanding the decision-making process of assessing the consciousness or purity of motives. Mills acknowledges that the power elite are not totally free agents, subject to no controls. A chief executive officer of a major corporation is answerable to unions, OSHA, the FDA, or other regulatory bodies. Pentagon officials are subject to congressional investigations and budget constraints. Defense contractors are liable to the Federal False Claims Act, which gives a share in the settlement to any employee who can prove that the contractor has defrauded the government (Stevenson 1991). The president of the United States is constrained by bureaucratic red tape and by a sometimes slow-moving, politically oriented Congress. Mills questions, however, whether these constraints on the power elite have "much significance when weighed against the areas of unrestricted action open to [them]" (Hacker 1971, p. 136).

In nuclear weapons production, testing, and waste disposal, the overlapping interests of the political, military, and corporate (defense contractors)

elite are particularly evident. Over the past 50 years or so, a handful of people (all men) have decided where to locate nuclear production, testing, and waste disposal sites without consulting local populations and/or explaining accurately the risks involved (see Figures 15.1 and 15.2). Their decisions have affected countless lives and destinies, and will continue to do so for as long as 240,000 years. During the course of the Cold War, hundreds of thousands of American soldiers, uranium miners, workers in weapons facilities, and residents of surrounding communities known as "downwinders" were exposed to great amounts of radiation or radioactive fallout. For example, an estimated 220,000 American soldiers were exposed to varying degrees of radiation from above-ground nuclear testing. Approximately 160,000 U.S. troops, known as "atomic soldiers," occupied Nagasaki and Hiroshima in 1945 and 1946. The U.S. Department of Defense is only beginning to release documents that reveal the extent and the consequences of these exposures.

Pacific Islanders and Nuclear Testing In *Day of Two Suns* Jane Dibblin (1988) describes how decisions by the power elite affected the lives and destinies of one group of people—the residents of the Marshall Islands, a chain of 34 small atolls 2,500 miles off the coast of New Zealand. The United States used these islands for above-ground nuclear weapons tests between 1948 and 1958. Since that time, the islands have been used to test long-range missiles and weapons for the Strategic Defense Initiative program (SDI, referred to colloquially as Star Wars).

The United States conducted 66 above-ground nuclear weapons tests on Bikini Island and Eniwetok Island. Before testing, U.S. military personnel evacuated the inhabitants of these islands to an uninhabited island (Rongerik), promising them that eventually they would return. Rongerik was uninhabited for several reasons: the land was not very fertile, there were few edible fish, and there was not enough fresh water. Consequently, many of the evacuees died or suffered from malnutrition. Meanwhile the inhabitants of nearby islands (Rongelap and Utrik) were showered with radioactive fallout.

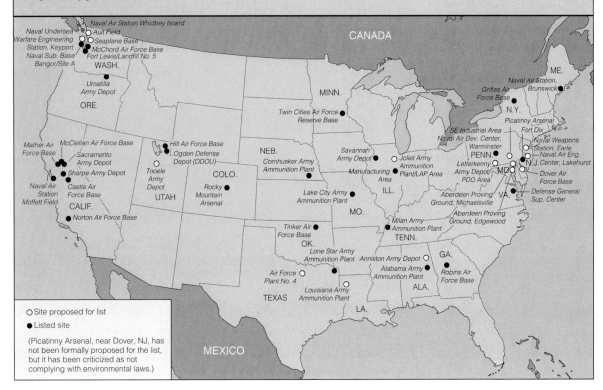

FIGURE 15.1 **Some of the Most Dangerous Toxic Waste Sites in the United States**

This map shows the military installations with abandoned toxic waste dumps or spill sites that are on the Environmental Protection Agency's "national priority list" or that have been formally proposed for the list. Listing means a site warrants immediate attention.

SOURCE: "Some of the Most Dangerous Waste Sites." Copyright © 1988 by The New York Times Company. Reprinted by permission.

One islander interviewed by Dibblin recalls:

I was 14 at the time and my sister Roko was 12. . . . We saw the bright light and heard a sound—boom—we were really scared. At that time we had no idea what it was. After noon, something powdery fell from the sky. Only later were we told it was fallout. . . .

That night we couldn't sleep, our skin itched so much. On our feet were burns, as if from hot water. Our hair fell out. We'd look at each other and laugh—you're bald, you look like an old man. But really we were frightened and sad.
(DIBBLIN 1988, PP. 24–25)

The U.S. military evacuated the exposed islanders, first to Kwajalein Island and then to Mejato Island, where they were to live until it was safe for them to return:

After two days on Kwajalein, a group of military doctors began studying the victims. Nausea, skin burns, diarrhea, headaches, eye pain, hair fall-out, numbness, skin discoloration were among common complaints. It had been so for quite a while. The children were more critical. My 10-year-old adopted son had severe burns in his body, feet, head, neck and ears. I cannot help remembering those sleepless nights we had

FIGURE 15.2 Map of Nuclear Production Sites in the Soviet Union

SOURCE: From "The Nuclear Epidemic," by Carla Anne Robbins et al., p. 49. Copyright © March 16, 1992 by *U.S. News & World Report.* Reprinted by permission. Original sources for basic data: Arms Control Association, Natural Resources Defense Council, Emerging Nuclear Suppliers Project, and Monterey Institute of International Studies.

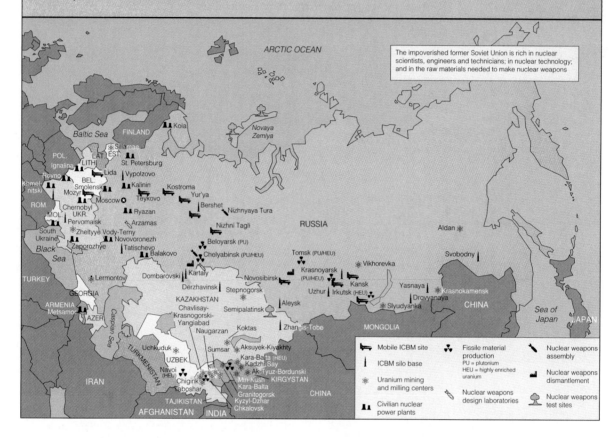

to hold him down onto his bed as he would have jumped up and down, scratching, rolling, as though insane.

Although I had also some burns on my back, feet and hands, and my hair was falling off, I knew I had been the least affected and I deeply felt pity about those who suffered the most. . . .

I saw three different women give birth to strange things after the "bomb." One was like the bark of a coconut tree. One was like a watery mass that was not humanlike. Another was again like a watery mass of grapes or something like that. I believe that these things are all caused by "the bomb." (PP. 27, 36)

In 1988, U.S. Department of Energy officials announced that the adults could return to Rongelap but that it was unsafe for their children. The adults did not return.

The United States is not the only country that has conducted above-ground and below-ground nuclear tests in the Pacific Ocean. The French tested on the Polynesian Islands; the British tested on Christmas Island and several islands off the coast of Australia. There, too, many islanders were exposed to radioactive fallout. The Chinese tested in Lop Nor, a desert in the northwest region of China, exposing Moslems of central Asia who lived downwind from the sites. The Soviets conducted their tests in Soviet central Asia.

Children at a Bikini cemetery pay their final respects to loved ones they must leave behind as they are forced to evacuate their homes to make way for atomic bomb tests.
UPI/Bettmann

Soviet Nuclear Testing One description of Soviet nuclear bomb testing appeared in *Letters to Gorbachev*, an anthology of some of the 5,000 to 7,000 letters that the editor of the Soviet newspaper *Argumenty I Fakty* received each day from an estimated 33 million readers. L. Boikava, a builder, wrote:

> On those days, we were herded into a deep ravine and told to lie on the ground, face down, with the mouth wide open (the latter was supposed to protect our eardrums from bursting). But of course we children wanted to see what was happening. So we watched. We saw three nuclear bomb carriers circle the sky, drop the bomb, and promptly fly away. The bomb would be slowly descending, in the glow of a blinding white light. When it fell, a cloud would rise up in the air, assuming the characteristic mushroom shape. The mushroom cloud was whitish- or greyish-pink, and it billowed and whirled and grew across the sky, eventually breaking into smaller clouds, before falling upon the earth, upon the fields and the River Irtysh. It all depended on the wind.
>
> Sometimes, it would blow towards Abolsk region, at other times towards us. There was also the sound wave. It would come more or less immediately, knocking people off their feet.
>
> Cats, dogs, cattle and other domestic animals would run away in panic, meowing, barking and lowing, adding to the noise of whining sirens. It was terrible, like the end of the world.
>
> Our school was the only two-storey building in our village. During one of the exercises, the top storey was sliced off, like with a knife. Many houses collapsed. Later, soldiers built a new school for us, and rebuilt the houses. After tests carried out in winter, the weather would change and it would be warm, like summer, with the ice on the River Irtysh melting completely. Sometimes the tests would be carried out at night-time. Children were woken up and dragged out of bed in haste, dressed warmly, if it was winter, and taken to the ravine. At night, the effect was truly spectacular: it was brighter than full sunshine; one could have picked needles out of a haystack.
>
> I have so many memories from those days, it would take too much time to tell them all. We used to hear a lot about Hiroshima and Nagasaki and weren't Americans awful, testing their nuclear weapons on humans. Yet no one ever mentioned that in the Soviet Union nuclear bombs were being tested on humans, i.e. on us. Clearly, we were not considered humans then— or now, for that matter. There were never any medical check-ups, in spite of the radiation we were exposed to. People in our village began to die of leukaemia, but for some reason it had to be kept quiet, and we children knew nothing about the mysterious illness. In 1963 my family moved to Urdshar, a village still in the same province, but 560 km, south of Semipalatinsk. (MCKAY 1991, PP. 80–81)

To this point we have examined two important agents of change: innovations and the actions of people in positions of authority (charismatic leaders and the power elite). We turn now to a third agent of change, conflict, which is intertwined with these two agents. In its most basic form, conflict involves clashes between groups over their shares of wealth, prestige, and other valued resources. Recall that the introduction of innovations (discoveries, inventions, and paradigms) disrupts the balance of power, causing conflict between those who stand to benefit and those who stand to lose from the widespread acceptance of an innovation. Recall also that charismatic leaders and the power elite possess the authority to impose their will (for better or worse), despite resistance by others.

Conflict

Sociologist Lewis Coser points out in his essay "Social Conflict and the Theory of Social Change" that conflict will always exist, if only because there is never a perfect "concordance between what individuals and groups within a system consider their just due and the system of allocation" (1973, p. 120). Conflict occurs whenever a group takes action to increase its share of or control over wealth, power, prestige, or some other valued resource and when these demands are resisted by those who benefit from the current distribution system.

Conflict, whether it involves violent clashes or public debate, is both a consequence and a cause of change. In general, we can say that any kind of change has the potential to trigger conflict between those who benefit from the change and those who stand to lose because of it. When the bicycle was invented in the 1840s, for example, horse dealers organized against it because it threatened their livelihoods. Some physicians declared that people who rode bicycles risked getting "cyclist sore throat" and "bicycle stoop." Church groups protested that bicycles would swell the ranks of "reckless" women (because bicycles could not be ridden sidesaddle).[9]

Conflict can lead to change as well. It can be a constructive and invigorating force that prevents a social system from becoming stagnant, unresponsive, or inefficient. Conflict such as that resulting from the anti-nuclear, civil rights, and women's movements can create new norms, relationships, and ways of thinking. Conflict also can generate new and efficient technologies.

Coser notes that employers have responded to workers' demands for higher wages by investing in technologies that simplify the work task or cut the number of workers. Ironically, the most destructive form of conflict—war—has generated advances in life-saving medical technologies. During World War I many soldiers fighting on manure-covered farmlands contracted tetanus.[10] In addition, large numbers of soldiers were injured by machine gun shrapnel and bombshells. Physicians experimenting with antitoxins eventually found a cure for tetanus and also made considerable advances in reconstructive surgery. Similarly, the kinds of injuries incurred during World War II motivated doctors to create a system of collecting and preserving blood plasma and to mass produce an effective drug, penicillin, to treat wound infections (Colihan and Joy 1984). More recently, the Pentagon's Advanced Research Project Agency [ARPA] funds civilian projects that it believes will have military applications. For example, the agency funds a project for electric-vehicle systems that would allow military vehicles to travel silently over enemy-controlled terrain (Wald 1994).

Whether conflict will lead to reform (improvements or alterations in current practices) or to revolution (complete and drastic change) depends on a broad range of factors and contingencies. Sociologist Ralf Dahrendorf has identified and described some of these factors in his essay "Toward a Theory of Social Conflict" (1973).

Structural Origins of Conflict

In his essay, Ralf Dahrendorf asks two questions: (1) What is the structural source of conflict? and (2) What forms can conflict take? Dahrendorf's answers rest on the following assumptions. First, in every organization that has a formal authority structure (a state, a corporation, the military, the judicial system, a school system), clear dichotomies exist between those who control the formal system of rewards and punishment (and thus have the authority to issue commands) and those who must obey those commands or face the consequences (loss of job, jail, low grades, and so on). Second, a distinction between "us" and "them" arises naturally from the unequal distribution of power. In view of these assumptions, the structural origins of conflict can be traced to the nature of authority relations. That conflict can assume many forms. It can be mild or severe; "it can even disappear for limited periods from the field of vision of a superficial observer" (p. 111). As long as an authority structure exists, however, conflict cannot be abolished.

Dahrendorf outlined a three-stage model of conflict in which progression from one stage to another depends on many things. He does not claim to give an exhaustive list with regard to the possible course of a conflict. In fact, he reminds readers that the conflicts he names are some of the most obvious. The point is that conflict—its course and its resolution—is a complicated phenomenon in which many elements must be considered. Dahrendorf described the three stages of conflict as follows:

Conflict Model Stage 1 Every authority structure contains at least two groups with opposite and latent interests. Those with power have an interest in preserving the system; those without power have an interest in changing it. These opposing interests, however, remain latent or below the surface until the groups (especially those without power) organize. Adam Michnik, one of the founders of Solidarity,[11] observes that "it is immeasurably difficult to trace the path on which a person . . . encounters other people just like himself, and at a certain point . . . [says] 'Let us join hands, friends, so that they will not pick us off one by one'" (1990, p. 240). In the uranium mine town of Marysvale, Utah, for example, the miners who worked between 1948 and 1966 hauling the uranium ore that the Atomic Energy Commission purchased for nuclear weapons production started to become aware of their shared fate when the first miners began dying in the early 1960s (Schneider 1990a).

Often a significant event makes powerless people aware that they share an interest in seeing the system changed. Vaclav Havel, the former president of Czechoslovakia and now the president of the Czech Republic, believes that Chernobyl may have played an important role in bringing about the revolutions in central Europe. Chernobyl is located in Ukraine and was the site of a nuclear power plant meltdown in 1986, the most serious kind of accident that can occur at a power plant. Although Ukraine, Russia, and Belarus suffered the most radioactive contamination, areas as far away as Sweden were also affected. After Chernobyl, people in Czechoslovakia dared to complain openly and loudly to one another (Ash 1989).

Sometimes, too, people organize because they have nothing left to lose. As one East German scientist explained, "You don't need courage to speak out against a regime. You just need not to care anymore—not to care about being punished or beaten. I don't know why it all happened this year [in 1989]. We finally reached the point where enough people didn't care anymore what would happen if they spoke out" (Reich 1989, p. 20).

Conflict Model Stage 2 If those without authority have opportunities to communicate with one another, the freedom to organize, the necessary resources, and a leader, then they will organize. In communist countries, public protest usually is suppressed by force. Before *perestroika*,[12] for example, when people in the Warsaw Pact countries demanded political freedoms, the Soviets sent in soldiers and tanks. But when the Soviet leaders announced that they would not intervene in the internal affairs of other communist countries and condemned past interventions, people quickly mobilized to topple the pro-Soviet regimes. The speed of these revolutions suggests that the political strength of the communist governments rested on a monopoly on physical force, which, until 1989, they made clear they would use.

This example shows that people in positions of authority often use their positions to keep potentially damaging information from those who might use it to organize to change the system. The hallmarks of the nuclear weapons program in the United States (as well as the former Soviet Union) were an obsessive and unchallenged secrecy and an absence of outside oversight that allowed scientists, government leaders, and weapon contractors to emphasize production over the environment and human health and safety. As a result information about radioactivity and its health effects was kept from workers, soldiers, and residents living in affected communities. U.S. government leaders censored, dismissed, and/or withdrew funds of researchers who sought to make public such information. However, we know that thousands of people persisted in their effort to study the effects or to gain access to government records (see "Scientist Who Managed to 'Shock the World' on Atomic Workers' Health").

Scientist Who Managed to "Shock the World" on Atomic Workers' Health

It was in September 1975 that Dr. Alice Stewart, a 68-year-old British epidemiologist, first saw statistical evidence that radiation was killing workers at the Hanford nuclear weapons plant in Washington State.

She was astonished. It was the first time researchers had found elevated levels of cancer in weapons workers exposed to low levels of radiation. "It came at me all of a sudden that what we were going to say would shock the world," she said in a recent interview. "Not even I had thought that the effect of such a small dose on an adult would be as great as it was."

Dr. Stewart and two colleagues published the finding, which was based on Government-funded research, in the journal *Health Physics* in 1977. It drew immediate and vitriolic criticism from other scientists, many of them under contract to the weapons industry.

A Policy of Secrecy

That study and the ensuing furor drew Dr. Stewart into a long struggle with the American Government and its nuclear weapons industry. For 14 years, this exquisitely courteous British scientist . . . battled a policy of scientific secrecy that she said threatened the lives of thousands of workers in weapons plants and of people living near them. Even before the Hanford finding was published, the Government cut off funds for the study, prompting a Congressional committee to investigate allegations of a scientific cover-up.

In March, in a change in policy on research and radiation protection that goes to the heart of Dr. Stewart's criticism, Energy Secretary James D. Watkins said he would open records on worker health that

the Government has long kept secret. Mr. Watkins also said he would end the Energy Department's control of the Government's most important program for studying the effects of radiation on workers in the weapons industry; by August, he said, the Department of Health and Human Services would manage the studies.

Those orders strengthened Dr. Stewart's reputation as perhaps the Energy Department's most influential and feared scientific critic, and they enhanced her credibility as the leading technical expert on worker safety for the department's opponents in Congress and for citizen groups.

Still, Much Dispute

"When I first started in this area, I didn't know what a nuclear reprocessing plant was," she said in an interview in Denver, where she was testifying at a worker's compensation hearing in behalf of a widow who said her husband had died from exposure to radiation at the nearby Rocky Flats weapons plant. "I was just an epidemiologist prepared to take up a subject and go after it.

"I have formulated views since then on the strength of my data. And that is, this industry is a great deal more dangerous than you are being told."

Her conclusions about the hazards of low-level radiation are by no means shared by all experts, particularly in the Department of Energy. Among her most ardent critics is Dr. Shirley A. Fry, the leader of the department's epidemiology group at the Oak Ridge atomic weapons plant in Tennessee. "The data do not support such statements," said

Dr. Fry, whose group may be substantially changed by Mr. Watkins's orders. "Dr. Stewart's work has been widely criticized for technical problems."

But with increasing frequency, scientists in and out of the Government have been confirming Dr. Stewart's views about the dangers of low-level radiation. Last December, in the most authoritative affirmation yet of her theories, the National Academy of Sciences said low-level radiation was a much more potent carcinogen than it had reported nine years before. . . .

Though she was regarded as one of Britain's best and most creative epidemiologists, she was unknown abroad until 1956, when she first tangled with the atomic industry over low-level radiation.

That year, in a letter to *The Lancet,* the British medical journal, Dr. Stewart reported on a disturbing finding she had made about prenatal X-rays: children who had died of cancer in England from 1953 to 1955 had received twice as many X-rays before birth as those without cancer.

It was the first time low levels of radiation had been shown to have any effect on health, and the finding was met with outrage from physicians and the nuclear industry. They rejected the idea that millions of people might be at risk not only from a common medical procedure but also from radiation produced by the nuclear power and weapons industries.

Scientists had known that high doses of radiation could cause illnesses; a dose of 500 rems or more, comparable to the levels received by some rescuers in the Chernobyl nuclear accident in 1986, was fatal.

But what was the biological effect of doses of 1 rem or lower? And was there any danger from years of exposure to low levels of radiation? In the 1950's, scientists agreed that there was not.

By the mid-1970's, Dr. Stewart's findings had been duplicated by other scientists, and the conclusion that X-rays were hazardous to unborn children had gained worldwide acceptance.

"We benefited from all the opposition, really," Dr. Stewart said recently. "If everybody accepted the finding we laid out in the 1950's, we'd have no reason to go on and collect more and more data. We shouldn't know one-quarter of what we know now."

Dr. Stewart retired from Oxford in 1974; she is now a senior research fellow at the University of Birmingham in the English Midlands. In 1975, she was asked by Thomas Mancuso, a researcher at the University of Pittsburgh, to help in a study for the Atomic Energy Commission of workers at the Hanford Reservation.

A few days after Easter that year, she arrived in Pittsburgh with her collaborator George W. Kneale, a statistician. By the fall, it was clear to the three researchers that Hanford workers exposed to radiation levels less than half the Federal safety limit of 5 rems a year suffered from at least a third more than the expected levels of pancreatic cancer, lung cancer and multiple myeloma, a rare bone marrow cancer.

"We knew this was going to be inflammatory," she said. "I persuaded Mancuso that he must warn his people. It wouldn't be fair to spring this on them. What I never counted on was their opposition."

U.S. Grant Is Cut Off

In early 1976, Dr. Mancuso presented preliminary results to the Atomic Energy Commission, the predecessor of the Energy Department. That March, the A.E.C. notified him that his 13-year research grant had been terminated. In Congressional hearings in 1978, Government officials denied that he had been dismissed because of the cancer findings, and later made public an inspector general's report that said the dismissal was not improper.

But Representative Paul Rogers of Florida, the chairman of the House subcommittee that looked into the incident, disputed the inspector general's conclusion. "We regret that the shortcomings with the report may only fuel charges of 'cover-up' already voiced," he wrote in a letter to James R. Schlesinger, Secretary of the newly formed Department of Energy, in August 1978.

In the decade that followed, Dr. Stewart often came to the United States to speak to community groups, Congressional committees and scientific conferences. At hearings in 1988 and 1989, she told Senate and House committees that the Energy Department's program for assessing radiation hazards at nuclear weapons plants was badly flawed, hindering the free exchange of scientific ideas.

First, she argued, it was scientifically indefensible for the department to lock in its own files the medical records of 600,000 American nuclear weapons employees that had been collected by the Government since the industry started in 1942. Those records, she said, were the best source of raw data on the effects of low-level radiation, yet they were available only to researchers under contract to the Energy Department.

Second, Dr. Stewart argued for a halt to what she called a troubling conflict of interest in American radiation research. Not only was the Energy Department the owner and operator of the vast weapons industry, she said at a Senate hearing last August, but it was also the Government's principal source of funds to study the health effects of radiation.

In March, facing growing Congressional pressure and public embarrassment, the department moved to address both problems. The studies of health effects would be taken over by the Health and Human Services Department, Mr. Watkins said, and the secret medical records would be made available to independent scientists as quickly as possible, though he declined to set a date. The first of the independent scientists to receive the material will be Alice Stewart, as principal investigator for the Three Mile Island Public Health Fund, a private group in Pennsylvania that studies the effects of radiation.

"Many times I've been asked why I didn't follow my friends into quiet retirement," she said. "If I was a coward and afraid of my job, I wouldn't say a thing. But I am retired. I have no department that anybody's dependent upon for work. I speak out because I think there are not a lot of people in such a good position. I have nothing to lose. A lot of people do. This area of research can be shut down. I've watched it happen."

SOURCE: "The Scientist Who Managed to 'Shock the World' on Atomic Workers' Health," by Keith Schneider. Copyright © 1990 by The New York Times Company. Reprinted by permission.

Six Stories in Search of an Author

Donald N. Michael and Walter Truett Anderson identify what they call six major stories—competing world views or paradigms—that have emerged in reaction to the Western-style mystique of progress that the economic system of capitalism embodies. An excerpt from their article follows.

First among them, from the American perspective, is the progress story. Faltering a bit now, it still commands the allegiance of the American mainstream and the political establishment—both Republican and Democratic. It has been mythologized in the arts and literature, articulated in the works of many theorists; Herman Kahn has been perhaps the most dogged spokesman in recent decades. Its main prop in domestic policy is the GNP and the ethic of growth; its main contribution to foreign policy is the theory of development, with all its elaborate concepts about how primitive societies become Western-style industrialized democracies.

A competitor to the progress story in American society is the fundamentalist story of return to a society governed by Christian values. The true fundamentalist is suspicious of foreign influences, fiercely protective of national sovereignty, and deeply pessimistic about rapid change. Although fundamentalism is not new, its current manifestation as a potent political force does have a definite late twentieth century character: It is largely a reaction to the increasing pace of global connectedness. And although many conservative Americans manage to fit both the myth of progress and Christian fundamentalism into their personal worldviews, there is a basic conflict between the two. The Republican party's internal schism between progressive internationalists and Christian fundamentalists shows how serious this conflict can become—serious enough that many political analysts believe it to be an insurmountable obstacle to a period of Republican domination of American politics.

There are other fundamentalist stories that have the same emotional dynamic, even though their outward form is entirely different. The most prominent of these in contemporary global politics is Islam. In many parts of the Third World, most notably Africa, it is Islam that stands as the real adversary of the Western-style story of progress. Islam offers to oppressed peoples a compelling political and economic ideology rooted in cosmology, and the basis for a pox-on-both-your-houses rejection of both capitalism and Marxism.

Meanwhile the classic Marxist story of revolution still gives form and purpose to political life for many people. It holds out the promise of the overthrow of the capitalist state, the creation of the socialist state, all wealth reallocated, a final end to class conflict. The Marxist story is the official ideology of a large part of the world; it is also the inspiration of guerrilla groups, a strong, although somewhat waning, force in the intellectual world, and (in one of the more historically ironic developments of our time) part of a radical new religious movement, liberation theology.

Most of the peoples of the world live out their lives according to one or another of the above four stories.

Conflict Model Stage 3 Once organized, those without power enter a state of conflict with those in power. The speed and the depth of change depend on the capacity of those who rule to stay in power and on the kind and degree of pressure exerted from below. The intensity of the conflict can range from heated debate to violent civil war, but it is always contingent on many factors, including opportunities for mobility within the organization and the ability of those in power to control the conflict. If those who do not have authority are confident that eventually they will achieve such a position, the conflict is unlikely to become violent or revolutionary. If those in power decide that they cannot afford to compromise, and mobilize all their resources to thwart protests, two results are possible. First, the protesters may believe the sacrifices are too great and may then withdraw. Apparently this was the case with the democracy movement in the People's Republic of China. Once the demonstrators were fired on in Tiananmen Square, the movement ended. Alternatively, the protesters may decide to meet the

Each story has its "establishment," its official version and revealed truths—but there are many variations, sects, and combinations. People are highly creative about such things.

All four of the major stories took form during the modern era. Both the myth of progress and the ideology of Marxism are products of the Enlightenment and the Industrial Revolution, and contemporary fundamentalist stories (whether Christian or Islamic or any of the other versions) are reactions to them. Fundamentalist stories are never merely simple returns to the beliefs of the past, but rather new ideologies created out of old material in response to current conditions. And now postmodern stories are emerging: New candidates are in the field, and others will probably appear in the decades ahead. Here we will take note of two: the green story and the new paradigm story. Many people see these as one, but we think there are significant differences that make it worthwhile to distinguish between them.

The green story is identified with environmental values, with the mystique of solar energy and organic farming, with the animal rights movement, and, of course, with the Green party of West Germany and similar political groups in other countries. It is, like fundamentalism, a reaction against the Industrial Revolution's myth of progress—but a postmodern and highly sophisticated one that finds adherents among idealistic young people, intellectuals, and some feminists. The green story tells of bringing a halt to the destructiveness of industrialized progress and returning to a simpler time of reverence for Earth and life focused on the local bioregion.

The new paradigm story is superprogress: a sudden leap forward to an entirely new way of being, and a new way of understanding the world. Although some new-paradigm thinkers draw heavily on the green mystique, others embrace big business and high technology and are enthusiastic about future exploration of space. The new paradigm story is a postmodern version of ancient millennarian cults that predicted the imminent coming of a new order, a paradise on Earth. But where old millennarian stories were derived from religious prophecy, new paradigm advocates like to support their expectations by reference to scientific theory such as Ilya Prigogine's work on dissipative structures. Evolutionary ideas, usually more of the Teilhardian than the Darwinian variety, also figure prominently. New paradigm thinkers tend to be convinced that the bright future they predict will inevitably come to pass: this is another point of distinction from the green movement, which is often pessimistic, alarmed, and militant.

All of these stories—and others as well—compete for credibility in the postmodern world. All of their adherents say: This is where the world is going; this is how it must be. Clearly we are in for much conflict—not only the bipolar conflict between West and East, the myth of progress vs. the ideology of revolution—but also fundamentalists against Marxists, Moslems against Christians, Greens against industrial progress, and many other permutations.

SOURCE: From "Norms in Conflict and Confusion: Six Stories in Search of an Author," by Donald N. Michael and Walter Truett Anderson. Pp. 107–115 in *Technological Forecasting and Social Change.* Copyright © 1987 by the authors. Reprinted by permission.

"enemy" directly, in which case the conflict becomes bloody.

To this point we have discussed conflict in very general terms. The fourth important agent of change represents a specific kind of conflict, motivated by the pursuit of profit. This agent is capitalism, an economic system whose modern origins can be traced back 500 years. It is significant in promoting social change in general and in understanding the dynamics of the Cold War. Capitalism promoted social change in that socialism as an organizing principle arose theoretically as a more humane alternative to capitalism. "The classic Marxist story of revolution still gives form and purpose to political life for many people. It holds out the promise of the overthrow of the capitalist system, the creation of the socialist state, all wealth reallocated, a final end to class conflict" (Michael and Anderson 1987, pp. 110–11; see "Six Stories in Search of an Author").

Capitalism

Karl Marx believed that an economic system—capitalism—ultimately caused the explosion of technological innovation and the enormous and unprecedented increase in the amount of goods and services produced during the Industrial Revolution. In a capitalist system, profit is the most important measure of success.[13] To maximize profit, the successful entrepreneur reinvests profits in order to expand consumer markets and to obtain technologies that allow products and services to be produced at the highest quality and the greatest cost-effectiveness.[14]

The capitalist system is a vehicle of change in that it requires the instruments of production to be revolutionized constantly. Marx believed that capitalism was the first economic system capable of maximizing the immense productive potential of human labor and ingenuity. He also believed, however, that capitalism ignored too many human needs, and that too many people could not afford to buy the products of their labor. Marx stated that capitalism already had unleashed "wonders far surpassing Egyptian pyramids, Roman aqueducts, and Gothic cathedrals . . . [and] expeditions that put in the shade all former Exoduses of nations and crusades" ([1881] 1965, p. 531). He believed that if this economic system were in the right hands—those of socially conscious people motivated not by a desire for profit or by self-interest but by an interest in the greatest benefit to society—public wealth would be more than abundant and would be distributed according to need.

Instead, according to Marx, capitalism survived and flourished by sucking the blood of living labor. The drive for profit (which Marx maintained is derived from the labor of those directly involved in the production process) is a "boundless thirst . . . [a] werewolflike hunger . . . [that] takes no account of the health and the length of life of the worker unless society forces it to do so" (Marx 1987, p. 142). The thirst for profit "chases the bourgeoisie over the whole surface of the globe" ([1881] 1965, p. 531). Marx's theories influenced a group of contemporary sociologists—world system theorists—to write about capitalism as the agent of change underlying global interdependence.

World System Theory

Immanuel Wallerstein (1984) is the sociologist most frequently associated with **world system theory,** a modern theory concerned with capitalism. Since the early 1970s he has been writing a four-volume work (three volumes of which have been published) about the ceaseless expansion, over the past 500 years, of a single market force—capitalism. According to Wallerstein, although stagnant periods have occurred and some countries (the communist countries, for example) have tried to withdraw from the capitalist economy, no real contraction has occurred. "Hence, by the late nineteenth century, the capitalist world-economy included virtually the whole inhabited earth and it is presently striving to overcome the technological limits to cultivating the remaining corners; the deserts, the jungles, the seas, and indeed the other planets of the solar system" (p. 165).

Wallerstein distinguishes between the terms *world economy* and *world-economy.* People who use the term *world economy* (without the hyphen) envision the world as consisting of 160 or so national economies that have established trade relationships with one another. In this vision, globalization is portrayed as a relatively new phenomenon and as a process in which the countries of the world have moved from relatively isolated, self-sufficient economies to economies that trade with one another to varying degrees. Although popular, this conception of global interdependence is not very accurate. The more accurate term (and conception) is **world-economy.** The world-economy is not recent; it has been evolving for at least 500 years and is still evolving. People who use the hyphenated term envision a world (encompassing hundreds of countries and thousands of cultures) interconnected by a single division of labor. In the world-economy, economic transactions transcend national boundaries. Although each government seeks to shape the global market in ways that benefit its "own" corporations and national interests, no single political structure (world government) or national government has authority over the system of production and distribution.

One way in which capitalism has become global is by aggressively seeking out markets for its products in every part of the world, including this Kuna Indian village near Panama City, Panama.

Wim Van Cappellen/Impact Visuals

Wallerstein argues that the world-economy is capitalist because "its economy has been dominated by those who operate on the primacy of endless accumulation, . . . driving from the arena those who seek to operate on other premises" (1984, p. 15). Critics counter that this is an exaggeration, that there are many countries with economies that are not capitalist and that no country in the world has an economy that runs on purely capitalist principles.[15]

Wallerstein counters this criticism in part with the argument that the communist countries are the equivalent of huge state-owned capitalist corporations. All depend on the world-economy, and all trade on some levels with countries that are bitterly opposed to their political and economic system. Even before the collapse of the Soviet Union, the United States, for example, exported corn and wheat there and the Soviets exported chemicals, fuels, and minerals to the United States. Likewise, even before the Soviet Union disbanded, the Soviets were exporting natural and enriched uranium to Western countries (Broad 1991). Another feature that makes the world-economy capitalist is the fact that profits from goods and services are distributed unevenly through the global market to a network of beneficiaries, most of whom live in the mechanized-rich countries.

The Role of Capitalism in the Global Economy

How has capitalism come to dominate the global network of economic relationships? One answer lies in the ways in which capitalists respond to changes in the economy, especially to economic stagnation. Historically there have been five important responses, all designed to create economic growth:

1. Lowering production costs by hiring employees who will work for lower wages (for example, by busting unions, buying out workers' contracts, or offering early retirement plans), by introducing labor-saving technologies (such as computerizing the production process), or by moving production facilities out of high-wage zones and into lower-wage zones inside or outside the country.

2. Creating a new product that consumers "need" to buy, such as the videocassette recorder, the computer, or the fax machine (products introduced in the late 1970s and early 1980s).

3. Improving on an existing product and thus making previous versions obsolete (see "Capitalism in Action: The Case of Nike"). Fax machines are a good example of this strategy: "About 475,000 FAX machines were installed [in 1987], up from 190,000 in 1986. But while most companies are still deciding whether to purchase their first FAX machines, a new generation is already on the horizon" (Furchgott 1988, p. 484).

Capitalism in Action: The Case of Nike

"Just do it!" the TV ads admonish, and Nike has. In a market as packed with competitors as the New York City Marathon, the Beaverton, Oregon, firm has become the world's top designer and marketer of sports shoes. Last year sales jumped 31% to $2.2 billion, about $75 million higher than archrival Reebok International. Nike's lead in profits was almost as great. It made $243 million, vs. $177 million for Reebok.

Consumers know Nike for attention-grabbing commercials that feature athletes like Bo Jackson and Michael Jordan. Shoe sellers like the company for programs such as Nike Next Day—which fills merchants' orders within 24 hours. Says a distributor in North Carolina: "I deal with 100 other vendors. Nike is *the* standard in the industry." But from the standpoint of Nike executives, the company's main advantage is new shoes.

Its catalogue lists more than 800 models for use in 25 or so sports. Nike has led the industry in breaking the market into lucrative subsegments. It makes three lines of basketball shoes, each expressing what Nike calls a different attitude. The Áir Jordan (retail price: $125) is for consumers who want to follow in the footsteps of the Chicago Bulls superstar. The Flight (up to $115) is for players who value the lightest Nikes, while the Force (up to $150) incorporates the latest designs—like a customized-fit air bladder, Nike's answer to the Reebok Pump.

By updating each shoe at least every six months, the company tempts customers to lace on new pairs before last year's wear out. Says vice president Andrew Mooney: "We're like the auto industry, constantly looking for new technologies to pep up the product." Nike contracts the manufacture of its shoes to foreign factories, as do most sneaker makers. That saves capital, says CEO Philip Knight, and leaves Nike free to have its cobbling done wherever labor costs are lowest—currently Korea, Chile, and Thailand.

The designs, though, are home grown. Nike took off in 1979 by introducing the Tailwind, a shoe whose sole incorporated a gas-filled sac to cushion the foot. The so-called air technology revolutionized running shoes. A recent embodiment is the Air Huarache, a $110 model that features a Neoprene/Lycra bootie bonded to a thermo-plastic footbridge, Phylon midsole, and Waffle outsole. (Translation: It is made of synthetic elastic material attached to a cushy but tough plastic sole that leaves footprints with the pattern of a waffle.) It looks like a shoe from Mars—an aspect heightened by its Royale Blue and Scream Green decoration.

Such creations emanate from the Michael Jordan building, a glass and maple R&D center on Nike's 74-acre campus, where exercise physiologists and mechanical engineers study the stresses on athletes' feet and collaborate with stylists on new shoe ideas. The team publishes research in scientific journals such as *Medicine and Science in Sports and Exercise*. Knight boasts that Nike has stockpiled so many shoe ideas that, introduced into the market in rapid-fire succession, they will keep the competition on the run for at least three years.

SOURCE: From "Smart Moves by Quality Champs," by Erik Calonius, pp. 26–27. Copyright © 1991 by *Fortune*. Reprinted by permission.

---

4. Expanding the outer boundaries of the world-economy and creating new markets. Since the fall of the Berlin Wall in 1989, American, Western European, and Japanese corporations have been expanding their markets into Eastern Europe, Russia, and the new states of Eurasia. Procter & Gamble, for example, already produces and markets detergent, toothpaste, shampoo, and diapers in Czechoslovakia, Hungary, and Poland (Rawe 1991). The Coca-Cola Company moved particularly quickly into Eastern Europe. Almost immediately after the Berlin Wall fell and East Germans started to visit West Germany and West Berlin, Coca-Cola was there, handing out free Coca-Cola. This event was very popular in West Berlin; people sought out the Coca-Cola vendor for samples. Within weeks, the U.S. corporation was in East Germany discussing the distribution of Coca-Cola there:

Almost at the same time we reorganized: As soon as we saw the changes, we moved East

Germany into the West German and E. C. group, and transferred the infrastructure, talent and technology of our West German operations into East Germany. Within weeks, we were shipping Coca-Cola in cans into East Germany to distributors with whom we had made agreements. Within a month or two, we were selling a million cases a month in East Germany. . . .

By July, we had put our first East German production facility in place. Now we're no longer solely importing into East Germany, we're producing and distributing in the country. Next year we expect to sell 30 million cases and, by 1995, 100 million cases a year in East Germany. (GUTTMAN 1990, P. 16)

5. Redistributing wealth to enable more people to purchase products and services. Henry Ford was the first to do this on a large scale; in 1908 he came up with the revolutionary concept of paying workers a wage (five dollars per day) large enough to allow them to purchase the products of their labor (Halberstam 1986). Because of the shortage of hard currency in Russia, the new states of Eurasia, and Eastern Europe, American, Western European, and Japanese executives have set up barter systems there (Holusha 1989). "Pepsico, for example, exports wooden chairs from Poland to its Pizza Hut franchises in the United States, and sells its soft drink to the Soviet Union in exchange for old submarines" (O'Sullivan 1990, p. 22).

As a result of these responses to economic stagnation, capitalism has spread steadily to encompass the globe. In addition, every country of the world has come to play one of three different and unequal roles in the global economy: core, peripheral, and semiperipheral.

The Roles of Core, Peripheral, and Semiperipheral Economies

Core economies include those of the mechanized rich nations—nations characterized by strong stable governments. Core economies tend to be highly diversified. Mechanized rich nations absorb nearly two-thirds of developing countries' exports. They import raw materials from labor-intensive poor countries and make use of free trade zones around the world. The overwhelming majority of "the great global enterprises that make the key decisions—about what people eat and drink, what they read and hear, what sort of air they breathe and water they drink, and, ultimately, which societies will flourish and which city blocks will decay" (Barnet 1990, p. 59)—have their headquarters in countries with core economies. The sales of these corporations exceed the gross national products of many countries (Currie and Skolnick 1988; see Table 15.4). When economic activity weakens in the industrial world, the labor-intensive poor countries suffer because the amount of exports declines and price levels fall.

Labor-intensive poor countries have **peripheral economies**, which are not highly diversified; most of the jobs are low-paying and require few skills. Peripheral societies depend disproportionately on a single commodity such as coffee, peanuts, or tobacco or a single mineral resource such as tin, copper, or zinc. The aggregate GNP of the peripheral economies is less than that of the European Economic Community (Van Evera 1990). Peripheral economies have a dependent relationship with core economies that is rooted in colonialism. Peripheral economies operate on the fringes of the world-economy. In the midst of widespread and chronic poverty, there are islands of economic activity including off-shore manufacturing zones, highly vulnerable extractive and single-commodity economies, and tourist zones.

Between the core and the periphery are the **semiperipheral economies**, characterized by moderate wealth (but extreme inequality) and moderate diversification. Taiwan, Brazil, South Korea, and Mexico fall into this category. Semiperipheral economies exploit peripheral economies and are exploited by core economies. By this definition, Iraq and Kuwait were semiperipheral economies before the Gulf War: the core economies relied on them for cheap oil, and they relied on peripheral economies for cheap labor. The extent of that reliance became clear after Iraq invaded Kuwait: at that point millions of migrant workers fled Kuwait, including Iranians (70,000), Iraqis (2.2 million), Yemenis (45,000), Sudanese (21,800), Egyptians

TABLE 15.4 Corporate Sales and Gross National Products of Selected Countries, 1993

The sales of the 10 largest U.S.-headquartered industrial corporations combined exceed the GNP of many countries. The sales of some industrial corporations such as General Motors, Exxon, and Philip Morris are larger than the GNP of many countries.

Country or Corporation	Amount ($ billion)	Ranking
United States	5,686.0	1
Japan	3,337.1	2
Unified Germany	1,516.7	3
France	1,167.7	4
Italy	1,072.2	5
United Kingdom	963.6	6
Top 10 U.S. Industrial Corporations*	681.0	7
Canada	568.8	8
Spain	486.6	9
Russia	479.5	10
Brazil	447.3	11
Australia	287.8	12
India	284.7	13
Netherlands	278.8	14
South Korea	274.5	15
Mexico	252.4	16
Sweden	218.9	17
General Motors	133.6	18
Saudia Arabia	105.1	19
Exxon	97.8	20
Poland	70.6	21
Philip Morris	50.6	22
Romania	31.1	23

*The ten largest industrial corporations in 1993 were General Motors, Ford Motor, Exxon, IBM, General Electric, Mobil, Philip Morris, Chrysler, Texaco, E.I. du Pont de Nemours.

SOURCES: Compiled by Renée D. Johnson. Corporate Sales from *Fortune*, p. 220 (1994); GNP estimates *The 1994 Canadian Global Almanac*, pp. 307–503 (1993).

(700,000), Jordanians (220,000), and Palestinians (30,000). Similarly, Pakistanis (67,600), Indians (150,000), Bangladeshis (85,000), Vietnamese (16,000), and Filipinos (30,000) fled Iraq (Miller 1991). According to Wallerstein, semiperipheral economies play an important role in the world-economy because they are politically stable enough to provide useful places for capitalist investment if wage and benefit demands become too great in core economies.

World system theorists maintain that political upheavals are caused by uneven and unequal integration into the world-economy. Corporations take actions and governments make trade decisions that leave out some groups. Eventually, oppressed groups may organize around one of two main themes: "class" (disadvantaged position in the labor force) and "nation" (common community, culture, language, territory, and ethnicity).

According to this viewpoint it would be a mistake to believe that conflicts in Russia, the new states of Eurasia and central Europe, and Africa are simply ethnic in origin. They center around deeply rooted, chronic, persistent inequalities and conflicts over scarce and valued resources (territory, minerals, and so on).

Discussion

When sociologists study a social change in the organization and operation of social life they ask at least two key questions: what factors are causing that change? And what are the consequences of that change for society? The first question is difficult to answer because it is not usually possible to identify a single factor as a cause of a change because it is usually related to a seemingly endless sequence of events. Likewise the second question is difficult to answer because it is difficult to predict how a change will affect society. Yet such difficulties should not lead us to conclude that change is impossible to study.

With regard to the first question sociologists have identified at least four main agents of change under which the various interactions and reactions that bring about change can be classified. These agents are the creation of and reactions to innovation, the actions of leaders, conflict over scarce and valued resources, and behaviors and policies shaped by capitalist principles. In this chapter we have considered a major change: the end of the Cold War and the rise of what some call the New World Order, as reflected in the shift on the part of the former Soviet Union and the United States away from nuclear weapons proliferation to reduction. We used the four agents of change to organize our discussion of the many factors responsible for this shift which include (1) the basic innovations that led to the creation of an atomic bomb; (2) the series of improving innovations on the part of the United States and Soviet Union that followed the first bomb's creation; (3) the issue of cultural lag (specifically that both countries put weapons production ahead of environmental, health, and safety concerns to maintain their military edge); (4) Soviet

and American leaders' decisions to mobilize people and resources to participate in an arms race and decisions about where to locate nuclear production, testing, and waste disposal sites without consulting local populations and/or explaining accurately the risks involved; (5) the mobilization of people affected to challenge the actions of government leaders and contractors; and (6) the rise of socialism as a theoretically more human alternative to capitalism and its collapse in Russia and the new states of Eurasia. All these events (and many more) have pushed the United States and Russia to the point where each country is compelled to deal with the Cold War's effects on the environment and on the psyche of its citizens.

The second question—what are the consequences of a change for society?—is also difficult to answer for several reasons. It is difficult to predict exactly how a change will affect society, if only because it is difficult to pinpoint a time in the future when it will cease to have an impact. Yet this should not deter people from thinking about consequences. In this regard, sociology offers a rich vocabulary of concepts and some powerful theoretical perspectives for framing and clarifying thought and discussion about the consequences of change. A second reason is that people react to change, and their reactions play a role in shaping the consequences. The element of choice (that people decide *how* to react to a change) is the hopeful element in the study of change because it suggests that people are not passive agents. Sociology offers a unique perspective to those who are interested in making responsible and constructive responses—that perspective is the sociological imagination.

Epilogue: The Sociological Imagination Revisited

In Chapter 1 we learned that perceiving the connection between troubles and issues is fundamental to the meaning of sociology. Understanding this connection is central to the sociological imagination—the ability to link seemingly impersonal and remote historical experiences to an individual's life. To those who possess it, the sociological imagination provides a quality of mind or a framework that helps them think about "what is going on in the world and of what may be happening within themselves" (Mills 1959, p. 5).

The payoff for those who possess this quality of mind is that they can understand their own experiences and "fate" by locating themselves in history; they can recognize the variety of responses available to them by becoming aware of all the individuals who share the same situation as themselves. By understanding that a variety of responses exist for every situation, people can see that they need not be passive figures who respond in predictable and unthinking ways to the larger issues that affect their lives. Thus they can come to see that people shape a society, however minutely, if only by living in that society, even as that society shapes them.

Such a hopeful perspective on the role of the individual in society seems to be out of place when we consider that people may have to live with the consequences of the Cold War for as long as 240,000 years. In fact, upon writing this last sentence, I asked Renée Johnson, my research assistant, what I could say at this point to convince student readers that an informed individual response makes a difference. She replied, "That's a tough question. I don't know how to answer it, but in light of the material you've just presented, you would have to say something very profound to make me believe in the so-called promise of the sociological imagination."

Renée's comments spoke to my apprehension of ending this book without convincing readers that there is a reward for making the mental effort required to develop the sociological imagination. As I thought about this problem, I was reminded of another conversation that I had had many years ago

with Dr. Horatio C Wood, author of the Focus essay "Reflections on Change," which concludes this chapter. He quoted from "The Future of an Illusion," by Sigmund Freud. "The voice of the intellect is a soft one but it does not rest until it has gained a hearing." The voice of the intellect, informed by an ability to connect history with biography, is soft, but it is a compelling source of hope.

As disturbing as the seemingly endless consequences of the Cold War may be, I am struck by the always-present voice of the intellect over the past five decades of bomb building—by the large numbers of people all over the world who persistently questioned the government and defense contractors' decisions about how and where to build and test nuclear weapons and how to dispose of the radioactive waste, even before the first bomb was tested in 1945. Even Andrei Sakharov, the physicist who is considered the father of the Soviet H-bomb and Robert Oppenheimer, the U.S. physicist who directed the atomic energy research project (the "Manhattan Project") at Los Alamos, New Mexico, could not rest with regard to making the world aware of the moral implications and technical dangers of testing and producing nuclear weapons. Andrei Sakharov (1992) wrote in *Memoirs*:

> *I have read that on August 6, 1945, Robert Oppenheimer locked himself in his office while his younger colleagues ran around the Los Alamos laboratory shouting Indian war whoops, and that he also wept at his meeting with President Truman. Oppenheimer's personal tragedy disturbs me deeply, all the more so because I believe [that in helping to create the bomb] he was acting in good faith, for reasons of principle. Of course, the whole sad story of Hiroshima and Nagasaki which so affected his soul was even more troubling.* (P. 97)

Oppenheimer was chairman of the Atomic Energy Commission from 1946 to 1953, but was relieved of his post in 1953 on the grounds that he was a security risk to the United States because he had allegedly associated with persons known to be

Teenage members of Youth for Environmental Sanity (YES) exhibit their belief in taking a global perspective—and remaining optimistic that individuals working together can effect social change.

J.B. Diederich/Contact Press Images

communists and had obstructed the development of the hydrogen bomb. For his part, Andrei Sakharov "campaigned against nuclear testing and ideological distortions of science" (Kline 1989, p. xv). Nikita Khrushchev, premier of the Soviet Union in the late 1950s and early 1960s, wrote in his memoirs that in 1958

> *Sakharov called on our government to cancel a scheduled nuclear explosion and not to engage in any further testing, at least of the hydrogen bomb: "As a scientist and as the designer of the hydrogen bomb, I know what harm these explosions can bring down on the head of mankind."*
>
> *He was devoted to the idea that science should bring peace and prosperity to the world, that it should help preserve and improve the conditions for human life. However, he went too far in thinking that he had the right to decide whether the bomb he had developed could ever be used in the future.* (KLINE 1989, P. XV)

In 1979 Sakharov was arrested, banned from Moscow, and exiled to Gorky, a site off-limits to foreigners, for opposing the Soviet invasion of Afghanistan. The international community viewed the invasion as an example of Soviet expansionism; among other things, it resulted in the U.S. Senate's refusal to ratify the SALT II Treaty.

One can argue that the payoffs for Oppenheimer, Sakharov, and the countless people who persisted in placing the facts before politicians and the people are the 1993 SALT II treaty, the first treaty between Russia and the United States to reduce (dismantle) the number of nuclear weapons in each country; Hazel O'Leary's announcement that the U.S. Department of Energy must be forthcoming with the American people about events during the Cold War; and the joint efforts of scientists from the United States, Russia, and the new states of Eurasia to find ways to dismantle nuclear weapons.

Oppenheimer, Sakharov, and O'Leary were not passive agents to the social forces of their day. They were not simply spectators with regard to the larger forces affecting their lives. Knowing their responses helps us to see that people have choices. The promise of sociology is that it offers people who take its message seriously a perspective that enables them to assess the larger social and historical forces affecting their lives and, with that knowledge, gives them the ability to gauge the variety of responses open to them. We close with an essay by Dr. Horatio C Wood IV, M.D., who outlines some of the social factors that can intrude on people's capacity to respond in constructive ways to significant events in their lives. On the other hand, he points to our source of hope: the often soft voice of the intellect, which does not rest until it gains a hearing.

FOCUS
Reflections on Change
Horatio C Wood IV, M.D.

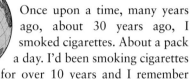

Once upon a time, many years ago, about 30 years ago, I smoked cigarettes. About a pack a day. I'd been smoking cigarettes for over 10 years and I remember clearly how much I enjoyed the process: the ritual of lighting up, the first "drag." I was sobered when the hospital I worked in developed a big ward for patients with lung cancer; I worked on the ward, and they were all heavy cigarette smokers, and all men. I decided that it was time to lessen my risks of developing lung cancer, to increase my chances of living out my allotted three score and ten and dying of something I couldn't predict.

Over the years I had managed to quit a few times, usually in the fall when I played soccer. This time the decision to quit was to be permanent. I was fortunate to have my wife quit, too, and to have two young children, a daughter and a son, who seemed to be encouraging both of us. I say "seemed" because I'm really not sure what their reasons were for encouraging us to stop. It might simply have been that cigarettes were not something little kids could have, and so neither could we. Or perhaps cigarettes got in the way and were smoky and unpleasant. I was fortunate, too, to have the then recently published research studies linking cigarette smoking and lung cancer. In any event the acute phase of the quitting process lasted about a month. Then there were the uncomfortable moments—sharp pangs—for a few years, when it would have been extremely nice to light up a cigarette.

I did manage to change with respect to cigarettes. Cigarette smoking is a very strong and compelling addiction to nicotine, reinforced by habits that build up over years, as well as by fellow smokers and by advertising. When I quit it was not well understood that cigarette smoking involved a nico-

tine addiction—we have since learned that. And over the past 30 years, thousands of people have managed to give up cigarettes, even while tens of thousands more have died of lung cancer acquired through years of cigarette smoking. Why should that be? How can people continue to kill themselves with cigarettes? Why don't people change? Why can't people change?

Human beings are endowed with a highly complex body and brain. Our search to know more, whether of our internal universe or our external universe, advances each week in a steady, inexorable march. For example, we are engaged in a project to map the human genome, to spell out precisely the amino acid makeup of every one of the 24 different chromosomes, to chart the composition of all the DNA sequences that comprise and govern the genetic determinants of the human being. Furthermore, current technology is enabling us to map activity areas in the human brain. We are beginning to understand a lot more about how the human brain works. For instance, we now know that memories of recent events are stored at first in one part of the brain, the hippocampus, but then eventually stored elsewhere in the brain (Zola-Morgan and Squire 1990).

What does this expansion of what we know about ourselves imply for our understanding of our ability to change? We do know from the studies of monozygotic twins reared apart that genetic factors "exert a pronounced and pervasive influence on behavioral variability" (Bouchard 1990). Does this mean we are all behaving in genetically predetermined ways? Are we predestined to feel, think, and act in pre-programmed fashion? Or do we have an ability to change? Chances are our increased knowledge will lead to our eventually expanding our options: further knowledge will give us new, now unforeseen, not yet appreciated, choices.

In psychiatry most of our knowledge was originally based on the individual. This has been the method of medicine from its inception: to study the individual patient and to see how that person's illness, disease, or disability differed from the healthy

SOURCE: "Reflections on Change," by Horatio C Wood IV, M.D., Federal Medical Center, Lexington, KY. (1991, revised 1994).

person. After World War II, however, our field of vision began to broaden. For example, various studies began to coalesce and culminate in our understanding the powerful interactions of maternal care and child mental and medical health. In 1945 the work of Rene Spitz cast a sharp light on the effects of hospitals in creating grief, depression, and marasmus (progressive emaciation) in very young children separated from their mothers or from any other consistent caregiver. John Bowlby also began studying the impact of separations and losses around that time. He focused on children at various early ages who had been separated from their parents during the wartime bombing of London and found that prolonged separations could have profound and lasting adverse effects (for example, nail-biting, bed-wetting, anxieties) on the children. His findings led him to want to know why. And so he worked for the next thirty years on elucidating the nature of bonds between parent and child, and the effects of disrupting those bonds.

I cite John Bowlby's work for its all-important finding that the emotional environment can have an important effect on people. Children exposed to losses and separations may come to adolescence, youth, and young adulthood lacking the ability to make choices. Worries preoccupy such people, intruding on their capacity to attend to significant matters. Considering a situation from many angles, looking at the many possible outcomes that would derive from whatever course is selected, becomes impossible when anxiety and tension flood one's thinking.

What are some of the other eventualities that operate to limit the chances of change in people? I speak here of things that affect all people, but that affect children most strongly. Listing these things is quite distressing to me because we have the knowledge and the means to eliminate them but have not yet been able to do so. I am talking about hunger. I am talking about endemic disease, polluted water supplies, and unsanitary sewage systems. And I am talking about basic universal education, which would give people access through literacy to knowledge extending beyond the household and the village, when, and if, they wanted to find it. Without the compass- and map-reading ability that literacy offers, the trails to further knowledge cannot be ap-

preciated, let alone followed. Without the elimination of endemic disease, illness and weakness dictate actions; there can be very few choices. And chronic hunger and thirst produce apathy and desperation.

What takes people so long to change? If we know that people would be better off by not being hungry, or by not having disease, why don't we who know how take the necessary steps to change things? Here the matter gets complicated.

Consider the effects of scurvy (a disease caused by lack of ascorbic acid) on seamen of the Royal Navy in the eighteenth century. James Lind, a Scotchman, was a physician to the Naval Hospital at Haslar. In 1754 he wrote a treatise on the scourge of scurvy (75 percent of the sailors on a 1740 sailing expedition had scurvy), a disease that caused people to become debilitated, weakened, and hemorrhagic, and often to die. The condition had been observed after sieges of cities (Breda [1625], Nuremberg [1631], Augsburg [1632]) and in a number of long sea voyages in the seventeenth century. Orange and lemon juice, as well as fresh vegetables, were discovered to be quickly curative. But not until 1795 was an admiralty order issued on the regular use of lime and lemon juice in the Royal Navy, after which scurvy disappeared. Today the only remnant of the terrible disease is that British sailors are still called "limeys" (Garrison 1929).

Why didn't Dr. Lind's treatise on scurvy in 1754 convince the important people in the Admiralty to issue the orders for change? Why did it take 45 years after Dr. Lind's treatise? Discounting the Peter Principle, we do find that the majority of people, some of whom are in leadership positions in institutions, tend to maintain the status quo. This means that to bring about change in most institutions, it is necessary to persist, to pound away with the hammer of knowledge, until the change occurs. When enough people are convinced, conditions are optimal for change. Sigmund Freud stated the matter well in his paper on "The Future of an Illusion": "The voice of the intellect is a soft one but it does not rest until it has gained a hearing." The voice of the intellect is soft, but it is one of our sources of hope. The lag time between the discovery of the cause of something and the application of the cure or preventative measure (vitamin C for scurvy, quitting smoking for lung cancer, seat belts/body restraints for auto accident injuries and fatalities) is about one to two generations.

Looking further at considerations of change, what can we understand about people who appear to be trapped in particular situations? What about a Native American trapped on the reservation, an African-American trapped in an inner-city housing project, a white American trapped in a lock-step, culturally self-limiting and self-deceiving existence? Or a black South African trapped in an hostile white minority culture? Or an Indian untouchable trapped in an entrenched socio-politico-religious swamp? Or a Palestinian or Israeli trapped in an ongoing, no-way-out conflict? Or a prisoner trapped in a jail? What choices can these people have? How can these people change?

The list of people throughout history who made a choice and a change is impressive. The list includes Rosa Parks, who on December 12, 1955, refused the bus driver's order to move from her seat in the middle of the bus and stand at the rear so that a white man could sit in her seat. She was arrested, booked, fingerprinted, and incarcerated. Her choice not to vacate her seat led to the famous Montgomery, Alabama, bus boycott. She did not come to her point of change suddenly; she was the secretary of the local chapter of the National Association for the Advancement of Colored People and had experienced the insults of segregation for 50 years. But she made a courageous and critical choice not to move from her seat. Her action is considered to have been an important catalyst in the civil rights movement (Branch 1988).

Another example is Kenyan biologist, environmentalist, and human rights activist Wangari Maathai who founded and launched one of the most effective environmental movements in the world in 1975—the Green Belt Movement—a nutrition, family planning, and reforestation project to restore a devastated natural environment. From Maathai's perspective the ills of the natural environment reflected the human unhappiness within it. Since 1975 Maathai mobilized more than 50,000 women (or "foresters without diplomas") to plant more than 10 million trees and she has led many protests against the destruction and inefficient use of the environment (Library of Congress 1993).

In 1993 Maathai decided that she had to do something about the government-incited ethnic violence taking place in the western part of her country. She and a group of friends traveled to the area of strife to assist and organize those who were under attack. She explained her reasons for choosing to get involved and risking her personal safety:

> To me, here is a people who are displaced. I read about them. I see that their houses have been burned. I hear that their children are not going to school. I hear that they have to go back to their ancestral land. My response is to say, "This is wrong. And the politicians must be stopped." But I don't stop there. . . . I say I'll stop them. If nobody's stopping them, I'll stop them. (1993, p. 353)

The list also includes U.S. Surgeon General C. Everett Koop who worked to keep his personal beliefs separate from his decisions about what was best for public health. Koop's 1986 report on AIDS, which was mailed to every American household, is credited as opening a responsible public dialogue about the disease at a time when AIDS was considered a homosexual disease and conservative supporters of the Reagan administration hoped the report would be used to condemn homosexual behavior.

Allen Wheelis in his book *How People Change* points out that we are what we do. If I am an engineer working with computers and I want to be an airline pilot, I need to do some things differently from what I am doing—I need to take time away from my computers and learn to fly an airplane. If I learn to do enough things differently, and do them satisfactorily, I will become a licensed pilot. You are what you do (Wheelis 1973). I may say I can't do it. I may insist that I don't have any choice, that my responsibilities in my present position prevent me from doing what I say I want to do. I may think that external demands and forces keep me from making the change. But, finally, the choice is really mine.

Wherever people exist, there are choices, even for those who are "trapped." There are courses of action to choose among. Exercising those choices will make us all the agents of our own fate.

Key Concepts

Adaptive Culture 573

Anomaly 574

Authority 575

Basic Innovations 570

Capitalism 566

Charismatic Authority 575

Core Economies 593

Cultural Base 570

Cultural Lag 571

Great Changes 576

Improving Innovations 570

Innovation 566

Legal-Rational Authority 576

Paradigms 574

Peripheral Economies 593

Power Elite 576

Routinized Charisma 576

Scientific Revolution 574

Semiperipheral Economies 593

Simultaneous-Independent
  Inventions 571

Social Change 564

Technological Determinist 573

World-Economy 590

World System Theory 590

Notes

1. On October 3, 1990, East Germany and West Germany were reunified. The fall of the Berlin Wall and the surprisingly swift German reunification were capstones to the largely peaceful revolutions that had taken place first in Poland, then in Hungary, Czechoslovakia, and East Germany. These revolutions were followed by declarations of independence by six Soviet republics (Azerbaijan, Estonia, Georgia, Latvia, Lithuania, and Moldava). By August 1991, 14 of the 15 Soviet republics had declared independence at some level (McKay 1991).

2. In Romania the revolution was violent. On December 16, 1989, Nicolai Ceausescu, then president of Romania, ordered security forces to fire on demonstrators. Hundreds were killed, but protests continued to spread throughout Romania. Army units joined the protesters, and a group known as the Council of National Salvation announced it had overthrown the Ceausescu regime. Fierce fighting broke out between forces loyal to Ceausescu and those who backed the new government. Eventually Ceausescu and his wife were captured, tried, found guilty of genocide against the Romanian people, and executed by a firing squad.

3. After World War II, the United States emerged as the sole possessor of the atomic bomb, and with an intact economy and infrastructure (roads, factories, communication networks). In contrast, its allies and enemies were devastated. Consequently the United States, although a powerful country in its own right, assumed global power by default.

4. An atomic bomb uses the energy of nuclear fission—a process in which the nucleus of a heavy atom is split, resulting in a tremendous release of energy. Uranium and plutonium are two heavy chemical elements commonly used in the fission process. A hydrogen bomb, on the other hand, uses the energy of nuclear fusion—a process in which the nuclei of two light atoms are fused together. The process results in a far greater, more violently explosive release of energy that converts hydrogen to helium. A hydrogen bomb is much more devastating than an atomic bomb.

5. Since the end of World War II, the foreign and domestic policies of the United States have been shaped by Cold War dynamics. Virtually every policy, from the 1949 Marshall Plan to covert aid to the Contras during the Reagan administration, was influenced in some way by America's professed desire to protect the world from the Soviet influence and the spread of communism. Robert S. McNamara, U.S. secretary of defense under presidents Kennedy and Johnson, remarked that "on occasion after occasion, when confronted with a choice between support of democratic governments and support of anti-Soviet dictatorships, we have turned our backs on our traditional values and have supported the antidemocratic" and brutally repressive and totalitarian regimes (McNamara 1989, p. 96).

6. Science correspondents for *The New York Times* consulted with experts to find out which technologies are expected to be the most important in the coming decades (Broad et al. 1991). Many of

the basic innovations they named will increase even further the level of global interdependence. These include micromotors (devices so small they fit on the head of a pin), teraflops (personal computers that can calculate a trillion problems every second), genetic redesign (techniques for altering genetic makeup and for transferring genes from one species to another), digital imaging (a merger of television with computers), mighty chips (silicon chips that hold 100 million transistors), and fiber optics networks capable of transmitting and receiving massive amounts of information. Although all of these technologies exist, a number of difficulties must be worked out before they are ready for widespread application. For example, the current obstacle in the field of micromotors is finding a material less brittle than silicon to make the tiny gears. Currently, researchers at the University of Michigan are experimenting with nickel.

7. Weber also identified a third type of authority, *traditional authority*. It rests on the sanctity of time-honored norms that govern the selection of someone to a powerful position (chief, king, queen) and specify responsibilities and appropriate conduct for those selected. People comply because they believe they are accountable to the past and have an obligation to perpetuate it. (They are apt to reason, "it has always been like that.") To give up past ways of doing things is to renounce a heritage and an identity (Boudon and Bourricaud 1989).

8. It is difficult to comprehend how millions of people could have suffered and died under Stalin's system without rising up to stop him. Even the best investigators offer no satisfactory explanation. In a review of *The Great Terror: A Reassessment* by Robert Conquest, Tatyana Tolstaya wrote:

This book is not a storeroom of facts, but a profoundly analytical investigation. Instead of getting tangled up in the abundance of information, you untangle the knots of the Soviet nightmare under the author's patient direction. Having finished this book, no one can ever again say: "I didn't know." Now we all know.

But the question Why? remains unanswered. Perhaps the only answer is "Because." Period. (TOLSTAYA 1991, P. 4)

9. I read about this resistance to the bicycle many years ago while teaching a course titled Sociology of Sport. This example is quite memorable.

10. Tetanus is an infectious disease characterized by muscular spasms and difficulty in opening the mouth (lockjaw). The tetanus bacterium is dangerous because it produces a toxin that affects the heart and the breathing muscles.

11. Solidarity is a federation of independent trade unions in Poland. Lech Walesa, who had been an unemployed electrician, is one of the union's best-known founders. After its formation in 1980, Solidarity grew to include more than 12 million members, a substantial proportion of Poland's population of 31 million.

12. One architect of *perestroika* defined it as the Soviet "quest to achieve a qualitatively new condition of Soviet society—politically, economically, culturally, morally, spiritually. . . . It is the renewal of socialism, striving to reveal the truly democratic and human face of socialism" (Yakolev 1989, pp. 38–39).

13. As a system, capitalism depends on a free market—a situation in which (1) no single company has exclusive control over a particular commercial activity, (2) the driving force is supply and demand, and (3) government does not interfere with the wage and price structure or with the production and distribution process. Ideally, private ownership in conjunction with these three qualities ensures that the goods and services that consumers demand will be of the highest quality and will be produced at the lowest cost.

14. Although pursuit of profit is a largely self-centered activity, advocates of capitalism consider it socially legitimate because they assume that what is profitable for entrepreneurs and corporations and their stockholders also benefits society as a whole. Quality and cost-effectiveness are enhanced when competition exists among those who produce or distribute the same goods and services. Ideally, well-informed consumers "vote" for products with their purchases. Manufacturers and providers who cannot sustain consumers' interest and who cannot match or surpass their competitors in cost and quality will be forced to improve, or eventually they will go out of business.

15. For example, although most Americans consider the United States to be a capitalist country, capi-

talist principles do not apply to the way the defense industry operates. In the defense market, a single customer (the Pentagon) describes the weapons or other technologies it desires and contracts with a company to make them. If the industry operated on capitalist principles, the supplier would invest its own money and resources to create and develop the product and then would compete with other companies to sell the product to the buyer (Lambert 1992). Because capitalist principles do not apply to defense contractors, they do not have to control costs, understand customer needs, or be familiar with marketing or sales strategies.

APPENDIX

References for Interconnection Data
That Start Each Chapter

The Chronicle of Higher Education Almanac. 1993. "Characteristics of Recipients of Doctorates 1991" (August 25):16.

The College Handbook. 1993. "Study Abroad." New York: Guidance. U.S. Bureau of the Census. 1978. "Foreign Student Enrollment." *Statistical Abstract of the United States 1978.* Washington, DC: U.S. Government Printing Office.

———. 1982. "Certificated Route Air Carriers Summary: 1960–81." *National Data Book and Guide to Sources.* Washington, DC: U.S. Government Printing Office.

———. 1993. "Earned Degrees Conferred by Field of Study"; "Foreign (Nonimmigrant) Student Enrollment"; "Higher Education Registration in Foreign Languages"; "Military Personnel on Active Duty by Location: 1970–1992." *Statistical Abstract of the United States 1993.* Washington, DC: U.S. Government Printing Office.

U.S. Department of Commerce. 1981. "Selected Data of U.S. Affiliates 1977, Table 8." *Foreign Direct Investment in the United States.* Washington, DC: U.S. Government Printing Office.

———. 1992. "Table 12.2" *Survey of Current Business Vol 72 (5)* Washington, DC: U.S. Government Printing Office.

———. 1992. "Table A-4, Utility Patent Applications from Foreign Inventors by Country of Origin." *Highlights in Patent Activity.* Washington, DC: U.S. Government Printing Office.

U.S. Department of Defense. 1993. *Worldwide Military Strength as of June 30, 1993.* Washington, DC: U.S. Government Printing Office.

U.S. Department of Education. 1988. "Doctor's Degrees Conferred." *Digest of Educational Statistics 1988.* Washington, DC: U.S. Government Printing Office.

———. 1993. "Doctor's Degrees Conferred." *Digest of Educational Statistics 1993.* Washington, DC: U.S. Government Printing Office.

U.S. Federal Communication Commission. 1976. "Table 13." *Statistics of Communication Common Carriers.* Washington, DC: U.S. Government Printing Office.

———. 1991. "Table 4.1." *Statistics of Communication Common Carriers.* Washington, DC: U.S. Government Printing Office.

———. 1992. "Table 4.1." *Statistics of Communication Common Carriers.* Washington, DC: U.S. Government Printing Office.

U.S. Department of Immigration and Naturalization. 1992. "Table 5." *Commissioner's Fact Book.* Washington, DC: U.S. Government Printing Office.

———. 1992. "Table 36." *1991 Statistical Yearbook.* Washington, DC: U.S. Government Printing Office.

U.S. Department of Transportation. 1993. "Passenger Travel between U.S. and Foreign Countries, Table IIa and Table IId." *U.S. International Air Travel Statistics 1991.* Washington, DC: U.S. Government Printing Office.

KEY CONCEPTS

Ability Grouping the arranging of elementary school students into instructional groups according to similarities in past academic performance and/or on standardized test scores.

Achieved Characteristics attributes acquired through some combination of choice, effort, and ability. In other words, people must act in some way to acquire the attribute. Examples include occupation, marital status, level of education, and income.

Achieved Status positions acquired through effort and ability.

Active Adaptation a biologically based tendency to adjust to and resolve environmental challenges.

Adaptive Culture the portion of nonmaterial culture (norms, values, and beliefs) that adjusts to material innovations.

Advanced Market Economics an economic arrangement that offers widespread employment opportunities to women as well as to men.

Alienation a state in which human life is dominated by the forces of human inventions.

Annual Per Capita Consumption of Energy the average amount of energy each person in a nation consumes over a year.

Anomaly an observation or observations that a paradigm cannot explain, and that threaten the paradigm's explanatory value and hence its status. (See also *paradigms.)*

Anomie see *structural strain.*

Apartheid a former policy in South Africa that was a rigid system of racial classification designed to promote and maintain white supremacy.

Ascribed Characteristics attributes that people (1) have at birth (such as skin color, gender, or hair color), (2) develop over time (such as baldness, gray hair, wrinkles, retirement, or reproductive capacity), or (3) possess through no effort or fault of their own (national origin or religious affiliation that was "inherited" from parents).

Ascribed Status positions that people are born into, grow into, or otherwise acquire through no fault or virtue of their own.

Assimilation a process by which ethnic and racial distinctions between groups disappear.

Authority legitimate power in which people believe that the differences in power are just and proper and that a leader is entitled to give orders.

Automate to use the computer to increase workers' speed and consistency, as a source of surveillance, and to maintain divisions of knowledge and thus a hierarchical arrangement between management and workers.

Back Stage the region out of sight where individuals can do things that would be inappropriate or unexpected on the front stage.

Basic Innovations revolutionary, unprecedented, or ground-breaking ideas, practices, and tools that are the cornerstones for a wide range of applications.

Beliefs conceptions that people accept as true about how the world operates and about the place of the individual in the world.

Bourgeoisie the owners of the means of production. (See *means of production.)*

Bureaucracy in theory, a completely rational organization—one that uses the most efficient means to achieve a valued goal.

Capitalism an economic system in which natural resources and the means of producing and distributing goods and services are privately owned.

Caste System any scheme of social stratification in which people are ranked on the basis of physical or cultural traits over which they have no control and that they usually cannot change.

Catechisms short books covering religious principles written in question-and-answer format.

Charismatic Authority authority that rests on the exceptional and exemplary qualities of the person issuing the commands.

Church according to Durkheim, a group whose members hold the same beliefs with regard to the sacred and the profane, who behave in the same way in the presence of the sacred, and who gather together in body and spirit at agreed-upon times to reaffirm their commitment to those beliefs and practices.

Civil Religion "any set of beliefs and rituals, related to the past, present and/or future of a people (nation), which are understood in some transcendental fashion" (Hammond 1976, p. 171).

Claims Makers people who articulate and promote claims and who tend to gain if the targeted audience accepts their claims as true.

Classless Society a propertyless society providing equal access to the means of production.

Class System any scheme of social stratification in which people are ranked on the basis of merit, talent, ability, or past performance.

Cohort a group of people sharing a common characteristic or life event.

Collective Memory experiences shared and recalled by significant numbers of people.

Concepts powerful thinking and communication tools that enable us to give and receive complex information in an efficient manner.

Confederate someone who works in cooperation with the experimenter.

Conformists people who have not violated the rules of a group and are treated accordingly.

Conformity 1. behavior and appearance that follow and maintain standards set by a group. 2. the acceptance of the cultural goals and the pursuit of these goals through legitimate means.

Constrictive Pyramids population pyramids that are characteristic of some European societies, most notably Switzerland and western Germany, and that are narrower at the base than in the middle. This shape shows that the population is composed disproportionately of middle-aged and older people. (See also *population pyramid*.)

Constructionist Approach a sociological approach that focuses on the process by which some groups, activities, conditions, or artifacts become defined as social problems.

Content (of interaction) the cultural factors (norms, values, beliefs, material culture) that guide interpretations, behavior, and dialogue during interaction.

Context (of interaction) the larger historical circumstances that bring people together.

Control Variables variables suspected of causing spurious correlations.

Core Economies the highly diversified economies of mechanized rich nations characterized by strong, stable governments.

Corporate Crime crime committed by a corporation as it competes with other companies for market share and profits.

Correlation a relationship between two variables such that a change in one variable is associated with a change in another.

Counterculture a subculture that conspicuously challenges, rejects, or clashes with the central norms and values of the dominant culture.

Crime deviance that is punished by formal sanctions.

Cults generally very small, loosely organized groups, usually founded by a charismatic leader who attracts people by virtue of his or her personal qualities.

Cultural Base the number of existing inventions.

Cultural Genocide the outcome of a situation in which people of one society define the culture of another society not only as offensive but as so intolerable that they attempt to destroy it.

Cultural Lag the failure to adapt to a new invention; a situation in which adaptive culture fails to adjust in necessary ways to a material innovation. (See *adaptive culture*.)

Cultural Relativism a perspective in which elements of foreign cultures are viewed in their cultural context, not in isolation or by standards of a home culture.

Culture a society's distinctive and complete design for living.

Culture Shock the physical and mental strain that people from one culture experience when they must reorient themselves to the ways of a new culture.

Data printed, visual, and spoken materials. Data becomes information after someone reads it, listens to it, or views it.

Dearth of Feedback a factor in creating data that is poor in quality. Much of the data that is televised and published is not subject to honest, constructive feedback because there are too many messages and not enough critical readers and listeners to evaluate the data before it is released or picked up by the popular media. Without feedback the creators cannot correct their mistakes; thus, the data they produce becomes poor in quality.

Demographic Gap the difference between birthrates and death rates.

Demographic Trap the point at which population growth overwhelms the environment's carrying capacity.

Demography a subdiscipline within sociology that studies population trends.

Denomination a formal, hierarchical, well-integrated organization in a society in which church and state are usually separate.

Dependent Variable the variable affected by a change in the independent variable. (See also *variable*.)

Deviance any behavior or physical appearance that is socially challenged and condemned because it departs from the norms and expectations of a group.

Deviant Subcultures groups that are part of the larger society but whose members adhere to norms and values that favor violation of the larger society's laws.

Differential Association a theory of socialization that explains the origins of delinquent behavior. It refers to the idea that "when persons become criminal, they do so because of contacts with criminal patterns and also because of isolation from anti-criminal patterns" (Sutherland and Cressey 1978, p. 78).

Diffusion the process by which an idea, an invention, or some other item is borrowed from a foreign source.

Discoveries the uncovering of something that had existed before but had remained hidden, unnoticed, or undescribed.

Discrimination the intentionally or unintentionally unequal treatment of individuals or groups on the basis of attributes unrelated to merit, ability, or past performance. The treatment may be based on such attributes as skin color, weight, religion, ethnicity, or social class. Discrimination is behavior aimed at denying members of minority groups equal opportunities to achieve valued social goals (education, health care, long life) and/or blocking their access to valued goods and services.

Disenchantment a great spiritual void accompanied by a crisis of meaning. It occurs when people focus so uncritically on the ways they go about achieving a valued goal that they lose sight of that goal.

Disenchantment of the World Max Weber's phrase for a great spiritual void accompanied by a crisis of meaning.

Dispositional Traits personal or group traits such as motivation level, mood, and inherent ability.

Division of Labor work broken down into specialized tasks, with each task performed by a different set of persons.

Documents written or printed material.

Dominant Group the ethnic and racial group at the top of the hierarchy.

Doubling Time the estimated number of years required for a country's population to double in size.

Downward Mobility a change in social class that corresponds to a loss in rank or prestige

Dramaturgical Model a model in which interaction is viewed as though it were theater, people as though they were actors, and roles as though they were performances presented before an audience in a particular setting.

Dysfunctions parts that have disruptive consequences to the system or to some segments of society.

Ecclesiae a formal and well-integrated religious organization led by a hierarchy of leaders that claims as its members everyone in a society.

Emigration the departure of individuals from a country.

Engram physical traces formed by chemicals produced in the brain that store the recollections of experiences.

Established Sects religious organizations that share characteristics of denominations and sects. They are renegades from denominations or ecclesiae but have existed long enough to acquire a significantly large membership and to achieve respectability.

Ethgender related to people who share (or are believed by themselves or others to share) the same sex and race and ethnicity.

Ethnicity a shared national origin, ancestry, distinctive and visible cultural traits (religious practice, dietary habits, style of dress, body ornaments, or language), and/or socially important physical characteristics.

Ethnocentrism a viewpoint in which one's home culture is used as the standard for judging the worth of foreign ways.

Expansive Pyramids population pyramids that are characteristic of labor-intensive poor nations. They are triangular in shape, broadest at the base, and each successive bar is smaller than the one below it. The relative sizes of the age cohorts in expansive pyramids show that the population is increasing in size and that it is composed disproportionately of young people. (See also *population pyramid*.)

Externality Costs costs that are not figured into the price of a product but that are nevertheless a price we pay for using or creating a product. An example of externality costs is the cost of restoring contaminated and barren environments and of assisting people to cope.

Facade of Legitimacy an explanation that members in dominant groups give to justify exploitive practices; a justifying ideology.

Falsely Accused people who have not broken the rules but who are treated as if they have done so.

Family two or more people related to one another by blood, marriage, adoption, or some other socially recognized criteria.

Feeling Rules norms specifying appropriate ways to express the bodily sensations that we experience in relationships with other people. (See *social emotions*.) These rules are so powerful that they affect how people solve problems.

Femininity (or feminine characteristics) a term that signifies the physical, behavioral, and mental or emotional traits believed to be characteristics of females.

Feminist a man or woman who actively opposes gender scripts and believes that men's and women's self-image, aspirations, and life chances should not be constrained by those scripts.

Folkways norms that apply to routine matters.

Formal Curriculum the various subjects such as mathematics, science, English, reading, physical education, and so on. (See *hidden curriculum*.)

Formal Dimension (of an organization) the official, written guidelines, rules, regulations, and policies that define the goals of the organization and its relationship to other organizations and integral parties. This term also applies to the roles, the nature of the relationships among roles, and the way in which tasks should be carried out to realize the goals.

Formal Education a systematic, purposeful, and planned effort intended to impart specific skills and modes of thought. (See *informal education* and *schooling* also.)

Formal Sanctions definite and systematic laws, rules, regulations, and policies that specify (usually in writing) the conditions under which people should be rewarded or punished, and that define the procedures for allocating rewards and imposing punishments. Examples of formal sanctions include medals, cash bonuses, diplomas, fines, prison sentences, and the death penalty. (See *informal sanctions*.)

Fortified Households preindustrial arrangements in which there is no police force, militia, national guard, or other peacekeeping organization. The household is an armed unit, and the head of the household is its military commander.

Front Stage the region where people take care to create and maintain expected images and behavior.

Function the contribution of a part to the larger system and its effect on other parts in the system.

Fundamentalism a complex religious phenomenon that involves a belief in the timeless nature of sacred writings, and a belief that such writings are applicable to all kinds of environments including high-technology societies.

Games structured and organized activities that almost always involve more than one person.

Gender social distinctions based on culturally conceived and learned ideas about appropriate behavior, appearance, and mental or emotional characteristics for males and females.

Gender Nonconformists (1) persons whose primary characteristics are not clear-cut (the intersexed); (2) those whose secondary characteristics depart from the ideal conceptions of masculinity and femininity; (3) those whose interests, feelings, sexual orientation, choice of occupation, or academic major do not match gender-polarized scripts; and (4) those "who actively oppose the gender scripts of the culture" (Bem 1993, p. 167).

Gender Polarization as defined by Sandra Lipsitz Bem, "the organizing of social life around the male-female distinction, so that people's sex is connected to virtually every other aspect of human experience, including modes of dress, social roles, and even ways of expressing emotion and experiencing sexual desire" (1993, p. 192).

Gender Schematic Decisions choices related to any aspect of life that are influenced by society's polarized definitions of masculinity and femininity rather than on the basis of

other criteria such as self-fulfillment, interest, ability, or personal comfort.

Generalizability the extent to which the findings of a research project can be applied to the population from which the sample is drawn.

Generalized Other a system of expected behaviors, meanings, and points of view.

Global Interdependence a situation in which the lives of people around the world are intertwined closely and in which any one nation's problems—the search for employment, drug abuse, environmental pollution, the acquisition and allocation of scarce resources—are part of a larger global problem.

Great Changes events whose causes lie outside ordinary people's characters or their immediate environments but profoundly affect their life chances.

Hate Crimes actions aimed at humiliating members of a minority group and destroying their property or lives.

Hawthorne Effect a phenomenon whereby observed persons alter their behavior when they learn they are being observed.

Hidden Curriculum all the things that students learn along with the subject matter.

Hidden Rape rape that goes unreported.

Household all related and unrelated persons who share the same dwelling.

Hypothesis a trial explanation put forward as the focus of research that predicts how the independent and dependent variables are related. (See also *independent variable* and *dependent variable*.)

Ideal Type a standard against which real cases can be compared.

Ideologies fundamental ideas that support the interests of dominant groups.

Ideology a set of ideas that do not hold up under the rigors of scientific investigation and that support the interests of dominant groups.

Illiteracy the inability to understand and use a symbol system, whether it is based on sounds, letters, numbers, pictographs, or other type of character.

Immigration the entrance of individuals into a new country.

Impression Management the process by which people in social situations manage the setting, their dress, their words, and their gestures to correspond to the impressions they are trying to make or the image they are trying to project.

Improving Innovations modifications of basic inventions in order to improve on them—that is, to make them smaller, faster, less complicated, or more efficient, attractive, durable, or profitable. (See also *basic innovations.)*

Independent Variable the variable of cause; a change in this variable brings about a change in the dependent variable. (See also *variable.)*

Individual Discrimination any overt action on the part of an individual that depreciates minority persons, denies members of minorities opportunities to participate, or does violence to minority members' lives and property.

Infant Mortality the number of deaths in the first year of life for every 1,000 live births.

Informal Dimension (of an organization) those dimensions of organizational life that include worker-generated norms that evade, bypass, do not correspond with, or are not systematically stated in official policies, rules, and regulations.

Informal Education education that occurs in a spontaneous, unplanned way.

Informal Sanctions spontaneous and unofficial expressions of approval or disapproval; they are not backed by the force of law.

Informate to use the computer to empower workers with knowledge of the overall production process, with the expectation that they will make critical and collaborative judgments about production tasks.

Information data that someone has read, listened to, or viewed.

Information Explosion an unprecedented rate of increase in the volume of information due to the revolution in voice, data, and image processing, transmission, and storage that resulted from the development of the computer and telecommunications. These two inventions combine to create an intricate maze of electronic pathways that can move new and old information around the world in seconds.

Ingroup those groups with which people identify and feel closely attached, particularly when that attachment is founded on hatred for another group.

In-Migration the movement of people into a designated area. (See *out-migration.)*

Innovation (as a response to structural strain) the acceptance of the cultural goals but the rejection of legitimate means to obtain these goals. For the innovator, success means winning the game rather than playing by the rules of the game.

Innovation the development of something new—an idea, a practice, or a tool.

Institutionalized Discrimination the established and customary ways of doing things in society—the rules, policies, and day-to-day practices that we take for granted and do not challenge, which impede or limit minority members' achievements and keep them in a subordinate and disadvantaged position.

Institutionally Complete a term that describes a subculture whose members do not interact with anyone outside the subculture.

Intergenerational Mobility a change in social class over two or more generations.

Internalization to take as one's own and accept as binding the norms, values, beliefs, and language of one's culture.

Internal Migration movement within the boundaries of a single nation—from one state, region, or city to another.

International Migration the movement of people between countries.

Intersexed a term used to classify people with some mixture of male and female biological characteristics.

Interviews face-to-face sessions or telephone conversations between an interviewer and a respondent in which the interviewer asks the respondent questions and records his or her answers.

Intragenerational Mobility a change in social class during an individual's lifetime.

Involuntary Minorities ethnic and racial groups that do not choose to be a part of a country.

Issues public matters that can be explained by factors outside an individual's control and immediate environment.

Latent Dysfunctions the unintended, unanticipated negative consequences that a part causes in some segment of society.

Latent Functions the unintended, unrecognized, and unanticipated or unpredicted consequences that contribute to the smooth operation of the system.

Legal-Rational Authority power that rests on a system of impersonal rules that formally specifies the qualifications for occupying a powerful position.

Liberation Theology an approach to the role of religion in society maintaining that organized religions have a responsibility to demand social justice for the marginalized peoples of the world, especially landless peasants and the urban poor, and to take an active role at the grass-roots level in order to bring about political and economic justice.

Life Chances opportunities that include "everything from the chance to stay alive during the first year after birth to the chance to view fine art, the chance to remain healthy and grow tall, and if sick to get well again quickly, the chance to avoid becoming a juvenile delinquent—and very crucially, the chance to complete an intermediary or higher educational grade" (Gerth and Mills 1954, p. 313).

Looking-Glass Self a phrase coined by Charles Horton Cooley to describe the way in which a sense of self develops: we visualize how we appear to others, we imagine a judgment of that appearance, and we develop a feeling somewhere between pride and shame.

Low-Technology Tribal Societies hunting-and-gathering societies with technologies that do not permit the creation of surplus wealth, or wealth beyond what is needed to meet the basic needs (food and shelter).

Manifest Dysfunctions the expected or anticipated disruptions that a part causes in some segment of the system.

Manifest Functions the intended, recognized, expected, or predictable consequences of a part for society.

Masculinity (or masculine characteristics) a term that signifies physical, behavioral, and mental or emotional traits believed to be characteristics of males.

Material Culture the objects or physical substances available to the people of a society that have been changed by human intervention.

Means of Production the resources—land, tools, equipment, factories, transportation, and labor— essential to the production and distribution of goods and services.

Mechanical Solidarity societal order based on a common conscience or uniform thinking, and characteristic of preindustrial societies.

Mechanisms of Social Control all of the methods that people employ to teach, persuade, or force others to conform.

Mechanization the addition of external sources of power such as oil or steam to hand tools and to modes of transportation.

Melting Pot Assimilation a process of cultural blending in which the groups involved accept many new behaviors and values from one another.

Methods of Data Collection the research procedure used to gather relevant data to test hypotheses.

Migration the movement of people from one area to another.

Minorities the ethnic and racial groups at the bottom of the hierarchy.

Minority Groups subgroups within a society that can be distinguished from members of the dominant groups by visible and identifying characteristics, including physical and cultural attributes.

Mixed Contacts social situations in which the stigmatized and normals are in each other's company. (See also *normals.*)

Modern Capitalism "a form of economic life which involved the careful calculation of costs and profits, the borrowing and lending of money, the accumulation of capital in the form of money and material assets, investment, private property, and the employment of laborers and employees in a more or less unrestricted labor market" (Robertson 1987, p. 6).

Mores norms that people consider pivotal to the well-being of the group.

Mortality Crises frequent and violent fluctuations in the death rate caused by war, famine, and epidemics, during which time the death rate has no limit.

Multinational Corporation an enterprise that owns or controls production or service facilities in countries outside the one in which it is headquartered.

Mystical Religions religions in which the sacred is sought in states of being that, at their peaks, can exclude all awareness of one's existence, sensations, thoughts, and surroundings.

Nature human genetic makeup or biological inheritance.

Negative Sanction an expression of disapproval for noncompliance; the punishment may be withdrawal of affection, ridicule, ostracism, banishment, physical harm, imprisonment, solitary confinement, or even death. (See *positive sanction.*)

Nonhouseholder Class propertyless laborers and servants usually residing within fortified households. (See *fortified households.*)

Nonmaterial Culture intangible creations that cannot be identified directly through the senses but that exert considerable influence over people's behavior.

Nonparticipant Observation a research procedure that involves detached watching and listening; the researcher does not interact or become involved with those being studied.

Nonprejudiced Nondiscriminators (all-weather liberals) persons who accept the creed of equal opportunity and whose conduct conforms to that creed.

Normals those people who are in the majority or those who possess no discrediting attributes.

Norms the written and unwritten rules that specify the behavior appropriate to specific situations.

Nurture the environment or the interaction experiences that make up every individual's life.

Objective Secularization a twofold process: the decline in the control of religion over education, medicine, law, and politics, and the emergence of an environment in which people are free to choose from many equally valid religions the one to which they wish to belong.

Objectivity a position taken by researchers in which they do not let personal and subjective views about the topic influence the outcome of the research.

Obligations the relationship and behavior that a person enacting a role must assume toward others in a particular status.

Observation method of data gathering in which the researcher not only watches and listens but remains open to other considerations. Success results from identifying what is worth observing. (See

also *nonparticipant observation* and *participant observation*.)

Oligarchy rule by the few, or the concentration of decision-making power in the hands of a few persons who hold the top positions in an organization's hierarchy.

Operational Definitions clear and precise definitions and instructions about how to observe and measure the variables being studied.

Organic Solidarity order based on interdependence and cooperation among a wide range of diverse and specialized tasks; this type of solidarity characterizes industrial societies.

Organization a coordinating mechanism created by people to achieve stated objectives—whether to maintain order; to challenge an established order; to keep track of people; to grow, harvest, or process food; to produce goods; or to provide a service.

Outgroup a group of individuals toward which members of an ingroup feel separateness, opposition, or even hatred. (See also *ingroup*.)

Out-Migration movement out of a designated area. (See *in-migration*.)

Paradigms the dominant and widely accepted theories and concepts in a particular field of study.

Participant Observation a research procedure in which a researcher does one or more of the following: joins a group and participates as a member, interacts directly with those whom he or she is studying, assumes a position critical to the outcome of the study, or lives in a community under study.

Per Capita Income the average share of income that each person in a country would receive if the country's gross national product were divided evenly.

Peripheral Economies the economies of labor-intensive poor countries, where most of the jobs are low-paying and require few skills. Peripheral economies are not highly diversified.

Play a voluntary and often spontaneous activity with few or no formal rules, which is not subject to constraints of time.

Population the total number of individuals, groups, households, items, or entities that could be studied.

Population (study of) a specialty within sociology that focuses on the number of people in and composition of various social groupings that live within specified boundaries and the factors that lead to changes in that social grouping's size and composition.

Population Pyramid a series of horizontal bar graphs each of which represents a different five-year age cohort. (See *cohort*.) Two bar graphs are constructed for each cohort, one for males and another for females; the bars are placed end to end, separated by a line that represents zero. Usually, the left-hand side of the pyramid depicts the number or percentage of males that make up each age cohort and the right-hand side depicts the number or percentage of females. The graphs are stacked according to age; the age 0–4 cohort forms the base of the pyramid and the 80+ cohort is at the apex of the pyramid. The population pyramid allows us to view the relative sizes of the age cohorts and to compare the relative number of males and females.

Positive Checks events that increase mortality, including epidemics of infectious and parasitic disease, war, and famine.

Positive Sanction an expression of approval and a reward for compliance. Such a sanction may take the form of applause, an approving smile, or a pat on the back.

Power Elite those few people positioned so high in the social structure of leading institutions that their decisions have consequences that affect millions of people worldwide.

Predestination the belief that God has foreordained all things, including the salvation or damnation of individual souls.

Prejudice a rigid judgment about an outgroup, usually unfavorable, that does not change in the face of contradictory evidence and that applies to anyone who shares the distinguishing characteristics of that group.

Prejudiced Discriminators (active bigots) persons who reject the "American creed" and profess a right, even a duty, to discriminate.

Prejudiced Nondiscriminators (timid bigots) persons who do not accept the "American creed" but who refrain from discriminatory actions primarily because they fear the sanctions they may encounter if they are caught.

Primary Groups major socializing agents, especially in the early years, because they give newcomers their first exposure to the "rules of life." These groups are characterized by face-to-face contact and strong ties among members.

Primary Sex Characteristics the anatomical traits essential to reproduction. Most cultures divide the population into two categories—male and female—largely on the basis of what most people consider to be clear anatomical distinctions.

Private Households an arrangement that exists when the workplace is separate from the home, where men are heads of households and assume a breadwinner role, and where women remain responsible for housekeeping and childrearing.

Probabilistic Model a model in which the hypothesized effect does not always result from a hypothesized cause.

Profane everything that is not sacred, including those things opposed to the sacred (the unholy, the irreverent, the contemptuous, the blasphemous) and those things that, although not opposed to the sacred, stand apart from it (the ordinary, the commonplace, the unconsecrated, the temporal, the bodily).

Professionalization (within organizations) a hiring trend in organizations in which experts are hired who have formal training in a particular subject or activity that is essential to achieving organizational goals.

Proletariat those who must sell their labor to the bourgeoisie.

Prophetic Religions religions in which conceptions of the sacred revolve around items that symbolize significant historical events or around the lives, teachings, and writings of great people.

Pull Factors the conditions that encourage people to move into a particular area. Some of the most common pull factors are employment opportunities, favorable climate, and the relative absence of discrimination.

Pure Deviants people who have broken the rules and are caught, punished, and labeled as outsiders.

Push Factors the conditions that encourage people to move out of an area. Some of the most common push factors include religious or political persecution, discrimination, depletion of natural resources, lack of employment opportunities, and natural disasters.

Race 1. a biological term that refers to a group of people who possess certain distinctive physical characteristics. 2. a group of people who possess certain distinctive and conspicuous physical characteristics.

Racism an ideology that maintains that something in the biological makeup of a specific racial or ethnic group explains its subordinate or superior status.

Random Sample a sample drawn in such a way that every case in the population has an equal chance of being selected.

Rationalization as defined by Max Weber, a process whereby thought and action rooted in emotion (love, hatred, revenge, joy), in superstition, in respect for mysterious forces, and in tradition are replaced by thought and action grounded in the logical assessment of cause and effect or means and ends.

Rebellion the full or partial denunciation of both goals and means and the introduction of a new set of goals and means.

Reflexive Thinking stepping outside the self to observe and evaluate it from another's viewpoint.

Refugee "someone who has a well-founded fear of persecution on the basis of his or her race, religion, nationality, political opinion or membership in a particular social group" (Walsh 1993, p. A6).

Reliability the extent to which the operational definition gives consistent results.

Religion according to Émile Durkheim, a system of shared beliefs and rituals about the sacred that bind together a community of worshipers. (See also *sacred*.)

Representative Sample a sample with the same distribution of characteristics as the population from which it was selected.

Research a fact-gathering and fact-explaining enterprise governed by strict rules.

Research Design a plan for gathering data to test hypotheses.

Research Methods the procedure that sociologists and other investigators use to formulate meaningful research questions and to collect, analyze, and interpret facts in ways that other researchers can duplicate.

Research Methods Literate the ability to know how to collect data that is worth putting into the computer and to know how to interpret the data that comes out of it.

Resocialization the process of discarding values and behaviors unsuited to new circumstances and replacing them with new, more appropriate values and standards of behavior.

Retreatism the rejection of both cultural goals and the means of achieving these goals.

Reverse Ethnocentrism the tendency to see the home culture as inferior to a foreign culture.

Rights the behaviors that a person assuming a role can demand or expect from others.

Ritualism the abandonment of cultural goals but a rigid adherence to the legitimate means of those goals. It is the opposite of innovation (as a response to structural strain); the game is played according to the rules despite defeat.

Rituals rules that govern how people must behave when in the presence of the sacred. These rules may take the form of instructions detailing the appropriate context, the roles of various participants, ac-

ceptable attire, and the precise wording of chants, songs, and prayers.

Role the behavior expected of a status in relation to another status.

Role Conflict a predicament in which the expectations associated with two or more roles in a role set are contradictory.

Role Set the array of roles associated with every status.

Role Strain a predicament in which contradictory or conflicting expectations are associated with the role that a person is occupying.

Role-Taking stepping outside the self and viewing its appearance and behavior imaginatively from an outsider's perspective.

Routinized Charisma a situation in which the community must establish procedures, rules, and traditions to regulate the members' conduct, to recruit new members, and to ensure the orderly transfer of power.

Sacramental Religion religions in which the sacred is sought in places, objects, and actions believed to house a god or a spirit.

Sacred all phenomena that are regarded as extraordinary and that inspire in believers deep and absorbing sentiments of awe, respect, mystery, and reverence.

Sample a portion of cases from a particular population.

Sampling Frame a complete list of every case in the population.

Sanctions reactions of approval and disapproval to behavior and appearances. Sanctions can be positive or negative, formal or informal.

Scapegoat a person or a group that is assigned blame for conditions that cannot be controlled, that threaten a community's sense of

well-being, or that shake the foundations of a trusted institution.

Schooling a program of formal and systematic instruction that takes place primarily in a classroom but also includes extracurricular activities and out-of-classroom assignments.

Scientific Method an approach to data collection guided by the assumptions that knowledge about the world is acquired through the senses (is observable) and that the truth of the knowledge is confirmed by other persons making the same observations.

Scientific Revolution a condition that occurs when enough people in the community break with an old paradigm and change the nature of their research in favor of the incompatible new paradigm.

Secondary Sex Characteristics physical traits not essential to reproduction (breast development, quality of voice, distribution of facial and body hair, and skeletal form) that result from the action of so-called male hormones (androgen) and female hormones (estrogen). Although testes produce androgen and ovaries estrogen, the adrenal cortex produces androgen and estrogen in both sexes.

Secondary Sources data that has been collected by other researchers for some other purpose.

Secret Deviants people who have broken the rules, but whose violation goes unnoticed, or, if it is noticed, no one reacts to enforce the law.

Sect a small community of believers led by a lay ministry, with no formal hierarchy or official governing body to oversee the various religious gatherings and activities.

Sectarian Community a geographically distinct group of people who profess allegiance to a particular religion, such as Islam or Christianity, and who also have strong ties to a powerful family, clan, or ethnic group.

Secularization a process by which religious influences on thought and behavior are reduced.

Selective Perception the process whereby prejudiced persons notice only those behaviors or events that support their stereotypes about an outgroup.

Self-Administered Questionnaire a set of questions given or mailed to respondents, who read the instructions and fill in the answers themselves.

Self-Fulfilling Prophecy a concept that begins with a false definition of a situation. The false definition is assumed to be accurate, and people behave as if the definition were true. In the end the misguided behavior produces responses that confirm the false definition.

Semiperipheral Economies the moderately diversified economies of moderately wealthy countries.

Sexist Ideologies the ideologies that justify one sex's social, economic, and political dominance over the other.

Sexual Property "the relatively permanent claim to exclusive sexual rights over a particular person" (Collins 1971, p. 7).

Sick Role a term coined by sociologist Talcott Parsons to represent the rights and obligations accorded people when they are sick.

Significant Others people or characters who are important in a person's life—important in the sense that they have considerable influence on self-evaluation and encourage a person to behave in a certain manner.

Significant Symbols words, gestures, and other learned signs that are used to convey a meaning that is the same for both the communicator and the recipient. Particularly important significant symbols are language and symbolic gestures.

Simultaneous-Independent Inventions situations in which the same invention is created by two or more persons working independently of one another at about the same time (sometimes within a few days or months).

Situational Factors forces outside an individual's control, such as environmental conditions or bad luck.

Small Groups groups of 2 to about 20 people who interact with one another in meaningful ways.

Social Action behavior or actions that people take in response to others.

Social Change any alteration, modification, or transformation of social phenomena over a specified period.

Social Emotions internal bodily sensations that people experience in relationships with others.

Social Identity the category to which a person belongs and the qualities that others believe, rightly or wrongly, to be "ordinary and natural" (Goffman 1963, p. 2) for a member of that category.

Social Interactions events involving at least two people who communicate through language and symbolic gestures to affect one another's behavior and thinking.

Socialization a complex, lifelong process of learning about the social world. Socialization begins immediately after birth and continues throughout life. It is a process through which newcomers develop their human capacities, acquire a

unique personality and identity, and internalize the norms, values, beliefs, and language needed to participate in the larger society.

Social Promotion passing students from one grade to another on the basis of age rather than academic competency.

Social Relativity the view that ideas, beliefs, and behavior are, to a large extent, a product of time and place.

Social Status a position in a system of social relationships or a social structure.

Social Stratification a systematic, nonrandom process by which people in a society are ranked according to a scale of social worth and are awarded unequal amounts of income, wealth, prestige, and power on the basis of that ranking.

Social Structure two or more people interacting and interrelating in expected ways, regardless of the unique personalities involved.

Society a group of people living in a given territory who share a culture and who interact with people of that territory more than with people of another territory.

Sociological Imagination the ability to connect seemingly impersonal and remote historical forces to the most basic incidents of an individual's life. The sociological imagination enables people to distinguish between personal troubles and public issues.

Sociological Theory a set of principles and definitions that tell how societies operate and how people relate to one another.

Sociology the scientific study of the causes and consequences of human interaction.

Solidarity the ties that bind people to one another.

Spurious Correlation a correlation that is coincidental or accidental.

State a governing body organized to manage and control specified activities of people living in a given territory.

Stationary Pyramids population pyramids that are characteristic of most developed nations and that are similar to constrictive pyramids except that all of the age cohorts in the population are roughly the same size, and fertility is at replacement level. (See also *population pyramid* and *constrictive pyramids*.)

Status Group a plurality of persons held together by virtue of a common lifestyle, formal education, family background, or occupation and "by level of social esteem and honor accorded them by others" (Coser 1977, p. 229).

Status System classification of achievements resulting in popularity, respect, and acceptance into the crowd versus disdain, discouragement, and disrespect.

Status Value a situation in which persons who possess one category of a characteristic (white skin versus brown skin, blond hair versus dark hair) are believed to be and are treated as more valuable or worthy than persons who possess other categories (Ridgeway 1991).

Stereotypes exaggerated and inaccurate generalizations about people who are members of an outgroup.

Stigmas statuses that are deeply discrediting in the sense that they overshadow all other statuses that a person occupies.

Streaming the arranging of middle school and high school students into instructional groups according to similarities in past academic performance and/or on standardized test scores.

Structural Strain (or anomie) a condition that occurs when the valued goals have no clear boundaries, when it is not clear whether the legitimate means that society provides will lead to the goals, and when the legitimate opportunities for meeting the goals are closed to a significant portion of the population.

Structured Interview an interview in which the wording and the sequence of questions are predetermined and cannot be altered during the course of the interview.

Subcultures groups that share in some parts of the dominant culture but have their own distinctive values, norms, language, or material culture.

Subjective Secularization a decrease in the number of people who view the world and their place in it from a religious perspective.

Symbol any kind of physical phenomenon—a word, a taste—to which people assign a meaning or value.

Symbolic Gestures extraverbal cues that include tone of voice, inflection, facial expression, posture, and other body movements or positions that convey meaning from one person to another.

Technological Determinist someone who believes that human beings have no free will and are controlled entirely by their material innovations.

Technology the knowledge, skills, and tools used to transform resources into forms with specific purposes, and the skills and knowledge required to use them.

Territories settings that have borders or that are set aside for particular activities.

Theory a framework that can be used to comprehend and explain events.

Theory of the Demographic Transition a model that outlines historical changes in birth- and death rates among the mechanized rich countries and the factors underlying those changes. Some demographers have theorized that a country's birth- and death rates are linked to its level of industrial or economic development.

This-Worldly Asceticism a belief that people are instruments of divine will and that their activities are determined and directed by God.

Total Fertility the average number of children women bear over their lifetime.

Total Institutions settings in which people surrender control of their lives, voluntarily or involuntarily, to an administrative staff and in which they (as inmates) carry out daily activities in the presence of other inmates.

Traces materials or physical transformations that yield information about human activity, such as the items that people throw away, the number of lights on in a house, or changes in water pressure.

Tracking the arranging of middle school and high school students into instructional groups according to similarities in past academic performance and/or on standardized test scores.

Trained Incapacity the inability to respond to new and unusual circumstances or to recognize when official rules and procedures are outmoded or no longer applicable.

Transformative Powers of History the dramatic consequences of important historical events on people's thinking and behavior.

Troubles private matters that can be explained in terms of personal characteristics (motivation level, mood, personality, or ability) or immediate relationships with others (family members, friends, acquaintances, or co-workers). The resolution of a trouble, if it can be resolved, lies in changing an individual's character or immediate relationships.

Unit of Analysis who or what is to be studied in a research project.

Unprejudiced Discriminators (fair-weather liberals) persons who believe in equal opportunity but engage in discriminatory behaviors because it is to their advantage to do so or because they fail to consider the discriminatory consequences of some of their actions.

Unstructured Interview an interview that is flexible and open-ended; the question–answer sequence is spontaneous and like a conversation.

Upward Mobility a change in social class that corresponds to a gain in rank or prestige.

Urbanization an increase in the number of cities and in the proportion of the population living in cities.

Urban Underclass the ghetto poor, or a "heterogeneous grouping of families and individuals in the inner city that are outside the mainstream of the American occupational system and that consequently represent the very bottom of the economic hierarchy" (Wilson 1983, p. 80).

Validity the degree to which an operational definition measures what it claims to measure.

Values general conceptions about what is good, right, appropriate, worthwhile, and important with regard to modes of conduct and states of existence.

Variable any trait or characteristic that can vary or have more than one category. (For example, gender is a variable with two categories: male and female.)

Vertical Mobility a change in class status that corresponds to a gain or loss in rank or prestige.

Voluntary Minorities racial and ethnic groups that come to a country expecting to improve their way of life.

White-Collar Crime "crime committed by persons of respectability and high social status in the course of their occupations" (Sutherland and Cressey 1978, p. 44).

Witch-Hunt a campaign or purge launched to reach subversive elements on the pretext of investigating and correcting activities that undermine a group or a country.

World-Economy an economy in which economic transactions transcend national boundaries.

REFERENCES

Chapter 1

Abercrombie, Nicholas, Stephen Hill, and Bryan S. Turner. 1988. *The Penguin Dictionary of Sociology*. New York: Penguin.

Alger, C. E. and J. E. Harf. 1985. *Global Education: Why? About What?* Washington, DC: American Association of Colleges for Teacher Education.

Bardis, Panos D. 1980. "Sociology As a Science." *Social Science Journals* 55(3): 141–80.

Berger, Peter L. 1963. *Invitation to Sociology: A Humanistic Perspective*. New York: Anchor.

Boden, Deirdre, Anthony Giddens, and Harvey L. Molotch. 1990. "Sociology's Role in Addressing Society's Problems Is Undervalued and Misunderstood in Academe." *The Chronicle of Higher Education* (February 21):B1, B3.

Charyn, Jerome. 1978. "Black Diamond." *The New York Review of Books* (August 17):41.

Cortes, Carlos E. 1983. "Multiethnic and Global Education: Partners for the Eighties?" *Phi Delta Kappan* (April): 568–71.

Coser, Lewis A. 1977. *Masters of Sociological Thought*, edited by R. K. Merton. New York: Harcourt Brace Jovanovich.

Craig, Ben T. 1987. "Financial Globalization: The American Consumer." *Vital Speeches of the Day* (February 1):230–31.

Edensword, Diana and Gary Milhollin. 1993. "Iraq's Bomb—An Update." *The New York Times* (April 26):A15.

Fanning, Deirdre. 1990. "The Executive Life: Making Japanese Bosses Stand Up and Applaud." *The New York Times* (July 1):F21.

Fletcher, Max E. 1974. "Harriet Martineau and Ayn Rand: Economics in the Guise of Fiction." *American Journal of Economics and Sociology* 33(4):367–79.

Freund, Julien. 1968. *The Sociology of Max Weber*. New York: Random House.

Gordon, John Steele. 1989. "When Our Ancestors Became Us." *American Heritage* (December):106–21.

Gould, Stephen Jay. 1981. *The Mismeasure of Man*. New York: Norton.

Harries, Owen. 1990. "Students Should Know More About the Real World Than Past History and Geography." *NASSP Bulletin* (January):16–20.

Henslin, James M. 1993. *Down to Earth Sociology: Introductory Readings*. New York: Free Press.

Isaacson, Walter. 1992. "Is There Room for Morality in a World Governed by Power?" *Los Angeles Times* (October 4):M3.

Jehl, Douglas. 1993. "Iraq's Purchases in the A-Bomb Supermarket." *The New York Times* (July 18):E5.

Lengermann, Patricia M. 1974. *Definitions of Sociology: A Historical Approach*. Columbus, OH: Merrill.

Marshall, Bruce and Philip Boys. 1991. *The Real World: Understanding the Modern World Through the New Geography*. Boston: Houghton Mifflin.

Martineau, Harriet. [1837] 1968. *Society in America*, edited and abridged by S. M. Lipset. Gloucester, MA: Peter Smith.

Miller, S. M. 1963. *Max Weber: Selections from His Work*. New York: Crowell.

Mills, C. Wright. 1959. *The Sociological Imagination*. New York: Oxford University Press.

Moffat, Susan. 1992. "An American Dilemma: Typewriter Wars Point Up Problem of Defining U.S. Goods." *Los Angeles Times* (July 20):D1,2.

The New Columbia Encyclopedia. 1975. "Steamship." New York: Columbia University Press.

Newhouse News Service. 1990. "Congressman: No Parking for Foreign Cars." *The Cincinnati Post* (December 12):7A.

Ornstein, Robert and Paul Ehrlich. 1989. *New World New Mind*. New York: Touchstone Books.

Ostar, Allan. 1988. "Entering the Global Village." *Vital Speeches of the Day* (May 5):459–61.

Reich, Robert B. 1988. "Corporation and Nation." *The Atlantic Monthly* (May):76–81.

Terry, James L. 1983. "Bringing Women . . . In: A Modest Proposal." *Teaching Sociology* 10(2):251–61.

Webb, R. K. 1960. *Harriet Martineau, A Radical Victorian*. New York: Columbia University Press.

Zuboff, Shoshana. 1988. *In the Age of the Smart Machine*. New York: Basic Books.

Chapter 2

Andrews, Edmund L. 1990. "Patents: The Illogical Process of Invention." *The New York Times* (May 9):13.

Applebome, Peter. 1986. "U.S. Goods Made in Mexico Raise Concerns on Job Losses." *The New York Times* (October 29):Y1+.

Barrio, Federico. 1988. "History and Perspectives of the Maquiladora Industry in Mexico." Pp. 7–13 in *Mexico: In-Bond Industry*, edited by T. P. Lee. Tiber, Mexico: ASI.

Bearden, Tom. 1990. "Finally—Bus Stop (Report on Greyhound Bus Strike That Began Two Months Ago)." "MacNeil/ Lehrer Newshour" (transcript #3723). New York: WNET.

———. 1993. "Focus: Help Wanted." Interview with anonymous Denver woman on her use of undocumented worker for child care. "MacNeil/Lehrer Newshour" (transcript #4548). New York: WNET.

Behar, Richard. 1990. "The Price of Freedom." *Time* (May 14):70–72.

Blumer, Herbert. 1962. "Society as Symbolic Interaction." In *Human Behavior and Social Processes*, edited by A. Rose. Boston: Houghton Mifflin.

Carver, Terrell. 1987. *A Marx Dictionary*. Totowa, NJ: Barnes and Noble.

Chast, Roz. 1990. "Why Oil Spills Are Good." *The New Yorker* (May 14):88.

Cornejo, José A. Perez. 1988. "The Implications for the U.S. Economy of Tariff Schedule Item 807 and Mexico's Maquila Program." *Maquiladora Newsletter* 15(5):2–5.

Dodge, David. 1988. "Insights into the Mexicans." Pp. 46–55 in *Fodor's Mexico*, edited by A. Beresky. New York: Fodor's Travel Publications.

Donato, Katharine M., Jorge Durand, and Douglas S. Massey. 1992. "Stemming the Tide? Assessing the Deterrent Effects of the Immigration Reform and Control Act." *Demography* 29(2): 139–57.

Dupont, David. 1989. Quoted in "As Bitter Coal Strike Nears End, Ripples from the Dispute Are Widely Felt." *The New York Times* (December 23):16Y.

Ehrenreich, Barbara and Annette Fuentes. 1985. "Life on the Global Assembly Line." Pp. 373–88 in *Crisis in American Social Institutions*, 6th ed., edited by J. H. Skolnick and E. Currie. Boston: Little, Brown.

Excelsior. 1994. "Advertisement Circular." (February 25).

Flanigan, James. 1993. "Job Site, Sweet Job Site." *Los Angeles Times* (September 20):3+.

"FRONTLINE." 1990. "New Harvest, Old Shame." Boston: WGBH Educational Foundation.

Gans, Herbert. 1972. "The Positive Functions of Poverty." *American Journal of Sociology* 78: 275–89.

Garcia, Juan Ramon. 1980. *Operation Wetback: The Mass Deportation of Mexican Undocumented Workers in 1954.* Westwood, CT: Greenwood.

Gault, Charlayne Hunter. 1990. "Focus: Crossed Lines (Gulf Crisis and Communication Gap)." "MacNeil/Lehrer Newshour" (transcript #3909). New York: WNET.

Gergen, David. 1989. "MacNeil/Lehrer Newshour" (transcript #3630). New York: WNET.

Glionna, John M. 1992. "The Paper Chase." *Los Angeles Times* (November 8):E1.

Golden, Tim. 1993. "After the Deluge, Life in Tijuana Is Grimmer." *The New York Times* (February 1):A1.

Grunwald, Joseph. 1985. "Internationalization of Industry: U.S.–Mexican Linkages." Pp. 110–38 in *The U.S. and Mexico: Borderland Development and the National Economies,* edited by L. J. Gibson and A. C. Renteria. Boulder, CO: Westview.

Hall, Edward T. 1959. *The Silent Language.* New York: Doubleday.

———. 1992. *An Anthropology of Everyday Life.* New York: Doubleday.

Hamilton, Virginia. 1988. *In the Beginning: Creation Stories from Around the World.* New York: Harcourt Brace Jovanovich.

Hansen, Nils. 1985. "The Nature and Significance of Border Development Patterns." Pp. 3–17 in *The U.S. and Mexico: Borderland Development and the National Economies,* edited by L. J. Gibson and A. C. Renteria. Boulder, CO: Westview.

Herzog, Lawrence A. 1985. "The Cross-Cultural Dimensions of Urban Land Use Policy on the U.S.–Mexico Border: A San Diego–Tijuana Case Study." *Social Science Journal* (July):29–46.

Hinds, Michael deCourcy. 1989. "As Bitter Coal Strike Nears End, Ripples from the Dispute Are Widely Felt." *The New York Times* (December 23):16Y.

———. 1990. "Strike Leaves Deep Vein of Anger in Coal Country." *The New York Times* (January 3):Y9.

Hurlbert, Lynn. 1987. "Maquiladoras: Logistics Tips for Supporting a Maquiladora." *Distribution* (October):53–58.

Hussein, Saddam. 1990. "Focus: Crossed Lines (Gulf Crisis and Communication Gap)." "MacNeil/Lehrer Newshour" (transcript #3909). New York: WNET.

Jacobs, Gary. 1986. Interview with Bill Moyers in "One River, One Country: The U.S.–Mexican Border" (documentary). New York: CBS.

Jacobson, Gary. 1988. "The Boom on Mexico's Border." *Management Review* (July):21–24.

Kilborn, Peter T. 1990a. "Replacement Workers: Management's Big Stick." *The New York Times* (March 13):A11.

———. 1990b. "When Plant Shuts Down, Retraining Laid-Off Workers Is Toughest Job Yet." *The New York Times* (April 23):A12.

———. 1992. "Tide of Migrant Labor Tells of a Law's Failure." *The New York Times* (November 4):A9.

Koenenn, Connie. 1992. "The Power of Pulling Purse Strings." *Los Angeles Times* (December 1):E1.

Langley, Lester D. 1988. *MexAmerica: Two Countries, One Future.* New York: Crown.

Lekachman, Robert. 1985. "The Specter of Full Employment." Pp. 74–80 in *Crisis in American Institutions,* edited by J. H. Skolnick and E. Currie. Boston: Little, Brown.

Madden, Steve. 1990. Quoted in "Finally—Bus Stop (Report on Greyhound Bus Strike That Began Two Months Ago)." "MacNeil/Lehrer Newshour" (transcript #3723). New York: WNET.

Magaziner, Ira C. and Mark Patinkin. 1989. *The Silent War: Inside the Global Business Battles Shaping America's Future.* New York: Random House.

Marx, Karl. [1888] 1961. "The Class Struggle." Pp. 529–35 in *Theories of Society,* edited by T. Parsons, E. Shils, K. D. Naegele, and J. R. Pitts. New York: Free Press.

Mead, George H. 1934. *Mind, Self and Society.* Chicago: University of Chicago Press.

Merton, Robert K. 1967. "Manifest and Latent Functions." Pp. 73–137 in *On Theoretical Sociology: Five Essays, Old and New.* New York: Free Press.

Moyers, Bill. 1986. "One River, One Country: The U.S.–Mexican Border" (documentary). New York: CBS.

Murray, Alan. 1988. "Out of Luck: Joblessness Tops 25 Percent in Some Areas Despite National Drop to Lowest in Nearly a Decade." *Wall Street Journal* (April 21):1+.

Myerson, Allen R. 1994. "Greyhound: The Airline of the Road." *The New York Times* (January 18):C1.

National Public Radio/"Morning Edition." 1993. "Commentator James Fallows" transcript (September 17):3–4.

O'Hare, William P. 1988. "The Rise of Poverty in Rural America." *Population Trends and Public Policy* (July). Washington, DC: Population Reference Bureau.

Oster, Patrick. 1989. *The Mexicans: A Personal Portrait of a People.* New York: Harper & Row.

Partida, Gilbert A. and Cesar Ochoa. 1990. "Border Waste Program." *Twin Plant News: The Magazine of the Maquiladora and Mexican Industries* (April):29–31.

Pearce, Jean. 1987. "Mexico Holds Special Place in the Heart of Japan: An Interview with Sergio Gonzalez Galvez, Ambassador of Mexico." *Business Japan* (November/December):47–49.

Pranis, Peter P., Jr. 1989. (Vice president, COSTEP) Personal correspondence, October 3.

———. 1990. Personal correspondence, August 16.

Roberts, Steven V. with Ann E. Andrews. 1990. "The Hunt for New Americans." *U.S. News & World Report* (May 14):35–36.

Rodale, Jerome Irving. 1986. *The Synonym Finder,* revised by L. Urdang and N. LaRoche. New York: Warner.

Roderick, Larry M. and J. Rene Villalobos. 1992. "Pollution at the Border." *Twin Plant News* (November):64–67.

Rodriguez, Richard. 1992. *Days of Obligation: An Argument with My Mexican Father.* New York: Viking.

Rose, Kenneth J. 1988. *The Body in Time.* New York: Wiley.

Sanchez, Jesus and Juanita Darling. 1992. "Border Can Drive Them Up the Wall." *Los Angeles Times* (July 14):D1.

Sanders, Thomas G. 1986. "Maquiladoras: Mexico's In-bond Industries." *UFSI Reports.* No. 15:1–8.

———. 1987. "Tijuana, Mexico's Pacific Coast Metropolis." *UFSI Reports.* No. 38:1–8.

Sharp, Kathleen. 1989. "For Migrant Workers, Legality Lowers Wages." *The New York Times* (December 3):Y12.

Silverstein, Stuart. 1993. "Survey: Good News for Working Parents." *Los Angeles Times* (September 16):D1.

Suro, Roberto. 1991. "Border Boom's Dirty Residue Imperils U.S.–Mexico Trade." *The New York Times* (March 31):Y1.

Tuleja, Tad. 1987. *Curious Customs: The Stories Behind 296 Popular American Rituals.* New York: Harmony Books.

Tumin, Melvin. 1964. "The Functionalist Approach to Social Problems." *Social Problems* 12:379–88.

Twin Plant News. 1990a. Advertisement: "Shopping the Interior? Picture This. . . ." (February):7.

———. 1990b. Advertisement: "Maquila's Multi-Billion Dollar Market: At Your Fingertips!" (July):53.

———. 1992a. Advertisement: "What Industrial City Is 3 Times the Size of

Dallas and 45 Times Closer to You Than Taiwan?" (July):2.

———. 1992b. "Maquila Scoreboard" (November):74.

Uchitelle, Louis. 1993. "Those High-Tech Jobs Can Cross the Border, Too." *The New York Times* (March 28):4E.

Uhlig, Mark A. 1990. "In Panama, Counting the Invasion Dead Is a Matter of Dispute." *The New York Times* (October 28):2E.

U.S. Bureau of Labor Statistics. 1993. *Major Work Stoppages: 1947 to 1992.* Washington, DC: U.S. Government Printing Office.

U.S. General Accounting Office. 1990. *Immigration Reform: Employer Sanctions and the Question of Discrimination.* Washington, DC: U.S. Government Printing Office.

———. 1992. *Hired Farmworkers: Health and Well-Being at Risk.* GAO/HRD–92–46 (February 14). Washington, DC: U.S. Government Printing Office.

U.S. Immigration and Naturalization Service. 1990. *Statistical Yearbook of the Immigration and Naturalization Service 1989.* Washington, DC: U.S. Government Printing Office.

U.S. Immigration Reform and Control Act of 1986. 1986. Publication 99-603 [S. 1200]. Washington, DC: U.S. Government Printing Office.

Weisman, Alan. 1986. *La Frontera: The United States Border with Mexico.* New York: Harcourt Brace Jovanovich.

White, Leslie A. 1949. *The Science of Culture: A Study of Man and Civilization.* New York: Farrar, Straus.

Yim, Yong Soon. 1989. "American Perceptions of Korean-Americans: An Analytical Study of a 1988 Survey." *Korean and World Affairs* 13(Fall): 519–42.

Chapter 3

Allison, Anne. 1991. "Japanese Mothers and *Obentōs*: The Lunch-Box as Ideological State Apparatus." *Anthropological Quarterly* 64:195–208.

Bailey, William T. and Wade C. Mackey. 1989. "Observations of Japanese Men and Children in Public Places: A Comparative Study." *Psychological Reports* 65:731–34.

Barnlund, Dean C. 1989. *Communicative Styles of Japanese and Americans: Images and Realities.* Belmont, CA: Wadsworth.

Barnlund, Dean C. and Shoko Araki. 1985. "Intercultural Encounters: The Management of Compliments by Japanese and Americans." *Journal of Cross-Cultural Psychology* 16(1):9–26.

Barringer, Felicity. 1990. "The Census, in One Not-So-Easy Lesson." *The New York Times* (April 22):E1+.

Barzun, Jacques. 1987. "Doing Research—Should the Sport Be Regulated?" *Columbia* (February): 18–22.

Bell, Daniel. 1979. "Communication Technology—for Better or for Worse?" *Harvard Business Review* (May/June): 34–41.

Branscomb, Lewis M. 1981. "The Electronic Library." *Journal of Communication* (Winter):143–50.

Cameron, William B. 1963. *Informal Sociology.* New York: Random House.

Caplan, Lincoln. 1990. "A Reporter at Large (Open Adoption—Part II)." *The New Yorker* (May 28): 73–95.

Compaine, Benjamin M. 1981. "Shifting Boundaries in the Information Marketplace." *Journal of Communication* (Winter):132–47.

Durkheim, Emile. [1924] 1953. *Sociology and Philosophy.* Glencoe, IL: Free Press.

———. 1951. *Suicide: A Study in Sociology.* Translated by J. A. Spaulding and G. Simpson. New York: Free Press.

Fallows, James. 1989. "The Real Japan." *The New York Review of Books* (July 20):23–28.

Goffman, Erving. 1963. *Behavior in Public Places.* New York: Free Press.

Gregg, Alan. 1989. Quoted on pp. 48–55 in *Science and the Human Spirit,* edited by R. D. White. Belmont, CA: Wadsworth.

Hagan, Frank E. 1989. *Research Methods in Criminal Justice and Criminology.* New York: Macmillan.

Herskovits, Melville J. 1948. *Man and His Works: The Science of Cultural Anthropology.* New York: Knopf.

Hirsch, E. D., Jr., Joseph F. Kett, and James Trefil. 1988. *The Dictionary of Cultural Literacy (What Every American Needs to Know).* Boston: Houghton Mifflin.

Ishii-Kuntz, Masako. 1992. "Are Japanese Families 'Fatherless'?" *Sociology and Social Research* 76 (3):105–10.

Joseph, Michael. 1982. *The Timetable of Technology.* London: Marshal Editions.

Jussaume, Ramond A., Jr., and Yoshiharu Yamada. 1990. "A Comparison of the Viability of Mail Surveys in Japan and the United States." *Public Opinion Quarterly* 54:219–28.

Kagay, Michael R. 1990. "Polls Try to Account for the Many Uncounted by the Census." *The New York Times* (April 26):C24.

Katzer, Jeffrey, Kenneth H. Cook, and Wayne W. Crouch. 1991. *Evaluating Information: A Guide for Users of Social Science Research,* 3rd edition. New York: McGraw-Hill.

Klapp, Orrin E. 1986. *Overload and Boredom: Essays on the Quality of Life*

in the Information Society. New York: Greenwood.

Knepper, Paul. 1992. "How to Find Sources for Research Papers." Pp. 247–56 in *Study Guide to Accompany Sociology: A Global Perspective* by J. Ferrante and M. Murray. Belmont, CA: Wadsworth.

Kōji, Kata. 1983. "Pachinko." P. 143 in *Kodansha Encyclopedia of Japan.* New York: Kodansha International.

Kurosu, Satomi. 1991. "Suicide in Rural Areas: The Case of Japan 1960–1980." *Rural Sociology* 56 (4):603–18.

Levin, Doron P. 1990. "Wheeling Out Iacocca for a Long-Shot Attack." *The New York Times* (May 13).

Lewin, Kurt. 1948. *Resolving Social Conflicts.* New York: Harper & Row.

Lewis, Catherine C. 1988. "Japanese First-Grade Classrooms: Implications for U.S. Theory and Research." *Comparative Education Review* 32(2): 159–72.

Lewontin, R. C. 1990. "Fallen Angels." *The New York Review of Books* (June 14):3–7.

Lofland, Lyn. 1973. *A World of Strangers.* New York: Basic Books.

Lucky, Robert W. 1985. "Message by Light Wave." *Science* (November): 112–13.

Lynd, Robert S. and Helen M. Lynd. [1929] 1956. *Middletown: A Study in Modern American Culture.* New York: Harcourt, Brace & World.

Machlup, Fritz. 1988. "Are the Social Sciences Really Inferior?" *Society* (May/June):57–65.

Marsa, Linda. 1992. "Scientific Fraud." *Omni* (June):39+.

Matsuo, Hisako. 1992. "Identificational Assimilation of Japanese Americans: A Reassessment of Primordialism and Circumstantialism." *Sociological Perspectives* 35(3):505–23.

Michael, Donald. 1984. "Too Much of a Good Thing?: Dilemmas of an Information Society." *Technological Forecasting and Social Change* 25(4):347–54.

Navarro, Mireya. 1990. "The Census Knocks, But Many Refuse to Answer." *The New York Times* (May 14):E4.

Ozaki, Robert. 1978. *The Japanese.* Tokyo: Tuttle.

Paulos, John A. 1988. *Innumeracy: Mathematical Illiteracy and Its Consequences.* New York: Hill and Wang.

Rathje, William L. and Cullen Murphy. 1992. *Rubbish: The Archaeology of Garbage.* New York: HarperCollins.

Rennie, Drummond. 1986. "Guarding the Guardians: A Conference on Editorial Peer Review." *Journal of the American Medical Association* 256(17): 2391–392.

Robinson, Richard D. 1985. "Another Look at Japanese–United States Trade Relations." *UFSI Reports* No. 19:1–7.

Roethlisberger, F. J. and William J. Dickson. 1939. *Management and the Worker.* Cambridge, MA: Harvard University Press.

Rossi, Peter H. 1988. "On Sociological Data." Pp. 131–54 in *Handbook of Sociology,* edited by N. Smelser. Newberry Park, CA: Sage.

Ryūzō, Satō. 1991. "Maturing from Weism to Global You-ism." *Japan Quarterly* (July/September):273–82.

Sanger, David E. 1989. "IBM, in Surprise, Sees a Vital Edge over Japan." *The New York Times* (June 23):1+.

———. 1990. "Hitachi's Quest for Super Chips." *The New York Times* (September 10):C1+.

———. 1992. "A Defiant Detroit Still Depends on Japan." *The New York Times* (February 27):A1+.

Saxonhouse, Gary R. 1991. "Sony Side Up: Japan's Contribution to the U.S. Economy." *Policy Review* (Spring):60–64.

Schonberg, Harold C. 1981. "Sumō Embodies Ancient Rituals." *The New York Times*:B9.

Shotola, Robert W. 1992. "Small Groups." Pp. 1796–1806 in *Encyclopedia of Sociology. Vol. 4,* edited by E. F. Borgatta and M. L. Borgatta. New York: Macmillan.

Singleton, Royce A., Jr., Bruce C. Straits, and Margaret Miller Straits. 1993. *Approaches to Social Research,* 2nd ed. New York: Oxford.

Smith, Joel. 1991. "A Methodology for the Twenty-First Century Sociology." *Social Forces,* 70(1): 1–17.

Thayer, John E. III. 1983. "Sumō." Pp. 270–74 in *Kodansha Encyclopedia of Japan, Vol. 7.* Tokyo: Kodansha.

Thayer, Nathaniel B. and Stephen E. Weiss. 1987. "The Changing Logic of a Former Minor Power." Pp. 45–74 in *National Negotiating Styles,* edited by H. Binnendijk. Washington, DC: U.S. Government Printing Office.

Thomas, Bill. 1992. "King Stacks." *Los Angeles Times Magazine* (November 15):31+.

Tocqueville, Alexis de. [1835] 1956. *Democracy in America.* New York: Mentor.

Totten, Bill. 1990. "Japan's Mythical Trade Surplus." *The New York Times* (December 9):F13.

Tsutomu, Kano. 1976. *The Silent Power.* Tokyo: Simul.

U.S. Department of Commerce. 1985. *United States Trade Performance in 1984 and Outlook.* Washington, DC: U.S. Government Printing Office.

———. 1989a. *United States Trade Performance in 1988.* Washington, DC: U.S. Government Printing Office.

———. 1989b. *Foreign Direct Investment in the United States.* Washington, DC: U.S. Government Printing Office.

———. 1992. *U.S. Foreign Trade Highlights 1991.* Washington, DC: U.S. Government Printing Office.

White, Lynn K. and Yuko Matsumato. 1992. "Why Are U.S. and Japanese Divorce Rates So Different?: An Aggregate-Level Analysis." Pp. 85–92 in *Families: East and West, Vol. 1.* Indianapolis: University of Indianapolis Press.

Whyte, William H. 1988. *City: Rediscovering the Center.* New York: Doubleday.

Wood, Margaret. 1934. *The Stranger.* New York: Columbia University Press.

Chapter 4

Benedict, Ruth. 1976. Quoted on p. 14 in *The Person: His and Her Development Throughout the Life Cycle,* by Theodor Lidz. New York: Basic Books.

Berger, Peter. 1989. Conversation on pp. 484–93 in *A World of Ideas,* with Bill Moyers. New York: Doubleday.

Berkhofer, Robert F., Jr. 1978. *The White Man's Indian: Images of the American Indian from Columbus to the Present.* New York: Knopf.

Bok, Lee Suk. 1987. *The Impact of U.S. Forces in Korea.* Washington, DC: National Defense University Press.

Braunschweig, Robert. 1990. "A Revolution on Wheels." *UNESCO Courier* (October):22–29.

Breton, Raymond. 1967. "Institutional Completeness of Ethnic Communities and the Personal Relations of Immigrants." *American Journal of Sociology* 70:193–205.

Brown, Rita Mae. 1988. *Rubyfruit Jungle.* New York: Bantam Books.

Chinoy, Ely. 1961. *Society: An Introduction to Sociology.* New York: Random House.

Clifton, James A. 1989. *Being and Becoming Indian: Biographical Studies of North American Frontiers.* Chicago: Dorsey.

Critchfield, Richard. 1985. "Science, Villagers, and Us." *UFSI Reports* No. 3:1–17.

Doi, Takeo, M.D. 1986. *The Anatomy of Dependence,* translated by J. Bester. Tokyo: Kodansha International.

Fallows, James. 1988. "Trade: Korea Is Not Japan." *The Atlantic Monthly* (October):22–33.

———. 1990. "Wake Up, America!" *The New York Review of Books* (March 1):14–19.

Frank, Lawrence. 1948. "World Order and Cultural Diversity." Pp. 389–95 in *Society as the Patient.* New Brunswick, NJ: Rutgers University Press.

"FRONTLINE." 1988. "American Game, Japanese Rules" (transcript #611). Boston: WGBH Educational Foundation.

Gelb, Leslie H. 1990. "The Stuff That Makes the World Go Round." *The New York Times Book Review* (December 9):1, 4–35.

Gordon, Steven L. 1981. "The Sociology of Sentiments and Emotion." Pp. 562–92 in *Social Psychology Sociological Perspectives,* edited by M. Rosenberg and R. H. Turner. New York: Basic Books.

Halberstam, David. 1986. *The Reckoning.* New York: Morrow.

Hall, Edward T. 1959. *The Silent Language.* Garden City, NY: Doubleday.

Henry, William A. 1988. "No Time for the Poetry: NBC's Cool Coverage Stints on the Drama." *Time* (October 3):80.

Henslin, James M. 1985. *Down to Earth Sociology: Introductory Readings.* New York: Free Press.

Herskovits, Melville J. 1948. *Man and His Works: The Science of Cultural Anthropology.* New York: Knopf.

Hochschild, Arlie R. 1976. "The Sociology of Feeling and Emotion: Selected Possibilities." Pp. 280–307 in *Another Voice,* edited by M. Millman and R. Kanter. New York: Octagon.

———. 1979. "Emotion Work, Feeling Rules, and Social Structure." *American Journal of Sociology* 85:551–75.

Horton, Paul B. and Chester L. Hunt. 1984. *Sociology.* New York: McGraw-Hill.

Hughes, Everett C. 1984. *The Sociological Eye: Selected Papers.* New Brunswick, NJ: Transaction Books.

Hurst, G. Cameron. 1984a. "Getting a Piece of the R.O.K.: American Problems of Doing Business in Korea." *UFSI Reports* No. 19.

———. 1984b. "Japanese Education: Trouble in Paradise?" *UFSI Reports* No. 40.

Ingram, Erik. 1992. "Water Use Continues to Decline: Bay Area Districts Report Record Savings." *San Francisco Chronicle* (July):A15.

International Baseball Association. 1993. "International Baseball Association Membership List" (July). Indianapolis.

Iyer, Pico. 1988. "The Yin and Yang of Paradoxical, Prosperous Korea." *Smithsonian* (August):47–58.

Jun, Suk-ho and Daniel Dayan. 1986. "An Interactive Media Event: South Korea's Televised 'Family Reunion.'" *Journal of Communication* (Spring):73–82.

Kephart, William M. 1987. *Extraordinary Groups*. New York: St. Martin's Press.

Kim, Choong Soon. 1989. "Attribute of 'Asexuality' in Korean Kinship and Sundered Koreans during the Korean War." *Journal of Comparative Family Studies* 20(3):309–25.

Kim, Eun Mee. 1993. "Contradictions and Limits of a Developmental State: With Illustrations from the South Korean Case." *Social Problems* 40(2): 228–49.

Kim, W. Chan and R. A. Mauborgne. 1987. "Cross-Cultural Strategies." *Journal of Business Strategy* 7(Spring):28–35.

Kluckhohn, Clyde. 1949. *Mirror for Man: Anthropology and Modern Life*. New York: McGraw-Hill.

———. 1964. "The Study of Culture." Pp. 40–54 in *Sociological Theory: A Book of Readings*, edited by L. A. Coser and B. Rosenberg. New York: Macmillan.

Lamb, David. 1987. *The Arabs: Journeys Beyond the Mirage*. New York: Random House.

Larson, James F. and Nancy K. Rivenburgh. 1991. "A Comparative Analysis of Australian, U.S., and British Telecasts of the Seoul Olympic Opening Ceremony." *Journal of Broadcasting and Electronic Media* 35(1):75–94.

"Letters." 1988. "Reading for Meaning, and Not Just for Sounds." *The New York Times* (November 12):Y4.

Lidz, Theodore. 1976. *The Person: His and Her Development Throughout the Life Cycle*. New York: Basic Books.

Linton, Ralph. 1936. *The Study of Man: An Introduction*. New York: Appleton-Century-Crofts.

Magaziner, Ira C. and Mark Patinkin. 1989. *The Silent War: Inside the Global Business Battles Shaping America's Future*. New York: Random House.

Moran, Robert T. 1987. "Cross-Cultural Contact: What's Funny to You May Not Be Funny to Other Cultures." *International Management* 42 (July/August):74.

Park, Myung-Seok. 1979. *Communication Styles in Two Different Cultures: Korean and American*. Seoul: Han Shin.

Peterson, Mark. 1977. "Some Korean Attitudes Toward Adoption." *Korea Journal* 17(12):28–31.

Protzman, Ferdinand. 1991. "As Marriage Nears, Germans in the Wealthy West Fear Cost in Billions." *The New York Times* (September 24):A6.

Reader, John. 1988. *Man on Earth*. Austin: University of Texas Press.

Reid, Daniel P. 1988. *Korea: The Land of the Morning Calm*. Lincolnwood, IL: Passport Books.

Reinhold, Robert. 1989. "The Koreans' Big Entry into Business." *The New York Times* (September 24):4E.

Rohner, Ronald P. 1984. "Toward a Conception of Culture for Cross-Cultural Psychology." *Journal of Cross-Cultural Psychology* 15(2):111–38.

Rosenfeld, Jeffrey P. 1987. "Barking up the Right Tree." *American Demographics* (May):40–43.

Sanchez, Jesus. 1987. "L.A. Top Choice for Koreans in Business in U.S." *Los Angeles Times* (June 6):IV, 1.

Sapir, Edward. 1949. "Selected Writings of Edward Sapir," in *Language, Culture and Personality*, edited by D. G. Mandelbaum. Berkeley: University of California Press.

Segall, Marshall H. 1984. "More Than We Need to Know About Culture, But Are Afraid Not to Ask." *Journal of Cross-Cultural Psychology* 15(2): 153–62.

Sterngold, James. 1991. "New Doubts on Uniting 2 Koreas." *The New York Times* (May 30):C1.

Sumner, William Graham. 1907. *Folkways*. Boston: Ginn.

Temple, Robert. 1986. *The Genius of China: 3,000 Years of Science, Discovery, and Inventions*. London: Multimedia.

Tuleja, Tad. 1987. *Curious Customs: The Stories Behind 296 Popular American Rituals*. New York: Harmony Books.

Underwood, Horace G. 1977. "Foundations in Thought and Values: How We Differ." *Korea Journal* 17(12):6–9.

Visser, Margaret. 1988. *Much Depends on Dinner*. New York: Grove.

———. 1989. "A Meditation on the Microwave." *Psychology Today* (December):38–42.

Vogt, Evon Z. and John M. Roberts. 1960. "A Study of Values." Pp. 109–17 in *Readings in Sociology*, 2nd edition, edited by E. A. Schuler, T. A. Hoult, D. L. Gibson, M. L. Fiero, and W. B. Brookover. New York: Crowell.

Wald, Matthew L. 1989. "Why Americans Consume More Energy to Produce Less." *The New York Times* (March 12):6E.

Welch, Wilford H. 1991. "Regional Powerhouse Reshapes World." *The World Paper* (November):3.

White, Leslie A. 1949. *The Science of Culture*. New York: Farrar, Straus & Cudahy.

Williams, Robin M., Jr. 1970. *American Society: A Sociological Interpretation*. New York: Knopf.

Winchester, Simon. 1988. *Korea: A Walk Through the Land of Miracles*. New York: Prentice Hall.

World Monitor. 1992. "The Map: Batters Up!" (April):11.

———. 1993. "The Map: Hoop-la" (February): 10–11.

Yeh, May. 1991. "A Letter." *Amerasia* 17(2):1–7.

Yim, Yong Soon. 1989. "American Perceptions of Korean-Americans: An Analytical Study of a 1988 Survey." *Korea and World Affairs* (Fall) Vol. 13(3): 519–42.

Yoo, Yushin. 1987. *Korea the Beautiful: Treasures of the Hermit Kingdom*. Los Angeles: Golden Pond.

Chapter 5

Al-Batrawi, Khaled and Mouin Rabbani. 1991. "Break Up of Families: A Case Study in Creeping Transfer." *Race and Class* 32(4):35–44.

Atta, Nasser. 1989. "Letters to the Editor: Israel a Military Power, Violates Human Rights." *The News Record* (February 8):5.

Ben-David, Amith and Yoav Lavee. 1992. "Families in the Sealed Room: Interaction Patterns of Israeli Families During SCUD Missile Attacks." *Family Process* 31(1):35–44.

Berman, Morris. 1991. "Interview: Morris Berman." *Omni* (August 19):60–91.

Bourne, Jenny. 1990. "The Rending Pain of Reenactment." *Race and Class* 32(2):67–72.

Broder, John M. and Norman Kempster. 1993. "'Enough of Blood and Tears': Israel and P.L.O. Adopt Framework for Peace." *Los Angeles Times* (September 14):A1+.

Bunuel, Luis. 1985. Quoted on p. 22 in *The Man Who Mistook His Wife for a Hat and Other Clinical Tales*, by Oliver Sacks. New York: Summit Books.

Charney, Marc D. 1988. "The Battleground for the Jordan to the Sea." *The New York Times* (February 28):C1.

Cooley, Charles Horton. 1909. *Social Organization*. New York: Scribners.

———. 1961. "The Social Self." Pp. 822–28 in *Theories of Society: Foundations of Modern Sociological Theory*, edited by T. Parsons, E. Shils, K. D. Naegele, and J. R. Pitts. New York: Free Press.

Corsaro, William A. 1985. *Friendship and Peer Culture in the Early Years*. Norwood, NJ: Ablex.

Coser, Lewis A. 1992. "The Revival of the Sociology of Culture: The Case of Collective Memory." *Sociological Forum* 7(2):365–73.

Davis, Kingsley. 1940. "Extreme Isolation of a Child." *American Journal of Sociology* 45:554–65.

———. 1947. "Final Note on a Case of Extreme Isolation." *American Journal of Sociology* 3(5):432–37.

Delgado, Jose M. R. 1970. Quoted on p. 170 in "Brain Researcher Jose Delgado Asks—'What Kind of Humans Would

We Like to Construct?'" *The New York Times Magazine* (November 15):46+.

Dyer, Gwynne. 1985. *War*. New York: Crown.

Elon, Amos. 1993. "The Jews' Jews." *The New York Review of Books* (June 10):14–18.

Encyclopedia Judaica Yearbook. 1987. Jerusalem: Keter.

Facts About Israel. 1985. Ministry of Foreign Affairs, Information Division. Jerusalem: Israel Information Centre.

Faris, Ellsworth. 1964. "The Primary Group: Essence and Accident." Pp. 314–19 in *Sociological Theory: A Book of Readings*, edited by L. A. Coser and B. Rosenberg. New York: Macmillan.

Fields, Rona M. 1979. "Child Terror Victims and Adult Terrorists." *Journal of Psychohistory* 7(1):71–75.

Figler, Stephen K. and Gail Whitaker. 1991. *Sport and Play in American Life*. Dubuque, IA: Brown.

Freeman, Norman H. 1987. "Children's Drawings of Human Figures." Pp. 135–39 in *The Oxford Companion to the Mind*, edited by R. L. Gregory. Oxford: Oxford University Press.

Freud, Anna and Dorothy T. Burlingham. 1943. *War and Children*. New York: Ernst Willard.

Freud, Anna and Sophie Dann. 1958. "An Experiment in Group Upbringing." Pp. 127–68 in *The Psychoanalytic Study of the Child*, Vol. 6, edited by R. S. Eissler, A. Freud, H. Hartmann, and E. Kris. New York: Quadrangle.

Friedman, Robert I. 1992. *Zealots for Zion: Inside Israel's West Bank Settlement Movement*. New York: Random House.

Friedman, Thomas L. 1989. *From Beirut to Jerusalem*. New York: Farrar, Straus & Giroux.

Goffman, Erving. 1961. *Asylums: Essays on the Social Situation of Mental Patients and Other Inmates*. New York: Anchor.

Goldman, Ari L. 1989. "Mementos to Preserve the Record of Anguish." *The New York Times* (February 29):14Y.

Gorkin, Michael. 1986. "Countertransference in Cross-Cultural Psychotherapy: The Example of Jewish Therapist and Arab Patient." *Psychiatry* 49:69–79.

Gould, Stephen Jay. 1990. Interview on "MacNeil/Lehrer Newshour." (January 1). New York: WNET.

Griffith, Marlin S. 1977. "The Influences of Race on the Psychotherapeutic Relationship." *Psychiatry* 40:27–40.

Grossman, David. 1988. *The Yellow Wind*, translated from Hebrew by H. Watzman. New York: Farrar, Straus & Giroux.

Halbwachs, Maurice. 1980. *The Collective Memory*, translated from French by F. J. Ditter, Jr., and V. Y. Ditter. New York: Harper & Row.

Hellerstein, David. 1988. "Plotting a Theory of the Brain." *The New York Times Magazine*. (May 22):17+.

Hull, Jon D. 1991. "The Good Life in Gaza." *Time* (July 1):40.

Jacob, Mira. 1989a. "Letters to the Editor: U.S. Should Realize Truth About Israel." *The News Record* (February 6):5.

———. 1989b. "Letters to the Editor: Israel Modernized Country Deserving Trust, Sympathy." *The News Record* (February 22):5.

Jacobs, Dan. 1989. "Most in U.S. Don't Grasp Complex Israeli Situation." *The News Record* (January 30):5.

Kagan, Jerome. 1988a. Interview "The Mind." PBS.

———. 1988b. Quoted on pp. 21–22 in *The Mind*, by R. M. Restak. New York: Bantam Books.

———. 1989. *Unstable Ideas: Temperament, Cognition, and Self*. Cambridge: Harvard University Press.

Kotowitz, Victor and Matt Moody. 1993. "A World of Civil Strife." *Los Angeles Times* (June 8):H8.

Lesch, Ann Mosely. 1989. "Anatomy of an Uprising: The Palestinian *Intifadah*." Pp. 87–110 in *Palestinians Under Occupation*. Washington, DC: Georgetown University Press.

Mannheim, Karl. 1952. "The Problem of Generations." Pp. 276–322 in *Essays on the Sociology of Knowledge*, edited by P. Kecskemeti. New York: Oxford University Press.

Matar, Ibrahim. 1983. "Israeli Settlements and Palestinian Rights." Pp. 117–41 in *Occupation: Israel over Palestine*, edited by N. H. Aruri. Belmont, MA: AAUG.

Mathabane, Mark. 1993. "Appearances Are Destructive." *The New York Times* (August 28):A11.

Mead, George Herbert. 1934. *Mind, Self and Society*. Chicago: University of Chicago Press.

Merton, Robert K. 1976. *Sociological Ambivalence and Other Essays*. New York: Free Press.

Miller, Donald E. and Lorna Touryan Miller. 1991. "Memory and Identity Across the Generations: A Case Study of Armenian Survivors and Their Progeny." *Qualitative Sociology* 14(1):13–38.

Montgomery, Geoffrey. 1989. "Molecules of Memory." *Discover* (December):46–55.

Morrow, Lance. 1988. "Israel: At 40, the Dream Confronts Palestinian Fury and a Crisis of Identity." *Time* (April 4):36–50.

Nesvisky, Matthew. 1986. "Culture Shock: Thousands of Ethiopian Jews Are Now in Israel. How Are They Doing?" *Present Tense* 14(1):12–19.

The New York Times. 1993a. "Where the Israeli Settlements Are" (September 8):A8.

———. 1993b. "P.L.O. Accepts Israel, Disavows Force; Attains Recognition After Three Decades" (September 10):A1.

"Nova." 1986. "Life's First Feelings" (February 11).

Ornstein, Robert and Richard F. Thompson. 1984. *The Amazing Brain*. Boston: Houghton Mifflin.

Patai, Raphael. 1971. "Zionism." P. 1262 in *Encyclopedia of Zionism and Israel*. New York: McGraw-Hill.

Pawel, Ernst. 1989. *The Labyrinth of Exile: A Life of Theodor Herzl*. New York: Farrar, Straus & Giroux.

Penfield, Wilder and P. Perot. 1963. "The Brain's Record of Auditory and Visual Experience: A Final Summary and Discussion." *Brain* 86:595–696.

Piaget, Jean. 1923. *The Language and Thought of the Child*, translated by M. Worden. New York: Harcourt, Brace & World.

———. 1929. *The Child's Conception of the World*, translated by J. Tomlinson and A. Tomlinson. Savage, MD: Rowan & Littlefield.

———. 1932. *The Moral Judgement of the Child*, translated by M. Worden. New York: Harcourt, Brace & World.

———. 1946. *The Child's Conception of Time*, translated by A. J. Pomerans. London: Routledge & Kegan Paul.

———. 1967. *On the Development of Memory and Identity*. Worchester, MA: Clarks University Press.

Rabin, Yitzhak. 1993. "Making a New Middle East. 'Shalom, Salaam, Peace.' Views of 3 Leaders." *Los Angeles Times* (September 14):A7.

Restak, Richard M. 1988. *The Mind*. New York: Bantam Books.

Rose, Peter I., Myron Glazer, and Penina M. Glazer. 1979. "In Controlled Environments: Four Cases of Intensive Resocialization." Pp. 320–38 in *Socialization and the Life Cycle*, edited by P. I. Rose. New York: St. Martin's Press.

Rosenthal, Elisabeth. 1989. "Mystery on Arrival." *Discover* (December):78–82.

Rubinstein, Danny. 1988. "The Uprising: Reporter's Notebook." *Present Tense* 15:22–25.

———. 1991. *The People of Nowhere: The Palestinian Vision of Home*, translated by T. Friedman. New York: Random House.

Sacks, Oliver. 1989. *Seeing Voices: A Journey into the World of the Deaf*. Los Angeles: University of California Press.

Satterly, D. J. 1987. "Jean Piaget (1896–1980)." Pp. 621–22 in *The Oxford Companion to the Mind*, edited by R. I. Gregory. Oxford: Oxford University Press.

Schiff, Ze'ev and Ehud Ya'ari. 1991. *Intifada: The Palestinian Uprising—Israel's Third Front*. New York: Touchstone.

Shadid, Mohammed and Rick Seltzer. 1988. "Political Attitudes of Palestinians in the West Bank and Gaza Strip." *The Middle East Journal* 42:16–32.

Shipler, David. 1986. *Arab and Jew: Wounded Spirits in a Promised Land*. New York: Times Books.

Sichrovsky, Peter. 1991. *Abraham's Children: Israel's Young Generation*. New York: Pantheon.

Smooha, Sammy. 1980. "Control of Minorities in Israel and Northern Ireland." *Society for Comparative Study of Society and History* 10:256–80.

Spitz, Rene A. 1951. "The Psychogenic Diseases in Infancy: An Attempt at Their Etiological Classification." Pp. 255–78 in *The Psychoanalytic Study of the Child*, Vol. 27, edited by R. S. Eissler and A. Freud. New York: Quadrangle.

Steiner, George. 1967. *Language and Silence: Essays on Language, Literature, and the Inhuman*. New York: Atheneum.

Theodorson, George A. and Achilles G. Theodorson. 1979. *A Modern Dictionary of Sociology*. New York: Barnes & Noble.

Townsend, Peter. 1962. Quoted on pp. 146–47 in *The Last Frontier: The Social Meaning of Growing Old*, by Andrea Fontana. Beverly Hills, CA: Sage.

U.S. Bureau for Refugee Programs. 1988. *World Refugee Report*. Washington, DC: U.S. Government Printing Office.

Usher, Graham. 1991. "Children of Palestine." *Race and Class* 32(4):1–18.

Yishai, Yael. 1985. "Hawkish Proletariat: The Case of Israel." *Journal of Political and Military Sociology* 13:53–73.

Chapter 6

Aina, Tade. 1988. "The Myth of African Promiscuity." Pp. 78–80 in *Blaming Others: Prejudice, Race and Worldwide AIDS*, edited by R. Sabatier. Philadelphia: New Society Publishers.

Altman, Lawrence K. 1986. "Anxiety on Transfusions." *The New York Times* (July 18):A1, B4.

Anderson, Elijah. 1990. "The Police and the Black Male." Pp. 190–206 in *Streetwise: Race, Class, and Change in an Urban Community*. Chicago: University of Chicago Press.

Barr, David. 1990. "What Is AIDS? Think Again." *The New York Times* (December 1):Y15.

Bloor, Michael, David Goldberg, and John Emslie. 1991. "Research Note: Ethnostatics [*sic*] and the AIDS Epidemic." *The British Journal of Sociology* 42(1):131–38.

Brooke, James. 1987. "In Cradle of AIDS Theory, a Defensive Africa Sees a Disguise for Racism." *The New York Times* (November 19):B13.

———. 1988a. "In Africa, Tribal Hatreds Defy the Borders of State." *The New York Times* (August 28):E1.

———. 1988b. "Mobutu's Village Basks in His Glory." *The New York Times* (September 9):Y4.

Carovano, Kathryn. 1990. "Point of View: Women Are More Than Mothers." *WorldAIDS* (8):2.

Chin, James and Jonathan M. Mann. 1988. "The Global Patterns and Prevalence of AIDS and HIV Infection." *AIDS 2* (supplement 1):S247–52.

Clark, Matt with Stryker McGuire. 1980. "Blood Across the Border." *Newsweek* (December 29):61.

Clarke, Thurston. 1988. *Equator: A Journey*. New York: Morrow.

Colby, Ron. 1986. Quoted in "Did Media Sensationalize Student AIDS Case?" by John McGauley. *Editor and Publisher* 119:19.

Conrad, Joseph. 1971. *Heart of Darkness*, revised and edited by R. Kimhough. New York: Norton.

De Cock, Kevin M. and Joseph B. McCormick. 1988. "Correspondence: Reply to HIV Infection in Zaire." *New England Journal of Medicine* 319(5):309.

Doyal, Lesley with Imogen Pennell. 1981. *The Political Economy of Health*. Boston: South End Press.

Durkheim, Emile. [1933] 1964. *The Division of Labor in Society*, translated by G. Simpson. New York: Free Press.

The Economist. 1981. "America the Blood Bank" (October 17):87.

———. 1983. "Vein Hopes, Mainline Profits" (January 22):63–64.

Fox, Renée. 1988. *Essays in Medical Sociology: Journeys into the Field*. New Brunswick, NJ: Transaction Books.

"FRONTLINE." 1993. "AIDS, Blood, and Politics." Boston: WGBH Educational Foundation and *Health Quarterly*.

Giese, Jo. 1987. "Sexual Landscape: On the Difficulty of Asking a Man to Wear a Condom." *Vogue* 177 (June):227+.

Goffman, Erving. 1959. *The Presentation of Self in Everyday Life*. New York: Anchor.

———. 1963. *Stigma: Notes on the Management of Spoiled Identity*. Englewood Cliffs, NJ: Prentice-Hall.

Grmek, Mirkod. 1990. *History of AIDS: Emergence and Origin of a Modern Pandemic*, translated by R. C. Maulitz and J. Duffin. Princeton, NJ: Princeton University Press.

Grover, Jan Zita. 1987. "AIDS: Keywords." *October* 43:17–30.

Harris, Robert and Jeremy Paxman. 1982. *A Higher Form of Killing*. New York: Hill and Wang.

Henig, Robin Marantz. 1993. *A Dancing Matrix: Voyages Along the Viral Frontier*. New York: Knopf.

Hiatt, Fred. 1988. "Tainted U.S. Blood Blamed for AIDS' Spread in Japan." *The Washington Post* (June 23):A29.

Hilts, Philip J. 1988. "Dispelling Myths About AIDS in Africa." *Africa Report* 33:27–31.

Hunt, Charles W. 1989. "Migrant Labor and Sexually Transmitted Disease: AIDS in Africa." *Journal of Health and Social Behavior* 30:353–73.

Hurley, Peter and Glenn Pinder. 1992. "Ethics, Social Forces, and Politics in AIDS-Related Research: Experience in Planning and Implementing a Household HIV Seroprevalence Survey." *The Milbank Quarterly* 70(4):605–28.

International Federation of Pharmaceutical Manufacturers Associations. 1981. Personal correspondence.

Irwin, Kathleen. 1991. "Knowledge, Attitudes and Beliefs About HIV Infection and AIDS Among Healthy Factory Workers and Their Wives, Kinshasa, Zaire." *Social Science and Medicine* 32(8):917–30.

Johnson, Diane and John F. Murray, M.D. 1988. "AIDS Without End." *The New York Review of Books* (August 18):57–63.

Kaptchuk, Ted and Michael Croucher, with the BBC. 1986. *The Healing Arts: Exploring the Medical Ways of the World*. New York: Summit.

Kerr, Dianne L. 1990. "AIDS Update: Ryan White's Death." *Journal of School Health* 60(5):237–38.

Kolata, Gina. 1989. "AIDS Test May Fail to Detect Virus for Years, Study Finds." *The New York Times* (June 1):1Y.

Kornfield, Ruth. 1986. "Dr., Teacher, or Comforter?: Medical Consultation in a Zairian Pediatrics Clinic." *Culture, Medicine and Psychiatry* 10:367–87.

Kramer, Reed. 1993. "Ties That Bind: Pressure Points Considered to Back Change in Zaire." *Africa News* (March 8–21):2.

Kramer, Staci D. 1988. "The Media and AIDS." *Editor and Publisher* 121:10–11, 43.

Krause, Richard. 1993. Quoted on p. xii in Robin Marantz Henig's *A Dancing Matrix: Voyage Along the Viral Frontier*. New York: Knopf.

Lamb, David. 1987. *The Africans*. New York: Random House.

Lasker, Judith N. 1977. "The Role of Health Services in Colonial Rule: The Case of the Ivory Coast." *Culture, Medicine and Psychiatry* 1:277–97.

Lippmann, Walter. 1976. "The World Outside and the Pictures in Our Heads." Pp. 174–81 in *Drama in Life: The Uses of Communication in Society,* edited by J. E. Combs and M. W. Mansfield. New York: Hastings House.

Liversidge, Anthony. 1993. "Heresy!: 3 Modern Galileos." *Omni* (June):43–51.

Mahler, Halfdan. 1989. Quoted on p. 91 in *AIDS and Its Metaphors,* by Susan Sontag. New York: Farrar, Straus & Giroux.

Mannheim, Karl. 1952. "The Problem of Generations." Pp. 276–322 in *Essays on the Sociology of Knowledge,* edited by P. Kecskemeti. New York: Oxford University Press.

McNeill, William H. 1976. *Plagues and People*. New York: Anchor.

Meltzer, Milton. 1960. *Mark Twain: A Pictorial Biography*. New York: Bonanza.

Merton, Robert K. 1957. *Social Theory and Social Structure*. Glencoe, IL: Free Press.

Naisbitt, John. 1984. *Megatrends: Ten New Directions Transforming Our Lives*. New York: Warner.

Noble, Kenneth B. 1989. "More Zaire AIDS Cases Show Less Underreporting." *The New York Times* (December 26):J4.

———. 1992. "As the Nation's Economy Collapses, Zairians Squirm Under Mobutu's Heel." *The New York Times* (August 30):4Y.

The Panos Institute. 1989. *AIDS and the Third World*. Philadelphia: New Society Publishers.

Parsons, Talcott. 1975. "The Sick Role and the Role of the Physician Reconsidered." *Milbank Memorial Fund Quarterly: Health and Society* 53(1):257–78.

Peretz, S. Michael. 1984. "Providing Drugs to the Third World: An Industry View." *Multinational Business* 84 (Spring):20–30.

Postman, Neil. 1985. *Amusing Ourselves to Death*. New York: Penguin.

Rock, Andrea. 1986. "Inside the Billion-Dollar Business of Blood." *Money* (March):153–72.

Rozenbaum, Willy. 1989. Quoted on p. 91 in *AIDS and Its Metaphors,* by Susan Sontag. New York: Farrar, Straus & Giroux.

Shilts, Randy. 1987. *And the Band Played On: Politics, People, and the AIDS Epidemic*. New York: St. Martin's Press.

Sontag, Susan. 1989. *AIDS and Its Metaphors*. New York: Farrar, Straus & Giroux.

Stolberg, Sheryl. 1992. "New AIDS Definition to Increase Tally." *Los Angeles Times* (December 31):A1+.

Swenson, Robert M. 1988. "Plagues, History, and AIDS." *The American Scholar* 57:183–200.

Thomas, William I. and Dorothy Swain Thomas. [1928] 1970. *The Child in America*. New York: Johnson Reprint.

The Times Atlas of World History. 1984. Maplewood, NJ: Hammond.

Treichler, Paula A. 1989. "Seduced and Terrorized: AIDS and Network Television." *ARTFORUM* 28:147–51.

Tuchman, Barbara W. 1981. *Practicing History: Selected Essays*. New York: Ballantine.

Turnbull, Colin M. 1961. *The Forest People*. New York: Simon & Schuster.

———. 1962. *The Lonely African*. New York: Simon & Schuster.

———. 1965. *Wayward Servants*. New York: Doubleday.

———. 1983. *The Human Cycle*. New York: Simon & Schuster.

Twain, Mark. 1973. "King Leopold's Soliloquy on the Belgian Congo." Pp. 41–60 in *Mark Twain and the Three R's,* edited by Maxwell Geismar. New York: International Publishers.

U.S. Bureau for Refugee Programs. 1988. *World Refugee Report*. Washington, DC: U.S. Government Printing Office.

U.S. Bureau of the Census. 1992a. Report FT947/91-A. *U.S. Exports and General Imports by Hormonized Commodity by Country*. Washington, DC: U.S. Government Printing Office.

———. 1992b. *Statistical Abstract of the United States*, 112th edition. Washington, DC: U.S. Government Printing Office.

U.S. Department of Health and Human Services. 1990. *HIV/AIDS Surveillance Report*. Washington, DC: U.S. Government Printing Office.

———. 1992. "AIDS Knowledge and the Attitudes for January–March 1991: Provisional Data from the National Health Interview Survey." No. 216. Washington, DC: U.S. Government Printing Office.

U.S. General Accounting Office (GAO). 1987. *AIDS: Information of Global Dimensions and Possible Impacts*. Washington, DC: U. S. Government Printing Office.

Watson, William. 1970. "Migrant Labor and Detribalization." Pp. 38–48 in *Black Africa: Its Peoples and Their Cultures Today,* edited by J. Middleton. London: Collier-Macmillan.

Whitaker, Jennifer Seymour. 1988. *How Can Africa Survive?* New York: Harper & Row.

World Health Organization (WHO). 1988. "A Global Response to AIDS." *Africa Report* (November/December):13–16.

Zuck, Thomas F. 1988. "Transfusion-Transmitted AIDS Reassessed." *New England Journal of Medicine* 318: 511–12.

Chapter 7

Aldrich, Howard E. and Peter V. Marsden. 1988. "Environments and Organizations." Pp. 361–92 in *Handbook of Sociology,* edited by N. J. Smelser. Newbury Park, CA: Sage.

Alvares, Claude. 1987. "A Walk Through Bhopal." Pp. 159–82 in *The Bhopal Syndrome,* by David Weir. San Francisco: Sierra Club Books.

Asimov, Isaac and Frederik Pohl. 1991. *Our Angry Earth: A Ticking Time Bomb*. New York: Doherty.

Barnet, Richard J. and Ronald E. Müller. 1974. *Global Reach: The Power of the Multinational Corporations*. New York: Simon & Schuster.

Blau, Peter M. 1974. *On the Nature of Organizations*. New York: Wiley.

Blau, Peter M. and Richard A. Schoenherr. 1973. *The Structure of Organizations*. White Plains, NY: Longman.

Bleiberg, Robert M. 1987. "Thought for Labor Day: In the Bhopal Disaster, There's Plenty of Blame to Go 'Round." *Barron's* 67 (September 7):9.

Castleman, Barry. 1986. Quoted in "The Dilemmas of Advanced Technology for the Third World," by Rashid A. Shaikh. *Technology Review* 89 (April):62.

Chenevière, Alain. 1987. *Vanishing Tribes*. Garden City, NY: Doubleday.

Clark, Andrew. 1993. "Learning the Rules of Global Citizenship: Transnationals Need to Meet Their Challenges, or Be Overwhelmed by Them." *The World Paper* (February):5.

Crossette, Barbara. 1989. "New Delhi Prepares Attempt to Control Pervasive Pollution." *The New York Times* (July 4):24Y.

———. 1991. "India's Descent." *The New York Times Magazine* (May 19): 26–67.

Derdak, Thomas. 1988. *International Directory of Company Histories,* Vol. 1. Chicago: St. James Press.

Diamond, Stuart. 1985a. "The Bhopal Disaster: How It Happened." *The New York Times* (January 28):A1+.

———. 1985b. "The Disaster in Bhopal: Workers Recall Horror." *The New York Times* (January 30):A1+.

———. 1985c. "Problems at Chemical Plants Raise Broad Safety Concerns." *The New York Times* (November 25):A1, D11.

Directory of American Firms Operating in Foreign Countries, 11th ed., Vol. 3. 1987. New York: World Trade Academy Press.

Ember, Lois R. 1985. "Technology in India: An Uneasy Balance of Progress and Tradition." *Chemical and Engineering News* 62: 61–65.

Engler, Robert. 1985. "Many Bhopals: Technology Out of Control." *The Nation* 240 (April 27):488–500.

Everest, Larry. 1986. *Behind the Poison Cloud: Union Carbide's Bhopal Massacre.* Chicago: Banner.

Faux, Jeff. 1990. "Labor in the New Global Economy: Mobile Capital, Multinationals, Working Standards." *Dissent* (Summer):376–82.

Fortune. 1989. "The Fortune 500: The Largest U.S. Industrial Corporations" (April 24):354.

———. 1992. "The World's Largest Industrial Corporations" (July 27):179.

Franklin, Ben A. 1985. "Toxic Cloud Leaks at Carbide Plant in West Virginia." *The New York Times* (August 12):1+.

Freund, Julien. 1968. *The Sociology of Max Weber.* New York: Random House.

Gross, Edward and Amitai Etzioni. 1985. *Organizations in Society.* Englewood Cliffs, NJ: Prentice-Hall.

Hazarika, Sanjoy. 1984. "Stores Reopen in Indian City as Poison Disposal Proceeds." *The New York Times* (December 19):A10.

———. 1993. "Settlement Slow in India Gas Disaster Claims." *The New York Times* (March 25):A6.

Holmes, Bob. 1992. "The Joy Ride Is Over: Farmers Are Discovering That Pesticides Increasingly Don't Kill Pests." *U.S. News & World Report* (September 14):72–73.

Hutchison, Robert A. 1989. "A Tree Hugger Stirs Villagers in India to Save Their Forests." *Smithsonian* (May): 185–96.

International Directory of Corporate Affiliations 1988/1989. 1988. Wilmette, IL: National Register.

Jasanoff, Sheila. 1988. "The Bhopal Disaster and the Right to Know." *Social Science and Medicine* 27(10):1113–123.

Keller, George M. 1986. "International Business and the National Interest." *Vital Speeches of the Day* (December 1):124–28.

Kennedy, Paul. 1993. *Preparing for the Twenty-First Century.* New York: Random House.

Khan, Rahat Nabi. 1986. "Multinational Companies and the World Economy: Economic and Technological Impact." *Impact of Science on Society* 36(141):15–25.

Kurzman, Dan. 1987. *A Killing Wind: Inside Union Carbide and the Bhopal Catastrophe.* New York: McGraw-Hill.

Kusy, Frank. 1989. *Cadogan Guides: India,* 2nd ed., edited by R. Fielding and P. Levey. Old Saybrook, CT: Globe Pequot Press.

Leisinger, Klaus M. 1988. "Multinationals and The Third World Sell Solutions, Not Just Products." *The New York Times* (February 21):K4.

Lengermann, Patricia M. 1974. *Definitions of Sociology: A Historical Approach.* Columbus, OH: Merrill.

Lepkowski, Wil. 1985. "Chemical Safety in Developing Countries: The Lessons of Bhopal." *Chemical and Engineering News* 63:9–14.

———. 1992. "Union Carbide–Bhopal Saga Continues as Criminal Proceedings Begin in India." *Chemical and Engineering News* 70(11):7–14.

Lydenberg, Steven D., Alice Tepper Marlin, Sean O'Brien Strub, and the Council on Economic Priorities. 1986. *Rating America's Corporate Conscience.* Reading, MA: Addison-Wesley.

McIntosh-Fletcher, W. Thomas. 1990. "When Two Cultures Meet." *Twin Plant News: The Magazine of the Maquiladora Industry* 5(April):32–33.

Michels, Robert. 1962. *Political Parties,* translated by E. Paul and C. Paul. New York: Dover.

Moskowitz, Milton. 1987. *The Global Marketplace.* New York: Macmillan.

National Public Radio /"Morning Edition." 1990. "Global Corporations." Tape/cassette #900820 (August 20).

The New York Times. 1985a. "Most at Plant Thought Poison Was Chiefly Skin-Eye Irritant" (January 1):A6.

———. 1985b. "Strong Odor Alerted Residents to Chemical Plant Problem" (August 12):A12.

———. 1989. "Rain Forest Worth More If Uncut, Study Says" (July 4):24Y.

———. 1992. "Compensation for Bhopal Set" (June 22):C5.

Nikore, Monika and Marian Leahy. 1993. *World Monitor* (February): 46–52.

Organization for Economic Cooperation and Development (OECD). 1987. *OECD Environmental Data: Compendium 1987.* Paris: Author.

Passell, Peter. 1990. "Rebel Economists Add Ecological Cost to Price of Progress." *The New York Times* (November 27):B5, B6.

Prewitt, Kenneth. 1983. "Scientific Illiteracy and Democratic Theory." *Daedalus* 112 (Spring):49–64.

Reich, Robert B. 1988. "Corporation and Nation." *The Atlantic Monthly* (May):76–81.

———. 1990. Quoted on p. 84 in "Calhoun County Goes Global," by Jan Bowermaster. *The New York Times Magazine* (December 2):58+.

Schumacher, E. F. 1985. Quoted in "Technology Out of Control," by Robert Engler. *Nation* 240 (April 27):490.

Sekulic, Dusko. 1978. "Approaches to the Study of Informal Organization." *Sociologija* 20(1):27–43.

Shabecoff, Philip. 1985. "Tangled Rules on Toxic Hazards Hamper Efforts to Protect Public." *The New York Times* (November 27):A1+.

———. 1988a. "Military Is Accused of Ignoring Rules on Hazardous Waste." *The New York Times* (June 14):C4.

———. 1988b. "A Guide to Some of the Scariest Things on Earth." *The New York Times* (December 25):E3.

———. 1989. "U.S. Only Narrowly Avoided 17 Bhopal-Like Disasters, Study Says." *The New York Times* (April 30):A16.

Shaikh, Rashid A. 1986. "The Dilemmas of Advanced Technology for the Third World." *Technology Review* 89:56–64.

Sharma, Ravi. 1987. "Assessing Development Costs in India." *Environment* 29(3):6–38.

Shrivastava, Paul. 1992. *Bhopal: Anatomy of a Crisis.* London: Chapman.

Snow, Charles P. 1961. *Science and Government.* Cambridge, MA: Harvard University Press.

Standke, Klaus-Heinrich. 1986. "Technology Assessment: An Essentially Political Process." *Impact of Science on Society* 36(141):65–76.

Stevens, William K. 1984. "Workers at Site of Leak Described as Unskilled." *The New York Times* (December 6): A10.

Union Carbide Annual Report. 1984. "After Bhopal." Danbury, CT: Union Carbide.

U.S. Department of Labor. 1992. *Valuing Cultural Diversity* (self-instructional package). Washington, DC: U.S. Government Printing Office.

U.S. General Accounting Office. 1978. *U.S. Foreign Relations and Multinational Corporations: What's the Connection?* Washington, DC: U.S. Government Printing Office.

Veblen, Thorstein. 1933. *The Engineers and the Price System.* New York: Viking.

Wald, Matthew L. 1990. "Where All That Gas Goes: Drivers' Thirst for Power." *The New York Times* (November 21):A1, C17.

Weber, Max. 1947. *The Theory of Social and Economic Organization,* edited and translated by A. M. Henderson and T. Parsons. New York: Macmillan.

Weir, David. 1987. *The Bhopal Syndrome: Pesticides, Environment, and Health.* San Francisco: Sierra Club Books.

Wexler, Mark N. 1989. "Learning from Bhopal." *The Midwest Quarterly* 31(1):106–29.

Young, T. R. 1975. "Karl Marx and Alienation: The Contributions of Karl Marx to Social Psychology." *Humboldt Journal of Social Relations* 2(2):26–33.

Zuboff, Shoshana. 1988. *In the Age of the Smart Machine: The Future of Work and Power.* New York: Basic Books.

FOCUS Chapter 7

Boly, William. 1990. "Downwind from the Cold War." *In Health* (July/August):58–69.

Nevin, David. 1988. "Millennial Dreams." *New Scientist* 118 (June): 31–32.

Newsweek. 1961. "Survival: Are Shelters the Answer?" (November 6): 19–23.

Norris, Robert S., Thomas B. Cochran, and William M. Arkin. 1985. "History of the Nuclear Stockpile." *Bulletin of the Atomic Scientist* 41 (August): 104–108.

Time. 1961a. "The Minutemen" (November 3):18–19.

———. 1961b. "Gun Thy Neighbor" (August 18):58.

Chapter 8

Author X. 1992. "Mao Fever—Why Now?" translated and adapted from the Chinese by R. Terrill. *World Monitor* (December): 22–25.

Becker, Howard S. 1963. *Outsiders: Studies in the Sociology of Deviance.* New York: Free Press.

———. 1973. "Labelling Theory Reconsidered" in *Outsiders: Studies in the Sociology of Deviance.* New York: Free Press.

Belkin, Lisa. 1990. "Airport Anti-Drug Nets Snare Many People Fitting 'Profiles.'" *The New York Times* (March 20):A1, A11.

Bem, Sandra. 1990. Quoted on pp. 83–84 in "Blurring the Line: Androgyny on Trial," by Don Monkerud. *Omni* (October):81–86+.

Bernstein, Richard. 1982. *From the Center of the Earth: The Search for the Truth About China.* Boston: Little, Brown.

Bernstein, Thomas P. 1983. "Starving to Death in China." *The New York Review of Books* (June 16):36–38.

Best, Joel. 1989. *Images of Issues: Typifying Contemporary Social Problems.* New York: Aldine de Gruyter.

Bonavia, David. 1989. *The Chinese.* London: Penguin.

Bracey, Dorothy H. 1985. "The System of Justice and the Concept of Human Nature in the People's Republic of China." *Justice Quarterly* 2(1):139–44.

Broadfoot, Robert. 1993. Quoted in "Ancient Power Steps into Asian Spotlight," by D. Holley. (A special Pacific Rim edition of World Report) *Los Angeles Times* (June 15):H15.

Burns, John F. 1986. "When Peking Fights Crime, News Is on the Wall." *The New York Times* (January 25):Y3.

Butterfield, Fox. 1976. "Mao Tse-Tung: Father of Chinese Revolution." *The New York Times* (September 10):A13+.

———. 1980. "The Pragmatists Take China's Helm." *The New York Times Magazine* (December 28):22–35.

———. 1982. *China: Alive in the Bitter Sea.* New York: Times Books.

Calhoun, Craig. 1989. "Revolution and Repression in Tiananmen Square." *Society.* (September/October):21–38.

Chambliss, William. 1974. "The State, the Law, and the Definition of Behavior as Criminal or Delinquent." Pp. 7–44 in *Handbook of Criminology,* edited by D. Glaser. Chicago: Rand McNally.

Chang Jung. 1991. *Wild Swans: Three Daughters of China.* New York: Simon and Schuster.

———. 1992. Quoted in "Literature of the Wounded," by Jonathan Mirsky. *The New York Review of Books* (March 5): 6.

Chiu Hungdah. 1988. "China's Changing Criminal Justice System." *Current History* (September):265–72.

Clark, John P. and Shirley M. Clark. 1985. "Crime in China—As We Saw It." *Justice Quarterly* 2(1):103–10.

Collins, Randall. 1982. *Sociological Insight: An Introduction to Nonobvious Sociology.* New York: Oxford University Press.

Cooley, Charles Horton. [1902] 1964. *Human Nature and the Social Order.* New York: Schocken.

Cowell, Alan. 1992. "Strike Hits Tobacco Industry, and 13 Million Italians Suffer." *The New York Times* (November 19):A5.

Durkheim, Emile. [1901] 1982. *The Rules of Sociological Method and Selected Texts on Sociology and Its Method,* edited by S. Lukes and translated by W. D. Halls. New York: Free Press.

Erikson, Kai T. 1966. *Wayward Puritans.* New York: Wiley.

Fairbank, John King. 1987. *The Great Chinese Revolution 1800–1985.* New York: Harper & Row.

———. 1989. "Why China's Rulers Fear Democracy." *The New York Review of Books* (September 28):32–33.

Fang Lizhi. 1990. "The Chinese Amnesia." *The New York Review of Books* (September 27):30–31.

Feng Jicai. 1991. *Voices from the Whirlwind: An Oral History of the Chinese Cultural Revolution.* New York: Pantheon.

Fox, James Alan and Jack Levin. 1990. "Inside the Mind of Charles Stuart." *Boston Magazine* (April):66–70.

Gabriel, Trip. 1988. "China Strains for Olympic Glory." *The New York Times Magazine* (April 24):30–35.

Goldman, Merle. 1989. "Vengeance in China." *The New York Review of Books* (November 9):5–9.

Gorriti, Gustavo A. 1989. "How to Fight the Drug War." *The Atlantic Monthly* (July):70–76.

Gould, Stephen Jay. 1990. "Taxonomy as Politics: The Harm of False Classification." *Dissent* (Winter):73–78.

Han Xu. 1989. "The Chinese Ambassador's Version. . . ." *The New York Times* (August 21):Y19.

Hareven, Tamara K. 1987. "Divorce, Chinese Style." *The Atlantic Monthly* (April):70–76.

Haub, Carl and Machiko Yanagishita. 1994. *1994 World Population Data Sheet.* Washington, DC: Population Reference Bureau.

Henriques, Diana B. 1993. "Great Men and Tiny Bubbles: For God, Country and Coca-Cola." *The New York Times Book Review* (May 23):13.

Holley, David. 1993. "Ancient Power Steps into Asian Spotlight." (A special Pacific Rim edition of World Report) *Los Angeles Times* (June 15): H1+.

Holley, David and Christine Courtney. 1993. "View from Hong Kong: Colony Looks to South China After '97." (A special Pacific Rim edition of World Report) *Los Angeles Times* (June 15): H12.

Ignatius, Adi. 1988. "China's Birthrate Is Out of Control Again as One-Child Policy Fails in Rural Areas." *Asian Wall Street Journal Weekly* (July 18): 18.

Jacobs, James B. 1983. "Smashed." *The New York Review of Books* (April 14):36–38.

Jao, Y. C. and C. K. Leung. 1986. *China's Special Economic Zones: Policies, Problems and Prospects.* Hong Kong: Oxford University Press.

Kitsuse, John I. 1962. "Societal Reaction to Deviant Behavior: Problems of Theory and Method." *Social Problems* 9 (Winter):247–56.

Kometani, Foumiko. 1987. "Pictures from Their Nightmare." *The New York Times Book Review* (July 19):9–10.

Kristof, Nicholas D. 1989a. "China Erupts." *The New York Times Magazine* (June 4):27+.

———. 1989b. "China Is Planning 2 Years of Labor for Its Graduates." *The New York Times* (August 13):Y1.

———. 1992. "Chinese Police Halt Tiananmen Square Memorial." *The New York Times* (June 4):A5.

Kwong, Julia. 1988. "The 1986 Student Demonstrations in China." *Asian Survey* 28(9):970–85.

Lemert, Edwin M. 1951. *Social Pathology.* New York: McGraw-Hill.

Leys, Simon. 1989. "After the Massacres." *The New York Review of Books* (October 12):17–19.

———. 1990. "The Art of Interpreting Nonexistent Inscriptions Written in Invisible Ink on a Blank Page." *The New York Review of Books* (October 11): 8–13.

Lilly, J. Robert. 1991. "Prisons: How to Help Eastern Europe." *The Angolite* 16(1):1.

Link, Perry. 1989. "The Chinese Intellectuals and the Revolt." *The New York Review of Books* (June 29):38–41.

Liu Binyan. 1993. "An Unnatural Disaster," translated by P. Link. *The New York Review of Books* (April 8):3–6.

Liu Zaifu. 1989. Quoted in "The Chinese Intellectuals and the Revolt," by Perry Link. *The New York Review of Books* (June 29):40.

Lubman, Stanley. 1983. "Comparative Criminal Law and Enforcement: China." Pp. 182–93 in *Encyclopedia of Crime and Justice*, edited by S. H. Kadish. New York: Free Press.

Mao Tse-tung (Mao Zedong). 1965. "Report on an Investigation of the Peasant Movement in Hunan (March 1927)." In *Selected Works of Mao Tse-tung*. Peking: Foreign Language Press.

Mathews, Jay and Linda Mathews. 1983. *One Billion: A China Chronicle.* New York: Random House.

Merton, Robert K. 1957. *Social Theory and Social Structure.* Glencoe, IL: Free Press.

Milgram, Stanley. 1974. *Obedience to Authority: An Experimental View.* New York: Harper & Row.

———. 1987. "Obedience." Pp. 566–68 in *The Oxford Companion to the Mind*, edited by R. L. Gregory. Oxford: Oxford University Press.

Monkerud, Don. 1990. "Blurring the Lines: Androgyny on Trial." *Omni* (October):81–86+.

Montalbano, William D. 1993. "Lifetime of Change in a Dozen Years." (A special Pacific Rim edition of World Report) *Los Angeles Times* (June 15):H2.

Mosher, Steven W. 1991. "Chinese Prison Labor." *Society* (November/December): 49–59.

Nadelmann, Ethan. 1989. Cited in "How to Fight the Drug War," by G. A. Goriti. *The Atlantic Monthly* (July):71.

National Council for Crime Prevention in Sweden. 1985. *Crime and Criminal Policy in Sweden* (Report No. 19). Stockholm, Sweden: Liber Distribution.

Oxman, Robert. 1993a. "Focus, Olympic Hurdle." Interview on "MacNeil/Lehrer Newshour," September 21 (transcript #4759). WNET.

———. 1993b. "China in Transition." Interview on "MacNeil/Lehrer Newshour," December 27 (transcript #4828). WNET.

———. 1993c. "China in Transition: Mao to Markets." Interview on "MacNeil/Lehrer Newshour," December 28 (transcript #4829). WNET.

———. 1993d. "China in Transition: Status Report (Chinese Women)." Interview on "MacNeil/Lehrer Newshour," December 29 (transcript #4830). WNET.

———. 1994a. "China in Transition: Taking the Plunge (Education in China; Going into Business in China)." Interview on "MacNeil/Lehrer Newshour," January 4 (transcript #4834). WNET.

———. 1994b. "China in Transition: Up on the Farm." Interview on "MacNeil/Lehrer Newshour," January 6 (transcript #4836). WNET.

———. 1994c. "China in Transition: Any Progress? (Human Rights)." Interview on MacNeil/Lehrer Newshour," January 31 (transcript #4853). WNET.

Pannell, Clifton W. 1987. Review of *China's Special Economic Zones: Policies, Problems, and Prospects*, edited by Y. C. Jao and C. K. Leung. *Economic Geography* 63(3):277–78.

Personal correspondence, 1993. Comments by an anonymous reviewer.

Ramos, Francisco Martins. 1993. "My American Glasses." Pp. 1–10 in *Distant Mirrors: America as a Foreign Culture*, by Philip R. DeVita and James D. Armstrong. Belmont, CA: Wadsworth.

Reinarman, Craig and Harry G. Levine. 1989. "The Crack Attack: Politics and Media in America's Latest Drug Scare." Pp. 115–38 in *Images of Issues: Typifying Contemporary Social Problems*, edited by J. Best. New York: Aldine de Gruyter.

Rojek, Dean G. 1985. "The Criminal Process in the People's Republic of China." *Justice Quarterly* 2(1):117–25.

Rorty, Amelie Oksenberg. 1982. "Western Philosophy in China." *Yale Review* 72(1):141–60.

Ross, Madelyn C. 1984. "China's New and Old Investment Zones." *China Business Review* (November/December):14–18.

Shipp, E. R., Dean Baquet, and Martin Gottlieb. 1991. "Slaying Casts a New Glare on Law's Uncertain Path." *The New York Times* (June 23):A1+.

Simmons, J. L., with Hazel Chambers. 1965. "Public Stereotypes of Deviants." *Social Problems* 3(2):223–32.

Spector, Malcolm and J. I. Kitsuse. 1977. *Constructing Social Problems.* Menlo Park, CA: Cummings.

Stepanek, James B. 1982. "China's SEZs." *China Business Review* (March/April):38–39.

Strebeigh, Fred. 1989. "Training China's New Elite." *The Atlantic Monthly* (April):72–80.

Sumner, William Graham. 1907. *Folkways.* Boston: Ginn.

Sutherland, Edwin H. and Donald R. Cressey. 1978. *Principles of Criminology*, 10th ed. Philadelphia: Lippincott.

Tannenbaum, Frank. 1938. *Crime and the Community.* New York: Ginn.

Theroux, Paul. 1989. "Travel Writing: Why I Bother." *The New York Times Book Review* (July 30):7–8.

Tien H. Yuan. 1990. "Demographer's Page: China's Population Planning After Tiananmen." *Population Today* 18(9):6–8.

Tien H. Yuan, Zhang Tianlu, Ping Yu, Li Jingneng, and Liang Zhongtang. 1992. "China's Demographic Dilemmas." *Population Bulletin* 47(1):1–44.

Tobin, Joseph J., David Y. H. Wu, and Dana H. Davidson. 1989. *Preschool in Three Cultures: Japan, China and the United States.* New Haven, CT: Yale University Press.

U.S. Bureau of Justice Statistics. 1992. *Criminal Victimization in the United States, 1991.* Washington, DC: U.S. Government Printing Office.

Wang Ruowang. 1989. Quoted in "The Chinese Intellectuals and the Revolt," by Perry Link. *The New York Review of Books* (June 29):40.

Williams, Terry. 1989. *The Cocaine Kids: The Inside Story of a Teenage Drug Ring.* Reading, MA: Addison-Wesley.

Wu Han. 1981. Quoted on p. 39 in *Coming Alive: China After Mao*, by Roger Garside. New York: McGraw-Hill.

Wu Yimei. 1988. Quoted on p. 31 in "China Strains for Olympic Glory," by Trip Gabriel. *The New York Times Magazine* (April 24):31–32.

WuDunn, Sheryl. 1993. "Booming China Is a Dream Market for West." *The New York Times* (February 15):A1+.

Chapter 9

Angelou, Maya. 1987. "Intra-Racism." Interview on the "Oprah Winfrey Show" (Journal Graphics transcript #W172):2.

Berger, Joseph. 1992. "South African Students Learn to Learn Together." *The New York Times* (May 12):A4.

Berger, Peter L. and Brigitte Berger. 1983. *Sociology: A Biological Approach*. New York: Penguin Books.

Berreman, Gerald D. 1972. "Race, Caste, and Other Invidious Distinctions in Social Stratification." *Race* 13(4): 385–414.

Boudon, Raymond and François Bourricaud. 1989. *A Critical Dictionary of Sociology*, selected and translated by P. Hamilton. Chicago: University of Chicago Press.

Bratton, Michael and Nicolas van de Walle. 1993. "Neopatrimonial Regimes and Political Transactions in Africa." Paper presented at the annual meeting of the American Sociological Association, Miami, FL.

Chapkis, Wendy. 1986. *Beauty Secrets: Women and the Politics of Appearance*. Boston: South End Press.

Chass, Murray. 1992. "A Zillionaire at the Bat." *International Herald Tribune* (February 28):16.

———. 1993. "25 Men on a Team and 7 Figures Per Man." *The New York Times* (April 11):4S.

Coser, Lewis A. 1977. *Masters of Sociological Thought*, 2nd ed., edited by R. K. Merton. New York: Harcourt Brace Jovanovich.

Coser, Lewis A. and Bernard Rosenberg. 1976. *Sociological Theory: A Book of Readings*, 4th ed. New York: Macmillan.

Crapanzano, Vincent. 1985. *Waiting: The Whites of South Africa*. New York: Random House.

Davis, Kingsley and Wilbert E. Moore. 1945. "Some Principles of Stratification." Pp. 413–25 in *Sociological Theory: A Book of Readings*, edited by L. A. Coser and B. Rosenberg. New York: Macmillan.

Eiseley, Loren. 1990. "Man: Prejudice and Personal Choice." Pp. 640–943 in *The Random House Encyclopedia*, 3rd ed. New York: Random House.

Finnegan, William. 1986. *Crossing the Line: A Year in the Land of Apartheid*. New York: Harper & Row.

Franklin, John Hope. 1990. Quoted in "That's History, Not Black History," by Mark Mcgurl. *The New York Times Book Review* (June 3):13.

"The Freedom Charter." 1990. *One Nation, One Country*, by Nelson Mandela. The Phelps-Stokes Fund 4 (May):47–51.

"FRONTLINE."1985. "A Class Divided" (transcript #309). Boston: WGBH Educational Foundation.

Gerth, Hans and C. Wright Mills. 1954. *Character and Social Structure: The Psychology of Social Institutions*. London: Routledge & Kegan Paul.

Hacker, Andrew. 1991. "Class Dismissed." *The New York Review of Books* (March 7):44–46.

Jencks, Christopher. 1990. Quoted in "The Rise of the 'HyperPoor,'" by David Whitman. *U.S. News & World Report* (October 15):40–42.

Johnson, Allan G. 1989. *Human Arrangements*, 2nd ed., edited by R. K. Merton. New York: Harcourt Brace Jovanovich.

Jones, Arthur. 1994. "Reality Check." *World Trade* (February):34–38.

Keller, Bill. 1994. "Mandela's Party Publishes Plan to Redistribute Wealth." *The New York Times* (January 15):Y3.

Lamb, David. 1987. *The Africans*. New York: Vintage.

Lambert, Father Rollins. 1988. "A Day in the Life of Apartheid: The Editors Interview Father Rollins Lambert." *U.S. Catholic* 53:26–32.

Lee, Spike. 1989. Quoted in "He's Got to Have It His Way," by Jeanne McDowell. *Time* (July 17):92, 94.

Lelyveld, Joseph. 1985. *Move Your Shadow: South Africa, Black and White*. New York: Penguin Books.

Liebenow, J. Gus. 1986. "South Africa: Home, 'Not-So-Sweet,' Homelands." *UFSI Reports* No. 23.

Loy, John W. and Joseph F. Elvogue. 1971. "Racial Segregation in American Sport." *International Review of Sport Sociology* 5:5–24.

Mabuza, Lindiwe. 1990. "Apartheid: Far from Over." *The New York Times* (June 20):A15.

Mallaby, Sebastian. 1992. *After Apartheid: The Future of South Africa*. New York: Times Books.

Mandela, Nelson. 1990. "I Am the First Accused" (Rivonia Trial Statement 1964). *One Nation, One Country*. The Phelps-Stokes Fund 4 (May):17–45.

———. 1991. Quoted in "The World: Apartheid's Laws Are Dismantled, but Not Its Cages," by Christopher S. Wren. *The New York Times* (February 10):4E.

———. 1994. "Inaugural Address." *Los Angeles Times* (May 11):A6.

Marx, Karl. 1909. *Capital: A Critique of Political Economy* Vol. III, edited by F. Engles, translated by E. Untermann. Chicago: Kerr.

———. [1895] 1976. *The Class Struggles in France 1848–1850*. New York: International Publishers.

Medoff, Marshall H. 1977. "Positional Segregation and Professional Baseball."

International Review of Sport Sociology 12:49–56.

Merton, Robert K. 1958. *Social Theory and Social Structure*. New York: Free Press.

Mukherjee, Bharati. 1990. Quoted on pp. 3–10 in *Bill Moyers, A World of Ideas II: Public Opinion from Private Citizens*, edited by A. Tucher. New York: Doubleday.

Myrdal, Gunnar. 1944. *An American Dilemma: The Negro Problem and Modern Democracy*. New York: Harper & Brothers.

Nasar, Sylvia. 1992. "Those Born Wealthy or Poor Usually Stay So, Studies Say." *The New York Times* (May 18):A1, C7.

O'Hare, William P. 1992. "America's Minorities—The Demographics of Diversity." *Population Today* 47(4):1–47.

O'Hare, William P. and Brenda Curry-White. 1992. "Demographer's Page: Is There a Rural Underclass?" *Population Today* 20(3):6–8.

Ohene, Elizabeth. 1993. "Tales from South Africa: A Beauty for Our Times." *BBC Focus on Africa* Magazine 4(4):23–25.

Poussaint, Alvin. 1987. "Intra-Racism." Interview on "Oprah Winfrey Show" (Journal Graphics transcript #W172):7.

Ridgeway, Cecilia. 1991. "The Social Construction of Status Value: Gender and Other Nominal Characteristics." *Social Forces* 70(2):367–86.

Ross, Edward Alsworth. [1908] 1929. *Social Psychology: An Outline and Source Book*. New York: Macmillan.

Russell, Diana E. H. 1989. *Lives of Courage: Women for a New South Africa*. New York: Basic Books.

Seidman, Judy. 1978. *Ba Ye Zwa: The People Live*. Boston: South End Press.

Simpson, Richard L. 1956. "A Modification of the Functional Theory of Social Stratification." *Social Forces* 35:132–37.

Sparks, Allister. 1990. *The Mind of South Africa*. New York: Knopf.

Thomas, Isaiah. 1987. Quoted in "The Coloring of Bird," by Ira Berkow. *The New York Times* (June 2):D27.

Tumin, Melvin M. 1953. "Some Principles of Stratification: A Critical Analysis." *American Sociological Review* 18:387–94.

U.S. Bureau of the Census. 1990. *Statistical Abstract of the United States: 1990*. Washington, DC: U.S. Government Printing Office.

———. 1992a. *Statistical Abstract of the United States: 1992*. Washington, DC: U.S. Government Printing Office.

———. 1992b. *Money Income of Households, Families, and Persons in the United States: 1991*. Washington DC: U.S. Government Printing Office.

Wacquant, Loic J. D. 1989. "The Ghetto, the State, and the New Capitalist Economy." *Dissent* (Fall):508–20.

Weber, Max. 1982. "Status Groups and Classes." Pp. 69–73 in *Classes, Power, and Conflict: Classical and Contemporary Debates,* edited by A. Giddens and D. Held. Los Angeles: University of California.

———. [1947] 1985. "Social Stratification and Class Structure." Pp. 573–76 in *Theories of Society: Foundations of Modern Sociological Theory,* edited by T. Parsons, E. Shils, K. D. Naegele, and J. R. Pitts. New York: Free Press.

Weekend World: Johannesburg. 1977. "Lesson 8: Basic Economics at People's College." (April 24).

Wilson, Francis and Mamphela Ramphele. 1989. *Uprooting Poverty: The South African Challenge.* New York: Norton.

Wilson, William Julius. 1983. "The Urban Underclass: Inner-City Dislocations." *Society* 21:80–86.

———. 1987. *The Truly Disadvantaged: The Inner City, the Underclass, and Public Policy.* Chicago: University of Chicago Press.

Wirth, Louis. [1945] 1985. "The Problem of Minority Groups." Pp. 309–15 in *Theories of Society: Foundations of Modern Sociological Theory,* edited by T. Parsons, E. Shils, K. D. Naegele, and J. R. Pitts. New York: Free Press.

The World Almanac and Book of Facts 1991. 1990. New York: Pharos.

The World Almanac and Book of Facts 1994. 1993. Mahwah, NJ: World Almanac.

World Development Report. 1990. "Poverty." New York: Oxford University Press.

Wren, Christopher S. 1990a. "De Klerk Hopes to Show Bush the Change Is Real." *The New York Times* (September 23):8Y.

———. 1990b. "A South Africa Color Bar Falls Quietly." *The New York Times* (October 16):A3.

———. 1991. "South Africans Desegregate Some White Public Schools." *The New York Times* (January 10):A1.

Yeutter, Clayton. 1992. "When 'Fairness' Isn't Fair." *The New York Times* (March 24):A13.

Chapter 10

Adams, Anne V. 1992. "Translator's Afterword." Pp. 234–37 in *Showing Our Colors: Afro-German Women Speak Out,* edited by M. Opitz, K. Oguntoye, and D. Schultz. Amherst: University of Massachusetts Press.

Adomako, Abena. 1992. "Mother: Afro-German Father: Ghanian." Pp. 199–203 in *Showing Our Colors: Afro-German Women Speak Out,* edited by M.

Opitz, K. Oguntoye, and D. Schultz. Amherst: University of Massachusetts Press.

Alba, Richard D. 1992. "Ethnicity." Pp. 575–84 in *Encyclopedia of Sociology* Vol. 2, edited by E. F. Borgatta and M. L. Borgatta. New York: Macmillan.

Anson, Robert Sam. 1987. *Best Intentions: The Education and Killing of Edmund Perry.* New York: Random House.

Ashe, Arthur and Arnold Rampersad. 1993. *Days of Grace: A Memoir.* New York: Knopf.

Atkins, Elizabeth. 1991. "For Many Mixed-Race Americans, Life Isn't Simply Black or White." *The New York Times* (June 5):B8.

Barrins, Adeline. 1992. Quoted in "The Tallest Fence: Feelings on Race in a White Neighborhood." *The New York Times* (June 21):12Y.

Böhning, W. R. 1992. "Integration and Immigration Pressures in Western Europe." *International Labour Review* 130(4):445–58.

Breton, Raymond, Wsevolod W. Isajiw, Warren E. Kalbach, and Jeffrey G. Reitz. 1990. *Ethnic Identity and Equality: Varieties of Experience in a Canadian City.* Toronto: University of Toronto.

Buckley, Jerry. 1991. "Mt. Airy, Philadelphia." *U.S. News and World Report* (July 22):22–28.

Bustamante, Jorge A. 1993. "Mexico-Bashing: A Case Where Words Can Hurt." *Los Angeles Times* (August 13):B7.

Carver, Terrell. 1987. *A Marx Dictionary.* Totowa, NJ: Barnes and Noble.

Castles, Stephen. 1985. "The Guests Who Stayed—The Debate on 'Foreigners Policy' in the German Federal Republic." *International Migration Review* 19(3):517–34.

———. 1986. "The Guest-Worker in Western Europe—An Obituary." *International Migration Review* 20(4):761–78.

Castles, Stephen and Godula Kosack. 1985. *Immigrant Workers and Class Structure in Western Europe,* 2nd ed. New York: Oxford University.

Cornell, Stephen. 1990. "Land, Labour and Group Formation: Blacks and Indians in the United States." *Ethnic and Racial Studies* 13(3):368–88.

Crapanzano, Vincent. 1985. *Waiting: The Whites of South Africa.* New York: Random House.

Darnton, John. 1993. "Western Europe Is Ending Its Welcome to Immigrants." *The New York Times* (August 10):A1+.

Davis, F. James. 1978. *Minority–Dominant Relations: A Sociological Analysis.* Arlington Heights, IL: AHM.

De Witt, Karen. 1993. "Conversations/ Elaine R. Jones: In a Color-Conscious Society, She Challenges the 'Color Blind.'" *The New York Times* (July 18):E9.

Dunne, John Gregory. 1991. "Law and Disorder in Los Angeles." *New York Review of Books* (October 10):26.

Emde, Helga. 1992. "An 'Occupation Baby' in Postwar Germany." Pp. 101–11 in *Showing Our Colors: Afro-German Women Speak Out,* edited by M. Opitz, K. Oguntoye, and D. Schultz. Amherst: University of Massachusetts Press.

Erlanger, Steven. 1993. "Germany Pays to Keep Ethnic Germans in Russia." *The New York Times* (May 9):1A+.

Feen, Richard. 1993. "The Never Ending Story: The Haitian Boat People." *Migration World* 21(1):13–15.

Fein, Helen. 1978. "A Formula for Genocide: Comparison of the Turkish Genocide (1915) and the German Holocaust (1939–1945)." *Comparative Studies in Sociology* 1:271–94.

Frontline. 1988. "Racism 101." Boston: WGBH Educational Foundation.

General Assembly of the State of North Carolina. 1831. "An Act to Prevent All Persons from Teaching Slaves to Read or Write, the Use of Figures Excepted." Raleigh, NC.

German Information Center. 1993. "Focus On . . . German Citizenship and Naturalization." New York.

Goffman, Erving. 1963. *Stigma: Notes on the Management of Spoiled Identity.* Englewood Cliffs, NJ: Prentice-Hall.

Gordon, Milton M. 1978. *Human Nature, Class, and Ethnicity.* New York: Oxford University Press.

Gould, Stephen Jay. 1981a. "The Politics of Census." *Natural History* 90(1):20–24.

———. 1981b. *The Mismeasure of Man.* New York: Norton.

Green, Peter S. 1992. "The Nomads of Eastern Europe." *U.S. News and World Report* (October 26):31.

Hacker, Andrew. 1992. *Two Nations: Black and White, Separate, Hostile, Unequal.* New York: Scribner.

Halberstam, David. 1986. *The Reckoning.* New York: Morrow.

Heilig, Gerhard, Thomas Büttner, and Wolfgang Lutz. 1990. "Germany's Population: Turbulent Past, Uncertain Future." *Population Bulletin* 45(4):1–46.

Hoffmann, A. 1992. *Facts About Germany,* translated by G. Finan. Braunschweig: Westermann.

Holmes, Steven A. 1994. "Behind a Dark Mirror: Traditional Victims Give Vent to Racism." *The New York Times* (February 15):4E.

Holzner, Lutz. 1982. "The Myth of Turkish Ghettos: A Geographic Case Study of West German Responses Towards a Foreign Minority." *Journal of Ethnic Studies* 9(4):65–85.

Hopkins, Mary-Carol. 1993. "Prospectus" for *Don't Give Me Your Tired, Your Poor: A Cambodian (Khmer) Community in an American City*. Department of Sociology, Anthropology, and Philosophy, Northern Kentucky University.

Horton, Paul B., Gerald R. Leslie and Richard F. Larson. 1988. *The Sociology of Social Problems*, 9th ed. Englewood Cliffs, NJ: Prentice-Hall.

Houston, Velin Hasu. 1991. "The Past Meets the Future: A Cultural Essay." *Amerasia Journal* 17(1):53–56.

Jones, Tamara and Hugh Pope. 1993. "Kurds Raid Turk Offices in Europe." *The New York Times* (June 25):A1+.

Keely, Charles B. 1993. "The Politics of Migration Policy in the United States." *Migration World* 21(1):20–23.

King, Lloyd. 1992. "Lloyd King." Pp. 397–401 in *Race: How Blacks and Whites Think and Feel About the American Obsession*, edited by S. Terkel. New York: New Press.

Kinzer, Stephen. 1992. "Vietnamese, Easy Target, Fear Ouster by Germany." *The New York Times* (December 6):Y3.

———. 1993. "Right Groups Attack Plan in Germany to Limit Asylum." *The New York Times* (February 7):84.

Kramer, Jane. 1993. "Letter from Europe: Neo-Nazis: A Chaos in the Head." *The New Yorker* (June 14): 52–70.

Lacayo, Richard. 1991. "Give Me Your Rich, Your Lucky. . . ." *Time* (October 14):26–27.

Lee, Patrick. 1993. "Studies Challenge View That Immigrants Harm Economy." *Los Angeles Times* (August 13): A1+.

Lee, Spike. 1989. Quoted in "He's Got to Have It His Way," by Jeanne McDowell. *Time* (July 17): 92–94.

Levin, Jack and Jack McDevitt. 1993. *Hate Crimes: The Rising Tide of Bigotry and Bloodshed*. New York: Plenum.

Lieberman, Leonard. 1968. "The Debate Over Race: A Study in the Sociology of Knowledge." *Phylon* 39 (Summer):127–41.

Light, Ivan. 1990. Quoted in "As Bias Crime Seems to Rise, Scientists Study Roots of Racism," by Daniel Goleman. *The New York Times* (May 29):B5+.

Los Angeles Times. 1992. "Probe Finds Pattern of Excessive Force, Brutality by Deputies" (July 21):A18.

Mandel, Ruth. 1989. "Turkish Headscarves and the 'Foreigner Problem': Constructing Difference Through Em-

blems of Identity." *New German Critique* 46(Winter):27–46.

Marshall, Tyler. 1992a. "German Opposition Party Backs Immigration Limits." *Los Angeles Times* (November 17):A1+.

———. 1992b. "Saying 'No' to Nazis in Germany." *Los Angeles Times* (December 5):A1, 13.

———. 1993. "Arson Attacks on Foreigners No Longer Big News in Germany." *Los Angeles Times* (July 17): A8.

Martin, Philip L. and Mark J. Miller. 1990. "Guests or Immigrants?: Contradiction and Change in the German Immigration Policy Debate Since the Recruitment Stop." *Migration World* 15(1):8–13.

McClain, Leanita. 1986. *A Foot in Each World: Essays and Articles*, edited by C. Page. Evanston, IL: Northwestern University Press.

McDowell, Jeanne. 1989. "He's Got to Have It His Way." *Time* (July 17): 92–94.

McIntosh, Peggy. 1992. "White Privilege and Male Privilege: A Personal Account of Coming to See Correspondences Through Work in Women's Studies." Pp. 70–81 in *Race, Class, and Gender: An Anthology*, edited by M. L. Andersen and P. H. Collins. Belmont, CA: Wadsworth.

Meddis, Sam V. 1993. "Stereotypes Fuel Cycle of Suspicion, Arrest." *USA Today* (July 23):6A.

Merton, Robert K. 1957. *Social Theory and Social Structure*. New York: Free Press.

———. 1976. "Discrimination and the American Creed." Pp. 189–216 in *Sociological Ambivalence and Other Essays*. New York: Free Press.

Miller, Alan C. and Ronald J. Ostrow. 1993. "Immigration Policy Failures Invite Overhaul." *Los Angeles Times* (July 11):A1+.

Mydans, Seth. 1991. "40,000 Aliens to Win Legal Status in Lottery." *The New York Times* (September 25):A1+.

National Public Radio/"All Things Considered." 1990. "Prejudice Puzzle" (September 13).

National Public Radio/"Morning Edition." 1993. "Immigration."

The New York Times. 1990. "Advertising: New Group Makes the Case for Black Agencies and Media" (October 26):C18.

"Nightline." 1987. "Prejudice in Baseball." *Journal Graphics* (transcript #1532), April 8.

O'Connor, Peggy. 1992. Quoted in "The Tallest Fence: Feelings on Race in a White Neighborhood." *The New York Times* (June 21):12Y.

Ogbu, John U. 1990. "Minority Status and Literacy in Comparative Perspective." *Daedalus* 119(2): 141–68.

Opitz, May. 1992a. "Racism Here and Now." Pp. 125–44 in *Showing Our Colors: Afro-German Women Speak Out*, edited by M. Opitz, K. Oguntoye, and D. Schultz. Amherst: University of Massachusetts Press.

———. 1992b. "Three Afro-German Women in Conversation with Dagmar Schultz: The First Exchange for This Book." Pp. 145–64 in *Showing Our Colors: Afro-German Women Speak Out*, edited by M. Opitz, K. Oguntoye, and D. Schultz. Amherst: University of Massachusetts Press.

———. 1992c. "In Search of My Father (From a Conversation with Ellen Wiedenroth." Pp. 172–77 in *Showing Our Colors: Afro-German Women Speak Out*, edited by M. Opitz, K. Oguntoye, and D. Schultz. Amherst: University of Massachusetts Press.

———. 1992d. "Recapitulation and Outlook." Pp. 228–33 in *Showing Our Colors: Afro-German Women Speak Out*, edited by M. Opitz, K. Oguntoye, and D. Schultz. Amherst: University of Massachusetts Press.

Page, Clarence. 1990. "Black Youth Need More Help, Not More Scorn." *The Cincinnati Post* (March 9):A10.

Pitman, Paul M., III. 1988. *Turkey: A Country Study*. Washington, DC: U.S. Government Printing Office.

Portes, Alejandro and Cynthia Truelove. 1987. "Making Sense of Diversity: Recent Research on Hispanic Minorities in the United States." *Annual Review of Sociology* 13:359–85.

Proffitt, Steve. 1993. "Anna Deavere Smith: Finding a Voice for the Cacophony That Is Los Angeles." *Los Angeles Times* (July 11):M3.

Protzman, Ferdinand. 1993. "Germany Moves to Make First Cut in Its Generous Social Safety Net." *The New York Times* (August 12):A1.

Raspberry, William. 1991. "MacNeil/Lehrer Newshour" (June) WNET.

Rawley, James A. 1981. *The Transatlantic Slave Trade: A History*. New York: Norton.

Reynolds, Larry T. 1992. "A Retrospective on 'Race': The Career of a Concept." *Sociological Focus* 25(1):1–14.

Riding, Alan. 1991. "Europe's Growing Debate Over Whom to Let Inside." *The New York Times* (December 4):2E.

Rogers, Rosemarie. 1992. "The Future of Refugee Flows and Policies." *International Migration Review* 26(4):1112–43.

Safran, William. 1986. "Islamization in Western Europe: Political Consequences and Historical Parallels." *Annals* 485(May):98–112.

Salt, John. 1992. "The Future of International Labor Migration." *International Migration Review* 26(4): 1077–111.

Sayari, Sabri. 1986. "Migration Policies of Sending Countries: Perspectives on the Turkish Experience." *Annals* 485 (May):87–97.

Schneider, Peter. 1989. "If the Wall Came Tumbling Down." *The New York Times Magazine* (June 25):22+.

Schulz, Peter. 1975. "Turks and Yugoslavs: Guests or New Berlinein." *International Migration Review* 13:53–59.

Shearer, Derek. 1993. "The 'German Models' Loses Its Punch." *Los Angeles Times* (June 28):B7.

Smokes, Saundra. 1992. "A Lifetime of Racial Rage Control Snaps with a Telephone Call." *The Cincinnati Post* (May 13):14A.

Stark, Evan. 1990. "The Myth of Black Violence." *The New York Times* (July 18):A21.

Starr, Paul D. 1978. "Ethnic Categories and Identification in Lebanon." *Urban Life* 7(1):111–42.

Steele, Shelby. 1990. "A Negative Vote on Affirmative Action." *The New York Times Magazine* (May 13):46–49+.

Strasser, Hermann. 1993. "The German Debate Over Multicultural Society: Climax or Test of Organized Capitalism?" Paper presented at American Sociological Association Annual Meeting. Miami, FL.

Teraoka, Arlene Akiko. 1989. "Talking 'Turk': On Narrative Strategies and Cultural Stereotypes." *New German Critique* 46:104–28.

Terkel, Studs. 1992. *Race: How Blacks and Whites Think and Feel About the American Obsession.* New York: New Press.

Thränhardt, Dietrich. 1989. "Patterns of Organization Among Different Ethnic Minorities." *New German Critique* 46:10–26.

The Times Atlas of World History. 1984. Maplewood, NJ:Hammond.

U.S. Bureau for Refugee Programs. 1992. *World Refugee Report: A Report Submitted to the Congress as Part of the Consultations on FY 1993 Refugee Admission to the United States.* Washington, DC: Government Printing Office.

U.S. Commission on Civil Rights. 1981. *Affirmative Action in the 1980s: Dismantling the Process of Discrimination (A Proposed Statement).* Clearinghouse Publication 65.

Walsh, Mary Williams. 1993. "Battered Women as Refugees." *Los Angeles Times* (January 23):A1+.

Walton, Anthony. 1989. "Willie Horton and Me." *The New York Times Magazine* (August 20):52+.

Waters, Mary C. 1991. "The Role of Lineage in Identity Formation Among Black Americans." *Qualitative Sociology* 14(1):57–76.

Weiner, Myron. 1990. "Immigration: Perspectives from Receiving Countries." *Third World Quarterly* 12(1): 140–65.

———. 1992. "People and States in a New Ethnic Order?" *Third World Quarterly* 13(2):317–33.

Weiner, Tim. 1993. "Pleas for Asylum Inundate System for Immigration." *The New York Times* (April 25):A1+.

Wiedenroth, Ellen. 1992. "What Makes Me So Different in the Eyes of Others?" Pp. 165–77 in *Showing Our Colors: Afro-German Women Speak Out,* edited by M. Opitz, K. Oguntoye, and D. Schultz. Amherst: University of Massachusetts Press.

Wilpert, Czarina. 1991. "Migration and Ethnicity in a Non-Immigration Country: Foreigners in a United Germany." *New Community* 18(1):49–62.

Wirth, Louis. 1945. "The Problem of Minority Groups." Pp. 347–72 in *The Science of Man,* edited by R. Linton. New York: Columbia.

Zinn, Howard. 1980. *A People's History of the United States.* New York: Harper & Row.

FOCUS Chapter 10

Campbell, Mavis. 1990. *The Maroons of Jamaica 1655–1796.* Trenton NJ: Africa World Press.

CKSSG Chodorow, S., M. Knox, C. Schirokauer, J. Strayer, and H. Gatzke. 1989. *The Mainstream of Civilization to 1715.* Orlando, FL: Harcourt Brace Jovanovich.

Forbes, Jack D. 1993. *Africans and Native Americans.* Urbana and Chicago: University of Illinois Press.

hooks, bell. 1992. *Black Looks: Race and Representation.* Boston: South End Press.

Johnston, James H. 1929. "Documentary Evidence of the Relations of Negroes and Indians." *Journal of Negro History* Vol. 14 (1).

Katz, William L. 1986. *Black Indians.* New York: Macmillan.

———. 1987. *The Black West.* Seattle, WA: Open Hand Publishing.

Levin, Michael D. 1991. "Population Differentiation and Racial Classification." *Encyclopedia of Human Biology* Vol. 6. Academic Press.

Mullin, Michael. 1992. *Africa in America.* Chicago: University of Illinois Press.

Porter, Kenneth W. 1932. "Relations Between Negroes and Indians Within the Present Limits of the United States." *Journal of Negro History* Vol. XVII, No. 3 (July).

Rogers, J. A. 1984. *Sex and Race,* Vol. 2. St. Petersburg FL: Helga M. Rogers.

Strickland, Rennard. 1980. *The Indians in Oklahoma.* Norman, OK, and London: University of Oklahoma Press.

Woodson, Carter G. 1920. "The Relations of Negroes and Indians in Massachusetts." *Journal of Negro History* Vol. 5 (1).

Chapter 11

Ahmed, Syed Zubair. 1994. "What Do Men Want?" *The Times of India* (January 28).

Alderman, Craig, Jr. 1990. "10 February 1989 Memo for Mr. Peter Nelson." P. 108 in *Gays in Uniform: The Pentagon's Secret Reports,* edited by K. Dyer. Boston: Alyson.

Almquist, Elizabeth M. 1992. Review of "Gender, Family, and Economy: The Triple Overlap." *Contemporary Sociology* 21(3):331–32.

American Medical Association Bureau of Investigation. 1929. "The Tricho System: Albert C. Gryser X-ray Method of Depilation." *Journal of the American Medical Association* 92:252.

Anspach, Renee R. 1987. "Prognostic Conflict in Life-and-Death Decisions: The Organization as an Ecology of Knowledge." *Journal of Health and Social Behavior* 28(3):215–31.

Anthias, Floya and Nira Yuval-Davis. 1989. "Introduction." Pp. 1–15 in *Woman-Nation-State,* edited by N. Yuval-Davis and F. Anthias. New York: St. Martin's Press.

Baumgartner-Papageorgiou, Alice. 1982. *My Daddy Might Have Loved Me: Student Perceptions of Differences Between Being Male and Being Female.* Denver: Institute for Equality in Education.

Bem, Sandra Lipsitz. 1993. *The Lenses of Gender: Transforming the Debate on Sexual Inequality.* Binghamton, NY: Vail-Ballou.

Borden, Anthony. 1992. "The Yugoslav Conflict." *The European Security Network* (1):1–8.

Boroughs, Don L. 1990. "Valley of the Doll?" *U.S. News & World Report* (December 3):56–59.

Brew, Jo. 1994. "European Feminists Meet." *Off Our Backs* 24(1):1.

Brownmiller, Susan. 1975. *Against Our Will: Men, Women and Rape.* New York: Simon & Schuster.

Brownstein, Ronald. 1993. "Identifying Fathers Called Crucial to Welfare Reform." *Los Angeles Times* (December 16):A1+.

Burns, John F. 1992. "Canada Moves to Strengthen Sexual Assault Law." *The New York Times* (February 21):B9.

Canadian Broadcasting Corporation. 1994. "Sunday Morning" (January 9).

Collins, Randall. 1971. "A Conflict Theory of Sexual Stratification." *Social Problems* 19(1):3–21.

Cordes, Helen. 1992. "What a Doll! Barbie: Materialistic Bimbo or Feminist Trailblazer." *Utne Reader* (March/April):46, 50.

Curtis, Glenn E. 1992. *Yugoslavia: A Country Study*, 3rd ed. Washington, DC: U.S. Government Printing Office.

Darville, Ray L. and Joy B. Reeves. 1992. "Social Inequality Among Yugoslav Women in Directoral Positions." *Sociological Spectrum* 12(3):279–92.

Davis, F. James. 1979. *Understanding Minority-Dominant Relations: Sociological Contributions*. Arlington Heights, IL: AHM.

Dewhurst, Christopher J. and Ronald R. Gordon. 1993. Quoted in "How Many Sexes Are There?" *The New York Times* (March 12):A15.

Doherty, Jake. 1993. "Conference to Focus on Plight of Wartime 'Comfort Women.'" *Los Angeles Times* (February 20):B3.

Drakulić, Slavenka. 1990. "Women of Eastern Europe." *Ms.* (July/August):36–47.

———. 1992. *How We Survived Communism and Even Laughed*. New York: Norton.

——— 1993a. *The Balkan Express: Fragments from the Other Side of War*. New York: Norton.

———. 1993b. "Women and the New Democracy in the Former Yugoslavia." Pp. 123–30 in *Gender Politics and Post-Communism: Reflections from Eastern Europe and the Former Soviet Union*, edited by N. Funk and M. Mueller. New York: Routledge.

Enloe, Cynthia. 1993. *The Morning After: Sexual Politics at the End of the Cold War*. Los Angeles, CA: University of California Press.

Fagot, Beverly, Richard Hagan, Mary Driver Leinbach, and Sandra Kronsberg. 1985. "Differential Reactions to Assertive and Communicative Acts of Toddler Boys and Girls." *Child Development* 56(6):1499–505.

Fairstein, Linda A. 1993. *Sexual Violence: Our War Against Rape*. New York: Morrow.

Fausto-Sterling, Anne. 1993. "How Many Sexes Are There?" *The New York Times* (March 12):A15.

Ferrante, Joan. 1988. "Biomedical Versus Cultural Constructions of Abnormality: The Case of Idiopathic Hirsutism in the United States." *Culture, Medicine and Psychiatry* 12:219–38.

Garb, Frances. 1991. "Secondary Sex Characteristics." Pp. 326–27 in *Women's Studies Encyclopedia Volume 1: Views from the Sciences*, edited by H. Tierney. New York: Bedrick.

Gauguin, Paul. [1919] 1985. *Noa Noa: The Tahitian Journal*, translated by O. F. Theis. New York: Dover.

Geschwender, James A. 1992. "Ethgender, Women's Waged Labor, and Economic Mobility." *Social Problems* 39(1):1–16.

Glenny, Mirsha. 1992. "Yugoslavia: The Revenger's Tragedy." *The New York Review of Books* (August 13):37–43.

Grady, Denise. 1992. "Sex Test of Champions." *Discover* (June):78–82.

Gross, Jane. 1993. "Big Grocery Chain Reaches Landmark Sex-Bias Accord." *The New York Times* (December 17): A1; B11.

Hall, Edward T. 1959. *The Silent Language*. New York: Doubleday.

Harper's. 1990. "Employment History: Zeitgeist Barbie" (August):20.

Henry, Jules. 1963. *Culture Against Men*. New York: Vintage.

Hoon, Shim Jae. 1992. "Haunted by the Past." *Far Eastern Economic Review* (February 6):20.

Ignatieff, Michael. 1993. *Blood and Belonging: Journeys into the New Nationalism*. Toronto: Viking.

Johnson, G. David, Gloria J. Palileo, and Norma B. Gray. 1992. "'Date Rape' on a Southern Campus: Reports from 1991." *SSR* 76(2):37–44.

Jones, Ann. 1994. "Change from Within." *The Women's Review of Books* 11(4):14.

Josefowitz, Natasha. 1980. *Paths to Power*. New York: Addison-Wesley.

Katzarova, Mariana. 1993. "Opening the Door." *The Nation* (July 26): 148–50.

Kifner, John. 1994. "Bosnian Serbs Order General Mobilization for 'Conclusion of War.'" *The New York Times* (February 1):A4.

Kinzer, Stephen. 1993. "Feminist Gadfly Unappreciated in Her Own Land." *The New York Times* (December 11):4Y.

Kirka, Danica. 1993. "Slavenka Drakulić: Out of Sync in a Country Unaccustomed to Democracy." *Los Angeles Times* (December 19):M3.

Kolata, Gina. 1992. "Track Federation Urges End to Gene Test for Femaleness." *The New York Times* (February 12):A1; B11.

Komarovsky, Mirna. 1991. "Some Reflections on the Feminist Scholarship in Sociology." Pp. 1–25 in *Annual Review of Sociology Vol. 17*, edited by W. R. Scott and J. Blake. Palo Alto, CA: Annual Reviews.

Koss, Mary P., Christine A. Gidycz, and Nadine Wisniewski. 1987. "The Scope of Rape: Incidence and Prevalence of Sexual Aggression and Victimization in a National Sample of Higher Education Students." *Journal of Consulting and Clinical Psychology* 55(2):162–70.

Laber, Jeri. 1993. "Bosnia: Questions About Rape." *The New York Review of Books* (March 25):3–6.

Lambert, Bruce. 1993. "Abandoned Filipinas Sue U.S. Over Child Support." *The New York Times* (June 21):A3.

Lemonick, Michael D. 1992. "Genetic Tests Under Fire." *Time* (February 24):65.

Lewin, Tamar. 1993. "At Bases, Debate Rages Over Impact of New Gay Policy." *The New York Times* (December 24):A1+.

Milic, Andgelka. 1993. "Women and Nationalism in the Former Yugoslavia." Pp. 109–22 in *Gender Politics and Post-Communism: Reflections from Eastern Europe and the Former Soviet Union*, edited by N. Funk and M. Mueller. New York: Routledge.

Mills, Janet Lee. 1985. "Body Language Speaks Louder Than Words." *Horizons* (February):8–12.

Mirsada. 1993. "Testimony." *Los Angeles Times Magazine* (January 31): 28–32.

Mladjenovic, Lepa. 1993. "Universal Soldier: Rape Is War by a Feminist in Serbia." *Off Our Backs* (March): 14–15.

Morawski, Jill G. 1991. "Femininity." Pp. 136–39 in *Women's Studies Encyclopedia Volume 1: Views from the Sciences*, edited by H. Tierney. New York: Bedrick.

Morgan, Robin. 1993. "Editorial: Isolated Incidents." *Ms.* (March/April):1.

Morgenson, Gretchen. 1991. "Barbie Does Budapest." *Forbes* (January 7):66–69.

Pion, Alison. 1993. "Accessorizing Ken." *Origins* (November):8.

Ramet, Sabrina P. 1991. *Social Currents in Eastern Europe*. Durham, NC: Duke University Press.

Rank, Mark R. 1989. "Fertility Among Women on Welfare: Incidence and Determinants," in *American Sociological Review* 54(4):296–304.

Rieff, David. 1992. "Letter from Bosnia: Original Virtue, Original Sin." *The New Yorker* (November 23):82–95.

Roiphe, Katie. 1993. *The Morning After: Sex, Fear, and Feminism on Campus*. New York: Little, Brown.

Sarbin, Theodore R. and Kenneth E. Karols. 1990. "Nonconforming Sexual Orientations and Military Suitability." Pp. 6–49 in *Gays in Uniform: The Pentagon's Secret Reports*, edited by K. Dyer. Boston: Alyson.

Schaller, Jane Green and Elena O. Nightingale. 1992. "Children and Childhoods: Hidden Casualties of War and Civil Unrest." *Journal of the American Medical Association* 268(5):642–44.

Schmalz, Jeffrey. 1993. "From Midshipman to Gay-Rights Advocate." *The New York Times* (February 4):B1+.

Segal, Lynne. 1990. *Slow Motion: Changing Masculinities, Changing Men*. London: Virago.

Shweder, Richard A. 1994. "What Do Men Want? A Reading List for the Male Identity Crisis." *The New York Times Book Review* (January 9):3, 24.

Solinger, Rickie. 1992. *Wake Up Little Susie: Single Pregnancy and Race Before Roe v. Wade*. New York: Routledge.

Stevenson, Samantha. 1994. "Sister Strikeout Does the Job Well." *The New York Times* (February 23):B10.

Sturdevant, Saundra Pollock and Brenda Stoltzfus. 1992. *Let the Good Times Roll: Prostitution and the U.S. Military in Asia*. New York: New Press.

Swiss, Shana and Joan E. Giller. 1993. "Rape as a Crime of War." *Journal of the American Medical Association* 270(5):612–15.

Tattersall, Ian. 1993. "Focus—All in the Family (Homosapien Exhibit at New York's American Museum of Natural History)." Interview on "MacNeil/Lehrer Newshour," July 12 (transcript #4708). WNET.

Tierney, Helen. 1991. "Gender/Sex." P. 153 in *Women's Studies Encyclopedia Volume 1: Views from the Sciences*, edited by H. Tierney. New York: Bedrick.

United Nations. 1991. *The World's Women 1970–1990: Trends and Statistics*. New York: UN.

U.S. Bureau of the Census. 1990.

U.S. Department of Defense. 1990. "DOD Directive 1332.14." P. 19 in *Gays in Uniform: The Pentagon's Secret Reports*, edited by K. Dyer. Boston: Alyson.

Weitzman, Lenore J. 1985. *The Divorce Revolution: The Unexpected Social and Economic Consequences for Women and Children in America*. New York: Free Press.

Williams, Carol J. 1993. "Postscript: 'Ethnic Cleansing' Threatens to Wipe Away Memory of Tito." *Los Angeles Times* (October 26):H1+.

Williams, Lena. 1993. "Pregnant Teenagers Are Outcasts No Longer." *The New York Times* (December 2):B1+.

Chapter 12

Adams, Bert. 1968. *Kinship in an Urban Setting*. Chicago: Markham.

Allan, Graham. 1977. "Sibling Solidarity." *Journal of Marriage and the Family* 9(1):177–83.

Bass, Thomas A. 1990. "A New Life Begins for the Island of Hope and Tears." *Smithsonian* (June): 89–97.

Behnam, Djamshid. 1990. "An International Inquiry into the Future of the Family: A UNESCO Project." *International Social Science Journal* 42: 547–52.

Bell, Daniel. 1989. "The Third Technological Revolution." *Dissent* (Spring): 164–76.

Berelson, Bernard. 1978. "Prospects and Programs for Fertility Reduction: What? Where?" *Population and Development Review* 4:579–616.

Beresky, Andrew E., ed. 1991. *Fodor's Brazil: Including Bahia and Adventures in the Amazon*. New York: Fodor's Travel Publications.

Bernardo, Felix M. 1967. "Kinship Interaction and Communications Among Space-Age Migrants." *Journal of Marriage and the Family* 29(3):541–54.

Boccaccio, Giovanni. (1353) 1984. "The Black Death." Pp. 728–40 in *The Norton Reader: An Anthology of Expository Prose*, 6th ed., edited by A. M. Eastman. New York: Norton.

Brooke, James. 1989. "Decline in Births in Brazil Lessens Population Fears." *The New York Times* (August 8):Y1+.

Brown, Christy. 1992. "The Letter 'A.'" Pp. 85–90 in *One World Many Cultures*, edited by S. Hirschberg. New York: Macmillan.

Brown, Lester R. 1987. "Analyzing the Demographic Trap." *State of the World 1987: A Worldwatch Institute Report on Progress Toward a Sustainable Society*. New York: Norton.

Burke, B. Meredith. 1989. "Ceausescu's Main Victims: Women and Children." *The New York Times* (January 16): Y15.

Butts, Yolanda and Donald J. Bogue. 1989. *International Amazonia: Its Human Side*. Chicago: Social Development Center.

Calsing, Elizeu Francisco. 1985. "Extent and Characteristics of Poverty in Brazil. Estimation of Social Inequalities." *Revista Paraguaya de Sociología* 22: 29–53.

Caufield, Catherine. 1985. "A Reporter At Large: The Rain Forests." *The New Yorker* (January 14):41+.

Cowell, Adrian. 1990. *The Decade of Destruction: The Crusade to Save the Amazon Rain Forest*. New York: Henry Holt.

Davis, Kingsley. 1984. "Wives and Work: The Sex Role Revolution and Its Consequences." *Population and Development Review* 10(3):397–417.

Dean, Warren. 1991. Review of *Coffee, Contention, and Change: In the Making of Modern Brazil*, edited by Mauricio A. Font and Charles Tilly. *Journal of Latin American Studies* 23(3): 649–50.

Durning, Alan B. 1990. "Ending Poverty." Pp. 135–53 in *State of the World 1990: A Worldwatch Institute Report on Progress Toward a Sustainable Society*, edited by L. Starke. New York: Norton.

Dychtwald, Ken and Joe Flower. 1989. *Age Wave: The Challenges and Opportunities of an Aging America*. Los Angeles: Tarcher.

Eckholm, Erik. 1990. "An Aging Nation Grapples with Caring for the Frail." *The New York Times* (March 27):A1+.

Feder, Ernest. 1971. *The Rape of the Peasantry: Latin America's Landholding System*. Garden City, NY: Anchor Books.

Fonseca, Claudia. 1986. "Orphanages, Foundlings, and Foster Mothers: The System of Child Circulation in a Brazilian Squatter Settlement." *Anthropological Quarterly* 59:15–27.

Glascock, Anthony P. 1982. "Decrepitude and Death Hastening: The Nature of Old Age in Third World Societies (Part I)." *Studies in Third World Societies* 22:43–66.

Goldani, Ana Maria. 1990. "Changing Brazilian Families and the Consequent Need for Public Policy." *International Social Science Journal* 42(4): 523–38.

Goldenberg, Sheldon. 1987. *Thinking Sociologically*. Belmont, CA: Wadsworth.

Goode, Judith. 1987. "Gaminismo: The Changing Nature of the Street Child Phenomenon in Colombia." *UFSI Reports*, No. 28.

Gutis, Philip S. 1989a. "Family Redefines Itself, and Now the Law Follows." *The New York Times* (May 28):B1.

———. 1989b. "What Makes a Family? Traditional Limits Are Challenged." *The New York Times* (August 31): Y15+.

Hall, Michael. 1992. "Mission to Somalia: The Anatomy of Starvation." *Los Angeles Times* (December 12):A8.

Halberstam, David. 1993. *The Fifties*. New York: Villard Books.

Harrison, Paul. 1987. *Inside the Third World: The Anatomy of Poverty*, 2nd. ed. New York: Viking Penguin.

Haub, Carl, Mary Mederias Kent, and Machiko Yanagishita. 1991. *1991 World Population Data Sheet*. Washington, DC: Population Reference Bureau.

Haub, Carl and Machiko Yanagishita. 1992. *1992 World Population Data Sheet*. Washington, DC: Population Reference Bureau.

Human Development Report, 1993. 1993. New York: Oxford University Press.

Johansson, S. Ryan. 1987. "Status Anxiety and Demographic Contraction of Privileged Populations." *Population and Development Review* 13(3): 439–70.

Leigh, Geoffrey. 1982. "Kinship Interaction over the Family Life Span." *Journal of Marriage and the Family* 41(1): 197–208.

Lengermann, Patricia M. 1974. *Definitions of Sociology: A Historical Approach*. Columbus, OH: Merrill.

Lewin, Tamar. 1990. "Strategies to Let Elderly Keep Some Control." *The New York Times* (March 28):A1, A11.

Light, Ivan. 1983. *Cities in World Perspective*. New York: Macmillan.

Litwak, Eugene. 1960. "Geographic Mobility and Extended Family Cohesion." *American Sociological Review* 25:385–94.

Malthus, Thomas R. (1798) 1965. *First Essay on Population*. New York: Augustus Kelley.

Marciano, Teresa and Marvin B. Sussman. 1991. *Wider Families: New Traditional Family Forms*. Binghamton, NY: Haworth.

Margolick, David. 1990. "Lesbians' Custody Fights Test Family Law Frontier." *The New York Times* (July 4): Y1+.

Molano, Alfredo. 1993. Quoted on p. 43 in "Colombia's Vanishing Forests," *World Press Review* (June):43.

Nations, Marilyn K. and Mara Lucia Amaral. 1991. "Flesh, Blood, Souls, and Households: Cultural Validity in Morality Inquiry." *Medical Anthropology Quarterly* 5(3):204–20.

Nolty, Denise, ed. 1990. *Fodor's '90 Brazil: Including the Amazon and Bahia*. New York: Fodor's Travel Publications.

Olshansky, S. Jay and A. Brian Ault. 1986. "The Fourth Stage of the Epidemiologic Transition: The Age of Delayed Degenerative Diseases." *The Milbank Quarterly* 64(3):355–91.

Omran, Abdel R. 1971. "The Epidemiologic Transition: A Theory of the Epidemiology of Population Change." *The Milbank Quarterly* 49(4):509–38.

Parsons, Talcott. 1966. "The Kinship System of the Contemporary United States." Pp. 177–96 in *Essays in Sociological Theory,* revised ed. New York: Free Press.

Perlman, Janice. 1967. *The Myth of Marginality*. Berkeley: University of California Press.

Pullman, Thomas W. 1992. "Population." Pp. 1499–1507 in *Encyclopedia of Sociology,* Vol. 3, edited by E. F. Borgatta and M. L. Borgatta. New York: Macmillan.

Ramirez, Anthony. 1990. "Lessons in the Cracker Market." *The New York Times* (July 5):C1+.

Revkin, Andrew. 1990. *The Burning Season: The Murder of Chico Mendes and the Fight for the Amazon Rain Forest*. Boston: Houghton Mifflin.

Riesman, David, with Nathan Glazer and Reuel Denney. 1977. *The Lonely Crowd: A Study of the Changing American Character,* abridged ed. New Haven, CT: Yale University Press.

Rock, Andrea. 1990. "Can You Afford Your Kids?" *Money* (July):88–99.

Rusinow, Dennison. 1986. "Mega-Cities Today and Tomorrow: Is the Cup Half Full or Half Empty?" *UFSI Reports,* No. 12.

Sagan, Carl. 1978. *Murmurs of Earth: The Voyager Interstellar Record*. New York: Random House.

Sanders, Thomas G. 1986. "The Politics of Agrarian Reform in Brazil." *UFSI Reports,* No. 32.

———. 1987a. "Brazilian Street Children, Part I: Who They Are." *UFSI Reports,* No. 17.

———. 1987b. "Brazilian Street Children, Part II: The Public and Political Response." *UFSI Reports,* No. 18.

———. 1988. "Happiness Also Rises Up There: The Favelas of Rio." *UFSI Reports,* No. 2.

Sayre, Robert F. 1983. "The Parents' Last Lessons." Pp. 124–42 in *Life Studies: A Thematic Reader,* edited by D. Cavitch. New York: St. Martin's Press.

Semana. 1993. "Colombia's Vanishing Forests." *World Press Review* (June):43.

Simons, Marlise. 1988. "Man-Made Amazon Fires Tied to Global Warming." *The New York Times* (August 12):Y1+.

———. 1989. "Brazil, Smarting from the Outcry over the Amazon, Charges Foreign Plot." *The New York Times* (March 23):Y6.

Skidmore, Thomas E. 1993. "Bi-racial U.S.A. vs. Multi-racial Brazil: Is the Contrast Still Valid?" *Journal of Latin American Studies* 25:373–86.

Soldo, Beth J. and Emily M. Agree. 1988. "America's Elderly." *Population Bulletin* 43(3):5+.

Sorel, Nancy Caldwell. 1984. *Ever Since Eve: Personal Reflections on Childbirth*. New York: Oxford.

Statistical Abstract of Latin America. 1989. Vol. 27, edited by James W. Wilkie and Enrique Ochoa. Los Angeles: UCLA.

Stockwell, Edward G. and H. Theodore Groat. 1984. *World Population: An Introduction to Demography*. New York: Franklin Watts.

Stockwell, Edward G. and Karen A. Laidlaw. 1981. *Third World Development: Problems and Prospects*. Chicago: Nelson-Hall.

Stone, Robyn, Gail Lee Cafferata, and Judith Sangl. 1987. "Caregivers of the Frail Elderly: A National Profile." *The Gerontologist* 27(5):616–26.

Stub, Holger R. 1982. *The Social Consequences of Long Life*. Springfield, IL: Thomas.

Targ, Dena B. 1989. "Feminist Family Sociology: Some Reflections." *Sociological Focus* 22(3): 151–60.

Tremblay, Hélène. 1988. *Families of the World: Family Life at the Close of the Twentieth Century, Vol. 1: The Americas and the Caribbean*. New York: Farrar, Straus & Giroux.

Turner, Jonathan H. 1978. *Sociology: Studying the Human System*. Santa Monica, CA: Goodyear.

United Nations. 1983. *World Population Trends and Policies: 1983 Monitoring Report*. Vol. 1. New York.

U.S. Bureau of the Census. 1947. *Statistical Abstract of the United States, 1947*. Washington, DC: U.S. Government Printing Office.

———. 1989. *World Population Profile: 1989*. Washington, DC: U.S. Government Printing Office.

———. 1991. *World Population Profile: 1991*. Washington, DC: U.S. Government Printing Office.

———. 1993. *Statistical Abstract of the United States, 1993*. Washington, DC: U.S. Government Printing Office.

U.S. Department of the Army. 1983. *Brazil, A Country Study*, 4th ed., edited by Richard F. Nyrop. Washington, DC: U.S. Government Printing Office.

van de Kaa, Dirk J. 1987. "Europe's Second Demographic Transition." *Population Bulletin* 42(1):1–59.

Watkins, Susan C. and Jane Menken. 1985. "Famines in Historical Perspective." *Population and Development Review* 11(4):647–75.

The World Bank. 1990. *World Development Report, 1990*. New York: Oxford University Press.

The World Factbook 1992. 1992. Washington, DC: Central Intelligence Agency.

Zelditch, Morris. 1964. "Family, Marriage, and Kinship." Pp. 680–733 in *Handbook of Modern Sociology,* edited by R. E. L. Faris. Chicago: Rand McNally.

Chapter 13

Alpert, Bracha. 1991. "Students' Resistance in the Classroom." *Anthropology and Education Quarterly* 22(4): 350–66.

Barrett, Michael J. 1990. "The Case for More School Days." *The Atlantic Monthly* (November):78–106.

Bennett, William J. 1987. "Epilogue: Implications for American Education." Pp. 69–71 in *Japanese Education Today*. Washington, DC: U.S. Government Printing Office.

Bettelheim, Bruno and Karen Zeland. 1981. *On Learning to Read: The*

Child's Fascination with Meaning. New York: Knopf.

Blakeslee, Sandra. 1989. "Crack's Toll Among Babies: A Joyless View of Even Toys." *The New York Times* (September 17):Y1+.

Bloom, Benjamin S. 1981. *All Our Children Learning: A Primer for Parents, Teachers and Other Educators.* New York: McGraw-Hill.

The Book of the States, 1992–93 Edition. 1992. Lexington, KY: Council of State Governments.

Botstein, Leon. 1990. "Damaged Literacy: Illiteracies and American Democracy." *Daedalus* 119 (2): 55–84.

Boyer, Ernest. 1986. "Forum: How Not to Fix the Schools." *Harper's* (February):39–51.

Celis, William III. 1992. "A Texas-Size Battle to Teach Rich and Poor Alike." *The New York Times* (February 12):B6.

———. 1993a. "International Report Card Shows U.S. Schools Work." *The New York Times* (December 9):A1+.

———. 1993b. "Study Finds Rising Concentration of Black and Hispanic Students." *The New York Times* (December 14):A1+.

Cetron, Marvin. 1988. "Forum: Teach Our Children Well." *Omni* 10(11):14.

Chira, Susan. 1991. "Student Tests in Other Nations Offer U.S. Hints, Study Says." *The New York Times* (May 20):A1+.

Clements, Marcelle. 1992. "Fear of Reading." *The New York Times* (May 18):A11.

Cohen, David K. and Barbara Neufeld. 1981. "The Failure of High Schools and the Progress of Education." *Daedalus* (Summer):69–89.

Coleman, James S. 1960. "The Adolescent Subculture and Academic Achievement." *American Journal of Sociology* 65:337–47.

———. 1966. *Equality of Educational Opportunity.* Washington, DC: U.S. Government Printing Office.

———. 1977. "Choice in American Education." Pp. 1–12 in *Parents, Teachers, and Children: Prospects for Choice in American Education.* San Francisco: Institute for Contemporary Studies.

Coleman, James S., John W. C. Johnstone, and Kurt Jonassohn. 1961. *The Adolescent Society.* New York: Free Press.

Cooke, Charles, Alan Ginsburg, and Marshall Smith. 1985. "The Sorry State of Education Statistics." *The Education Digest* 51(4):28–30.

Csikszentmihalyi, Mihaly. 1990. "Literacy and Intrinsic Motivation." *Daedalus* 119(2):115–40.

Cummings, William K. 1989. "The American Perception of Japanese Edu-

cation." *Comparative Education* 25(3):293–302.

Currie, Elliott and Jerome H. Skolnick. 1988. *America's Problems: Social Issues and Public Policy,* 2nd ed. Boston: Little, Brown.

Danner, Mark D. 1986. "Forum: How Not to Fix the Schools." *Harper's* (February):39–51.

De Icaza, Maria A. 1991. "Letters: U.S. Students Memorize, but Don't Understand." *The New York Times* (November 6):A13.

Dorfman, Andrea. 1989. "Alcohol's Youngest Victims." *Time* (August 28):60.

Dorris, Michael. 1989. *The Broken Cord.* New York: Harper & Row.

Durkheim, Emile. 1961. "On the Learning of Discipline." Pp. 860–65 in *Theories of Society: Foundations of Modern Sociological Theory* Vol. 2, edited by T. Parsons, E. Shils, K. D. Naegele, and J. R. Pitts. New York: Free Press.

———. 1968. *Education and Sociology,* translated by S. D. Fox. New York: Free Press.

Early, Margaret. 1987. *Streamers Workbook.* Chicago: Harcourt Brace Jovanovich.

Elam, Stanley M. 1989. "The Second Gallup/Phi Delta Kappa Poll of Teachers' Attitudes Toward the Public Schools." *Phi Delta Kappan* (June): 785–98.

Erickson, Frederick. 1984. "School Literacy, Reasoning, and Civility: An Anthropologist's Perspective." *Review of Educational Research* 54(4): 525–46.

Folbre, Nancy. 1987. *A Field Guide to the U.S. Economy: 160 Graphic Keys to How the System Works.* New York: Pantheon Books.

Foster, Jack D. 1991. "The Role of Accountability in Kentucky's Education Reform Act of 1990." *Education Leadership*: 34–36.

Gardner, John W. 1984. *Excellence: Can We Be Equal and Excellent Too?* New York: Norton.

Gisi, Lynn Grover. 1985. "How States Are Reforming Public Education." *USA Today* 113(2478):76–78.

Guzzardi, Walter, Jr. 1976. "The Uncertain Passage from College to Job." *Fortune* (January):126–29, 168–72.

Hahn, Andrew. 1987. "Reaching Out to America's Dropouts: What to Do?" *Phi Delta Kappan* 69:256–63.

Hakim, Joy. 1993. Interview on *National Public Radio/* "Morning Edition." (June 2):4–7.

Hallinan, M. T. 1988. "Equality of Educational Opportunity." Pp. 249–68 in *Annual Review of Sociology* Vol. 14, edited by W. R. Scott and J. Blake. Palo Alto, CA: Annual Reviews.

Harrison, Charles. 1988. *Public Schools USA: A Comparative Guide to School Districts.* Charlotte, VT: Williamson.

Hawkins, David. 1990. "The Roots of Literacy." *Daedalus* 119(2):1–14.

Hechinger, Fred M. 1990. "About Education: Why France Outstrips the United States in Nurturing Its Children." *The New York Times* (August 1):B8.

Henry, Jules. 1965. *Culture Against Man.* New York: Random House.

Hirsch, E. D., Jr. 1989. "The Primal Scene of Education." *The New York Review of Books* (March 2):29–35.

Horn, Miriam. 1987. "The Burgeoning Educational Underclass." *U.S. News & World Report* (May 18):66–67.

Lapointe, Archie E., Nancy A. Mead, and Gary W. Phillips. 1989. *A World of Differences: An International Assessment of Mathematics and Science.* Princeton, NJ: Educational Testing Service.

Lefkowitz, Bernard. 1987. "Who Cares About Eddie?" *Present Tense* 14(July/August):15–21.

Lightfoot, Sara Lawrence. 1988. "Bill Moyers' World of Ideas" (transcript #123). New York: Public Affairs Television.

Limage, Leslie. 1990. "Reports from Around the World—The Industrialized Countries: Questions and Answers." *Unesco Courier* (July):16–20.

Los Angeles Times. 1993. "Gifted Students Found Unchallenged." (November 5):A39.

Luria, A. R. 1979. *The Making of Mind: A Personal Account of Soviet Psychology,* edited by M. Cole and S. Cole. Cambridge, MA: Harvard University Press.

Lynn, Richard. 1988. *Educational Achievement in Japan: Lessons for the West.* London: Macmillian.

Merton, Robert K. 1957. *Social Theory and Social Structure.* Glencoe, IL: Free Press.

The National Commission on Excellence in Education. 1983. *A Nation at Risk: The Imperative for Educational Reform.* Washington, DC: U.S. Government Printing Office.

National Endowment for the Humanities. 1991. *National Tests: What Other Countries Expect Their Students to Know.* Washington, DC: U.S. Government Printing Office.

National Science Board. 1991. *Science and Engineering Indicators 1991.* Washington, DC: U.S. Government Printing Office.

Oakes, Jeannie. 1985. *Keeping Track: How Schools Structure Inequality.* Binghamton, NY: Vail-Ballou.

———. 1986a. "Keeping Track, Part 1: The Policy and Practice of Curriculum Inequality." *Phi Delta Kappan* 67(September):12–17.

———. 1986b. "Keeping Track, Part 2: Curriculum Inequality and School Reform." *Phi Delta Kappan* 67(October): 148–54.

O'Connor, John J. 1990. "Critic's Notebook: How TV Sends Mixed Messages About Education." *The New York Times* (September 13):B1.

Ouane, Adama. 1990. "National Languages and Mother Tongues." *Unesco Courier* (July):27–29.

Phelan, Patricia and Ann Locke Davidson. 1994. "Looking Across Borders: Students' Investigations of Family, Peer, and School Worlds as Cultural Therapy." Pp. 35–59 in *Pathways to Cultural Awareness: Cultural Therapy With Teachers and Students*, edited by George and Louise Spindler. Thousand Oaks, CA.: Corwin Press, Inc.

Phelan, Patricia, Ann Locke Davidson, and Hanh Cao Yu. 1993. "Students' Multiple Worlds: Navigating the Borders of Family, Peer, and School Cultures." Pp. 89–107 in *Renegotiating Cultural Diversity in American Schools*, edited by Patrica Phelan and Ann Locke Davidson. New York: Teachers College Press.

Phelan, Patricia, Ann Locke Davidson, and Hanh Thanh Cao. 1991. "Students' Multiple Worlds: Negotiating the Boundaries of Family, Peer, and School Cultures." *Anthropology and Education Quarterly* 22(3):224–50.

Potter, J. Hasloch and A. E. W. Sheard. 1918. *Catechizings for the Church and Sunday Schools*, 2nd series. London: Skeffington.

Ramirez, Francisco and John W. Meyer. 1980. "Comparative Education: The Social Construction of the Modern World System." Pp. 369–99 in *Annual Review of Sociology* Vol. 6, edited by A. Inkeles, N. J. Smelser, and R. H. Turner. Palo Alto, CA: Annual Reviews.

Reich, Robert. 1993. Quoted on p. E1 in "Skipping School," by Bettijane Levine. *Los Angeles Times* (July 3):E1, 5.

Remlinger, Connie and Debra Vance. 1989. "Business Enlisting in War on Illiteracy." *The Kentucky Post*. Special Supplement on Literacy. (October 19):2.

Resnick, Daniel P. 1990. "Historical Perspectives on Literacy and Schooling." *Daedalus* 119(2):15–32.

Resnick, Daniel P. and Lauren B. Resnick. 1989. "Varieties of Literacy." Pp. 171–206 in *Social History and Issues in Human Consciousness*, edited by A. E. Barnes and P. N. Stearns.

Resnick, Lauren B. 1990. "Literacy In School and Out." *Daedalus* 119(2): 169–86.

Richardson, Lynda. 1994. "More Schools Are Trying to Write Textbooks Out of the Curriculum." *The New York Times* (January 31):A1+.

Rodriguez, Richard. 1975. *On Becoming a Chicano*. New York: Georges Borchardt.

Rohlen, Thomas P. 1986. "Japanese Education: If They Can Do It, Should We?" *The American Scholar* 55:29–43.

Rosenthal, Robert and Lenore Jacobson. 1968. *Pygmalion in the Classroom*. New York: Holt, Rinehart & Winston.

Rush, Benjamin. 1966. Quoted on p. 34 in "Forming the National Character," by David Tyack. *Harvard Educational Review* 36:29–41.

Sanchez, Claudio. 1993. Interview on *National Public Radio*/"Morning Edition" (December 8):11–13.

Schlack, Lawrence B. 1992. "Letter: School Days." *World Monitor* (November):5.

Shelley, Kristina J. 1992. "The Future of Jobs for College Graduates." *Monthly Labor Review* (July): 13–21.

Sowell, Thomas. 1981. *Ethnic America: A History*. New York: Basic Books.

Stevenson, Harold. 1992. Interview on *National Public Radio*/"Morning Edition" (December 10):8–10.

Stevenson, Harold W., Shin-ying Lee, and James W. Stigler. 1986. "Mathematics Achievement of Chinese, Japanese, and American Children." *Science* 231:693–99.

Sweetland, Bill. 1989. "Boredom, Banality, and Bad Design Characterize Text for Slow Readers." *Curriculum Review* (October):3–7.

Thomas, William I. and Dorothy Swain Thomas. [1928] 1970. *The Child in America*. New York: Johnson Reprint.

Thomson, Scott D. 1989. "Report Card USA: How Much Do Americans Value Schooling?" *NASSP Bulletin* 73(519): 51–67.

Time. 1990. "Reading, Writing and Rhetoric." (February 12):54.

Tyack, David and Elisabeth Hansot. 1981. "Conflict and Consensus in American Public Education." *Daedalus* (Summer):1–43.

Tyler, Ralph. 1974. Quoted in "Divergent Views on the Schools: Some Optimism Justified," by Alan C. Purves. *The New York Times* (January 16):C74.

U.S. Bureau of the Census. 1982. *Illiteracy: 1982*. Washington, DC: U.S. Government Printing Office.

———. 1993. *Statistical Abstract of the United States* 113th ed. Washington, DC: U.S. Government Printing Office.

U.S. Bureau of Indian Affairs. 1988. *Report on BIA Education: Excellence in Indian Education Through the Effective School Process*. Washington, DC: U.S. Government Printing Office.

U.S. Department of Education. 1987. *Japanese Education Today*. Washington, DC: U.S. Government Printing Office.

———. 1991. *Occupational and Educational Outcomes of Recent College Graduates One Year After Graduation: 1989*. Washington, DC: U.S. Government Printing Office.

———. 1992a. *National Adult Literacy Survey 1992*. Washington, DC: U.S. Government Printing Office.

———. 1992b. *International Mathematics and Science Assessments: What Have We Learned?* Washington, DC: U.S. Government Printing Office.

———. 1992c. *International Education Comparisons*. Washington, DC: U.S. Government Printing Office.

———. 1993a. *Occupational and Educational Outcomes of Recent College Graduates 1 Year After Graduation: 1991*. Washington, DC: U.S. Government Printing Office.

———. 1993b. *Adult Literacy in America: A First Look at the Results of the National Literacy Survey*. Washington, DC: U.S. Government Printing Office.

Vélis, Jean-Pierre. 1990. "Waste." *Unesco Courier* (July):30–32.

Webster, Noah. 1966. Quoted on pp. 32–33 in "Forming the National Character," by David Tyack. *Harvard Educational Review* 36:29–41.

White, Robert. 1989. "Building a Better Classroom." *The Cincinnati Post* (September 18):1A+.

The World Almanac and Book of Facts 1994. 1993. New York: World Almanac.

Chapter 14

Abercrombie, Nicholas and Bryan S. Turner. 1978. "The Dominant Ideology Thesis." *British Journal of Sociology* 29(2):149–70.

Aikman, David. 1992. "Interview: The World Is Fresh and Bright and Beautiful." *Time* (May 18):57–58.

Alston, William P. 1972. "Religion." Pp. 140–45 in *The Encyclopedia of Philosophy* Vol. 7, edited by P. Edwards. New York: Macmillan.

Aron, R. 1969. Quoted on p. 204 in *The Sociology of Max Weber* by Julien Freund. New York: Random House.

Barakat, Halim. 1979. "The Social Context." Pp. 3–20 in *Lebanon in Crisis*, edited by P. E. Haley and L. W. Snider. New York: Syracuse University Press.

Berger, Peter L. 1967. *The Sacred Canopy: Elements of a Sociological Theory of Religion*. New York: Doubleday.

Boling, Rick. 1985. "The Ecology of War: Earth." *Omni* (November):14.

Bush, George H. 1991. State of the Union Address by the President of the United States (January 29).

Caplan, Lionel. 1987. "Introduction: Popular Conceptions of Fundamentalism." Pp. 1–24 in *Studies in Religious*

Fundamentalism, edited by L. Caplan. Albany: State University of New York Press.

Chinoy, Ely. 1961. *Society: An Introduction to Sociology.* New York: Random House.

Christianity Today. 1986. "Letters." (October 17):6.

Coles, Robert. 1990. *The Spiritual Life of Children.* Boston: Houghton Mifflin.

Cryderman, Lyn. 1990. "Dance of Faith and Death: The Church in Lebanon." *Christianity Today* (November 5): 41–49.

Doerner, William R. 1985. "Movements Within Movements." *Time* (July 1):24.

Durkheim, Emile. [1915] 1964. *The Elementary Forms of the Religious Life,* 5th ed., translated by J. W. Swain. New York: Macmillan.

———. 1951. *Suicide: A Study in Sociology,* translated by J. A. Spaulding and G. Simpson. New York: Free Press.

———. 1984. Quoted in *Durkheim's Sociology of Religion* by W. S. F. Pickering. London: Routledge & Kegan Paul.

Ebersole, Luke. 1967. "Sacred." P. 613 in *A Dictionary of the Social Sciences,* edited by J. Gould and W. L. Kolb. New York: UNESCO.

Esposito, John L. 1986. "Islam in the Politics of the Middle East." *Current History* (February):53–57, 81.

———. 1992. *The Islamic Threat: Myth or Reality?* New York: Oxford University Press.

Farrell, Thomas. 1945. Quoted on p. 65 in *In a Dark Time: Images for Survival,* edited by R. J. Lifton and N. Humphrey. Cambridge, MA: Harvard University Press.

Fisk, Robert. 1990. *Pity the Nation: The Abduction of Lebanon.* New York: Atheneum.

Forbes, James. 1990. "Up from Invisibility." Review of *The Black Church in the African American Experience* by C. Eric Lincoln and Lawrence H. Mamiya. *The New York Times Book Review* (December 23):1–2.

Friedman, Thomas L. 1989. *From Beirut to Jerusalem.* New York: Farrar, Straus & Giroux.

Gallup, George, Jr., and Jim Castelli. 1989. *The People's Religion: American Faith in the 90s.* New York: Macmillan.

Genet, Harry. 1984. "Why It Is Easy to Be Confused About Lebanon." *Christianity Today* (February 17):10.

Goodman, Walter. 1991. "The Rite of Exorcism on '20/20.'" *The New York Times* (April 5):B5.

Haddad, Yvonne. 1991. Interview with Bill Moyers on "Images of God in the Arab World." PBS television.

Hammond, Phillip E. 1976. "The Sociology of American Civil Religion: A Bib-

liographic Essay." *Sociological Analysis* 37(2):169–82.

Hedges, Chris. 1993. "Israel Keeps Pounding South Lebanon." *The New York Times* (July 29):A6.

Heiman, Andrea. 1992. "Spiritual Journey." *Los Angeles Times* (August 8): B4–5.

Hourani, Albert. 1991. *A History of the Arab Peoples.* Cambridge, MA: Belknap Press.

Ibrahim, Youssef M. 1991. "In Kuwait, Ramadan Has a Bitter Taste." *The New York Times* (March 19):A1, A7.

Lamb, David. 1987. *The Arabs: Journeys Beyond the Mirage.* New York: Random House.

Lechner, Frank J. 1989. "Fundamentalism Revisited." *Society* (January/February):51–59.

LeMoyne, James. 1990. "Fuel for Saudi Debate: Opening Society Without Causing Strife." *The New York Times* (October 31):A7.

Leo, John. 1991. "Lesson from a Sanitized War." *U.S. News & World Report* (March 18):26.

Lifton, Robert Jay. 1969. *Boundaries: Psychological Man in Revolution.* New York: Touchstone.

Lifton, Robert Jay and Nicholas Humphrey. 1984. *In a Dark Time: Images of Survival.* Cambridge, MA: Harvard University Press.

Lincoln, C. Eric and Lawrence H. Mamiya. 1990. *The Black Church in the African American Experience.* Durham, NC: Duke University Press.

Mahjubah: The Magazine for Moslem Women. 1984 (July).

Marx, Karl. [1844] 1990. "Religion, the Opium of the People." Pp. 79–91 in *The World Treasury of Modern Religious Thought,* edited by Jaroslav Pelikan and Clifton Fadiman. Boston: Little, Brown.

Mead, George Herbert. 1940. *Mind, Self and Society,* 3rd ed. Chicago: University of Chicago Press.

Merton, Robert K. 1957. *Social Theory and Social Structure.* Glencoe, IL: Free Press.

Murphy, Kim. 1993a. "Ethnic Discord: What About the . . . " *Los Angeles Times* (November 23):H2.

———. 1993b. "Lebanon: Tiny Land Still Caught in Middle." *Los Angeles Times* (December 4):A1, 10.

National Public Radio. 1984a. "Black Islam." *The World of Islam.* Washington, DC.

———. 1984b. "Decay or Rebirth: The Plight of Islamic Art." *The World of Islam.* Washington, DC.

———. 1984c. "Voices of Resurgence." *The World of Islam.* Washington, DC.

The New Columbia Encyclopedia. 1975. Edited by William H. Harris and Judith

S. Levey. New York: Columbia University Press.

The New York Public Library Desk Reference. 1989. New York: Simon & Schuster.

Norton, Augustus R. 1986. "Estrangement and Fragmentation in Lebanon." *Current History* (February):58–62+.

Norton, Augustus R. and Jillian Schwendler. 1993. "Swiss Soldiers, Ta'if Clocks, and Early Elections." *Middle East Insight* (Nov.–Dec.):45–54.

Nottingham, Elizabeth K. 1971. *Religion: A Sociological View.* New York: Random House.

Pickering, W. S. F. 1984. *Durkheim's Sociology of Religion.* London: Routledge & Kegan Paul.

Roberts, Keith A. 1990. *Religion in Sociological Perspective.* Belmont, CA: Wadsworth.

Robertson, Roland. 1987. "Economics and Religion." Pp. 1–11 in *The Encyclopedia of Religion.*

Shames, Stephen. 1980. "Thoughts of a Teenage War Veteran in Lebanon." *Senior Scholastic* (October 31):6.

Shulman, Ken. 1991. "A Man of Principle Pays the Price." *The New York Times* (March 3):Y27.

Smart, Ninian. 1976. *The Religious Experience of Mankind.* New York: Charles Scribner.

Spencer, Herbert. 1972. Quoted on p. 140 in "Religion." Pp. 140–45 in *The Encyclopedia of Philosophy* Vol. 7, edited by P. Edwards. New York: Macmillan.

Spokesman for the Iraqi government. 1991. Quoted in "Iraqi Message: 'Duty' Fulfilled." *The New York Times* (February 26):Y1.

Stark, Rodney and William S. Bainbridge. 1985. *The Future of Religion: Secularization, Revival and Cult Formation.* Berkeley, CA: University of California Press.

Stavenhagen, Rodolfo. 1991. "Ethnic Conflicts and Their Impact on International Society." *International Social Science Journal* (February):117–32.

Steinfels, Peter. 1991. "Priest Allows ABC to Film an Exorcism on a 16-Year-Old Girl." *The New York Times* (April 4):B1.

Turner, Bryan S. 1974. *Weber and Islam: A Critical Study.* Boston: Routledge & Kegan Paul.

Turner, Jonathan H. 1978. *Sociology: Studying the Human System.* Santa Monica, CA: Goodyear.

U.S. Department of the Army. 1989. *Lebanon: A Country Study,* edited by Thomas Collelo. Washington, DC: U.S. Government Printing Office.

Van Doren, Charles L. 1991. *A History of Knowledge: Past, Present, and Future.* New York: Carol.

Vecsey, George. 1991. "Sports of the Times: Lokar's Last Point Was His Best." *The New York Times* (February 15):B12.

Watchtower Bible and Tract Society. 1987. "Life in a Peaceful New World." Brooklyn: Watchtower.

Weber, Max. 1922. *The Sociology of Religion,* translated by E. Fischoff. Boston: Beacon Press.

———. 1958. *The Protestant Ethic and the Spirit of Capitalism,* 5th ed., translated by T. Parsons. New York: Scribner.

Wenger, Martha with Julie Denny. 1990. "Primer: Lebanon's Fifteen-Year War 1975–1990." *Middle East Report* (January/February):23–25.

Whetten, Lawrence L. 1979. "The Military Dimension." Pp. 75–90 in *Lebanon in Crisis,* edited by P. E. Haley and L. W. Snider. Syracuse: Syracuse University Press.

Wills, Garry. 1990. *Under God: Religion and American Politics.* New York: Simon & Schuster.

Wilmore, Gayraud S. 1972. *Black Religion and Black Radicalism.* Garden City, NY: Doubleday.

The World Almanac and Book of Facts 1991. 1990. New York: Pharos Books.

The World Treasury of Modern Religious Thought. 1990. Edited by Jaroslav Pelikan and Clifton Fadiman. Boston: Little, Brown.

Yinger, J. Milton. 1971. *The Scientific Study of Religion.* New York: Macmillan.

Zangwill, O. L. 1987. "Isolation Experiments." Pp. 393–94 in *The Oxford Companion to the Mind,* edited by R. L. Gregory. New York: Oxford University Press.

Chapter 15

Ash, Timothy Garton. 1989. *The Uses of Adversity: Essays on the Fate of Central Europe.* New York: Random House.

Barnet, Richard J. 1990. "Reflections: Defining the Moment." *The New Yorker* (July 16):45–60.

———. 1991. "Reflections: The Uses of Force." *The New Yorker* (April 29): 82–95.

Barron, James. 1991. "Base Panel Spreads Glee and Gloom." *The New York Times* (July 1):A8.

Bauerlein, Monika. 1992. "Plutonium Is Forever: Is There Any Sane Place to Put Nuclear Waste?" *Utne Reader* (July/August):34–37.

Bell Telephone System. 1945. "Advertisement: Millions of Military Telephones." *Life* (August 10):3.

Boudon, Raymond and François Bourricaud. 1989. *A Critical Dictionary of Sociology,* selected and translated by P. Hamilton. Chicago: University of Chicago Press.

Broad, William J. 1991. "Nuclear Designers from East and West Plan Bomb Disposal." *The New York Times* (December 17):B5+.

———. 1992. "Present Since Atom Was Split, Physicist Reflects on a Turbulent Era." *The New York Times* (December 1):B5+.

———. 1994. "Book Says Britain Bluffed About Its H-Bomb." *The New York Times* (March 24):A4.

Broad, William J., Barnaby J. Feder, John Holusha, John Markoff, Andrew Pollack, and Matthew L. Wald. 1991. "Transforming the Decade: 10 Critical Technologies." *The New York Times* (January 1): Y15+.

Burdick, Alan. 1992. "The Last Cold-War Monument: Designing the 'Keep Out' Sign for a Nuclear Waste Site." *Harper's* (August):62–67.

Calonius, Erik. 1991. "Smart Moves by Quality Champs." *Fortune* (Spring/Summer):24–28.

Cobb, Charles E. Jr. 1989. "Living with Radiation." *National Geographic* (April):403–37.

Colihan, Jane and Robert J. T. Joy. 1984. "Military Medicine." *American Heritage* (October/November):65.

Coser, Lewis A. 1973. "Social Conflict and the Theory of Social Change." Pp. 114–22 in *Social Change: Sources, Patterns, and Consequences,* edited by E. Etzioni-Halevy and A. Etzioni. New York: Basic Books.

Currie, Elliott and Jerome H. Skolnick. 1988. *America's Problems: Social Issues and Public Policy,* 2nd ed. Boston: Little, Brown.

Dahrendorf, Ralf. 1973. "Toward a Theory of Social Conflict." Pp. 100–13 in *Social Change: Sources, Patterns, and Consequences,* 2nd ed., edited by E. Ezioni-Halevy and A. Etzioni. New York: Basic Books.

D'Antonio, Michael. 1994. "Scars and Secrets: The Atomic Trail." *Los Angeles Times* (March 20): 14–20+.

Darnton, Robert. 1990. "The Writing on the Wall." *UNESCO Courier* (June): 12–17.

Dawisha, Karen and Bruce Parrott. 1994. *Russia and the New States of Eurasia: The Politics of Upheaval.* New York: Cambridge.

Dibblin, Jane. 1988. *Day of Two Suns: U.S. Nuclear Testing and the Pacific Islanders.* New York: New Amsterdam.

Dyer, Gwynne. 1985. *War.* New York: Crown Publishers.

Erikson, Kai T. 1971. "Sociology and the Historical Perspective." Pp. 61–77 in *The Sociology of the Future,* edited by W. Bell and J. A. Mau. New York: Russell Sage Foundation.

Facts on File. 1993. "Law Mandating Layoff Notices Diluted." (June 24):470.

Fortune. 1994. "The Fortune 500: The Largest Industrial Corporations." (April 18):220–300.

Furchgott, Roy. 1988. "Management's High-Tech Challenge." *Editorial Research Reports* (September 30):482–91.

Galbraith, John K. 1958. *The Affluent Society.* Boston: Houghton Mifflin.

Gould, Stephen Jay. 1990. Interview on "MacNeil/ Lehrer Newshour," January 1. New York: WNET.

Griffin, Rodman D. 1992. "Nuclear Proliferation." *The CQ Researcher* (June 5):483–501.

Guttman, Robert J. 1990. "Interview with John Georgas: The Coca-Cola Company." *Europe* (July/August):16.

Hacker, Andrew. 1971. "Power to Do What?" Pp. 134–46 in *The New Sociology: Essays in Social Science and Social Theory in Honor of C. Wright Mills,* edited by I. L. Horowitz. New York: Oxford University Press.

Halberstam, David. 1986. *The Reckoning.* New York: William Morrow.

Heilbroner, Robert. 1990. "Reflections: After Communism." *The New Yorker* (September 10): 91–100.

Holusha, John. 1989. "Eastern Europe: Its Lure and Hurdles." *The New York Times* (December 18): Y25+.

Idelson, Holly. 1991. "Energy: Watkins' Seizure of Waste Site Meets with Mixed Reviews." *Congressional Quarterly* 49 (41):2945.

———. 1992. "Environment: House Votes to Begin Testing New Mexico Nuclear Dump." *Congressional Quarterly* 50 (30):2171.

Ifill, Gwen. 1991. "Final Cut Is Made on Military Bases That Should Close." *The New York Times* (July 1):A1+.

Jerome, Richard. 1994. "Bombs Away!" *The New York Times Magazine* (April 3):46–47.

Johnson, Lyndon B. 1987. Quoted on p. 907 in *America's History Since 1865,* by James A. Henretta, W. Elliott Brownlee, David Brody, and Susan Ware. Chicago: Dorsey.

Kline, Edward. 1989. "Foreword." Pp. xiii–xviii in *Memoirs* by Andrei Sakharov (1992). New York: Vintage.

Kuhn, Thomas S. 1975. *The Structure of Scientific Revolutions.* Chicago: University of Chicago Press.

Lambert, Michael. 1992. "Defense Contractors Use Down-Sizing, Not Diversification, to Maintain Profits." *Aviation Week and Space Technology* (March 9):61–63.

Lenssen, Nicholas. 1991. *Nuclear Waste: The Problem That Won't Go Away.* Washington, DC: Worldwatch Institute.

Life. 1945. "The Jap Homeland: Allies Give It a Terrific Beating." (August 13): 32–33.

Mandelbaum, Maurice. 1977. *The Anatomy of Historical Knowledge.* Baltimore: Johns Hopkins University Press.

Martel, Leon. 1986. *Mastering Change: The Key to Business Success.* New York: Simon & Schuster.

Marx, Karl. [1881] 1965. "The Class Struggle." Pp. 529–35 in *Theories of Society,* edited by T. Parsons, E. Shils, K. D. Naegele, and J. R. Pitts. New York: Free Press.

———. 1987. Quoted in *A Marx Dictionary,* by Terrell Carver. Totowa, NJ: Barnes & Noble.

McKay, Ron. 1991. *Letters to Gorbachev: Life in Russia Through the Postbag of* Argumenty I Fakty. London: Michael Joseph.

McNamara, Robert S. 1989. *Out of the Cold: New Thinking for American Foreign and Defense Policy in the 21st Century.* New York: Simon & Schuster.

Michael, Donald N. and Walter Truett Anderson. 1987. "Norms in Conflict and Confusion: Six Stories in Search of an Author." *Technological Forecasting and Social Change* 31:107–15.

Michnik, Adam. 1990. "The Moral and Spiritual Origins of Solidarity." Pp. 239–50 in *Without Force or Lies: Voices from the Revolution of Central Europe in 1989–90.* San Francisco: Mercury House.

Miller, Judith. 1991. "Displaced in the Gulf War: 5 Million Refugees." *The New York Times* (June 16):E3.

Mills, C. Wright. 1959. *The Sociological Imagination.* New York: Oxford University Press.

———. 1963. "The Structure of Power in American Society." Pp. 23–38 in *Power, Politics and People: The Collected Essays of C. Wright Mills,* edited by I. L. Horowitz. New York: Oxford University Press.

———. 1973. "The Sources of Societal Power." Pp. 123–30 in *Social Change: Sources, Patterns, and Consequences,* 2nd ed., edited by E. Etzioni-Halevy and A. Etzioni. New York: Basic Books.

The New York Times. 1988. "Some of the Most Dangerous Waste Sites." (June 14):16Y.

The 1994 Canadian Global Almanac. 1994. Edited by John Robert Colombo. Toronto: Macmillan Canada.

Norman, Donald A. 1988. Quoted on pp. 483, 488 of "Management's High-Tech Challenge." *Editorial Research Report* (September 30):482–91.

Ogburn, William F. 1968. "Cultural Lag as Theory." Pp. 86–95 in *William F. Ogburn on Culture and Social Change,* 2nd ed., edited by O. D. Duncan. Chicago: University of Chicago Press.

O'Leary, Hazel. 1993. Interview on "MacNeil/Lehrer Newshour," December 7. New York: WNET.

Oppenheimer, Robert. 1986. Quoted on p. 11 *The Making of the Bomb* by Richard Rhodes. New York: Touchstone.

O'Sullivan, Anthony. 1990. "Eastern Europe." *Europe* (September):21–22.

Parrott, Bruce. 1983. *Politics and Technology in the Soviet Union.* Cambridge, MA: MIT.

Rabi, Isidor I. 1969. "The Revolution in Science." Pp. 28–66 in *The Environment of Change,* edited by A. W. Warner, D. Morse, and T. E. Cooney. New York: Columbia.

Rawe, Dick. 1991. "P&G Expands Entry in Eastern Europe." *The Cincinnati Post* (June 20):10B.

Reich, Jens. 1989. Quoted on p. 20 of "People of the Year." *Newsweek* (December 25):18–25.

Rhodes, Richard. 1986. *The Making of the Atomic Bomb.* New York: Touchstone.

Robbins, Carla Anne with Douglas Pasternak, Bruce B. Auster, Julie Corwin, and Douglas Stanglin. 1992. "The Nuclear Epidemic." *U.S. News & World Report* (March 16):40–51.

Sakharov, Andrei. 1992. *Memoirs: Andrei Sakharov.* New York: Vintage.

Schmemann, Serge. 1993. "Bush and Yeltsin Sign Pact Making Deep Missile Cuts." *The New York Times* (January 4):A1+.

Schneider, Keith. 1990a. "Uranium Miners Inherit Dispute's Sad Legacy." *The New York Times* (January 9):Y1+.

———. 1990b. "Scientist Who Managed to 'Shock the World' on Atomic Workers' Health." *The New York Times* (May 3):A11.

———. 1990c. "U.S. Releases Health Data on 44,000 at Nuclear Plant." *The New York Times* (July 18):A9.

———. 1991. "Military Has New Strategic Goal in Cleanup of Vast Toxic Waste." *The New York Times* (August 5):A1, C3.

———. 1994. "Redressing the Harms of the Nuclear Age May Not Be Cheap." *The New York Times* (January 9):E3.

Seltzer, Richard. 1991. "Move to Test Nuclear Waste Site Draws Fire." *Chemical and Engineering News* 69 (41):5–6.

Steele, Karen. 1993. "Cleaning Up After the Cold War." Pp. 129–46 in *The 1993 Environmental Almanac,* compiled by World Resources Institute. New York: Houghton Mifflin.

Stevenson, Richard W. 1991. "Northrop Settles Workers' Suit on False Missile Tests for $8 Million." *The New York Times* (June 25):A7.

Tibbett, Paul. 1985. Quoted on p. 96 in *War* by Gwynne Dyer. New York: Crown.

Tolstaya, Tatyana. 1991. "In Cannibalistic Times," translated by J. Gambrell. *The New York Review of Books* (April 11):3–6.

U.S. Department of Defense. 1991. *100 Companies Receiving the Largest Dollar Volume of Prime Contract Awards: Fiscal Year 1991.* Washington, DC: U.S. Government Printing Office.

U.S. Department of Energy. 1988. *Waste Isolation Pilot Plant—WIPP.* Produced by Westinghouse Electric Corporation.

Van Evera, Stephen. 1990. "The Case Against Intervention." *The Atlantic Monthly* (July):72–80.

Wald, Matthew L. 1989. "Finding a Burial Place for Nuclear Wastes Grows More Difficult." *The New York Times* (December 5):Y19+.

———. 1994. "A Military-Industrial Alliance Turns Plowshares to Swords." *The New York Times* (March 16):A1+.

Wallerstein, Immanuel. 1984. *The Politics of the World-Economy: The States, the Movements and the Civilizations.* New York: Cambridge University Press.

Weber, Max. 1947. *The Theory of Social and Economic Organization,* translated by A. M. Henderson and T. Parsons. Glencoe, IL: Free Press.

White, Leslie A. 1949. *The Science of Culture: A Study of Man and Civilization.* New York: Grove.

The World Almanac and Book of Facts 1991. 1990. New York: Pharos Books.

Yakolev, Aleksandr. 1989. "Perestroika or the 'Death of Socialism.'" Pp. 33–75 in *Voices of Glasnost: Interviews with Gorbachev's Reformers,* edited by S. F. Cohen and K. vanden Heuvel. New York: Norton.

Zickel, Raymond E. 1991. *Soviet Union: A Country Study,* 2nd ed. Washington, DC: U.S. Government Printing Office.

FOCUS Chapter 15

Bowlby, John. 1988. *A Secure Base.* New York: Basic Books.

Bouchard, Thomas, Jr. 1990. "Sources of Human Psychological Differences: The Minnesota Study of Twins Reared Apart." *Science* 250 (October 12): 223–28.

Branch, Taylor. 1988. Pp. 124–42 in *Parting the Waters.* New York: Simon & Schuster.

Garrison, Fielding. 1929. Pp. 307, 362–63 in *An Introduction to the History of Medicine,* revised ed. by W. B. Saunders.

Library of Congress. 1933. *1994 Women Who Dare.* Rohnert Park, CA: Pomegranate.

Maathai, Wangari. 1993. Quoted on p. 353 of *Current Biography Yearbook 1993,* edited by Judith Graham. New York: H. W. Wilson Co.

Watson, James. 1990. "The Human Genome Project: Past, Present and Future." *Science* 248 (April 6): 44–49.

Wheelis, Allen. 1973. *How People Change.* New York: Harper & Row.

Zola-Morgan, Stuart and Larry Squire. 1990. "The Primate Hippocampal Formation: Evidence for a Time-Limited Role in Memory Storage." *Science* 250 (October 12): 288–89.

CREDITS

From *Culture Against Man* by Jules Henry. Copyright © 1963 Random House, Inc. Reprinted by permission of the publisher. (498–499, 500)

From *Keeping Track: How Schools Structure Inequality* by Jeannie Oakes. Copyright © 1985 Jeannie Oakes. Reprinted by permission of Yale University Press. (503, 504)

From *The Making of Mind: A Personal Account of Soviet Psychology* by A. R. Luria. Copyright © 1979 the President and Fellows of Harvard College. Reprinted by permission of Harvard University Press. (506)

From *The Adolescent Society* by James S. Coleman. Copyright © 1961 The Free Press, a Division of Macmillan, Inc. Reprinted by permission of the publisher. (510, 511)

From "The Roots of Literacy" by David Hawkins, *Daedalus*, Spring 1990. Copyright © 1990 the American Academy of Arts and Sciences. Reprinted by permission. (516)

Chapter 14

From *The Spiritual Life of Children* by Robert Coles. Copyright © 1990 Robert Coles. Reprinted by permission of Houghton Mifflin Company. (532)

Chapter 15

From *Day of Two Suns: U.S. Nuclear Testing and the Pacific Islanders* by Jane Dibblin, New Amsterdam, 1988. Reprinted by permission of New Amsterdam Books and Peters, Fraser & Dunlop. (581–582)

INDEX

bureaucracy in, 239–240
expert knowledge and responsibility in, 249–251
explanation of, 227, 229
formal vs. informal dimensions of, 241–243
problems of oligarchy in, 250–252
pros and cons of, 257
rationalization as tool in modern, 234–236
records of performance in, 247–249
relationship between environment and, 256–259
trained incapacity in, 243–247
Outgroups, 162–164
Out-migration, 454
Oxman, Robert, 265–266, 290

P

Page, Clarence, 363
Palestine, 147–148
Palestinians. *See also* Arabs
conflict between Israelis and, 146–150, 152
displacement of, 148–149
images of Israelis, 168
ingroup-outgroup relations and, 162–164
in Lebanon, 528–530
memories held by, 157
military training for, 161–162
origin of, 147
peace negotiations with, 152
primary group experiences and views of, 159–160
in refuge camps, 151
role-taking among, 166–167
Palileo, Gloria J., 416
Paradigms, 574
Park, Myung-Seok, 128
Parks, Rosa, 375, 600
Parsons, Talcott, 200
Participant observation, 85–86
Passaris, Solange, 494
Patinkin, Mark, 127
Patino, Mario José Martinez, 394
Peer pressure, 177–179

People's Republic of China
Cultural Revolution in, 265, 267–268
democracy movement in, 588–589. *See also* Tiananmen Square incident
differential association in, 290, 292
early intervention and rehabilitation for criminals in, 291
ideological commitment in, 268
legal system in, 284
mores and folkways in, 274
overview of, 268–270, 292
population growth in, 450, 452
social control in, 274–276, 278–279, 292–293
structural strain in, 289
ties between United States and, 264
West's debt to, 124–125
Per capita, 438
Performance measures, 247–249
Perlman, Janice, 462–463
Persian Gulf War, 158–159
Phelan, Patricia, 511–513
Piaget, Jean, 168–170
Pickering, W.S.F., 545
Play, 166
Police, African-American males and, 217–220
Population. *See also* Demographic transition
doubling time of, 438
explanation of, 83
in labor-intensive poor countries, 450–452
study of, 431
Population pyramids, 450, 451
Positive sanctions, 276
Postman, Neil, 214
Poverty
among African-Americans, 361
functionalist view of, 35–36
mobility and, 334–335, 337
Power elite
decision-making process for, 580
explanation of, 576–577

Predestination, 549
Prejudice
explanation of, 363
reinforcement of, 364
relationship between discrimination and, 364–365
Prejudiced discriminators, 366–367
Prejudiced nondiscriminators, 366
Preoperational stage of cognitive development, 169
Prewitt, Kenneth, 257
Primary groups, 158
Primary sex characteristics, 393, 394
Private households, 402
Probabilistic models, 94–95
Production, means of, 16
Profane
distinction between sacred and, 536, 556, 558
explanation of, 533
Professionalization, of organizations, 249, 251
Proletariat, 16
Prophetic religions, 533
Pull factos, 452
Punishment, 279
Pure deviants, 280–281
Push factors, 452

Q

Questionnaires
self-administered, 84
United States vs. Japanese versions of, 86–87

R

Rabi, Isador I., 573
Rabin, Yitzhak, 146, 152, 177
Race
as burden, 343
explanation of, 305, 308, 350
ways of viewing, 372–373
Racial classification
explanation of, 305
hierarchical ranking used in, 359
problematic nature of, 351–354